T0251320

A Numerical Library in

for Scientists and Engineers

H.T. Lau, Ph.D.

The software mentioned in this book is now available for download on our Web site at: http://www.crcpress.com/e_products/downloads/default.asp

CRC Press
Boca Raton Ann Arbor London Tokyo

LIMITED WARRANTY

CRC Press warrants the physical diskette(s) enclosed herein to be free of defects in materials and workmanship for a period of thirty days from the date of purchase. If within the warranty period CRC Press receives written notification of defects in materials or workmanship, and such notification is determined by CRC Press to be correct, CRC Press will replace the defective diskette(s).

The entire and exclusive liability and remedy for breach of this Limited Warranty shall be limited to replacement of defective diskette(s) and shall not include or extend to any claim for or right to cover any other damages, including but not limited to, loss of profit, data, or use of the software, or special, incidental, or consequential damages or other similar claims, even if CRC Press has been specifically advised of the possibility of such damages. In no event will the liability of CRC Press for any damages to you or any other person ever exceed the lower suggested list price or actual price paid for the software, regardless of any form of the claim.

CRC Press SPECIFICALLY DISCLAIMS ALL OTHER WARRANTIES, EXPRESS OR IMPLIED, INCLUDING BUT NOT LIMITED TO, ANY IMPLIED WARRANTY OF MERCHANTABILITY OR FITNESS FOR A PARTICULAR PURPOSE. Specifically, CRC Press makes no representation or warranty that the software is fit for any particular purpose and any implied warranty of merchantability is limited to the thirty-day duration of the Limited Warranty covering the physical diskette(s) only (and not the software) and is otherwise expressly and specifically disclaimed.

Since some states do not allow the exclusion of incidental or consequential damages, or the limitation on how long an implied warranty lasts, some of the above may not apply to you.

DISCLAIMER OF WARRANTY AND LIMITS OF LIABILITY: The author(s) of this book have used their best efforts in preparing this material. These efforts include the development, research, and testing of the theories and programs to determine their effectiveness. NEITHER THE AUTHOR(S) NOR THE PUBLISHER MAKE WARRANTIES OF ANY KIND, EXPRESS OR IMPLIED, WITH REGARD TO THESE PROGRAMS OR THE DOCUMENTATION CONTAINED IN THIS BOOK, INCLUDING WITHOUT LIMITATION WARRANTIES OF MERCHANTABILITY OR FITNESS FOR A PARTICULAR PURPOSE. NO LIABILITY IS ACCEPTED IN ANY EVENT FOR ANY DAMAGES, INCLUDING INCIDENTAL OR CONSEQUENTIAL DAMAGES, LOST PROFITS, COSTS OF LOST DATA OR PROGRAM MATERIAL, OR OTHERWISE IN CONNECTION WITH OR ARISING OUT OF THE FURNISHING, PERFORMANCE, OR USE OF THE PROGRAMS IN THIS BOOK.

Library of Congress Cataloging-in-Publication Data

Lau, H. T. (Hang Tong), 1952–
Numerical library in C for scientists and engineers / Hang-Tong Lau.
p. cm.
Includes bibliographical references and index.
ISBN 0-8493-7376-X
1. C (Computer program language) I. Title.
QA76.73.C15L38 1994
519.4'0285'53—dc20 94-37928
CIP

International Standard Book Number 0-8493-7376-X
Library of Congress Card Number 94-37928
2 3 4 5 6 7 8 9 0

To my wife, Helen,
and the children Matthew, Lawrence and Tabia
for their love and support

Contents

Introduction

The following collection of procedures is a translation into C of the library *NUMAL* (*NUM*erical procedures in *AL*gol 60) [He80] developed at the Mathematical Centre, Amsterdam. The scope of the library is indicated by the list of procedure descriptions which follows this introduction and refers to linear algebra, the numerical solution of ordinary and partial differential equations, optimisation, parameter estimation, special functions of mathematical physics and indeed most concerns of an active research group. Each main procedure is assembled from auxiliary submodules and for this reason the library is highly compact. The submodules implement routine numerical processes and are of use in the development of new methods.

The expert is offered both a collection of procedures for implementing numerical methods which are by now standard and with whose theory he will be perfectly familiar and a powerful research tool. The novice is offered a prodigious source of entertainment and instruction: he may run the test programs which carry out the worked examples provided with each critical procedure or group of procedures, he is free to construct other examples and apply the procedures to them, and he is even at liberty to devise examples which provoke the procedures into emitting one of the failure indications consistently provided.

The use of some Algol procedures in the NUMAL library was explained in detail in a seminar on numerical software. Furthermore, the commercially available program libraries IMSL (IMSL Inc., Houston, Texas, USA) and NAG (Numerical Algorithms Group, Oxford, United Kingdom) were compared with the Algol NUMAL library. The proceedings of that seminar appeared (in Dutch) in [Bus76, Rie77].

NUMAL was originally constructed with the aid of a CDC-Cyber 70 system. However, personal computers, most of them equipped with C-compilers, are now within the purchasing capability of the professionally active. The C version of NUMAL, being highly compact, is peculiarly suited to implementation upon a personal computer.

The original order of the NUMAL procedures has been preserved and is indicated by a list which also follows this introduction. The documentations are taken partly from the original NUMAL source and partly from extended forms given in a FORTRAN version of NUMAL [Wy81] constructed, with considerable help from the author, by P. Wynn for his own use. To ensure maximal compactness of the main library, the worked examples have been grouped together at the end of the C version library. The references are collected in Appendix A. The prototype declarations for all the procedures of NUMAL are listed in Appendix B. Appendix C contains a capsule statement of what each procedure does in the library.

While dynamic allocation of memory is one of the fundamental features of Algol 60, such allocation is less direct in C and must be given special treatment. Storage manipulation procedures, similar to those given in [Pr92], are contained in Appendix D.

Each procedure in the library is described using the following scheme:

Top line	Each procedure in the library is identified by its name as the top line.
Module description	Following the top line are a description of the procedure and a brief discussion on the methods used. References are given for further details of the underlying mathematical method.
Function parameters	This is the function specification of the procedure, followed by the description of each parameter of the procedure.
Functions used	List of other functions in the library used by this procedure.
Method	Supplementary remarks on the method and performance of the module.
Program listing	The C code listing of the procedure.

To ensure portability, the ANSI C standard is used throughout. The programs have been tested using the Turbo C (version 2.0 from Borland International) on an IBM-AT personal computer using the MS-DOS operation system.

In conclusion it should perhaps be emphasized that the library is offered as a research tool. The documentation to the library procedures briefly describes the algorithms used; it is in no way intended as a treatise upon numerical methods. For fuller theory the user is directed to the references given and to further research texts - in particular to the relevant Mathematical Centre colloquium publications. Reference [Pr92] also serves as an excellent elementary introduction to part of the library.

1. Elementary Procedures

This chapter contains elementary operations for vectors and matrices such as the assignment of initial values to vectors and matrix slices, duplication and interchange of such slices, rotations and reflections. The procedures are taken from [Dek68, DekHo68]. Most of them are used in subsequent chapters. The elementary procedures are all quite short; prepared versions may be re-coded, making optimal use of machine capabilities, in assembly language.

1.1 Real vector and matrix - Initialization

A. inivec

Initializes part of a vector a by setting $a_i=x$ $(i=l,l+1,...,u)$.

Function Parameters:

void inivec (l,u,a,x)

l,u: int; lower and upper index of the vector a, respectively;
a: float $a[l:u]$; the array to be initialized;
x: float; initialization constant.

```
void inivec(int l, int u, float a[], float x)
{
    for (; l<=u; l++) a[l]=x;
}
```

B. inimat

Initializes a rectangular submatrix a by setting

$$a_{i,j}=x \quad (i=lr,lr+1,...,ur;\ j=lc,lc+1,...,uc).$$

Function Parameters:

void inimat (lr,ur,lc,uc,a,x)

lr,ur,lc,uc: int; lower and upper row-index, and lower and upper column-index of the matrix a, respectively;
a: float $a[lr:ur,lc:uc]$; the matrix to be initialized;
x: float; initialization constant.

```
void inimat(int lr, int ur, int lc, int uc, float **a, float x)
{
    int j;

    for (; lr<=ur; lr++)
        for (j=lc; j<=uc; j++) a[lr][j]=x;
}
```

C. inimatd

Initializes a forward diagonal sequence of elements of a rectangular matrix *a* by setting

$$a_{i,i+shift}=x \quad (i=lr,lr+1,...,ur).$$

Function Parameters:

void inimatd (*lr,ur,shift,a,x*)

lr,ur: int; lower and upper row-index of the codiagonal to be initialized;
shift: int; distance between diagonal and codiagonal;
a: float *a[lr:ur,lr+shift:ur+shift]*; the array to be initialized;
x: float; initialization constant.

```
void inimatd(int lr, int ur, int shift, float **a, float x)
{
    for (; lr<=ur; lr++) a[lr][lr+shift]=x;
}
```

D. inisymd

Initializes a forward diagonal sequence of elements of a symmetric square matrix *b*. The upper triangle of *b* is stored columnwise in a linear array *a* such that $b_{i,j}$ is $a_{(j-1)j/2+i}$.

Function Parameters:

void inisymd (*lr,ur,shift,a,x*)

lr,ur: int; lower and upper row-index of a codiagonal of a symmetric matrix of order n to be initialized, *lr* and *ur* should satisfy: $lr \geq 1$, $ur \leq n$;
shift: distance between diagonal and codiagonal, $-n < shift < n$;
a: float *a*[1:n(n+1)/2]; the linear array containing the columnwise stored upper triangle of a symmetric matrix;
x: float; initialization constant.

```
#include <math.h>

void inisymd(int lr, int ur, int shift, float a[], float x)
{
    shift=abs(shift);
    ur += shift+1;
    shift += lr;
    lr += ((shift-3)*shift)/2;
    lr += shift;
    while (shift < ur) {
        a[lr]=x;
        shift++;
        lr += shift;
    }
}
```

E. inisymrow

Initializes part of a row of a symmetric square matrix *b*. The upper triangle of *b* is stored columnwise in a linear array *a* such that $b_{i,j}$ is $a_{(j-1)j/2+i}$.

Function Parameters:

void inisymrow (*l,u,i,a,x*)

l,u: int; lower and upper index of row-element of a codiagonal of a symmetric matrix of

order n to be initialized, *l* and *u* should satisfy: $1 \le l \le n$, $1 \le u \le n$;
i: row index, $1 \le i \le n$;
a: float *a*[1:n(n+1)/2]; the linear array containing the columnwise stored upper triangle of a symmetric matrix;
x: float; initialization constant.

```
void inisymrow(int l, int u, int i, float a[], float x)
{
    int k;

    if (l <= i) {
        k=((i-1)*i)/2;
        l += k;
        k += (u<i) ? u : i;
        for (; l<=k; l++) a[l]=x;
        l=i+1;
    }
    if (u > i) {
        k=((l-1)*l)/2+i;
        do {
            a[k]=x;
            l++;
            k += l-1;
        } while (l <= u);
    }
}
```

1.2 Real vector and matrix - Duplication

A. dupvec

Duplicates part of a vector *b* by setting

$$a_i = b_{i,i+shift} \quad (i=l,l+1,\ldots,u).$$

Function Parameters:

void dupvec (*l,u,shift,a,b*)

l,u: int; lower and upper vector-index, respectively;
shift: int; index-shifting parameter;
a,b: float *a[l:u]*, *b[l+shift:u+shift]*; the array *b* is copied into *a*.

```
void dupvec(int l, int u, int shift, float a[], float b[])
{
    for (; l<=u; l++) a[l]=b[l+shift];
}
```

B. dupvecrow

Duplicates part of a row vector of a rectangular matrix *b* by setting

$$a_j = b_{i,j} \quad (j=l,l+1,\ldots,u).$$

Function Parameters:

void dupvecrow (*l,u,i,a,b*)

l,u: int; lower and upper vector-index, respectively;
i: int; row-index of the matrix *b*;
a,b: float *a[l:u]*, *b[i:i,l:u]*; *b* is copied into *a*.

```
void dupvecrow(int l, int u, int i, float a[], float **b)
{
    for (; l<=u; l++) a[l]=b[i][l];
}
```

C. duprowvec

Replaces part of a row vector of a rectangular matrix *a* by a corresponding part of a vector *b* by setting

$$a_{i,j}=b_j \ (j=l,l+1,...,u).$$

Function Parameters:

void duprowvec (*l,u,i,a,b*)

l,u: int; lower and upper vector-index, respectively;
i: int; row-index of the matrix *a*;
a,b: float *a[i:i,l:u], b[l:u]*; *b* is copied into *a*.

```
void duprowvec(int l, int u, int i, float **a, float b[])
{
    for (; l<=u; l++) a[i][l]=b[l];
}
```

D. dupveccol

Duplicates part of a column vector of a rectangular matrix *b* by setting

$$a_i=b_{i,j} \ (i=l,l+1,...,u).$$

Function Parameters:

void dupveccol (*l,u,j,a,b*)

l,u: int; lower and upper vector-index, respectively;
j: int; column-index of the matrix *b*;
a,b: float *a[l:u], b[l:u,j:j]*; *b* is copied into *a*.

```
void dupveccol(int l, int u, int j, float a[], float **b)
{
    for (; l<=u; l++) a[l]=b[l][j];
}
```

E. dupcolvec

Replaces part of a column vector of a rectangular matrix *a* by a corresponding part of a vector *b* by setting

$$a_{i,j}=b_i \ (i=l,l+1,...,u).$$

Function Parameters

void dupcolvec (*l,u,j,a,b*)

l,u: int; lower and upper vector-index, respectively;
j: int; column-index of the matrix *a*;
a,b: float *a[l:u,i:i], b[l:u]*; *b* is copied into *a*.

```
void dupcolvec(int l, int u, int j, float **a, float b[])
{
    for (; l<=u; l++) a[l][j]=b[l];
}
```

F. dupmat

Replaces a submatrix of the rectangular matrix *a* by a corresponding submatrix of the rectangular matrix *b* by setting

$$a_{m,n}=b_{m,n} \ (m=l,l+1,...,u; \ n=i,i+1,...,j).$$

Function Parameters:

void dupmat (l,u,i,j,a,b)

l,u: int; lower and upper row-index, respectively;
i,j: int; lower and upper column-index, respectively;
a,b: float *a[l:u,i:j], b[l:u,i:j]*; *b* is copied into *a*.

```
void dupmat(int l, int u, int i, int j, float **a, float **b)
{
    int k;

    for (; l<=u; l++)
        for (k=i; k<=j; k++) a[l][k]=b[l][k];
}
```

1.3 Real vector and matrix - Multiplication

A. mulvec

Forms a constant multiple of part of a vector by setting

$$a_i=xb_{i+shift} \ (i=l,l+1,...,u).$$

Function Parameters:

void mulvec $(l,u,shift,a,b,x)$

l,u: int; lower and upper vector-index, respectively;
shift: int; subscript-shifting parameter;
a,b: float *a[l:u], b[l+shift:u+shift]*; the product of the contents of *b* are stored in *a*;
x: multiplication factor.

```
void mulvec(int l, int u, int shift, float a[], float b[], float x)
{
    for (; l<=u; l++) a[l]=b[l+shift]*x;
}
```

B. mulrow

Replaces a row sequence of elements of a rectangular matrix *a* by a constant multiple of a row sequence of elements of a rectangular matrix *b* by setting

$$a_{i,k}=xb_{j,k} \ (k=l,l+1,...,u).$$

Function Parameters:

void mulrow (*l,u,i,j,a,b,x*)

l,u: int; lower and upper column-index, respectively;
i,j: int; row indices of *a* and *b*;
a,b: float *a[i:i,l:u], b[j:j,l:u]*; the contents of *b* multiplied by *x* are stored into *a*;
x: multiplication factor.

```
void mulrow(int l, int u, int i, int j, float **a, float **b, float x)
{
    for (; l<=u; l++) a[i][l]=b[j][l]*x;
}
```

C. mulcol

Replaces a column sequence of elements of a rectangular matrix *a* by a constant multiple of a column sequence of elements of a rectangular matrix *b* by setting

$$a_{k,i}=xb_{k,j} \; (k=l,l+1,...,u).$$

Function Parameters:

void mulcol (*l,u,i,j,a,b,x*)

l,u: int; lower and upper row-index, respectively;
i,j: int; column indices of *a* and *b*;
a,b: float *a[l:u,i:i], b[l:u,j:j]*; the contents of *b* multiplied by *x* are stored into *a*;
x: multiplication factor.

```
void mulcol(int l, int u, int i, int j, float **a, float **b, float x)
{
    for (; l<=u; l++) a[l][i]=b[l][j]*x;
}
```

D. colcst

Replaces the elements of a column sequence taken from a rectangular matrix *a* by constant multiples of the same elements, by setting

$$a_{i,j}=xa_{i,j} \; (i=l,l+1,...,u).$$

Function Parameters:

void colcst (*l,u,j,a,x*)

l,u: int; lower and upper row-index, respectively;
j: int; column index;
a: float *a[l:u,j:j]*; the rectangular matrix;
x: multiplication factor.

```
void colcst(int l, int u, int j, float **a, float x)
{
    for (; l<=u; l++) a[l][j] *= x;
}
```

E. rowcst

Replaces the elements of a row sequence taken from a rectangular matrix *a* by constant multiples of the same elements, by setting

$$a_{i,j}=xa_{i,j} \quad (j=l,l+1,...,u).$$

Function Parameters:

void rowcst (l,u,i,a,x)

l,u: int; lower and upper column-index, respectively;
i: int; row index;
a: float *a[i:i,l:u]*; the rectangular matrix;
x: multiplication factor.

```
void rowcst(int l, int u, int i, float **a, float x)
{
    for (; l<=u; l++) a[i][l] *= x;
}
```

1.4 Real vector vector products

A. vecvec

Forms the inner product s of part of a vector a and part of a vector b by setting

$$s = \sum_{i=l}^{u} a_i * b_{i+shift}.$$

Function Parameters:

float vecvec $(l,u,shift,a,b)$

vecvec: given the value of s in the above;
l,u: int; lower and upper bound of the running subscript;
shift: int; index-shifting parameter of the vector b;
a,b: float *a[l:u], b[l+shift:u+shift]*.

```
float vecvec(int l, int u, int shift, float a[], float b[])
{
    int k;
    float s;

    s=0.0;
    for (k=l; k<=u; k++) s += a[k]*b[k+shift];
    return (s);
}
```

B. matvec

Forms the inner product s of part of a row of a rectangular matrix a and the corresponding part of a vector b by setting

$$s = \sum_{k=l}^{u} a_{i,k} * b_k.$$

Function Parameters:

float matvec (l,u,i,a,b)

matvec: given the value of s in the above;
l,u: int; lower and upper bound of the running subscript;
i: int; row-index of a;
a,b: float $a[i:i,l:u]$, $b[l:u]$.

```
float matvec(int l, int u, int i, float **a, float b[])
{
    int k;
    float s;

    s=0.0;
    for (k=l; k<=u; k++) s += a[i][k]*b[k];
    return (s);
}
```

C. tamvec

Forms the inner product s of part of a column of a rectangular matrix a and the corresponding part of a vector b by setting

$$s = \sum_{k=l}^{u} a_{k,i} * b_k.$$

Function Parameters:

float tamvec (l,u,i,a,b)

tamvec: given the value of s in the above;
l,u: int; lower and upper bound of the running subscript;
i: int; column-index of a;
a,b: float $a[l:u,i:i]$, $b[l:u]$.

```
float tamvec(int l, int u, int i, float **a, float b[])
{
    int k;
    float s;

    s=0.0;
    for (k=l; k<=u; k++) s += a[k][i]*b[k];
    return (s);
}
```

D. matmat

Forms the inner product s of part of a row of a rectangular matrix a and the corresponding part of a column of a rectangular matrix b by setting

$$s = \sum_{k=l}^{u} a_{i,k} * b_{k,j}.$$

Function Parameters:

float matmat (l,u,i,j,a,b)

matmat: given the value of s in the above;
l,u: int; lower and upper bound of the running subscript;
i,j: int; row-index of a and column-index of b;
a,b: float $a[i:i,l:u]$, $b[l:u,j:j]$.

```
float matmat(int l, int u, int i, int j, float **a, float **b)
{
    int k;
    float s;

    s=0.0;
    for (k=l; k<=ü; k++) s += a[i][k]*b[k][j];
    return (s);
}
```

E. tammat

Forms the inner product s of part of a column of a rectangular matrix a and the corresponding part of a column of a rectangular matrix b by setting

$$s = \sum_{k=l}^{u} a_{k,i} * b_{k,j}.$$

Function Parameters:

float tammat (l,u,i,j,a,b)

tammat: given the value of s in the above;
l,u: int; lower and upper bound of the running subscript;
i,j: int; column-indices of a and b, respectively;
a,b: float $a[l:u,i:i]$, $b[l:u,j:j]$.

```
float tammat(int l, int u, int i, int j, float **a, float **b)
{
    int k;
    float s;

    s=0.0;
    for (k=l; k<=u; k++) s += a[k][i]*b[k][j];
    return (s);
}
```

F. mattam

Forms the inner product s of part of a row of a rectangular matrix a and the corresponding part of a row of a rectangular matrix b by setting

$$s = \sum_{k=l}^{u} a_{i,k} * b_{j,k}.$$

Function Parameters:

float mattam (l,u,i,j,a,b)

mattam: given the value of s in the above;
l,u: int; lower and upper bound of the running subscript;
i,j: int; row-indices of a and b, respectively;
a,b: float *a[i:i,l:u], b[j:j,l:u]*.

```
float mattam(int l, int u, int i, int j, float **a, float **b)
{
    int k;
    float s;

    s=0.0;
    for (k=l; k<=u; k++) s += a[i][k]*b[j][k];
    return (s);
}
```

G. seqvec

Forms the sum

$$s = \sum_{j=l}^{u} a_{il+(j+l-1)*(j-l)/2} * b_{j+shift}.$$

Function Parameters:

float seqvec $(l,u,il,shift,a,b)$

seqvec: given the value of s in the above;
l,u: int; lower and upper bound of the running subscript;
il: int; lower bound of the vector a;
shift: int; index-shifting parameter of the vector b;
a,b: float *a[p:q], b[l+shift:u+shift]*; where $p \le il$ and $q \ge il+(u+l-1)(u-l)/2$.

```
float seqvec(int l, int u, int il, int shift, float a[], float b[])
{
    float s;

    s=0.0;
    for (; l<=u; l++) {
        s += a[il]*b[l+shift];
        il += l;
    }
    return (s);
}
```

H. scaprd1

Forms the stepped inner product

$$s = \sum_{j=1}^{n} a_{la+(j-1)*sa} * b_{lb+(j-1)*sb}$$

from the elements of two vectors a and b.

Function Parameters:

float scaprd1 (*la,sa,lb,sb,n,a,b*)

scaprd1: given the value of s in the above;
la,lb: int; lower bounds of the vectors a and b, respectively;
sa,sb: int; index_shifting parameters of the vectors a and b, respectively;
a,b: float *a[p:q]*, *b[r:s]*; the subscripts above and the values of $la+(j-1)*sa$ and $lb+(j-1)*sb$, $j=1,..,n$ should not contradict each other.

```
float scaprd1(int la, int sa, int lb, int sb, int n, float a[], float b[])
{
    int k;
    float s;

    s=0.0;
    for (k=1; k<=n; k++) {
        s += a[la]*b[lb];
        la += sa;
        lb += sb;
    }
    return (s);
}
```

I. symmatvec

Forms the inner product of part of a row of a symmetric square matrix c, whose elements are stored columnwise in the linear array a, and the corresponding part of a vector b, by setting

$$s = \sum_{k=l}^{u} c_{i,k} * b_k.$$

Function Parameters:

float symmatvec (*l,u,i,a,b*)

symmatvec: given the value of s in the above;
l,u: int; lower and upper bound of the vector b, $l \geq 1$;
i: int; row index of a, $i \geq 1$;
a: float *a[p:q]*; array with: if $i > l$ then $p=(i-1)i/2+l$ else $p=(l-1)l/2+i$ and if $i > u$ then $q=(i-1)i/2+u$ else $q=(u-1)u/2+i$;
b: float b[l:u].

Functions used: vecvec, seqvec.

```
float symmatvec(int l, int u, int i, float a[], float b[])
{
    int k, m;
    float vecvec(int, int, int, float [], float []);
    float seqvec(int, int, int, int, float [], float []);

    m=(l>i) ? l : i;
    k=(m*(m-1))/2;
    return (vecvec(l, (i<=u) ? i-1 : u, k,b,a) + seqvec(m,u,k+i,0,a,b));
}
```

1.5 Real matrix vector products

A. fulmatvec

Premultiplies part of a vector by a submatrix by setting

$$c_i = \sum_{j=lc}^{uc} a_{i,j} * b_j, \quad i=lr,...,ur.$$

Function Parameters:

void fulmatvec (*lr,ur,lc,uc,a,b,c*)

lr,ur: int; lower and upper bound of the row-index;
lc,uc: int; lower and upper bound of the column-index;
a: float *a[lr:ur,lc:uc]*; the matrix;
b: float *b[lc:uc]*; the vector;
c: float *c[lr:ur]*; the result *a*b* is delivered in *c*.

```
void fulmatvec(int lr, int ur, int lc, int uc,
                     float **a, float b[], float c[])
{
    float matvec(int, int, int, float **, float []);

    for (; lr<=ur; lr++) c[lr]=matvec(lc,uc,lr,a,b);
}
```

B. fultamvec

Premultiplies part of a vector by a transposed submatrix by setting

$$c_j = \sum_{i=lr}^{ur} a_{i,j} * b_i, \quad j=lc,...,uc.$$

Function Parameters:

void fultamvec (*lr,ur,lc,uc,a,b,c*)

lr,ur: int; lower and upper bound of the row-index;
lc,uc: int; lower and upper bound of the column-index;
a: float *a[lr:ur,lc:uc]*; the matrix;
b: float *b[lr:ur]*; the vector;

c: float *c[lc:uc]*; the result $a'*b$ is delivered in *c*, a' denotes the transposed of the matrix *a*.

```
void fultamvec(int lr, int ur, int lc, int uc,
                    float **a, float b[], float c[])
{
   float tamvec(int, int, int, float **, float []);

   for (; lc<=uc; lc++) c[lc]=tamvec(lr,ur,lc,a,b);
}
```

C. fulsymmatvec

Premultiplies part of a vector by a submatrix of a linearly stored symmetric matrix by setting

$$c_i = \sum_{j=lc}^{uc} a_{i,j}*b_j, \quad i=lr,...,ur.$$

Function Parameters:

void fulsymmatvec (*lr,ur,lc,uc,a,b,c*)

lr,ur: int; lower and upper bound of the row-index, $lr \geq 1$;
lc,uc: int; lower and upper bound of the column-index, $lc \geq 1$;
a: float *a[l:u]*; where
$l = min\ (lr(lr-1)/2+lc,\ lc(lc-1)/2+lr)$,
$u = max\ (ur(ur-1)/2+uc,\ uc(uc-1)/2+ur)$
and the (i,j)-th element of the symmetric matrix should be given in $a_{j(j-1)/2+i}$;
b: float *b[lc:uc]*; the vector;
c: float *c[lr:ur]*; the result $a*b$ is delivered in *c*.

```
void fulsymmatvec(int lr, int ur, int lc, int uc,
                        float a[], float b[], float c[])
{
   float symmatvec(int, int, int, float [], float []);

   for (; lr<=ur; lr++) c[lr]=symmatvec(lc,uc,lr,a,b);
}
```

D. resvec

Determines a residual vector by setting

$$c_i = \sum_{j=lc}^{uc} a_{i,j}*b_j + xc_i, \quad i=lr,...,ur.$$

Function Parameters:

void resvec (*lr,ur,lc,uc,a,b,c,x*)

lr,ur: int; lower and upper bound of the row-index;

lc,uc: int; lower and upper bound of the column-index;
a: float *a[lr:ur,lc:uc]*; the matrix;
b: float *b[lc:uc]*; the vector;
x: float; the value of the multiplying scalar;
c: float *c[lr:ur]*; the result *a*b+x*c* is overwritten on *c*.

```
void resvec(int lr, int ur, int lc, int uc,
                  float **a, float b[], float c[], float x)
{
    float matvec(int, int, int, float **, float []);

    for (; lr<=ur; lr++) c[lr]=matvec(lc,uc,lr,a,b)+c[lr]*x;
}
```

E. symresvec

Determines a residual vector from a submatrix of a linearly stored symmetric matrix by setting

$$c_i = \sum_{j=lc}^{uc} a_{i,j}*b_j + xc_i, \quad i=lr,...,ur.$$

Function Parameters:

void symresvec (*lr,ur,lc,uc,a,b,c,x*)

lr,ur: int; lower and upper bound of the row-index, $lr \geq 1$;
lc,uc: int; lower and upper bound of the column-index, $lc \geq 1$;
a: float *a[l:u]*; where
$l = min\ (lr(lr-1)/2+lc,\ lc(lc-1)/2+lr)$,
$u = max\ (ur(ur-1)/2+uc,\ uc(uc-1)/2+ur)$
and the (i,j)-th element of the symmetric matrix should be given in $a_{j(j-1)/2+i}$;
b: float *b[lc:uc]*; the vector;
x: float; the value of the multiplying scalar;
c: float *c[lr:ur]*; the result *a*b+x*c* is overwritten on *c*.

```
void symresvec(int lr, int ur, int lc, int uc,
                     float a[], float b[], float c[], float x)
{
    float symmatvec(int, int, int, float [], float []);

    for (; lr<=ur; lr++) c[lr]=symmatvec(lc,uc,lr,a,b)+c[lr]*x;
}
```

1.6 Real matrix matrix products

A. hshvecmat

Premultiplies a submatrix by an elementary reflector: $a=(I-xu^Tu)a$ where

$$a_{i-lr+1,j-lc+1}=a_{i,j}\ (i=lr,...,ur;\ j=lc,...,uc),\ u_{i-lr+1}=u_i\ (i=lr,...,ur).$$

Function Parameters:

void hshvecmat (*lr,ur,lc,uc,x,u,a*)

lr,ur: int; lower and upper row indices;
lc,uc: int; lower and upper column indices;
x: float; the Householder constant;
u: float *u[lr:ur]*; the Householder vector;
a: float *a[lr:ur,lc:uc]*; the matrix to be premultiplied by the Householder matrix.

Functions used: tamvec, elmcolvec.

```
void hshvecmat(int lr, int ur, int lc, int uc,
                  float x, float u[], float **a)
{
   void elmcolvec(int, int, int, float **, float [], float);
   float tamvec(int, int, int, float **, float []);

   for (; lc<=uc; lc++) elmcolvec(lr,ur,lc,a,u,tamvec(lr,ur,lc,a,u)*x);
}
```

B. hshcolmat

Premultiplies one submatrix by an elementary reflector formed from a column of another: $a=(I-xu^Tu)a$ where $a_{k-lr+1,j-lc+1}=a_{k,j}$ $(k=lr,...,ur;$ $j=lc,...,uc)$, $u_{k-lr+1}=U_{k,i}$ $(k=lr,...,ur)$.

Function Parameters:

void hshcolmat (*lr,ur,lc,uc,i,x,u,a*)

lr,ur: int; lower and upper row indices;
lc,uc: int; lower and upper column indices;
i: int; the column index of the Householder vector;
x: float; the Householder constant;
u: float *u[lr:ur,i:i]*; the Householder vector;
a: float *a[lr:ur,lc:uc]*; the matrix to be premultiplied by the Householder matrix.

Functions used: tammat, elmcol.

```
void hshcolmat(int lr, int ur, int lc, int uc, int i,
                  float x, float **u, float **a)
{
   void elmcol(int, int, int, int, float **, float **, float);
   float tammat(int, int, int, int, float **, float **);

   for (; lc<=uc; lc++) elmcol(lr,ur,lc,i,a,u,tammat(lr,ur,lc,i,a,u)*x);
}
```

C. hshrowmat

Premultiplies one submatrix by an elementary reflector formed from a row of another: $a=(I-xu^Tu)a$ where $a_{k-lr+1,j-lc+1}=a_{k,j}$ $(k=lr,...,ur;$ $j=lc,...,uc)$, $u_{k-lr+1}=U_{i,k}$ $(k=lr,...,ur)$.

Function Parameters:

void hshrowmat (*lr,ur,lc,uc,i,x,u,a*)

lr,ur: int; lower and upper row indices;
lc,uc: int; lower and upper column indices;
i: int; the row index of the Householder vector;
x: float; the Householder constant;
u: float *u[i:i,lr:ur]*; the Householder vector;
a: float *a[lr:ur,lc:uc]*; the matrix to be premultiplied by the Householder matrix.

Functions used: matmat, elmcolrow.

```
void hshrowmat(int lr, int ur, int lc, int uc, int i,
                    float x, float **u, float **a)
{
    float matmat(int, int, int, int, float **, float **);
    void elmcolrow(int, int, int, int, float **, float **, float);

    for (; lc<=uc; lc++) elmcolrow(lr,ur,lc,i,a,u,matmat(lr,ur,i,lc,u,a)*x);
}
```

D. hshvectam

Postmultiplies a submatrix by an elementary reflector: $a=(I-xu^Tu)a$ where

$$a_{i-lr+1,j-lc+1}=a_{i,j}\ (i=lr,...,ur;\ j=lc,...,uc),\ u_{i-lc+1}=u_i\ (i=lc,...,uc).$$

Function Parameters:

void hshvectam (*lr,ur,lc,uc,x,u,a*)

lr,ur: int; lower and upper row indices;
lc,uc: int; lower and upper column indices;
x: float; the Householder constant;
u: float *u[lc:uc]*; the Householder vector;
a: float *a[lr:ur,lc:uc]*; the matrix to be postmultiplied by the Householder matrix.

Functions used: matvec, elmrowvec.

```
void hshvectam(int lr, int ur, int lc, int uc,
                    float x, float u[], float **a)
{
    float matvec(int, int, int, float **, float []);
    void elmrowvec(int, int, int, float **, float [], float);

    for (; lr<=ur; lr++) elmrowvec(lc,uc,lr,a,u,matvec(lc,uc,lr,a,u)*x);
}
```

E. hshcoltam

Postmultiplies one submatrix by an elementary reflector formed from a column of another: $a=(I-xu^Tu)a$ where $a_{k-lr+1,j-lc+1}=a_{k,j}$ $(k=lr,...,ur;$ $j=lc,...,uc),$ $u_{k-lc+1}=U_{k,i}$ $(k=lc,...,uc).$

Function Parameters:

void hshcoltam (*lr,ur,lc,uc,i,x,u,a*)

lr,ur: int; lower and upper row indices;

lc,uc: int; lower and upper column indices;
i: int; the column index of the Householder vector;
x: float; the Householder constant;
u: float *u[lc:uc,i:i]*; the Householder vector;
a: float *a[lr:ur,lc:uc]*; the matrix to be postmultiplied by the Householder matrix.

Functions used: matmat, elmrowcol.

```
void hshcoltam(int lr, int ur, int lc, int uc, int i,
                float x, float **u, float **a)
{
   float matmat(int, int, int, int, float **, float **);
   void elmrowcol(int, int, int, int, float **, float **, float);

   for (; lr<=ur; lr++) elmrowcol(lc,uc,lr,i,a,u,matmat(lc,uc,lr,i,a,u)*x);
}
```

F. hshrowtam

Postmultiplies one submatrix by an elementary reflector formed from a row of another: $a=(I-xu^Tu)a$ where $a_{k-lr+1,j-lc+1}=a_{k,j}$ $(k=lr,...,ur;\ \ j=lc,...,uc)$, $u_{k-lc+1}=U_{i,k}$ $(k=lc,...,uc)$.

Function Parameters:

void hshrowtam (*lr,ur,lc,uc,i,x,u,a*)

lr,ur: int; lower and upper row indices;
lc,uc: int; lower and upper column indices;
i: int; the row index of the Householder vector;
x: float; the Householder constant;
u: float *u[i:i,lc:uc]*; the Householder vector;
a: float *a[lr:ur,lc:uc]*; the matrix to be postmultiplied by the Householder matrix.

Functions used: mattam, elmrow.

```
void hshrowtam(int lr, int ur, int lc, int uc, int i,
                float x, float **u, float **a)
{
   float mattam(int, int, int, int, float **, float **);
   void elmrow(int, int, int, int, float **, float **, float);

   for (; lr<=ur; lr++) elmrow(lc,uc,lr,i,a,u,mattam(lc,uc,lr,i,a,u)*x);
}
```

1.7 Real vector and matrix - Elimination

A. elmvec

Adds a constant multiple of part of a vector *b* to part of a vector *a* by setting

$$a_i = a_i+xb_{i+shift} \quad (i=l,l+1,...,u).$$

Function Parameters:

void elmvec (*l,u,shift,a,b,x*)

l,u: int; lower and upper bound of the running subscript;
shift: int; index-shifting parameter of the vector *b*;
a,b: float *a[l:u], b[l+shift:u+shift]*;
x: float; elimination factor.

```
void elmvec(int l, int u, int shift, float a[], float b[], float x)
{
    for (; l<=u; l++) a[l] += b[l+shift]*x;
}
```

B. elmcol

Adds a constant multiple of part of a column of a rectangular matrix *b* to part of a column of a rectangular matrix *a* by setting

$$a_{k,i} = a_{k,i}+xb_{k,j} \quad (k=l,l+1,...,u).$$

Function Parameters:

void elmcol (*l,u,i,j,a,b,x*)

l,u: int; lower and upper bound of the running subscript;
i: int; column-index of *a*;
j: int; column-index of *b*;
a,b: float *a[l:u,i:i], b[l:u,j:j]*;
x: float; elimination factor.

```
void elmcol(int l, int u, int i, int j, float **a, float **b, float x)
{
    for (; l<=u; l++) a[l][i] += b[l][j]*x;
}
```

C. elmrow

Adds a constant multiple of part of a row of a rectangular matrix *b* to part of a row of a rectangular matrix *a* by setting

$$a_{i,k} = a_{i,k}+xb_{j,k} \quad (k=l,l+1,...,u).$$

Function Parameters:

void elmrow (*l,u,i,j,a,b,x*)

l,u: int; lower and upper bound of the running subscript;
i: int; row-index of *a*;
j: int; row-index of *b*;
a,b: float *a[i:i,l:u], b[j:j,l:u]*;
x: float; elimination factor.

```
void elmrow(int l, int u, int i, int j, float **a, float **b, float x)
{
    for (; l<=u; l++) a[i][l] += b[j][l]*x;
}
```

D. elmveccol

Adds a constant multiple of part of a column of a rectangular matrix b to part of a vector a by setting

$$a_k = a_k + x b_{k,i} \quad (k=l,l+1,...,u).$$

Function Parameters:

void elmveccol (l,u,i,a,b,x)

l,u: int; lower and upper bound of the running subscript;
i: int; column-index of b;
a,b: float $a[l:u]$, $b[l:u,i:i]$;
x: float; elimination factor.

```
void elmveccol(int l, int u, int i, float a[], float **b, float x)
{
    for (; l<=u; l++) a[l] += b[l][i]*x;
}
```

E. elmcolvec

Adds a constant multiple of part of a vector b to part of a column of a rectangular matrix a by setting

$$a_{k,i} = a_{k,i} + x b_k \quad (k=l,l+1,...,u).$$

Function Parameters:

void elmcolvec (l,u,i,a,b,x)

l,u: int; lower and upper bound of the running subscript;
i: int; column-index of a;
a,b: float $a[l:u,i:i]$, $b[l:u]$;
x: float; elimination factor.

```
void elmcolvec(int l, int u, int i, float **a, float b[], float x)
{
    for (; l<=u; l++) a[l][i] += b[l]*x;
}
```

F. elmvecrow

Adds a constant multiple of part of a row of a rectangular matrix b to the corresponding part of a vector a by setting

$$a_k = a_k + x b_{i,k} \quad (k=l,l+1,...,u).$$

Function Parameters:

void elmvecrow (l,u,i,a,b,x)

l,u: int; lower and upper bound of the running subscript;
i: int; row-index of b;
a,b: float $a[l:u]$, $b[i:i,l:u]$;
x: float; elimination factor.

```
void elmvecrow(int l, int u, int i, float a[], float **b, float x)
{
    for (; l<=u; l++) a[l] += b[i][l]*x;
}
```

G. elmrowvec

Adds a constant multiple of part of a vector b to the corresponding part of a row of a rectangular matrix a by setting

$$a_{i,k} = a_{i,k}+xb_k \quad (k=l,l+1,...,u).$$

Function Parameters:

void elmrowvec (l,u,i,a,b,x)

l,u: int; lower and upper bound of the running subscript;
i: int; row-index of a;
a,b: float $a[i:i,l:u]$, $b[l:u]$;
x: float; elimination factor.

```
void elmrowvec(int l, int u, int i, float **a, float b[], float x)
{
	for (; l<=u; l++) a[i][l] += b[l]*x;
}
```

H. elmcolrow

Adds a constant multiple of part of a row of a rectangular matrix b to the corresponding part of a column of a rectangular matrix a by setting

$$a_{k,i} = a_{k,i}+xb_{j,k} \quad (k=l,l+1,...,u).$$

Function Parameters:

void elmcolrow (l,u,i,j,a,b,x)

l,u: int; lower and upper bound of the running subscript;
i: int; column-index of a;
j: int; row-index of b;
a,b: float $a[l:u,i:i]$, $b[j:j,l:u]$, when $a=b$ then correct elimination is guaranteed only when the row and column are disjunct;
x: float; elimination factor.

```
void elmcolrow(int l, int u, int i, int j, float **a, float **b, float x)
{
	for (; l<=u; l++) a[l][i] += b[j][l]*x;
}
```

I. elmrowcol

Adds a constant multiple of part of a column of a rectangular matrix b to the corresponding part of a row of a rectangular matrix a by setting

$$a_{i,k} = a_{i,k}+xb_{k,j} \quad (k=l,l+1,...,u).$$

Function Parameters:

void elmrowcol (l,u,i,j,a,b,x)

l,u: int; lower and upper bound of the running subscript;
i: int; row-index of a;

j: int; column-index of *b*;
a,b: float *a[i:i,l:u], b[l:u,j:j]*, when *a=b* then correct elimination is guaranteed only when the row and column are disjunct;
x: float; elimination factor.

```
void elmrowcol(int l, int u, int i, int j, float **a, float **b, float x)
{
    for (; l<=u; l++) a[i][l] += b[l][j]*x;
}
```

J. maxelmrow

Performs the same operations as *elmrow* and, in addition, sets *maxelmrow=l* if $l > u$ and otherwise *maxelmrow=m* where *m* is that value of *k* for which $|a_{i,k}|$ is a maximum over $k=l,l+1,...,u$.

Function Parameters:

int maxelmrow (*l,u,i,j,a,b,x*)

maxelmrow: delivers the index of the maximal element after elimination step upon *a*;
l,u: int; lower and upper bound of the running subscript;
i: int; row-index of *a*;
j: int; row-index of *b*;
a,b: float *a[i:i,l:u], b[i:i,l:u]*;
x: float; elimination factor.

```
#include <math.h>

int maxelmrow(int l, int u, int i, int j, float **a, float **b, float x)
{
    int k;
    float r, s;

    s=0.0;
    for (k=l; k<=u; k++) {
        r=(a[i][k] += b[j][k]*x);
        if (fabs(r) > s) {
            s=fabs(r);
            l=k;
        }
    }
    return (l);
}
```

1.8 Real vector and matrix - Interchanging

A. ichvec

Interchanges two parts of the same vector *a* by setting

$$a_i \leftrightarrow a_{i+shift} \quad (i=l,l+1,...,u).$$

Function Parameters:

void ichvec (*l,u,shift,a*)

l,u: int; lower and upper bound of the running subscript;
shift: int; index-shifting parameter;

a: float $a[p:q]$; p and q should satisfy: $p \le l$, $q \ge u$, $p \le l+shift$, $q \ge u+shift$.

```
void ichvec(int l, int u, int shift, float a[])
{
    float r;

    for (; l<=u; l++) {
        r=a[l];
        a[l]=a[l+shift];
        a[l+shift]=r;
    }
}
```

B. ichcol

Interchanges corresponding parts of two columns of the same rectangular matrix a by setting

$$a_{k,i} \leftrightarrow a_{k,j} \ (k=l,l+1,...,u).$$

Function Parameters:

void ichcol (l,u,i,j,a)

l,u: int; lower and upper bound of the running subscript;
i,j: int; column-indices of a;
a: float $a[l:u,p:q]$; p and q should satisfy: $p \le i$, $p \le j$, $q \ge i$, $q \ge j$.

```
void ichcol(int l, int u, int i, int j, float **a)
{
    float r;

    for (; l<=u; l++) {
        r=a[l][i];
        a[l][i]=a[l][j];
        a[l][j]=r;
    }
}
```

C. ichrow

Interchanges corresponding parts of two rows of the same rectangular matrix a by setting

$$a_{i,k} \leftrightarrow a_{j,k} \ (k=l,l+1,...,u).$$

Function Parameters:

void ichrow (l,u,i,j,a)

l,u: int; lower and upper bound of the running subscript;
i,j: int; row-indices of a;
a: float $a[p:q,l:u]$; p and q should satisfy: $p \le i$, $p \le j$, $q \ge i$, $q \ge j$;

```
void ichrow(int l, int u, int i, int j, float **a)
{
    float r;

    for (; l<=u; l++) {
        r=a[i][l];
        a[i][l]=a[j][l];
        a[j][l]=r;
    }
}
```

D. ichrowcol

Interchanges corresponding parts of a row and a column of the same rectangular matrix a by setting

$$a_{i,k} \leftrightarrow a_{k,j} \quad (k=l,l+1,...,u).$$

Function Parameters:

void ichrowcol (l,u,i,j,a)

l,u: int; lower and upper bound of the running subscript;
i: int; row-index of a;
j: int; column-index of a;
a: float $a[p:q,r:s]$; p, q, r and s should satisfy: $p \leq i$, $p \leq l$, $q \geq i$, $q \geq u$, $r \leq j$, $r \leq l$, $s \geq j$ and $s \geq u$, furthermore the row and column to be interchanged should be disjoint.

```
void ichrowcol(int l, int u, int i, int j, float **a)
{
    float r;

    for (; l<=u; l++) {
        r=a[i][l];
        a[i][l]=a[l][j];
        a[l][j]=r;
    }
}
```

E. ichseqvec

Interchanges two parts of the same vector by setting

$$a_{il+(j+l-1)(j-l)/2} \leftrightarrow a_{j+shift} \quad (j=l,l+1,...,u).$$

Function Parameters:

void ichseqvec $(l,u,il,shift,a)$

l,u: int; lower and upper bound of the running subscript;
il: int; lower bound of the vector a;
$shift$: int; index-shifting parameter;
a: float $a[p:q]$; it is assumed that the values of $l+shift$, $u+shift$ and $il+(u+l-1)(u-l)/2$ are not out of range.

```
void ichseqvec(int l, int u, int il, int shift, float a[])
{
    float r;

    for (; l<=u; l++) {
        r=a[il];
        a[il]=a[l+shift];
        a[l+shift]=r;
        il += l;
    }
}
```

F. ichseq

Interchanges two parts of the same vector by setting

$$a_{il+(j+l-1)(j-l)/2} \leftrightarrow a_{shift+il+(j+l-1)(j-l)/2} \quad (j=l,l+1,...,u).$$

Function Parameters:

void ichseq $(l,u,il,shift,a)$

l,u: int; lower and upper bound of the running subscript;
il: int; lower bound of the vector a;
$shift$: int; index-shifting parameter;
a: float $a[p:q]$; it is assumed that the values of $il+shift+(j+l-1)(j-l)/2$, $j=l,...,u$, do not go out of range.

```
void ichseq(int l, int u, int il, int shift, float a[])
{
    float r;

    for (; l<=u; l++) {
        r=a[il];
        a[il]=a[il+shift];
        a[il+shift]=r;
        il += l;
    }
}
```

1.9 Real vector and matrix - Rotation

A. rotcol

Rotates two columns of a rectangular matrix by setting

$$a_{k,i} = ca_{k,i} + sa_{k,j}, \qquad a_{k,j} = ca_{k,j} - sa_{k,i} \qquad (k=l,l+1,...,u).$$

Function Parameters:

void rotcol (l,u,i,j,a,c,s)

l,u: int; lower and upper bound of the running subscript;
i,j: int; column-indices of the array a;
a: float $a[l:u,p:q]$; p and q should satisfy: $p \le i$, $p \le j$, $q \ge i$ and $q \ge j$;
c,s: float; rotation factors.

```
void rotcol(int l, int u, int i, int j, float **a, float c, float s)
{
    float x, y;

    for (; l<=u; l++) {
        x=a[l][i];
        y=a[l][j];
        a[l][i]=x*c+y*s;
        a[l][j]=y*c-x*s;
    }
}
```

B. rotrow

Rotates two rows of a rectangular matrix by setting

$$a_{i,k} = ca_{i,k} + sa_{j,k}, \qquad a_{j,k} = ca_{j,k} - sa_{i,k} \qquad (k=l,l+1,...,u).$$

Function Parameters:

void rotrow (l,u,i,j,a,c,s)

l,u: int; lower and upper bound of the running subscript;

i,j: int; row-indices of the array *a*;
a: float *a[p:q,l:u]*; *p* and *q* should satisfy: $p \leq i$, $p \leq j$, $q \geq i$ and $q \geq j$;
c,s: float; rotation factors.

```
void rotrow(int l, int u, int i, int j, float **a, float c, float s)
{
    float x, y;

    for (; l<=u; l++) {
        x=a[i][l];
        y=a[j][l];
        a[i][l]=x*c+y*s;
        a[j][l]=y*c-x*s;
    }
}
```

1.10 Real vector and matrix - Norms

A. infnrmvec

Computes the ∞-norm of part of a vector and determines the index of the element with largest absolute value ($\rho = \max|a_i|$ $(l \leq i \leq u)$ and the smallest k such that $|a_k|=\rho$ $(l \leq k \leq u)$).

Function Parameters:

float infnrmvec (*l,u,k,a*)

infnrmvec: delivers the ∞-norm of the vector *a*, the value of ρ above;
l,u: int; lower and upper bound of the index of the vector *a*, respectively;
k: int *; exit: the value of *k* above;
a: float *a[l:u]*.

```
#include <math.h>

float infnrmvec(int l, int u, int *k, float a[])
{
    float r, max;

    max=0.0;
    *k=l;
    for (; l<=u; l++) {
        r=fabs(a[l]);
        if (r > max) {
            max=r;
            *k=l;
        }
    }
    return (max);
}
```

B. infnrmrow

Computes the ∞-norm of part of a row of a rectangular matrix, and determines the index of the element with largest absolute value ($\rho = \max|a_{i,j}|$ $(l \leq j \leq u)$ and the smallest k such that $|a_{i,k}|=\rho$ $(l \leq k \leq u)$).

Function Parameters:

float infnrmrow (*l,u,i,k,a*)

infnrmrow: delivers the ∞-norm of the row vector of *a*, the value of ρ above;

l,u: int; lower and upper bound of the column index of *a*, respectively;
i: int; the row index;
k: int *; exit: the value of *k* above;
a: float *a[i:i,l:u]*.

```
#include <math.h>

float infnrmrow(int l, int u, int i, int *k, float **a)
{
    float r, max;

    max=0.0;
    *k=l;
    for (; l<=u; l++) {
        r=fabs(a[i][l]);
        if (r > max) {
            max=r;
            *k=l;
        }
    }
    return (max);
}
```

C. infnrmcol

Computes the ∞-norm of part of a column of a rectangular matrix, and determines the index of the element with largest absolute value ($\rho = \max|a_{i,j}|$ $(l \le i \le u)$ and the smallest k such that $|a_{k,j}| = \rho$ $(l \le k \le u)$).

Function Parameters:

float infnrmcol (*l,u,j,k,a*)

infnrmcol: delivers the ∞-norm of the column vector of *a*, the value of ρ above;
l,u: int; lower and upper bound of the row index of *a*, respectively;
j: int; the column index;
k: int *; exit: the value of *k* above;
a: float *a[l:u,j:j]*.

```
#include <math.h>

float infnrmcol(int l, int u, int j, int *k, float **a)
{
    float r, max;

    max=0.0;
    *k=l;
    for (; l<=u; l++) {
        r=fabs(a[l][j]);
        if (r > max) {
            max=r;
            *k=l;
        }
    }
    return (max);
}
```

D. infnrmmat

Computes the ∞-norm of a submatrix and determines the index of the row whose 1-norm is equal to the ∞-norm of the submatrix:

$$\rho = \max \sum_{j=lc}^{uc} |a_{i,j}|, \quad lr \le i \le ur,$$

and the smallest *kr* such that

$$\rho = \sum_{j=lc}^{uc} |a_{kr,j}|.$$

Function Parameters:

float infnrmmat (*lr,ur,lc,uc,kr,a*)

infnrmmat: delivers the ∞-norm of the matrix *a*, the value of ρ above;
lr,ur: int; lower and upper bound of the row index, respectively;
lc,uc: int; lower and upper bound of the column index, respectively;
kr: int *; exit: the value of *kr* above;
a: float *a[lr:ur,lc:uc]*.

Function used: onenrmrow.

```
float infnrmmat(int lr, int ur, int lc, int uc, int *kr, float **a)
{
    float onenrmrow(int, int, int, float **);
    float r, max;

    max=0.0;
    *kr=lr;
    for (; lr<=ur; lr++) {
        r=onenrmrow(lc,uc,lr,a);
        if (r > max) {
            max=r;
            *kr=lr;
        }
    }
    return (max);
}
```

E. onenrmvec

Computes the 1-norm of part of a vector:

$$\rho = \sum_{i=l}^{u} |a_i|.$$

Function Parameters:

float onenrmvec (*l,u,a*)

onenrmvec: delivers the 1-norm of the vector *a*, the value of ρ above;
l,u: int; lower and upper bound of the index of the vector *a*, respectively;
a: float *a[l:u]*.

```
#include <math.h>

float onenrmvec(int l, int u, float a[])
{
    float sum;

    sum=0.0;
    for (; l<=u; l++) sum += fabs(a[l]);
    return (sum);
}
```

F. onenrmrow

Computes the 1-norm of part of a row of a rectangular matrix:

$$\rho = \sum_{j=l}^{u} |a_{i,j}|.$$

Function Parameters:

float onenrmrow (l,u,i,a)

onenrmrow: delivers the 1-norm of a row vector, the value of ρ above;
l,u: int; lower and upper bound of the column index of a, respectively;
i: int; the row index;
a: float $a[i:i,l:u]$.

```
#include <math.h>

float onenrmrow(int l, int u, int i, float **a)
{
    float sum;

    sum=0.0;
    for (; l<=u; l++) sum += fabs(a[i][l]);
    return (sum);
}
```

G. onenrmcol

Computes the 1-norm of part of a column of a rectangular matrix:

$$\rho = \sum_{i=l}^{u} |a_{i,j}|.$$

Function Parameters:

float onenrmcol (l,u,j,a)

onenrmcol: delivers the 1-norm of a column vector, the value of ρ above;
l,u: int; lower and upper bound of the row index of a, respectively;
j: int; the column index;
a: float $a[l:u,j:j]$.

```
#include <math.h>

float onenrmcol(int l, int u, int j, float **a)
{
    float sum;

    sum=0.0;
    for (; l<=u; l++) sum += fabs(a[l][j]);
    return (sum);
}
```

H. onenrmmat

Computes the 1-norm of a submatrix and determines the index of the column whose 1-norm is that of the submatrix:

$$\rho = \max \sum_{i=lr}^{ur} |a_{i,j}|, \quad lc \le j \le uc,$$

and the smallest *kc* such that

$$\rho = \sum_{i=lr}^{ur} |a_{i,kc}|.$$

Function Parameters:

float onenrmmat (lr,ur,lc,uc,kc,a)

onenrmmat: delivers the 1-norm of the matrix a, the value of ρ above;
lr,ur: int; lower and upper bound of the row index, respectively;
lc,uc: int; lower and upper bound of the column index, respectively;
kc: int *; exit: the value of *kc* above;
a: float $a[lr:ur,lc:uc]$.

Function used: onenrmcol.

```
float onenrmmat(int lr, int ur, int lc, int uc, int *kc, float **a)
{
    float onenrmcol(int l, int u, int j, float **a);
    float r, max;

    max=0.0;
    *kc=lc;
    for (; lc<=uc; lc++) {
        r=onenrmcol(lr,ur,lc,a);
        if (r > max) {
            max=r;
            *kc=lc;
        }
    }
    return (max);
}
```

I. absmaxmat

Computes the absolute value of an element with largest absolute value belonging to a submatrix, and the row and column indices of the element in question ($\rho = \max|a_{k,h}|$, ($lr \leq k \leq ur$, $lc \leq h \leq uc$) and the smallest i for the smallest j such that $\rho = |a_{i,j}|$).

Function Parameters:

float absmaxmat (*lr,ur,lc,uc,i,j,a*)

absmaxmat: delivers for a given matrix *a* the modulus of an element which is of maximum absolute value, the value of ρ above;

lr,ur: int; lower and upper bound of the row index, respectively;

lc,uc: int; lower and upper bound of the column index, respectively;

i,j: int *; exit: the row and column index of an element for which the modulus is maximal;

a: float *a[lr:ur,lc:uc]*.

Function used: infnrmcol.

```
float absmaxmat(int lr, int ur, int lc, int uc, int *i, int *j, float **a)
{
    float infnrmcol(int, int, int, int *, float **);
    int ii;
    float r, max;

    max=0.0;
    *i=lr;
    *j=lc;
    for (; lc<=uc; lc++) {
        r=infnrmcol(lr,ur,lc,&ii,a);
        if (r > max) {
            max=r;
            *i=ii;
            *j=lc;
        }
    }
    return (max);
}
```

1.11 Real vector and matrix - Scaling

reascl

Normalizes the (non-null) columns of a matrix in such a way that, in each column, an element of maximum absolute value equals 1. The normalized vectors are delivered in the corresponding columns of the matrix.

Function Parameters:

void reascl (*a,n,n1,n2*)

a: float *a[1:n,n1:n2]*;

entry: the *n2-n1+1* column vectors must be given in *a*;

exit: the normalized vectors (i.e. in each vector an element of maximum absolute value equals 1) are delivered in the corresponding columns of *a*;

n: int; the number of rows of *a*;

n1,n2: int; the lower and upper bound of the column indices of *a*.

```
#include <math.h>

void reascl(float **a, int n, int n1, int n2)
{
```

```
    int i, j;
    float s;

    for (j=n1; j<=n2; j++) {
        s=0.0;
        for (i=1; i<=n; i++)
            if (fabs(a[i][j]) > fabs(s)) s=a[i][j];
        if (s != 0.0)
            for (i=1; i<=n; i++) a[i][j] /= s;
    }
}
```

1.12 Complex vector and matrix - Multiplication

A. comcolcst

Replaces part of a column of a matrix with complex elements by a constant complex multiple of that part:

$$a_{i,j}=xa_{i,j},\ i=l,...,u.$$

Function Parameters:

void comcolcst (l,u,j,ar,ai,xr,xi)

l,u: int; lower and upper bound of the column vector;
j: int; column-index of the column vector;
ar,ai: float *ar[l:u,j:j], ai[l:u,j:j]*;
entry: ar: real part;
ai: imaginary part of the column vector;
exit: the transformed complex column;
xr,xi: float;
entry: xr: real part of the multiplication factor;
xi: imaginary part of the multiplication factor.

Function used: commul.

```
void comcolcst(int l, int u, int j,
               float **ar, float **ai, float xr, float xi)
{
    void commul(float, float, float, float, float *, float *);

    for (; l<=u; l++)
        commul(ar[l][j],ai[l][j],xr,xi,&ar[l][j],&ai[l][j]);
}
```

B. comrowcst

Replaces part of a row of a matrix with complex elements by a constant complex multiple of that part:

$$a_{i,j}=xa_{i,j},\ j=l,...,u.$$

Function Parameters:

void comrowcst (l,u,i,ar,ai,xr,xi)

l,u: int; lower and upper bound of the row vector;
j: int; row-index of the row vector;

ar,ai: float *ar[i:i,l:u], ai[i:i,l:u]*;
entry: ar: real part;
ai: imaginary part of the row vector;
exit: the transformed complex row;
xr,xi: float;
entry: xr: real part of the multiplication factor;
xi: imaginary part of the multiplication factor.

Function used: commul.

```
void comrowcst(int l, int u, int i,
                    float **ar, float **ai, float xr, float xi)
{
    void commul(float, float, float, float, float *, float *);

    for (; l<=u; l++)
        commul(ar[i][l],ai[i][l],xr,xi,&ar[i][l],&ai[i][l]);
}
```

1.13 Complex vector and matrix - Scalar products

A. commatvec

Forms the inner product of part of a row of a matrix and part of a vector, both having complex elements:

$$r = \sum_{j=l}^{u} a_{i,j} b_j.$$

Function Parameters:

void commatvec (*l,u,i,ar,ai,br,bi,rr,ri*)

l,u: int; lower and upper bound of the vectors;
i: int; the row index of *ar* and *ai*;
ar,ai: float *ar[i:i,l:u],ai[i:i,l:u]*;
entry: ar: real part and
ai: imaginary part of the matrix;
br,bi: float *br[l:u],bi[l:u]*;
entry: br: real part and
bi: imaginary part of the vector;
rr,ri: float *;
exit: rr: the real part and
ri: the imaginary part of the scalar product.

Function used: matvec.

```
void commatvec(int l, int u, int i, float **ar, float **ai,
                    float br[], float bi[], float *rr, float *ri)
{
    float matvec(int, int, int, float **, float []);
    float mv;
```

```
    mv=matvec(l,u,i,ar,br)-matvec(l,u,i,ai,bi);
    *ri=matvec(l,u,i,ai,br)+matvec(l,u,i,ar,bi);
    *rr=mv;
}
```

B. hshcomcol

Given the components $a_{i,j}$ $(i=l+1,...,n)$ of the j-th column of the complex valued matrix a, and a real tolerance tol, and if

$$\sum_{i=l+1}^{u} |a_{i,j}|^2 > tol \qquad (1)$$

this function determines a complex vector c such that, the j-th column of

$$a = (I-2c\bar{c}^T/(\bar{c}^Tc))a \qquad \text{(where } \bar{c}^T \text{ denotes conjugating and transposing)}$$

has zero elements in the second,..., last positions, and computes

$$t = (\bar{c}^Tc)/2$$

and the first component of c.

Function Parameters:

int hshcomcol $(l,u,j,ar,ai,tol,k,c,s,t)$

hshcomcol: if the condition (1) above is satisfied then a transformation is performed and *hshcomcol*=1, otherwise *hshcomcol*=0.

l,u,j: int; the complex vector to be transformed, must be given in the *j*-th column from row *l* until row *u* of a complex matrix;

ar,ai: float *ar[l:u,j:j],ai[l:u,j:j]*;
- entry: the real part and the imaginary part of the vector to be transformed must be given in the arrays *ar* and *ai*, respectively;
- exit: if the condition (1) above is satisfied then the real part and the imaginary part of the vector *c* of the Householder matrix are delivered in the arrays *ar* and *ai*, respectively, otherwise the arrays *ar* and *ai* are unaltered;

tol: float;
- entry: a tolerance (for example, the square of the machine precision times a norm of the matrix under consideration);

k,c,s: float *;
- exit: if the condition (1) above is satisfied then the modulus, cosine and sine of the argument of the first element of the transformed vector are delivered in *k*, *c* and *s*, respectively, otherwise the modulus, cosine and sine of the argument of the complex number *ar[l,j]+ai[l,j]*I* are delivered;

t: float *;
- exit: if the condition (1) above is satisfied then it has the value *t* in the above, otherwise it has the value -1.

Functions used: carpol, tammat.

```
#include <math.h>

int hshcomcol(int l, int u, int j, float **ar, float **ai, float tol,
                    float *k, float *c, float *s, float *t)
{
    void carpol(float, float, float *, float *, float *);
    float tammat(int, int, int, int, float **, float **);
    float vr, mod, h, arlj, ailj;

    vr=tammat(l+1,u,j,j,ar,ar)+tammat(l+1,u,j,j,ai,ai);
    arlj=ar[l][j];
    ailj=ai[l][j];
    carpol(arlj,ailj,&mod,c,s);
    if (vr > tol) {
        vr += arlj*arlj+ailj*ailj;
        h = *k = sqrt(vr);
        *t=vr+mod*h;
        if (arlj == 0.0 && ailj == 0.0)
            ar[l][j]=h;
        else {
            ar[l][j]=arlj + *c * *k;
            ai[l][j]=ailj + *s * *k;
            *s = - *s;
        }
        *c = - *c;
        return (1);
    } else {
        *k=mod;
        *t = -1.0;
        return (0);
    }
}
```

C. hshcomprd

Given the components of the complex vector c, and

$$t = (c\bar{c}^T)/2 \qquad \text{(where } \bar{c}^T \text{ denotes conjugating and transposing)}$$

and the submatrix a composed of the elements $a_{k,h}$ ($k=i,...,ii;\ h=l,...,u$) of the complex matrix a, forms the product

$$a = (I - t\bar{c}^T c)a \ ,$$

it being known that c is such that the elements in rows $l+1,...,u$ of the j-th column of a are zero.

Function Parameters:

void hshcomprd (*i,ii,l,u,j,ar,ai,br,bi,t*)

i,ii,l,u: int; the complex matrix to be premultiplied must be given in the l-th to u-th column from row i to row ii of a complex matrix;

j: int; the complex vector c of the Householder matrix must be given in the j-th column from row i to row ii of a complex matrix given in (*br,bi*).

ar,ai: float *ar[i:ii,l:u],ai[i:ii,l:u]*;
- entry: the real part and the imaginary part of the vector to be premultiplied must be given in the arrays *ar* and *ai*, respectively;
- exit: the real part and the imaginary part of the resulting matrix are delivered in the arrays *ar* and *ai*, respectively, otherwise the arrays *ar* and *ai* are unaltered;

br,bi: float *br[i:ii,j:j], bi[i:ii,j:j]*;

entry: the real part and the imaginary part of the complex vector *c* of the Householder matrix must be given in the arrays *br* and *bi*, respectively;

t: float;

entry: the scalar *t* above of the Householder matrix (for example, as delivered by *hshcomcol*).

Functions used: tammat, elmcomcol.

```
void hshcomprd(int i, int ii, int l, int u, int j, float **ar,
               float **ai, float **br, float **bi, float t)
{
   void elmcomcol(int, int, int, int, float **, float **,
                  float **, float **, float, float);
   float tammat(int, int, int, int, float **, float **);

   for (; l<=u; l++)
      elmcomcol(i,ii,l,j,ar,ai,br,bi,
                (-tammat(i,ii,j,l,br,ar)-tammat(i,ii,j,l,bi,ai))/t,
                (tammat(i,ii,j,l,bi,ar)-tammat(i,ii,j,l,br,ai))/t);
}
```

1.14 Complex vector and matrix - Elimination

A. elmcomveccol

Adds a constant complex multiple of part of a column of a rectangular matrix to part of a vector, both having complex elements: $a_i=a_i+xb_{i,j}$, $i=l,...,u$.

Function Parameters:

void elmcomveccol (*l,u,j,ar,ai,br,bi,xr,xi*)

l,u: int; lower and upper bounds of the vectors;

j: int; column-index of *br* and *bi*;

ar,ai: float *ar[l:u], ai[l:u]*;

entry: ar: real part and
ai: imaginary part of the vector;

exit: the resulting vector (adds *xr+xi*i* times the complex column vector given in arrays *br* and *bi* to the complex vector given in arrays *ar* and *ai*);

br,bi: float *br[l:u,j:j], bi[l:u,j:j]*;

entry: br: real part and
bi: imaginary part of the column vector;

xr,xi: float;

entry: xr: real part and
xi: imaginary part of the elimination factor.

Function used: elmveccol.

```
void elmcomveccol(int l, int u, int j, float ar[], float ai[],
                  float **br, float **bi, float xr, float xi)
{
   void elmveccol(int, int, int, float [], float **, float);

   elmveccol(l,u,j,ar,br,xr);
   elmveccol(l,u,j,ar,bi,-xi);
   elmveccol(l,u,j,ai,br,xi);
   elmveccol(l,u,j,ai,bi,xr);
```

}

B. elmcomcol

Adds a constant complex multiple of part of a column of one complex matrix to part of a column of another: $a_{k,i}=a_{k,i}+(xr+xi*i)b_{k,j},\ k=l,...,u.$

Function Parameters:

void elmcomcol (*l,u,i,j,ar,ai,br,bi,xr,xi*)

l,u: int; lower and upper bounds of the vectors;
i,j: int;
i: column-index of *ar* and *ai*;
j: column-index of *br* and *bi*;
ar,ai: float *ar[l:u,i:i], ai[l:u,i:i]*;
entry: ar: real part and
ai: imaginary part of the column vector;
exit: the resulting vector (adds *xr+xi*i* times the complex column vector given in arrays *br* and *bi* to the complex column vector given in arrays *ar* and *ai*);
br,bi: float *br[l:u,j:j], bi[l:u,j:j]*;
entry: br: real part and
bi: imaginary part of the column vector;
xr,xi: float;
entry: xr: real part and
xi: imaginary part of the elimination factor.

Function used: elmcol.

```
void elmcomcol(int l, int u, int i, int j, float **ar, float **ai,
               float **br, float **bi, float xr, float xi)
{
   void elmcol(int, int, int, int, float **, float **, float);

   elmcol(l,u,i,j,ar,br,xr);
   elmcol(l,u,i,j,ar,bi,-xi);
   elmcol(l,u,i,j,ai,br,xi);
   elmcol(l,u,i,j,ai,bi,xr);
}
```

C. elmcomrowvec

Adds a constant complex multiple of part of a vector to part of a row of a rectangular matrix, both having complex elements: $a_{i,j}=a_{i,j}+xb_j,\ j=l,...,u.$

Function Parameters:

void elmcomrowvec (*l,u,i,ar,ai,br,bi,xr,xi*)

l,u: int; lower and upper bounds of the vectors;
i: int; row-index of *ar* and *ai*;
ar,ai: float *ar[i:i,l:u], ai[i:i,l:u]*;
entry: ar: real part and
ai: imaginary part of the row vector;
exit: the resulting vector (adds *xr+xi*i* times the complex vector given in arrays *br*

and *bi* to the complex row vector given in arrays *ar* and *ai*);

br,bi: float *br[l:u], bi[l:u]*;
entry: br: real part and
bi: imaginary part of the vector;

xr,xi: float;
entry: xr: real part and
xi: imaginary part of the elimination factor.

Function used: elmrowvec.

```
void elmcomrowvec(int l, int u, int i, float **ar, float **ai,
                  float br[], float bi[], float xr, float xi)
{
    void elmrowvec(int, int, int, float **, float [], float);

    elmrowvec(l,u,i,ar,br,xr);
    elmrowvec(l,u,i,ar,bi,-xi);
    elmrowvec(l,u,i,ai,br,xi);
    elmrowvec(l,u,i,ai,bi,xr);
}
```

1.15 Complex vector and matrix - Rotation

A. rotcomcol

Rotates two columns of a matrix with complex elements:

$$a_{k,i} = (cr+ci*i)a_{k,i} + sa_{k,j}, \qquad a_{k,j} = (cr+ci*i)a_{k,i} - sa_{k,j} \qquad (k=l,...,u).$$

Function Parameters:

void rotcomcol (*l,u,i,j,ar,ai,cr,ci,s*)

l,ui,j: int; the rotation is performed on the column vectors *ar[l:u,i:i], ai[l:u,i:i]* and *ar[l:u,j:j], ai[l:u,j:j]*;

ar,ai: float *ar[l:u,i:j], ai[l:u,i:j]*;
entry: ar: real part and
ai: imaginary part of the column vectors;
exit: the resulting vectors;

cr,ci,s: float; rotation factors.

```
void rotcomcol(int l, int u, int i, int j, float **ar, float **ai,
               float cr, float ci, float s)
{
    float arli,aili,arlj,ailj;

    for (; l<=u; l++) {
        arli=ar[l][i];
        aili=ai[l][i];
        arlj=ar[l][j];
        ailj=ai[l][j];
        ar[l][i]=cr*arli+ci*aili-s*arlj;
        ai[l][i]=cr*aili-ci*arli-s*ailj;
        ar[l][j]=cr*arlj-ci*ailj+s*arli;
        ai[l][j]=cr*ailj+ci*arlj+s*aili;
    }
}
```

B. rotcomrow

Rotates two rows of a matrix with complex elements:

$$a_{i,k} = (cr+ci*i)a_{i,k} + sa_{j,k}, \qquad a_{j,k} = (cr+ci*i)a_{j,k} - sa_{i,k} \qquad (k=l,...,u).$$

Function Parameters:

void rotcomrow (*l,u,i,j,ar,ai,cr,ci,s*)

l,ui,j: int; the rotation is performed on the row vectors *ar[i:i,l:u]*, *ai[i:i,l:u]* and *ar[j:j,l:u]*, *ai[j:j,l:u]*;

ar,ai: float *ar[i:j,l:u]*, *ai[i:j,l:u]*;
 entry: ar: real part and
 ai: imaginary part of the row vectors;
 exit: the resulting vectors;

cr,ci,s: float; rotation factors.

```
void rotcomrow(int l, int u, int i, int j, float **ar, float **ai,
                    float cr, float ci, float s)
{
    float aril,aiil,arjl,aijl;

    for (; l<=u; l++) {
        aril=ar[i][l];
        aiil=ai[i][l];
        arjl=ar[j][l];
        aijl=ai[j][l];
        ar[i][l]=cr*aril+ci*aiil+s*arjl;
        ai[i][l]=cr*aiil-ci*aril+s*aijl;
        ar[j][l]=cr*arjl-ci*aijl-s*aril;
        ai[j][l]=cr*aijl+ci*arjl-s*aiil;
    }
}
```

C. chsh2

Computes the complex Householder matrix that maps the complex vector (a_1,a_2) into the direction (1,0): given complex a_1 and a_2, it determines real c and complex s such that

$$\begin{pmatrix} -c & s \\ s & c \end{pmatrix}\begin{pmatrix} a_1 \\ a_2 \end{pmatrix} = \begin{pmatrix} z \\ 0 \end{pmatrix}$$

where z is nonzero if $|a_1|+|a_2|\neq 0$.

Function Parameters:

void chsh2 (*a1r,a1i,a2r,a2i,c,sr,si*)

a1r: float; entry: the real part of the first vector component a_1 above;

a1i: float; entry: the imaginary part of the first vector component a_1 above;

a2r: float; entry: the real part of the second vector component a_2 above;

a2i: float; entry: the imaginary part of the second vector component a_2 above;

c: float *; exit: the value of *c* above;

sr,si: float *; exit: the real part and imaginary part of *s* above.

```
#include <math.h>

void chsh2(float a1r, float a1i, float a2r, float a2i,
               float *c, float *sr, float *si)
{
```

```
    float r;

    if (a2r != 0.0 || a2i != 0.0) {
        if (a1r != 0.0 || a1i != 0.0) {
            r=sqrt(a1r*a1r+a1i*a1i);
            *c=r;
            *sr=(a1r*a2r+a1i*a2i)/r;
            *si=(a1r*a2i-a1i*a2r)/r;
            r=sqrt(*c * *c + *sr * *sr + *si * *si);
            *c /= r;
            *sr /= r;
            *si /= r;
        } else {
            *si = *c = 0.0;
            *sr=1.0;
        }
    } else {
        *c=1.0;
        *sr = *si = 0.0;
    }
}
```

1.16 Complex vector and matrix - Norms

comeucnrm

Computes the Euclidean norm

$$\rho = \sqrt{\sum_{i=1}^{n}\sum_{j=1}^{n} |a_{i,j}|^2}$$

of the $n \times n$ complex band matrix a for which $a_{i,j}=0$ when $|i-j|>lw$.

Function Parameters:

float comeucnrm (*ar,ai,lw,n*)

comeucnrm: float; delivers the Euclidean norm of matrix a with lw codiagonals, value of ρ above;

n: int; the order of the matrix;

lw: int; the number of lower codiagonals;

ar,ai: float *ar[1:n,1:n],ai[1:n,1:n]*; the real part and imaginary part of the complex matrix with lw lower codiagonals.

Function used: mattam.

```
#include <math.h>

float comeucnrm(float **ar, float **ai, int lw, int n)
{
    float mattam(int, int, int, int, float **, float **);
    int i,l;
    float r;

    r=0.0;
    for (i=1; i<=n; i++) {
        l=(i>lw) ? i-lw : 1;
        r += mattam(l,n,i,i,ar,ar)+mattam(l,n,i,i,ai,ai);
    }
    return (sqrt(r));
}
```

1.17 Complex vector and matrix - Scaling

A. comscl

Scales a sequence of columns, some complex, others pure real of an n-rowed matrix a. The real and imaginary components of the complex columns, and the real elements of the pure real columns are stored in succession in columns $n1$ to $n2$ of the matrix a. The complex and pure real columns of the original matrix may be distinguished by inspection of the values stored in locations $n1$ to $n2$ of the real vector im. In succession, starting with $j_1=n1$, if $im(j_k)=0$, column j_k of a contains the elements of a real column of the input a; if $im(j_k)\neq 0$ then $im(j_k+1)=-im(j_k)$, and columns j_k and j_k+1 of a contain the real and imaginary parts respectively of a complex column of the input a; in the first case $j_{k+1}=j_k+1$, in the second $j_{k+1}=j_k+2$. The columns are scaled by taking $\mu_j=a_{k,j}$ where $|\mu_j|=\max|a_{i,j}|$ $(1\leq i\leq n)$ and $a_{i,j}=a_{i,j}/\mu_j$ $(i=1,...,n)$ where j runs through the column suffices of the stored columns of a.

Function Parameters:

void comscl (*a,n,n1,n2,im*)

a: float *a[1:n,n1:n2]*;
  - entry: each real eigenvector must be given in a column of array *a* whose corresponding element of array *im* equals 0; the real and imaginary part of each complex eigenvector must be given in consecutive columns of array *a* whose corresponding elements of array *im* are not equal to 0;
  - exit: the normalized eigenvectors (i.e. in each eigenvector an element of maximum modulus equals 1) are delivered in the corresponding columns of *a*;

n: int; the number of rows of array *a*;

n1,n2: int; the lower and upper bound of the column indices of array *a*;

im: float *im[n1:n2]*; the imaginary parts of the eigenvalues of which the eigenvectors are given in the corresponding columns of array *a* must be given in array *im*.

```
#include <math.h>

void comscl(float **a, int n, int n1, int n2, float im[])
{
    int i,j,k;
    float s,u,v,w,aij,aij1;

    for (j=n1; j<=n2; j++) {
        s=0.0;
        if (im[j] != 0.0) {
            for (i=1; i<=n; i++) {
                aij=a[i][j];
                aij1=a[i][j+1];
                u=aij*aij+aij1*aij1;
                if (u > s) {
                    s=u;
                    k=i;
                }
            }
            if (s != 0.0) {
                v=a[k][j]/s;
                w = -a[k][j+1]/s;
                for (i=1; i<=n; i++) {
                    u=a[i][j];
                    s=a[i][j+1];
                    a[i][j]=u*v-s*w;
                    a[i][j+1]=u*w+s*v;
                }
```

```
                }
                j++;
            } else {
                for (i=1; i<=n; i++)
                    if (fabs(a[i][j]) > fabs(s)) s=a[i][j];
                if (s != 0.0)
                    for (i=1; i<=n; i++)
                        a[i][j] /= s;
            }
        }
}
```

B. sclcom

Scales the columns *n1* to *n2* of an *n*-rowed matrix *a* with complex elements by dividing the elements in each column by the element of largest modulus in that column (with $\mu_j=a_{k,j}$ where $|\mu_j|=\max|a_{i,j}|$ $(1 \leq i \leq n)$, $a_{i,j}=a_{i,j}/\mu_j$ $(i=1,...,n)$ when $\mu_j \neq 0$; $j=n1,...,n2$)).

Function Parameters:

void sclcom (*ar,ai,n,n1,n2*)

ar,ai: float *ar[1:n,n1:n2], ai[1:n,n1:n2]*;
- entry: the real part and the imaginary part of the matrix of which the columns are to be scaled must be given in the arrays *ar* and *ai*, respectively;
- exit: the real part and imaginary part of the matrix with scaled columns are delivered in the arrays *ar* and *ai*;

n: int; the order of the matrix;

n1,n2: int; the *n1*-th to *n2*-th column vectors are to be scaled.

Function used: comcolcst.

```
void sclcom(float **ar, float **ai, int n, int n1, int n2)
{
    void comcolcst(int, int, int, float **, float **, float, float);
    int i,j,k;
    float s,r,arij,aiij;

    for (j=n1; j<=n2; j++) {
        s=0.0;
        for (i=1; i<=n; i++) {
            arij=ar[i][j];
            aiij=ai[i][j];
            r=arij*arij+aiij*aiij;
            if (r > s) {
                s=r;
                k=i;
            }
        }
        if (s != 0.0) comcolcst(1,n,j,ar,ai,ar[k][j]/s,-ai[k][j]/s);
    }
}
```

1.18 Complex monadic operations

A. comabs

Computes the modulus of a complex number: $\rho = |xr+xi*\mathrm{i}|$.

Function Parameters:

float comabs (*xr,xi*)

comabs: float; delivers the modulus of the complex number, value of ρ above;
xr,xi: float; the real part and imaginary part of the complex number, respectively.

```
#include <math.h>

float comabs(float xr, float xi)
{
    float temp;

    xr=fabs(xr);
    xi=fabs(xi);
    if (xi > xr) {
        temp=xr/xi;
        return (sqrt(temp*temp+1.0)*xi);
    }
    if (xi == 0.0)
        return (xr);
    else {
        temp=xi/xr;
        return (sqrt(temp*temp+1.0)*xr);
    }
}
```

B. comsqrt

Forms the square root of a complex number.

Function Parameters:

void comsqrt (*ar,ai,pr,pi*)

xr,xi: float;
entry: the real part and imaginary part of the complex number, respectively;
pr,pi: float *;
exit: the real part and imaginary part of the square root, respectively.

```
#include <math.h>

void comsqrt(float ar, float ai, float *pr, float *pi)
{
    float br,bi,h,temp;

    if (ar == 0.0 && ai == 0.0)
        *pr = *pi = 0.0;
    else {
        br=fabs(ar);
        bi=fabs(ai);
        if (bi < br) {
            temp=bi/br;
            if (br < 1.0)
                h=sqrt((sqrt(temp*temp+1.0)*0.5+0.5)*br);
            else
                h=sqrt((sqrt(temp*temp+1.0)*0.125+0.125)*br)*2;
        } else {
            if (bi < 1.0) {
                temp=br/bi;
                h=sqrt((sqrt(temp*temp+1.0)*bi+br)*2)*0.5;
            } else {
                if (br+1.0 == 1.0)
                    h=sqrt(bi*0.5);
                else {
                    temp=br/bi;
                    h=sqrt(sqrt(temp*temp+1.0)*bi*0.125+br*0.125)*2;
                }
            }
        }
```

```
        if (ar >= 0.0) {
            *pr=h;
            *pi=ai/h*0.5;
        } else {
            *pi = (ai >= 0.0) ? h : -h;
            *pr = bi/h*0.5;
        }
    }
}
```

C. carpol

Determines the polar form of a complex number: obtains r, $c=\cos\phi$ and $s=\sin\phi$ where

$$ar+ai*\mathrm{i}=re^{\mathrm{i}\phi}.$$

Function Parameters:

void carpol (*ar,ai,r,c,s*)

ar,ai: float;
  entry: the real part and imaginary part of the complex number, respectively;
r,c,s: float *;
  exit: the modulus of the complex number is delivered in *r* and the cosine and the sine of the argument are delivered in *c* and *s*, respectively; when $ar=ai=0$ then $c=1$ and $r=s=0$.

```
#include <math.h>

void carpol(float ar, float ai, float *r, float *c, float *s)
{
    float temp;

    if (ar == 0.0 && ai == 0.0) {
        *c = 1.0;
        *r = *s = 0.0;
    } else {
        if (fabs(ar) > fabs(ai)) {
            temp=ai/ar;
            *r = fabs(ar)*sqrt(1.0+temp*temp);
        } else {
            temp=ar/ai;
            *r = fabs(ai)*sqrt(1.0+temp*temp);
        }
        *c = ar / *r;
        *s = ai / *r;
    }
}
```

1.19 Complex dyadic operations

A. commul

Forms the product of two complex numbers: $rr+ri*\mathrm{i} = (ar+ai*\mathrm{i})(br+bi*\mathrm{i})$.

Function Parameters:

void commul (*ar,ai,br,bi,rr,ri*)

ar,ai: float;
  entry: the real part and imaginary part of the first complex number, respectively;
br,bi: float;

entry: the real part and imaginary part of the second complex number, respectively;
rr,ri: float *;
exit: the real part and imaginary part of the product of the two complex numbers, respectively.

```
void commul(float ar, float ai, float br, float bi, float *rr, float *ri)
{
    *rr=ar*br-ai*bi;
    *ri=ar*bi+ai*br;
}
```

B. comdiv

Forms the quotient of two complex numbers $zr+zi*i = (xr+xi*i)/(yr+yi*i)$. It is assumed that $yr+yi*i \neq 0$.

Function Parameters:

void comdiv (*xr,xi,yr,yi,zr,zi*)

xr,xi: float;
entry: the real part and imaginary part of the first complex number, respectively;
yr,yi: float;
entry: the real part and imaginary part of the second complex number, respectively;
zr,zi: float *;
exit: the real part and imaginary part of the quotient of the two complex numbers, respectively.

```
#include <math.h>

void comdiv(float xr, float xi, float yr, float yi, float *zr, float *zi)
{
    float h,d;

    if (fabs(yi) < fabs(yr)) {
        if (yi == 0.0) {
            *zr=xr/yr;
            *zi=xi/yr;
        } else {
            h=yi/yr;
            d=h*yi+yr;
            *zr=(xr+h*xi)/d;
            *zi=(xi-h*xr)/d;
        }
    } else {
        h=yr/yi;
        d=h*yr+yi;
        *zr=(xr*h+xi)/d;
        *zi=(xi*h-xr)/d;
    }
}
```

1.20 Long integer arithmetic

A. lngintadd

Forms the sum of two multilength integers, each expressed in the form

$$\sum_{j=0}^{i_0} i_j B^j$$

where the i_j are single length nonnegative integers, and B is a single length positive integer:

$$\sum_{j=0}^{s_0} s_j B^j = \sum_{j=0}^{u_0} u_j B^j + \sum_{j=0}^{v_0} v_j B^j,$$

$s_0 \le \max\{u_0,v_0\}+1$; B^2+B not greater than the largest integer having a machine representation.

Function Parameters:

void lngintadd (*u,v,sum*)

u,v: int *u[0:u[0]], v[0:v[0]]*;
  entry: the long integers to be added, values of *u* and *v* above;
sum: int *sum[0,*max(*u[0],v[0]*)+1*]*;
  exit: the multilength sum of *u* and *v*, while *u* and *v* remain unchanged.

Method: see the function *lngintpower*.

```
#define BASE 100        /* value of B in the above */

void lngintadd(int u[], int v[], int sum[])
{
    int lu,lv,diff,carry,i,t,max;

    lu=u[0];
    lv=v[0];
    if (lu >= lv) {
        max=lu;
        diff=lu-lv+1;
        carry=0;
        for (i=lu; i>=diff; i--) {
            t=u[i]+v[i-diff+1]+carry;
            carry = (t < BASE) ? 0 : 1;
            sum[i]=t-carry*BASE;
        }
        for (i=diff-1; i>=1; i--) {
            t=u[i]+carry;
            carry = (t < BASE) ? 0 : 1;
            sum[i]=t-carry*BASE;
        }
    } else {
        max=lv;
        diff=lv-lu+1;
        carry=0;
        for (i=lv; i>=diff; i--) {
            t=v[i]+u[i-diff+1]+carry;
            carry = (t < BASE) ? 0 : 1;
            sum[i]=t-carry*BASE;
        }
        for (i=diff-1; i>=1; i--) {
            t=v[i]+carry;
            carry = (t < BASE) ? 0 : 1;
            sum[i]=t-carry*BASE;
        }
    }
    if (carry == 1) {
        for (i=max; i>=1; i--) sum[i+1]=sum[i];
        sum[1]=1;
        max=max+1;
    }
    sum[0]=max;
}
```

B. lngintsubtract

Forms the difference of two multilength integers, each expressed in the form

$$\sum_{j=1}^{i_0} i_j B^{j-1}$$

where the i_j are single length nonnegative integers, and B is a single length positive integer:

$$\sum_{j=1}^{d_0} d_j B^{j-1} = \sum_{j=1}^{u_0} u_j B^{j-1} - \sum_{j=1}^{v_0} v_j B^{j-1},$$

$d_0 \leq u_0$; B^2+B not greater than the largest integer having a machine representation.

Function Parameters:

void lngintsubtract (*u,v,difference*)

u,v: int *u[0:u[0]], v[0:v[0]]*;
entry: the long integers to be subtracted, values of *u* and *v* above;

difference: int *difference[0,u[0]]*;
exit: the multilength difference *u-v*, if *u*<*v* then *difference[0]=0*; *u* and *v* remain unchanged.

Method: see the function *lngintpower*.

```
#define BASE 100        /* value of B in the above */

void lngintsubtract(int u[], int v[], int difference[])
{
    int lu,lv,diff,i,t,j,carry;

    lu=u[0];
    lv=v[0];
    if ((lu < lv) || ((lu == lv) && (u[1] < v[1]))) {
        difference[0]=0;
        return;
    }
    diff=lu-lv+1;
    carry=0;
    for (i=lu; i>=diff; i--) {
        t=u[i]-v[i-diff+1]+carry;
        carry = (t < 0) ? -1 : 0;
        difference[i]=t-carry*BASE;
    }
    for (i=diff-1; i>=1; i--) {
        t=u[i]+carry;
        carry = (t < 0) ? -1 : 0;
        difference[i]=t-carry*BASE;
    }
    if (carry == -1) {
        difference[0]=0;
        return;
    }
    i=1;
    j=lu;
    while ((difference[i] == 0) && (j > 1)) {
        j--;
        i++;
```

```
    }
    difference[0]=j;
    if (j < lu)
        for (i=1; i<=j; i++) difference[i]=difference[lu+i-j];
}
```

C. lngintmult

Forms the difference of two multilength integers, each expressed in the form

$$\sum_{j=1}^{i_0} i_j B^{j-1}$$

where the i_j are single length nonnegative integers, and B is a single length positive integer:

$$\sum_{j=1}^{d_0} d_j B^{j-1} = \sum_{j=1}^{u_0} u_j B^{j-1} - \sum_{j=1}^{v_0} v_j B^{j-1},$$

$d_0 \leq u_0$; B^2+B not greater than the largest integer having a machine representation.

Function Parameters:

void lngintsubtract (*u,v,difference*)

u,v: int *u[0:u[0]], v[0:v[0]]*;
entry: the long integers to be subtracted, values of *u* and *v* above;

difference: int *difference[0,u[0]]*;
exit: the multilength difference *u-v*, if *u<v* then *difference[0]=0*; *u* and *v* remain unchanged.

Method: see the function *lngintpower*.

```
#define BASE 100        /* value of B in the above */

void lngintmult(int u[], int v[], int product[])
{
    int lu,lv,luv,i,j,carry,t;

    lu=u[0];
    lv=v[0];
    luv=lu+lv;
    for (i=lu+1; i<=luv; i++) product[i]=0;
    for (j=lu; j>=1; j--) {
        carry=0;
        for (i=lv; i>=1; i--) {
            t=u[j]*v[i]+product[j+i]+carry;
            carry=t/BASE;
            product[j+i]=t-carry*BASE;
        }
        product[j]=carry;
    }
    if (product[1] == 0) {
        for (i=2; i<=luv; i++) product[i-1]=product[i];
        luv--;
    }
    product[0]=luv;
}
```

D. lngintdivide

Forms the quotient and remainder from two multilength integers, each expressed in the form

$$\sum_{j=1}^{i_0} i_j B^{j-1}$$

where the i_j are single length nonnegative integers, and B is a single length positive integer:

$$\sum_{j=1}^{u_0} u_j B^{j-1} = \sum_{j=1}^{q_0} q_j B^{j-1} \times \sum_{j=1}^{v_0} v_j B^{j-1} + \sum_{j=1}^{r_0} r_j B^{j-1} ,$$

$q_0=u_0-v_0+1$; $r_0 \le v_0$; B^2+B not greater than the largest integer having a machine representation.

Function Parameters:

void lngintdivide (*u,v,quotient,remainder*)

u,v: int *u[0:u[0]], v[0:v[0]]*;
entry: *u* contains the dividend, *v* the divisor ($v \ne 0$);

quotient,remainder: int *quotient[0:u[0]-v[0]+1], remainder[0:v[0]]*;
exit: results of the division, *u* and *v* remain unchanged.

Method: see the function *lngintpower*.

```
#define BASE 100          /* value of B in the above */

void lngintdivide(int u[], int v[], int quotient[], int remainder[])
{
    int *allocate_integer_vector(int, int);
    void free_integer_vector(int *, int);
    int lu,lv,v1,diff,i,t,scale,d,q1,j,carry,*uu,*a;

    lu=u[0];
    lv=v[0];
    v1=v[1];
    diff=lu-lv;

    if (lv == 1) {
        carry=0;
        for (i=1; i<=lu; i++) {
            t=carry*BASE+u[i];
            quotient[i]=t/v1;
            carry=t-quotient[i]*v1;
        }
        remainder[0]=1;
        remainder[1]=carry;
        if (quotient[1] == 0) {
            for (i=2; i<=lu; i++) quotient[i-1]=quotient[i];
            quotient[0]=lu - ((lu == 1) ? 0 : 1);
        } else
            quotient[0]=lu;
        return;
    }

    if (lu < lv) {
        quotient[0]=1;
        quotient[1]=0;
        for (i=0; i<=lu; i++) remainder[i]=u[i];
        return;
    }
```

```
uu=allocate_integer_vector(0,lu);
a=allocate_integer_vector(0,lv);
for (i=0; i<=lu; i++) uu[i]=u[i];
scale=BASE/(v1+1);
if (scale > 1) {
    /* normalize u */
    carry=0;
    for (i=lu; i>=1; i--) {
        t=scale*uu[i]+carry;
        carry=t/BASE;
        uu[i]=t-carry*BASE;
    }
    uu[0]=carry;
    /* normalize v */
    carry=0;
    for (i=lv; i>=1; i--) {
        t=scale*v[i]+carry;
        carry=t/BASE;
        v[i]=t-carry*BASE;
    }
    v1=v[1];
} else
    uu[0]=0;

/* compute quotient and remainder */
for (i=0; i<=diff; i++) {
    d=uu[i]*BASE+uu[i+1];
    q1 = (uu[i] == v1) ? BASE-1 : d/v1;
    if (v[2]*q1 > (d-q1*v1)*BASE+uu[i+2]) {
        q1--;
        if (v[2]*q1 > (d-q1*v1)*BASE+uu[i+2]) q1--;
    }
    /* uu[i:i+lv]=u[i:i+lv]-q1*v[1:lv] */
    carry=0;
    for (j=lv; j>=1; j--) {
        t=q1*v[j]+carry;
        carry=t/BASE;
        a[j]=t-carry*BASE;
    }
    a[0]=carry;
    carry=0;
    for (j=lv; j>=0; j--) {
        t=uu[i+j]-a[j]+carry;
        carry = (t < 0) ? -1 : 0;
        uu[i+j]=t-carry*BASE;
    }
    /* if carry=-1 then q1 is one too large,
       and v must be added back to uu[i:i+lv] */
    if (carry == -1) {
        q1--;
        carry=0;
        for (j=lv; j>=1; j--) {
            t=uu[i+j]+v[j]+carry;
            carry = (t < BASE) ? 0 :1;
            uu[i+j]=t-carry*BASE;
        }
    }
    quotient[i]=q1;
}

/* correct storage of quotient */
if (quotient[0] != 0) {
    for (i=diff; i>=0; i--) quotient[i+1]=quotient[i];
    quotient[0]=diff+1;
} else
    if (quotient[1] != 0)
        quotient[0]=diff;
    else {
        for (i=1; i<=diff-1; i++) quotient[i]=quotient[i+1];
        quotient[0]=diff-1;
    }

/* remainder=uu[diff+1:lu]/scale */
if (scale > 1) {
```

```
            carry=0;
            for (i=1; i<=lv; i++) {
                t=carry*BASE+uu[diff+i];
                remainder[i]=t/scale;
                carry=t-remainder[i]*scale;
            }
        } else
            for (i=1; i<=lv; i++) remainder[i]=uu[diff+i];

        /* correct storage of remainder */
        i=1;
        j=lv;
        while (remainder[i] == 0 && j > 1) {
            j--;
            i++;
        }
        remainder[0]=j;
        if (j < lv)
            for (i=1; i<=j; i++) remainder[i]=remainder[lv+i-j];

        /* unnormalize the divisor v */
        if (scale > 1) {
            carry=0;
            for (i=1; i<=lv; i++) {
                t=carry*BASE+v[i];
                v[i]=t/scale;
                carry=t-v[i]*scale;
            }
        }

        free_integer_vector(uu,0);
        free_integer_vector(a,0);
}
```

E. lngintpower

Forms the single length positive integer power of a multilength integer expressed in the form

$$\sum_{j=1}^{i_0} i_j B^{j-1}$$

where the i_j are single length nonnegative integers, and B is a single length positive integer:

$$\sum_{j=1}^{r_0} r_j B^{j-1} = \left(\sum_{j=1}^{u_0} u_j B^{j-1} \right)^e$$

$r_0 \leq eu_0$; B^2+B not greater than the largest integer having a machine representation.

Function Parameters:

void lngintpower (*u,exponent,result*)

u: int *u[0:u[0]]*;
entry: *u* must contain the long integer which has to be raised to the power exponent;

exponent: int;
entry: the (positive) power to which the long integer *u* will be raised;

result: int *result[0,u[0]*exponent]*;
exit: contains the value of *u* raised to the power of *exponent*, *u* remains unchanged.

exit: contains the value of u raised to the power of *exponent*, u remains unchanged.

Function used: lngintmult.

Method: For the method of the functions *lngintadd, lngintsubtract, lngintmult* and *lngintdivide*, see [K69]; the function *lngintpower* uses the binary method for exponentiation [K69].

```
void lngintpower(int u[], int exponent, int result[])
{
    int *allocate_integer_vector(int, int);
    void free_integer_vector(int *, int);
    void lngintmult(int [], int [], int []);
    int max,i,n,exp,*y,*z,*h;

    exp=exponent;
    max=u[0]*exp;
    y=allocate_integer_vector(0,max);
    z=allocate_integer_vector(0,max);
    h=allocate_integer_vector(0,max);

    y[0]=y[1]=1;
    for (i=u[0]; i>=0; i--) z[i]=u[i];
    while (1) {
        n=exp/2;
        if (n+n != exp) {
            lngintmult(y,z,h);
            for (i=h[0]; i>=0; i--) y[i]=h[i];
            if (n == 0) {
                for (i=y[0]; i>=0; i--) result[i]=y[i];
                free_integer_vector(y,0);
                free_integer_vector(z,0);
                free_integer_vector(h,0);
                return;
            }
        }
        lngintmult(z,z,h);
        for (i=h[0]; i>=0; i--) z[i]=h[i];
        exp=n;
    }
}
```

2. Algebraic Evaluations

2.1 Evaluation of polynomials in Grunert form

A. pol

Computes the sum of the polynomial

$$p(x) = \sum_{i=0}^{n} a_i x^i$$

using Horner's rule. The error growth is given by a linear function of the degree of the polynomial [Wi63].

Function Parameters:

float pol (n,x,a)

pol: given the value of $p(x)$ above;
n: int; the degree of the polynomial;
x: float; the argument of the polynomial;
a: float $a[0:n]$;
entry: the coefficients of the polynomial.

```
float pol(int n, float x, float a[])
{
    float r;

    r=0.0;
    for (; n>=0; n--) r=r*x+a[n];
    return (r);
}
```

B. taypol

Computes the values of the terms $x^j D^j p(x)/j!$ $(j=0,...,k \leq n)$ where

$$p(x) = \sum_{i=0}^{n} a_i x^i.$$

Function Parameters:

void taypol (n,k,x,a)

n: int; the degree of the polynomial;
k: int; the first *k* terms of the above are to be calculated;
x: float; the argument of the polynomial;
a: float $a[0:n]$;
entry: the coefficients of the polynomial;
exit: the *j*-th term x^j*(*j*-th derivative)/*j*! is delivered in $a[j]$, $j=0,1,...,k \leq n$,

the other elements of a are generally altered.

Method: The method of evaluation is given in [ShT74]. The more sophisticated algorithm based on divisors of n+1 in [ShT74] was not implemented because of the more complex appearance of the implementation and because of the difficulty in choosing the most efficient divisor. In this implementation of the one-parameter family of algorithms, the linear number of multiplications is preserved. See [Wo74] for the k-th normalized derivative.

```
void taypol(int n, int k, float x, float a[])
{
    int i,j,nm1;
    float xj,aa,h;

    if (x != 0.0) {
        xj=1;
        for (j=1; j<=n; j++) {
            xj *= x;
            a[j] *= xj;
        }
        aa=a[n];
        nm1=n-1;
        for (j=0; j<=k; j++) {
            h=aa;
            for (i=nm1; i>=j; i--) h = a[i] += h;
        }
    } else {
        for (; k>=1; n--) a[k]=0;
    }
}
```

C. norderpol

Computes the first k normalized derivatives $D^j p(x)/j!$, $(j=0,...,k \leq n)$ of the polynomial

$$p(x) = \sum_{i=0}^{n} a_i x^i.$$

Function Parameters:

void norderpol (n,k,x,a)

n: int; the degree of the polynomial;

k: int; the first k normalized derivatives (j-th derivative / j factorial) are to be calculated;

x: float; the argument of the polynomial;

a: float $a[0:n]$;
entry: the coefficients of the polynomial;
exit: the j-th normalized derivative is delivered in $a[j]$, $j=0,1,...,k \leq n$, the other elements of a are generally altered.

Method: see the function *taypol*.

```
void norderpol(int n, int k, float x, float a[])
{
    float *allocate_real_vector(int, int);
    void free_real_vector(float *, int);
    int i,j,nm1;
```

```
    float xj,aa,h,*xx;

    if (x != 0.0) {
        xx=allocate_real_vector(0,n);
        xj=1;
        for (j=1; j<=n; j++) {
            xx[j] = xj *= x;
            a[j] *= xj;
        }
        h=aa=a[n];
        nm1=n-1;
        for (i=nm1; i>=0; i--) h = a[i] += h;
        for (j=1; j<=k; j++) {
            h=aa;
            for (i=nm1; i>=j; i--) h = a[i] += h;
            a[j]=h/xx[j];
        }
        free_real_vector(xx,0);
    }
}
```

D. derpol

Computes the first k derivatives $D^j p(x)$ $(j=0,...,k \leq n)$ of the polynomial

$$p(x) = \sum_{i=0}^{n} a_i x^i.$$

Function Parameters:

void derpol (n,k,x,a)

- n: int; the degree of the polynomial;
- k: int; the first k derivatives are to be calculated;
- x: float; the argument of the polynomial;
- a: float $a[0:n]$;
  - entry: the coefficients of the polynomial;
  - exit: the j-th derivative is delivered in $a[j]$, $j=0,1,...,k \leq n$, the other elements of a are generally altered.

Function used: norderpol.

Method: see the function *taypol*.

```
void derpol(int n, int k, float x, float a[])
{
    void norderpol(int, int, float, float []);
    int j;
    float fac;

    fac=1.0;
    norderpol(n,k,x,a);
    for (j=2; j<=k; j++) {
        fac *= j;
        a[j] *=fac;
    }
}
```

2.2 Evaluation of general orthogonal polynomials

This section contains six procedures for evaluating orthogonal polynomials [AbS65, Lu69]. The procedures ending with *sym* are versions for symmetric polynomials.

A. ortpol

Evaluates the single polynomial $p_n(x)$, where

$$p_0(x) = 1, \quad p_1(x) = x-b_0,$$
$$p_{k+1}(x) = (x-b_k)p_k(x) - c_k p_{k-1}(x), \qquad k=1,...,n-1$$

with given b_k and c_k.

Function Parameters:

float ortpol (n,x,b,c)

ortpol: given the value of the orthogonal polynomial in the above;
n: int; the degree of the polynomial;
x: float; the argument of the orthogonal polynomial;
b,c: float *b[0:n-1], c[1:n-1]*;
entry: the recurrence coefficients of the polynomial.

```
float ortpol(int n, float x, float b[], float c[])
{
    int k,l;
    float r,s,h;

    if (n == 0) return (1.0);
    r=x-b[0];
    s=1.0;
    l=n-1;
    for (k=1; k<=l; k++) {
        h=r;
        r=(x-b[k])*r-c[k]*s;
        s=h;
    }
    return (r);
}
```

B. ortpolsym

Given the c_k and x, evaluates the single polynomial $p_n(x)$, where

$$p_0(x) = 1, \quad p_1(x) = x,$$
$$p_{k+1}(x) = xp_k(x) - c_k p_{k-1}(x), \qquad k=1,...,n-1.$$

Function Parameters:

float ortpolsym (n,x,c)

ortpolsym: given the value of the orthogonal polynomial in the above;
n: int; the degree of the polynomial;
x: float; the argument of the orthogonal polynomial;
c: float *c[1:n-1]*;
entry: the recurrence coefficients of the polynomial.

```
float ortpolsym(int n, float x, float c[])
{
    int k,l;
    float r,s,h;

    if (n == 0) return (1.0);
    r=x;
```

```
    s=1.0;
    l=n-1;
    for (k=1; k<=l; k++) {
        h=r;
        r=x*r-c[k]*s;
        s=h;
    }
    return (r);
}
```

C. allortpol

Evaluates the sequence of polynomials $p_k(x)$ for $k=0,...,n$, where

$$p_0(x) = 1, \quad p_1(x) = x-b_0,$$
$$p_{k+1}(x) = (x-b_k)p_k(x) - c_k p_{k-1}(x), \quad k=1,...,n-1$$

with given b_k and c_k.

Function Parameters:

void allortpol (n,x,b,c,p)

n: int; the degree of the polynomial;
x: float; the argument of the orthogonal polynomial;
b,c: float $b[0:n-1]$, $c[1:n-1]$;
entry: the recurrence coefficients of the polynomial;
p: float $p[0:n]$;
exit: $p[k]$ contains, for the argument, the value of the k-th orthogonal polynomial as defined by the recurrence coefficients.

```
void allortpol(int n, float x, float b[], float c[], float p[])
{
    int k,k1;
    float r,s,h;

    if (n == 0) {
        p[0]=1.0;
        return;
    }
    r=p[1]=x-b[0];
    s=p[0]=1.0;
    k=1;
    for (k1=2; k1<=n; k1++) {
        h=r;
        p[k1]=r=(x-b[k])*r-c[k]*s;
        s=h;
        k=k1;
    }
}
```

D. allortpolsym

Given the c_k and x, evaluates the sequence of polynomials $p_k(x)$ for $k=0,...,n$, where

$$p_0(x) = 1, \quad p_1(x) = x,$$
$$p_{k+1}(x) = xp_k(x) - c_k p_{k-1}(x), \quad k=1,...,n-1.$$

Function Parameters:

void allortpolsym (n,x,c,p)

n: int; the degree of the polynomial;
x: float; the argument of the orthogonal polynomial;

c: float *c[1:n-1]*;
entry: the recurrence coefficients of the polynomial;
p: float *p[0:n]*;
exit: *p[k]* contains, for the argument, the value of the *k*-th orthogonal polynomial as defined by the recurrence coefficients.

```
void allortpolsym(int n, float x, float c[], float p[])
{
    int k;
    float r,s,h;

    if (n == 0) {
        p[0]=1.0;
        return;
    }
    r=p[1]=x;
    s=p[0]=1.0;
    for (k=2; k<=n; k++) {
        h=r;
        p[k]=r=x*r-c[k-1]*s;
        s=h;
    }
}
```

E. sumortpol

Evaluates the sum

$$s = \sum_{k=0}^{n} a_k p_k(x),$$

where $p_0(x) = 1, \quad p_1(x) = x-b_0,$
$p_{k+1}(x) = (x-b_k)p_k(x) - c_k p_{k-1}(x), \quad k=1,...,n-1,$
with given a_k, b_k and c_k.

Function Parameters:

float sumortpol (n,x,b,c,a)

sumortpol: given the value of the sum *s* in the above;
n: int; the degree of the polynomial;
x: float; the argument of the orthogonal polynomial;
b,c: float *b[0:n-1]*, *c[1:n-1]*;
entry: the recurrence coefficients of the polynomial;
a: float *a[0:n]*;
entry: the value of a_k in the above.

```
float sumortpol(int n, float x, float b[], float c[], float a[])
{
    int k;
    float h,r,s;

    if (n == 0) return (a[0]);
    r=a[n];
    s=0.0;
    for (k=n-1; k>=1; k--) {
        h=r;
        r=a[k]+(x-b[k])*r+s;
        s = -c[k]*h;
```

```
    }
    return (a[0]+(x-b[0])*r+s);
}
```

F. sumortpolsym

Given the a_k, c_k and x, evaluates the sum

$$s = \sum_{k=0}^{n} a_k p_k(x),$$

where $p_0(x) = 1, \quad p_1(x) = x,$

$$p_{k+1}(x) = xp_k(x) - c_k p_{k-1}(x), \qquad k=1,...,n-1.$$

Function Parameters:

float sumortpolsym (*n,x,c,a*)

sumortpolsym: given the value of the sum s in the above;
n: int; the degree of the polynomial;
x: float; the argument of the orthogonal polynomial;
c: float *c[1:n-1]*;
entry: the recurrence coefficients of the polynomial;
a: float *a[0:n]*;
entry: the values of a_k in the above.

```
float sumortpolsym(int n, float x, float c[], float a[])
{
    int k;
    float h,r,s;

    if (n == 0) return (a[0]);
    r=a[n];
    s=0.0;
    for (k=n-1; k>=1; k--) {
        h=r;
        r=a[k]+x*r+s;
        s = -c[k]*h;
    }
    return (a[0]+x*r+s);
}
```

2.3 Evaluation of Chebyshev polynomials

This section contains four procedures for evaluating Chebyshev polynomials of the first kind [FoP68, Riv74].

A. chepolsum

Uses the Clenshaw or generalized Horner algorithm [Cle62] to evaluate a sum of Chebyshev polynomials

$$S = \sum_{i=0}^{n} a_i T_i(x)$$

where $T_i(x)$ is a Chebyshev polynomial, by use of the backward recursion

$$S_{n+1} = 0, \quad S_n = a_n,$$
$$S_k = a_k + 2xS_{k+1} - S_{k+2}, \qquad k=n-2,n-3,...,1;$$
$$S = a_0 + S_1x - S_2$$

for $n \geq 2$.

Function Parameters:

float chepolsum (n,x,a)

chepolsum: given the value of the Chebyshev sum S in the above;
n: int; the degree of the polynomial represented by the Chebyshev sum, $n \geq 0$;
x: float; the argument of the Chebyshev polynomials, $|x| \leq 1$;
a: float $a[0:n]$;
entry: $a[k]$ is the coefficient of the Chebyshev polynomial of degree k, $0 \leq k \leq n$.

```
float chepolsum(int n, float x, float a[])
{
    int k;
    float h,r,s,tx;

    if (n == 0) return (a[0]);
    if (n == 1) return (a[0]+a[1]*x);
    tx=x+x;
    r=a[n];
    h=a[n-1]+r*tx;
    for (k=n-2; k>=1; k--) {
        s=r;
        r=h;
        h=a[k]+r*tx-s;
    }
    return (a[0]-r+h*x);
}
```

B. oddchepolsum

Given the a_i and x, uses the Clenshaw or generalized Horner algorithm [Cle62] to evaluate a sum of Chebyshev polynomials of odd degree

$$S = \sum_{i=0}^{n} a_i T_{2i+1}(x)$$

the T_i being Chebyshev polynomials.

Function Parameters:

float oddchepolsum (n,x,a)

oddchepolsum: given the value of the Chebyshev sum S in the above;
n: int; the degree of the polynomial represented by the Chebyshev sum is $2n+1$, $n \geq 0$;
x: float; the argument of the Chebyshev polynomials, $|x| \leq 1$;
a: float $a[0:n]$;
entry: $a[k]$ is the coefficient of the Chebyshev polynomial of degree $2k+1$, $0 \leq k \leq n$.

```
float oddchepolsum(int n, float x, float a[])
{
    int k;
    float h,r,s,y;
```

```
	if (n == 0) return (x*a[0]);
	if (n == 1) return (x*(a[0]+a[1]*(4.0*x*x-3.0)));
	y=4.0*x*x-2.0;
	r=a[n];
	h=a[n-1]+r*y;
	for (k=n-2; k>=0; k--) {
		s=r;
		r=h;
		h=a[k]+r*y-s;
	}
	return (x*(h-r));
}
```

C. chepol

Computes the value of the Chebyshev polynomial $T_n(x)$ by use of the recursion [St72]

$$T_0(x) = 1, \quad T_1(x) = x,$$
$$T_i(x) = 2xT_{i-1}(x) - T_{i-2}(x), \qquad i=2,...,n.$$

Function Parameters:

float chepol (n,x)

chepol: given the value of the Chebyshev polynomial of the first kind, $T_n(x)$ above;
n: int; the degree of the polynomial, $n \geq 0$;
x: float; the argument of the Chebyshev polynomial, $|x| \leq 1$.

```
float chepol(int n, float x)
{
	int i;
	float t1,t2,h,x2;

	if (n == 0) return (1.0);
	if (n == 1) return (x);
	t2=x;
	t1=1.0;
	x2=x+x;
	for (i=2; i<=n; i++) {
		h=x2*t2-t1;
		t1=t2;
		t2=h;;
	}
	return (h);
}
```

D. allchepol

Computes the values of the Chebyshev polynomials $T_i(x)$, $i=1,...,n$, by use of the recursion [St72]

$$T_0(x) = 1, \quad T_1(x) = x,$$
$$T_i(x) = 2xT_{i-1}(x) - T_{i-2}(x), \qquad i=2,...,n.$$

Function Parameters:

void allchepol (n,x,t)

n: int; the degree of the last polynomial, $n \geq 0$;
x: float; the argument of the Chebyshev polynomials, $|x| \leq 1$;
t: float $t[0:n]$;
exit: the value of the Chebyshev polynomial of the first kind of degree k, for the argument x, is delivered in $t[k]$, $k=0,...,n$.

```
void allchepol(int n, float x, float t[])
{
    int i;
    float t1,t2,h,x2;

    if (n == 0) {
        t[0]=1.0;
        return;
    }
    if (n == 1) {
        t[0]=1.0;
        t[1]=x;
        return;
    }
    t[0]=t1=1.0;
    t[1]=t2=x;
    x2=x+x;
    for (i=2; i<=n; i++) {
        t[i]=h=x2*t2-t1;
        t1=t2;
        t2=h;;
    }
}
```

2.4 Evaluation of Fourier series

A. sinser

Computes the value of a sine series

$$s = \sum_{j=1}^{n} b_j \sin(j\theta)$$

where θ and the b_j are real. When $-½ \le \cos(\theta) \le ½$, the recursion

$$u_{n+2} = u_{n+1} = 0,$$
$$u_k = 2\cos(\theta)u_{k+1} - u_{k+2} + b_k, \qquad k=n,...,1$$

is used. When $\cos(\theta) < -½$, the equivalent recursion

$$u_{n+1} = d_{n+1} = 0,$$
$$d_k = 2(1 + \cos(\theta))u_{k+1} - d_{k+1} + b_k, \qquad k=n,...,1$$
$$u_k = d_k - u_{k+1},$$

and when $\cos(\theta) > ½$, the further recursion

$$u_{n+1} = e_{n+1} = 0,$$
$$e_k = 2(\cos(\theta) - 1)u_{k+1} - e_{k+1} + b_k, \qquad k=n,...,1$$
$$u_k = e_k - u_{k+1}$$

are used. In each case $s = u_1 \sin(\theta)$.

Function Parameters:

float sinser (*n,theta,b*)

sinser: given the value of the sine series, the value of *s* above;

n: int;
    entry: the number of terms in the sine series;

theta: float;
    entry: the argument of the sine series;

b: float *b[1:n]*;
    entry: the coefficients of the sine series.

Method: see the function *comfouser2*.

```
#include <math.h>

float sinser(int n, float theta, float b[])
{
    int k;
    float c,cc,lambda,h,dun,un,un1,temp;

    c=cos(theta);
    if (c < -0.5) {
        temp=cos(theta/2.0);
        lambda=4.0*temp*temp;
        un=dun=0.0;
        for (k=n; k>=1; k--) {
            dun=lambda*un-dun+b[k];
            un=dun-un;
        }
    } else {
        if (c > 0.5) {
            temp=sin(theta/2.0);
            lambda = -4.0*temp*temp;
            un=dun=0.0;
            for (k=n; k>=1; k--) {
                dun += lambda*un+b[k];
                un += dun;
            }
        } else {
            cc=c+c;
            un=un1=0.0;
            for (k=n; k>=1; k--) {
                h=cc*un-un1+b[k];
                un1=un;
                un=h;
            }
        }
    }
    return (un*sin(theta));
}
```

B. cosser

Computes the value of a cosine series

$$s = \sum_{j=0}^{n} a_j \cos(j\theta)$$

where θ and the a_j are real. The method used is based upon recursion of the documentation to *sinser*, where now $s = a_0 + 2u_1 cos(\theta) - u_2$,
and upon further recursions similar to those given in that documentation.

Function Parameters:

float cosser (*n,theta,a*)

cosser: given the value of the cosine series, value of *s* above;

n: int;
 entry: the degree of the trigonometric polynomial;

theta: float;
 entry: the argument of the cosine series;

a: float *a[0:n]*;
 entry: the coefficients of the cosine series.

Method: see the function *comfouser2*.

```
#include <math.h>

float cosser(int n, float theta, float a[])
{
    int k;
    float c,cc,lambda,h,dun,un,un1,temp;

    c=cos(theta);
    if (c < -0.5) {
        temp=cos(theta/2.0);
        lambda=4.0*temp*temp;
        un=dun=0.0;
        for (k=n; k>=0; k--) {
            un=dun-un;
            dun=lambda*un-dun+a[k];
        }
        return (dun-lambda/2.0*un);
    } else {
        if (c > 0.5) {
            temp=sin(theta/2.0);
            lambda = -4.0*temp*temp;
            un=dun=0.0;
            for (k=n; k>=0; k--) {
                un += dun;
                dun += lambda*un+a[k];
            }
            return (dun-lambda/2.0*un);
        } else {
            cc=c+c;
            un=un1=0.0;
            for (k=n; k>=1; k--) {
                h=cc*un-un1+a[k];
                un1=un;
                un=h;
            }
            return (a[0]+un*c-un1);
        }
    }
}
```

C. fouser

Computes the value of

$$s = \sum_{j=0}^{n} a_j \cos(j\theta) + \sum_{j=1}^{n} a_j \sin(j\theta)$$

where θ and the a_j are real, by methods similar to those described in the documentations to *sinser* and *cosser*.

Function Parameters:

float fouser $(n,theta,a)$

fouser: given the value of the fourier series, value of s above;

n: int;
entry: the degree of the trigonometric polynomial;

theta: float;
entry: the argument of the fourier series;

a: float *a[0:n]*;

entry: the coefficients of the (finite) fourier series.

```
#include <math.h>

float fouser(int n, float theta, float a[])
{
    int k;
    float c,cc,lambda,h,dun,un,un1,c2,s2;

    c=cos(theta);
    if (c < -0.5) {
        c2=cos(theta/2.0);
        lambda=4.0*c2*c2;
        un=dun=0.0;
        for (k=n; k>=0; k--) {
            un=dun-un;
            dun=lambda*un-dun+a[k];
        }
        return (dun+2.0*c2*(sin(theta/2.0)-c2)*un);
    } else {
        if (c > 0.5) {
            s2=sin(theta/2.0);
            lambda = -4.0*s2*s2;
            un=dun=0.0;
            for (k=n; k>=0; k--) {
                un += dun;
                dun += lambda*un+a[k];
            }
            return (dun+2.0*s2*(s2+cos(theta/2.0))*un);
        } else {
            cc=c+c;
            un=un1=0.0;
            for (k=n; k>=1; k--) {
                h=cc*un-un1+a[k];
                un1=un;
                un=h;
            }
            return (a[0]-un1+(c+sin(theta))*un);
        }
    }
}
```

D. fouser1

Computes the value of

$$s = \sum_{j=0}^{n} a_j \cos(j\theta) + \sum_{j=1}^{n} b_j \sin(j\theta)$$

where θ, a_j and b_j are real, using a Horner scheme.

Function Parameters:

float fouser1 (*n,theta,a,b*)

fouser1: given the value of the fourier series, value of *s* above;

n: int;

entry: the degree of the trigonometric polynomial;

theta: float;

entry: the argument of the fourier series;

a,b: float *a[0:n], b[1:n]*;

entry: the coefficients of the (finite) fourier series,

with a_k coefficient of cos(k*θ), k=0,...,n,
and b_k coefficient of sin(k*θ), k=1,...,n.

Method: see the function *comfouser2*.

```
#include <math.h>

float fouser1(int n, float theta, float a[], float b[])
{
    int i;
    float r,s,h,co,si;

    r=s=0.0;
    co=cos(theta);
    si=sin(theta);
    for (i=n; i>=1; i--) {
        h=co*r+si*s+a[i];
        s=co*s-si*r+b[i];
        r=h;
    }
    return (co*r+si*s+a[0]);
}
```

E. fouser2

Computes the value of

$$s = \sum_{j=0}^{n} a_j \cos(j\theta) + \sum_{j=1}^{n} b_j \sin(j\theta)$$

where θ, a_j and b_j are real, by methods similar to those described in the documentations to *sinser* and *cosser*.

Function Parameters:

float fouser2 (*n,theta,a,b*)

fouser2: given the value of the fourier series, value of *s* above;
n: int;
entry: the degree of the trigonometric polynomial;
theta: float;
entry: the argument of the fourier series;
a,b: float *a[0:n], b[1:n]*;
entry: the coefficients of the (finite) fourier series,
with a_k coefficient of cos(k*θ), k=0,...,n,
and b_k coefficient of sin(k*θ), k=1,...,n.

Functions used: sinser, cosser.

Method: see the function *comfouser2*.

```
#include <math.h>

float fouser2(int n, float theta, float a[], float b[])
{
    float sinser(int, float, float []);
    float cosser(int, float, float []);
```

```
    return (cosser(n,theta,a)+sinser(n,theta,b));
}
```

F. comfouser

Computes the value of

$$s = \sum_{j=0}^{n} a_j e^{ij\theta}$$

where θ and the a_j are real, by methods similar to those described in the documentations to *sinser* and *cosser*.

Function Parameters:

void comfouser (*n,theta,a,rr,ri*)

n: int;
entry: the degree of the polynomial in $e^{i\theta}$;

theta: float;
entry: the argument of the fourier series;

a: float *a[0:n]*;
entry: the real coefficients in the series;

rr,ri: float *;
exit: the real part and the imaginary part of *s* above are delivered in *rr* and *ri*, respectively.

```
#include <math.h>

void comfouser(int n, float theta, float a[], float *rr, float *ri)
{
    int k;
    float c,cc,lambda,h,dun,un,un1,temp;

    c=cos(theta);
    if (c < -0.5) {
        temp=cos(theta/2.0);
        lambda=4.0*temp*temp;
        un=dun=0.0;
        for (k=n; k>=0; k--) {
            un=dun-un;
            dun=lambda*un-dun+a[k];
        }
        *rr=dun-lambda/2.0*un;
    } else {
        if (c > 0.5) {
            temp=sin(theta/2.0);
            lambda = -4.0*temp*temp;
            un=dun=0.0;
            for (k=n; k>=0; k--) {
                un += dun;
                dun += lambda*un+a[k];
            }
            *rr=dun-lambda/2.0*un;
        } else {
            cc=c+c;
            un=un1=0.0;
            for (k=n; k>=1; k--) {
                h=cc*un-un1+a[k];
                un1=un;
                un=h;
            }
```

```
            *rr=a[0]+un*c-un1;
        }
    }
    *ri=un*sin(theta);
}
```

G. comfouser1

Computes the value of

$$s = \sum_{j=0}^{n} a_j e^{ij\theta}$$

where θ is real and the a_j are complex, using a Horner scheme.

Function Parameters:

void comfouser1 (*n,theta,ar,ai,rr,ri*)

n: int;
 entry: the degree of the polynomial in $e^{i\theta}$;
theta: float;
 entry: the argument of the fourier series;
ar,ai: float *ar[0:n], ai[0:n]*;
 entry: the real part and the imaginary part of the complex coefficients in the series must be given in arrays *ar* and *ai*, respectively;
rr,ri: float *;
 exit: the real part and the imaginary part of *s* above are delivered in *rr* and *ri*, respectively.

Method: see the function *comfouser2*.

```
#include <math.h>

void comfouser1(int n, float theta, float ar[], float ai[],
                float *rr, float *ri)
{
    int k;
    float h,hr,hi,co,si;

    hr=hi=0.0;
    co=cos(theta);
    si=sin(theta);
    for (k=n; k>=1; k--) {
        h=co*hr-si*hi+ar[k];
        hi=co*hi+si*hr+ai[k];
        hr=h;
    }
    *rr=co*hr-si*hi+ar[0];
    *ri=co*hi+si*hr+ai[0];
}
```

H. comfouser2

Computes the value of

$$s = \sum_{j=0}^{n} a_j e^{ij\theta}$$

where θ is real and the a_j are complex, by methods similar to those described in the documentations to *sinser* and *cosser*.

Function Parameters:

void comfouser2 (*n,theta,ar,ai,rr,ri*)

n: int;
entry: the degree of the polynomial in $e^{i\theta}$;

theta: float;
entry: the argument of the fourier series;

ar,ai: float *ar[0:n], ai[0:n]*;
entry: the real part and the imaginary part of the complex coefficients in the series must be given in arrays *ar* and *ai*, respectively;

rr,ri: float *;
exit: the real part and the imaginary part of *s* above are delivered in *rr* and *ri*, respectively.

Function used: comfouser.

Method: For the evaluation of a finite fourier series two algorithms are used: (a) the Horner scheme in the function fouser1, and (b) a combination of the Clenshaw algorithm [Ge69, Lu69, St72] and the modification of Reinsch [Re67, St72] in the functions *sinser*, *cosser*, *fouser* and *fouser2*. A modification of the idea of Newbery is not implemented because of the introduction of sine (cosine) terms in a cosine (sine) series and the inefficiency of the algorithm [N73]. For the evaluation of a finite complex fourier series, two algorithms, in real arithmetic, are used: (a) the Horner scheme in the function *comfouser1*, and (b) a combination of the Clenshaw algorithm and the modification of Reinsch in functions *comfouser* and *comfouser2*. The Horner scheme is implemented because of the simplicity of the algorithm (although this algorithm is less efficient than the Goertzel-Watt-Clenshaw-Reinsch algorithm) and the stable nature of orthogonal transformations. A combination of the algorithm of Goertzel-Watt-Clensaw and the modification of Reinsch is implemented because of the efficiency of the Goertzel-Watt-Clensaw algorithm and the stability of the modification of Reinsch, especially for small values of the argument. An upper bound for the error growth is given by a linear function of the degree for both implemented algorithms.

```
#include <math.h>

void comfouser2(int n, float theta, float ar[], float ai[],
                float *rr, float *ri)
{
   void comfouser(int, float, float [], float *, float *);
   float car,cai,sar,sai;

   comfouser(n,theta,ar,&car,&sar);
   comfouser(n,theta,ai,&cai,&sai);
   *rr=car-sai;
   *ri=cai+sar;
}
```

2.5 Evaluation of continued fractions

jfrac

Computes the value of the convergent

$$C_n = b_0 + \frac{a_1}{b_1+}\frac{a_2}{b_2+}\cdots\frac{a_n}{b_n}$$

by use of the recursion

$$D_1 = b_n,$$
$$D_{i+1} = b_{n-i} + a_{n-i+1}/D_i, \qquad i=1,...,n,$$

when $C_n = D_{n+1}$.

Function Parameters:

float jfrac (n,a,b)

jfrac: delivers the value of the terminating continued fraction, the value of C_n above;
n: int; the upper index of the arrays *a* and *b*;
a,b: float *a[1:n],b[0:n]*; the elements of the continued fraction, a_i and b_i above.

```
float jfrac(int n, float a[], float b[])
{
    int i;
    float d;

    d=0.0;
    for (i=n; i>=1; i--) d=a[i]/(b[i]+d);
    return (d+b[0]);
}
```

2.6 Transformation of polynomial representation

A. polchs

Given the a_k, derives the b_k occurring in the relationship

$$\sum_{k=0}^{n} a_k x^k = \sum_{k=0}^{n} b_k T_k(x)$$

the $T_k(x)$ being Chebyshev polynomials.

Function Parameters:

void polchs (n,a)

n: int; the degree of the polynomial;
a: float *a[0:n]*;
 entry: the coefficients of the power sum;
 exit: the coefficients of the Chebyshev sum.

Method: Although the transformation of representations of polynomials could have been obtained by fast evaluation and fast interpolation, the algorithm of Hamming [H73] was implemented here because of its simple appearance.

```
void polchs(int n, float a[])
{
    int k,l,twopow;

    if (n > 1) {
        twopow=2;
        for (k=1; k<=n-2; k++) {
            a[k] /= twopow;
            twopow *= 2;
        }
        a[n-1]=2.0*a[n-1]/twopow;
        a[n] /= twopow;
        a[n-2] += a[n];
        for (k=n-2; k>=1; k--) {
            a[k-1] += a[k+1];
            a[k]=2.0*a[k]+a[k+2];
            for (l=k+1; l<=n-2; l++) a[l] += a[l+2];
        }
    }
}
```

B. chspol

Given the b_k, derives the a_k occurring in the relationship

$$\sum_{k=0}^{n} b_k T_k(x) = \sum_{k=0}^{n} a_k x^k$$

the $T_k(x)$ being Chebyshev polynomials.

Function Parameters:

void chspol (n,a)

n: int; the degree of the polynomial;

a: float $a[0:n]$;
 entry: the coefficients of the Chebyshev sum;
 exit: the coefficients of the power sum.

Method: see the function *polchs*.

```
void chspol(int n, float a[])
{
    int k,l,twopow;

    if (n > 1) {
        for (k=0; k<=n-2; k++) {
            for (l=n-2; l>=k; l--) a[l] -= a[l+2];
            a[k+1] /= 2.0;
        }
        twopow=2;
        for (k=1; k<=n-2; k++) {
            a[k] *= twopow;
            twopow *= 2;
        }
        a[n-1] *= twopow;
        a[n] *= twopow;
    }
```

```
}
```

C. polshtchs

Given the a_k, derives the b_k occurring in the relationship

$$\sum_{k=0}^{n} a_k x^k = \sum_{k=0}^{n} b_k S_k(x)$$

the $S_k(x)$ being shifted Chebyshev polynomials defined by $S_k(x)=T_k(2x-1)$, $T_k(x)$ being a Chebyshev polynomial.

Function Parameters:

void polshtchs (n,a)

n: int; the degree of the polynomial;
a: float $a[0:n]$;
entry: the coefficients of the power sum;
exit: the coefficients of the shifted Chebyshev sum.

Functions used: lintfmpol, polchs.

Method: see the function *polchs*.

```
void polshtchs(int n, float a[])
{
    void lintfmpol(float, float, int, float []);
    void polchs(int, float []);

    lintfmpol(0.5,0.5,n,a);
    polchs(n,a);
}
```

D. shtchspol

Given the b_k, derives the a_k occurring in the relationship

$$\sum_{k=0}^{n} b_k S_k(x) = \sum_{k=0}^{n} a_k x^k$$

the $S_k(x)$ being shifted Chebyshev polynomials defined by $S_k(x)=T_k(2x-1)$, $T_k(x)$ being a Chebyshev polynomial.

Function Parameters:

void shtchspol (n,a)

n: int; the degree of the polynomial;
a: float $a[0:n]$;
entry: the coefficients of the shifted Chebyshev sum;
exit: the coefficients of the power sum.

Functions used: lintfmpol, chspol.

Method: see the function *polchs*.

```
void shtchspol(int n, float a[])
{
    void chspol(int, float []);
    void lintfmpol(float, float, int, float []);

    chspol(n,a);
    lintfmpol(2.0,-1.0,n,a);
}
```

E. grnnew

Given the coefficients a_i occurring in the polynomial

$$P(x) = \sum_{i=0}^{n} a_i x^i$$

and the tabulation points x_ν, computes the divided differences δ_j in the equivalent Newton series representation

$$P(x) = \sum_{j=0}^{n} \left(\prod_{\nu=0}^{j-1} (x-x_\nu) \right) \delta_j.$$

Function Parameters:

void grnnew (n,x,a)

n: int; the degree of the polynomial;

x: float *x[0:n-1]*;
  entry: the interpolation points, values of x_ν above;

a: float *a[0:n]*;
  entry: the coefficients of the power sum;
  exit: the coefficients of the Newton sum, values of δ_j above.

Method: see the function *polchs*.

```
void grnnew(int n, float x[], float a[])
{
    int k,l;

    for (k=n-1; k>=0; k--)
        for (l=n-1; l>=n-1-k; l--) a[l] += a[l+1]*x[n-1-k];
}
```

F. newgrn

Given the coefficients $\delta_j f(x_0)$, together with the values of the arguments x_i from which they are formed, in the truncated Newton interpolation series

$$f_n(x) = \sum_{j=0}^{n} \left(\prod_{\nu=0}^{j-1} (x - x_\nu) \right) \delta_j f(x_0),$$

computes the coefficients c_i, $i=0,...,n$, in the equivalent polynomial form

$$f_n(x) = \sum_{i=0}^{n} c_i x^i.$$

Function Parameters:

void newgrn (n,x,a)

n: int; the degree of the polynomial;
x: float $x[0:n-1]$;
  entry: the interpolation points, values of x_i above;
a: float $a[0:n]$;
  entry: the coefficients of the Newton sum, values of $\delta_i f(x_0)$;
  exit: the coefficients of the power sum, values of c_i above.

Function used: elmvec.

Method: see the function *polchs*.

```
void newgrn(int n, float x[], float a[])
{
    void elmvec(int, int, int, float [], float [], float);
    int k;

    for (k=n-1; k>=0; k--)
        elmvec(k,n-1,1,a,a,-x[k]);
}
```

G. lintfmpol

Given the a_i occurring in the polynomial expression

$$P(x) = \sum_{i=0}^{n} a_i x^i$$

and p, q, derives the b_j occurring in the equivalent expression

$$P(x) = \sum_{j=0}^{n} b_j y^j.$$

where $x=py+q$.

Function Parameters:

void lintfmpol (p,q,n,a)

n: int; the degree of the polynomial;

p,q: float;

entry: defining the linear transformation of the independent variable $x=py+q$; $p=0$ gives the value of the polynomial with argument q;

a: float $a[0:n]$;

entry: the coefficients of the power sum in x, values of a_i above;

exit: the coefficients of the power sum in y, values of b_j above.

Function used: norderpol.

Method: see the function *polchs*.

```
void lintfmpol(float p, float q, int n, float a[])
{
    void norderpol(int, int, float, float []);
    int k;
    float ppower;

    norderpol(n,n,q,a);
    ppower=p;
    for (k=1; k<=n; k++) {
        a[k] *= ppower;
        ppower *= p;
    }
}
```

2.7 Operations on orthogonal polynomials

intchs

Given the real coefficients a_j in the expansion

$$f(x) = \sum_{j=0}^{n} a_j T_j(x)$$

where $T_j(x)$ is the Chebyshev polynomial of the first kind of degree j, those in the expansion

$$\int_0^x f(x)dx = \sum_{j=1}^{n+1} b_j T_j(x)$$

are derived.

Function Parameters:

void intchs (n,a,b)

n: int; the degree of the polynomial represented by the Chebyshev series;

a,b: float $a[0:n]$, $b[1:n+1]$;

entry: the coefficients of the Chebyshev series, values of a_j above;

exit: the coefficients of the integral Chebyshev series, values of b_j above.

Method: For a description of the algorithm see [Cle62, FoP68].

```
void intchs(int n, float a[], float b[])
{
    int i;
    float h,l,dum;

    if (n == 0) {
        b[1]=a[0];
        return;
    }
    if (n == 1) {
        b[2]=a[1]/4.0;
        b[1]=a[0];
        return;
    }
    h=a[n];
    dum=a[n-1];
    b[n+1]=h/((n+1)*2);
    b[n]=dum/(n*2);
    for (i=n-1; i>=2; i--) {
        l=a[i-1];
        b[i]=(l-h)/(2*i);
        h=dum;
        dum=l;
    }
    b[1]=a[0]-h/2.0;
}
```

3. Linear Algebra

3.1 Full real general matrices

3.1.1 Preparatory procedures

A. dec

Decomposes the $n \times n$ matrix A in the form $LU=PA$, where L is lower triangular, U is unit upper triangular and P is a permutation matrix.

Function Parameters:

void dec (*a,n,aux,p*)

a: float *a[1:n,1:n]*;
  entry: the matrix to be decomposed;
  exit: the calculated lower triangular matrix and unit upper triangular matrix with its unit diagonal omitted;

n: int; the order of the matrix;

aux: float *aux[1:3]*;
  entry:
  aux[2]: a relative tolerance: a reasonable choice for this value is an estimate of the relative precision of the matrix elements; however, it should not be chosen smaller than the machine precision;
  exit:
  aux[1]: if R is the number of elimination steps performed (see *aux[3]*), then *aux[1]* equals 1 if the determinant of the principal submatrix of order R is positive, else *aux[1]* equals -1;
  aux[3]: the number of elimination steps performed; if *aux[3]*<*n* then the process has been broken off because the selected pivot is too small relative to the maximum of the Euclidean norms of the rows of the given matrix;

p: int *p[1:n]*;
  exit: the pivot indices; row i and row *p[i]* are interchanged in the i-th iteration.

Functions used: matmat, mattam, ichrow.

Method: The decomposition uses only partial pivoting [Dek68, Wi63, Wi65]. Since, in exceptional cases, partial pivoting may yield useless results, even for well-conditioned matrices, the user is advised to use the function gsselm. However, if the number of variables is small relative to the number of binary digits in the mantissa of the number representation then the function *dec* may also be used. Refer to the function *gsselm* for more details.

```
#include <math.h>

void dec(float **a, int n, float aux[], int p[])
{
    float *allocate_real_vector(int, int);
    void free_real_vector(float *, int);
    float matmat(int, int, int, int, float **, float **);
```

```
    float mattam(int, int, int, int, float **, float **);
    void ichrow(int, int, int, int, float **);
    int i,k,k1,pk,d;
    float r,s,eps,*v;

    v=allocate_real_vector(1,n);
    r = -1.0;
    for (i=1; i<=n; i++) {
        s=sqrt(mattam(1,n,i,i,a,a));
        if (s > r) r=s;
        v[i]=1.0/s;
    }
    eps=aux[2]*r;
    d=1.0;
    for (k=1; k<=n; k++) {
        r = -1.0;
        k1=k-1;
        for (i=k; i<=n; i++) {
            a[i][k] -= matmat(1,k1,i,k,a,a);
            s=fabs(a[i][k])*v[i];
            if (s > r) {
                r=s;
                pk=i;
            }
        }
        p[k]=pk;
        v[pk]=v[k];
        s=a[pk][k];
        if (fabs(s) < eps) break;
        if (s < 0.0) d = -d;
        if (pk != k) {
            d = -d;
            ichrow(1,n,k,pk,a);
        }
        for (i=k+1; i<=n; i++) a[k][i]=(a[k][i]-matmat(1,k1,k,i,a,a))/s;
    }
    aux[1]=d;
    aux[3]=k-1;
    free_real_vector(v,1);
}
```

B. gsselm

Decomposes the $n \times n$ matrix A in the form $LU=P_1AP_2$, where L is lower triangular, U is unit upper triangular and P_1 and P_2 are permutation matrices. This function uses partial pivoting, with complete pivoting if the former does not yield stable results.

Function Parameters:

void gsselm (*a,n,aux,ri,ci*)

a: float *a[1:n,1:n]*;
  entry: the matrix to be decomposed;
  exit: the calculated lower triangular matrix and unit upper triangular matrix with its unit diagonal omitted;

n: int; the order of the matrix;

aux: float *aux[1:7]*;
  entry:
  aux[2]: a relative tolerance: a reasonable choice for this value is an estimate of the relative precision of the matrix elements; however, it should not be chosen smaller than the machine precision;
  aux[4]: a value which is used for controlling pivoting, usually *aux[4]=8* will give good results;
  exit:

aux[1]: if *R* is the number of elimination steps performed (see *aux[3]*), then *aux[1]* equals 1 if the determinant of the principal submatrix of order *R* is positive, else *aux[1]* equals -1;

aux[3]: the number of elimination steps performed; if *aux[3]*<*n* then the process has been broken off because the selected pivot is too small relative to the maximum of the moduli of elements of the given matrix;

aux[5]: the modulus of an element which is of maximum absolute value for the matrix which had been given in *a*;

aux[7]: an upper bound for the growth (the modulus of an element which is of maximum absolute value for the matrices occurring during elimination);

ri: int *ri[1:n]*;
exit: the pivotal row-indices; see *ci* below;

ci: int *ci[1:n]*;
exit: the pivotal column-indices; in the *i*-th iteration, row *i* is interchanged with row *ri[i]* and column *j* is interchanged with column *ci[j]*.

Functions used: rowcst, elmrow, maxelmrow, ichcol, ichrow, absmaxmat.

Method: The process of Gaussian elimination is performed in at most *n* steps, where *n* denotes the order of the matrix. Partial pivoting is used as long as the calculated upper bound for the growth [Busi71] is less than a critical value that equals *aux[4]*n* times the modulus of an element which is of maximum absolute value for the given matrix. In the partial pivoting strategy, that element is chosen as pivot in the k-th step whose absolute value is maximal for the k-th column of the lower triangular matrix *L*. However, if the upper bound for the growth exceeds this critical value in the k-th step then a pivot is selected in the j-th step, j=k,...,n, in such a way that its absolute value is maximal for the remaining submatrix of order n-k+1 (complete pivoting). Since in practice, if we choose *aux[4]* properly, the upper bound for the growth rarely exceeds this critical value [Busi71, Wi63], we usually take advantage of the greater speed of partial pivoting (order n-k+1 in the k-th step), while in a few doubtful cases numerical difficulties will be recognized and the process will switch to complete pivoting (order $(n-k+1)^2$ in the k-th step). Using the function gsselm, the upper bound for the relative error in the solution of a linear system [Wi63, Wi65] will be at most *aux[4]*n* times the upper bound using Gaussian elimination with complete pivoting only. Usually this will be a crude overestimate. The choice *aux[4]* < *1/n* will result in complete pivoting only, while partial pivoting will be used in every step if we choose *aux[4]* > $2^{n-1}/n$. Usually *aux[4]=8* will give good results [Busi71]. The process will also switch to complete pivoting if the modulus of the pivot obtained with partial pivoting is less than a certain tolerance which equals the given relative tolerance *aux[2]* times the modulus of an element which is of maximum absolute value for the given matrix. If all elements in the remaining submatrix are smaller in absolute value than this tolerance then the process is broken off and the previous step number is delivered in *aux[3]*. In contrast with the method used in the function dec, no equilibrating is done in this pivoting strategy. The user has to take care for a reasonable scaling of the matrix elements.

```
#include <math.h>

void gsselm(float **a, int n, float aux[], int ri[], int ci[])
{
    void elmrow(int, int, int, int, float **, float **, float);
    int maxelmrow(int, int, int, int, float **, float **, float);
    void ichcol(int, int, int, int, float **);
    void ichrow(int, int, int, int, float **);
    void rowcst(int, int, int, float **, float);
    float absmaxmat(int, int, int, int, int *, int *, float **);
    int i,j,p,q,r,r1,jpiv,rank,signdet,partial;
    float crit,pivot,rgrow,max,aid,max1,eps;

    aux[5]=rgrow=absmaxmat(1,n,1,n,&i,&j,a);
    crit=n*rgrow*aux[4];
    eps=rgrow*aux[2];
    max=0.0;
    rank=n;
    signdet=1;
    partial = rgrow != 0;
    for (q=1; q<=n; q++)
        if (q != j) {
            aid=fabs(a[i][q]);
            if (aid > max) max=aid;
        }
    rgrow += max;
    for (r=1; r<=n; r++) {
        r1=r+1;
        if (i != r) {
            signdet = -signdet;
            ichrow(1,n,r,i,a);
        }
        if (j != r) {
            signdet = -signdet;
            ichcol(1,n,r,j,a);
        }
        ri[r]=i;
        ci[r]=j;
        pivot=a[r][r];
        if (pivot < 0.0) signdet = -signdet;
        if (partial) {
            max=max1=0.0;
            j=r1;
            rowcst(r1,n,r,a,1.0/pivot);
            for (p=r1; p<=n; p++) {
                elmrow(r1,n,p,r,a,a,-a[p][r]);
                aid=fabs(a[p][r1]);
                if (max < aid) {
                    max=aid;
                    i=p;
                }
            }
            for (q=r1+1; q<=n; q++) {
                aid=fabs(a[i][q]);
                if (max1 < aid) max1=aid;
            }
            aid=rgrow;
            rgrow += max1;
            if ((rgrow > crit) || (max < eps)) {
                partial=0;
                rgrow=aid;
                max=absmaxmat(r1,n,r1,n,&i,&j,a);
            }
        } else {
            if (max <= eps) {
                rank=r-1;
                if (pivot < 0.0) signdet = -signdet;
                break;
            }
            max = -1.0;
            rowcst(r1,n,r,a,1.0/pivot);
            for (p=r1; p<=n; p++) {
                jpiv=maxelmrow(r1,n,p,r,a,a,-a[p][r]);
                aid=fabs(a[p][jpiv]);
```

```
                if (max < aid) {
                    max=aid;
                    i=p;
                    j=jpiv;
                }
            }
            if (rgrow < max) rgrow=max;
        }
    }
    aux[1]=signdet;
    aux[3]=rank;
    aux[7]=rgrow;
}
```

C. onenrminv

Computes $\|a^{-1}\|_1$, a being an $n \times n$ matrix assumed to have been triangularly decomposed by the function dec or gsselm. a^{-1} is first computed by forward and back substitution [Dek68, Wi63, Wi65], and $\|a^{-1}\|_1$ is then determined.

Function Parameters:

float onenrminv *(a,n)*

onenrminv: given the 1-norm of the calculated inverse of the matrix whose triangularly decomposed form is given in array *a*;

a: float *a[1:n,1:n]*;
  entry: the triangularly decomposed form of a matrix, as delivered by the function dec or gsselm; the elements of *a* remain unaltered on exit of the function;

n: int; the order of the matrix.

Function used: matvec.

```
#include <math.h>

float onenrminv(float **a, int n)
{
    float *allocate_real_vector(int, int);
    void free_real_vector(float *, int);
    float matvec(int, int, int, float **, float []);
    int i,j;
    float norm,max,aid,*y;

    y=allocate_real_vector(1,n);
    norm=0.0;
    for (j=1; j<=n; j++) {
        for (i=1; i<=n; i++)
            y[i]=(i < j) ? 0 :
                    ((i == j) ? 1.0/a[i][i] : -matvec(j,i-1,i,a,y)/a[i][i]);
        max=0.0;
        for (i=n; i>=1; i--) {
            aid = y[i] -= matvec(i+1,n,i,a,y);
            max += fabs(aid);
        }
        if (norm < max) norm=max;
    }
    free_real_vector(y,1);
    return (norm);
}
```

D. erbelm

Computes a rough error bound r for the relative error $\|\Delta x\|_1 / \|x\|$ in the solution of a linear system of n-equations $Ax = b$; it is assumed that A has been triangularly decomposed by the function *gsselm* [Dek68, Wi63, Wi65], and that an upper bound for $\|A^{-1}\|_1$ (possibly determined by the function *onenrminv*) is available (r is a function of A alone). With *eps* the machine precision, n the order of A, *epsa* an upper bound for the relative error in the elements of A, g the upper bound for the growth of auxiliary numbers produced during Gauss elimination (see the function *gsselm*: *aux[7]*), C the calculated inverse of A^{-1}, and

$$q = g * (0.75*n^3 + 4.5*n^2) * eps + epsa$$

$$p = q * \|C\|_1 / (1 - q*\|C\|_1)$$

r is given by $r = p/(1-p)$. If A is very badly conditioned, it may occur that the above formula becomes unusable.

Function Parameters:

void erbelm (*n,aux,nrminv*)

n: int; the order of the linear system in consideration;

aux: float *aux[0:11]*;

entry:

aux[0]: the machine precision;

aux[5]: the modulus of an element which is of maximum absolute value for the matrix of the linear system, this value is delivered by the function gsselm;

aux[6]: an upper bound for the relative error in the elements of the matrix of the linear system;

aux[7]: an upper bound for the growth during Gaussian elimination, this value is delivered in *aux[7]* by the function gsselm;

exit:

aux[9]: the value of *nrminv*;

aux[11]: a rough upper bound for the relative error in the solution of a linear system when Gaussian elimination is used for the calculation of this solution, if no use can be made of the formula for the error bound because of a very bad condition of the matrix then *aux[11]* = *-1*;

nrminv: float;

entry: the 1-norm of the inverse of the matrix of the linear system must be given in *nrminv*, this value may be obtained by the function *onenrminv*.

```
#include <math.h>

void erbelm(int n, float aux[], float nrminv)
{
    float aid,eps;

    eps=aux[0];
    aid=(1.06*eps*(0.75*n+4.5)*(n*n)*aux[7]+aux[5]*aux[6])*nrminv;
    aux[11]=(2.0*aid >= (1.0-eps)) ? -1.0 : aid/(1.0-2.0*aid);
    aux[9]=nrminv;
}
```

E. gsserb

Performs a triangular decomposition of the $n \times n$ matrix A and calculates an upper bound for the relative error in the solution of linear systems of the form $Ax = b$.

Function Parameters:

void gsserb (*a,n,aux,ri,ci*)

a: float *a[1:n,1:n]*;
entry: the matrix to be decomposed;
exit: the calculated lower triangular matrix and unit upper triangular matrix with its unit diagonal omitted;

n: int; the order of the matrix;

aux: float *aux[0:11]*;
entry:
aux[0]: the machine precision;
aux[2]: a relative tolerance;
aux[4]: a value which is used for controlling pivoting;
aux[6]: an upper bound for the relative precision of the matrix elements;
exit:
aux[1]: if *R* is the number of elimination steps performed, then *aux[1]* equals 1 if the determinant of the principal submatrix of order *R* is positive, otherwise *aux[1]* = *-1*;
aux[3]: the number of elimination steps performed;
aux[5]: the modulus of an element which is of maximum absolute value for the matrix which has been given in array *a*;
aux[7]: an upper bound for the growth;
aux[9]: if *aux[3]=n* then *aux[9]* will be equal to the 1-norm of the inverse matrix, else *aux[9]* will be undefined;
aux[11]: if *aux[3]=n* then the value of *aux[11]* will be a rough upper bound for the relative error in the solution of linear systems with a matrix as given in array *a*, else *aux[11]* will be undefined. If no use can be made of the formula for the error bound as given above because of a very bad condition of the matrix then *aux[11]* = *-1*;

ri: int *ri[1:n]*; exit: the pivotal row-indices; see *gsselm*;

ci: int *ci[1:n]*; exit: the pivotal column-indices.

Functions used: gsselm, onenrminv, erbelm.

```
void gsserb(float **a, int n, float aux[], int ri[], int ci[])
{
    void gsselm(float **, int, float [], int [], int []);
    float onenrminv(float **, int);
    void erbelm(int, float [], float);

    gsselm(a,n,aux,ri,ci);
    if (aux[3] == n) erbelm(n,aux,onenrminv(a,n));
}
```

F. gssnri

Performs a triangular decomposition of the *n*×*n* matrix *A* and calculates $\|A^{-1}\|_1$.

Function Parameters:

void gssnri (*a,n,aux,ri,ci*)

a: float *a[1:n,1:n]*;
entry: the matrix to be decomposed;
exit: the calculated lower triangular matrix and unit upper triangular matrix with its

unit diagonal omitted;

n: int; the order of the matrix;

aux: float *aux[1:9]*;

entry:

aux[2]: a relative tolerance;

aux[4]: a value used for controlling pivoting;

exit:

aux[1]: if *R* is the number of elimination steps performed, then *aux[1]* equals 1 if the determinant of the principal submatrix of order *R* is positive, otherwise *aux[1]* = *-1*;

aux[3]: the number of elimination steps performed;

aux[5]: the modulus of an element which is of maximum absolute value for the matrix which has been given in array *a*;

aux[7]: an upper bound for the growth;

aux[9]: if *aux[3]=n* then *aux[9]* will be equal to the 1-norm of the inverse matrix, else *aux[9]* will be undefined;

ri: int *ri[1:n]*;

exit: the pivotal row-indices; see *gsselm*;

ci: int *ci[1:n]*;

exit: the pivotal column-indices.

Functions used: gsselm, onenrminv.

```
void gssnri(float **a, int n, float aux[], int ri[], int ci[])
{
    void gsselm(float **, int, float [], int [], int []);
    float onenrminv(float **, int);

    gsselm(a,n,aux,ri,ci);
    if (aux[3] == n) aux[9]=onenrminv(a,n);
}
```

3.1.2 Calculation of determinant

determ

Calculates the determinant of a triangularly decomposed matrix. The calculation of the determinant is done directly by calculating the product of the diagonal elements of the lower triangular matrix given in array *a*. The user is warned that overflow may occur if the order of the matrix is large.

Function Parameters:

float determ (*a,n,sign*)

determ: delivers the calculated value of the determinant of the matrix;

a: float *a[1:n,1:n]*;

entry: the diagonal elements of the lower triangular matrix, obtained by triangular decomposition of the matrix, has to be given in $a_{i,i}$, $i=1,...,n$;

n: int; the order of the matrix whose determinant has to be calculated;

sign: int;

entry: if the determinant of the matrix is positive then the value of *sign* should be +1, else -1; this value is delivered by the function *gsselm* or *dec* in *aux[1]*.

```
#include <math.h>

float determ(float **a, int n, int sign)
{
    int i;
    float det;

    det=1.0;
    for (i=1; i<=n; i++) det *= a[i][i];
    return (sign*fabs(det));
}
```

3.1.3 Solution of linear equations

A. sol

Solves the linear system whose matrix has been triangularly decomposed. *sol* should be called after the function *dec* and solves the linear system with a matrix whose triangularly decomposed form as produced by *dec* is given in array *a* and a right hand side as given in array *b*. *sol* leaves the array *a* and the permutation array *p* unaltered. After one call of *dec*, several calls of *sol* may follow for solving several systems having the same matrix but different right hand sides.

Function Parameters:

void sol (*a,n,p,b*)

a: float *a[1:n,1:n]*;
  entry: the triangularly decomposed form of the matrix of the linear system as produced by the function *dec*;
n: int; the order of the matrix;
p: int *p[1:n]*;
  exit: the pivotal indices, as produced by *dec*;
b: float *b[1:n]*;
  entry: the right hand side of the linear system;
  exit: the solution of the linear system.

Function used: matvec.

```
void sol(float **a, int n, int p[], float b[])
{
    float matvec(int, int, int, float **, float []);
    int k,pk;
    float r;

    for (k=1; k<=n; k++) {
        r=b[k];
        pk=p[k];
        b[k]=(b[pk]-matvec(1,k-1,k,a,b))/a[k][k];
        if (pk != k) b[pk]=r;
    }
    for (k=n; k>=1; k--) b[k] -= matvec(k+1,n,k,a,b);
}
```

B. decsol

Solves a well-conditioned linear system of equations $Ax = b$ whose order is small relative to the number of binary digits in the number representation.

Function Parameters:

void decsol (*a,n,aux,b*)

a: float *a[1:n,1:n]*;
  entry: the *n*-th order matrix;
  exit: the calculated lower triangular matrix and unit upper triangular matrix with its unit diagonal omitted;
n: int; the order of the matrix;
aux: float *aux[1:3]*;
  entry:
  aux[2]: a relative tolerance; a reasonable choice for this value is an estimate of the relative precision of the matrix elements; however, it should not be chosen smaller than the machine precision;
  exit:
  aux[1]: if *R* is the number of elimination steps performed (see *aux[3]*), then *aux[1]* equals 1 if the determinant of the principal submatrix of order *R* is positive, else *aux[1]* = *-1*;
  aux[3]: the number of elimination steps performed, if *aux[3]* < *n* then the process is terminated and no solution will be calculated;
b: float *b[1:n]*;
  entry: the right hand side of the linear system;
  exit: if *aux[3]* = *n*, then the calculated solution of the linear system is overwritten on *b*, else *b* remains unaltered.

Functions used: dec, sol.

```
void decsol(float **a, int n, float aux[], float b[])
{
    int *allocate_integer_vector(int, int);
    void free_integer_vector(int *, int);
    void sol(float **, int, int [], float []);
    void dec(float **, int, float [], int []);
    int *p;

    p=allocate_integer_vector(1,n);
    dec(a,n,aux,p);
    if (aux[3] == n) sol(a,n,p,b);
    free_integer_vector(p,1);
}
```

C. solelm

Solves the $n \times n$ system of equations $Ax = b$ whose matrix has been triangularly decomposed by the function *gsselm* or *gsserb*. *solelm* leaves the matrix *a* and permutation arrays *ri* and *ci* unaltered. After one call of *gsselm* or *gsserb*, several calls of *solelm* may follow for solving several systems having the same matrix but different right hand sides.

Function Parameters:

void solelm (*a,n,ri,ci,b*)

a: float *a[1:n,1:n]*;

entry: the triangularly decomposed form of the matrix of the linear system as produced by *gsselm*;

n: int; the order of the matrix;

ri: int *ri[1:n]*;
entry: the pivotal row indices, as produced by *gsselm*;

ci: int *ci[1:n]*;
entry: the pivotal column indices, as produced by *gsselm*;

b: float *b[1:n]*;
entry: the right hand side of the linear system;
exit: the solution of the linear system.

Function used: sol.

```
void solelm(float **a, int n, int ri[], int ci[], float b[])
{
    void sol(float **, int, int [], float []);
    int r,cir;
    float w;

    sol(a,n,ri,b);
    for (r=n; r>=1; r--) {
        cir=ci[r];
        if (cir != r) {
            w=b[r];
            b[r]=b[cir];
            b[cir]=w;
        }
    }
}
```

D. gsssol

Solves a linear system. *gsssol* first calls *gsselm* to decompose the matrix and then *solelm* to solve the linear system.

Function Parameters:

void gsssol (*a,n,aux,b*)

a: float *a[1:n,1:n]*;
entry: the *n*-th order matrix;
exit: the calculated lower triangular matrix and unit upper triangular matrix with its unit diagonal omitted;

n: int; the order of the matrix;

aux: float *aux[1:7]*;
entry:
aux[2]: a relative tolerance; a reasonable choice for this value is an estimate of the relative precision of the matrix elements; however, it should not be chosen smaller than the machine precision;
aux[4]: a value used for controlling pivoting, see *gsselm*;
exit:
aux[1]: if *R* is the number of elimination steps performed (see *aux[3]*), then *aux[1]* equals 1 if the determinant of the principal submatrix of order *R* is positive, else *aux[1]* = *-1*;
aux[3]: the number of elimination steps performed, if *aux[3]* < *n* then the process is

terminated and no solution will be calculated;

aux[5]: the modulus of an element which is of maximum absolute value for the matrix given in array *a*;

aux[7]: an upper bound for the growth, see *gsselm*;

b: float *b[1:n]*;

entry: the right hand side of the linear system;

exit: if *aux[3]* = *n*, then the calculated solution of the linear system is overwritten on *b*, else *b* remains unaltered.

Functions used: solelm, gsselm.

```
void gsssol(float **a, int n, float aux[], float b[])
{
    int *allocate_integer_vector(int, int);
    void free_integer_vector(int *, int);
    void solelm(float **, int, int [], int [], float []);
    void gsselm(float **, int, float [], int [], int []);
    int *ri,*ci;

    ri=allocate_integer_vector(1,n);
    ci=allocate_integer_vector(1,n);
    gsselm(a,n,aux,ri,ci);
    if (aux[3] == n) solelm(a,n,ri,ci,b);
    free_integer_vector(ri,1);
    free_integer_vector(ci,1);
}
```

E. gsssolerb

Solves the $n \times n$ system of equation $Ax = b$, and provides an upper bound for the relative error in x. *gsssolerb* calls *gsserb* to perform the triangular decomposition and to calculate an upper bound for the relative error, and then calls *solelm* to calculate x.

Function Parameters:

void gsssolerb (*a,n,aux,b*)

a: float *a[1:n,1:n]*;

entry: the *n*-th order matrix;

exit: the calculated lower triangular matrix and unit upper triangular matrix with its unit diagonal omitted;

n: int;

entry: the order of the matrix;

aux: float *aux[0:11]*;

entry:

aux[0]: the machine precision;

aux[2]: a relative tolerance; a reasonable choice for this value is an estimate of the relative precision of the matrix elements; however, it should not be chosen smaller than the machine precision;

aux[4]: a value used for controlling pivoting, see *gsselm*;

aux[6]: an upper bound for the relative precision of the given matrix elements;

exit:

aux[1]: if *R* is the number of elimination steps performed (see *aux[3]*), then *aux[1]* equals 1 if the determinant of the principal submatrix of order *R* is positive, else *aux[1]* = *-1*;

aux[3]: the number of elimination steps performed, if *aux[3]* < *n* then the process is terminated and no solution will be calculated;
aux[5]: the modulus of an element which is of maximum absolute value for the matrix given in array *a*;
aux[7]: an upper bound for the growth, see *gsselm*;
aux[9]: if *aux[3]* = *n* then *aux[9]* will be equal to the 1-norm of the inverse matrix, else *aux[9]* will be undefined;
aux[11]: if *aux[3]* = *n* then the value of *aux[11]* will be a rough upper bound for the relative error in the calculated solution of the given linear system, else *aux[11]* will be undefined; if no use can be made of the formula for the error bound because of a very bad condition of the matrix, otherwise *aux[11]* = *-1*.

b: float *b[1:n]*;
entry: the right hand side of the linear system;
exit: if *aux[3]* = *n*, then the calculated solution of the linear system is overwritten on *b*, else *b* remains unaltered.

Functions used: solelm, gsserb.

```
void gsssolerb(float **a, int n, float aux[], float b[])
{
	int *allocate_integer_vector(int, int);
	void free_integer_vector(int *, int);
	void solelm(float **, int, int [], int [], float []);
	void gsserb(float **, int, float [], int [], int []);
	int *ri,*ci;

	ri=allocate_integer_vector(1,n);
	ci=allocate_integer_vector(1,n);
	gsserb(a,n,aux,ri,ci);
	if (aux[3] == n) solelm(a,n,ri,ci,b);
	free_integer_vector(ri,1);
	free_integer_vector(ci,1);
}
```

3.1.4 Matrix inversion

A. inv

Calculates the inverse of a matrix that has been triangularly decomposed by *dec*.

Function Parameters:

void inv (*a,n,p*)

a: float *a[1:n,1:n]*;
entry: the triangularly decomposed form of the matrix as produced by the function *dec*;
exit: the calculated inverse matrix;
n: int; the order of the matrix;
p: int *p[1:n]*;
entry: the pivotal indices, as produced by *dec*.

Functions used: matmat, ichcol, dupcolvec.

```
void inv(float **a, int n, int p[])
{
    float *allocate_real_vector(int, int);
    void free_real_vector(float *, int);
    float matmat(int, int, int, int, float **, float **);
    void ichcol(int, int, int, int, float **);
    void dupcolvec(int, int, int, float **, float []);
    int j,k,k1;
    float r,*v;

    v=allocate_real_vector(1,n);
    for (k=n; k>=1; k--) {
        k1=k+1;
        for (j=n; j>=k1; j--) {
            a[j][k1]=v[j];
            v[j] = -matmat(k1,n,k,j,a,a);
        }
        r=a[k][k];
        for (j=n; j>=k1; j--) {
            a[k][j]=v[j];
            v[j] = -matmat(k1,n,j,k,a,a)/r;
        }
        v[k]=(1.0-matmat(k1,n,k,k,a,a))/r;
    }
    dupcolvec(1,n,1,a,v);
    for (k=n-1; k>=1; k--) {
        k1=p[k];
        if (k1 != k) ichcol(1,n,k,k1,a);
    }
    free_real_vector(v,1);
}
```

B. decinv

Obtains the inverse of the $n \times n$ matrix by partial pivoting using successive calls of the functions *dec* and *inv*.

Function Parameters:

void decinv (*a,n,aux*)

a: float *a[1:n,1:n]*;
  entry: the matrix whose inverse has to be calculated;
  exit: if *aux[3]* = *n*, then the calculated inverse matrix;
n: int; the order of the matrix;
aux: float *aux[1:3]*;
  entry:
  aux[2]: a relative tolerance; a reasonable choice for this value is an estimate of the relative precision of the matrix elements; however, it should not be chosen smaller than the machine precision;
  exit:
  aux[1]: if *R* is the number of elimination steps performed (see *aux[3]*), then *aux[1]* equals 1 if the determinant of the principal submatrix of order *R* is positive, else *aux[1]* = *-1*;
  aux[3]: the number of elimination steps performed, if *aux[3]* < *n* then the process is terminated and no inverse will be calculated.

Functions used: dec, inv.

```
void decinv(float **a, int n, float aux[])
{
    int *allocate_integer_vector(int, int);
```

```
    void free_integer_vector(int *, int);
    void dec(float **, int, float [], int []);
    void inv(float **, int, int []);
    int *p;

    p=allocate_integer_vector(1,n);
    dec(a,n,aux,p);
    if (aux[3] == n) inv(a,n,p);
    free_integer_vector(p,1);
}
```

C. inv1

Calculates the inverse of a matrix that has been triangularly decomposed by the function *gsselm* or *gsserb*. The 1-norm of the inverse matrix might also be calculated.

Function Parameters:

float inv1 (*a,n,ri,ci,withnorm*)

inv1: if the value of *withnorm* is nonzero then the value of *inv1* will be equal to the 1-norm of the calculated inverse matrix, else *inv1* = 0;

a: float *a[1:n,1:n]*;
entry: the triangularly decomposed form of the matrix as produced by *gsselm*;
exit: the calculated inverse matrix;

n: int; the order of the matrix;

ri: int *ri[1:n]*;
entry: the pivotal row indices, as produced by *gsselm*;

ci: int *ci[1:n]*;
entry: the pivotal column indices, as produced by *gsselm*;

withnorm: int;
entry: if the value of *withnorm* is nonzero then the 1-norm of the inverse matrix will be calculated and assigned to *inv1*, else *inv1* = 0.

Functions used: ichrow, inv.

```
#include <math.h>

float inv1(float **a, int n, int ri[], int ci[], int withnorm)
{
    void ichrow(int, int, int, int, float **);
    void inv(float **, int, int []);
    int l,k,k1;
    float aid,nrminv;

    inv(a,n,ri);
    nrminv=0.0;
    if (withnorm)
        for (l=1; l<=n; l++) nrminv += fabs(a[l][n]);
    for (k=n-1; k>=1; k--) {
        if (withnorm) {
            aid=0.0;
            for (l=1; l<=n; l++) aid += fabs(a[l][k]);
            if (nrminv < aid) nrminv=aid;
        }
        k1=ci[k];
        if (k1 != k) ichrow(1,n,k,k1,a);
    }
    return (nrminv);
}
```

D. gssinv

Uses the function *gsselm* to perform a triangular decomposition of the matrix and the function *inv1* to calculate the inverse matrix.

Function Parameters:

void gssinv (*a,n,aux*)

a: float *a[1:n,1:n]*;
entry: the matrix whose inverse has to be calculated;
exit: if *aux[3]* = *n* then the calculated inverse matrix;

n: int;
entry: the order of the matrix;

aux: float *aux[1:9]*;
entry:
aux[2]: a relative tolerance; a reasonable choice for this value is an estimate of the relative precision of the matrix elements; however, it should not be chosen smaller than the machine precision;
aux[4]: a value used for controlling pivoting, see *gsselm*;
exit:
aux[1]: if *R* is the number of elimination steps performed (see *aux[3]*), then *aux[1]* equals 1 if the determinant of the principal submatrix of order *R* is positive, else *aux[1]* = *-1*;
aux[3]: the number of elimination steps performed, if *aux[3]* < *n* then the process is terminated and no solution will be calculated;
aux[5]: the modulus of an element which is of maximum absolute value for the matrix given in array *a*;
aux[7]: an upper bound for the growth, see *gsselm*;
aux[9]: if *aux[3]* = *n* then *aux[9]* will be equal to the 1-norm of the calculated inverse matrix, else *aux[9]* will be undefined.

Functions used: inv1, gsselm.

```
void gssinv(float **a, int n, float aux[])
{
    int *allocate_integer_vector(int, int);
    void free_integer_vector(int *, int);
    void gsselm(float **, int, float [], int [], int []);
    float inv1(float **, int, int [], int [], int);
    int *ri,*ci;

    ri=allocate_integer_vector(1,n);
    ci=allocate_integer_vector(1,n);
    gsselm(a,n,aux,ri,ci);
    if (aux[3] == n) aux[9]=inv1(a,n,ri,ci,1);
    free_integer_vector(ri,1);
    free_integer_vector(ci,1);
}
```

E. gssinverb

Uses the function *gsselm* to perform the triangular decomposition of the matrix *a*, the function *inv1* to calculate the inverse matrix and its 1-norm and the function *erbelm* to calculate an upper bound for the relative error in the calculated inverse.

Function Parameters:

void gssinverb (*a,n,aux*)

a: float *a[1:n,1:n]*;
entry: the matrix whose inverse has to be calculated;
exit: if *aux[3]* = *n* then the calculated inverse matrix;

n: int;
entry: the order of the matrix;

aux: float *aux[0:11]*;
entry:
aux[0]: the machine precision;
aux[2]: a relative tolerance; a reasonable choice for this value is an estimate of the relative precision of the matrix elements; however, it should not be chosen smaller than the machine precision;
aux[4]: a value used for controlling pivoting, see *gsselm*;
aux[6]: an upper bound for the relative precision of the given matrix elements;
exit:
aux[1]: if *R* is the number of elimination steps performed (see *aux[3]*), then *aux[1]* equals 1 if the determinant of the principal submatrix of order *R* is positive, else *aux[1]* = *-1*;
aux[3]: the number of elimination steps performed, if *aux[3]* < *n* then the process is terminated and no solution will be calculated;
aux[5]: the modulus of an element which is of maximum absolute value for the matrix given in array *a*;
aux[7]: an upper bound for the growth, see *gsselm*;
aux[9]: if *aux[3]* = *n* then *aux[9]* will be equal to the 1-norm of the inverse matrix, else *aux[9]* will be undefined;
aux[11]: if *aux[3]* = *n* then the value of *aux[11]* will be a rough upper bound for the relative error in the calculated inverse matrix, else *aux[11]* will be undefined if no use can be made of the formula for the error bound because of a very bad condition of the matrix, otherwise *aux[11]* = *-1*.

Functions used: inv1, gsselm, erbelm.

```
void gssinverb(float **a, int n, float aux[])
{
    int *allocate_integer_vector(int, int);
    void free_integer_vector(int *, int);
    void gsselm(float **, int, float [], int [], int []);
    float inv1(float **, int, int [], int [], int);
    void erbelm(int, float [], float);
    int *ri,*ci;

    ri=allocate_integer_vector(1,n);
    ci=allocate_integer_vector(1,n);
    gsselm(a,n,aux,ri,ci);
    if (aux[3] == n) erbelm(n,aux,inv1(a,n,ri,ci,1));
    free_integer_vector(ri,1);
    free_integer_vector(ci,1);
}
```

3.1.5 Iteratively improved solution

A. itisol

Solves a linear system $Ax = b$ whose matrix has been triangularly decomposed by the function *gsselm* or *gsserb*. This solution will be refined iteratively until the calculated relative correction to this solution will be less than a prescribed value (see *aux[10]*). Each iteration of the refinement process consists of the following three steps [Bus72a, Dek71, Wi65]:

(a) calculate in double precision, the residual vector r defined by $r = Ax^k - b$, where x^k denotes the solution obtained in the k-th iteration;
(b) calculate the solution c of the linear system $Ac = r$, with the aid of the triangularly decomposed matrix as given in the array *lu*;
(c) calculate the new solution $x^{k+1} = x^k - c$.

Function Parameters:

void itisol (*a,lu,n,aux,ri,ci,b*)

a: float *a[1:n,1:n]*;
entry: the matrix of the linear system;
lu: float *lu[1:n,1:n]*;
entry: the triangularly decomposed form of the matrix given in *a*, as delivered by *gsselm*;
n: int; the order of the matrix;
aux: float *aux[10:13]*;
entry:
aux[10]: a relative tolerance for the solution vector; if the 1-norm of the vector of corrections to the solution divided by the 1-norm of the calculated solution is smaller than *aux[10]*, then the process will stop; the user should not choose the value of *aux[10]* smaller than the relative precision of the elements of the matrix and the right hand side of the linear system;
aux[12]: the maximum number of iterations allowed for the refinement of the solution; if the number of iterations exceeds the value of *aux[12]* then the process will be broken off; usually *aux[12]* = 5 will give good results;
exit:
aux[11]: the 1-norm of the vector of corrections to the solution in the last iteration step divided by the 1-norm of the calculated solution;
if *aux[11]* > *aux[10]* then the process has been broken off because the number of iterations exceeded the value given in *aux[12]*;
aux[13]: the 1-norm of the residual vector r above;
ri: int *ri[1:n]*;
entry: the pivotal row-indices, as produced by *gsselm*;
ci: int *ci[1:n]*;
entry: the pivotal column-indices, as produced by *gsselm*;
b: float *b[1:n]*;
entry: the right hand side of the linear system;
exit: the calculated solution of the linear system.

Functions used: solelm, inivec, dupvec.

Method: If the condition of the matrix is not too bad then the precision of the calculated solution will be of the order of the precision asked for in *aux[10]*. If the condition of the matrix is very bad then this process will possibly not converge or, in exceptional cases, converge to a useless result. If the user wants to make certain about the precision of the calculated solution then the function *itisolerb* should be used. *itisol* leaves *a, lu, ri* and *ci* unaltered, so after one call of *gsselm* several calls of *itisol* may follow to calculate the solution of several linear systems with the same matrix but different right hand sides.

```
#include <math.h>

void itisol(float **a, float **lu, int n, float aux[],
            int ri[], int ci[], float b[])
{
    float *allocate_real_vector(int, int);
    void free_real_vector(float *, int);
    void solelm(float **, int, int [], int [], float []);
    void inivec(int, int, float [], float);
    void dupvec(int, int, int, float [], float []);
    int i,j,iter,maxiter;
    float maxerx,erx,nrmres,nrmsol,r,rr,*res,*sol;
    double dtemp;

    res=allocate_real_vector(1,n);
    sol=allocate_real_vector(1,n);
    maxerx=erx=aux[10];
    maxiter=aux[12];
    inivec(1,n,sol,0.0);
    dupvec(1,n,0,res,b);
    iter=1;
    do {
        solelm(lu,n,ri,ci,res);
        erx=nrmsol=nrmres=0.0;
        for (i=1; i<=n; i++) {
            r=res[i];
            erx += fabs(r);
            rr=sol[i]+r;
            sol[i]=rr;
            nrmsol += fabs(rr);
        }
        erx /= nrmsol;
        for (i=1; i<=n; i++) {
            dtemp = -(double)b[i];
            for (j=1; j<=n; j++)
                dtemp += (double)a[i][j]*(double)sol[j];
            r = -dtemp;
            res[i]=r;
            nrmres += fabs(r);
        }
        iter++;
    } while ((iter <= maxiter) && (maxerx < erx));
    dupvec(1,n,0,b,sol);
    aux[11]=erx;
    aux[13]=nrmres;
    free_real_vector(res,1);
    free_real_vector(sol,1);
}
```

B. gssitisol

Uses the function *gsselm* to perform a triangular decomposition of the matrix and the function *itisol* to calculate an iteratively refined solution of the given linear system.

Function Parameters:

void gssitisol (*a,n,aux,b*)

a: float *a[1:n,1:n]*;
entry: the *n*-th order matrix;

exit: the calculated lower triangular matrix and unit upper triangular matrix with its unit diagonal omitted;

n: int; entry: the order of the matrix;

aux: float *aux[1:13]*;

entry:

aux[2]: a relative tolerance for the process of triangular decomposition; a reasonable choice for this value is an estimate of the relative precision of the matrix elements; however, it should not be chosen smaller than the machine precision;

aux[4]: a value used for controlling pivoting, see *gsselm*;

aux[10]: a relative tolerance for the solution vector; if the 1-norm of the vector of corrections to the solution divided by the 1-norm of the calculated solution is smaller than *aux[10]*, then the process will stop; the user should not choose the value of *aux[10]* smaller than the relative precision of the elements of the matrix and the right hand side of the linear system;

aux[12]: the maximum number of iterations allowed for the refinement of the solution; if the number of iterations exceeds the value of *aux[12]* then the process will be broken off; usually *aux[12]* = 5 will give good results;

exit:

aux[1]: if *R* is the number of elimination steps performed (see *aux[3]*), then *aux[1]* equals 1 if the determinant of the principal submatrix of order *R* is positive, else *aux[1]* = *-1*;

aux[3]: the number of elimination steps performed, if *aux[3]* < *n* then the process has been broken off and no solution will have been calculated;

aux[5]: modulus of an element which is of maximum absolute value for array *a*;

aux[7]: an upper bound for the growth, see *gsselm*;

aux[11]: if *aux[3]* < *n* then *aux[11]* will be undefined, else *aux[11]* will be equal to the 1-norm of the vector of corrections to the solution in the last step divided by the 1-norm of the calculated solution; if *aux[11]* > *aux[10]* then the process has been broken off because the number of iterations exceeded the value given in *aux[12]*;

aux[13]: if *aux[3]* = *n* then the value of *aux[13]* will be equal to the 1-norm of the residual vector (see *itisol*), else *aux[13]* will be undefined;

b: float *b[1:n]*;

entry: the right hand side of the linear system;

exit: if *aux[3]* = *n*, then the calculated solution of the linear system is overwritten on *b*, else *b* remains unaltered.

Functions used: dupmat, gsselm, itisol.

```
void gssitisol(float **a, int n, float aux[], float b[])
{
    int *allocate_integer_vector(int, int);
    float **allocate_real_matrix(int, int, int, int);
    void free_integer_vector(int *, int);
    void free_real_matrix(float **, int, int, int);
    void gsselm(float **, int, float [], int [], int []);
    void itisol(float **, float **, int, float [],
                int [], int [], float []);
    void dupmat(int, int, int, int, float **, float **);
    int *ri,*ci;
    float **aa;

    ri=allocate_integer_vector(1,n);
    ci=allocate_integer_vector(1,n);
```

```
    aa=allocate_real_matrix(1,n,1,n);

    dupmat(1,n,1,n,aa,a);
    gsselm(a,n,aux,ri,ci);
    if (aux[3] == n) itisol(aa,a,n,aux,ri,ci,b);

    free_integer_vector(ri,1);
    free_integer_vector(ci,1);
    free_real_matrix(aa,1,n,1);
}
```

C. itisolerb

Solves a linear system $Ax = b$ whose matrix has been triangularly decomposed by the function *gssnri* (the function *gssnri* also delivers the proper values for the odd elements of the array *aux*). *itisolerb* calculates, with the use of the function *itisol*, an iteratively improved solution of the linear system. Moreover, with values of the machine precision ε and of $\|A^{-1}\|_1$ available, and also with upper bounds η_A for $\|\Delta A\|_1/\|A\|_1$, ΔA being the matrix of errors in A, and η_b for $\|\Delta b\|_1/\|b\|_1$, Δb being the vector of errors in b available, *itisolerb* calculates a realistic upper bound for the relative error $\|\Delta x\|_1/\|x\|_1$ in the calculated solution [Bus72a, Dek71]. The latter bound is given by

$$\frac{\|\Delta x\|_1}{\|x\|_1} = \frac{p}{1-p}$$

where

$$p = \left(\frac{\|Ax-b\|_1}{\|x\|_1} + \frac{\eta_b}{\|x\|_1} + \eta_A\right)\frac{\|C\|_1}{1-q\|C\|_1}$$

C being the computed inverse of A, and

$$q = \alpha(0.75n^3 + 4.5n^2)\varepsilon + \eta_A$$

α being the upper bound for the growth of auxiliary numbers during Gauss elimination (the value allocated to *aux[7]* at exit from the function *gssnri*). *itisolerb* leaves *a, lu, ri* and *ci* unaltered, so after one call of *gssnri* several calls of *itisolerb* may follow, to calculate the solution of several linear systems with the same matrix but different right hand sides.

Function Parameters:

void itisolerb (*a,lu,n,aux,ri,ci,b*)

a: float *a[1:n,1:n]*;
entry: the matrix of the linear system;

lu: float *lu[1:n,1:n]*;
entry: the triangularly decomposed form of the matrix given in *a*, as delivered by *gssnri*;

n: int;
entry: the order of the matrix;

aux: float *aux[0:13]*;
entry:

aux[0]: the machine precision;
aux[5]: the modulus of an element which is of maximum absolute value for the matrix of the linear system, this value is delivered by *gssnri* in *aux[5]*;
aux[6]: an upper bound for the relative error in the elements of the matrix of the linear system; the value of η_A above;
aux[7]: an upper bound for the growth during Gaussian elimination, this value is delivered by *gssnri* in *aux[7]*; the value of α above;
aux[8]: an upper bound for the relative error in the elements of the right hand side of the linear system; the value of η_b above;
aux[9]: the 1-norm of the inverse matrix, this value is delivered by *gssnri* in *aux[9]*; the value of $\|C\|_1$ above;
aux[10]: a relative tolerance for the solution vector; if the 1-norm of the vector of corrections to the solution divided by the 1-norm of the calculated solution is smaller than *aux[10]*, then the process will stop; the user should not choose the value of *aux[10]* smaller than the relative precision of the elements of the matrix and the right hand side of the linear system, given in *aux[6]* and *aux[8]*;
aux[12]: the maximum number of iterations allowed for the refinement of the solution; if the number of iterations exceeds the value of *aux[12]* then the process will be broken off; usually *aux[12]* = 5 will give good results;

exit:
aux[11]: a realistic upper bound (the value of *p/(1-p)* above) for the relative error in the calculated solution; if no use can be made of the error formula then *aux[11]* = -1;
aux[13]: the 1-norm of the residual vector; the value of $\|Ax-b\|_1$ for which the last iterative scheme operates.

ri: int *ri[1:n]*;
entry: the pivotal row-indices, as produced by *gssnri*;
ci: int *ci[1:n]*;
entry: the pivotal column-indices, as produced by *gssnri*;
b: float *b[1:n]*;
entry: the right hand side of the linear system;
exit: the calculated solution of the linear system.

Function used: itisol.

```
#include <math.h>

void itisolerb(float **a, float **lu, int n, float aux[],
               int ri[], int ci[], float b[])
{
   void itisol(float **, float **, int, float [],
               int [], int [], float []);
   int i;
   float nrmsol,nrminv,nrmb,alfa,tola,eps;

   eps=aux[0];
   nrminv=aux[9];
   tola=aux[5]*aux[6];
   nrmb=nrmsol=0.0;
   for (i=1; i<=n; i++) nrmb += fabs(b[i]);
   itisol(a,lu,n,aux,ri,ci,b);
   for (i=1; i<=n; i++) nrmsol += fabs(b[i]);
   alfa=1.0-(1.06*eps*aux[7]*(0.75*n+4.5)*n*n+tola)*nrminv;
   if (alfa < eps)
```

```
            aux[11] = -1.0;
        else {
            alfa=((aux[13]+aux[8]*nrmb)/nrmsol+tola)*nrminv/alfa;
            aux[11]=(1.0-alfa < eps) ? -1.0 : alfa/(1.0-alfa);
        }
}
```

D. gssitisolerb

Uses the function *gssnri* to perform a triangular decomposition of the matrix and the function *itisolerb* to calculate an iteratively refined solution of the given linear system and a realistic upper bound for the relative error in the solution.

Function Parameters:

void gssitisolerb (*a,n,aux,b*)

a: float *a[1:n,1:n]*;
- entry: the *n*-th order matrix;
- exit: the calculated lower triangular matrix and unit upper triangular matrix with its unit diagonal omitted;

n: int;
- entry: the order of the matrix;

aux: float *aux[0:13]*;
- entry:
  - *aux[0]*: the machine precision;
  - *aux[2]*: a relative tolerance; a reasonable choice for this value is an estimate of the relative precision of the matrix elements; however, it should not be chosen smaller than the machine precision;
  - *aux[4]*: a value used for controlling pivoting, see *gsselm*;
  - *aux[6]*: an upper bound for the relative error in the matrix elements of the linear system;
  - *aux[8]*: an upper bound for the relative error in the elements of the right hand side;
  - *aux[10]*: a relative tolerance for the solution vector; if the 1-norm of the vector of corrections to the solution divided by the 1-norm of the calculated solution is smaller than *aux[10]*, then the process will stop; the user should not choose the value of *aux[10]* smaller than the relative precision of the elements of the matrix and the right hand side of the linear system (*aux[10]* $\geq$ *aux[2]*);
  - *aux[12]*: the maximum number of iterations allowed for the refinement of the solution; if the number of iterations exceeds the value of *aux[12]* then the process will be broken off; usually *aux[12]* = 5 will give good results;
- exit:
  - *aux[1]*: if R is the number of elimination steps performed (see *aux[3]*), then *aux[1]* equals 1 if the determinant of the principal submatrix of order R is positive, else *aux[1]* = *-1*;
  - *aux[3]*: the number of elimination steps performed, if *aux[3]* < n then the process has been broken off and no solution will have been calculated;
  - *aux[5]*: the modulus of an element which is of maximum absolute value for the matrix which had been given in array a;
  - *aux[7]*: an upper bound for the growth, see *gsselm*;
  - *aux[9]*: if *aux[3]* = n then *aux[9]* will be equal to the 1-norm of the calculated inverse

matrix, else *aux[9]* will be undefined;

aux[11]: if *aux[3]* < *n* then *aux[11]* will be undefined, else *aux[11]* will be equal to a realistic upper bound for the relative error in the calculated solution; however, if no use can be made of the error formula (see *itisolerb*) then *aux[11]* = -1;

aux[13]: if *aux[3]* = *n* then the value of *aux[13]* will be equal to the 1-norm of the residual vector (see *itisol*), else *aux[13]* will be undefined;

b: float *b[1:n]*;

entry: the right hand side of the linear system;

exit: if *aux[3]* = *n*, then the calculated solution of the linear system is overwritten on *b*, else *b* remains unaltered.

Functions used: dupmat, gssnri, itisolerb.

```
void gssitisolerb(float **a, int n, float aux[], float b[])
{
    int *allocate_integer_vector(int, int);
    float **allocate_real_matrix(int, int, int, int);
    void free_integer_vector(int *, int);
    void free_real_matrix(float **, int, int, int);
    void gssnri(float **, int, float [], int [], int []);
    void itisolerb(float **, float **, int, float [],
                        int [], int [], float []);
    void dupmat(int, int, int, int, float **, float **);
    int *ri,*ci;
    float **aa;

    ri=allocate_integer_vector(1,n);
    ci=allocate_integer_vector(1,n);
    aa=allocate_real_matrix(1,n,1,n);

    dupmat(1,n,1,n,aa,a);
    gssnri(a,n,aux,ri,ci);
    if (aux[3] == n) itisolerb(aa,a,n,aux,ri,ci,b);

    free_integer_vector(ri,1);
    free_integer_vector(ci,1);
    free_real_matrix(aa,1,n,1);
}
```

3.2 Real Symmetric positive definite matrices

3.2.1 Preparatory procedures

A. chldec2

Calculates the Cholesky decomposition of a positive definite symmetric matrix whose upper triangle is given in a two-dimensional array. For a given symmetric positive definite matrix A, it computes the upper triangular matrix U for which $U^TU=A$. The columns $u_{(k)}$ $(k=1,...,n)$ of U are determined in succession. The decomposition process is broken off at stage k if

$$a_{k,k} - u_{(k)}^T u_{(k)} < \tau * B, \qquad \text{with } B = \max a_{i,i}, \quad 1 \le i \le n,$$

where τ is a tolerance supplied by the user. In this case the matrix, possibly modified by rounding errors, is not positive definite.

Function Parameters:

void chldec2 (*a,n,aux*)

a: float *a[1:n,1:n]*;
entry: the upper triangle of the positive definite symmetric matrix must be given in the upper triangular part of *a* (the elements *a[i,j]*, $i \leq j$);
exit: the Cholesky decomposition of the matrix is delivered in the upper triangle of *a*;
n: int;
entry: the order of the matrix;
aux: float *aux[2:3]*;
entry:
aux[2]: a relative tolerance used to control the calculation of the diagonal elements;
exit:
aux[3]: normal exit: *aux[3]* = *n*;
abnormal exit: if the decomposition fails because the matrix is not positive definite then *aux[3]* = *k-1*, where *k* is the last stage number.

Function used: tammat.

Method: The method used is Cholesky's square root method without pivoting [Dek68, Wi65].

```
#include <math.h>

void chldec2(float **a, int n, float aux[])
{
    float tammat(int, int, int, int, float **, float **);
    int k,j;
    float r,epsnorm;

    r=0.0;
    for (k=1; k<=n; k++)
        if (a[k][k] > r) r=a[k][k];
    epsnorm=aux[2]*r;
    for (k=1; k<=n; k++) {
        r=a[k][k]-tammat(1,k-1,k,k,a,a);
        if (r <= epsnorm) {
            aux[3]=k-1;
            return;
        }
        a[k][k]=r=sqrt(r);
        for (j=k+1; j<=n; j++)
            a[k][j]=(a[k][j]-tammat(1,k-1,j,k,a,a))/r;
    }
    aux[3]=n;
}
```

B. chldec1

Performs the Cholesky decomposition of a positive definite symmetric matrix whose upper triangle is given columnwise in a one-dimensional array, by Cholesky's square root method without pivoting (see the description of the function *chldec2*).

Function Parameters:

void chldec1 (*a,n,aux*)

a: float *a[1:(n+1)n/2]*;

entry: the upper triangular part of the positive definite symmetric matrix must be given columnwise in array *a* (the (i,j)-th element of the matrix should be given in array $a[(j-1)j/2+i]$, $1 \le i \le j \le n$);
exit: the Cholesky decomposition of the matrix is delivered columnwise in *a*;

n: int;
entry: the order of the matrix;

aux: float *aux[2:3]*;
entry:
aux[2]: a relative tolerance used to control the calculation of the diagonal elements;
exit:
aux[3]: normal exit: *aux[3]* = *n*;
abnormal exit: if the decomposition fails because the matrix is not positive definite then *aux[3]* = *k-1*, where *k* is the last stage number.

Function used: vecvec.

```
#include <math.h>

void chldec1(float a[], int n, float aux[])
{
    float vecvec(int, int, int, float [], float []);
    int j,k,kk,kj,low,up;
    float r,epsnorm;

    r=0.0;
    kk=0;
    for (k=1; k<=n; k++) {
        kk += k;
        if (a[kk] > r) r=a[kk];
    }
    epsnorm=aux[2]*r;
    kk=0;
    for (k=1; k<=n; k++) {
        kk += k;
        low=kk-k+1;
        up=kk-1;
        r=a[kk]-vecvec(low,up,0,a,a);
        if (r <= epsnorm) {
            aux[3]=k-1;
            return;
        }
        a[kk]=r=sqrt(r);
        kj=kk+k;
        for (j=k+1; j<=n; j++) {
            a[kj]=(a[kj]-vecvec(low,up,kj-kk,a,a))/r;
            kj +=j;
        }
    }
    aux[3]=n;
}
```

3.2.2 Calculation of determinant

A. chldeterm2

Calculates the determinant of a symmetric positive definite matrix whose Cholesky matrix is given in the upper triangle of a two-dimensional array. *chldeterm2* should be called after a successful call of *chldec2* or *chldecsol2*. *chldeterm2* should not be called if overflow is to be expected.

Function Parameters:

float chldeterm2 *(a,n)*

a: float *a[1:n,1:n]*;
entry: the upper triangular part of the Cholesky matrix as produced by *chldec2* or *chldecsol2* must be given in the upper triangle of *a*;
exit: the contents of *a* are not changed;

n: int;
entry: the order of the matrix.

```
float chldeterm2(float **a, int n)
{
    int k;
    float d;

    d=1.0;
    for (k=1; k<=n; k++) d *= a[k][k];
    return (d*d);
}
```

B. chldeterm1

Calculates the determinant of a symmetric positive definite matrix whose Cholesky matrix is given columnwise in a one-dimensional array. *chldeterm1* should be called after a successful call of *chldec1* or *chldecsol1*. *chldeterm1* should not be called if overflow is to be expected.

Function Parameters:

float chldeterm1 *(a,n)*

a: float *a[1:(n+1)n/2]*;
entry: the upper triangular part of the Cholesky matrix as produced by *chldec1* or *chldecsol1* must be given columnwise in array *a*;
exit: the contents of *a* are not changed;

n: int;
entry: the order of the matrix.

```
float chldeterm1(float a[], int n)
{
    int k,kk;
    float d;

    d=1.0;
    kk=0;
    for (k=1; k<=n; k++) {
        kk += k;
        d *= a[kk];
    }
    return (d*d);
}
```

3.2.3 Solution of linear equations

A. chlsol2

Calculates the solution of a system of linear equations, provided that the coefficient matrix has been decomposed by a successful call of the function *chldec2* or *chldecsol2*. The solution is obtained by carrying out the forward and back substitution with the Cholesky matrix and the right hand side. The elements of the Cholesky matrix are not changed. Several systems of linear equations with the same coefficient matrix but different right hand sides can be solved by successive calls of *chlsol2* [Dek68].

Function Parameters:

void chlsol2 (*a,n,b*)

a: float *a[1:n,1:n]*;
 entry: the upper triangular part of the Cholesky matrix as produced by *chldec2* or *chldecsol2* must be given in the upper triangle of *a*;
 exit: the contents of *a* are not changed;
n: int;
 entry: the order of the matrix;
b: float *b[1:n]*;
 entry: the right hand side of the system of linear equations;
 exit: the solution of the system.

Functions used: matvec, tamvec.

```
void chlsol2(float **a, int n, float b[])
{
    float matvec(int, int, int, float **, float []);
    float tamvec(int, int, int, float **, float []);
    int i;

    for (i=1; i<=n; i++) b[i]=(b[i]-tamvec(1,i-1,i,a,b))/a[i][i];
    for (i=n; i>=1; i--) b[i]=(b[i]-matvec(i+1,n,i,a,b))/a[i][i];
}
```

B. chlsol1

Calculates the solution of a system of linear equations, provided that the coefficient matrix has been decomposed by a successful call of the function *chldec1* or *chldecsol1*. The solution is obtained by carrying out the forward and back substitution with the Cholesky matrix and the right hand side. The elements of the Cholesky matrix are not changed. Several systems of linear equations with the same coefficient matrix but different right hand sides can be solved by successive calls of *chlsol1*.

Function Parameters:

void chlsol1 (*a,n,b*)

a: float *a[1:(n+1)n/2]*;
 entry: the upper triangular part of the Cholesky matrix as produced by *chldec1* or *chldecsol1* must be given columnwise in array *a*;
 exit: the contents of *a* are not changed;
n: int;
 entry: the order of the matrix;
b: float *b[1:n]*;
 entry: the right hand side of the system of linear equations;
 exit: the solution of the system.

Functions used: vecvec, seqvec.

```
void chlsol1(float a[], int n, float b[])
{
    float vecvec(int, int, int, float [], float []);
    float seqvec(int, int, int, int, float [], float []);
    int i,ii;

    ii=0;
    for (i=1; i<=n; i++) {
        ii += i;
        b[i]=(b[i]-vecvec(1,i-1,ii-i,b,a))/a[ii];
    }
    for (i=n; i>=1; i--) {
        b[i]=(b[i]-seqvec(i+1,n,ii+i,0,a,b))/a[ii];
        ii -= i;
    }
}
```

C. chldecsol2

Solves a system of linear equations with a symmetric positive definite coefficient matrix by calling *chldec2* and, if the call is successful, *chlsol2*. The coefficient matrix must be given in the upper triangle of a two-dimensional array.

Function Parameters:

void chldecsol2 (a,n,aux,b)

a: float *a[1:n,1:n]*;
  entry: the upper triangle of the symmetric positive definite matrix must be given in the upper triangular part of *a* (the elements *a[i,j]*, $i \leq j$);
  exit: the Cholesky decomposition of the matrix is delivered in the upper triangle of *a*;
n: int;
  entry: the order of the matrix;
aux: float *aux[2:3]*;
  entry:
  aux[2]: a relative tolerance used to control the calculation of the diagonal elements;
  exit:
  aux[3]: normal exit: *aux[3]* = *n*;
    abnormal exit: if the decomposition fails because the matrix is not positive definite then *aux[3]* = *k-1*, where *k* is the last stage number.
b: float *b[1:n]*;
  entry: the right hand side of the system of linear equations;
  exit: the solution of the system.

Functions used: chldec2, chlsol2.

```
void chldecsol2(float **a, int n, float aux[], float b[])
{
    void chldec2(float **, int, float []);
    void chlsol2(float **, int, float []);

    chldec2(a,n,aux);
    if (aux[3] == n) chlsol2(a,n,b);
}
```

D. chldecsol1

Solves a system of linear equations with a symmetric positive definite coefficient matrix by calling *chldec1* and, if the call is successful, *chlsol1*. The upper triangle of the coefficient matrix must be stored columnwise in a one-dimensional array.

Function Parameters:

void chldecsol1 (*a,n,aux,b*)

a: float *a[1:(n+1)n/2]*;
- entry: the upper triangular part of the symmetric positive definite matrix must be given columnwise in array *a* (the elements (i,j)-th element of the matrix must be given in *a[(j-1)*j/2+i]*, $1 \leq i \leq j \leq n$);
- exit: the Cholesky decomposition of the matrix is delivered columnwise in *a*;

n: int;
- entry: the order of the matrix;

aux: float *aux[2:3]*;
- entry:
  - *aux[2]*: a relative tolerance used to control the calculation of the diagonal elements;
- exit:
  - *aux[3]*: normal exit: *aux[3]* = *n*;
    abnormal exit: if the decomposition fails because the matrix is not positive definite then *aux[3]* = *k-1*, where *k* is the last stage number.

b: float *b[1:n]*;
- entry: the right hand side of the system of linear equations;
- exit: the solution of the system.

Functions used: chldec1, chlsol1.

```
void chldecsol1(float a[], int n, float aux[], float b[])
{
   void chldec1(float [], int, float []);
   void chlsol1(float [], int, float []);

   chldec1(a,n,aux);
   if (aux[3] == n) chlsol1(a,n,b);
}
```

3.2.4 Matrix inversion

A. chlinv2

Calculates the inverse *X* of a symmetric positive definite matrix *A*, provided that the matrix has been decomposed ($A = U^T U$, where *U* is the Cholesky matrix) by a successful call of *chldec2* or *chldecsol2*. The Cholesky matrix must be given in the upper triangle of a two-dimensional array. The inverse *X* is obtained from the conditions that *X* be symmetric and *UX* be a lower triangular matrix whose main diagonal elements are the reciprocals of the diagonal elements of *U*. The upper triangular elements of *X* are calculated by back substitution [Dek68].

Function Parameters:

void chlinv2 (*a,n*)

a: float *a[1:n,1:n]*;
 entry: the upper triangular part of the Cholesky matrix as produced by *chldec2* or *chldecsol2* must be given in the upper triangle of *a*;
 exit: the upper triangular part of the inverse matrix is delivered in the upper triangle of *a*;

n: int;
 entry: the order of the matrix.

Functions used: matvec, tamvec, dupvecrow.

```
void chlinv2(float **a, int n)
{
    float *allocate_real_vector(int, int);
    void free_real_vector(float *, int);
    void dupvecrow(int, int, int, float [], float **);
    float matvec(int, int, int, float **, float []);
    float tamvec(int, int, int, float **, float []);
    int i,j,i1;
    float r,*u;

    u=allocate_real_vector(1,n);
    for (i=n; i>=1; i--) {
        r=1.0/a[i][i];
        i1=i+1;
        dupvecrow(i1,n,i,u,a);
        for (j=n; j>=i1; j--)
            a[i][j] = -(tamvec(i1,j,j,a,u)+matvec(j+1,n,j,a,u))*r;
        a[i][i]=(r-matvec(i1,n,i,a,u))*r;
    }
    free_real_vector(u,1);
}
```

B. chlinv1

Calculates the inverse X of a symmetric positive definite matrix A, provided that the matrix has been decomposed ($A = U^TU$, where U is the Cholesky matrix) by a successful call of *chldec1* or *chldecsol1*. The upper triangular part of the Cholesky matrix must be given columnwise in a one-dimensional array. The inverse X is obtained from the conditions that X be symmetric and UX be a lower triangular matrix whose main diagonal elements are the reciprocals of the diagonal elements of U. The upper triangular elements of X are calculated by back substitution.

Function Parameters:

void chlinv1 (*a,n*)

a: float *a[1:(n+1)n/2]*;
 entry: the upper triangular part of the Cholesky matrix as produced by *chldec1* or *chldecsol1* must be given columnwise in array *a*;
 exit: the upper triangular part of the inverse matrix is delivered columnwise in array *a*;

n: int;
 entry: the order of the matrix.

Functions used: seqvec, symmatvec.

```
void chlinv1(float a[], int n)
{
    float *allocate_real_vector(int, int);
    void free_real_vector(float *, int);
    float seqvec(int, int, int, int, float [], float []);
    float symmatvec(int, int, int, float [], float []);
    int i,ii,i1,j,ij;
    float r,*u;

    u=allocate_real_vector(1,n);
    ii=((n+1)*n)/2;
    for (i=n; i>=1; i--) {
        r=1.0/a[ii];
        i1=i+1;
        ij=ii+i;
        for (j=i1; j<=n; j++) {
            u[j]=a[ij];
            ij += j;
        }
        for (j=n; j>=i1; j--) {
            ij -= j;
            a[ij] = -symmatvec(i1,n,j,a,u)*r;
        }
        a[ii]=(r-seqvec(i1,n,ii+i,0,a,u))*r;
        ii -= i;
    }
    free_real_vector(u,1);
}
```

C. chldecinv2

Calculates the inverse of a symmetric positive definite coefficient matrix by calling *chldec2* and, if the call is successful, *chlinv2*. The coefficient matrix must be given in the upper triangle of a two-dimensional array.

Function Parameters:

void chldecinv2 (*a,n,aux*)

a: float *a[1:n,1:n]*;
- entry: the upper triangle of the symmetric positive definite matrix must be given in the upper triangular part of *a* (the elements *a[i,j]*, $i \leq j$);
- exit: the upper triangular part of the inverse matrix is delivered in the upper triangle of *a*;

n: int;
- entry: the order of the matrix;

aux: float *aux[2:3]*;
- entry:
- *aux[2]*: a relative tolerance used to control the calculation of the diagonal elements;
- exit:
- *aux[3]*: normal exit: *aux[3]* = *n*;
  abnormal exit: if the decomposition fails because the matrix is not positive definite then *aux[3]* = *k-1*, where *k* is the last stage number.

Functions used: chldec2, chlinv2.

```
void chldecinv2(float **a, int n, float aux[])
{
    void chldec2(float **, int, float []);
    void chlinv2(float **, int);
```

```
    chldec2(a,n,aux);
    if (aux[3] == n) chlinv2(a,n);
}
```

D. chldecinv1

Calculates the inverse of a symmetric positive definite matrix by calling *chldec1* and, if the call is successful, *chlinv1*. The upper triangle of the coefficient matrix must be stored columnwise in a one-dimensional array.

Function Parameters:

void chldecinv1 (*a,n,aux*)

a: float *a[1:(n+1)n/2]*;
entry: the upper triangular part of the symmetric positive definite matrix must be given columnwise in array *a* (the elements (i,j)-th element of the matrix must be given in $a[(j-1)j/2+i]$, $1 \le i \le j \le n$);
exit: the upper triangular part of the inverse matrix is delivered columnwise in *a*;

n: int;
entry: the order of the matrix;

aux: float *aux[2:3]*;
entry:
aux[2]: a relative tolerance used to control the calculation of the diagonal elements;
exit:
aux[3]: normal exit: $aux[3] = n$;
abnormal exit: if the decomposition fails because the matrix is not positive definite then $aux[3] = k-1$, where k is the last stage number.

Functions used: chldec1, chlinv1.

```
void chldecinv1(float a[], int n, float aux[])
{
    void chldec1(float [], int, float []);
    void chlinv1(float [], int);

    chldec1(a,n,aux);
    if (aux[3] == n) chlinv1(a,n);
}
```

3.3 General real symmetric matrices

3.3.1 Preparatory procedure

decsym2

Calculates the LDL^T decomposition of a symmetric matrix which may be indefinite and/or singular. Before the decomposition is performed, a check is made to see whether the matrix is symmetric. If the matrix is asymmetric then no decomposition is performed.
Given a real symmetric $n \times n$ matrix A, it attempts to obtain the pivot reference integers $p(i)$ associated with a permutation matrix P, a unit lower triangular matrix L and a block diagonal matrix D with 1×1 or 2×2 blocks of elements on its principal diagonal and zeros

elsewhere, such that $P^TAP = LDL^T$. The decomposition process used involves the recursive construction of a system of $i{\times}i$ matrices $A^{(i)}$ $(i=n,...,k)$ with $A^{(n)} = AA^{(i-s)}$, where $s=1$ or $s=2$ is determined from $A^{(i)}$, by use of the decomposition

$$P^{(i)T}A^{(i)}P^{(i)} = \begin{pmatrix} E & C^T \\ C & Y \end{pmatrix}$$

where $P^{(i)}$ is an $i{\times}i$ permutation matrix and E is $s{\times}s$ nonsingular, and of the reduction formula

$$A^{(i-s)} = Y - CE^{-1}C^T .$$

In the following $I_{m,r}$ is the permutation matrix obtained by interchanging rows m and r of the $i{\times}i$ identity matrix I; $\alpha = (1 + 17^{1/2})/8 \approx 0.6404$ is a fixed parameter; the determinations of $P^{(i)}$ and s at each stage result from computations of a sequence of steps which may terminate at the stages a2, b2, c1 or d1 (thus if termination at stage a2 does not take place, b1 and b2 are embarked upon, etc.).

a1: obtain $\lambda = \max|A^{(i)}_{k,1}|$ $(2 \le k \le i)$ and r for which $\lambda = |A^{(i)}_{r,1}|$;

a2: if $|A^{(i)}_{1,1}| \ge \alpha\lambda$, set $P^{(i)} = I$, $s = 1$.

b1: obtain $\sigma = \max|A^{(i)}_{k,r}|$ $(1 \le k,r \le i;\ k \ne r)$;

b2: if $|A^{(i)}_{1,1}|\sigma \ge \alpha\lambda^2$, set $P^{(i)} = I$, $s=1$.

c1: if $|A^{(i)}_{r,r}| \ge \alpha\sigma$, set $P^{(i)} = I_{1,r}$, $s = 1$.

d1: set $P^{(i)} = I_{2,r}$, $s = 2$.

With

$$\mu^{(i)} = \max|A_{j,k}^{(i)}| \quad (1 \le j,k \le i) \; ,$$

the choice of $\alpha = (1 + 17^{1/2})/8$ above leads to the element growth inequality

$$\max(\mu^{(1)},\mu^{(2)}) \le 2.57^{n-1}\mu^{(n)} \; .$$

When $D_{i,i+1}=0$, $L_{i+1,i}=0$, so that if the block structure of D is known, the elements of D and L may be stored in the closed upper triangular part of the two dimensional array in which the elements of A are stored at entry. At exit from *decsym2*, the successive locations of the one-dimensional integer array p in the parameter list of *decsym2* contain not only the pivot reference integers associated with the $P^{(i)}$ above, but also information concerning the block structure of D: if $p[i]>0$ and $p[i+1]=0$, $D_{i,i+1}\neq 0$ and $L_{i+1,i}=0$. Upon successful exit from *decsym2*, the successive locations of the one-dimensional real array *detaux* contains numbers which are of use in computing the determinant of A. If $p[i]$, $p[i+1] > 0$, $detaux[i] = D_{i,i}$; if $p[i] > 0$, $p[i+1] = 0$, $detaux[i] = 1$ and $detaux[i+1]$ is the determinant of

$$\begin{pmatrix} D_{i,i} & D_{i,i+1} \\ D_{i+1,i} & D_{i+1,i+1} \end{pmatrix} .$$

If g is the number of 1×1 blocks in D, so that $h=(n-g)/2$ is the number of 2×2 blocks, and u, v and w are respectively the numbers of positive, negative and zero 1×1 blocks, then $u+h$ is the number of positive eigenvalues of A, $v+h$ the number of negative eigenvalues, and w the number of zero eigenvalues (these numbers are delivered in successive locations of *aux* at exit from *decsym2*). The decision as to whether a 1×1 block $D_{i,i}$ is zero is governed by the small real number τ allocated to the real variable *tol* upon call of *decsym2*: if $|D_{i,i}| < \tau$, $D_{i,i}$ is taken to be zero.

Function Parameters:

void decsym2 (*a,n,tol,aux,p,detaux*)

a: float *a[1:n,1:n]*;
  entry: the symmetric coefficient matrix;
  exit: the elements of the LDL^T decomposition of a are stored in the upper triangular part of a; D is a block diagonal matrix with blocks of order 1 or 2, for a block of order 2 we always have $D_{i,i+1}\neq 0$ and $L_{i+1,i}=0$, so that D and L^T fit in the upper triangular part of a, the strictly lower triangular part of a is left undisturbed;

n: int;
  entry: the order of the matrix;

tol: float;
  entry: a relative tolerance used to control the calculation of the block diagonal elements, the value of τ above;

aux: int *aux[2:5]*;
  exit:
  aux[2]: if the matrix is symmetric then 1; otherwise 0, and no decomposition of a is performed;
  aux[3]: if the matrix is symmetric then the number of its positive eigenvalues,

otherwise 0; if *aux[3]=n* then the matrix is positive definite;

aux[4]: if the matrix is symmetric then the number of its negative eigenvalues, otherwise 0; if *aux[4]=n* then the matrix is negative definite;

aux[5]: if the matrix is symmetric then the number of its zero eigenvalues, otherwise 0; if *aux[5]=0* then the matrix is symmetric and non-singular;

p: int *p[1:n]*;

exit: a vector recording (1) the interchanges performed on array *a* during the computation of the decomposition and (2) the block structure of *D*; if *p[i] > 0* and *p[i+1] = 0* a 2×2 block has been found ($D_{i,i+1}\neq 0$ and $L_{i+1,i}=0$);

detaux: float *detaux[1:n]*;

exit: if *p[i]>0* and *p[i+1]>0* then *detaux[i]* equals the exit value of *a[i,i]*; if *p[i]>0* and *p[i+1]=0* then *detaux[i]=1* and *detaux[i+1]* equals the value of the determinant of the corresponding 2×2 diagonal block as determined by *decsym2*.

Functions used: elmrow, ichrow, ichrowcol.

Method: The function *decsym2* computes the LDL^T decomposition of a symmetric matrix according to a method due to Bunch, Kaufman and Parlett [BunK77, BunKP76]. For the inertia problem it is important that *decsym2* can accept singular matrices. However, in order to find the number of zero eigenvalues of singular matrices, the singular value decomposition might be preferred.

```
#include <math.h>

void decsym2(float **a, int n, float tol,
               int aux[], int p[], float detaux[])
{
   void elmrow(int, int, int, int, float **, float **, float);
   void ichrow(int, int, int, int, float **);
   void ichrowcol(int, int, int, int, float **);
   int i,j,m,ip1,ip2,onebyone,sym;
   float det,s,t,alpha,lambda,sigma,aii,aip1,aip1i,temp;

   aux[3]=aux[4]=0;
   sym=1;
   i=0;
   while (sym && (i < n)) {
      i++;
      j=i;
      while (sym && (j < n)) {
         j++;
         sym = sym && (a[i][j] == a[j][i]);
      }
   }
   if (sym)
      aux[2]=1;
   else {
      aux[2]=0;
      aux[5]=n;
      return;
   }
   alpha=(1.0+sqrt(17.0))/8.0;
   p[n]=n;
   i=1;
   while (i < n) {
      ip1=i+1;
      ip2=i+2;
      aii=fabs(a[i][i]);
      p[i]=i;
      lambda=fabs(a[i][ip1]);
      j=ip1;
```

```
        for (m=ip2; m<=n; m++)
            if (fabs(a[i][m]) > lambda) {
                j=m;
                lambda=fabs(a[i][m]);
            }
        t=alpha*lambda;
        onebyone=1;
        if (aii < t) {
            sigma=lambda;
            for (m=ip1; m<=j-1; m++)
                if (fabs(a[m][j]) > sigma) sigma=fabs(a[m][j]);
            for (m=j+1; m<=n; m++)
                if (fabs(a[j][m]) > sigma) sigma=fabs(a[j][m]);
            if (sigma*aii < lambda) {
                if (alpha*sigma < fabs(a[j][j])) {
                    ichrow(j+1,n,i,j,a);
                    ichrowcol(ip1,j-1,i,j,a);
                    t=a[i][i];
                    a[i][i]=a[j][j];
                    a[j][j]=t;
                    p[i]=j;
                } else {
                    if (j > ip1) {
                        ichrow(j+1,n,ip1,j,a);
                        ichrowcol(ip2,j-1,ip1,j,a);
                        t=a[i][i];
                        a[i][i]=a[j][j];
                        a[j][j]=t;
                        t=a[i][j];
                        a[i][j]=a[i][ip1];
                        a[i][ip1]=t;
                    }
                    temp=a[i][ip1];
                    det=a[i][i]*a[ip1][ip1]-temp*temp;
                    aip1i=a[i][ip1]/det;
                    aii=a[i][i]/det;
                    aip1=a[ip1][ip1]/det;
                    p[i]=j;
                    p[ip1]=0;
                    detaux[i]=1.0;
                    detaux[ip1]=det;
                    for (j=ip2; j<=n; j++) {
                        s=aip1i*a[ip1][j]-aip1*a[i][j];
                        t=aip1i*a[i][j]-aii*a[ip1][j];
                        elmrow(j,n,j,i,a,a,s);
                        elmrow(j,n,j,ip1,a,a,t);
                        a[i][j]=s;
                        a[ip1][j]=t;
                    }
                    aux[3]++;
                    aux[4]++;
                    i=ip2;
                    onebyone=0;
                }
            }
        }
        if (onebyone) {
            if (tol < fabs(a[i][i])) {
                aii=a[i][i];
                detaux[i]=a[i][i];
                if (aii > 0.0)
                    aux[3]++;
                else
                    aux[4]++;
                for (j=ip1; j<=n; j++) {
                    s = -a[i][j]/aii;
                    elmrow(j,n,j,i,a,a,s);
                    a[i][j]=s;
                }
            }
            i=ip1;
        }
    }
    if (i == n) {
```

```
        if (tol < fabs(a[n][n])) {
            if (a[n][n] > 0.0)
                aux[3]++;
            else
                aux[4]++;
        }
        detaux[n]=a[n][n];
    }
    aux[5]=n-aux[3]-aux[4];
}
```

3.3.2 Calculation of determinant

determsym2

Calculates the determinant of a symmetric matrix A, det(A). The function *decsym2* should be called to perform the LDL^T decomposition of the symmetric matrix. Given the values of the determinants of the blocks of the $n \times n$ block diagonal matrix D (which has 1×1 or 2×2 blocks of elements on its principal diagonal and zeros elsewhere) and the number m of zero eigenvalues of A, *determsym2* evaluates det(A). If $m \neq 0$ then det(A)=0, otherwise det(A) is simply the product of the above determinants of blocks.

Function Parameters:

float determsym2 (*detaux,n,aux*)

determsym2: delivers the calculated value of the determinant of the matrix;

detaux: float *detaux[1:n]*;

entry: the array *detaux* as produced by *decsym2*;

n: int; the order of the array *detaux* (= the order of the matrix);

aux: int *aux[2:5]*;

entry: the array *aux* as produced by *decsym2*.

```
float determsym2(float detaux[], int n, int aux[])
{
    int i;
    float det;

    if (aux[5] > 0)
        det=0.0;
    else {
        det=1.0;
        for (i=1; i<=n; i++) det *= detaux[i];
    }
    return (det);
}
```

3.3.3 Solution of linear equations

A. solsym2

Solves a symmetric system of linear equations, assuming that the matrix has been decomposed into LDL^T form by a call of *decsym2*.

Function Parameters:

void solsym2 (*a,n,b,p,detaux*)

a: float *a[1:n,1:n]*;

entry: the LDL^T decomposition of A as produced by *decsym2*;

n: int;
entry: the order of the matrix;

b: float *b[1:n]*;
entry: the right hand side of the system of linear equations;
exit: the solution of the system;

p: int *p[1:n]*;
entry: a vector recording the interchanges performed on the matrix by *decsym2*, *p* also contains information on the block structure of the matrix as decomposed by *decsym2*;

detaux: float *detaux[1:n]*;
entry: the array *detaux* as produced by *decsym2*.

Functions used: matvec, elmvecrow.

```
void solsym2(float **a, int n, float b[], int p[], float detaux[])
{
    float matvec(int, int, int, float **, float []);
    void elmvecrow(int, int, int, float [], float **, float);
    int i,ii,k,ip1,pi,pii;
    float det,temp,save;

    i=1;
    while (i < n) {
        ip1=i+1;
        pi=p[i];
        save=b[pi];
        if (p[ip1] > 0) {
            b[pi]=b[i];
            b[i]=save/a[i][i];
            elmvecrow(ip1,n,i,b,a,save);
            i=ip1;
        } else {
            temp=b[i];
            b[pi]=b[ip1];
            det=detaux[ip1];
            b[i]=(temp*a[ip1][ip1]-save*a[i][ip1])/det;
            b[ip1]=(save*a[i][i]-temp*a[i][ip1])/det;
            elmvecrow(i+2,n,i,b,a,temp);
            elmvecrow(i+2,n,ip1,b,a,save);
            i += 2;
        }
    }
    if (i == n) {
        b[i] /= a[i][i];
        i=n-1;
    } else
        i=n-2;
    while (i > 0) {
        if (p[i] == 0)
            ii=i-1;
        else
            ii=i;
        for (k=ii; k<=i; k++) {
            save=b[k];
            save += matvec(i+1,n,k,a,b);
            b[k]=save;
        }
        pii=p[ii];
        b[i]=b[pii];
        b[pii]=save;
        i=ii-1;
    }
}
```

B. decsolsym2

Computes the solution of a symmetric system of linear equations by first calling *decsym2* to compute the LDL^T decomposition of the symmetric matrix. If the matrix is found to be non-singular then the function *solsym2* is called to compute the solution vector, otherwise the function *solsym2* is not called.

Function Parameters:

void decsolsym2 (*a,n,b,tol,aux*)

a: float *a[1:n,1:n]*;
  entry: see *decsym2*;
  exit: see *decsym2*;

n: int;
  entry: the order of the matrix;

b: float *b[1:n]*;
  entry: the right hand side of the system of linear equations;
  exit: if the matrix *a* is non-singular then *b* contains the solution of the system, otherwise *b* is left undisturbed;

tol: float;
  entry: see *decsym2*;

aux: int *aux[2:5]*;
  exit: see *decsym2*.

Functions used: decsym2, solsym2.

```
void decsolsym2(float **a, int n, float b[], float tol, int aux[])
{
    int *allocate_integer_vector(int, int);
    float *allocate_real_vector(int, int);
    void free_integer_vector(int *, int);
    void free_real_vector(float *, int);
    void decsym2(float **, int, float, int [], int [], float []);
    void solsym2(float **, int, float [], int [], float []);
    int *p;
    float *detaux;

    p=allocate_integer_vector(1,n);
    detaux=allocate_real_vector(1,n);
    decsym2(a,n,tol,aux,p,detaux);
    if (aux[5] == 0) solsym2(a,n,b,p,detaux);
    free_integer_vector(p,1);
    free_real_vector(detaux,1);
}
```

3.4 Real full rank overdetermined systems

3.4.1 Preparatory procedures

A. lsqortdec

Reduces an $n{\times}m$ matrix A ($n \geq m$) to the column permuted form A', where $R = QA'$, R being an $n{\times}m$ upper triangular matrix, and Q being an $n{\times}m$ orthogonal matrix having the form

$$\prod_{\nu=1}^{p} (1 - 2u^{(\nu)T}u^{(\nu)}/u^{(\nu)}u^{(\nu)T}) \qquad (p \leq m) .$$

Q is the product of at most m Householder matrices which are represented by their generating vectors. A is reduced to R in at most m stages. At the k-th stage the k-th column of the (already modified) matrix is interchanged with the column of maximum Euclidean norm (the pivot column). Then the matrix is multiplied by a Householder matrix such that the subdiagonal elements of the k-th column become zero while the first k-1 columns remain unchanged. The process terminates prematurely if at some stage the Euclidean norm of the pivotal column is less than some tolerance (a given tolerance *aux[2]* times the maximum of the Euclidean norms of the columns of the given matrix).

Function Parameters:

void lsqortdec (*a,n,m,aux,aid,ci*)

a: float *a[1:n,1:m]*;
- entry: the coefficient matrix of the linear least square problem;
- exit: in the upper triangle of *a* (the elements *a[i,j]* with *i*<*j*) the superdiagonal elements of the upper triangular matrix (matrix *R* above) produced by the Householder transformation; in the other part of the columns of *a* the significant elements of the generating vectors of the Householder matrices used for the Householder triangularization (values of $u^{(\nu)}$ above);

n: int;
- entry: the number of rows of the matrix;

m: int;
- entry: the number of columns of the matrix ($n \geq m$);

aux: float *aux[2:5]*;
- entry:
  - *aux[2]*: contains a relative tolerance used for calculating the diagonal elements of the upper triangular matrix;
- exit:
  - *aux[3]*: delivers the number of the diagonal elements of the upper triangular matrix which are found not negligible, normal exit *aux[3]* = *m*;
  - *aux[5]*: the maximum of the Euclidean norms of the columns of the given matrix;

aid: float *aid[1:m]*;
- exit: normal exit (*aux[3]* = *m*) *aid* contains the diagonal elements of the upper triangular matrix produced by the Householder triangularization;

ci: int *ci[1:m]*;
- exit: contains the pivotal indices of the interchanges of the columns of the given matrix.

Functions used: tammat, elmcol, ichcol.

Method: The method is Householder triangularization with column interchanges. It is a modification of [Dek68] where a derivation is given by [BusiG65].

```
#include <math.h>

void lsqortdec(float **a, int n, int m, float aux[],
```

```
              float aid[], int ci[])
{
    float *allocate_real_vector(int, int);
    void free_real_vector(float *, int);
    float tammat(int, int, int, int, float **, float **);
    void elmcol(int, int, int, int, float **, float **, float);
    void ichcol(int, int, int, int, float **);
    int j,k,kpiv;
    float beta,sigma,norm,w,eps,akk,aidk,temp,*sum;

    sum=allocate_real_vector(1,m);
    norm=0.0;
    aux[3]=m;
    for (k=1; k<=m; k++) {
        w=sum[k]=tammat(1,n,k,k,a,a);
        if (w > norm) norm=w;
    }
    w=aux[5]=sqrt(norm);
    eps=aux[2]*w;
    for (k=1; k<=m; k++) {
        sigma=sum[k];
        kpiv=k;
        for (j=k+1; j<=m; j++)
            if (sum[j] > sigma) {
                sigma=sum[j];
                kpiv=j;
            }
        if (kpiv != k) {
            sum[kpiv]=sum[k];
            ichcol(1,n,k,kpiv,a);
        }
        ci[k]=kpiv;
        akk=a[k][k];
        sigma=tammat(k,n,k,k,a,a);
        w=sqrt(sigma);
        aidk=aid[k]=((akk < 0.0) ? w : -w);
        if (w < eps) {
            aux[3]=k-1;
            break;
        }
        beta=1.0/(sigma-akk*aidk);
        a[k][k]=akk-aidk;
        for (j=k+1; j<=m; j++) {
            elmcol(k,n,j,k,a,a,-beta*tammat(k,n,k,j,a,a));
            temp=a[k][j];
            sum[j] -= temp*temp;
        }
    }
    free_real_vector(sum,1);
}
```

B. lsqdglinv

Computes the principal diagonal elements of the inverse of A^TA, where A is the coefficient matrix of a linear least squares problem. It is assumed that A has been decomposed after calling *lsqortdec* successfully. These values can be used for the computation of the standard deviations of least squares solutions.

Function Parameters:

void lsqdglinv (*a,m,aux,ci,diag*)

a,m,aid,ci: see *lsqortdec*; the contents of *a, aid* and *ci* should be produced by a successful call of *lsqortdec*;

diag: float *diag[1:m]*;

exit: the diagonal elements of the inverse of A^TA, where A is the matrix of the linear least squares problem.

Functions used: vecvec, tamvec.

```
void lsqdglinv(float **a, int m, float aid[], int ci[], float diag[])
{
    float vecvec(int, int, int, float [], float []);
    float tamvec(int, int, int, float **, float []);
    int j,k,cik;
    float w;

    for (k=1; k<=m; k++) {
        diag[k]=1.0/aid[k];
        for (j=k+1; j<=m; j++) diag[j] = -tamvec(k,j-1,j,a,diag)/aid[j];
        diag[k]=vecvec(k,m,0,diag,diag);
    }
    for (k=m; k>=1; k--) {
        cik=ci[k];
        if (cik != k) {
            w=diag[k];
            diag[k]=diag[cik];
            diag[cik]=w;
        }
    }
}
```

3.4.2 Least squares solution

A. lsqsol

Determines the least squares solution of the overdetermined system $Ax = b$, where A is an $n \times m$ matrix ($n \geq m$). It is assumed that A has been decomposed by calling *lsqortdec* successfully. The least squares solutions of several overdetermined systems with the same coefficient matrix can be solved by successive calls of *lsqsol* with different right hand sides.

Function Parameters:

void lsqsol (*a,n,m,aid,ci,b*)

a,n,m,aid,ci: see *lsqortdec*; the contents of the arrays *a, aid* and *ci* should be produced by a successful call of *lsqortdec*;

b: float *b[1:n]*;
- entry: contains the right hand side of a linear least squares problem;
- exit: *b[1:m]* contains the solution of the problem;
  b[m+1:n] contains a vector with Euclidean length equal to the Euclidean length of the residual vector.

Functions used: matvec, tamvec, elmveccol.

```
void lsqsol(float **a, int n, int m, float aid[], int ci[], float b[])
{
    float matvec(int, int, int, float **, float []);
    float tamvec(int, int, int, float **, float []);
    void elmveccol(int, int, int, float [], float **, float);
    int k,cik;
    float w;

    for (k=1; k<=m; k++)
        elmveccol(k,n,k,b,a,tamvec(k,n,k,a,b)/(aid[k]*a[k][k]));
    for (k=m; k>=1; k--) b[k]=(b[k]-matvec(k+1,m,k,a,b))/aid[k];
    for (k=m; k>=1; k--) {
        cik=ci[k];
        if (cik != k) {
            w=b[k];
```

```
                b[k]=b[cik];
                b[cik]=w;
            }
        }
}
```

B. lsqortdecsol

Computes the least squares solution of an overdetermined system $Ax = b$ (n linear equations in m unknowns), and computes the principal diagonal elements of the inverse of A^TA. The matrix A is first reduced to the column permuted form A', where $R = QA'$ (see *lsqortdec*) by calling *lsqortdec* and, if this call is successful, the least squares solutions are determined by *lsqsol* and the principal diagonal elements are calculated by *lsqdglinv*.

Function Parameters:

void lsqortdecsol (*a,n,m,aux,diag,b*)

a: float *a[1:n,1:m]*;
  entry: the coefficient matrix of the linear least square problem;
  exit: in the upper triangle of *a* (the elements *a[i,j]* with $i<j$) the superdiagonal elements of the upper triangular matrix (matrix R above) produced by the Householder transformation; in the other part of the columns of *a* the significant elements of the generating vectors of the Householder matrices used for the Householder triangularization;

n: int;
  entry: the number of rows of the matrix;

m: int;
  entry: the number of columns of the matrix ($n \geq m$);

aux: float *aux[2:5]*;
  entry:
  aux[2]: contains a relative tolerance used for calculating the diagonal elements of the upper triangular matrix;
  exit:
  aux[3]: delivers the number of the diagonal elements of the upper triangular matrix which are found not negligible, normal exit *aux[3]* = *m*;
  aux[5]: the maximum of the Euclidean norms of the columns of the given matrix;

diag: float *diag[1:m]*;
  exit: the diagonal elements of the inverse of A^TA, where A is the matrix of the linear least squares problem;

b: float *b[1:n]*;
  entry: contains the right hand side of a linear least squares problem;
  exit: *b[1:m]* contains the solution of the problem;
    b[m+1:n] contains a vector with Euclidean length equal to the Euclidean length of the residual vector.

Functions used: lsqortdec, lsqdglinv, lsqsol.

```
void lsqortdecsol(float **a, int n, int m, float aux[],
                        float diag[], float b[])
{
    int *allocate_integer_vector(int, int);
    float *allocate_real_vector(int, int);
    void free_integer_vector(int *, int);
```

```
    void free_real_vector(float *, int);
    void lsqortdec(float **, int, int, float [], float [], int []);
    void lsqdglinv(float **, int, float [], int [], float []);
    void lsqsol(float **, int, int, float [], int [], float []);
    int *ci;
    float *aid;

    ci=allocate_integer_vector(1,m);
    aid=allocate_real_vector(1,m);
    lsqortdec(a,n,m,aux,aid,ci);
    if (aux[3] == m) {
        lsqdglinv(a,m,aid,ci,diag);
        lsqsol(a,n,m,aid,ci,b);
    }
    free_integer_vector(ci,1);
    free_real_vector(aid,1);
}
```

3.4.3 Inverse matrix of normal equations

lsqinv

Calculates the inverse of the matrix A^TA, where A is the coefficient matrix of a linear least squares problem. *lsqinv* can be used for the calculation of the covariance matrix of a linear least squares problem. Given the $m \times m$ upper triangular matrix U and the pivot reference integers p_i, $i=1,...,m$, associated with the permutation matrix P, both occurring in the decomposition $Q^TU=AP$ of the $n \times m$ matrix A ($n \geq m$), where Q is $n \times n$ orthogonal, *lsqinv* constructs the $m \times m$ inverse matrix $X = (A^TA)^{-1}$. *lsqinv* is to be called after a successful call of *lsqortdec* which obtains the decomposition in question. Since $U^TU = P^TA^TAP$, X may be obtained by inverting U^TU in a call of *chlinv2*. Afterwards the covariance matrix is obtained by interchanges of the columns and rows of the inverse matrix.

Function Parameters:

void lsqinv (*a,m,aid,c*)

a: float *a[1:m,1:m]*;
 entry: in the upper triangle of *a* (the elements *a[i,j]* with $1 \leq i<j \leq m$) the superdiagonal elements should be given of the upper triangular matrix (U above) that is produced by the Householder triangularization in a call of the function *lsqortdec* with normal exit (*aux[3]=m*);
 exit: the upper triangle of the symmetric inverse matrix is delivered in the upper triangular elements of the array *a* (*a[i,j]*, $1 \leq i \leq j \leq m$);

m: int;
 entry: the number of columns of the matrix of the linear least squares problem;

aid: float *aid[1:m]*;
 entry: contains the diagonal elements of the upper triangular matrix produced by *lsqortdec*;

ci: int *ci[1:m]*;
 entry: contains the pivotal indices produced by a call of *lsqortdec*.

Functions used: chlinv2, ichcol, ichrow, ichrowcol.

```
void lsqinv(float **a, int m, float aid[], int c[])
{
    void chlinv2(float **, int);
    void ichcol(int, int, int, int, float **);
```

```
    void ichrow(int, int, int, int, float **);
    void ichrowcol(int, int, int, int, float **);
    int i,ci;
    float w;

    for (i=1; i<=m; i++) a[i][i]=aid[i];
    chlinv2(a,m);
    for (i=m; i>=1; i--) {
        ci=c[i];
        if (ci != i) {
            ichcol(1,i-1,i,ci,a);
            ichrow(i+1,ci-1,i,ci,a);
            ichrow(ci+1,m,i,ci,a);
            w=a[i][i];
            a[i][i]=a[ci][ci];
            a[ci][ci]=w;
        }
    }
}
```

3.4.4 Least squares with linear constraints

A. lsqdecomp

Let A denote the given matrix. *lsqdecomp* produces an n-th order orthogonal matrix Q and an $n \times m$ upper triangular matrix U such that $U=QA$ with permuted columns.

The constrained least squares problem considered is that of determining $x \epsilon R^m$ which minimizes $\| r_2 \|_E$ where

$$r_2 = b_2 - A_2 x, \qquad b_2 \in R^{n_2}$$

and A_2 is an $n_2 \times m$ matrix, subject to the condition that

$$A_1 x = b \qquad \text{where} \qquad b_1 \in R^{n_1},$$

A_1 being $n_1 \times m$ $(n_1+n_2=n)$. The required solution satisfies the equation

$$\begin{pmatrix} 0 & 0 & A_1 \\ 0 & I & A_2 \\ A_1^T & A_2^T & 0 \end{pmatrix} \begin{pmatrix} \lambda \\ r_2 \\ x \end{pmatrix} = \begin{pmatrix} b_1 \\ b_2 \\ c \end{pmatrix} \qquad \text{where } \lambda \in R^{n_1}, \tag{1}$$

λ is a vector of Lagrange multipliers and $c \epsilon R^m$ is zero. In the following,

$(A,B)^t$ denotes the compound matrix $\begin{matrix} A \\ B \end{matrix}$.

Let P be an $m \times m$ permutation matrix such that $(A_1,A_2)^t = (E,F)^t$ where $E=(G_{1,1},G_{1,2})$, $F=(G_{2,1},G_{2,2})$, $G_{1,1}$ being an $n_1 \times n_1$ nonsingular matrix and, for example, $G_{2,2}$ being $(n-n_1) \times (m-n_1)$. Determine an $n_1 \times n_1$ orthogonal matrix Q_1 such that $Q_1 E=(U_1,R_1)$ where U_1 is an $n_1 \times n_1$ upper triangular matrix, and R_1 is $n_1 \times (m-n_1)$. Determine a further $(n-n_1) \times (n-n_1)$ orthogonal matrix Q_2 such that

$$Q_2(G_{2,2} - G_{2,1}U_1^{-1}R_1) = (U_2,0)^t$$

where U_2 is an $(m-n_1)\times(m-n_1)$ upper triangular matrix, and 0 represents an $(n-m)\times(m-n_1)$ matrix of zeros. With

$$B = U_1^{-T}G_{2,1} \quad \text{and} \quad R = (I,0^T)Q_2,$$

I representing the $(m-n_1)\times(m-n_1)$ unit matrix, the two formulae

$$R = \begin{pmatrix} U_1 & R_1 \\ 0 & U_2 \end{pmatrix}, \quad APR^{-1} = \begin{pmatrix} Q_1 & B \\ 0 & R_2 \end{pmatrix}^T$$

are valid. Thus, setting

$$y = (y_1,y_2)^t, \quad d = (d_1,d_2)^t \quad \text{where} \quad y_1,d_1 \in R^{n_1}, \quad y_2,d_2 \in R^{m-n_1}$$

with

$$x = PR^{-1}y, \qquad d = P^TR^{-T}c \tag{2}$$

equation (1) may be rewritten as

$$\begin{pmatrix} 0 & 0 & Q_1^T & 0 \\ 0 & I & B^T & R^T \\ Q_1 & B & 0 & 0 \\ 0 & R_2 & 0 & 0 \end{pmatrix} \begin{pmatrix} \lambda \\ r_2 \\ y_1 \\ y_2 \end{pmatrix} = \begin{pmatrix} b_1 \\ b_2 \\ d_1 \\ d_2 \end{pmatrix}.$$

These equations may be solved: successively

$$y_1 = Q_1b_1, \qquad y_2 = g_1 - d_1$$

where

$$g = (g_1,g_2)^t = Q_2(b_2 - B^Ty_1),$$

$$r_2 = Q_2^T(d_2,g_2)^t, \qquad \lambda = Q_1^T(d_1 - Br_2).$$

x may be obtained from y by use of the first of relationships (2).

The orthogonal matrices Q_1 and Q_2 and permutation matrix P are obtained recursively by means of a process which may be explained as follows. Starting with a matrix G and setting $G=A^{(1)}$, $A^{(k+1)}$ is obtained from $A^{(k)}$ by use of the relationship

$$A^{(k+1)} = (I - \beta_k u^{(k)} u^{(k)T}) A^{(k)}$$

where, with $n(k)=n_1$ if $k<n_1$ and $n(k)=n$ if $k>n_1$ and j so chosen that

$$s_j = \sum_{i=k}^{n(k)} \left(A_{i,j}^{(k)}\right)^2, \qquad k \le j \le n$$

is a maximum,

$$A_{i,j}^{(k+1)} = 0 \qquad \text{for} \quad i=k+1,\dots,n(k).$$

(The k-th pivot reference integer $p(k)$ associated with P is j.)

$A_{i,j}^{(k+1)}$ is rendered zero as desired by setting

$$\sigma_k = \sqrt{s_j}, \quad \beta_k = (\sigma_k(\sigma_k + |A_{k,j}^{(k)}|))^{-1}$$

$$u_i^{(k)} = 0 \quad \text{for} \quad i<k,\ i>n(k)$$

$$u_k^{(k)} = \operatorname{sign}(A_{k,j}^{(k)})\ (\sigma_k + |A_{k,j}^{(k)}|)$$

$$u_i^{(k)} = A_{i,j}^{(k)} \qquad k<i\le n(k)$$

With $G=A_I$ in the above, Q_2 is the product

$$(I - \beta_{n_1} u^{(n_1)} u^{(n_1)T}) ... (I - \beta_1 u^{(1)} u^{(1)T}) \quad \text{and} \quad A_1^{(n_1)} = (U_1, R_1).$$

Setting

$$A^{(n_1+1)} = \begin{pmatrix} U_1 & R_1 \\ 0 & R' \end{pmatrix}$$

where

$$R' = G_{2,2} - Q_1^T R_1, \qquad A^{(m+1)} = (R,0)^t.$$

The user must prescribe a small real tolerance *tol*. If in the above $(s_j)^{1/2}$ is less than *tol* multiplied by the maximum Euclidean norm over the columns of $A^{(k)}$, the decomposition process is broken off.

Function Parameters:

void lsqdecomp (*a,n,m,n1,aux,aid,ci*)

a: float *a[1:n,1:m]*;
- entry: the original least squares matrix, where the first *n1* rows should form the constraint matrix (i.e. the first *n1* equations are to be strictly satisfied);
- exit: in the upper triangle of *a* (the elements *a[i,j]* with *i*<*j*) the superdiagonal part of the upper triangular matrix produced by the Householder transformation; in the other part of the columns of *a* the significant elements of the generating vectors of the Householder matrices used for the Householder triangularization;

n: int;
- entry: the number of rows of the matrix;

m: int;
- entry: the number of columns of the matrix;

n1: int;
- entry: number of linear constraints, i.e. the first *n1* rows of *a* set up a system of *n1* linear equations that must be strictly satisfied (if there are no constraints, *n1* must be chosen zero);

aux: float *aux[2:7]*;
- entry:
- *aux[2]*: contains a relative tolerance used for calculating the diagonal elements of the upper triangular matrix;
- exit:
- *aux[3]*: delivers the number of the diagonal elements which are not negligible, normal exit *aux[3]* = *m*;
- *aux[5]*: the maximum of the Euclidean norms of the columns of the given matrix;

aid: float *aid[1:m]*;
- exit: normal exit (*aux[3]* = *m*) *aid* contains the diagonal elements of the upper triangular matrix produced by the Householder transformation;

ci: int *ci[1:m]*;

exit: contains the pivotal indices of the interchanges of the columns of the given matrix.

Functions used: matmat, tammat, elmcol, ichcol.

Method: See [BjG67] for the QR decomposition of a least squares matrix.

```
#include <math.h>

void lsqdecomp(float **a, int n, int m, int n1, float aux[],
               float aid[], int ci[])
{
    float *allocate_real_vector(int, int);
    void free_real_vector(float *, int);
    float matmat(int, int, int, int, float **, float **);
    float tammat(int, int, int, int, float **, float **);
    void elmcol(int, int, int, int, float **, float **, float);
    void ichcol(int, int, int, int, float **);
    int j,k,kpiv,nr,s,fsum;
    float beta,sigma,norm,aidk,akk,w,eps,temp,*sum;

    sum=allocate_real_vector(1,m);
    norm=0.0;
    aux[3]=m;
    nr=n1;
    fsum=1;
    for (k=1; k<=m; k++) {
        if (k == n1+1) {
            fsum=1;
            nr=n;
        }
        if (fsum)
            for (j=k; j<=m; j++) {
                w=sum[j]=tammat(k,nr,j,j,a,a);
                if (w > norm) norm=w;
            }
        fsum=0;
        eps=aux[2]*sqrt(norm);
        sigma=sum[k];
        kpiv=k;
        for (j=k+1; j<=m; j++)
            if (sum[j] > sigma) {
                sigma=sum[j];
                kpiv=j;
            }
        if (kpiv != k) {
            sum[kpiv]=sum[k];
            ichcol(1,n,k,kpiv,a);
        }
        ci[k]=kpiv;
        akk=a[k][k];
        sigma=tammat(k,nr,k,k,a,a);
        w=sqrt(sigma);
        aidk=aid[k]=((akk < 0.0) ? w : -w);
        if (w < eps) {
            aux[3]=k-1;
            break;
        }
        beta=1.0/(sigma-akk*aidk);
        a[k][k]=akk-aidk;
        for (j=k+1; j<=m; j++) {
            elmcol(k,nr,j,k,a,a,-beta*tammat(k,nr,k,j,a,a));
            temp=a[k][j];
            sum[j] -= temp*temp;
        }
        if (k == n1)
            for (j=n1+1; j<=n; j++)
                for (s=1; s<=m; s++) {
                    nr = (s > n1) ? n1 : s-1;
                    w=a[j][s]-matmat(1,nr,j,s,a,a);
```

```
                a[j][s] = (s > n1) ? w : w/aid[s];
            }
        }
        free_real_vector(sum,1);
}
```

B. lsqrefsol

Solves a constrained least squares problem consisting of the determination of that $x \epsilon R^m$ which minimizes

$$\|r_2\|_E \quad \text{where} \quad r_2 = b_2 - A_2 x, \quad b_2 \epsilon R^{n_2}$$

and A_2 is an $n_2 \times m$ matrix, subject to the condition that

$$A_1 x = b_1 \quad \text{where} \quad b_1 \epsilon R^{n_1},$$

A_1 being $n_1 \times m$ $(n_1+n_2=n)$. The required solution satisfies the equation $Bz=h$ where

$$B = \begin{pmatrix} 0 & 0 & A_1 \\ 0 & I & A_2 \\ A_1^T & A_2^T & 0 \end{pmatrix}$$

$$z = (\lambda, r_2, x)^T, \quad h = (b_1, b_2, 0)^T, \quad \lambda \epsilon R^{n_1},$$

λ being a vector of Lagrange multipliers. It is assumed that the components of the vectors $u^{(k)}$ associated with the elementary reflectors defining orthogonal matrices Q_1 and Q_2 and those of an upper triangular matrix R, together with pivot reference integers associated with a permutation matrix P, have all been obtained by means of a successful call of *lsqdecomp*.

lsqrefsol first obtains a numerical solution $z^{(1)}$ of the equation $Bz=h$, and then uses an iterative scheme of the form $f^{(s)}=h-Bz^{(s)}$, $B\delta z^{(s)}=f^{(s)}$, $z^{(s+1)}=z^{(s)}+\delta z^{(s)}$, $s=1,2,...$ to obtain a refined solution to this equation, and in so doing a refined estimate of x, $z^{(1)}$ and, at each state, $\delta z^{(s)}$ are derived by the solution process outlined in the documentation to *lsqdecomp*. The above iterative scheme [BjG67] is terminated if either (a) $\|\delta z^{(s)}\|_E \le \varepsilon \|z^{(s)}\|_E$ where ε is a small real tolerance prescribed by the user or (b) $s=smax$ where the integer *smax* is also prescribed by the user. The least squares solutions of several overdetermined systems with the same constraints and coefficient matrix can be solved by successive calls of *lsqrefsol* with different right hand sides.

Function Parameters:

void lsqrefsol (*a,qr,n,m,n1,aux,aid,ci,b,ldx,x,res*)

a: float *a[1:n,1:m]*;
entry: the original least squares matrix, where the first *n1* rows should form the constraint matrix (i.e. the first *n1* equations are to be strictly satisfied);

qr: float *qr[1:n,1:m]*;
entry: the QR decomposition of the original least squares matrix as delivered by a successful call of *lsqdecomp*;
n: int;
entry: the number of rows of the matrices *a* and *qr*;
m: int;
entry: the number of columns of the matrices *a* and *qr*;
n1: int;
entry: number of linear constraints;
aux: float *aux[2:7]*;
entry:
aux[2]: contains a relative tolerance (value of ε above) as a criterion to stop iterative refining, if the Euclidean norm of the correction is smaller than *aux[2]* times the current approximation of the solution then the iterative refining is stopped;
aux[6]: maximum number of iterations allowed (value of *smax* above), usually *aux[6]*=5 will be sufficient;
exit:
aux[7]: the number of iterations performed (the last value of *s* for which a correction term $\delta z^{(s)}$ is determined in the above);
aid: float *aid[1:m]*;
entry: the diagonal elements of the upper triangular matrix as delivered by a successful call of *lsqdecomp*;
ci: int *ci[1:m]*;
entry: the pivotal indices as produced by *lsqdecomp*;
b: float *b[1:n]*;
entry: the right hand side of the least squares problem; first *n1* elements form the right hand sides of the constraints;
ldx: float *;
the Euclidean norm of the last correction of the solution (the value of $\|\delta x^{(s)}\|_E$ for the last $\delta x^{(s)}$ determined in the above, $x^{(s)}$ being formed from the last *m* components of $z^{(s)}$);
x: float *x[1:m]*;
exit: the solution vector;
res: float *res[1:n]*;
exit: the residual vector ($f^{(s)}$ in the above) corresponding to the solution.

Functions used: vecvec, matvec, tamvec, elmveccol, ichcol.

```
#include <math.h>

void lsqrefsol(float **a, float **qr, int n, int m, int n1,
               float aux[], float aid[], int ci[], float b[],
               float *ldx, float x[], float res[])
{
   float *allocate_real_vector(int, int);
   void free_real_vector(float *, int);
   float vecvec(int, int, int, float [], float []);
   float matvec(int, int, int, float **, float []);
   float tamvec(int, int, int, float **, float []);
   void ichcol(int, int, int, int, float **);
   void elmveccol(int, int, int, float [], float **, float);
   int i,j,k,s,startup;
   float c1,nexve,ndx,ndr,d,corrnorm,*f,*g;
   double dtemp;

   f=allocate_real_vector(1,n);
```

```
    g=allocate_real_vector(1,m);
    for (j=1; j<=m; j++) {
        s=ci[j];
        if (s != j) ichcol(1,n,j,s,a);
    }
    for (j=1; j<=m; j++) x[j]=g[j]=0.0;
    for (i=1; i<=n; i++) {
        res[i]=0.0;
        f[i]=b[i];
    }
    k=0;
    do {
        startup = (k <= 1);
        ndx=ndr=0.0;
        if (k != 0) {
            for (i=1; i<=n; i++) res[i] += f[i];
            for (s=1; s<=m; s++) {
                x[s] += g[s];
                dtemp=0.0;
                for (i=1; i<=n; i++)
                    dtemp += (double)a[i][s]*(double)res[i];
                d=dtemp;
                g[s]=(-d-tamvec(1,s-1,s,qr,g))/aid[s];
            }
            for (i=1; i<=n; i++) {
                dtemp = (i > n1) ? res[i] : 0.0;
                for (s=1; s<=m; s++)
                    dtemp += (double)a[i][s]*(double)x[s];
                f[i]=(double)b[i]-dtemp;
            }
        }
        nexve=sqrt(vecvec(1,m,0,x,x)+vecvec(1,n,0,res,res));
        for (s=1; s<=n1; s++)
            elmveccol(s,n1,s,f,qr,tamvec(s,n1,s,qr,f)/(qr[s][s]*aid[s]));
        for (i=n1+1; i<=n; i++)
            f[i] -= matvec(1,n1,i,qr,f);
        for (s=n1+1; s<=m; s++)
            elmveccol(s,n,s,f,qr,tamvec(s,n,s,qr,f)/(qr[s][s]*aid[s]));
        for (i=1; i<=m; i++) {
            c1=f[i];
            f[i]=g[i];
            g[i] = (i > n1) ? c1-g[i] : c1;
        }
        for (s=m; s>=1; s--) {
            g[s]=(g[s]-matvec(s+1,m,s,qr,g))/aid[s];
            ndx += g[s]*g[s];
        }
        for (s=m; s>=n1+1; s--)
            elmveccol(s,n,s,f,qr,tamvec(s,n,s,qr,f)/(qr[s][s]*aid[s]));
        for (s=1; s<=n1; s++)
            f[s] -= tamvec(n1+1,n,s,qr,f);
        for (s=n1; s>=1; s--)
            elmveccol(s,n1,s,f,qr,tamvec(s,n1,s,qr,f)/(qr[s][s]*aid[s]));
        aux[7]=k;
        for (i=1; i<=n; i++) ndr += f[i]*f[i];
        corrnorm=sqrt(ndx+ndr);
        k++;
    } while (startup || (corrnorm>aux[2]*nexve && k<=aux[6]));
    *ldx=sqrt(ndx);
    for (s=m; s>=1; s--) {
        j=ci[s];
        if (j != s) {
            c1=x[j];
            x[j]=x[s];
            x[s]=c1;
            ichcol(1,n,j,s,a);
        }
    }
    free_real_vector(f,1);
    free_real_vector(g,1);
}
```

3.5 Other real matrix problems

3.5.1 Solution of overdetermined systems

A. solsvdovr

Solves an overdetermined system of linear equations. *solsvdovr* determines that $x \epsilon R^n$ with minimum $\|x\|_2$ which minimizes $\|Ax-b\|_2$, where A is a real $m \times n$ matrix, $b \epsilon R^m$ $(m \geq n)$, the matrices U, Λ, V occurring in the singular value decomposition $A=U\Lambda V^T$ being available, where U is an $m \times n$ column orthogonal matrix ($U^TU=I$), Λ is an $n \times n$ diagonal matrix whose diagonal elements are the singular values of A (λ_i, i=1,...,n), and V is an $n \times n$ orthogonal matrix ($V^TV=VV^T=I$). The analytic solution of the above problem is $x=A^+b$, where A^+ is the pseudo-inverse of A: numerically, $x=V\Lambda^{(-1)}U^Tb$, where $\Lambda^{(-1)}$ is a diagonal matrix whose successive diagonal elements are $(\lambda_i)^{-1}$ if $\lambda_i > \delta$, and 0 otherwise, δ being a small positive real number prescribed by the user. The two stages in the determination of x are the formation of
$b'=\Lambda^{(-1)}U^Tb'$ and that of $x=Vb'$.

Function Parameters:

void solsvdovr (*u,val,v,m,n,x,em*)

u: float *u[1:m,1:n]*;
  entry: the matrix U in the singular values decomposition $U\Lambda V^T$;
val: float *val[1:n]*;
  entry: the singular values (diagonal elements λ_i of Λ);
v: float *v[1:n,1:n]*;
  entry: the matrix V in the singular values decomposition;
m: int;
  entry: the length of the right hand side vector;
n: int;
  entry: the number of unknowns, n should satisfy $n \leq m$;
x: float *x[1:m]*;
  entry: the right hand side vector;
  exit: the solution vector;
em: float *em[6:6]*;
  entry: the minimal non-neglectable singular value (value of δ in the above).

Functions used: matvec, tamvec.

Method: See [WiR71] for the solution of an overdetermined system of linear equations.

```
void solsvdovr(float **u, float val[], float **v, int m, int n,
               float x[], float em[])
{
    float *allocate_real_vector(int, int);
    void free_real_vector(float *, int);
    float matvec(int, int, int, float **, float []);
    float tamvec(int, int, int, float **, float []);
    int i;
    float min,*x1;

    x1=allocate_real_vector(1,n);
    min=em[6];
```

```
    for (i=1; i<=n; i++)
        x1[i] = (val[i] <= min) ? 0.0 : tamvec(1,m,i,u,x)/val[i];
    for (i=1; i<=n; i++) x[i]=matvec(1,n,i,v,x1);
    free_real_vector(x1,1);
}
```

B. solovr

Solves an overdetermined system of linear equations. *solovr* determines that $x \epsilon R^n$ with minimum $\|x\|_2$ which minimizes $\|Ax-b\|_2$, where A is a real $m \times n$ matrix, $b \epsilon R^m$ $(m \geq n)$. *solovr* first calls *qrisngvaldec* to obtain the matrices U, Λ, V occurring in the singular value decomposition $A=U\Lambda V^T$ and, if all singular values of A can be determined by use of this function, calls *solsvdovr* to obtain x.

Function Parameters:

int solovr (*a,m,n,x,em*)

solovr: given the number of singular values of *a* which cannot be found by *qrisngvaldec*;
a: float *a[1:m,1:n]*;
entry: the matrix of the system;
m: int;
entry: the number of rows of *a*;
n: int;
entry: the number of columns of *a*, $n \leq m$;
x: float *x[1:m]*;
entry: the right hand side vector;
exit: the solution vector;
em: float *em[0:7]*;
entry:
em[0]: the machine precision;
em[2]: the relative precision of the singular values;
em[4]: the maximal number of iterations to be performed in the singular values decomposition;
em[6]: the minimal non-neglected singular value;
exit:
em[1]: the infinity norm of the matrix;
em[3]: the maximal neglected superdiagonal element;
em[5]: the number of iterations performed in the singular values decomposition;
em[7]: the numerical rank of the matrix,
i.e. the number of singular values ≥ *em[6]*.

Functions used: qrisngvaldec, solsvdovr.

```
int solovr(float **a, int m, int n, float x[], float em[])
{
    float *allocate_real_vector(int, int);
    float **allocate_real_matrix(int, int, int, int);
    void free_real_vector(float *, int);
    void free_real_matrix(float **, int, int, int);
    int qrisngvaldec(float **, int, int, float [], float **, float []);
    void solsvdovr(float **, float [], float **, int, int,
                        float [], float []);
    int i;
    float *val,**v;
```

```
    val=allocate_real_vector(1,n);
    v=allocate_real_matrix(1,n,1,n);
    i=qrisngvaldec(a,m,n,val,v,em);
    if (i == 0) solsvdovr(a,val,v,m,n,x,em);
    free_real_vector(val,1);
    free_real_matrix(v,1,n,1);
    return i;
}
```

3.5.2 Solution of underdetermined systems

A. solsvdund

Solves an underdetermined system of linear equations [WiR71]. *solsvdund* determines that $x \epsilon R^m$ with minimum $\|x\|_2$ which minimizes $\|A^Tx-b\|$, where A is a real $m \times n$ matrix, $b \epsilon R^n$ $(m \geq n)$, the matrices U, Λ, V as described in the documentation to *solsvdovr* being available. The analytic solution of the above problem is $x=(A^+)^Tb$, where A^+ is the pseudo-inverse of A. The two stages in the determination of x are the formation of $b'=\Lambda^{(-1)}V^Tb$ and that of $x=Ub'$.

Function Parameters:

void solsvdund (*u,val,v,m,n,x,em*)

u: float *u[1:m,1:n]*;
entry: the matrix U in the singular values decomposition $U\Lambda V^T$;
val: float *val[1:n]*;
entry: the singular values (diagonal elements of Λ);
v: float *v[1:n,1:n]*;
entry: the matrix V in the singular values decomposition;
m: int;
entry: the number of unknowns;
n: int;
entry: the length of the right hand side vector, n should satisfy $n \leq m$;
x: float *x[1:m]*;
entry: the right hand side vector in *x[1:n]*;
exit: the solution vector in *x[1:m]*;
em: float *em[6:6]*;
entry: the minimal non-neglectable singular value.

Functions used: matvec, tamvec.

```
void solsvdund(float **u, float val[], float **v, int m, int n,
                float x[], float em[])
{
    float *allocate_real_vector(int, int);
    void free_real_vector(float *, int);
    float matvec(int, int, int, float **, float []);
    float tamvec(int, int, int, float **, float []);
    int i;
    float min,*x1;

    x1=allocate_real_vector(1,n);
    min=em[6];
    for (i=1; i<=n; i++)
        x1[i] = (val[i] <= min) ? 0.0 : tamvec(1,n,i,v,x)/val[i];
    for (i=1; i<=m; i++)   x[i] = matvec(1,n,i,u,x1);
    free_real_vector(x1,1);
```

}

B. solund

Solves an underdetermined system of linear equations. *solund* determines that $x \epsilon R^m$ with minimum $\|x\|_2$ which minimizes $\|A^T x-b\|_2$, where A is a real $m \times n$ matrix, $b \epsilon R^n$ $(m \geq n)$. *solund* first calls *qrisngvaldec* to obtain the matrices U, Λ, V occurring in the singular value decomposition $A=U\Lambda V^T$ and, if all singular values of A may be obtained by use of this function, calls *solsvdund* to obtain x.

Function Parameters:

int solund (*a,m,n,x,em*)

solund: given the number of singular values of *a* which cannot be found by *qrisngvaldec*;
a: float *a[1:m,1:n]*;
entry: the transpose of the matrix;
m: int;
entry: the number of rows of *a*;
n: int;
entry: the number of columns of *a*, $n \leq m$;
x: float *x[1:m]*;
entry: the right hand side vector in *x[1:n]*;
exit: the solution vector;
em: float *em[0:7]*;
entry:
em[0]: the machine precision;
em[2]: the relative precision for the singular values;
em[4]: the maximal number of iterations to be performed in the singular values decomposition;
em[6]: the minimal non-neglected singular value;
exit:
em[1]: the infinity norm of the matrix;
em[3]: the maximal neglected superdiagonal element;
em[5]: the number of iterations performed in the singular values decomposition;
em[7]: the numerical rank of the matrix,
i.e. the number of singular values $\geq$ *em[6]*.

Functions used: qrisngvaldec, solsvdund.

```
int solund(float **a, int m, int n, float x[], float em[])
{
    float *allocate_real_vector(int, int);
    float **allocate_real_matrix(int, int, int, int);
    void free_real_vector(float *, int);
    void free_real_matrix(float **, int, int, int);
    int qrisngvaldec(float **, int, int, float [], float **, float []);
    void solsvdund(float **, float [], float **, int, int,
                        float [], float []);
    int i;
    float *val,**v;

    val=allocate_real_vector(1,n);
    v=allocate_real_matrix(1,n,1,n);
    i=qrisngvaldec(a,m,n,val,v,em);
    if (i == 0) solsvdund(a,val,v,m,n,x,em);
```

```
    free_real_vector(val,1);
    free_real_matrix(v,1,n,1);
    return i;
}
```

3.5.3 Solution of homogeneous equation

A. homsolsvd

Given the matrices U, Λ, V occurring in a real singular value decomposition $A=U\Lambda V^T$ where A is an $m \times n$ matrix ($m \geq n$), the diagonal elements λ_i of Λ are reordered in such a way that, for the new order, $\lambda_i \geq \lambda_{i+1}$; the columns of U and V are simultaneously reordered to preserve the correspondence between the λ_i and the columns.

With u_i being the successive columns of U, v_i those of V, and λ_i the diagonal elements of Λ (all in the new ordering)

$$Av_i = \lambda_i u_i \qquad u_i^T A = \lambda_i v_i^T \qquad i=1,...,n.$$

If, for some $r<n$, $\lambda_i<\delta$ $(i=r+1,...,n)$, δ being a small positive real number, then approximately

$$Av_i = u_i^T A = 0 \qquad\qquad i=r+1,...,n;$$

for any $x \epsilon R^n$,

$$x = \sum_{i=1}^{n} x_i v_i, \qquad\qquad Ax = \sum_{i=1}^{r} x_i \lambda_i u_i,$$

i.e. Ax lies in the column subspace in R^m spanned by u_i $(i=1,...,r)$; any vector x in the complementary column subspace in R^n spanned by $v_{r+1},...,v_n$ satisfies the equation $Ax \approx 0$; for any $x^T \epsilon R^m$, $x^T A$ lies in the column subspace in R^n spanned by v_i $(i=1,...,r)$; any vector x in the complementary row subspace in R^m spanned by $(u_{r+1})^T,...,(u_n)^T$ satisfies the equation $x^T A=0$.

After a call of *homsolsvd*, the reordered columns of U and V and reordered λ_i are available in the two dimensional arrays u and v, and the one dimensional array *val*, respectively. The latter elements λ_i may thus be inspected, and the vector sets $v_1,...,v_r$; $v_{r+1},...,v_n$; $u_1,...,u_r$ and $u_{r+1},...,u_n$ defining the above four subspaces may be extracted [WiR71].

Function Parameters:

void homsolsvd (*u,val,v,m,n*)

u: float *u[1:m,1:n]*;
entry: the matrix U in the singular values decomposition $U\Lambda V^T$;
exit: the components *u[i,j]* of U occurring in a reordered decomposition $A=U\Lambda V^T$ in which the diagonal elements of Λ are set in nonascending order;

val: float *val[1:n]*;
entry: the singular values;

exit: the array will be ordered in such a way that $val[i]<val[j]$ if $j<i$;
v: float $v[1:n,1:n]$;
entry: the matrix V in the singular values decomposition;
exit: the components $v[i,j]$ of V in the reordered decomposition above;
m: int;
entry: the number of rows of u;
n: int;
entry: the number of columns of u;

Function used: ichcol.

```
void homsolsvd(float **u, float val[], float **v, int m, int n)
{
    void ichcol(int, int, int, int, float **);
    int i,j;
    float x;

    for (i=n; i>=2; i--)
        for (j=i-1; j>=1; j--)
            if (val[i] > val[j]) {
                x=val[i];
                val[i]=val[j];
                val[j]=x;
                ichcol(1,m,i,j,u);
                ichcol(1,n,i,j,v);
            }
}
```

B. homsol

Given an $m \times n$ matrix A ($m \geq n$), capable of real singular value decomposition, *homsol* determines the vectors $u_j \epsilon R^m$ ($j=r,...,n$) defining the column subspace in which Ax lies for all $x \epsilon R^n$, the vectors $v_j \epsilon R^n$ ($j=r+1,...,n$) defining the complementary column subspace in which $x \epsilon R^n$ must lie for the condition $Ax=0$ to hold, the vectors v_j ($j=1,...,r$) defining the column subspace in which $x^T A$ lies for all $x \epsilon R^m$ and the vectors $(u_j)^T$ ($j=r+1,...,n$) defining the complementary row subspace in which $x^T \epsilon R^m$ must lie for the condition $x^T A=0$ to hold.

homsol first calls *qrisngvaldec* to obtain the matrices U, Λ, V occurring in the singular value decomposition $A=U \Lambda V^T$ and, if all singular values of A can be determined by use this function, calls *homsolsvd* to reorder the columns of U and V and the diagonal elements of Λ. The numerical rank (r in the above) of A is determined by the conditions $\lambda_i > \delta$ (i=1,...,r) $\lambda_i \leq \delta$ (i=r+1,...,n) for the reordered λ_i, where δ is a small positive real number supplied by the user.

Function Parameters:

int homsol (a,m,n,v,em)

homsol: given the number of singular values of a which cannot be found by *qrisngvaldec*;
a: float $a[1:m,1:n]$;
entry: the matrix of the system;
exit: the value of the components $(u_j)^{(i)}$ of u_j, $i=1,...,m$; $j=1,...,n$;
m: int;
entry: the number of rows of a;
n: int;

entry: the number of columns of *a*;

v: float *v[1:n,1:n]*;
exit: the value of the components $(v_j)^{(i)}$ of v_j, $i,j=1,...,n$;

em: float *em[0:7]*;
entry:
em[0]: the machine precision;
em[2]: the relative precision of the singular values;
em[4]: the maximal number of iterations to be performed in the singular values decomposition;
em[6]: the minimal non-neglected singular value;
exit:
em[1]: the infinity norm of the matrix;
em[3]: the maximal neglected superdiagonal element;
em[5]: the number of iterations performed in the singular values decomposition;
em[7]: the numerical rank of the matrix, i.e. the number of singular values $\geq$ *em[6]*.

Functions used: qrisngvaldec, homsolsvd.

```
int homsol(float **a, int m, int n, float **v, float em[])
{
    float *allocate_real_vector(int, int);
    void free_real_vector(float *, int);
    int qrisngvaldec(float **, int, int, float [], float **, float []);
    void homsolsvd(float **, float [], float **, int, int);
    int i;
    float *val;

    val=allocate_real_vector(1,n);
    i=qrisngvaldec(a,m,n,val,v,em);
    if (i == 0) homsolsvd(a,val,v,m,n);
    free_real_vector(val,1);
    return i;
}
```

3.5.4 Pseudo-inversion

A. psdinvsvd

Calculates the pseudo-inverse of a matrix [WiR71]. Given the matrices U,Λ,V occurring in a real singular value decomposition $A=U\Lambda V^T$ of the $m \times n$ matrix A ($m \geq n$), *psdinvsvd* determines the generalized inverse A^+ of A (which satisfies and is defined by the relationships $AA^+A=A$, $A^+AA^+=A^+$, AA^+ and A^+A symmetric in the real case). With λ_i (i=1,...,n) being the diagonal elements of Λ, the successive diagonal elements of the diagonal matrix $\Lambda^{(-1)}$ are determined by taking them to be $(\lambda_i)^{-1}$ if $\lambda_i > \delta$, and 0 otherwise, δ being a small positive real number prescribed by the user. Thereafter $A^+=V\Lambda^{(-1)}U^T$. The matrix $X=\Lambda^{(-1)}U^T$ is first constructed, and then $A^+=VX$.

Function Parameters:

void psdinvsvd (*u,val,v,m,n,em*)

u: float *u[1:m,1:n]*;
entry: the matrix U in the singular values decomposition $U\Lambda V^T$;
exit: the transpose of the pseudo-inverse;

val: float *val[1:n]*;
entry: the singular values;
v: float *v[1:n,1:n]*;
entry: the matrix V in the singular values decomposition;
m: int;
entry: the number of rows of U;
n: int;
entry: the number of columns of V;
em: float *em[6:6]*;
entry: the minimal non-neglectable singular value.

Function used: matvec.

```
void psdinvsvd(float **u, float val[], float **v, int m, int n,
               float em[])
{
    float *allocate_real_vector(int, int);
    void free_real_vector(float *, int);
    float matvec(int, int, int, float **, float []);
    int i,j;
    float min,vali,*x;

    x=allocate_real_vector(1,n);
    min=em[6];
    for (i=1; i<=n; i++)
        if (val[i] > min) {
            vali=1.0/val[i];
            for (j=1; j<=m; j++) u[j][i] *= vali;
        } else
            for (j=1; j<=m; j++) u[j][i]=0.0;
    for (i=1; i<=m; i++) {
        for (j=1; j<=n; j++) x[j]=u[i][j];
        for (j=1; j<=n; j++) u[i][j]=matvec(1,n,j,v,x);
    }
    free_real_vector(x,1);
}
```

B. psdinv

Determines the generalized inverse A^+ of the $m \times n$ matrix A ($m \geq n$). The singular value decomposition $A=U\Lambda V^T$ is first carried out by a call of *qrisngvaldec* and, if all singular values have been obtained, *psdinvsvd* is then called to construct A^+.

Function Parameters:

int psdinv (*a,m,n,em*)

psdinv: given the number of singular values of *a* which cannot be found by *qrisngvaldec*;
a: float *a[1:m,1:n]*;
entry: the given matrix;
exit: the transpose of the pseudo-inverse;
m: int;
entry: the number of rows of *a*;
n: int;
entry: the number of columns of *a*, $n \leq m$;
em: float *em[0:7]*;
entry:
em[0]: the machine precision;

em[2]: the relative precision of the singular values;
em[4]: the maximal number of iterations to be performed;
em[6]: the minimal non-neglected singular value;
exit:
em[1]: the infinity norm of the matrix;
em[3]: the maximal neglected superdiagonal element;
em[5]: the number of iterations performed in the singular values decomposition;
em[7]: the numerical rank of the matrix,
i.e. the number of singular values $\geq$ *em[6]*.

Functions used: qrisngvaldec, psdinvsvd.

```
int psdinv(float **a, int m, int n, float em[])
{
    float *allocate_real_vector(int, int);
    float **allocate_real_matrix(int, int, int, int);
    void free_real_vector(float *, int);
    void free_real_matrix(float **, int, int, int);
    int qrisngvaldec(float **, int, int, float [], float **, float []);
    void psdinvsvd(float **, float [], float **, int, int, float []);
    int i;
    float *val,**v;

    val=allocate_real_vector(1,n);
    v=allocate_real_matrix(1,n,1,n);
    i=qrisngvaldec(a,m,n,val,v,em);
    if (i == 0) psdinvsvd(a,val,v,m,n,em);
    free_real_vector(val,1);
    free_real_matrix(v,1,n,1);
    return i;
}
```

3.6 Real sparse non-symmetric band matrices

3.6.1 Preparatory procedure

decbnd

Performs the decomposition of a matrix whose non-zero elements are in band form, and whose band elements are stored rowwise in a one-dimensional array [Dek68].

Given an $n \times n$ matrix A for whose elements $A_{i,j}=0$ when $i>j+lw$ or $j>i+rw$ ($i,j=1,...,n$), *decbnd* obtains (a) a unit lower triangular band matrix M for whose elements $M_{i,j}=0$ when $i>j+lw$ or $j>i$, and $M_{i,i}=1$; (b) an upper triangular band matrix U with $U_{i,j}=0$ when $j>i+lw+rw$ or $i>j$; and (c) a sequence of pivot reference integers $p(i)$ ($i=1,...,n$) associated with a permutation matrix P, such that $MU=PA$, by Gauss elimination using partial pivoting.

The method used involves the recursive construction of matrices $A^{(i)}$, with $A^{(1)}=A$. At the i-th stage the elements $(A_{k,j})^{(i)}$ ($j=1,...,i-1$; $k=j+1,...,n$) are zero and

$$(A_{k,j})^{(i)}=(U_{k,j})^{(i)} \quad (k=1,...,i-1;\ j=k,k+1,...,n).$$

Then (a) the smallest integer I for which $|(A_{I,i})^{(i)}| \geq |(A_{k,i})^{(i)}|$ for $k \geq i$ is determined; (b) $p(i)$ is set equal to I; (c) rows i and I of $A^{(i)}$ are interchanged (the i-th row of $A^{(i+1)}$ has now been determined); (d) with $M_{k,i}=(A_{k,i})^{(i)}/(A_{i,i})^{(i+1)}$, row($k$) of $A^{(i)}$ is replaced by row(k) - $M_{k,i}$row(i) to form row(k) of $A^{(i+1)}$ ($k=i+1,...,\min(n,i+lw)$). (The elements $(A_{k,i})^{(i+1)}$, $k>i$, are thus zero.) The process is arrested if $\delta>\delta_i$, where $\delta_i=|(A_{i,i})^{(i+1)}|/\| i\text{-th row of } A^{(i)}\|_2$ and δ is a small positive real number prescribed by the user.

Function Parameters:

void decbnd (*a,n,lw,rw,aux,m,pi*)

a: float *a[1:(lw+rw)(n-1)+n]*;

entry: *a* contains rowwise the band elements of the band matrix in such a way that the (*i,j*)-th element of the matrix is given in *a[(lw+rw)(i-1)+j]*, *i=1,...,n* and *j*=max(*1,i-lw*),...,min(*n,i+rw*), the values of the remaining elements of *a* are irrelevant;

exit: the band elements of the Gaussian eliminated matrix, which is an upper triangular band matrix *U* with (*lw+rw*) codiagonals, are rowwise delivered in *a* as follows: the (*i,j*)-th element of *U* is *a[(lw+rw)(i-j)+j]*, *i=1,...,n* and *j=i*,...,min(*n,i+lw+rw*);

n: int;

entry: the order of the band matrix;

lw: int;

entry: number of left codiagonals of *a*;

rw: int;

entry: number of right codiagonals of *a*;

aux: float *aux[1:5]*;

entry:

aux[2]: a relative tolerance to control the elimination process (value of δ above);

exit:

aux[1]: if successful, given the sign of the determinant of the matrix (+1 or -1);

aux[3]: if successful then *aux[3]=n*; otherwise *aux[3]=i-1*, where the reduction process was terminated at stage *i*;

aux[5]: if successful, given the minimum absolute value of pivot(i) divided by the Euclidean norm of the i-th row (value of min δ_j, $1 \le j \le n$, in the above);

m: float *m[1:lw(n-2)+1]*;

exit: the Gaussian multipliers (values of $M_{i,j}$ above) of all eliminations in such a way that the *i*-th multiplier of the *j*-th step is *m[lw(j-1)+i-j]*;

p: int *p[1:n]*;

exit: the pivotal indices.

Functions used: vecvec, elmvec, ichvec.

```
#include <math.h>

void decbnd(float a[], int n, int lw, int rw, float aux[],
            float m[], int p[])
{
    float *allocate_real_vector(int, int);
    void free_real_vector(float *, int);
    float vecvec(int, int, int, float [], float []);
    void elmvec(int, int, int, float [], float [], float);
    void ichvec(int, int, int, float []);
    int i,j,k,kk,kk1,pk,mk,ik,lw1,f,q,w,w1,w2,nrw,iw,sdet;
    float r,s,eps,min,*v;

    v=allocate_real_vector(1,n);
    f=lw;
    w1=lw+rw;
    w=w1+1;
    w2=w-2;
    iw=0;
    sdet=1;
    nrw=n-rw;
    lw1=lw+1;
```

```
    q=lw-1;
    for (i=2; i<=lw; i++) {
        q--;
        iw += w1;
        for (j=iw-q; j<=iw; j++) a[j]=0.0;
    }
    iw = -w2;
    q = -lw;
    for (i=1; i<=n; i++) {
        iw += w;
        if (i <= lw1) iw--;
        q += w;
        if (i > nrw) q--;
        v[i]=sqrt(vecvec(iw,q,0,a,a));
    }
    eps=aux[2];
    min=1.0;
    kk = -w1;
    mk = -lw;
    if (f > nrw) w2 += nrw-f;
    for (k=1; k<=n; k++) {
        if (f < n) f++;
        ik = kk += w;
        mk += lw;
        s=fabs(a[kk])/v[k];
        pk=k;
        kk1=kk+1;
        for (i=k+1; i<=f; i++) {
            ik += w1;
            m[mk+i-k]=r=a[ik];
            a[ik]=0.0;
            r=fabs(r)/v[i];
            if (r > s) {
                s=r;
                pk=i;
            }
        }
        if (s < min) min=s;
        if (s < eps) {
            aux[3]=k-1;
            aux[5]=s;
            aux[1]=sdet;
            free_real_vector(v,1);
            return;
        }
        if (k+w2 >= n) w2--;
        p[k]=pk;
        if (pk != k) {
            v[pk]=v[k];
            pk -= k;
            ichvec(kk1,kk1+w2,pk*w1,a);
            sdet = -sdet;
            r=m[mk+pk];
            m[mk+pk]=a[kk];
            a[kk]=r;
        } else
            r=a[kk];
        if (r < 0.0) sdet = -sdet;
        iw=kk1;
        lw1=f-k+mk;
        for (i=mk+1; i<=lw1; i++) {
            s = m[i] /= r;
            iw += w1;
            elmvec(iw,iw+w2,kk1-iw,a,a,-s);
        }
    }
    aux[3]=n;
    aux[5]=min;
    aux[1]=sdet;
    free_real_vector(v,1);
}
```

3.6.2 Calculation of determinant

determbnd

Calculates the determinant of the Gaussian eliminated upper triangular matrix provided with the correct sign that is delivered by *decbnd* or *decsolbnd*. *determbnd* should not be called when overflow can be expected.

Function Parameters:

float determbnd (*a,n,lw,rw,sgndet*)

determbnd: delivers the determinant of the band matrix;
a: float *a[1:(lw+rw)*(n-1)+n]*;
    entry: the contents of *a* are produced by *decbnd* or *decsolbnd*;
n: int;
    entry: the order of the band matrix;
lw: number of left codiagonals of *a*;
rw: number of right codiagonals of *a*;
sgndet: int;
    entry: the sign of the determinant as delivered in *aux[1]* by *decbnd*, if the elimination was successful.

```
#include <math.h>

float determbnd(float a[], int n, int lw, int rw, int sgndet)
{
    int i,l;
    float p;

    l=1;
    p=1.0;
    lw += rw+1;
    for (i=1; i<=n; i++) {
        p=a[l]*p;
        l += lw;
    }
    return (fabs(p)*sgndet);
}
```

3.6.3 Solution of linear equations

A. solbnd

Calculates the solution of a system of linear equations, provided that the matrix has been decomposed by a successful call of *decbnd*. The solution of the linear system is obtained by carrying out the elimination, for which the Gaussian multipliers are saved, on the right hand side, and by solving the new system with the upper triangular band matrix, as produced by *decbnd*, by back substitution. The solutions of several systems with the same coefficient matrix can be obtained by successive calls of *solbnd*.

Function Parameters:

void solbnd (*a,n,lw,rw,m,p,b*)

a,n,lw,rw,m,p: see *decbnd*;
    entry: the contents of the arrays *a,m,p* are as produced by *decbnd*;

b: float *b[1:n]*;
entry: the right hand side of the system of linear equations.

Functions used: vecvec, elmvec.

```
void solbnd(float a[], int n, int lw, int rw, float m[],
            int p[], float b[])
{
    float vecvec(int, int, int, float [], float []);
    void elmvec(int, int, int, float [], float [], float);
    int f,i,k,kk,w,w1,w2,shift;
    float s;

    f=lw;
    shift = -lw;
    w1=lw-1;
    for (k=1; k<=n; k++) {
        if (f < n) f++;
        shift += w1;
        i=p[k];
        s=b[i];
        if (i != k) {
            b[i]=b[k];
            b[k]=s;
        }
        elmvec(k+1,f,shift,b,m,-s);
    }
    w1=lw+rw;
    w=w1+1;
    kk=(n+1)*w-w1;
    w2 = -1;
    shift=n*w1;
    for (k=n; k>=1; k--) {
        kk -= w;
        shift -= w1;
        if (w2 < w1) w2++;
        b[k]=(b[k]-vecvec(k+1,k+w2,shift,b,a))/a[kk];
    }
}
```

B. decsolbnd

Calculates the solution of a system of linear equations by Gaussian elimination with partial pivoting if the coefficient matrix is in band form and is stored rowwise in a one-dimensional array. *decsolbnd* performs Gaussian elimination in the same way as *decbnd*, meanwhile also carrying out the elimination with the given right hand side. The solution of the eliminated system is obtained by back substitution.

Function Parameters:

void decsolbnd (*a,n,lw,rw,aux,b*)

a,n,lw,rw,aux: see *decbnd*;
b: see *solbnd*.

Functions used: vecvec, elmvec, ichvec.

```
#include <math.h>

void decsolbnd(float a[], int n, int lw, int rw, float aux[], float b[])
{
    float *allocate_real_vector(int, int);
    void free_real_vector(float *, int);
    float vecvec(int, int, int, float [], float []);
```

```
void elmvec(int, int, int, float [], float [], float);
void ichvec(int, int, int, float []);
int i,j,k,kk,kk1,pk,ik,lw1,f,q,w,w1,w2,iw,nrw,shift,sdet;
float r,s,eps,min,*m,*v;

m=allocate_real_vector(0,lw);
v=allocate_real_vector(1,n);
f=lw;
sdet=1;
w1=lw+rw;
w=w1+1;
w2=w-2;
iw=0;
nrw=n-rw;
lw1=lw+1;
q=lw-1;
for (i=2; i<=lw; i++) {
    q--;
    iw += w1;
    for (j=iw-q; j<=iw; j++) a[j]=0.0;
}
iw = -w2;
q = -lw;
for (i=1; i<=n; i++) {
    iw += w;
    if (i <= lw1) iw--;
    q += w;
    if (i > nrw) q--;
    v[i]=sqrt(vecvec(iw,q,0,a,a));
}
eps=aux[2];
min=1.0;
kk = -w1;
if (f > nrw) w2 += nrw-f;
for (k=1; k<=n; k++) {
    if (f < n) f++;
    ik = kk += w;
    s=fabs(a[kk])/v[k];
    pk=k;
    kk1=kk+1;
    for (i=k+1; i<=f; i++) {
        ik += w1;
        m[i-k]=r=a[ik];
        a[ik]=0.0;
        r=fabs(r)/v[i];
        if (r > s) {
            s=r;
            pk=i;
        }
    }
    if (s < min) min=s;
    if (s < eps) {
        aux[3]=k-1;
        aux[5]=s;
        aux[1]=sdet;
        free_real_vector(m,0);
        free_real_vector(v,1);
        return;
    }
    if (k+w2 >= n) w2--;
    if (pk != k) {
        v[pk]=v[k];
        pk -= k;
        ichvec(kk1,kk1+w2,pk*w1,a);
        sdet = -sdet;
        r=b[k];
        b[k]=b[pk+k];
        b[pk+k]=r;
        r=m[pk];
        m[pk]=a[kk];
        a[kk]=r;
    } else
        r=a[kk];
    iw=kk1;
```

```
        lw1=f-k;
        if (r < 0.0) sdet = -sdet;
        for (i=1; i<=lw1; i++) {
            s = m[i] /= r;
            iw += w1;
            elmvec(iw,iw+w2,kk1-iw,a,a,-s);
            b[k+i] -= b[k]*s;
        }
    }
    aux[3]=n;
    aux[5]=min;
    kk=(n+1)*w-w1;
    w2 = -1;
    shift=n*w1;
    for (k=n; k>=1; k--) {
        kk -= w;
        shift -= w1;
        if (w2 < w1) w2++;
        b[k]=(b[k]-vecvec(k+1,k+w2,shift,b,a))/a[kk];
    }
    aux[1]=sdet;
    free_real_vector(m,0);
    free_real_vector(v,1);
}
```

3.7 Real sparse non-symmetric tridiagonal matrices

3.7.1 Preparatory procedures

A. dectri

Given the $n \times n$ tridiagonal matrix T ($T_{i,j}=0$ for $|i-j|>1$) obtains a lower bidiagonal matrix L ($L_{i,j}=0$ for $i>j+1$ and $j>i$) and unit upper bidiagonal matrix U ($U_{i,i}=1$, $U_{i,j}=0$ for $i>j$ and $j>i+1$) such that $T=LU$. The columns of L and rows of U are determined in succession. If at stage k, $|L_{k,k}|<tol*\|t_k\|_1$, where tol is a tolerance prescribed by the user and t_k is the k-th row of T, the decomposition process is discontinued, and an integer K is given the value $k-1$; otherwise K is given the value n at exit.

Function Parameters:

void dectri (*sub,diag,super,n,aux*)

sub: float *sub[1:n-1]*;
- entry: the subdiagonal of the given matrix T, $T_{i+1,i}$ should be given in *sub[i]*, $i=1,...,n-1$;
- exit: the lower bidiagonal matrix, the value $L_{i+1,i}$ will be delivered in *sub[i]*, $i=1,...,aux[3]-1$;

diag: float *diag[1:n]*;
- entry: the diagonal of T, the value of $T_{i,i}$ in *diag[i]*;
- exit: $L_{i,i}$ will be delivered in *diag[i]*, $i=1,...,aux[3]$;

super: float *super[1:n-1]*;
- entry: the superdiagonal of T, value of $T_{i,i+1}$ in *super[i]*, $i=1,...,n-1$;
- exit: $U_{i,i+1}$ will be delivered in *super[i]*, $i=1,...,aux[3]-1$;

n: int;
- entry: the order of the matrix;

aux: float *aux[2:5]*;
- entry:
- *aux[2]*: a relative tolerance; a reasonable choice for this value is an estimate of the

relative precision of the matrix elements; however, it should not be chosen smaller than the machine precision (the value of *tol* above);

exit:

aux[3]: the number of elimination steps performed (value of K above);

aux[5]: if *aux[3]=n* then *aux[5]* will be equal to the infinity norm of the matrix, else *aux[5]* is set equal to the value of that element which causes the breakdown of the decomposition (value of $T_{K+1,K+1}$).

```
#include <math.h>

void dectri(float sub[], float diag[], float super[],
            int n, float aux[])
{
    int i,n1;
    float d,r,s,u,norm,norm1,tol;

    tol=aux[2];
    d=diag[1];
    r=super[1];
    norm=norm1=fabs(d)+fabs(r);
    if (fabs(d) <= norm1*tol) {
        aux[3]=0.0;
        aux[5]=d;
        return;
    }
    u=super[1]=r/d;
    s=sub[1];
    n1=n-1;
    for (i=2; i<=n1; i++) {
        d=diag[i];
        r=super[i];
        norm1=fabs(s)+fabs(d)+fabs(r);
        diag[i] = d -= u*s;
        if (fabs(d) <= norm1*tol) {
            aux[3]=i-1;
            aux[5]=d;
            return;
        }
        u=super[i]=r/d;
        s=sub[i];
        if (norm1 > norm) norm=norm1;
    }
    d=diag[n];
    norm1=fabs(d)+fabs(s);
    diag[n] = d -= u*s;
    if (fabs(d) <= norm1*tol) {
        aux[3]=n1;
        aux[5]=d;
        return;
    }
    if (norm1 > norm) norm=norm1;
    aux[3]=n;
    aux[5]=norm;
}
```

B. dectripiv

Given the $n \times n$ tridiagonal matrix T ($T_{i,j}=0$ for $|i-j|>1$) attempts to obtain a lower bidiagonal matrix L ($L_{i,j}=0$ for $i>j+1$) and unit upper band matrix (with at most two codiagonals) U ($U_{i,i}=1$, $U_{i,j}=0$ for $i>j$ and $j>i+2$) such that $LU=PT$ where P is a permutation matrix resulting from either no interchange or interchange of consecutive rows at each stage in the LU decomposition. The columns of $l^{(k)}$ ($k=1,...,n$) of L are determined in succession. If $|(l_i)^{(k)}|<tol*\|t^{(i)}\|_1$, $(l_i)^{(k)}$ is the i-th component of $l^{(k)}$ and $t^{(i)}$ is the i-th row of T and tol is a tolerance prescribed by the user, then the decomposition process is discontinued, and an

integer K is given the value k-1; otherwise K is given the value n at exit.

Function Parameters:

void dectripiv (*sub,diag,super,n,aid,aux,piv*)

sub: float *sub[1:n-1]*;
entry: the subdiagonal of the given matrix T, $T_{i+1,i}$ should be given in *sub[i]*, $i=1,...,n-1$;
exit: the lower bidiagonal matrix, the value $L_{i+1,i}$ will be delivered in *sub[i]*, $i=1,...,aux[3]-1$;

diag: float *diag[1:n]*;
entry: the diagonal of T, value of $T_{i,i}$ in *diag[i]*;
exit: $L_{i,i}$ will be delivered in *diag[i]*, $i=1,...,aux[3]$;

super: float *super[1:n-1]*;
entry: the superdiagonal of T, value of $T_{i,i+1}$ in *super[i]*, $i=1,...,n-1$;
exit: $U_{i,i+1}$ will be delivered in *super[i]*, $i=1,...,aux[3]-1$;

n: int;
entry: the order of the matrix;

aid: float *aid[1:n-2]*;
exit: the value of $U_{i,i+2}$ will be delivered in *aid[i]*, $i=1,...,aux[3]-2$;

aux: float *aux[2:5]*;
entry:
aux[2]: a relative tolerance; a reasonable choice for this value is an estimate of the relative precision of the matrix elements, however, it should not be chosen smaller than the machine precision (value of *tol* above);
exit:
aux[3]: the number of elimination steps performed (value of K above);
aux[5]: if *aux[3]=n* then *aux[5]* will be equal to the infinity norm of the matrix, else *aux[5]* is set equal to the value of that element which causes the breakdown of the decomposition (value of $T_{K+1,K+1}$);

piv: int *piv[1:n-1]*;
exit: the value of *piv[i]* will be nonzero if the i-th and (i+1)-th row are interchanged, i=1,...,min(*aux[3],n-1*), else *piv[i]* will be zero.

```
#include <math.h>

void dectripiv(float sub[], float diag[], float super[], int n,
               float aid[], float aux[], int piv[])
{
    int i,i1,n1,n2;
    float d,r,s,u,t,q,v,w,norm,norm1,norm2,tol;

    tol=aux[2];
    d=diag[1];
    r=super[1];
    norm=norm2=fabs(d)+fabs(r);
    n2=n-2;
    for (i=1; i<=n2; i++) {
        i1=i+1;
        s=sub[i];
        t=diag[i1];
        q=super[i1];
        norm1=norm2;
        norm2=fabs(s)+fabs(t)+fabs(q);
        if (norm2 > norm) norm=norm2;
        if (fabs(d)*norm2 < fabs(s)*norm1) {
            if (fabs(s) <= tol*norm2) {
                aux[3]=i-1;
```

```
                aux[5]=s;
                return;
            }
            diag[i]=s;
            u=super[i]=t/s;
            v=aid[i]=q/s;
            sub[i]=d;
            w = super[i1] = -v*d;
            d=diag[i1]=r-u*d;
            r=w;
            norm2=norm1;
            piv[i]=1;
        } else {
            if (fabs(d) <= tol*norm1) {
                aux[3]=i-1;
                aux[5]=d;
                return;
            }
            u=super[i]=r/d;
            d=diag[i1]=t-u*s;
            aid[i]=0.0;
            piv[i]=0;
            r=q;
        }
    }
    n1=n-1;
    s=sub[n1];
    t=diag[n];
    norm1=norm2;
    norm2=fabs(s)+fabs(t);
    if (norm2 > norm) norm=norm2;
    if (fabs(d)*norm2 < fabs(s)*norm1) {
        if (fabs(s) <= tol*norm2) {
            aux[3]=n2;
            aux[5]=s;
            return;
        }
        diag[n1]=s;
        u=super[n1]=t/s;
        sub[n1]=d;
        d=diag[n]=r-u*d;
        norm2=norm1;
        piv[n1]=1;
    } else {
        if (fabs(d) <= tol*norm1) {
            aux[3]=n2;
            aux[5]=d;
            return;
        }
        u=super[n1]=r/d;
        d=diag[n]=t-u*s;
        piv[n1]=0;
    }
    if (fabs(d) <= tol*norm2) {
        aux[3]=n1;
        aux[5]=d;
        return;
    }
    aux[3]=n;
    aux[5]=norm;
}
```

3.7.2 Solution of linear equations

A. soltri

Given the $n \times n$ lower bidiagonal matrix L ($L_{i,j}=0$ for $i>j+1$ or $j>i$) and unit upper bidiagonal matrix U ($U_{i,i}=1$, $U_{i,j}=0$ for $i>j$ or $j>i+1$) solves the system of equations Tx-b by forward and back substitution, where $T=LU$ (this decomposition results from successful exit from *dectri*). One call of *dectri* followed by several calls of *soltri* may be used to solve several linear systems having the same tridiagonal matrix, but different right hand sides.

Function Parameters:

void soltri (*sub,diag,super,n,b*)

sub: float *sub[1:n-1]*;
entry: the subdiagonal of the lower bidiagonal matrix, as delivered by *dectri*;
diag: float *diag[1:n]*;
entry: the diagonal of the lower bidiagonal matrix, as delivered by *dectri*;
super: float *super[1:n-1]*;
entry: the superdiagonal of the upper bidiagonal matrix, as delivered by *dectri*;
n: int;
entry: the order of the matrix;
b: float *b[1:n]*;
entry: the right hand side of the linear system;
exit: the calculated solution of the linear system.

```
void soltri(float sub[], float diag[], float super[], int n, float b[])
{
    int i;
    float r;

    r = b[1] /= diag[1];
    for (i=2; i<=n; i++) r=b[i]=(b[i]-sub[i-1]*r)/diag[i];
    for (i=n-1; i>=1; i--) r = b[i] -= super[i]*r;
}
```

B. decsoltri

Given the $n \times n$ tridiagonal matrix T, attempts to obtain a lower bidiagonal matrix L and unit upper bidiagonal matrix U such that $T=LU$ and, if successful in so doing, solves the system of equations $Tx=b$. The triangular decomposition of the given matrix is done by calling *dectri* and the forward and back substitution by calling *soltri*.

Function Parameters:

void decsoltri (*sub,diag,super,n,aux,b*)

sub: float *sub[1:n-1]*;
entry: the subdiagonal of the given matrix T, $T_{i+1,i}$ should be given in *sub[i]*, $i=1,...,n-1$;
exit: the lower bidiagonal matrix, the value $L_{i+1,i}$ will be delivered in *sub[i]*, $i=1,...,aux[3]-1$;
diag: float *diag[1:n]*;
entry: the diagonal of T, value of $T_{i,i}$ in *diag[i]*;
exit: $L_{i,i}$ will be delivered in *diag[i]*, $i=1,...,aux[3]$;
super: float *super[1:n-1]*;
entry: the superdiagonal of T, value of $T_{i,i+1}$ in *super[i]*, $i=1,...,n-1$;
exit: $U_{i,i+1}$ will be delivered in *super[i]*, $i=1,...,aux[3]-1$;
n: int; entry: the order of the matrix;
aid: float *aid[1:n-2]*;
exit: the value of $U_{i,i+2}$ will be delivered in *aid[i]*, $i=1,...,aux[3]-2$;

aux: float *aux[2:5]*;
entry:
aux[2]: a relative tolerance; a reasonable choice for this value is an estimate of the relative precision of the matrix elements; however, it should not be chosen smaller than the machine precision;
exit:
aux[3]: the number of elimination steps performed;
aux[5]: if *aux[3]=n* then *aux[5]* will be equal to the infinity norm of the matrix, else *aux[5]* is set equal to the value of that element which causes the breakdown of the decomposition;
b: float *b[1:n]*;
entry: the right hand side of the linear system;
exit: if *aux[3]=n* then the solution of the linear system is overwritten on *b*, else *b* remains unaltered.

Functions used: dectri, soltri.

```
void decsoltri(float sub[], float diag[], float super[], int n,
               float aux[], float b[])
{
    void dectri(float [], float [], float [],    int, float []);
    void soltri(float [], float [], float [], int, float []);

    dectri(sub,diag,super,n,aux);
    if (aux[3] == n) soltri(sub,diag,super,n,b);
}
```

C. soltripiv

Given the $n \times n$ lower bidiagonal matrix L ($L_{i,j}=0$ for $i>j+1$ and $j>i$) and the unit upper band matrix (with at most two codiagonals) U ($U_{i,i}=1$, $U_{i,j}=0$ for $i>j$ and $j>i+2$) such that $LU=PT$ (where P is a permutation matrix resulting from either no interchange or interchange of consecutive rows at each stage in the LU decomposition of T), solves the system of equations $Tx=b$ (the above decomposition is produced at successful exit from *dectripiv*). One call of *dectripiv* followed by several calls of *soltripiv* may be used to solve several linear systems having the same tridiagonal matrix, but different right hand sides.

Function Parameters:

void soltripiv (*sub,diag,super,n,aid,piv,b*)

sub: float *sub[1:n-1]*;
entry: the subdiagonal of the lower-bidiagonal matrix, as delivered by *dectripiv*;
diag: float *diag[1:n]*;
entry: the diagonal of the lower bidiagonal matrix, as delivered by *dectripiv*;
super: float *super[1:n-1]*;
entry: the first codiagonal of the upper triangular matrix, as delivered by *dectripiv*;
n: int; entry: the order of the matrix;
aid: float *aid[1:n-2]*;
entry: the second codiagonal of the upper triangular matrix, as delivered by *dectripiv*;
piv: int *piv[1:n-1]*;
entry: the pivot information as delivered by *dectripiv*;
b: float *b[1:n]*;

entry: the right hand side of the linear system;
exit: the calculated solution of the linear system.

```
void soltripiv(float sub[], float diag[], float super[], int n,
               float aid[], int piv[], float b[])
{
    int i,n1;
    float bi,bi1,r,s,t;

    n1=n-1;
    for (i=1; i<=n1; i++) {
        if (piv[i]) {
            bi=b[i+1];
            bi1=b[i];
        } else {
            bi=b[i];
            bi1=b[i+1];
        }
        r=b[i]=bi/diag[i];
        b[i+1]=bi1-sub[i]*r;
    }
    r = b[n] /= diag[n];
    t = b[n1] -= super[n1]*r;
    for (i=n-2; i>=1; i--) {
        s=r;
        r=t;
        t = b[i] -= super[i]*r + ((piv[i]) ? aid[i]*s : 0.0);
    }
}
```

D. decsoltripiv

Given the $n \times n$ tridiagonal matrix T ($T_{i,j}$=0 for $|i-j|>1$) attempts to obtain a lower bidiagonal matrix L ($L_{i,j}$=0 for $i>j+1$ and $j>i$) and unit upper band matrix (with at most two codiagonals) U ($U_{i,i}$=1, $U_{i,j}$=0 for $i>j$ and $j>i+2$) such that $LU=PT$ where P is a permutation matrix resulting from either no interchange or interchange of consecutive rows at each stage in the LU decomposition and, if successful in so doing, solves the system of equations $Tx=b$. One call of *decsoltripiv* is equivalent to calling consecutively *dectripiv* and *soltripiv*. However, *decsoltripiv* does not make use of *dectripiv* and *soltripiv*, to save memory space and time. This is only true in the case that linear systems with different matrices have to be solved.

Function Parameters:

void decsoltripiv (*sub,diag,super,n,aux,b*)

sub: float *sub[1:n-1]*;
entry: the subdiagonal of the given matrix T, $T_{i+1,i}$ should be given in *sub[i]*, *i=1,...,n-1*;
exit: the elements of *sub* will be destroyed;

diag: float *diag[1:n]*;
entry: the diagonal of T, value of $T_{i,i}$ in *diag[i]*;
exit: the elements of *diag* will be destroyed;

super: float *super[1:n-1]*;
entry: the superdiagonal of T, value of $T_{i,i+1}$ in *super[i]*, *i=1,...,n-1*;
exit: the elements of *super* will be destroyed;

n: int;
entry: the order of the matrix;

aux: float *aux[2:5]*;

entry:
aux[2]: a relative tolerance; a reasonable choice for this value is an estimate of the relative precision of the matrix elements; however, it should not be chosen smaller than the machine precision;
exit:
aux[3]: the number of elimination steps performed;
aux[5]: if *aux[3]=n* then *aux[5]* will be equal to the infinity norm of the matrix, else *aux[5]* is set equal to the value of that element which causes the breakdown of the decomposition;

b: float *b[1:n]*;
entry: the right hand side of the linear system;
exit: if *aux[3]=n* then the solution of the linear system is overwritten on *b*, else *b* remains unaltered.

```
#include <math.h>

void decsoltripiv(float sub[], float diag[], float super[], int n,
                  float aux[], float b[])
{
    int *allocate_integer_vector(int, int);
    void free_integer_vector(int *, int);
    int i,i1,n1,n2,*piv;
    float d,r,s,u,t,q,v,w,norm,norm1,norm2,tol,bi,bi1,bi2;

    piv=allocate_integer_vector(1,n);
    tol=aux[2];
    d=diag[1];
    r=super[1];
    bi=b[1];
    norm=norm2=fabs(d)+fabs(r);
    n2=n-2;
    for (i=1; i<=n2; i++) {
        i1=i+1;
        s=sub[i];
        t=diag[i1];
        q=super[i1];
        bi1=b[i1];
        norm1=norm2;
        norm2=fabs(s)+fabs(t)+fabs(q);
        if (norm2 > norm) norm=norm2;
        if (fabs(d)*norm2 < fabs(s)*norm1) {
            if (fabs(s) <= tol*norm2) {
                aux[3]=i-1;
                aux[5]=s;
                free_integer_vector(piv,1);
                return;
            }
            u=super[i]=t/s;
            b[i] = bi1 /= s;
            bi -= bi1*d;
            v=sub[i]=q/s;
            w = super[i1] = -v*d;
            d=diag[i1]=r-u*d;
            r=w;
            norm2=norm1;
            piv[i]=1;
        } else {
            if (fabs(d) <= tol*norm1) {
                aux[3]=i-1;
                aux[5]=d;
                free_integer_vector(piv,1);
                return;
            }
            u=super[i]=r/d;
            b[i] = bi /= d;
            bi=bi1-bi*s;
            d=diag[i1]=t-u*s;
```

```
            piv[i]=0;
            r=q;
        }
    }
    n1=n-1;
    s=sub[n1];
    t=diag[n];
    norm1=norm2;
    bi1=b[n];
    norm2=fabs(s)+fabs(t);
    if (norm2 > norm) norm=norm2;
    if (fabs(d)*norm2 < fabs(s)*norm1) {
        if (fabs(s) <= tol*norm2) {
            aux[3]=n2;
            aux[5]=s;
            free_integer_vector(piv,1);
            return;
        }
        u=super[n1]=t/s;
        b[n1] = bi1 /= s;
        bi -= bi1*d;
        d=r-u*d;
        norm2=norm1;
    } else {
        if (fabs(d) <= tol*norm1) {
            aux[3]=n2;
            aux[5]=d;
            free_integer_vector(piv,1);
            return;
        }
        u=super[n1]=r/d;
        b[n1] = bi /= d;
        bi=bi1-bi*s;
        d=t-u*s;
    }
    if (fabs(d) <= tol*norm2) {
        aux[3]=n1;
        aux[5]=d;
        free_integer_vector(piv,1);
        return;
    }
    aux[3]=n;
    aux[5]=norm;
    bi1=b[n]=bi/d;
    bi = b[n1] -= super[n1]*bi1;
    for (i=n-2; i>=1; i--) {
        bi2=bi1;
        bi1=bi;
        bi = b[i] -= super[i]*bi1 + ((piv[i]) ? sub[i]*bi2 : 0.0);
    }
    free_integer_vector(piv,1);
}
```

3.8 Sparse symmetric positive definite band matrices

3.8.1 Preparatory procedure

chldecbnd

Decomposes a symmetric positive definite band matrix A ($A_{i,j}=0$ for $|i-j|>w$, $i,j=1,...,n$) into the form $A=U^TU$, where U is an upper triangular band matrix ($U_{i,j}=0$ for $j<i$ and $j>i+w$) by use of Cholesky's square root method. For $j=1,...,n$, with $j(w)=\max(i,j-w)$

$$\delta_j = A_{j,j} - \sum_{k=j(w)}^{j-1} U_{k,j}^2$$

$$U_{j,j} = \sqrt{\delta_j}$$

$$U_{i,j} = (A_{i,j} - \sum_{k=j(w)}^{i-1} U_{k,j}U_{k,i})/U_{i,i} \qquad i=j(w),...,j-1.$$

If when $j=k$, either $\delta_j<0$ or $\delta_j < \delta \max|a_{i,i}|$ for $1 \le i \le j$, δ being a small positive real number prescribed by the user, then the above process is terminated, and the value of $k-1$ is allocated to *aux[3]* as a failure indication [Dek68].

Function Parameters:

void chldecbnd *(a,n,w,aux)*

a: float *a[1:w(n-1)+n]*;
 entry: *a* contains columnwise the upper triangular band elements of the symmetric band matrix (the (i,j)-th element of the matrix is in location $a[(j-1)w+i]$, $j=1,...,n$, $i=\max(1,j-w),...,j$);
 exit: the band elements of the Cholesky matrix, which is an upper triangular band matrix with *w* superdiagonals, are delivered columnwise in *a*;

n: int;
 entry: the order of the band matrix;

w: int;
 entry: number of superdiagonals of the matrix;

aux: float *aux[2:3]*;
 entry:
 aux[2]: a relative tolerance to control the calculation of the diagonal elements of the Cholesky matrix (value of δ above);
 exit:
 aux[3]: if successful then *aux[3]=n*; otherwise *aux[3]=k-1*, where *k* is the index of the diagonal element of the Cholesky matrix that cannot be calculated.

Function used: vecvec.

```
#include <math.h>

void chldecbnd(float a[], int n, int w, float aux[])
{
    float vecvec(int, int, int, float [], float []);
    int j,k,jmax,kk,kj,w1,start;
    float r,eps,max;

    max=0.0;
    kk = -w;
    w1=w+1;
    for (j=1; j<=n; j++) {
        kk += w1;
        if (a[kk] > max) max=a[kk];
    }
    jmax=w;
    w1=w+1;
    kk = -w;
    eps=aux[2]*max;
    for (k=1; k<=n; k++) {
        if (k+w > n) jmax--;
        kk += w1;
        start=kk-k+1;
        r=a[kk]-vecvec(((k <= w1) ? start : kk-w),kk-1,0,a,a);
        if (r <= eps) {
```

```
                aux[3]=k-1;
                return;
            }
            a[kk]=r=sqrt(r);
            kj=kk;
            for (j=1; j<=jmax; j++) {
                kj += w;
                a[kj]=(a[kj]-vecvec(((k+j <= w1) ? start : kk-w+j),
                                kk-1,kj-kk,a,a))/r;
            }
        }
        aux[3]=n;
}
```

3.8.2 Calculation of determinant

chldetermbnd

With the Cholesky decomposition $A=U^TU$ of the symmetric positive definite band matrix A for whose elements $A_{i,j}=0$ for $|i-j|>w$, $i,j=1,...,n$) available, *chldetermbnd* computes the determinant Δ of A,

$$\Delta = \left(\prod_{j=1}^{n} U_{j,j}\right)^2 .$$

chldetermbnd should not be called when overflow can be expected.

Function Parameters:

float chldetermbnd *(a,n,w)*

chldetermbnd: delivers the determinant of symmetric positive definite band matrix whose Cholesky matrix is stored in *a* (value of Δ above);

a: float *a[1:w(n-1)+n]*;
  entry: the contents of *a* are as produced by *chldecbnd* or *chldecsolbnd*;

n: int;
  entry: the order of the band matrix;

w: int;
  entry: number of superdiagonals of the matrix.

```
float chldetermbnd(float a[], int n, int w)
{
    int j,kk,w1;
    float p;

    w1=w+1;
    kk = -w;
    p=1.0;
    for (j=1; j<=n; j++) {
        kk += w1;
        p *= a[kk];
    }
    return (p*p);
}
```

3.8.3 Solution of linear equations

A. chlsolbnd

With the Cholesky decomposition $A=U^TU$ of the symmetric positive definite band matrix A for whose elements $A_{i,j}=0$ for $|i-j|>w$, $i,j=1,...,n$) available, *chlsolbnd* carries out the solution of the systems of equations $Ax=b$. The relationships

$$y_i = (b_j - \sum_{k=j(w)}^{j-1} U_{k,j}y_k)/U_{j,j}$$

$$x_j = (y_j - \sum_{k=j+1}^{J(w)} U_{j,k}x_k)/U_{j,j}$$

for $j=1,...,n$, where $j(w)=\max(1,j-w)$, $J(w)=\min(n,j+w)$, are used (they follow from $U^Ty=b$, $Ux=y$).

Function Parameters:

void chlsolbnd (a,n,w,b)

a: float $a[1:w(n-1)+n]$;
entry: the contents of a are as produced by *chldecbnd*;
n: int;
entry: the order of the band matrix;
w: int;
entry: number of superdiagonals of the matrix;
b: float $b[1:n]$;
entry: the right hand side of the system of linear equations;
exit: the solution of the system.

Functions used: vecvec, scaprd1.

```
void chlsolbnd(float a[], int n, int w, float b[])
{
    float vecvec(int, int, int, float [], float []);
    float scaprd1(int, int, int, int, int, float [], float []);
    int k,imax,kk,w1;

    kk = -w;
    w1=w+1;
    for (k=1; k<=n; k++) {
        kk += w1;
        b[k]=(b[k]-vecvec(((k <= w1) ? 1 : k-w),k-1,kk-k,b,a))/a[kk];
    }
    imax = -1;
    for (k=n; k>=1; k--) {
        if (imax < w) imax++;
        b[k]=(b[k]-scaprd1(kk+w,w,k+1,1,imax,a,b))/a[kk];
        kk -= w1;
    }
}
```

B. chldecsolbnd

Solves the system of equations $Ax=b$ where A is an $n\times n$ symmetric positive definite band matrix for whose elements $A_{i,j}=0$ for $|i-j|>w$, $i,j=1,...,n$), by use of Cholesky's square root method. The decomposition $A=U^TU$ is first obtained by means of *chldecbnd* and, if this was successful, *chlsolbnd* is then called to obtain x.

Function Parameters:

void chldecsolbnd (a,n,w,aux,b)

a,n,w,aux: see *chldecbnd*;
b: float *b[1:n]*;
entry: the right hand side of the system of linear equations;
exit: the solution of the system.

Functions used: chldecbnd, chlsolbnd.

```
void chldecsolbnd(float a[], int n, int w, float aux[], float b[])
{
   void chldecbnd(float [], int, int, float []);
   void chlsolbnd(float [], int, int, float []);

   chldecbnd(a,n,w,aux);
   if (aux[3] == n) chlsolbnd(a,n,w,b);
}
```

3.9 Symmetric positive definite tridiagonal matrices

3.9.1 Preparatory procedure

decsymtri

Given the $n \times n$ symmetric tridiagonal matrix T ($T_{i,j}=0$ for $|i-j|>1$) attempts to obtain a unit upper diagonal matrix U ($U_{i,i}=1$, $U_{i,j}=0$ for $i>j$ and $j>i+1$) and unit upper bidiagonal matrix U ($U_{i,i}=1$, $U_{i,j}=0$ for $i>j$ and $j>i+1$) and a diagonal matrix D such that $U^TDU=T$. The rows of U and diagonal elements of D are determined in succession. If at stage k, $|T_{k,k}|<tol* \| t^{(k)} \|_1$, where *tol* is a tolerance prescribed by the user and $t^{(k)}$ is the k-th row of T, the decomposition process is discontinued, and an integer K is given the value $k-1$; otherwise K is given the value n at exit.

Function Parameters:

void decsymtri $(diag,co,n,aux)$

diag: float *diag[1:n]*;
entry: the diagonal of T, value of $T_{i,i}$ in *diag[i]*, $i=1,...,n$;
exit: $D_{i,i}$ will be delivered in *diag[i]*, $i=1,...,aux[3]$;
co: float *co[1:n-1]*;
entry: the codiagonal of T, value of $T_{i,i+1}$ in *co[i]*, $i=1,...,n-1$;
exit: $U_{i,i+1}$ will be delivered in *co[i]*, $i=1,...,aux[3]-1$;
n: int;
entry: the order of the matrix;
aux: float *aux[2:5]*;
entry:
aux[2]: a relative tolerance; a reasonable choice for this value is an estimate of the relative precision of the matrix elements; however, it should not be chosen smaller than the machine precision (the value of *tol* above);

exit:

aux[3]: the number of elimination steps performed (value of K above);

aux[5]: if *aux[3]=n* then *aux[5]* will be equal to the infinity norm of the matrix, else *aux[5]* is set equal to the value of that element which causes the breakdown of the decomposition (value of $T_{K+1,K+1}$).

```
#include <math.h>

void decsymtri(float diag[], float co[], int n, float aux[])
{
	int i,n1;
	float d,r,s,u,tol,norm,normr;

	tol=aux[2];
	d=diag[1];
	r=co[1];
	norm=normr=fabs(d)+fabs(r);
	if (fabs(d) <= normr*tol) {
		aux[3]=0.0;
		aux[5]=d;
		return;
	}
	u=co[1]=r/d;
	n1=n-1;
	for (i=2; i<=n1; i++) {
		s=r;
		r=co[i];
		d=diag[i];
		normr=fabs(s)+fabs(d)+fabs(r);
		diag[i] = d -= u*s;
		if (fabs(d) <= normr*tol) {
			aux[3]=i-1;
			aux[5]=d;
			return;
		}
		u=co[i]=r/d;
		if (normr > norm) norm=normr;
	}
	d=diag[n];
	normr=fabs(d)+fabs(r);
	diag[n] = d -= u*s;
	if (fabs(d) <= normr*tol) {
		aux[3]=n1;
		aux[5]=d;
		return;
	}
	if (normr > norm) norm=normr;
	aux[3]=n;
	aux[5]=norm;
}
```

3.9.2 Solution of linear equations

A. solsymtri

Given the $n \times n$ unit upper bidiagonal matrix U ($U_{i,i}=1$, $U_{i,j}=0$ for $i>j$ and $j>i+1$) and diagonal matrix D, solves the system of equations $Tx=b$, where $T=U^TDU$ (this decomposition results from successful exit from *decsymtri*). One call of *decsymtri* followed by several calls of *solsymtri* may be used to solve several linear systems having the same symmetric tridiagonal matrix, but different right hand sides.

Function Parameters:

void solsymtri (*diag,co,n,b*)

diag: float *diag[1:n]*;
entry: the diagonal matrix, as delivered by *decsymtri*;
co: float *co[1:n-1]*;
entry: the codiagonal of the unit upper bidiagonal matrix, as delivered by *decsymtri*;
n: int;
entry: the order of the matrix;
b: float *b[1:n]*;
entry: the right hand side of the linear system;
exit: the calculated solution of the linear system.

```
void solsymtri(float diag[], float co[], int n, float b[])
{
    int i;
    float r,s;

    r=b[1];
    b[1]=r/diag[1];
    for (i=2; i<=n; i++) {
        r=b[i]-co[i-1]*r;
        b[i]=r/diag[i];
    }
    s=b[n];
    for (i=n-1; i>=1; i--) s = b[i] -= co[i]*s;
}
```

B. decsolsymtri

Given the $n \times n$ symmetric tridiagonal matrix T ($T_{i,j}=0$ for $|i-j|>1$) attempts to obtain a unit bidiagonal matrix U ($U_{i,i}=1$, $U_{i,j}=0$ for $i>j$ and $j>i+1$) and a diagonal matrix D such that $U^T DU=T$ by calling *decsymtri* and, if successful in so doing, solves the system of equations $Tx=b$ by calling *solsymtri*.

Function Parameters:

void decsolsymtri (*diag,co,n,aux,b*)

diag: float *diag[1:n]*;
entry: the diagonal of T, value of $T_{i,i}$ in *diag[i]*, $i=1,...,n$;
exit: $D_{i,i}$ will be delivered in *diag[i]*, $i=1,...,aux[3]$;
co: float *co[1:n-1]*;
entry: the codiagonal of T, value of $T_{i,i+1}$ in *co[i]*, $i=1,...,n-1$;
exit: $U_{i,i+1}$ will be delivered in *co[i]*, $i=1,...,aux[3]-1$;
n: int;
entry: the order of the matrix;
aux: float *aux[2:5]*;
entry:
aux[2]: a relative tolerance; a reasonable choice for this value is an estimate of the relative precision of the matrix elements; however, it should not be chosen smaller than the machine precision;
exit:
aux[3]: the number of elimination steps performed;
aux[5]: if *aux[3]=n* then *aux[5]* will be equal to the infinity norm of the matrix, else *aux[5]* is set equal to the value of that element which causes the breakdown of the decomposition;
b: float *b[1:n]*;

entry: the right hand side of the linear system;
exit: if *aux[3]=n* then the solution of the linear system is overwritten on *b*, else *b* remains unaltered.

Functions used: decsymtri, solsymtri.

```
void decsolsymtri(float diag[], float co[], int n,
                  float aux[], float b[])
{
    void decsymtri(float [], float [], int, float []);
    void solsymtri(float [], float [], int, float []);

    decsymtri(diag,co,n,aux);
    if (aux[3] == n) solsymtri(diag,co,n,b);
}
```

3.10 Sparse real matrices - Iterative methods

conjgrad

Solves the system of equations $Ax=b$, where A is a $n{\times}n$ positive definite symmetric matrix, by the method of conjugate gradients [R71]. This method involves the construction of sequences of approximations x_k to the solution, differences $\Delta x_k=x_{k+1}-x_k$, and the associated residual vectors $r^{(k)}=b-Ax^{(k)}$. With x_0 an initial approximation to x, and

$$\rho_k = (r_k,r_k) \qquad \eta_k = (r_k,Ar_k) \qquad (k=0,1,...)$$

the recursion is

$$q_0 = \eta_0/\rho_0, \qquad \Delta x_0 = r_0/q_0$$

and for k=1,2,...

$$x_k = x_{k-1} + \Delta x_{k-1}$$
$$e_{k-1} = q_{k-1}\rho_k/\rho_{k-1}$$
$$q_k = (\eta_k/\rho_k) + e_{k-1}$$
$$\Delta x_k = (r_k + e_{k-1}\Delta x_{k-1})/q_k\ .$$

Setting $a_k=1/q_k$, $b_k=e_{k-1}/q_{k-1}$, $p_k=r_k+e_{k-1}\Delta x_{k-1}$, the above formulae may be written as

$$b_k = \rho_k/\rho_{k-1}$$
$$p_k = r_k + b_k p_{k-1}, \qquad q_k = Ap_k$$
$$a_k = \rho_k/(p_k,q_k)$$
$$x_{k+1} = x_k + a_k p_k, \qquad r_{k+1} = r_k - a_k q_k$$
$$\rho_{k+1} = (r_{k+1},r_{k+1}).$$

The computations are terminated if $\max(\|r_k\|_2, \|\Delta x_{k-1}\|_2) \le tol$, where tol is a tolerance prescribed by the user.

Function Parameters:

void conjgrad (*matvec,x,r,l,n,goon,iterate,norm2*)

matvec: void (**matvec*)(*p,q*);
this function defines the coefficient matrix A of the system as follows: at each call *matvec* delivers in q the matrix-vector product Ap, p and q are one-dimensional arrays, float *p[l:n], q[l:n]*;

x: float *x[l:n]*;
entry: an initial approximation to the solution;
exit: the solution;

r: float *r[l:n]*;

entry: the right hand side of the system;
exit: the residual b-Ax;

l,n: int;
entry: *l* and *n* are respectively the lower and upper bound of the arrays *x,r,p,q*;

goon: int (**goon*)(*iterate, norm2*);
this is a user supplied function, *goon* indicates the continuation of the process depending on the current values of *iterate* and *norm2*, if *goon* equals zero then the iteration process is stopped;

iterate: int *;
exit: delivers the number of iteration steps already performed;

norm2: float *;
exit: delivers the squared Euclidean norm of the residual.

Functions used: vecvec, elmvec.

```
void conjgrad(void (*matvec)(float [], float []), float x[],
              float r[], int l, int n, int (*goon)(int, float),
              int *iterate, float *norm2)
{
    float *allocate_real_vector(int, int);
    void free_real_vector(float *, int);
    float vecvec(int, int, int, float [], float []);
    void elmvec(int, int, int, float [], float [], float);
    int i;
    float a,b,prr,rrp,*p,*ap;

    p=allocate_real_vector(l,n);
    ap=allocate_real_vector(l,n);
    *iterate=0;
    do {
        if (*iterate == 0) {
            (*matvec)(x,p);
            for (i=l; i<=n; i++) p[i] = r[i] -= p[i];
            prr=vecvec(l,n,0,r,r);
        } else {
            b=rrp/prr;
            prr=rrp;
            for (i=l; i<=n; i++) p[i]=r[i]+b*p[i];
        }
        (*matvec)(p,ap);
        a=prr/vecvec(l,n,0,p,ap);
        elmvec(l,n,0,x,p,a);
        elmvec(l,n,0,r,ap,-a);
        *norm2=rrp=vecvec(l,n,0,r,r);
        (*iterate)++;
    } while ((*goon)(*iterate,*norm2));
    free_real_vector(p,l);
    free_real_vector(ap,l);
}
```

3.11 Similarity transformation

3.11.1 Equilibration - real matrices

A. eqilbr

Equilibrates an $n \times n$ real matrix A by determining a real diagonal matrix D whose diagonal elements are integer powers of 2, and a permutation matrix P such that the Euclidean norms of the k-th row and the k-th column of $A' = PDAD^{-1}P^{-1}$ are approximately equal ($k=1,...,n$).

With $j(k)=1,2,...,n,1,2,...$, (i.e. $j(k)=(k-1 \mod(n))+1$, $k=1,2,...$) a sequence of diagonal matrices D_k, where the $j(k)$-th diagonal element of D_k is ρ_k^{-1} (ρ_k being an integer power of 2) and all others 1, together with a sequence of matrices $A_k=D_kA_{k-1}D_k^{-1}$, $A_0=A$ are obtained. ρ_k is determined by the condition that the Euclidean norms of the $j(k)$-th column and $j(k)$-th row of A_k have approximately equal values. If all of diagonal elements of either the $j(k)$-th row or the $j(k)$-th column of A_{k-1} is nearly zero, the row and column in question are interchanged with the next pair for which this is not so. The process is terminated if either (a) $1/2 < \rho_k < 2$ for one whole cycle of the $j(k)$ or (b) $k=(n+1)n^2$. D is the product of the D_k.

Function Parameters:

void eqilbr (*a,n,em,d,inter*)

a: float *a[1:n,1:n]*;
  entry: the matrix to be equilibrated;
  exit: the equilibrated matrix;

n: int;
  entry: the order of the matrix;

em: float *em[0:0]*;
  entry: the machine precision;

d: float *d[1:n]*;
  exit: the main diagonal of the transforming diagonal matrix;

inter: int *inter[1:n]*;
  exit: information defining the possible interchanging of some rows and the corresponding columns.

Functions used: tammat, mattam, ichcol, ichrow.

Method: The method is equilibrated by means of Osborne's diagonal similarity transformation possibly with interchanges [DekHo68, Os60].

```
#include <math.h>

void eqilbr(float **a, int n, float em[], float d[], int inter[])
{
    float tammat(int, int, int, int, float **, float **);
    float mattam(int, int, int, int, float **, float **);
    void ichcol(int, int, int, int, float **);
    void ichrow(int, int, int, int, float **);
    int i,im,i1,p,q,j,t,count,exponent,ni;
    float c,r,eps,omega,factor,di;

    factor=1.0/(2.0*log(2.0));
    eps=em[0];
    omega=1.0/eps;
    t=p=1;
    q=ni=i=n;
    count=((n+1)*n)/2;
    for (j=1; j<=n; j++) {
        d[j]=1.0;
        inter[j]=0;
    }
    i = (i < q) ? i+1 : p;
    while (count > 0 && ni > 0) {
        count--;
        im=i-1;
        i1=i+1;
        c=sqrt(tammat(p,im,i,i,a,a)+tammat(i1,q,i,i,a,a));
        r=sqrt(mattam(p,im,i,i,a,a)+mattam(i1,q,i,i,a,a));
```

```
        if (c*omega <= r*eps) {
            inter[t]=i;
            ni=q-p;
            t++;
            if (p != i) {
                ichcol(1,n,p,i,a);
                ichrow(1,n,p,i,a);
                di=d[i];
                d[i]=d[p];
                d[p]=di;
            }
            p++;
        } else
            if (r*omega <= c*eps) {
                inter[t] = -i;
                ni=q-p;
                t++;
                if (q != i) {
                    ichcol(1,n,q,i,a);
                    ichrow(1,n,q,i,a);
                    di=d[i];
                    d[i]=d[q];
                    d[q]=di;
                }
                q--;
            } else {
                exponent=log(r/c)*factor;
                if (fabs(exponent) > 1.0) {
                    ni=q-p;
                    c=pow(2.0,exponent);
                    r=1.0/c;
                    d[i] *= c;
                    for (j=1; j<=im; j++) {
                        a[j][i] *=c;
                        a[i][j] *= r;
                    }
                    for (j=i1; j<=n; j++) {
                        a[j][i] *=c;
                        a[i][j] *= r;
                    }
                } else
                    ni--;
            }
        i = (i < q) ? i+1 : p;
    }
}
```

B. baklbr

Given the diagonal elements of the $n \times n$ diagonal matrix D, the pivot reference integers associated with a permutation matrix P, and a set of column vectors $v^{(j)}$ $(j=n1,...,n2)$, constructs the sequence of vectors $u^{(j)}=PDv^{(j)}$ $(j=n1,...,n2)$. (If $v^{(j)}$ is an eigenvector of $PDAD^{-1}P^{-1}$, $u^{(j)}$ is an eigenvector of A.)

Function Parameters:

void baklbr (*n,n1,n2,d,inter,vec*)

n: int; entry: the length of the vectors to be transformed;

n1,n2: int;
entry: the serial numbers of the first and last vector to be transformed;

d: float *d[1:n]*;
entry: the main diagonal of the transforming diagonal matrix of order *n*, as produced by *eqilbr*;

inter: int *inter[1:n]*;
entry: information defining the possible interchanging of some rows columns, as

produced by *eqilbr*;

vec: float *vec[1:n,n1:n2]*;

entry: the *n2-n1+1* vectors of length *n* to be transformed;

exit: the *n2-n1+1* vectors of length *n* resulting from the back transformation.

Function used: ichrow.

```
void baklbr(int n, int n1, int n2, float d[], int inter[], float **vec)
{
    void ichrow(int, int, int, int, float **);
    int i,j,k,p,q;
    float di;

    p=1;
    q=n;
    for (i=1; i<=n; i++) {
        di=d[i];
        if (di != 1)
            for (j=n1; j<=n2; j++) vec[i][j] *= di;
        k=inter[i];
        if (k > 0)
            p++;
        else
            if (k < 0) q--;
    }
    for (i=p-1+n-q; i>=1; i--) {
        k=inter[i];
        if (k > 0) {
            p--;
            if (k != p) ichrow(n1,n2,k,p,vec);
        } else {
            q++;
            if (-k != q) ichrow(n1,n2,-k,q,vec);
        }
    }
}
```

3.11.2 Equilibration - complex matrices

A. eqilbrcom

Equilibrates an $n \times n$ complex matrix A by determining a real diagonal matrix D whose diagonal elements are integer powers of 2, and a permutation matrix P such that if $C=PDAD^{-1}P^{-1}$, the diagonal elements of $C'C-CC'$ (C' denotes the conjugate transpose of C) are approximately zero. With $j(k)=1,2,...,n,1,2,...$, (i.e. $j(k)=(k-1 \bmod(n))+1$, $k=1,2,...$) a sequence of diagonal matrices D_k, where the $j(k)$-th diagonal element of D_k is ρ_k^{-1} (ρ_k being an integer power of 2) and all others 1, together with a sequence of matrices C_k for which $C_k=D_kC_{k-1}D_{k-1}^{-1}$, $C_0=A$ are determined. ρ_k is determined by the condition that the Euclidean norms of the $j(k)$-th column and $j(k)$-th row of C_k have approximately equal values. If all off-diagonal elements of either the $j(k)$-th column or the $j(k)$-th row of C_{k-1} are nearly zero, the row and column in question are interchanged with the next pair for which this is not so. The process is terminated if either (a) $1/2 < \rho_k < 2$ for one whole cycle of the $j(k)$ or (b) $k=kmax+1$, where *kmax* is an integer prescribed by the user [DekHo68, Os60, PaR69].

Function Parameters:

void eqilbrcom (*a1,a2,n,em,d,inter*)

a1,a2: float *a1[1:n,1:n], a2[1:n,1:n]*;

entry: the real part and imaginary part of the matrix to be equilibrated must be given

in the arrays *a1* and *a2*, respectively;
exit: the real part and the imaginary part of the equilibrated matrix are delivered in the arrays *a1* and *a2*, respectively;

n: int;
entry: the order of the given matrix;

em: float *em[0:7]*;
entry:
em[0]: the machine precision;
em[6]: the maximum allowed number of iterations (value of *kmax* above);
exit:
em[7]: the number of iterations performed;

d: float *d[1:n]*;
exit: the scaling factors of the diagonal similarity transformation;

inter: int *inter[1:n]*;
exit: information defining the possible interchanging of some rows and the corresponding columns.

Functions used: ichcol, ichrow, tammat, mattam.

```
#include <math.h>

void eqilbrcom(float **a1, float **a2, int n, float em[], float d[],
               int inter[])
{
    void ichcol(int, int, int, int, float **);
    void ichrow(int, int, int, int, float **);
    float tammat(int, int, int, int, float **, float **);
    float mattam(int, int, int, int, float **, float **);
    int i,p,q,j,t,count,exponent,ni,im,i1;
    float c,r,eps,di;

    eps=em[0]*em[0];
    t=p=1;
    q=ni=i=n;
    count=em[6];
    for (j=1; j<=n; j++) {
        d[j]=1.0;
        inter[j]=0;
    }
    i = (i < q) ? i+1 : p;
    while (count > 0 && ni > 0) {
        count--;
        im=i-1;
        i1=i+1;
        c=tammat(p,im,i,i,a1,a1)+tammat(i1,q,i,i,a1,a1)+
            tammat(p,im,i,i,a2,a2)+tammat(i1,q,i,i,a2,a2);
        r=mattam(p,im,i,i,a1,a1)+mattam(i1,q,i,i,a1,a1)+
            mattam(p,im,i,i,a2,a2)+mattam(i1,q,i,i,a2,a2);
        if (c/eps <= r) {
            inter[t]=i;
            ni=q-p;
            t++;
            if (p != i) {
                ichcol(1,n,p,i,a1);
                ichrow(1,n,p,i,a1);
                ichcol(1,n,p,i,a2);
                ichrow(1,n,p,i,a2);
                di=d[i];
                d[i]=d[p];
                d[p]=di;
            }
            p++;
        } else
            if (r/eps <= c) {
```

```
                inter[t] = -i;
                ni=q-p;
                t++;
                if (q != i) {
                    ichcol(1,n,q,i,a1);
                    ichrow(1,n,q,i,a1);
                    ichcol(1,n,q,i,a2);
                    ichrow(1,n,q,i,a2);
                    di=d[i];
                    d[i]=d[q];
                    d[q]=di;
                }
                q--;
            } else {
                exponent=ceil(log(r/c)*0.36067);
                if (abs(exponent) > 1) {
                    ni=q-p;
                    c=pow(2.0,exponent);
                    d[i] *= c;
                    for (j=1; j<=im; j++) {
                        a1[j][i] *= c;
                        a1[i][j] /= c;
                        a2[j][i] *= c;
                        a2[i][j] /= c;
                    }
                    for (j=i1; j<=n; j++) {
                        a1[j][i] *= c;
                        a1[i][j] /= c;
                        a2[j][i] *= c;
                        a2[i][j] /= c;
                    }
                } else
                    ni--;
            }
        i = (i < q) ? i+1 : p;
    }
    em[7]=em[6]-count;
}
```

B. baklbrcom

Given a real $n \times n$ diagonal matrix D, a permutation matrix P and a sequence of complex vectors $v^{(k)}$ $(k=n1,...,n2)$, constructs the vectors $u^{(k)}=DPv^{(k)}$ $(k=n1,...,n2)$. (If $v^{(k)}$, $k=n1,...,n2$, are eigenvectors of $P^{-1}D^{-1}ADP$, $u^{(k)}$, $k=n1,...,n2$, are corresponding eigenvectors of A.)

Function Parameters:

void baklbrcom *(n,n1,n2,d,inter,vr,vi)*

n: int;
  entry: the order of the matrix of which the eigenvectors are calculated;

n1,n2: int;
  entry: the eigenvectors corresponding to the eigenvalues with indices *n1,...,n2* are to be transformed;

d: float *d[1:n]*;
  entry: the scaling factors of the diagonal similarity transformation as delivered by *eqilbrcom*;

inter: int *inter[1:n]*;
  entry: information defining the interchanging of some rows and columns, as produced by *eqilbrcom*;

vr,vi: float *vr[1:n,n1:n2], vi[1:n,n1:n2]*;
  entry: the back transformation is performed on the eigenvectors with the real parts given in array *vr* and the imaginary parts given in array *vi*;

exit: the real parts and imaginary parts of the resulting eigenvectors are delivered in the columns of the arrays *vr* and *vi*, respectively.

Function used: baklbr.

```
void baklbrcom(int n, int n1, int n2, float d[], int inter[],
               float **vr, float **vi)
{
    void baklbr(int, int, int, float [], int [], float **);

    baklbr(n,n1,n2,d,inter,vr);
    baklbr(n,n1,n2,d,inter,vi);
}
```

3.11.3 To Hessenberg form - real symmetric

A. tfmsymtri2

Reduces an $n \times n$ symmetric matrix A to tridiagonal form T by setting $A^{(1)}=A$ and $A^{(i+1)}=P^{(i)}AP^{(i)}$ $(i=1,...,n-2)$ when $T=A^{(n-1)}$ (setting

$$(v^{(i)})^T = (A_{k,1}{}^{(i)},...,A_{k,k-2}{}^{(i)},A_{k,k-1}{}^{(i)},0,0,...,0),$$

and

$$(u^{(i)})^T=(v^{(i)})^T \pm ((v^{(i)})^T v^{(i)}) e_k,$$

where $k=n-i+1$ and e_k is the k-th unit vector,
and $\pm$ is the sign of $A_{k,k-1}$, $P^{(i)} = I - 2u^{(i)}(u^{(i)})^T/(u^{(i)})^T u^{(i)}$ in the above). The i-th transformation reduces the elements in the $(n-i+1)$-th row and column of $A^{(i)}$ which do not also belong to the principal diagonal and the two adjacent diagonals, to zero; if all of these elements are already less in absolute value than $\|A\|_\infty$ times the machine precision, the i-th transformation is skipped [DekHo68, Wi65].

Function Parameters:

void tfmsymtri2 (*a,n,d,b,bb,em*)

a: float *a[1:n,1:n]*;
  entry: the upper triangle of the symmetric matrix must be given in the upper triangular part of *a* (the elements *a[i,j]*, $i \leq j$);
  exit: the data for Householder's back transformation is delivered in the upper triangular part of *a*, the elements *a[i,j]*, $i>j$ are neither used nor changed;

n: int;
  entry: the order of the given matrix;

d: float *d[1:n]*;
  exit: the main diagonal of the symmetric tridiagonal matrix *T*, produced by Householder's transformation;

b: float *b[1:n]*;
  exit: the codiagonal elements of *T* are delivered in *b[1:n-1]*, *b[n]* is set equal to zero;

bb: float *bb[1:n]*;
  exit: the squares of the codiagonal elements of *T* are delivered in *bb[1:n-1]*, *bb[n]* is set equal to zero;

em: float *em[0:1]*;
  entry: *em[0]*: the machine precision;

exit: *em[1]*: the infinity norm of the original matrix.

Functions used: tamvec, matmat, tammat, elmveccol, elmcolvec, elmcol.

```
#include <math.h>

void tfmsymtri2(float **a, int n, float d[], float b[], float bb[],
                float em[])
{
    float tammat(int, int, int, int, float **, float **);
    float matmat(int, int, int, int, float **, float **);
    void elmveccol(int, int, int, float [], float **, float);
    float tamvec(int, int, int, float **, float []);
    void elmcol(int, int, int, int, float **, float **, float);
    void elmcolvec(int, int, int, float **, float [], float);
    int i,j,r,r1;
    float w,x,a1,b0,bb0,machtol,norm;

    norm=0.0;
    for (j=1; j<=n; j++) {
        w=0.0;
        for (i=1; i<=j; i++) w += fabs(a[i][j]);
        for (i=j+1; i<=n; i++) w += fabs(a[j][i]);
        if (w > norm) norm=w;
    }
    machtol=em[0]*norm;
    em[1]=norm;
    r=n;
    for (r1=n-1; r1>=1; r1--) {
        d[r]=a[r][r];
        x=tammat(1,r-2,r,r,a,a);
        a1=a[r1][r];
        if (sqrt(x) <= machtol) {
            b0=b[r1]=a1;
            bb[r1]=b0*b0;
            a[r][r]=1.0;
        } else {
            bb0=bb[r1]=a1*a1+x;
            b0 = (a1 > 0.0) ? -sqrt(bb0) : sqrt(bb0);
            a1=a[r1][r]=a1-b0;
            w=a[r][r]=1.0/(a1*b0);
            for (j=1; j<=r1; j++)
                b[j]=(tammat(1,j,j,r,a,a)+matmat(j+1,r1,j,r,a,a))*w;
            elmveccol(1,r1,r,b,a,tamvec(1,r1,r,a,b)*w*0.5);
            for (j=1; j<=r1; j++) {
                elmcol(1,j,j,r,a,a,b[j]);
                elmcolvec(1,j,j,a,b,a[j][r]);
            }
            b[r1]=b0;
        }
        r=r1;
    }
    d[1]=a[1][1];
    a[1][1]=1.0;
    b[n]=bb[n]=0.0;
}
```

B. baksymtri2

Performs the back substitutions upon the intermediate numbers produced by *tfmsymtri2*.

Function Parameters:

void baksymtri2 *(a,n,n1,n2,vec)*

a: float *a[1:n,1:n]*;
entry: the data for the back transformation, as produced by *tfmsymtri2*, must be given in the upper triangular part of *a*;

n: int;
entry: the order of the given matrix;

n1,n2: int;
entry: the lower and upper bound, respectively, of the column numbers of *vec*;

vec: float *vec[1:n, n1:n2]*;
entry: the vectors on which the back transformation has to be performed;
exit: the transformed vectors.

Functions used: tammat, elmcol.

```
void baksymtri2(float **a, int n, int n1, int n2, float **vec)
{
    float tammat(int, int, int, int, float **, float **);
    void elmcol(int, int, int, int, float **, float **, float);
    int j,k;
    float w;

    for (j=2; j<=n; j++) {
        w=a[j][j];
        if (w < 0.0)
            for (k=n1; k<=n2; k++)
                elmcol(1,j-1,k,j,vec,a,tammat(1,j-1,j,k,a,vec)*w);
    }
}
```

C. tfmprevec

Computes the matrix P for which $PAP^T=T$ where A is an $n \times n$ matrix, and the $n \times n$ matrix T is of tridiagonal form, using the intermediate results generated by *tfmsymtri2*.

Function Parameters:

void tfmprevec (a,n)

a: float *a[1:n,1:n]*;
entry: the data for the back transformation, as produced by *tfmsymtri2*, must be given in the upper triangular part of *a*;
exit: the matrix which transforms the original matrix into a similar tridiagonal one;

n: int;
entry: the order of the given matrix.

Functions used: tammat, elmcol.

```
void tfmprevec(float **a, int n)
{
    float tammat(int, int, int, int, float **, float **);
    void elmcol(int, int, int, int, float **, float **, float);
    int i,j,j1,k;
    float ab;

    j1=1;
    for (j=2; j<=n; j++) {
        for (i=1; i<=j1-1; i++) a[i][j1]=0.0;
        for (i=j; i<=n; i++) a[i][j1]=0.0;
        a[j1][j1]=1.0;
        ab=a[j][j];
        if (ab < 0)
            for (k=1; k<=j1; k++)
                elmcol(1,j1,k,j,a,a,tammat(1,j1,j,k,a,a)*ab);
        j1=j;
    }
```

```
    for (i=n-1; i>=1; i--) a[i][n]=0.0;
    a[n][n]=1.0;
}
```

D. tfmsymtri1

Reduces an $n \times n$ symmetric matrix A to tridiagonal form T as in the implementation of *tfmsymtri2*. It is assumed that the upper triangular elements of A are given in a linear array.

Function Parameters:

void tfmsymtri1 (*a,n,d,b,bb,em*)

a: float *a[1:(n+1)n/2]*;
  entry: the upper triangle of the given matrix must be given in such a way that the (i,j)-th element of the matrix is $a[(j-1)j/2+i]$, $1 \leq i \leq j \leq n$;
  exit: the data for Householder's back transformation as used by *baksymtri1*;

n: int;
  entry: the order of the given matrix;

d: float *d[1:n]*;
  exit: the main diagonal of the symmetric tridiagonal matrix T, produced by Householder's transformation;

b: float *b[1:n]*;
  exit: the codiagonal elements of T are delivered in *b[1:n-1]*, *b[n]* is set equal to zero;

bb: float *bb[1:n]*;
  exit: the squares of the codiagonal elements of T are delivered in *bb[1:n-1]*, *bb[n]* is set equal to zero;

em: float *em[0:1]*;
  entry: *em[0]*: the machine precision;
  exit: *em[1]*: the infinity norm of the original matrix.

Functions used: vecvec, seqvec, elmvec.

```
#include <math.h>

void tfmsymtri1(float a[], int n, float d[], float b[], float bb[],
                float em[])
{
    float vecvec(int, int, int, float [], float []);
    float seqvec(int, int, int, int, float [], float []);
    void elmvec(int, int, int, float [], float [], float);
    int i,j,r,r1,p,q,ti,tj;
    float s,w,x,a1,b0,bb0,norm,machtol;

    norm=0.0;
    tj=0;
    for (j=1; j<=n; j++) {
        w=0.0;
        for (i=1; i<=j; i++) w += fabs(a[i+tj]);
        tj += j;
        ti=tj+j;
        for (i=j+1; i<=n; i++) {
            w += fabs(a[ti]);
            ti += i;
        }
        if (w > norm) norm=w;
    }
    machtol=em[0]*norm;
    em[1]=norm;
```

```
    q=((n+1)*n)/2;
    r=n;
    for (r1=n-1; r1>=1; r1--) {
        p=q-r;
        d[r]=a[q];
        x=vecvec(p+1,q-2,0,a,a);
        a1=a[q-1];
        if (sqrt(x) <= machtol) {
            b0=b[r1]=a1;
            bb[r1]=b0*b0;
            a[q]=1.0;
        } else {
            bb0=bb[r1]=a1*a1+x;
            b0 = (a1 > 0.0) ? -sqrt(bb0) : sqrt(bb0);
            a1=a[q-1]=a1-b0;
            w=a[q]=1.0/(a1*b0);
            tj=0;
            for (j=1; j<=r1; j++) {
                ti=tj+j;
                s=vecvec(tj+1,ti,p-tj,a,a);
                tj=ti+j;
                b[j]=(seqvec(j+1,r1,tj,p,a,a)+s)*w;
                tj=ti;
            }
            elmvec(1,r1,p,b,a,vecvec(1,r1,p,b,a)*w*0.5);
            tj=0;
            for (j=1; j<=r1; j++) {
                ti=tj+j;
                elmvec(tj+1,ti,p-tj,a,a,b[j]);
                elmvec(tj+1,ti,-tj,a,b,a[j+p]);
                tj=ti;
            }
            b[r1]=b0;
        }
        q=p;
        r=r1;
    }
    d[1]=a[1];
    a[1]=1.0;
    b[n]=bb[n]=0.0;
}
```

E. baksymtri1

Performs the back substitutions upon the intermediate numbers produced by *tfmsymtri1*.

Function Parameters:

void baksymtri1 *(a,n,n1,n2,vec)*

a: float *a[1:(n+1)n/2]*;
  entry: the data for the back transformation, as produced by *tfmsymtri1*;
n: int;
  entry: the order of the given matrix;
n1,n2: int;
  entry: the lower and upper bound, respectively, of the column numbers of *vec*;
vec: float *vec[1:n, n1:n2]*;
  entry: the vectors on which the back transformation has to be performed;
  exit: the transformed vectors.

Functions used: vecvec, elmvec.

```
void baksymtri1(float a[], int n, int n1, int n2, float **vec)
{
    float *allocate_real_vector(int, int);
```

```
    void free_real_vector(float *, int);
    float vecvec(int, int, int, float [], float []);
    void elmvec(int, int, int, float [], float [], float);
    int j,j1,k,ti,tj;
    float w,*auxvec;

    auxvec=allocate_real_vector(1,n);
    for (k=n1; k<=n2; k++) {
        for (j=1; j<=n; j++) auxvec[j]=vec[j][k];
        tj=j1=1;
        for (j=2; j<=n; j++) {
            ti=tj+j;
            w=a[ti];
            if (w < 0.0)
                elmvec(1,j1,tj,auxvec,a,vecvec(1,j1,tj,auxvec,a)*w);
            j1=j;
            tj=ti;
        }
        for (j=1; j<=n; j++) vec[j][k]=auxvec[j];
    }
    free_real_vector(auxvec,1);
}
```

3.11.4 To Hessenberg form - real asymmetric

A. tfmreahes

Given a real $n \times n$ matrix A, obtains a sequence of pivot reference integers $1 \leq p(i) \leq n$ $(i=1,...,n)$ associated with a permutation matrix P, a unit lower triangular matrix L with $L_{i,1}=0$ $(i=2,...,n)$, $|L_{i,j}|<1$ $(j=2,...,n;\ i=j+1,...,n)$ such that $PLH=APL$, where H is an upper Hessenberg matrix ($H_{i,j}=0$ for $i>j+1$). For further details see [DekHof68, Wi65].

Function Parameters:

void tfmreahes (*a,n,em,index*)

a: float *a[1:n,1:n]*;
 entry: the matrix to be transformed;
 exit: the upper Hessenberg matrix is delivered in the upper triangle and the first subdiagonal of *a*, the transforming matrix *L*, in the remaining part of *a*, i.e. $a[i,j]=L_{i,j+1}$, for $i=3,...,n$ and $j=1,...,i-2$;

n: int;
 entry: the order of the given matrix;

em: float *em[0:1]*;
 entry: *em[0]*: the machine precision;
 exit: *em[1]*: the infinity norm of the original matrix;

index: int *inter[1:n]*;
 exit: the pivotal indices defining the stabilizing row and column interchanges.

Functions used: matvec, matmat, ichcol, ichrow.

```
#include <math.h>

void tfmreahes(float **a, int n, float em[], int index[])
{
    float *allocate_real_vector(int, int);
    void free_real_vector(float *, int);
    float matvec(int, int, int, float **, float []);
    float matmat(int, int, int, int, float **, float **);
    void ichcol(int, int, int, int, float **);
    void ichrow(int, int, int, int, float **);
```

```
	int i,j,j1,k,l;
	float s,t,machtol,macheps,norm,*b;

	b=allocate_real_vector(0,n-1);
	macheps=em[0];
	norm=0.0;
	for (i=1; i<=n; i++) {
		s=0.0;
		for (j=1; j<=n; j++) s += fabs(a[i][j]);
		if (s > norm) norm=s;
	}
	em[1]=norm;
	machtol=norm*macheps;
	index[1]=0;
	for (j=2; j<=n; j++) {
		j1=j-1;
		l=0;
		s=machtol;
		for (k=j+1; k<=n; k++) {
			t=fabs(a[k][j1]);
			if (t > s) {
				l=k;
				s=t;
			}
		}
		if (l != 0) {
			if (fabs(a[j][j1]) < s) {
				ichrow(1,n,j,l,a);
				ichcol(1,n,j,l,a);
			} else
				l=j;
			t=a[j][j1];
			for (k=j+1; k<=n; k++) a[k][j1] /=t;
		} else
			for (k=j+1; k<=n; k++) a[k][j1]=0.0;
		for (i=1; i<=n; i++)
			b[i-1] = a[i][j] +=
				((l == 0) ? 0.0 : matmat(j+1,n,i,j1,a,a)) -
				matvec(1,(j1 < i-2) ? j1 : i-2,i,a,b);
		index[j]=l;
	}
	free_real_vector(b,0);
}
```

B. bakreahes1

Given a sequence of pivot reference integers $p(i)$ ($1 \le p(i) \le n$, $i=1,...,n$) associated with an $n \times n$ permutation matrix P, a unit lower triangular matrix L with $L_{i,1}=0$ ($i=2,...,n$) and a single column vector v, forms the vector $u=PLv$. (If v is an eigenvector of H, where $PLH=APL$, u is an eigenvector of A.)

Function Parameters:

void bakreahes1 $(a,n,index,v)$

a: float *a[1:n,1:n]*;
entry: the transforming matrix L, as produced by *tfmreahes* must be given in the part below the first subdiagonal of a, i.e. $a[i,j]=L_{i,j+1}$, for $i=3,...,n$ and $j=1,...,i-2$;

n: int;
entry: the length of the vector to be transformed;

index: int *index[1:n]*;
entry: the pivotal indices defining the stabilizing row and column interchanges as produced by *tfmreahes*;

v: float *v[1:n]*;
entry: the vector to be transformed;

exit: the transformed vector.

Function used: matvec.

```
void bakreahes1(float **a, int n, int index[], float v[])
{
	float *allocate_real_vector(int, int);
	void free_real_vector(float *, int);
	float matvec(int, int, int, float **, float []);
	int i,l;
	float w,*x;

	x=allocate_real_vector(1,n);
	for (i=2; i<=n; i++) x[i-1]=v[i];
	for (i=n; i>=2; i--) {
		v[i] += matvec(1,i-2,i,a,x);
		l=index[i];
		if (l > i) {
			w=v[i];
			v[i]=v[l];
			v[l]=w;
		}
	}
	free_real_vector(x,1);
}
```

C. bakreahes2

Given a sequence of pivot reference integers $p(i)$ $(1 \leq p(i) \leq n,\ i=1,...,n)$ associated with an $n \times n$ permutation matrix P, a unit lower triangular matrix L with $L_{i,1}=0$ $(i=2,...,n)$ and a system of column vectors $b^{(k)}$ $(k=n1,...,n2)$ forms the vectors $v^{(k)}=PLb^{(k)}$ $(k=n1,...,n2)$. (If $b^{(k)}$ is an eigenvector of H, where $PLH=APL$, $v^{(k)}$ is an eigenvector of A.)

Function Parameters:

void bakreahes2 $(a,n,n1,n2,index,vec)$

a: float $a[1:n,1:n]$;
  entry: the transforming matrix L, as produced by *tfmreahes* must be given in the part below the first subdiagonal of a, i.e. $a[i,j]=L_{i,j+1}$, for $i=3,...,n$ and $j=1,...,i-2$;

n: int;
  entry: the length of the vectors to be transformed;

n1,n2: int;
  the column numbers of the first and last vector to be transformed;

index: int $index[1:n]$;
  entry: the pivotal indices defining the stabilizing row and column interchanges as produced by *tfmreahes*;

vec: float $vec[1:n,n1:n2]$;
  entry: the $n2-n1+1$ vectors of length n to be transformed;
  exit: the $n2-n1+1$ vectors of length n resulting from the back transformation.

Functions used: tamvec, ichrow.

```
void bakreahes2(float **a, int n, int n1, int n2, int index[],
				float **vec)
{
	float *allocate_real_vector(int, int);
	void free_real_vector(float *, int);
	float tamvec(int, int, int, float **, float []);
	void ichrow(int, int, int, int, float **);
```

```
    int i,l,k;
    float *u;

    u=allocate_real_vector(1,n);
    for (i=n; i>=2; i--) {
        for (k=i-2; k>=1; k--) u[k+1]=a[i][k];
        for (k=n1; k<=n2; k++) vec[i][k] += tamvec(2,i-1,k,vec,u);
        l=index[i];
        if (l > i) ichrow(n1,n2,i,l,vec);
    }
    free_real_vector(u,1);
}
```

3.11.5 To Hessenberg form - complex Hermitian

A. hshhrmtri

Given an $n \times n$ Hermitian matrix A, obtains a complex diagonal matrix C and k ($k \leq n-2$) vectors $u^{(j)}$ such that with

$$E = \prod_{j=1}^{k} (I - 2u^{(j)T}u^{(j)}/(u^{(j)}u^{(j)T})),$$

$S=C^{-1}E'AEC$ (E' denotes the conjugate transpose of E) is a real symmetric tridiagonal matrix ($S_{i,j}=0$ if $|i-j|>1$ and $S_{i,i+1}=S_{i+1,i}$).

Function Parameters:

void hshhrmtri *(a,n,d,b,bb,em,tr,ti)*

a: float *a[1:n,1:n]*;
- entry: the real part of the upper triangle of the Hermitian matrix must be given in the upper triangular part of *a* (the elements *a[i,j]*, $i \leq j$); the imaginary part of the strict lower triangle of the Hermitian matrix must be given in the strict lower part of *a* (the elements *a[i,j]*, $i>j$);
- exit: the components of the $u^{(j)}$ and C;

n: int;
- entry: the order of the matrix;

d: float *d[1:n]*;
- exit: the main diagonal of the resulting symmetric tridiagonal matrix ($S_{i,i}$ above);

b: float *b[1:n-1]*;
- exit: the codiagonal elements of the resulting symmetric tridiagonal matrix (the values $S_{i,i+1}$, $i=1,...,n-1$, in the above);

bb: float *bb[1:n-1]*;
- exit: the squares of the moduli of the codiagonal elements of the resulting symmetric tridiagonal matrix (the values $(S_{i,i+1})^2$, $i=1,...,n-1$, in the above);

em: float *em[0:1]*;
- entry: *em[0]*: the machine precision;
- exit: *em[1]*: an estimate for a norm of the original matrix;

tr,ti: float *tr[1:n-1]*, *ti[1:n-1]*;
- exit: data for subsequent back transformations.

Functions used: matvec, tamvec, matmat, tammat, mattam, elmveccol, elmcolvec, elmcol, elmrow, elmvecrow, elmrowvec, elmrowcol, elmcolrow, carpol.

Method: see *hshhrmtrival*.

```
#include <math.h>

void hshhrmtri(float **a, int n, float d[], float b[], float bb[],
               float em[], float tr[], float ti[])
{
    float matvec(int, int, int, float **, float []);
    float tamvec(int, int, int, float **, float []);
    float matmat(int, int, int, int, float **, float **);
    float tammat(int, int, int, int, float **, float **);
    float mattam(int, int, int, int, float **, float **);
    void elmveccol(int, int, int, float [], float **, float);
    void elmcolvec(int, int, int, float **, float [], float);
    void elmcol(int, int, int, int, float **, float **, float);
    void elmrow(int, int, int, int, float **, float **, float);
    void elmvecrow(int, int, int, float [], float **, float);
    void elmrowvec(int, int, int, float **, float [], float);
    void elmrowcol(int, int, int, int, float **, float **, float);
    void elmcolrow(int, int, int, int, float **, float **, float);
    void carpol(float, float, float *, float *, float *);
    int i,j,j1,jm1,r,rm1;
    float nrm,w,tol2,x,ar,ai,mod,c,s,h,k,t,q,ajr,arj,bj,bbj;

    nrm=0.0;
    for (i=1; i<=n; i++) {
        w=fabs(a[i][i]);
        for (j=i-1; j>=1; j--) w += fabs(a[i][j])+fabs(a[j][i]);
        for (j=i+1; j<=n; j++) w += fabs(a[i][j])+fabs(a[j][i]);
        if (w > nrm) nrm=w;
    }
    t=em[0]*nrm;
    tol2=t*t;
    em[1]=nrm;
    r=n;
    for (rm1=n-1; rm1>=1; rm1--) {
        x=tammat(1,r-2,r,r,a,a)+mattam(1,r-2,r,r,a,a);
        ar=a[rm1][r];
        ai = -a[r][rm1];
        d[r]=a[r][r];
        carpol(ar,ai,&mod,&c,&s);
        if (x < tol2) {
            a[r][r] = -1.0;
            b[rm1]=mod;
            bb[rm1]=mod*mod;
        } else {
            h=mod*mod+x;
            k=sqrt(h);
            t=a[r][r]=h+mod*k;
            if (ar == 0.0 && ai == 0.0)
                a[rm1][r]=k;
            else {
                a[rm1][r]=ar+c*k;
                a[r][rm1] = -ai-s*k;
                s = -s;
            }
            c = -c;
            j=1;
            jm1=0;
            for (j1=2; j1<=r; j1++) {
                b[j]=(tammat(1,j,j,r,a,a)+matmat(j1,rm1,j,r,a,a)+
                        mattam(1,jm1,j,r,a,a)-matmat(j1,rm1,r,j,a,a))/t;
                bb[j]=(matmat(1,jm1,j,r,a,a)-tammat(j1,rm1,j,r,a,a)-
                        matmat(1,j,r,j,a,a)-mattam(j1,rm1,j,r,a,a))/t;
                jm1=j;
                j=j1;
            }
            q=(tamvec(1,rm1,r,a,b)-matvec(1,rm1,r,a,bb))/t/2.0;
```

```
                elmveccol(1,rm1,r,b,a,-q);
                elmvecrow(1,rm1,r,bb,a,q);
                j=1;
                for (j1=2; j1<=r; j1++) {
                    ajr=a[j][r];
                    arj=a[r][j];
                    bj=b[j];
                    bbj=bb[j];
                    elmrowvec(j,rm1,j,a,b,-ajr);
                    elmrowvec(j,rm1,j,a,bb,arj);
                    elmrowcol(j,rm1,j,r,a,a,-bj);
                    elmrow(j,rm1,j,r,a,a,bbj);
                    elmcolvec(j1,rm1,j,a,b,-arj);
                    elmcolvec(j1,rm1,j,a,bb,-ajr);
                    elmcol(j1,rm1,j,r,a,a,bbj);
                    elmcolrow(j1,rm1,j,r,a,a,bj);
                    j=j1;
                }
                bb[rm1]=h;
                b[rm1]=k;
            }
            tr[rm1]=c;
            ti[rm1]=s;
            r=rm1;
        }
        d[1]=a[1][1];
}
```

B. hshhrmtrival

Given an $n \times n$ Hermitian matrix A, obtains m complex vectors $c^{(k)}$ $(m \leq n\text{-}2)$ such that with $E=E_m E_{m-1} \dots E_1$, where

$$E_k = I - 2c^{(k)}\bar{c}^{(k)T}/\bar{c}^{(k)}c^{(k)}, \qquad EA\bar{E}^T=S, \qquad (\bar{E}^T\text{: conjugate and transpose of E})$$

where S is a Hermitian tridiagonal matrix and delivers the values $S_{i,i}$ $(i=1,\dots,n)$ and of $|T_{i,i+1}|_2$ $(i=1,\dots,n\text{-}1)$.

Function Parameters:

void hshhrmtrival *(a,n,d,bb,em)*

a: float *a[1:n,1:n]*;
- entry: the real part of the upper triangle of the Hermitian matrix must be given in the upper triangular part of a (the elements $a[i,j]$, $i \leq j$); the imaginary part of the strict lower triangle of the Hermitian matrix must be given in the strict lower part of a (the elements $a[i,j]$, $i>j$);
- exit: the elements of a are altered;

n: int;
- entry: the order of the given matrix;

d: float *d[1:n]*;
- exit: the main diagonal of the resulting Hermitian tridiagonal matrix ($S_{i,i}$ above);

bb: float *bb[1:n-1]*;
- exit: the squares of the moduli of the codiagonal elements of the resulting Hermitian tridiagonal matrix (the values $(S_{i,i+1})^2$, $i=1,\dots,n\text{-}1$, in the above);

em: float *em[0:1]*;
- entry: *em[0]*: the machine precision;
- exit: *em[1]*: an estimate for a norm of the original matrix.

Functions used: matvec, tamvec, matmat, tammat, mattam, elmveccol, elmcolvec, elmcol, elmrow, elmvecrow, elmrowvec, elmrowcol, elmcolrow.

Method: *hshhrmtrival* transforms a Hermitian matrix into a similar Hermitian tridiagonal matrix by means of Householder's transformation. *hshhrmtri* transforms a Hermitian matrix into a similar real tridiagonal matrix by means of Householder's transformation followed by a complex diagonal unitary similarity transformation in order to make the resulting tridiagonal matrix real symmetric. Householder's transformation for complex Hermitian matrices is a unitary similarity transformation, transforming a Hermitian matrix into a similar complex tridiagonal one [Mu66, Wi65].

```
#include <math.h>

void hshhrmtrival(float **a, int n, float d[], float bb[], float em[])
{
    float matvec(int, int, int, float **, float []);
    float tamvec(int, int, int, float **, float []);
    float matmat(int, int, int, int, float **, float **);
    float tammat(int, int, int, int, float **, float **);
    float mattam(int, int, int, int, float **, float **);
    void elmveccol(int, int, int, float [], float **, float);
    void elmcolvec(int, int, int, float **, float [], float);
    void elmcol(int, int, int, int, float **, float **, float);
    void elmrow(int, int, int, int, float **, float **, float);
    void elmvecrow(int, int, int, float [], float **, float);
    void elmrowvec(int, int, int, float **, float [], float);
    void elmrowcol(int, int, int, int, float **, float **, float);
    void elmcolrow(int, int, int, int, float **, float **, float);
    int i,j,j1,jm1,r,rm1;
    float nrm,w,tol2,x,ar,ai,h,t,q,ajr,arj,dj,bbj,mod2;

    nrm=0.0;
    for (i=1; i<=n; i++) {
        w=fabs(a[i][i]);
        for (j=i-1; j>=1; j--) w += fabs(a[i][j])+fabs(a[j][i]);
        for (j=i+1; j<=n; j++) w += fabs(a[i][j])+fabs(a[j][i]);
        if (w > nrm) nrm=w;
    }
    t=em[0]*nrm;
    tol2=t*t;
    em[1]=nrm;
    r=n;
    for (rm1=n-1; rm1>=1; rm1--) {
        x=tammat(1,r-2,r,r,a,a)+mattam(1,r-2,r,r,a,a);
        ar=a[rm1][r];
        ai = -a[r][rm1];
        d[r]=a[r][r];
        if (x < tol2)
            bb[rm1]=ar*ar+ai*ai;
        else {
            mod2=ar*ar+ai*ai;
            if (mod2 == 0.0) {
                a[rm1][r]=sqrt(x);
                t=x;
            } else {
                x += mod2;
                h=sqrt(mod2*x);
                t=x+h;
                h=1.0+x/h;
                a[r][rm1] = -ai*h;
                a[rm1][r]=ar*h;
            }
            j=1;
            jm1=0;
            for (j1=2; j1<=r; j1++) {
                d[j]=(tammat(1,j,j,r,a,a)+matmat(j1,rm1,j,r,a,a)+
                        mattam(1,jm1,j,r,a,a)-matmat(j1,rm1,r,j,a,a))/t;
```

```
                bb[j]=(matmat(1,jm1,j,r,a,a)-tammat(j1,rm1,j,r,a,a)-
                        matmat(1,j,r,j,a,a)-mattam(j1,rm1,j,r,a,a))/t;
                jm1=j;
                j=j1;
            }
            q=(tamvec(1,rm1,r,a,d)-matvec(1,rm1,r,a,bb))/t/2.0;
            elmveccol(1,rm1,r,d,a,-q);
            elmvecrow(1,rm1,r,bb,a,q);
            j=1;
            for (j1=2; j1<=r; j1++) {
                ajr=a[j][r];
                arj=a[r][j];
                dj=d[j];
                bbj=bb[j];
                elmrowvec(j,rm1,j,a,d,-ajr);
                elmrowvec(j,rm1,j,a,bb,arj);
                elmrowcol(j,rm1,j,r,a,a,-dj);
                elmrow(j,rm1,j,r,a,a,bbj);
                elmcolvec(j1,rm1,j,a,d,-arj);
                elmcolvec(j1,rm1,j,a,bb,-ajr);
                elmcol(j1,rm1,j,r,a,a,bbj);
                elmcolrow(j1,rm1,j,r,a,a,dj);
                j=j1;
            }
            bb[rm1]=x;
        }
        r=rm1;
    }
    d[1]=a[1][1];
}
```

C. bakhrmtri

Given a complex $n \times n$ diagonal matrix C and $m \leq n\text{-}2$ complex vectors $u^{(k)}$ (as produced, for example, at exit from *hshhrmtri*) and a set of complex vectors $v^{(j)}$ ($j=n1,...,n2$) forms the vectors

$$u^{(j)} = \prod_{k=1}^{m} (I - 2u^{(k)T}u^{(k)}/u^{(k)}u^{(k)T})Cv^{(j)} \qquad (j=n1,...,n2).$$

Function Parameters:

void bakhrmtri (*a,n,n1,n2,vecr,veci,tr,ti*)

a: float *a[1:n,1:n]*;
entry: the data for the back transformation as produced by *hshhrmtri*;

n: int;
entry: the order of the matrix of which the eigenvectors are calculated;

n1,n2: int;
entry: the eigenvectors corresponding to the eigenvalues with indices *n1,...,n2* are to be transformed;

vecr,veci: float *vecr[1:n,n1:n2], veci[1:n,n1:n2]*;
entry: the back transformation is performed on the real eigenvectors given in the columns of array *vecr*;
exit: *vecr* and *veci* contain the real part and imaginary part of the transformed eigenvectors;

tr,ti: float *tr[1:n-1], ti[1:n-1]*;
entry: data at exit from *hshhrmtri*.

Functions used: matmat, tammat, elmcol, elmcolrow, commul, comrowcst.

Method: see hshhrmtrival.

```
void bakhrmtri(float **a, int n, int n1, int n2, float **vecr,
               float **veci, float tr[], float ti[])
{
    float matmat(int, int, int, int, float **, float **);
    float tammat(int, int, int, int, float **, float **);
    void elmcol(int, int, int, int, float **, float **, float);
    void elmcolrow(int, int, int, int, float **, float **, float);
    void commul(float, float, float, float, float *, float *);
    void comrowcst(int, int, int, float **, float **, float, float);
    int i,j,r,rm1;
    float c,s,t,qr,qi;

    for (i=1; i<=n; i++)
        for (j=n1; j<=n2; j++) veci[i][j]=0.0;
    c=1.0;
    s=0.0;
    for (j=n-1; j>=1; j--) {
        commul(c,s,tr[j],ti[j],&c,&s);
        comrowcst(n1,n2,j,vecr,veci,c,s);
    }
    rm1=2;
    for (r=3; r<=n; r++) {
        t=a[r][r];
        if (t > 0.0)
            for (j=n1; j<=n2; j++) {
                qr=(tammat(1,rm1,r,j,a,vecr)-matmat(1,rm1,r,j,a,veci))/t;
                qi=(tammat(1,rm1,r,j,a,veci)+matmat(1,rm1,r,j,a,vecr))/t;
                elmcol(1,rm1,j,r,vecr,a,-qr);
                elmcolrow(1,rm1,j,r,vecr,a,-qi);
                elmcolrow(1,rm1,j,r,veci,a,qr);
                elmcol(1,rm1,j,r,veci,a,-qi);
            }
        rm1=r;
    }
}
```

3.11.6 To Hessenberg form - complex non-Hermitian

A. hshcomhes

Given an $n \times n$ complex matrix A, determines m complex vectors $u^{(j)}$ ($m \leq n-2$) and an $n \times n$ complex diagonal matrix D such that with $E=E_m E_{m-1} \ldots E_1$, where

$$E_j = I - 2u^{(j)}u^{(j)T}/u^{(j)T}\bar{u}^{(j)},$$

$D^{-1}EAED=H$ where H is a complex upper Hessenberg matrix ($H_{i,j}=0$ for $i>j+1$) with real nonnegative subdiagonal elements $H_{j+1,j}$. The $u^{(j)}$ are determined by imposing the conditions that the first j elements $u^{(j)}$ are zero and that with $A^{(1)}=A$, $A^{(j+1)}A^{(j)}E^{(j)}$ has zeros in positions $i=j+2,\ldots,n$ of the j-th column ($j=1,\ldots,n-2$). D is determined by imposing the condition that the elements of the subdiagonal of $D^{-1}A^{(n-1)}D$ are absolute values of those of $A^{(n-1)}$. For further details see [Mu66, Wi65].

Function Parameters:

void hshcomhes (*ar,ai,n,em,b,tr,ti,del*)

ar,ai: float *ar[1:n,1:n], ai[1:n,1:n]*;

entry: the real part and the imaginary part of the matrix to be transformed must be

given in the arrays *ar* and *ai*, respectively;
exit: the real part and the imaginary part of the upper triangle of the resulting upper Hessenberg matrix are delivered in the corresponding parts of the arrays *ar* and *ai*, respectively; data for the Householder back transformation are delivered in the strict lower triangles of the arrays *ar* and *ai*;

n: int;
entry: the order of the given matrix;

em: float *em[0:1]*;
entry: *em[0]*: the machine precision;
em[1]: an estimate of the norm of the complex matrix (for example, the sum of the infinity norms of the real part and imaginary part of the matrix);

b: float *b[1:n-1]*;
exit: the real nonnegative subdiagonal of the resulting upper Hessenberg matrix;

tr,ti: float *tr[1:n]*, *ti[1:n]*;
exit: the real part and the imaginary part of the diagonal elements of a diagonal similarity transformation are delivered in the arrays *tr* and *ti*, respectively;

del: float *del[1:n-2]*;
exit: information concerning the sequence of Householder matrices.

Functions used: hshcomcol, matmat, elmrowcol, hshcomprd, carpol, commul, comcolcst, comrowcst.

```
void hshcomhes(float **ar, float **ai, int n, float em[], float b[],
               float tr[], float ti[], float del[])
{
   float matmat(int, int, int, int, float **, float **);
   void elmrowcol(int, int, int, int, float **, float **, float);
   void hshcomprd(int, int, int, int, int, float **,
                  float **, float **, float **, float);
   void comcolcst(int, int, int, float **, float **, float, float);
   void comrowcst(int, int, int, float **, float **, float, float);
   void carpol(float, float, float *, float *, float *);
   void commul(float, float, float, float, float *, float *);
   int hshcomcol(int, int, int, float **, float **, float,
                  float *, float *, float *, float *);
   int r,rm1,i,nm1;
   float tol,t,xr,xi;

   nm1=n-1;
   t=em[0]*em[1];
   tol=t*t;
   rm1=1;
   for (r=2; r<=nm1; r++) {
      if (hshcomcol(r,n,rm1,ar,ai,tol,&(b[rm1]),&(tr[r]),&(ti[r]),&t)) {
         for (i=1; i<=n; i++) {
            xr=(matmat(r,n,i,rm1,ai,ai)-matmat(r,n,i,rm1,ar,ar))/t;
            xi=(-matmat(r,n,i,rm1,ar,ai)-matmat(r,n,i,rm1,ai,ar))/t;
            elmrowcol(r,n,i,rm1,ar,ar,xr);
            elmrowcol(r,n,i,rm1,ar,ai,xi);
            elmrowcol(r,n,i,rm1,ai,ar,xi);
            elmrowcol(r,n,i,rm1,ai,ai,-xr);
         }
         hshcomprd(r,n,r,n,rm1,ar,ai,ar,ai,t);
      }
      del[rm1]=t;
      rm1=r;
   }
   if (n > 1) carpol(ar[n][nm1],ai[n][nm1],&(b[nm1]),&(tr[n]),&(ti[n]));
   rm1=1;
   tr[1]=1.0;
   ti[1]=0.0;
   for (r=2; r<=n; r++) {
```

```
        commul(tr[rm1],ti[rm1],tr[r],ti[r],&(tr[r]),&(ti[r]));
        comcolcst(1,rm1,r,ar,ai,tr[r],ti[r]);
        comrowcst(r+1,n,r,ar,ai,tr[r],-ti[r]);
        rm1=r;
    }
}
```

B. bakcomhes

Given m complex vectors $u^{(j)}$ ($m \leq n-2$), an $n \times n$ complex diagonal matrix D and a sequence of complex column vectors $v^{(k)}$ ($k=n1,...,n2$) computes the vectors $w^{(k)}=EDv^{(k)}$ ($k=n1,...,n2$) where

$$E = \prod_{j=1}^{m} (I - 2u^{(j)}\overline{u}^{(j)T}/\overline{u}^{(j)T}u^{(j)}).$$

(If $v^{(k)}$ ($k=n1,...,n2$) are eigenvectors of $H=D^{-1}EAED$, $w^{(k)}$ ($k=n1,...,n2$) are corresponding eigenvectors of A.) For further details see [Mu66, Wi65].

Function Parameters:

void bakcomhes (*ar,ai,tr,ti,del,vr,vi,n,n1,n2*)

ar,ai,tr,ti,del: float *ar[1:n,1:n], ai[1:n,1:n], tr[1:n], ti[1:n], del[1:n-2]*;
entry: the data for the back transformation as produced by *hshcomhes*;

vr,vi: float *vr[1:n,n1:n2]*;
entry: the back transformation is performed on the eigenvectors with the real parts given in *vr* and the imaginary parts given in *vi*;
exit: the real parts and imaginary parts of the resulting eigenvectors are delivered in the columns of the *vr* and *vi*, respectively;

n: int;
entry: the order of the matrix of which the eigenvectors are calculated;

n1,n2: int;
entry: the eigenvectors corresponding to the eigenvalues with indices *n1,...,n2* are to be transformed.

Functions used: comrowcst, hshcomprd.

```
void bakcomhes(float **ar, float **ai, float tr[], float ti[],
               float del[], float **vr, float **vi, int n,
               int n1, int n2)
{
    void hshcomprd(int, int, int, int, int, float **,
                   float **, float **, float **, float);
    void comrowcst(int, int, int, float **, float **, float, float);
    int i,r,rm1;
    float h;

    for (i=2; i<=n; i++) comrowcst(n1,n2,i,vr,vi,tr[i],ti[i]);
    r=n-1;
    for (rm1=n-2; rm1>=1; rm1--) {
        h=del[rm1];
        if (h > 0.0) hshcomprd(r,n,n1,n2,rm1,vr,vi,ar,ai,h);
        r=rm1;
    }
}
```

3.12 Other transformations

3.12.1 To bidiagonal form - real matrices

A. hshreabid

Reduces an $m \times n$ symmetric matrix A to bidiagonal form B. With $A=A_1$, $u^{(1)}$ is so chosen that all elements but the first in the first column of

$$A_1' = (I - 2u^{(1)}u^{(1)T}/u^{(1)T}u^{(1)})A_1$$

are zero; $v^{(1)}$ is so chosen that all elements but the first two of the first row in

$$A_1'' = A_1'(I - 2v^{(1)}v^{(1)T}/v^{(1)T}v^{(1)})$$

are zero; the first row and column are stripped from A_1'' to produce the $(m-1)\times(n-1)$ matrix A_2 and the process is repeated.

Function Parameters:

void hshreabid (*a,m,n,d,b,em*)

a: float *a[1:m,1:n]*;
  entry: the given matrix;
  exit: data concerning the premultiplying and postmultiplying matrices;
m: int;
  entry: the number of rows of the given matrix;
n: int;
  entry: the number of columns of the given matrix;
d: float *d[1:n]*;
  exit: the diagonal of the bidiagonal matrix (diagonal of B above);
b: float *b[1:n]*;
  exit: the superdiagonal of the bidiagonal matrix is delivered in *b[1:n-1]*;
em: float *em[0:1]*;
  entry: *em[0]*: the machine precision;
  exit: *em[1]*: the infinity norm of the original matrix.

Functions used: tammat, mattam, elmcol, elmrow.

Method: *hshreabid* slightly improves a part of a procedure (svd) of Golub and Reinsch [WiR71] by skipping a transformation if the column or row in already in the desired form, (i.e. if the sum of the squares of the elements that ought to be zero is smaller than a certain constant). In svd the transformation is skipped only if the norm of the full row or column is small enough. As a result, some ill-defined transformations are skipped in *hshreabid*. Moreover, if a transformation is skipped, a zero is not stored in the diagonal or superdiagonal, but the value that would have been found if the column or row were in the desired form already is stored.

```
#include <math.h>

void hshreabid(float **a, int m, int n, float d[], float b[],
               float em[])
{
    float tammat(int, int, int, int, float **, float **);
    float mattam(int, int, int, int, float **, float **);
    void elmcol(int, int, int, int, float **, float **, float);
    void elmrow(int, int, int, int, float **, float **, float);
    int i,j,i1;
    float norm,machtol,w,s,f,g,h;

    norm=0.0;
    for (i=1; i<=m; i++) {
        w=0.0;
        for (j=1; j<=n; j++) w += fabs(a[i][j]);
        if (w > norm) norm=w;
    }
    machtol=em[0]*norm;
    em[1]=norm;
    for (i=1; i<=n; i++) {
        i1=i+1;
        s=tammat(i1,m,i,i,a,a);
        if (s < machtol)
            d[i]=a[i][i];
        else {
            f=a[i][i];
            s += f*f;
            d[i] = g = (f < 0.0) ? sqrt(s) : -sqrt(s);
            h=f*g-s;
            a[i][i]=f-g;
            for (j=i1; j<=n; j++)
                elmcol(i,m,j,i,a,a,tammat(i,m,i,j,a,a)/h);
        }
        if (i < n) {
            s=mattam(i1+1,n,i,i,a,a);
            if (s < machtol)
                b[i]=a[i][i1];
            else {
                f=a[i][i1];
                s += f*f;
                b[i] = g = (f < 0.0) ? sqrt(s) : -sqrt(s);
                h=f*g-s;
                a[i][i1]=f-g;
                for (j=i1; j<=m; j++)
                    elmrow(i1,n,j,i,a,a,mattam(i1,n,i,j,a,a)/h);
            }
        }
    }
}
```

B. psttfmmat

Computes the postmultiplying matrix from the intermediate results generated by *hshreadbid.*

Function Parameters:

void psttfmmat (a,n,v,b)

a: float *a[1:n,1:n]*;
  entry: the data concerning the postmultiplying matrix, as generated by *hshreabid*;

n: int;
  entry: the number of columns and rows of *a*;

v: float *v[1:n,1:n]*;
  exit: the postmultiplying matrix;

b: float *b[1:n]*;
  exit: the superdiagonal as generated by *hshreabid.*

Functions used: matmat, elmcol.

```
void psttfmmat(float **a, int n, float **v, float b[])
{
    float matmat(int, int, int, int, float **, float **);
    void elmcol(int, int, int, int, float **, float **, float);
    int i,i1,j;
    float h;

    i1=n;
    v[n][n]=1.0;
    for (i=n-1; i>=1; i--) {
        h=b[i]*a[i][i1];
        if (h < 0.0) {
            for (j=i1; j<=n; j++) v[j][i]=a[i][j]/h;
            for (j=i1; j<=n; j++)
                elmcol(i1,n,j,i,v,v,matmat(i1,n,i,j,a,v));
        }
        for (j=i1; j<=n; j++) v[i][j]=v[j][i]=0.0;
        v[i][i]=1.0;
        i1=i;
    }
}
```

C. pretfmmat

Computes the premultiplying matrix from the intermediate results generated by *hshreadbid.*

Function Parameters:

void pretfmmat (a,m,n,d)

a: float $a[1:m,1:n]$;
entry: the data concerning the premultiplying matrix, as generated by *hshreabid*;
exit: the premultiplying matrix;

m: int;
entry: the number of rows of a;

n: int;
entry: the number of columns of a, n should satisfy $n \leq m$;

d: float $d[1:n]$;
entry: the diagonal as generated by *hshreabid.*

Functions used: tammat, elmcol.

```
void pretfmmat(float **a, int m, int n, float d[])
{
    float tammat(int, int, int, int, float **, float **);
    void elmcol(int, int, int, int, float **, float **, float);
    int i,i1,j;
    float g,h;

    for (i=n; i>=1; i--) {
        i1=i+1;
        g=d[i];
        h=g*a[i][i];
        for (j=i1; j<=n; j++) a[i][j]=0.0;
        if (h < 0.0) {
            for (j=i1; j<=n; j++)
                elmcol(i,m,j,i,a,a,tammat(i1,m,i,j,a,a)/h);
            for (j=i; j<=m; j++) a[j][i] /= g;
        } else
            for (j=i; j<=m; j++) a[j][i]=0.0;
        a[i][i] += 1.0;
    }
```

}

3.13 The (ordinary) eigenvalue problem

3.13.1 Real symmetric tridiagonal matrices

A. valsymtri

Calculates all or some consecutive eigenvalues in descending order of magnitude, of a symmetric $n \times n$ tridiagonal matrix T. With d_i ($i=1,...,n$) the values of the elements of the principal diagonal of T, b_i ($i=1,...,n-1$) the squares of the values of the codiagonal elements of T, δ the product of the machine precision and $\|T\|$, the sequence $f(k,x)$ given by

$$f(1,x)=d_1-x, \quad f(k,x)=(d_k-xb_{k-1})/v_k$$

where $v_k=f(k-1,x)$ if $|f(k-1,x)|>\delta$, $-\delta$ if $f(k-1,x)\le 0$, and δ otherwise ($k=2,...,n$), is determined,

$$f(n,x)=p(n,x)/p(n-1,x)$$

where $p(n,x)$ = determinant of $T-xI$, and the $f(k,x)$ form a Sturm sequence. The zeros of $f(n,x)$ (i.e. the eigenvalues of T) are computed by a sophisticated mixture of linear interpolation and bisection (see *zeroin*). It is assumed that the values of d_i, b_i and $\|T\|$ above are available; these are generated as output from the tridiagonalization of a symmetric $n \times n$ matrix by both *tfmsymtri2* and *tfmsymtri1*. For further details see [DekHo68, Wi65].

Function Parameters:

void valsymtri (*d,bb,n,n1,n2,val,em*)

d: float *d[1:n]*;
  entry: the main diagonal of the symmetric tridiagonal matrix (values of d_i above);

bb: float *bb[1:n-1]*;
  entry: the squares of the codiagonal elements of the symmetric tridiagonal matrix (the values of b_i in the above);

n: int;
  entry: the order of the given matrix;

n1,n2: int;
  entry: the serial number of the first and last eigenvalue to be calculated, respectively;

val: float *val[n1:n2]*;
  exit: the *n2-n1+1* calculated consecutive eigenvalues in nonincreasing order;

em: float *em[0:3]*;
  entry: *em[0]*: the machine precision;
    em[1]: an upper bound for the moduli of the eigenvalues of the given matrix;
    em[2]: a relative tolerance for the eigenvalues;
  exit: *em[3]*: the total number of iterations used for calculating the eigenvalues.

```
#include <float.h>
#include <math.h>

void valsymtri(float d[], float bb[], int n, int n1, int n2,
               float val[], float em[])
{
   float sturm(float [], float [], int, float, int, float,
                  float, int *, float *, float *);
   int k,count,ext,extrapolate;
   float max,x,y,macheps,norm,re,machtol,ub,lb,lambda,
         c,fc,b,fb,a,fa,dd,fd,fdb,fda,w,mb,tol,m,p,q;
```

```
macheps=em[0];
norm=em[1];
re=em[2];
machtol=norm*macheps;
max=norm/macheps;
count=0;
ub=1.1*norm;
lb = -ub;
lambda=ub;
for (k=n1; k<=n2; k++) {
    y=ub;
    lb = -1.1*norm;
    x=lb;

    /* look for the zero of the polynomial function */

    b=x;
    fb=sturm(d,bb,n,x,k,machtol,max,&count,&lb,&ub);
    a=x=y;
    fa=sturm(d,bb,n,x,k,machtol,max,&count,&lb,&ub);
    c=a;
    fc=fa;
    ext=0;
    extrapolate=1;
    while (extrapolate) {
        if (fabs(fc) < fabs(fb)) {
            if (c != a) {
                dd=a;
                fd=fa;
            }
            a=b;
            fa=fb;
            b=x=c;
            fb=fc;
            c=a;
            fc=fa;
        }
        tol=fabs(x)*re+machtol;
        m=(c+b)*0.5;
        mb=m-b;
        if (fabs(mb) > tol) {
            if (ext > 2)
                w=mb;
            else {
                if (mb == 0.0)
                    tol=0.0;
                else
                    if (mb < 0.0) tol = -tol;
                p=(b-a)*fb;
                if (ext <= 1)
                    q=fa-fb;
                else {
                    fdb=(fd-fb)/(dd-b);
                    fda=(fd-fa)/(dd-a);
                    p *= fda;
                    q=fdb*fa-fda*fb;
                }
                if (p < 0.0) {
                    p = -p;
                    q = -q;
                }
                w=(p<FLT_MIN || p<=q*tol) ? tol : ((p<mb*q) ? p/q : mb);
            }
            dd=a;
            fd=fa;
            a=b;
            fa=fb;
            x = b += w;
            fb=sturm(d,bb,n,x,k,machtol,max,&count,&lb,&ub);
            if ((fc >= 0.0) ? (fb >= 0.0) : (fb <= 0.0)) {
                c=a;
                fc=fa;
                ext=0;
            } else
```

```
                    ext = (w == mb) ? 0 : ext+1;
            } else
                break;
        }
        y=c;

        /* end of the zero finding procedure */

        val[k] = lambda = (x > lambda) ? lambda : x;
        if (ub > x)
            ub = (x > y) ? x : y;
    }
    em[3]=count;
}

float sturm(float d[], float bb[], int n, float x, int k, float machtol,
                float max, int *count, float *lb, float *ub)
{
    /* this sturm sequence procedure is used internally by VALSYMTRI */

    int p,i;
    float f;

    (*count)++;
    p=k;
    f=d[1]-x;
    for (i=2; i<=n; i++) {
        if (f <= 0.0) {
            p++;
            if (p > n) return ((p == n) ? f : (n-p)*max);
        } else
            if (p < i-1) {
                *lb = x;
                return ((p == n) ? f : (n-p)*max);
            }
        if (fabs(f) < machtol)
            f = (f <= 0.0) ? -machtol : machtol;
        f=d[i]-x-bb[i-1]/f;
    }
    if (p == n || f <= 0.0)
        if (x < *ub) *ub = x;
    else
        *lb = x;
    return ((p == n) ? f : (n-p)*max);
}
```

B. vecsymtri

Calculates, by use of inverse iteration, the eigenvectors x_i corresponding to the given consecutive eigenvalues λ_i (in nonincreasing order) ($i=n1,...,n2$) of the $n\times n$ symmetric tridiagonal matrix T. The inverse iteration procedure used involves the determination of the vectors $x_i^{(k)}$ from the relationships $(T-\lambda_i I)x_i^{(k+1)}=Y^{(k)}x_i^{(k)}$ ($k=0,1,...$), where $Y^{(k)}$ is so chosen that $\|x_i^{(k+1)}\|=1$. The values of the machine precision m_0, the norm of the given matrix m_1, an orthogonalization parameter m_4, the relative tolerance of the eigenvectors m_6, and the maximum number of iterations allowed for the calculation of each eigenvector m_8 must be provided. If $|\lambda_i-\lambda_j|<m_0*m_1$ for any j in $1\leq j\leq i-1$, then λ_i is modified so that $|\lambda_i-\lambda_j|=m_0*m_1$. If $|\lambda_i-\lambda_j|<m_4*m_1$ for any j in the range $1\leq j\leq i-1$, then a Gram-Schmidt process is carried out at each iteration step, so that the eigenvectors are orthogonal to within working precision. The iteration terminates when either $\|x_n\|_2<m_1*m_6$ or $k>m_8$. *vecsymtri* may be called with $n1$=1 (so that eigenvectors are already known). However, it may occur that from previous use of *vecsymtri* or some other procedure, the eigenvectors corresponding to the $n1$-1 largest eigenvalues $\lambda_1,...,\lambda_{m+1}$ are already known. In this case the eigenvalues $\lambda_1,...,\lambda_{n1-1}$ must be provided (in array *val*) as well as the corresponding eigenvector (in array *vec*). Moreover, if *vecsymtri* has been called previously, then the number m_5 of eigenvectors involved in the last Gram-Schmidt process used must be provided. For further details see [DekHo68].

Function Parameters:

void vecsymtri (*d,b,n,n1,n2,val,vec,em*)

d: float *d[1:n]*;
entry: the main diagonal of the symmetric tridiagonal matrix (values of $T_{i,i}$ above);

b: float *b[1:n]*;
entry: the codiagonal of the symmetric tridiagonal matrix (the values of $T_{i,i+1}=T_{i+1,i}$ in the above) followed by an additional element 0;

n: int;
entry: the order of the given matrix;

n1,n2: int;
entry: lower and upper bound of the array *val*;

val: float *val[n1:n2]*;
entry: a row of nonincreasing eigenvalues (values of $\lambda_1,\lambda_2,...,\lambda_{n2}$ above) as delivered by *valsymtri*;

vec: float *vec[1:n,n1:n2]*;
entry: if *n1*>1 then the components of the eigenvectors corresponding to $\lambda_1,...,\lambda_{n1-1}$ must be given in *vec[1:n,1:n1-1]*;
exit: the eigenvectors corresponding with the given eigenvalues $\lambda_1,...,\lambda_{n2}$;

em: float *em[0:9]*;
entry:
em[0]: the machine precision (value of m_0 above);
em[1]: a norm of the given matrix (value of m_1 above);
em[4]: the orthogonalization parameter (value of m_4 above);
em[5]: if *n1*>1 then *em[6]* should be given the value of the number of eigenvectors involved in the last Gram-Schmidt process used in the determination of the eigenvectors corresponding to $\lambda_1,...,\lambda_{m-1}$ (value of m_5 above);
em[6]: a relative tolerance for the eigenvectors;
em[8]: maximum number of iterations allowed for calculating each eigenvector;
exit:
em[5]: the number of eigenvectors involved in the last Gram-Schmidt orthogonalization;
em[7]: the maximum Euclidean norm of the residues;
em[9]: the largest number of iterations performed for the calculation of some eigenvector.

Functions used: vecvec, tamvec, elmveccol.

```
#include <math.h>

void vecsymtri(float d[], float b[], int n, int n1, int n2,
               float val[], float **vec, float em[])
{
   int *allocate_integer_vector(int, int);
   float *allocate_real_vector(int, int);
   void free_integer_vector(int *, int);
   void free_real_vector(float *, int);
   float vecvec(int, int, int, float [], float []);
   void elmveccol(int, int, int, float [], float **, float);
   float tamvec(int, int, int, float **, float []);
   int i,j,k,count,maxcount,countlim,orth,ind,iterate,*index;
```

```
float bi,bi1,u,w,y,mi1,lambda,oldlambda,ortheps,valspread,spr,
      res,maxres,norm,newnorm,oldnorm,machtol,vectol,
      *m,*p,*q,*r,*x;

index=allocate_integer_vector(1,n);
m=allocate_real_vector(1,n);
p=allocate_real_vector(1,n);
q=allocate_real_vector(1,n);
r=allocate_real_vector(1,n);
x=allocate_real_vector(1,n);
norm=em[1];
machtol=em[0]*norm;
valspread=em[4]*norm;
vectol=em[6]*norm;
countlim=em[8];
ortheps=sqrt(em[0]);
maxcount=ind=0;
maxres=0.0;
if (n1 > 1) {
    orth=em[5];
    oldlambda=val[n1-orth];
    for (k=n1-orth+1; k<=n1-1; k++) {
        lambda=val[k];
        spr=oldlambda-lambda;
        if (spr < machtol) lambda=oldlambda-machtol;
        oldlambda=lambda;
    }
} else
    orth=1;
for (k=n1; k<=n2; k++) {
    lambda=val[k];
    if (k > 1) {
        spr=oldlambda-lambda;
        if (spr < valspread) {
            if (spr < machtol) lambda=oldlambda-machtol;
            orth++;
        } else
            orth=1;
    }
    count=0;
    u=d[1]-lambda;
    bi=w=b[1];
    if (fabs(bi) < machtol) bi=machtol;
    for (i=1; i<=n-1; i++) {
        bi1=b[i+1];
        if (fabs(bi1) < machtol) bi1=machtol;
        if (fabs(bi) >= fabs(u)) {
            mi1=m[i+1]=u/bi;
            p[i]=bi;
            y=q[i]=d[i+1]-lambda;
            r[i]=bi1;
            u=w-mi1*y;
            w = -mi1*bi1;
            index[i]=1;
        } else {
            mi1=m[i+1]=bi/u;
            p[i]=u;
            q[i]=w;
            r[i]=0.0;
            u=d[i+1]-lambda-mi1*w;
            w=bi1;
            index[i]=0;
        }
        x[i]=1.0;
        bi=bi1;
    }   /* transform */
    p[n] = (fabs(u) < machtol) ? machtol : u;
    q[n]=r[n]=0.0;
    x[n]=1.0;
    iterate=1;
    while (iterate) {
        u=w=0.0;
        for (i=n; i>=1; i--) {
            y=u;
```

```
                u=x[i]=(x[i]-q[i]*u-r[i]*w)/p[i];
                w=y;
            }   /* next iteration */
            newnorm=sqrt(vecvec(1,n,0,x,x));
            if (orth > 1) {
                oldnorm=newnorm;
                for (j=k-orth+1; j<=k-1; j++)
                    elmveccol(1,n,j,x,vec,-tamvec(1,n,j,vec,x));
                newnorm=sqrt(vecvec(1,n,0,x,x));
                if (newnorm < ortheps*oldnorm) {
                    ind++;
                    count=1;
                    for (i=1; i<=ind-1; i++) x[i]=0.0;
                    for (i=ind+1; i<=n; i++) x[i]=0.0;
                    x[ind]=1.0;
                    if (ind == n) ind=0;
                    w=x[1];
                    for (i=2; i<=n; i++) {
                        if (index[i-1]) {
                            u=w;
                            w=x[i-1]=x[i];
                        } else
                            u=x[i];
                        w=x[i]=u-m[i]*w;
                    }
                    continue;   /* iterate on */
                }   /* new start */
            }   /* orthogonalization */
            res=1.0/newnorm;
            if (res > vectol || count == 0) {
                count++;
                if (count <= countlim) {
                    for (i=1; i<=n; i++) x[i] *= res;
                    w=x[1];
                    for (i=2; i<=n; i++) {
                        if (index[i-1]) {
                            u=w;
                            w=x[i-1]=x[i];
                        } else
                            u=x[i];
                        w=x[i]=u-m[i]*w;
                    }
                } else
                    break;
            } else
                break;
        }
        for (i=1; i<=n; i++) vec[i][k]=x[i]*res;
        if (count > maxcount) maxcount=count;
        if (res > maxres) maxres=res;
        oldlambda=lambda;
    }
    em[5]=orth;
    em[7]=maxres;
    em[9]=maxcount;
    free_integer_vector(index,1);
    free_real_vector(m,1);
    free_real_vector(p,1);
    free_real_vector(q,1);
    free_real_vector(r,1);
    free_real_vector(x,1);
}
```

C. qrivalsymtri

Calculates all eigenvalues of a symmetric tridiagonal matrix T using square-root-free QR iteration. Values of m_0, the machine precision, a relative tolerance m_2, and $\|T\|$ must be provided. The relative error in the computed eigenvalues is less than $m_2 * \|T\|$, the absolute error is less than $m_0 * \|T\|$. If some eigenvalues λ_i are very small, the absolute error in their determination may be made smaller by prescribing a value of m_0 which is less than its true value. When this is done, the calculation of b_i/m_0^2 should cause no overflow, and that of $(m_0 * \|T\|)^2$ no underflow, where b_i is the i-th element of the codiagonal of T. For further details see [Re71, Wi65].

Function Parameters:

int qrivalsymtri (*d,bb,n,em*)

qrivalsymtri: given the number of eigenvalues not calculated;

d: float *d[1:n]*;
- entry: the main diagonal of the symmetric tridiagonal matrix;

bb: float *bb[1:n]*;
- entry: the squares of the codiagonal elements of the symmetric tridiagonal matrix followed by an additional element 0;
- exit: the squares of the codiagonal elements of the symmetric tridiagonal matrix resulting from the QR iteration;

n: int; entry: the order of the given matrix;

em: float *em[0:5]*;
- entry: *em[0]*: the machine precision;
  - *em[1]*: a norm of the given matrix;
  - *em[2]*: a relative tolerance for the eigenvalues;
  - *em[4]*: the maximum allowed number of iterations;
- exit: *em[3]*: the maximum absolute value of the codiagonal elements neglected;
  - *em[5]*: the number of iterations performed.

```
#include <math.h>

int qrivalsymtri(float d[], float bb[], int n, float em[])
{
    int i,i1,low,oldlow,n1,count,max;
    float bbtol,bbmax,bbi,bbn1,machtol,dn,delta,f,num,shift,g,h,
          t,p,r,s,c,oldg;

    t=em[2]*em[1];
    bbtol=t*t;
    machtol=em[0]*em[1];
    max=em[4];
    bbmax=0.0;
    count=0;
    oldlow=n;
    n1=n-1;
    while (n > 0) {
        i=n;
        do {
            low=i;
            i--;
        } while ((i >= 1) ? bb[i] > bbtol : 0);
        if (low > 1)
            if (bb[low-1] > bbmax) bbmax=bb[low-1];
        if (low == n)
            n=n1;
        else {
            dn=d[n];
            delta=d[n1]-dn;
            bbn1=bb[n1];
            if (fabs(delta) < machtol)
                r=sqrt(bbn1);
            else {
                f=2.0/delta;
                num=bbn1*f;
                r = -num/(sqrt(num*f+1.0)+1.0);
            }
```

```
            if (low == n1) {
                d[n]=dn+r;
                d[n1] -= r;
                n -= 2;
            } else {
                count++;
                if (count > max) break;
                if (low < oldlow) {
                    shift=0.0;
                    oldlow=low;
                } else
                    shift=dn+r;
                h=d[low]-shift;
                if (fabs(h) < machtol)
                    h = (h <= 0.0) ? -machtol : machtol;
                g=h;
                t=g*h;
                bbi=bb[low];
                p=t+bbi;
                i1=low;
                for (i=low+1; i<=n; i++) {
                    s=bbi/p;
                    c=t/p;
                    h=d[i]-shift-bbi/h;
                    if (fabs(h) < machtol)
                        h = (h <= 0.0) ? -machtol : machtol;
                    oldg=g;
                    g=h*c;
                    t=g*h;
                    d[i1]=oldg-g+d[i];
                    bbi = (i == n) ? 0.0 : bb[i];
                    p=t+bbi;
                    bb[i1]=s*p;
                    i1=i;
                }
                d[n]=g+shift;
            }
        }
        n1=n-1;
    }
    em[3]=sqrt(bbmax);
    em[5]=count;
    return n;
}
```

D. qrisymtri

Calculates the eigenvalues and eigenvectors of a symmetric tridiagonal matrix T of order n simultaneously by QR iteration. The process requires input of an auxiliary matrix A. If T results from tridiagonalization by use of *tfmprevec* then A has precisely the form produced as output by that procedure. If T is simply given as a symmetric tridiagonal matrix then A must be set to be the unit matrix. The maximum allowed number of iterations must be prescribed. If this number is exceeded, and the eigenvalues $\lambda_1,...,\lambda_{n-k}$ and their corresponding eigenvectors have been determined, *qrisymtri* is given the value k, $\lambda_1,...,\lambda_{n-k}$ are to be found in locations $k+1,...,n$ of an output array D, their eigenvectors are to be found in columns $k+1,...,n$ of an output array A, and the numbers found in the remaining positions of these arrays are rough approximations only. For further details see [DekHo68, Wi65].

Function Parameters:

int qrisymtri (*a,n,d,b,bb,em*)

qrisymtri: given the number (k above) of eigenvalues and eigenvectors not calculated;
a: float *a[1:n,1:n]*;
entry: some input matrix, possibly the identity matrix;

exit: the eigenvectors of the original symmetric tridiagonal matrix, premultiplied by the input matrix *a*;

n: int; entry: the order of the given matrix;

d: float *d[1:n]*;
entry: the main diagonal of the symmetric tridiagonal matrix;
exit: the eigenvalues of the matrix in some arbitrary order;

b: float *b[1:n]*;
entry: the codiagonal of the symmetric tridiagonal matrix followed by an additional element 0;
exit: the codiagonal of the symmetric tridiagonal matrix resulting from the QR iteration, followed by an additional element 0;

bb: float *bb[1:n]*;
entry: the squared codiagonal elements of the symmetric tridiagonal matrix, followed by an additional element 0;
exit: the squared codiagonal elements of the symmetric tridiagonal matrix resulting from the QR iteration;

em: float *em[0:5]*;
entry:
em[0]: the machine precision;
em[1]: a norm of the given matrix;
em[2]: a relative tolerance for the QR iteration;
em[4]: the maximum allowed number of iterations;
exit:
em[3]: the maximum absolute value of the codiagonal elements neglected;
em[5]: the number of iterations performed.

Function used: rotcol.

```
#include <math.h>

int qrisymtri(float **a, int n, float d[], float b[], float bb[],
              float em[])
{
    void rotcol(int, int, int, int, float **, float, float);
    int j,j1,k,m,m1,count,max;
    float bbmax,r,s,sin,t,cos,oldcos,g,p,w,tol,tol2,lambda,dk1;

    tol=em[2]*em[1];
    tol2=tol*tol;
    count=0;
    bbmax=0.0;
    max=em[4];
    m=n;
    do {
        k=m;
        m1=m-1;
        while (1) {
            k--;
            if (k <= 0) break;
            if (bb[k] < tol2) {
                if (bb[k] > bbmax) bbmax=bb[k];
                break;
            }
        }
        if (k == m1)
            m=m1;
        else {
            t=d[m]-d[m1];
            r=bb[m1];
```

```
            if (fabs(t) < tol)
                s=sqrt(r);
            else {
                w=2.0/t;
                s=w*r/(sqrt(w*w*r+1.0)+1.0);
            }
            if (k == m-2) {
                d[m] += s;
                d[m1] -= s;
                t = -s/b[m1];
                r=sqrt(t*t+1.0);
                cos=1.0/r;
                sin=t/r;
                rotcol(1,n,m1,m,a,cos,sin);
                m -= 2;
            } else {
                count++;
                if (count > max) break;
                lambda=d[m]+s;
                if (fabs(t) < tol) {
                    w=d[m1]-s;
                    if (fabs(w) < fabs(lambda)) lambda=w;
                }
                k++;
                t=d[k]-lambda;
                cos=1.0;
                w=b[k];
                p=sqrt(t*t+w*w);
                j1=k;
                for (j=k+1; j<=m; j++) {
                    oldcos=cos;
                    cos=t/p;
                    sin=w/p;
                    dk1=d[j]-lambda;
                    t *= oldcos;
                    d[j1]=(t+dk1)*sin*sin+lambda+t;
                    t=cos*dk1-sin*w*oldcos;
                    w=b[j];
                    p=sqrt(t*t+w*w);
                    g=b[j1]=sin*p;
                    bb[j1]=g*g;
                    rotcol(1,n,j1,j,a,cos,sin);
                    j1=j;
                }
                d[m]=cos*t+lambda;
                if (t < 0.0) b[m1] = -g;
            }
        }
    } while (m > 0);
    em[3]=sqrt(bbmax);
    em[5]=count;
    return m;
}
```

3.13.2 Real symmetric full matrices

A. eigvalsym2

Computes the m largest eigenvalues of an $n \times n$ symmetric matrix A whose upper triangular part is stored in a two dimensional array, using linear interpolation on a function derived from a Sturm sequence. A is first reduced to similar tridiagonal form T by a call of *tfmsymtri2*, and *valsymtri* is then used to calculate the eigenvalues.

Function Parameters:

void eigvalsym2 (*a,n,numval,val,em*)

a: float *a[1:n,1:n]*;

entry: the upper triangle of the symmetric matrix must be given in the upper triangular part of a (the elements $a[i,j]$, $i \leq j$);
exit: the data for the Householder's back transformation (which is not used by this procedure) is delivered in the upper triangular part of a;

n: int; entry: the order of the given matrix;
numval: int; entry: the serial number of the last eigenvalue to be calculated;
val: float *val[1:numval]*;
exit: the *numval* largest eigenvalues in monotonically nonincreasing order;
em: float *em[0:3]*;
entry:
em[0]: the machine precision;
em[2]: a relative tolerance for the eigenvalues;
exit:
em[1]: the infinity norm of the original matrix;
em[3]: the number of iterations used for calculating the *numval* eigenvalues.

Functions used: tfmsymtri2, valsymtri.

```
void eigvalsym2(float **a, int n, int numval, float val[], float em[])
{
    float *allocate_real_vector(int, int);
    void free_real_vector(float *, int);
    void tfmsymtri2(float **, int, float [], float [], float [],
                    float []);
    void valsymtri(float [], float [], int, int, int,
                    float [], float []);
    float *b,*bb,*d;

    b=allocate_real_vector(1,n);
    bb=allocate_real_vector(1,n);
    d=allocate_real_vector(1,n);
    tfmsymtri2(a,n,d,b,bb,em);
    valsymtri(d,bb,n,1,numval,val,em);
    free_real_vector(b,1);
    free_real_vector(bb,1);
    free_real_vector(d,1);
}
```

B. eigsym2

Computes the m largest eigenvalues and corresponding eigenvectors of an $n \times n$ symmetric matrix A whose upper triangular part is stored in a two-dimensional array, by means of inverse iteration. A is first reduced to similar tridiagonal form T by a call of *tfmsymtri2*, *valsymtri* is used to calculate the eigenvalues, *vecsymtri* is used to calculate the corresponding eigenvectors, and *baksymtri2* is then used to perform the necessary back transformation. *eigsym2* requires, as does *vecsymtri*, the value of an orthogonalization parameter m_4. When the distance between two consecutive eigenvalues is less than m_4, the eigenvectors are orthogonalized by a Gram-Schmidt process. The number m_5 of eigenvectors involved in the last Gram-Schmidt process is given by *eigsym2* as part of its output. If *eigsym2* is called more than once (to compute successive sets of eigenvalues and eigenvectors), the value m_5 arising from the preceding call is required as input for the current call.

Function Parameters:

void eigsym2 (*a,n,numval,val,vec,em*)

a: float *a[1:n,1:n]*;
entry: the upper triangle of the symmetric matrix must be given in the upper triangular part of *a* (the elements *a[i,j]*, $i \leq j$);
exit: the data for Householder's back transformation is delivered in the upper triangular part of *a*; the elements *a[i,j]* for $i>j$ are neither used nor changed;

n: int;
entry: the order of the given matrix;

numval: int;
entry: the serial number of the last eigenvalue to be calculated;

val: float *val[1:numval]*;
exit: the *numval* largest eigenvalues in monotonically nonincreasing order;

vec: float *vec[1:n,1:numval]*;
exit: the *numval* calculated eigenvectors, stored columnwise, corresponding to the calculated eigenvalues;

em: float *em[0:9]*;
entry:
em[0]: the machine precision;
em[2]: the relative tolerance for the eigenvalues;
em[4]: the orthogonalization parameter (value of m_4 above);
em[6]: the relative tolerance for the eigenvectors;
em[8]: the maximum number of inverse iterations allowed for the calculation of each eigenvector;
exit:
em[1]: the infinity norm of the matrix;
em[3]: the number of iterations used for calculating the *numval* eigenvalues;
em[5]: the number of eigenvectors involved in the last Gram-Schmidt orthogonalization;
em[7]: the maximum Euclidean norm of the residues of the calculated eigenvectors;
em[9]: the largest number of inverse iterations performed for the calculation of some eigenvector.

Functions used: tfmsymtri2, valsymtri, vecsymtri, baksymtri2.

```
void eigsym2(float **a, int n, int numval, float val[],
             float **vec, float em[])
{
    float *allocate_real_vector(int, int);
    void free_real_vector(float *, int);
    void tfmsymtri2(float **, int, float [], float [], float [],
                    float []);
    void valsymtri(float [], float [], int, int, int,
                    float [], float []);
    void vecsymtri(float [], float [], int, int, int,
                    float [], float **, float []);
    void baksymtri2(float **, int, int, int, float **);
    float *b,*bb,*d;

    b=allocate_real_vector(1,n);
    bb=allocate_real_vector(1,n);
    d=allocate_real_vector(1,n);
    tfmsymtri2(a,n,d,b,bb,em);
    valsymtri(d,bb,n,1,numval,val,em);
    vecsymtri(d,b,n,1,numval,val,vec,em);
    baksymtri2(a,n,1,numval,vec);
    free_real_vector(b,1);
```

```
    free_real_vector(bb,1);
    free_real_vector(d,1);
}
```

C. eigvalsym1

Computes the m largest eigenvalues of an $n \times n$ symmetric matrix A whose upper triangular part is stored in a one dimensional array, using linear interpolation on a function derived from a Sturm sequence. A is first reduced to similar tridiagonal form T by a call of *tfmsymtri1*, and *valsymtri* is then used to calculate the eigenvalues.

Function Parameters:

void eigvalsym1 (*a,n,numval,val,em*)

a: float *a[1:(n+1)n/2]*;
entry: the upper triangle of the symmetric matrix must be given in such a way that the (i,j)-th element of the matrix is $a[(j-1)j/2+i]$, $1 \leq i \leq j \leq n$;
exit: the data for the Householder's back transformation (which is not used by this procedure) is delivered in *a*;

n: int;
entry: the order of the given matrix;

numval: int;
entry: the serial number of the last eigenvalue to be calculated;

val: float *val[1:numval]*;
exit: the *numval* largest eigenvalues in monotonically nonincreasing order;

em: float *em[0:3]*;
entry:
em[0]: the machine precision;
em[2]: a relative tolerance for the eigenvalues;
exit:
em[1]: the infinity norm of the original matrix;
em[3]: the number of iterations used for calculating the *numval* eigenvalues.

Functions used: tfmsymtri1, valsymtri.

```
void eigvalsym1(float a[], int n, int numval, float val[], float em[])
{
    float *allocate_real_vector(int, int);
    void free_real_vector(float *, int);
    void tfmsymtri1(float [], int, float [], float [], float [],
                    float []);
    void valsymtri(float [], float [], int, int, int,
                    float [], float []);
    float *b,*bb,*d;

    b=allocate_real_vector(1,n);
    bb=allocate_real_vector(1,n);
    d=allocate_real_vector(1,n);
    tfmsymtri1(a,n,d,b,bb,em);
    valsymtri(d,bb,n,1,numval,val,em);
    free_real_vector(b,1);
    free_real_vector(bb,1);
    free_real_vector(d,1);
}
```

D. eigsym1

Computes the m largest eigenvalues and corresponding eigenvectors of an $n \times n$ symmetric matrix A whose upper triangular part is stored in a one-dimensional array, by means of inverse iteration. A is first reduced to similar tridiagonal form T by a call of *tfmsymtri1*, *valsymtri* is used to calculate the eigenvalues, *vecsymtri* is used to calculate the corresponding eigenvectors, and *baksymtri1* is then used to calculate the necessary back transformation. The significances of an orthogonalization parameter m_4 and a number m_5 of eigenvectors involved in a Gram-Schmidt process are as for *eigsym2*.

Function Parameters:

void eigsym1 (*a,n,numval,val,vec,em*)

a: float *a[1:(n+1)n/2]*;
- entry: the upper triangle of the symmetric matrix must be given in such a way that the (i,j)-th element of the matrix is $a[(j-1)j/2+i]$, $1 \le i \le j \le n$;
- exit: the data for Householder's back transformation;

n: int;
- entry: the order of the given matrix;

numval: int;
- entry: the serial number of the last eigenvalue to be calculated;

val: float *val[1:numval]*;
- exit: the *numval* largest eigenvalues in monotonically nonincreasing order;

vec: float *vec[1:n,1:numval]*;
- exit: the *numval* calculated eigenvectors, stored columnwise, corresponding to the calculated eigenvalues;

em: float *em[0:9]*;
- entry:
  - *em[0]*: the machine precision;
  - *em[2]*: the relative tolerance for the eigenvalues;
  - *em[4]*: the orthogonalization parameter (value of m_4 above);
  - *em[6]*: the relative tolerance for the eigenvectors;
  - *em[8]*: the maximum number of inverse iterations allowed for the calculation of each eigenvector;
- exit:
  - *em[1]*: the infinity norm of the matrix;
  - *em[3]*: the number of iterations used for calculating the *numval* eigenvalues;
  - *em[5]*: the number of eigenvectors involved in the last Gram-Schmidt orthogonalization;
  - *em[7]*: the maximum Euclidean norm of the residues of the calculated eigenvectors;
  - *em[9]*: the largest number of inverse iterations performed for the calculation of some eigenvector.

Functions used: tfmsymtri1, valsymtri, vecsymtri, baksymtri1.

```
void eigsym1(float a[], int n, int numval, float val[],
          float **vec, float em[])
{
   float *allocate_real_vector(int, int);
   void free_real_vector(float *, int);
   void tfmsymtri1(float [], int, float [], float [], float [],
                   float []);
   void valsymtri(float [], float [], int, int, int,
                   float [], float []);
   void vecsymtri(float [], float [], int, int, int,
```

```
                    float [], float **, float []);
    void baksymtri1(float [], int, int, int, float **);
    float *b,*bb,*d;

    b=allocate_real_vector(1,n);
    bb=allocate_real_vector(1,n);
    d=allocate_real_vector(1,n);
    tfmsymtri1(a,n,d,b,bb,em);
    valsymtri(d,bb,n,1,numval,val,em);
    vecsymtri(d,b,n,1,numval,val,vec,em);
    baksymtri1(a,n,1,numval,vec);
    free_real_vector(b,1);
    free_real_vector(bb,1);
    free_real_vector(d,1);
}
```

E. qrivalsym2

Computes all eigenvalues of an $n \times n$ symmetric matrix A whose upper triangular part is stored in a two-dimensional array, by means of QR iteration. A is first reduced to similar tridiagonal form by a call of *tfmsymtri2*, and *qrivalsymtri* is then used to calculate the eigenvalues.

Function Parameters:

int qrivalsym2 (*a,n,val,em*)

qrivalsym2: given the number of eigenvalues not calculated;

a: float *a[1:n,1:n]*;
  entry: the upper triangle of the symmetric matrix must be given in the upper triangular part of *a* (the elements $a[i,j]$, $i \le j$);
  exit: the data for Householder's back transformation (which is not used by this procedure) is delivered in the upper triangular part of *a*; the elements $a[i,j]$ for $i>j$ are neither used nor changed;

n: int;
  entry: the order of the given matrix;

val: float *val[1:n]*;
  exit: the eigenvalues of the matrix in some arbitrary order;

em: float *em[0:5]*;
  entry:
  em[0]: the machine precision;
  em[2]: the relative tolerance for the eigenvalues;
  em[4]: the maximum allowed number of iterations;
  exit:
  em[1]: the infinity norm of the matrix;
  em[3]: the maximum absolute value of the codiagonal elements neglected;
  em[5]: the number of iterations performed.

Functions used: tfmsymtri2, qrivalsymtri.

```
int qrivalsym2(float **a, int n, float val[], float em[])
{
    float *allocate_real_vector(int, int);
    void free_real_vector(float *, int);
    void tfmsymtri2(float **, int, float [], float [], float [],
                    float []);
    int qrivalsymtri(float [], float [], int, float []);
    int i;
```

```
    float *b,*bb;

    b=allocate_real_vector(1,n);
    bb=allocate_real_vector(1,n);
    tfmsymtri2(a,n,val,b,bb,em);
    i=qrivalsymtri(val,bb,n,em);
    free_real_vector(b,1);
    free_real_vector(bb,1);
    return i;
}
```

F. qrisym

Computes all eigenvalues and corresponding eigenvectors of an $n \times n$ symmetric matrix A whose upper triangular part is stored in a two-dimensional array, by means of QR iteration. A is first reduced to similar tridiagonal form by a call of *tfmsymtri2*, the back transformation on the eigenvectors is prepared by a call of *tfmprevec*, and the eigenvalues and eigenvectors are then calculated by *qrisymtri*.

Function Parameters:

int qrisym (*a,n,val,em*)

qrisym: given the number of eigenvalues and eigenvectors not calculated;

a: float *a[1:n,1:n]*;
- entry: the upper triangle of the symmetric matrix must be given in the upper triangular part of *a* (the elements $a[i,j]$, $i \leq j$);
- exit: the eigenvectors of the symmetric matrix, stored columnwise;

n: int;
- entry: the order of the given matrix;

val: float *val[1:n]*;
- exit: the eigenvalues of the matrix corresponding to the calculated eigenvectors;

em: float *em[0:5]*;
- entry:
  - *em[0]*: the machine precision;
  - *em[2]*: the relative tolerance for the QR iteration;
  - *em[4]*: the maximum allowed number of iterations;
- exit:
  - *em[1]*: the infinity norm of the matrix;
  - *em[3]*: the maximum absolute value of the codiagonal elements neglected;
  - *em[5]*: the number of iterations performed.

Functions used: tfmsymtri2, tfmprevec, qrisymtri.

```
int qrisym(float **a, int n, float val[], float em[])
{
    float *allocate_real_vector(int, int);
    void free_real_vector(float *, int);
    void tfmsymtri2(float **, int, float [], float [], float [],
                    float []);
    void tfmprevec(float **, int);
    int qrisymtri(float **, int, float [], float [], float [], float []);
    int i;
    float *b,*bb;

    b=allocate_real_vector(1,n);
    bb=allocate_real_vector(1,n);
    tfmsymtri2(a,n,val,b,bb,em);
    tfmprevec(a,n);
```

```
    i=qrisymtri(a,n,val,b,bb,em);
    free_real_vector(b,1);
    free_real_vector(bb,1);
    return i;
}
```

G. qrivalsym1

Computes all eigenvalues of an $n \times n$ symmetric matrix A whose upper triangular part is stored in a one-dimensional array, by means of QR iteration. A is first reduced to similar tridiagonal form by a call of *tfmsymtri1*, and *qrivalsymtri* is then used to calculate the eigenvalues.

Function Parameters:

int qrivalsym1 *(a,n,val,em)*

qrivalsym1: given the number of eigenvalues not calculated;

a: float *a[1:(n+1)n/2]*;
- entry: the upper triangle of the symmetric matrix must be given in such a way that the *(i,j)*-th element of the matrix is $a[(j-1)j/2+i]$, $1 \leq i \leq j \leq n$;
- exit: the data for Householder's back transformation (which is not used by this procedure);

n: int;
- entry: the order of the given matrix;

val: float *val[1:n]*;
- exit: the eigenvalues of the matrix in some arbitrary order;

em: float *em[0:5]*;
- entry:
- *em[0]*: the machine precision;
- *em[2]*: the relative tolerance for the eigenvalues;
- *em[4]*: the maximum allowed number of iterations;
- exit:
- *em[1]*: the infinity norm of the matrix;
- *em[3]*: the maximum absolute value of the codiagonal elements neglected;
- *em[5]*: the number of iterations performed.

Functions used: tfmsymtri1, qrivalsymtri.

```
int qrivalsym1(float a[], int n, float val[], float em[])
{
    float *allocate_real_vector(int, int);
    void free_real_vector(float *, int);
    void tfmsymtri1(float [], int, float [], float [], float [],
                    float []);
    int qrivalsymtri(float [], float [], int, float []);
    int i;
    float *b,*bb;

    b=allocate_real_vector(1,n);
    bb=allocate_real_vector(1,n);
    tfmsymtri1(a,n,val,b,bb,em);
    i=qrivalsymtri(val,bb,n,em);
    free_real_vector(b,1);
    free_real_vector(bb,1);
    return i;
}
```

3.13.3 Symmetric matrices - Auxiliary procedures

A. mergesort

Determines the permutation of p integers in the range $l \leq i \leq u$, which rearranges the real numbers a_i $(i=l,...,u)$ in nonincreasing order of magnitude, so that $|a_{p(i)}| \geq |a_{p(i+1)}|$, $i=l,...,u-1$. For sorting by merging see [AHU74, K73].

Function Parameters:

void mergesort (*a,p,low,up*)

a: float *a[low:up]*;
 entry: the vector to be stored into nondecreasing order;
 exit: the contents of *a* are left invariant;
p: int *p[low:up]*;
 exit: the permutation of indices corresponding to sorting the elements of *vec1* into nondecreasing order;
low: int;
 entry: the lower index of the arrays *a* and *p* (value of *l* above);
up: int;
 entry: the upper index of the arrays *a* and *p* (value of *u* above).

```
void mergesort(float a[], int p[], int low, int up)
{
    int *allocate_integer_vector(int, int);
    void free_integer_vector(int *, int);
    void merge(int, int, int, int [], float [], int []);
    int i,lo,step,stap,umlp1,umsp1,rest,restv,*hp;

    hp=allocate_integer_vector(low,up);
    for (i=low; i<=up; i++) p[i]=i;
    restv=0;
    umlp1=up-low+1;
    step=1;
    do {
        stap=2*step;
        umsp1=up-stap+1;
        for (lo=low; lo<=umsp1; lo += stap)
            merge(lo,step,step,p,a,hp);
        rest=up-lo+1;
        if (rest > restv && restv > 0)
            merge(lo,rest-restv,restv,p,a,hp);
        restv=rest;
        step *= 2;
    } while (step < umlp1);
    free_integer_vector(hp,low);
}

void merge(int lo, int ls, int rs, int p[], float a[], int hp[])
{
    /* this procedure is used internally by MERGESORT */

    int l,r,lout,rout,i,pl,pr;

    l=lo;
    r=lo+ls;
    lout=rout=0;
    i=lo;
    do {
        pl=p[l];
        pr=p[r];
        if (a[pl] > a[pr]) {
            hp[i]=pr;
```

```
            r++;
            rout = (r == lo+ls+rs);
        } else {
            hp[i]=pl;
            l++;
            lout = (l == lo+ls);
        }
        i++;
    } while (!(lout || rout));
    if (rout) {
        for (i=lo+ls-1; i>=l; i--) p[i+rs]=p[i];
        r=l+rs;
    }
    for (i=r-1; i>=lo; i--) p[i]=hp[i];
}
```

B. vecperm

Given a sequence of distinct integers $p(i)$, $i=l,...,u$, in the range $l \leq p(i) \leq u$ and the real numbers a_i $(i=l,...,u)$, constructs the sequence of real numbers $a_{p(i)}$, $i=l,...,u$.

Function Parameters:

void vecperm (*perm,low,upp,vector*)

perm: int *perm[low:upp]*;
- entry: a given permutation (for example, as produced by *mergesort*) of the numbers in the array vector;
- exit: the contents of *a* are left invariant;

low: int;
- entry: the lower index of the arrays *perm* and *vector* (value of *l* above);

upp: int;
- entry: the upper index of the arrays *perm* and *vector* (value of *u* above).

vector: float *vector[low:upp]*;
- entry: the real vector to be permuted;
- exit: the permuted vector elements.

```
void vecperm(int perm[], int low, int upp, float vector[])
{
    int *allocate_integer_vector(int, int);
    void free_integer_vector(int *, int);
    int t,j,k,*todo;
    float a;

    todo=allocate_integer_vector(low,upp);
    for (t=low; t<=upp; t++) todo[t]=1;
    for (t=low; t<=upp; t++)
        if (todo[t]) {
            k=t;
            a=vector[k];
            j=perm[k];
            while (j != t) {
                vector[k]=vector[j];
                todo[k]=0;
                k=j;
                j=perm[k];
            }
            vector[k]=a;
            todo[k]=0;
        }
    free_integer_vector(todo,low);
}
```

C. rowperm

Given a sequence of distinct integers $p(j)$, $j=l,...,u$, in the range $l \leq p(j) \leq u$ and the elements $M_{i,j}$, $j=l,...,u$, belonging to the i-th row of a real matrix, constructs the corresponding row elements $M_{i,p(j)}$, $j=l,...,u$.

Function Parameters:

void rowperm (*perm,low,upp,i,mat*)

perm: int *perm[low:upp]*;
entry: a given permutation (for example, as produced by *mergesort*) of the numbers in the array vector;

low: int;
entry: the lower index of the arrays *perm*;

upp: int;
entry: the upper index of the arrays *perm*;

i: int;
entry: the row index of the matrix elements;

mat: float *mat[i:i,low:upp]*;
entry: *mat[i,low:upp]* should contain the elements to be permuted;
exit: *mat[i,low:upp]* contains the row of permuted elements.

```
void rowperm(int perm[], int low, int upp, int i, float **mat)
{
    int *allocate_integer_vector(int, int);
    void free_integer_vector(int *, int);
    int t,j,k,*todo;
    float a;

    todo=allocate_integer_vector(low,upp);
    for (t=low; t<=upp; t++) todo[t]=1;
    for (t=low; t<=upp; t++)
        if (todo[t]) {
            k=t;
            a=mat[i][k];
            j=perm[k];
            while (j != t) {
                mat[i][k]=mat[i][j];
                todo[k]=0;
                k=j;
                j=perm[k];
            }
            mat[i][k]=a;
            todo[k]=0;
        }
    free_integer_vector(todo,low);
}
```

3.13.4 Symmetric matrices - Orthogonalization

orthog

Given a system of linearly independent columns x_j, $j=l,...,u$, of an n-rowed matrix X, determines an orthonormal system of columns

$$x_j' = \sum_{k=l}^{u} c_{j,k} x_k, \qquad j=l,\dots,u$$

((x_i',x_j') equals 0 when $i \neq j$ and equals 1 with $i=j$) of a corresponding matrix X' by the modified Gram-Schmidt orthogonalization [Wi65].

Function Parameters:

void orthog (*n,lc,uc,x*)

n: int;
  entry: the order of the matrix *x*;
lc: int;
  entry: the lower column index of the matrix *x*;
uc: int;
  entry: the upper column index of the matrix *x*;
x: float *x[1:n,lc:uc]*;
  entry: the matrix columns to be orthogonalized;
  exit: the orthogonalized matrix columns.

Functions used: tammat, elmcol.

```
#include <math.h>

void orthog(int n, int lc, int uc, float **x)
{
    int *allocate_integer_vector(int, int);
    void free_integer_vector(int *, int);
    float tammat(int, int, int, int, float **, float **);
    void elmcol(int, int, int, int, float **, float **, float);
    int i,j,k;
    float normx;

    for (j=lc; j<=uc; j++) {
        normx=sqrt(tammat(1,n,j,j,x,x));
        for (i=1; i<=n; i++) x[i][j] /=normx;
        for (k=j+1; k<=uc; k++) elmcol(1,n,k,j,x,x,-tammat(1,n,k,j,x,x));
    }
}
```

3.13.5 Symmetric matrices - Iterative improvement

symeigimp

Given single precision estimates λ_j' of the eigenvalues λ_j and single precision estimates u_j' of the eigenvectors u_j (j=1,...,*n*) of an *n*×*n* real symmetric matrix *A*, derives (a) ranges of the form $[\lambda_j''-\delta_j,\lambda_n''+\eta_j]$ within which the λ_j lie, λ_j'' being an improved double precision estimate of λ_j and (b) improved single precision estimates u_j'' of the corresponding eigenvector u_j (j=1,...,*n*). For further details see [GrK69].

Function Parameters:

void symeigimp (*n,a,vec,val,lbound,ubound,aux*)

n: int;
  entry: the order of the matrix *a*;
a: float *a[1:n,1:n]*;
  entry: contains a real symmetric matrix whose eigensystem has to be improved;

vec: float *vec[1:n,1:n]*;
entry: contains a matrix whose columns are a system of approximate eigenvectors of matrix *a*; initial approximations;
exit: improved approximations;

val: float *val[1:n]*;
entry: initial approximations of the eigenvalues of *a*;
exit: improved eigenvalues of *a*;

lbound, ubound: float *lbound[1:n], ubound[1:n]*;
exit: the lower and upper error bounds respectively for the eigenvalue approximations in *val* such that the *i*-th exact eigenvalue lies between *val[i]-lbound[i]* and *val[i]+ubound[i]* (values of δ_j and η_j, j=1,...,*n*, respectively, in the above);

aux: float *aux[0:5]*;
entry:
aux[0]: the relative precision of the elements of *a*;
aux[2]: the relative tolerance for the residual matrix; the iteration ends when the maximum absolute value of the residual elements is smaller than *aux[2]*aux[1]*;
aux[4]: the maximum number of iterations allowed;
exit:
aux[1]: infinity norm of the matrix *a*;
aux[3]: maximum absolute element of the residual matrix;
aux[5]: number of iterations.

Functions used: vecvec, matmat, tammat, mergesort, vecperm, rowperm, orthog, qrisym, infnrmmat.

```
#include <float.h>
#include <math.h>

void symeigimp(int n, float **a, float **vec, float val[],
               float lbound[], float ubound[], float aux[])
{
    int *allocate_integer_vector(int, int);
    float *allocate_real_vector(int, int);
    float **allocate_real_matrix(int, int, int, int);
    void free_integer_vector(int *, int);
    void free_real_vector(float *, int);
    void free_real_matrix(float **, int, int, int);
    void orthog(int, int, int, float **);
    void mergesort(float [], int [], int, int);
    void vecperm(int [], int, int, float []);
    void rowperm(int [], int, int, int, float **);
    int qrisym(float **, int, float [], float []);
    float vecvec(int, int, int, float [], float []);
    float matmat(int, int, int, int, float **, float **);
    float tammat(int, int, int, int, float **, float **);
    float infnrmmat(int, int, int, int, int *, float **);
    int k,i,j,i0,i1,i01,iter,maxitp1,stop,n1,i0m1,i1p1,*perm;
    float s,max,tol,mateps,relerra,reltolr,norma,eps2,dl,dr,m1,
          **r,**p,**y,**pp,*rq,*eps,*z,*val3,*val4,*eta,em[6];
    double dtemp;

    perm=allocate_integer_vector(1,n);
    rq=allocate_real_vector(1,n);
    eps=allocate_real_vector(1,n);
    z=allocate_real_vector(1,n);
    val3=allocate_real_vector(1,n);
    eta=allocate_real_vector(1,n);
    r=allocate_real_matrix(1,n,1,n);
```

```
p=allocate_real_matrix(1,n,1,n);
y=allocate_real_matrix(1,n,1,n);
norma=infnrmmat(1,n,1,n,&i,a);
relerra=aux[0];
reltolr=aux[2];
maxitp1=aux[4]+1.0;
mateps=relerra*norma;
tol=reltolr*norma;
for (iter=1; iter<=maxitp1; iter++) {
    if (iter == 1)
        stop=0;
    else
        stop=1;
    max=0.0;
    for (j=1; j<=n; j++)
        for (i=1; i<=n; i++) {
            dtemp = -(double)(vec[i][j])*(double)(val[j]);
            for (k=1; k<=n; k++)
                dtemp += (double)(a[i][k])*(double)(vec[k][j]);
            r[i][j]=dtemp;
            if (fabs(r[i][j]) > max) max=fabs(r[i][j]);
        }
    if (max > tol) stop=0;
    if ((!stop) && (iter < maxitp1)) {
        for (i=1; i<=n; i++) {
            dtemp=(double)(val[i]);
            for (k=1; k<=n; k++)
                dtemp += (double)(vec[k][i])*(double)(r[k][i]);
            rq[i]=dtemp;
        }
        for (j=1; j<=n; j++) {
            for (i=1; i<=n; i++)
                eta[i]=r[i][j]-(rq[j]-val[j])*vec[i][j];
            z[j]=sqrt(vecvec(1,n,0,eta,eta));
        }
        mergesort(rq,perm,1,n);
        vecperm(perm,1,n,rq);
        for (i=1; i<=n; i++) {
            eps[i]=z[perm[i]];
            val3[i]=val[perm[i]];
            rowperm(perm,1,n,i,vec);
            rowperm(perm,1,n,i,r);
        }
        for (i=1; i<=n; i++)
            for (j=i; j<=n; j++)
                p[i][j]=p[j][i]=tammat(1,n,i,j,vec,r);
    }
    i0=1;
    do {
        j=i1=i0;
        j++;
        while ((j > n) ? 0 :
                  (rq[j]-rq[j-1] <= sqrt((eps[j]+eps[j-1])*norma))) {
            i1=j;
            j++;
        }
        if (stop || (iter == maxitp1)) {
            i=i0;
            do {
                j=i01=i;
                j++;
                while ((j>i1) ? 0 : rq[j]-rq[j-1] <= eps[j]+eps[j-1]) {
                    i01=j;
                    j++;
                }
                if (i == i01) {
                    if (i < n) {
                        if (i == 1)
                            dl=dr=rq[i+1]-rq[i]-eps[i+1];
                        else {
                            dl=rq[i]-rq[i-1]-eps[i-1];
                            dr=rq[i+1]-rq[i]-eps[i+1];
                        }
```

```
                } else
                   dl=dr=rq[i]-rq[i-1]-eps[i-1];
                eps2=eps[i]*eps[i];
                lbound[i]=eps2/dr+mateps;
                ubound[i]=eps2/dl+mateps;
             } else
                for (k=i; k<=i01; k++)
                   lbound[k]=ubound[k]=eps[k]+mateps;
             i01++;
             i=i01;
          } while (i <= i1);     /* bounds */
       } else {
          if (i0 == i1) {
             for (k=1; k<=n; k++)
                if (k == i0)
                   y[k][i0]=1.0;
                else
                   r[k][i0]=p[k][i0];
             val[i0]=rq[i0];
          } else {
             n1=i1-i0+1;
             em[0]=em[2]=FLT_EPSILON;
             em[4]=10*n1;
             val4=allocate_real_vector(1,n1);
             pp=allocate_real_matrix(1,n1,1,n1);
             m1=0.0;
             for (k=i0; k<=i1; k++) m1 += val3[k];
             m1 /= n1;
             for (i=1; i<=n1; i++)
                for (j=1; j<=n1; j++) {
                   pp[i][j]=p[i+i0-1][j+i0-1];
                   if (i == j) pp[i][j] += val3[j+i0-1]-m1;
                }
             for (i=i0; i<=i1; i++) {
                val3[i]=m1;
                val[i]=rq[i];
             }
             qrisym(pp,n1,val4,em);
             mergesort(val4,perm,1,n1);
             for (i=1; i<=n1; i++)
                for (j=1; j<=n1; j++)
                   p[i+i0-1][j+i0-1]=pp[i][perm[j]];
             i0m1=i0-1;
             i1p1=i1+1;
             for (j=i0; j<=i1; j++) {
                for (i=1; i<=i0m1; i++) {
                   s=0.0;
                   for (k=i0; k<=i1; k++) s += p[i][k]*p[k][j];
                   r[i][j]=s;
                }
                for (i=i1p1; i<=n; i++) {
                   s=0.0;
                   for (k=i0; k<=i1; k++) s += p[i][k]*p[k][j];
                   r[i][j]=s;
                }
                for (i=i0; i<=i1; i++) y[i][j]=p[i][j];
             }
             free_real_vector(val4,1);
             free_real_matrix(pp,1,n1,1);
          }  /* innerblock */
       }  /* not stop */
       i0=i1+1;
    } while (i0 <= n);     /* while i0 loop */
    if ((!stop) && (iter < maxitp1)) {
       for (j=1; j<=n; j++)
          for (i=1; i<=n; i++)
             if (val3[i] != val3[j])
                y[i][j]=r[i][j]/(val3[j]-val3[i]);
       for (i=1; i<=n; i++) {
          for (j=1; j<=n; j++) z[j]=matmat(1,n,i,j,vec,y);
          for (j=1; j<=n; j++) vec[i][j]=z[j];
       }
       orthog(n,1,n,vec);
    } else {
```

```
            aux[5]=iter-1;
            break;
        }
    }   /* for iter loop */
    aux[1]=norma;
    aux[3]=max;

    free_integer_vector(perm,1);
    free_real_vector(rq,1);
    free_real_vector(eps,1);
    free_real_vector(z,1);
    free_real_vector(val3,1);
    free_real_vector(eta,1);
    free_real_matrix(r,1,n,1);
    free_real_matrix(p,1,n,1);
    free_real_matrix(y,1,n,1);
}
```

3.13.6 Asymmetric matrices in Hessenberg form

A. reavalqri

Determines the eigenvalues λ_j (j=1,...,n) (all assumed to be real) of an $n \times n$ real upper Hessenberg matrix A ($A_{i,j}$=0 for $i>j+1$) by single QR iteration. The method used is the single QR iteration of Francis [Dek&Ho68, Fr61, Wi65]. The eigenvalues of a real upper-Hessenberg matrix are calculated provided that the matrix has real eigenvalues only.

Function Parameters:

int reavalqri (*a,n,em,val*)

reavalqri: given the value 0 provided that the process is completed within *em[4]* iterations; otherwise *reavalqri* is given the value *k*, of the number of eigenvalues not calculated;

a: float *a[1:n,1:n]*;
- entry: the elements of the real upper Hessenberg matrix must be given in the upper triangle and the first subdiagonal of array *a*;
- exit: the Hessenberg part of array *a* is altered;

n: int;
- entry: the order of the given matrix;

em: float *em[0:5]*;
- entry:
  - *em[0]*: the machine precision;
  - *em[1]*: a norm of the given matrix;
  - *em[2]*: the relative tolerance used for the QR iteration (*em[2]* > *em[0]*); if the absolute value of some subdiagonal element is smaller than *em[1]***em[2]* then this element is neglected and the matrix is partitioned;
  - *em[4]*: the maximum allowed number of iterations (for example, *em[4]*=10**n*);
- exit:
  - *em[3]*: the maximum absolute value of the subdiagonal elements neglected;
  - *em[5]*: the number of QR iterations performed; if the iteration process is not completed within *em[4]* iterations then the value *em[4]*+1 is delivered and in this case only the last *n-k* elements of *val* are approximate eigenvalues of the given matrix, where *k* is delivered in *reavalqri*;

val: float *val[1:n]*;

exit: the eigenvalues of the given matrix are delivered in *val*.

Functions used: rotcol, rotrow.

```
#include <math.h>

int reavalqri(float **a, int n, float em[], float val[])
{
    void rotcol(int, int, int, int, float **, float, float);
    void rotrow(int, int, int, int, float **, float, float);
    int n1,i,i1,q,max,count;
    float det,w,shift,kappa,nu,mu,r,tol,delta,machtol,s;

    machtol=em[0]*em[1];
    tol=em[1]*em[2];
    max=em[4];
    count=0;
    r=0.0;
    do {
        n1=n-1;
        i=n;
        do{
            q=i;
            i--;
        } while ((i >= 1) ? (fabs(a[i+1][i]) > tol) : 0);
        if (q > 1)
            if (fabs(a[q][q-1]) > r) r=fabs(a[q][q-1]);
        if (q == n) {
            val[n]=a[n][n];
            n=n1;
        } else {
            delta=a[n][n]-a[n1][n1];
            det=a[n][n1]*a[n1][n];
            if (fabs(delta) < machtol)
                s=sqrt(det);
            else {
                w=2.0/delta;
                s=w*w*det+1.0;
                s = (s <= 0.0) ? -delta*0.5 : w*det/(sqrt(s)+1.0);
            }
            if (q == n1) {
                val[n]=a[n][n]+s;
                val[n1]=a[n1][n1]-s;
                n -= 2;
            } else {
                count++;
                if (count > max) break;
                shift=a[n][n]+s;
                if (fabs(delta) < tol) {
                    w=a[n1][n1]-s;
                    if (fabs(w) < fabs(shift)) shift=w;
                }
                a[q][q] -= shift;
                for (i=q; i<=n-1; i++) {
                    i1=i+1;
                    a[i1][i1] -= shift;
                    kappa=sqrt(a[i][i]*a[i][i]+a[i1][i]*a[i1][i]);
                    if (i > q) {
                        a[i][i-1]=kappa*nu;
                        w=kappa*mu;
                    } else
                        w=kappa;
                    mu=a[i][i]/kappa;
                    nu=a[i1][i]/kappa;
                    a[i][i]=w;
                    rotrow(i1,n,i,i1,a,mu,nu);
                    rotcol(q,i,i,i1,a,mu,nu);
                    a[i][i] += shift;
                }
                a[n][n-1]=a[n][n]*nu;
                a[n][n]=a[n][n]*mu+shift;
            }
```

```
        }
    } while (n > 0);
    em[3]=r;
    em[5]=count;
    return n;
}
```

B. reaveches

Determines the eigenvector corresponding to a single real eigenvalue λ of an $n \times n$ real Hessenberg matrix A ($A_{i,j}=0$ for $i>j+1$) by inverse iteration [DekHo68, Vr66, Wi65]. Starting with $x^{(0)}=(1,1,...,1)^T$, vectors $y^{(k)}$ and $x^{(k)}$ are produced by use of the scheme

$$(A-\lambda I)y^{(k)}=x^{(k)},\ x^{(k+1)}=y^{(k)}/\|y^{(k)}\|_E\ (k=0,1,...).$$

The process is terminated if either

(a) $\|(A-\lambda I)x^{(k)}\|_E \leq \|A\| * tol$,

where *tol* is a tolerance prescribed by the user, or

(b) $k=kmax+1$,

where *kmax* is an integer prescribed by the user.

Function Parameters:

void reaveches (*a,n,lambda,em,v*)

a: float *a[1:n,1:n]*;
- entry: the elements of the real upper Hessenberg matrix must be given in the upper triangle and the first subdiagonal of array *a*;
- exit: the Hessenberg part of array *a* is altered;

n: int;
- entry: the order of the given matrix;

lambda: float;
- the given real eigenvalue of the upper Hessenberg matrix (value of λ above);

em: float *em[0:9]*;
- entry:
  - *em[0]*: the machine precision;
  - *em[1]*: a norm of the given matrix;
  - *em[6]*: the tolerance used for eigenvector (value of *tol* above, *em[6]* > *em[0]*); the inverse iteration ends if the Euclidean norm of the residue vector is smaller than *em[1]***em[6]*;
  - *em[8]*: the maximum allowed number of iterations (value of *kmax* above, for example, *em[8]*=5);
- exit:
  - *em[7]*: the Euclidean norm of the residue vector of the calculated eignenvector;
  - *em[9]*: the number of inverse iterations performed; if *em[7]* remains larger than *em[1]***em[6]* during *em[8]* iterations then the value *em[8]*+1 is delivered;

v: float *v[1:n]*;
- exit: the calculated eigenvector is delivered in *v*.

Functions used: vecvec, matvec.

```
#include <math.h>

void reaveches(float **a, int n, float lambda, float em[], float v[])
{
    int *allocate_integer_vector(int, int);
```

```
    void free_integer_vector(int *, int);
    float vecvec(int, int, int, float [], float []);
    float matvec(int, int, int, float **, float []);
    int i,i1,j,count,max,*p;
    float m,r,norm,machtol,tol;

    p=allocate_integer_vector(1,n);
    norm=em[1];
    machtol=em[0]*norm;
    tol=em[6]*norm;
    max=em[8];
    a[1][1] -= lambda;
    for (i=1; i<=n-1; i++) {
        i1=i+1;
        r=a[i][i];
        m=a[i1][i];
        if (fabs(m) < machtol) m=machtol;
        p[i] = (fabs(m) <= fabs(r));
        if (p[i]) {
            a[i1][i] = m /= r;
            for (j=i1; j<=n; j++)
                a[i1][j]=((j > i1) ? a[i1][j] : a[i1][j]-lambda)-m*a[i][j];
        } else {
            a[i][i]=m;
            a[i1][i] = m = r/m;
            for (j=i1; j<=n; j++) {
                r = (j > i1) ? a[i1][j] : a[i1][j]-lambda;
                a[i1][j]=a[i][j]-m*r;
                a[i][j]=r;
            }
        }
    }
    if (fabs(a[n][n]) < machtol) a[n][n]=machtol;
    for (j=1; j<=n; j++) v[j]=1.0;
    count=0;
    do {
        count++;
        if (count > max) break;
        for (i=1; i<=n-1; i++) {
            i1=i+1;
            if (p[i])
                v[i1] -= a[i1][i]*v[i];
            else {
                r=v[i1];
                v[i1]=v[i]-a[i1][i]*r;
                v[i]=r;
            }
        }
        for (i=n; i>=1; i--)
            v[i]=(v[i]-matvec(i+1,n,i,a,v))/a[i][i];
        r=1.0/sqrt(vecvec(1,n,0,v,v));
        for (j=1; j<=n; j++) v[j] *= r;
    } while (r > tol);
    em[7]=r;
    em[9]=count;
    free_integer_vector(p,1);
}
```

C. reaqri

Computes all eigenvalues λ_j (assumed to be real) and corresponding eigenvectors u_j (j=1,...,n) of the $n \times n$ real upper Hessenberg matrix A ($A_{i,j}=0$ for $i>j+1$) by single QR iteration. The eigenvectors are calculated by a direct method [see DekHo68], in contrast with *reaveches* which uses inverse iteration. If the Hessenberg matrix is not too ill-conditioned with respect to its eigenvalue problem then this method yields numerically independent eigenvectors and is competitive with inverse iteration as to accuracy and computation time.

Function Parameters:

int reaqri (*a,n,em,val,vec*)

reaqri: given the value 0 provided that the process is completed within *em[4]* iterations; otherwise *reaqri* is given the value *k*, of the number of eigenvalues and eigenvectors not calculated;

a: float *a[1:n,1:n]*;
entry: the elements of the real upper Hessenberg matrix must be given in the upper triangle and the first subdiagonal of array *a*;
exit: the Hessenberg part of array *a* is altered;

n: int;
entry: the order of the given matrix;

em: float *em[0:5]*;
entry:
em[0]: the machine precision;
em[1]: a norm of the given matrix;
em[2]: the relative tolerance used for the QR iteration (*em[2]* > *em[0]*); if the absolute value of some subdiagonal element is smaller than *em[1]*em[2]* then this element is neglected and the matrix is partitioned;
em[4]: the maximum allowed number of iterations (for example, *em[4]*=10**n*);
exit:
em[3]: the maximum absolute value of the subdiagonal elements neglected;
em[5]: the number of QR iterations performed; if the iteration process is not completed within *em[4]* iterations then the value *em[4]*+1 is delivered and in this case only the last *n-k* elements of *val* and the last *n-k* columns of *vec* are approximated eigenvalues and eigenvectors of the given matrix, where *k* is delivered in *reaqri*;

val: float *val[1:n]*;
exit: the eigenvalues of the given matrix are delivered in *val*;

vec: float *vec[1:n,1:n]*;
exit: the calculated eigenvectors corresponding to the eigenvalues in *val[1:n]* are delivered in the columns of *vec*.

Functions used: matvec, rotcol, rotrow.

```
#include <math.h>

int reaqri(float **a, int n, float em[], float val[], float **vec)
{
    float *allocate_real_vector(int, int);
    void free_real_vector(float *, int);
    float matvec(int, int, int, float **, float []);
    void rotcol(int, int, int, int, float **, float, float);
    void rotrow(int, int, int, int, float **, float, float);
    int m1,i,i1,m,j,q,max,count;
    float w,shift,kappa,nu,mu,r,tol,s,machtol,elmax,t,delta,det,*tf;

    tf=allocate_real_vector(1,n);
    machtol=em[0]*em[1];
    tol=em[1]*em[2];
    max=em[4];
    count=0;
    elmax=0.0;
    m=n;
    for (i=1; i<=n; i++) {
        vec[i][i]=1.0;
        for (j=i+1; j<=n; j++) vec[i][j]=vec[j][i]=0.0;
```

```
}
do {
    m1=m-1;
    i=m;
    do {
        q=i;
        i--;
    } while ((i >= 1) ? (fabs(a[i+1][i]) > tol) : 0);
    if (q > 1)
        if (fabs(a[q][q-1]) > elmax) elmax=fabs(a[q][q-1]);
    if (q == m) {
        val[m]=a[m][m];
        m=m1;
    } else {
        delta=a[m][m]-a[m1][m1];
        det=a[m][m1]*a[m1][m];
        if (fabs(delta) < machtol)
            s=sqrt(det);
        else {
            w=2.0/delta;
            s=w*w*det+1.0;
            s = (s <= 0.0) ? -delta*0.5 : w*det/(sqrt(s)+1.0);
        }
        if (q == m1) {
            val[m] = a[m][m] += s;
            val[q] = a[q][q] -= s;
            t = (fabs(s) < machtol) ? (s+delta)/a[m][q] : a[q][m]/s;
            r=sqrt(t*t+1.0);
            nu=1.0/r;
            mu = -t*nu;
            a[q][m] -= a[m][q];
            rotrow(q+2,n,q,m,a,mu,nu);
            rotcol(1,q-1,q,m,a,mu,nu);
            rotcol(1,n,q,m,vec,mu,nu);
            m -= 2;
        } else {
            count++;
            if (count > max) {
                em[3]=elmax;
                em[5]=count;
                free_real_vector(tf,1);
                return m;
            }
            shift=a[m][m]+s;
            if (fabs(delta) < tol) {
                w=a[m1][m1]-s;
                if (fabs(w) < fabs(shift)) shift=w;
            }
            a[q][q] -= shift;
            for (i=q; i<=m1; i++) {
                i1=i+1;
                a[i1][i1] -= shift;
                kappa=sqrt(a[i][i]*a[i][i]+a[i1][i]*a[i1][i]);
                if (i > q) {
                    a[i][i-1]=kappa*nu;
                    w=kappa*mu;
                } else
                    w=kappa;
                mu=a[i][i]/kappa;
                nu=a[i1][i]/kappa;
                a[i][i]=w;
                rotrow(i1,n,i,i1,a,mu,nu);
                rotcol(1,i,i,i1,a,mu,nu);
                a[i][i] += shift;
                rotcol(1,n,i,i1,vec,mu,nu);
            }
            a[m][m1]=a[m][m]*nu;
            a[m][m]=a[m][m]*mu+shift;
        }
    }
} while (m > 0);
for (j=n; j>=2; j--) {
    tf[j]=1.0;
    t=a[j][j];
```

```
            for (i=j-1; i>=1; i--) {
                delta=t-a[i][i];
                tf[i]=matvec(i+1,j,i,a,tf)/
                            ((fabs(delta) < machtol) ? machtol : delta);
            }
            for (i=1; i<=n; i++) vec[i][j]=matvec(1,j,i,vec,tf);
        }
        em[3]=elmax;
        em[5]=count;
        free_real_vector(tf,1);
        return m;
}
```

D. comvalqri

Determines the real and complex eigenvalues λ_j (j=1,...,*n*) of an *n*×*n* real upper Hessenberg matrix *A* ($A_{i,j}$=0 for *i*>*j*+*1*) by the double QR iteration of Francis [DekHo68, Fr61, Wi65].

Function Parameters:

int comvalqri (*a,n,em,re,im*)

comvalqri: given the value 0 provided that the process is completed within *em[4]* iterations; otherwise *comvalqri* is given the value *k*, of the number of eigenvalues not calculated;

a: float *a[1:n,1:n]*;
- entry: the elements of the real upper Hessenberg matrix must be given in the upper triangle and the first subdiagonal of array *a*;
- exit: the Hessenberg part of array *a* is altered;

n: int;
- entry: the order of the given matrix;

em: float *em[0:5]*;
- entry:
  - *em[0]*: the machine precision;
  - *em[1]*: a norm of the given matrix;
  - *em[2]*: the relative tolerance used for the QR iteration (*em[2]* > *em[0]*); if the absolute value of some subdiagonal element is smaller than *em[1]*em[2]* then this element is neglected and the matrix is partitioned;
  - *em[4]*: the maximum allowed number of iterations (for example, *em[4]*=10**n*);
- exit:
  - *em[3]*: the maximum absolute value of the subdiagonal elements neglected;
  - *em[5]*: the number of QR iterations performed; if the iteration process is not completed within *em[4]* iterations then the value *em[4]*+1 is delivered and in this case only the last *n-k* elements of *re* and *im* are approximate eigenvalues of the given matrix, where *k* is delivered in *comvalqri*;

re,im: float *re[1:n], im[1:n]*;
- exit: the real and imaginary parts of the calculated eigenvalues of the given matrix are delivered in *re, im[1:n]*, the members of each nonreal complex conjugate pair being consecutive.

```
#include <math.h>

int comvalqri(float **a, int n, float em[], float re[], float im[])
{
    int i,j,p,q,max,count,n1,p1,p2,imin1,i1,i2,i3,b;
    float disc,sigma,rho,g1,g2,g3,psi1,psi2,aa,e,k,s,norm,machtol2,
```

```
        tol,w;

    norm=em[1];
    w=em[0]*norm;
    machtol2=w*w;
    tol=em[2]*norm;
    max=em[4];
    count=0;
    w=0.0;
    do {
        i=n;
        do {
            q=i;
            i--;
        } while ((i >= 1) ? (fabs(a[i+1][i]) > tol) : 0);
        if (q > 1)
            if (fabs(a[q][q-1]) > w) w=fabs(a[q][q-1]);
        if (q >= n-1) {
            n1=n-1;
            if (q == n) {
                re[n]=a[n][n];
                im[n]=0.0;
                n=n1;
            } else {
                sigma=a[n][n]-a[n1][n1];
                rho = -a[n][n1]*a[n1][n];
                disc=sigma*sigma-4.0*rho;
                if (disc > 0.0) {
                    disc=sqrt(disc);
                    s = -2.0*rho/(sigma+((sigma >= 0.0) ? disc : -disc));
                    re[n]=a[n][n]+s;
                    re[n1]=a[n1][n1]-s;
                    im[n]=im[n1]=0.0;
                } else {
                    re[n]=re[n1]=(a[n1][n1]+a[n][n])/2.0;
                    im[n1]=sqrt(-disc)/2.0;
                    im[n] = -im[n1];
                }
                n -= 2;
            }
        } else {
            count++;
            if (count > max) break;
            n1=n-1;
            sigma=a[n][n]+a[n1][n1]+sqrt(fabs(a[n1][n-2]*a[n][n1])*em[0]);
            rho=a[n][n]*a[n1][n1]-a[n][n1]*a[n1][n];
            i=n-1;
            do {
                p1=i1=i;
                i--;
            } while ((i-1 >= q) ? (fabs(a[i][i-1]*a[i1][i]*
                    (fabs(a[i][i]+a[i1][i1]-sigma)+fabs(a[i+2][i1]))) >
                    fabs(a[i][i]*((a[i][i]-sigma)+
                        a[i][i1]*a[i1][i]+rho))*tol) : 0);
            p=p1-1;
            p2=p+2;
            for (i=p; i<=n-1; i++) {
                imin1=i-1;
                i1=i+1;
                i2=i+2;
                if (i == p) {
                    g1=a[p][p]*(a[p][p]-sigma)+a[p][p1]*a[p1][p]+rho;
                    g2=a[p1][p]*(a[p][p]+a[p1][p1]-sigma);
                    if (p1 <= n1) {
                        g3=a[p1][p]*a[p2][p1];
                        a[p2][p]=0.0;
                    } else
                        g3=0.0;
                } else {
                    g1=a[i][imin1];
                    g2=a[i1][imin1];
                    g3 = (i2 <= n) ? a[i2][imin1] : 0.0;
                }
                k = (g1 >= 0.0) ? sqrt(g1*g1+g2*g2+g3*g3) :
```

```
                                        -sqrt(g1*g1+g2*g2+g3*g3);
                b = (fabs(k) > machtol2);
                aa = (b ? g1/k+1.0 : 2.0);
                psi1 = (b ? g2/(g1+k) : 0.0);
                psi2 = (b ? g3/(g1+k) : 0.0);
                if (i != q)
                    a[i][imin1] = (i == p) ? -a[i][imin1] : -k;
                for (j=i; j<=n; j++) {
                    e=aa*(a[i][j]+psi1*a[i1][j]+
                            ((i2 <= n) ? psi2*a[i2][j] : 0.0));
                    a[i][j] -= e;
                    a[i1][j] -= psi1*e;
                    if (i2 <= n) a[i2][j] -= psi2*e;
                }
                for (j=q; j<=((i2 <= n) ? i2 : n); j++) {
                    e=aa*(a[j][i]+psi1*a[j][i1]+
                            ((i2 <= n) ? psi2*a[j][i2] : 0.0));
                    a[j][i] -= e;
                    a[j][i1] -= psi1*e;
                    if (i2 <= n) a[j][i2] -= psi2*e;
                }
                if (i2 <= n1) {
                    i3=i+3;
                    e=aa*psi2*a[i3][i2];
                    a[i3][i] = -e;
                    a[i3][i1] = -psi1*e;
                    a[i3][i2] -= psi2*e;
                }
            }
        }
    } while (n > 0);
    em[3]=w;
    em[5]=count;
    return n;
}
```

E. comveches

Determines the eigenvector corresponding to a single complex eigenvalue $\lambda+\mu i$ of an $n\times n$ real upper Hessenberg matrix A ($A_{i,j}=0$ for $i>j+1$) by inverse iteration [DekHo68, Vr66, Wi65]. Starting with $x^{(0)}=(1,1,...,1)^T$, vectors $y^{(k)}$ and $x^{(k)}$ are produced by use of the scheme

$$(A-\lambda I)y^{(k)}=x^{(k)},\ x^{(k+1)}=y^{(k)}/\|y^{(k)}\|_E\ (k=0,1,...).$$

The process is terminated if either (a) $\|(A-\lambda I)x^{(k)}\|_E\leq\|A\|*tol$, where tol is a tolerance prescribed by the user, or (b) $k=kmax+1$, where $kmax$ is an integer prescribed by the user.

Function Parameters:

void comveches (*a,n,lambda,mu,em,u,v*)

a: float *a[1:n,1:n]*;
- entry: the elements of the real upper Hessenberg matrix must be given in the upper triangle and the first subdiagonal of array *a*;
- exit: the Hessenberg part of array *a* is altered;

n: int;
- entry: the order of the given matrix;

lambda, mu: float; real and imaginary part of the given eigenvalue (values of λ and μ above);

em: float *em[0:9]*;
- entry:
- *em[0]*: the machine precision;
- *em[1]*: a norm of the given matrix;

em[6]: the tolerance used for eigenvector (value of *tol* above, *em[6]* > *em[0]*); the inverse iteration ends if the Euclidean norm of the residue vector is smaller than *em[1]*em[6]*;
em[8]: the maximum allowed number of iterations (value of *kmax* above, for example, *em[8]*=5);
exit:
em[7]: the Euclidean norm of the residue vector of the calculated eigenvector;
em[9]: the number of inverse iterations performed; if *em[7]* remains larger than *em[1]*em[6]* during *em[8]* iterations then the value *em[8]*+1 is delivered;

u,v: float *u[1:n]*, *v[1:n]*;

Functions used: vecvec, matvec, tamvec.

```
#include <math.h>

void comveches(float **a, int n, float lambda, float mu,
               float em[], float u[], float v[])
{
    float *allocate_real_vector(int, int);
    int *allocate_integer_vector(int, int);
    void free_integer_vector(int *, int);
    void free_real_vector(float *, int);
    float vecvec(int, int, int, float [], float []);
    float matvec(int, int, int, float **, float []);
    float tamvec(int, int, int, float **, float []);
    int i,i1,j,count,max,*p;
    float aa,bb,d,m,r,s,w,x,y,norm,machtol,tol,*g,*f;

    p=allocate_integer_vector(1,n);
    g=allocate_real_vector(1,n);
    f=allocate_real_vector(1,n);
    norm=em[1];
    machtol=em[0]*norm;
    tol=em[6]*norm;
    max=em[8];
    for (i=2; i<=n; i++) {
        f[i-1]=a[i][i-1];
        a[i][1]=0.0;
    }
    aa=a[1][1]-lambda;
    bb = -mu;
    for (i=1; i<=n-1; i++) {
        i1=i+1;
        m=f[i];
        if (fabs(m) < machtol) m=machtol;
        a[i][i]=m;
        d=aa*aa+bb*bb;
        p[i] = (fabs(m) < sqrt(d));
        if (p[i]) {
            f[i]=r=m*aa/d;
            g[i] = s = -m*bb/d;
            w=a[i1][i];
            x=a[i][i1];
            a[i1][i]=y=x*s+w*r;
            a[i][i1]=x=x*r-w*s;
            aa=a[i1][i1]-lambda-x;
            bb = -(mu+y);
            for (j=i+2; j<=n; j++) {
                w=a[j][i];
                x=a[i][j];
                a[j][i]=y=x*s+w*r;
                a[i][j]=x=x*r-w*s;
                a[j][i1] = -y;
                a[i1][j] -= x;
```

```
            }
        } else {
            f[i]=r=aa/m;
            g[i]=s=bb/m;
            w=a[i1][i1]-lambda;
            aa=a[i][i1]-r*w-s*mu;
            a[i][i1]=w;
            bb=a[i1][i]-s*w+r*mu;
            a[i1][i] = -mu;
            for (j=i+2; j<=n; j++) {
                w=a[i1][j];
                a[i1][j]=a[i][j]-r*w;
                a[i][j]=w;
                a[j][i1]=a[j][i]-s*w;
                a[j][i]=0.0;
            }
        }
    }
    p[n]=1;
    d=aa*aa+bb*bb;
    if (d < machtol*machtol) {
        aa=machtol;
        bb=0.0;
        d=machtol*machtol;
    }
    a[n][n]=d;
    f[n]=aa;
    g[n] = -bb;
    for (i=1; i<=n; i++) {
        u[i]=1.0;
        v[i]=0.0;
    }
    count=0;
    do {
        if (count > max) break;
        for (i=1; i<=n; i++)
            if (p[i]) {
                w=v[i];
                v[i]=g[i]*u[i]+f[i]*w;
                u[i]=f[i]*u[i]-g[i]*w;
                if (i < n) {
                    v[i+1] -= v[i];
                    u[i+1] -= u[i];
                }
            } else {
                aa=u[i+1];
                bb=v[i+1];
                u[i+1]=u[i]-(f[i]*aa-g[i]*bb);
                u[i]=aa;
                v[i+1]=v[i]-(g[i]*aa+f[i]*bb);
                v[i]=bb;
            }
        for (i=n; i>=1; i--) {
            i1=i+1;
            u[i]=(u[i]-matvec(i1,n,i,a,u)+
                    (p[i] ? tamvec(i1,n,i,a,v) : a[i1][i]*v[i1]))/a[i][i];
            v[i]=(v[i]-matvec(i1,n,i,a,v)-
                    (p[i] ? tamvec(i1,n,i,a,u) : a[i1][i]*u[i1]))/a[i][i];
        }
        w=1.0/sqrt(vecvec(1,n,0,u,u)+vecvec(1,n,0,v,v));
        for (j=1; j<=n; j++) {
            u[j] *= w;
            v[j] *= w;
        }
        count++;
    } while (w > tol);
    em[7]=w;
    em[9]=count;
    free_integer_vector(p,1);
    free_real_vector(g,1);
    free_real_vector(f,1);
}
```

3.13.7 Real asymmetric full matrices

A. reaeigval

Determines the eigenvalues λ_j (j=1,...,n) (all assumed to be real) of a real $n \times n$ matrix A. A is equilibrated to the form $A' = PDAD^{-1}P^{-1}$ (by means of a call of *eqilbr*) and transformed to similar real upper Hessenberg form H (by means of a call of *tfmreahes*). The eigenvalues of H are then computed by QR iteration (by means of a call of *reavalqri*). The procedure *reaeigval* should be used only if all eigenvalues are real. For further details see [DekHo68, Fr61, Wi65].

Function Parameters:

int **reaeigval** (*a,n,em,val*)

reaeigval: given the value 0 provided that the process is completed within *em[4]* iterations; otherwise *reaeigval* is given the value *k*, of the number of eigenvalues not calculated;

a: float *a[1:n,1:n]*;
- entry: the matrix whose eigenvalues are to be calculated;
- exit: the array elements are altered;

n: int;
- entry: the order of the given matrix;

em: float *em[0:5]*;
- entry:
  - *em[0]*: the machine precision;
  - *em[2]*: the relative tolerance used for the QR iteration (*em[2]* > *em[0]*);
  - *em[4]*: the maximum allowed number of iterations (for example, *em[4]*=10n);
- exit:
  - *em[1]*: the infinity norm of the equilibrated matrix;
  - *em[3]*: the maximum absolute value of the subdiagonal elements neglected;
  - *em[5]*: the number of QR iterations performed; if the iteration process is not completed within *em[4]* iterations then the value *em[4]*+1 is delivered and in this case only the last *n-k* elements of *val* are approximate eigenvalues of the given matrix, where *k* is delivered in *reaeigval*;

val: float *val[1:n]*;
- exit: the eigenvalues of the given matrix are delivered in monotonically nonincreasing order.

Functions used: eqilbr, tfmreahes, reavalqri.

```
#include <math.h>

int reaeigval(float **a, int n, float em[], float val[])
{
    int *allocate_integer_vector(int, int);
    float *allocate_real_vector(int, int);
    void free_integer_vector(int *, int);
    void free_real_vector(float *, int);
    void tfmreahes(float **, int, float [], int []);
    void eqilbr(float **, int, float [], float [], int []);
    int reavalqri(float **, int, float [], float []);
    int i,j,k,*ind,*ind0;
    float r,*d;
```

```
    ind=allocate_integer_vector(1,n);
    ind0=allocate_integer_vector(1,n);
    d=allocate_real_vector(1,n);
    eqilbr(a,n,em,d,ind0);
    tfmreahes(a,n,em,ind);
    k=reavalqri(a,n,em,val);
    for (i=k+1; i<=n; i++)
        for (j=i+1; j<=n; j++)
            if (val[j] > val[i]) {
                r=val[i];
                val[i]=val[j];
                val[j]=r;
            }
    free_integer_vector(ind,1);
    free_integer_vector(ind0,1);
    free_real_vector(d,1);
    return k;
}
```

B. reaeigl

Determines the eigenvalues λ_j (all assumed to be real) and corresponding eigenvectors $u^{(j)}$ ($j=1,...,n$) of a real $n \times n$ matrix A. A is equilibrated to the form $A'=PDAD^{-1}P^{-1}$ (by means of a call of *eqilbr*) and transformed to similar real upper Hessenberg form H (by means of a call of *tfmreahes*). The eigenvalues of H are then computed by single QR iteration (by means of a call of *reavalqri*). The eigenvectors of H are determined either by direct use of an iterative scheme of the form

$$(H - \lambda_j I)y_k^{(j)} = x_k^{(j)}, \qquad x_{k+1}^{(j)} = y_k^{(j)} / \|y_k^{(j)}\|_E \tag{1}$$

or by delayed application of such a scheme. If $\min|\lambda_j-\lambda_i| \le \varepsilon\|H\|$ $(j \ne i)$ for λ_i ranging over the previously determined eigenvalues, λ_j is replaced in (1) by μ_j, where $\min|\mu_j-\lambda_i|=\varepsilon\|H\|$, ε being the value of the machine precision supplied by the user. The inverse iteration scheme is terminated if (a) $\|(H-\lambda_j I)x_k^{(j)}\| \le \|H\|\tau$, where τ is a relative tolerance prescribed by the user, or (b) $k=kmax+1$, where the integer value of *kmax*, the maximum permitted number of inverse iterations, is also prescribed by the user. The above inverse iteration is performed by *reaveches*. The eigenvectors $v^{(j)}$ of the equilibrated matrix A' are then obtained from those, $x^{(j)}$, of H by back transformation (by means of a call of *bakreahes2*) and the eigenvectors $w^{(j)}$ of the original matrix A are recovered from the $v^{(j)}$ by means of a call of *baklbr*. Finally, the $w^{(j)}$ are scaled to $u^{(j)}$ by imposing the condition that the largest element of $u^{(j)}$ is 1 (by means of a call of *reascl*). The procedure *reaeigl* should be used only if all eigenvalues are real.

Function Parameters:

int reaeigl (*a,n,em,val,vec*)

reaeigl: given the value 0 provided that the process is completed within *em[4]* iterations; otherwise *reaeigl* is given the value *k*, of the number of eigenvalues and eigenvectors not calculated;

a: float *a[1:n,1:n]*;
- entry: the matrix whose eigenvalues and eigenvectors are to be calculated;
- exit: the array elements are altered;

n: int;
- entry: the order of the given matrix;

em: float *em[0:9]*;

entry:

em[0]: the machine precision (the value of ε above);

em[2]: the relative tolerance used for the QR iteration (*em[2]* > *em[0]*);

em[4]: the maximum allowed number of iterations (for example, *em[4]*=10*n*);

em[6]: the tolerance used for the eigenvectors (value of τ above; *em[6]* > *em[2]*); for each eigenvector the inverse iteration ends if the Euclidean norm of the residue vector is smaller than *em[1]*em[6]*;

em[8]: the maximum allowed number of inverse iterations for the calculation of each eigenvector (value of *kmax* above; for example, *em[8]*=5);

exit:

em[1]: the infinity norm of the equilibrated matrix;

em[3]: the maximum absolute value of the subdiagonal elements neglected;

em[5]: the number of QR iterations performed; if the iteration process is not completed within *em[4]* iterations then the value *em[4]*+1 is delivered and in this case only the last *n-k* elements of *val* and the last *n-k* columns of *vec* are approximate eigenvalues and eigenvectors of the given matrix, where *k* is delivered in *reaeig1*;

em[7]: the maximum Euclidean norm of the residues of the calculated eigenvectors of the transformed matrix;

em[9]: the largest number of inverse iterations performed for the calculation of some eigenvector; if, for some eigenvector the Euclidean norm of the residue remains larger than *em[1]*em[6]*, then the value *em[8]*+1 is delivered; nevertheless the eigenvectors may then very well be useful, this should be judged from the value delivered in *em[7]* or from some other test;

val: float *val[1:n]*;

exit: the eigenvalues of the given matrix are delivered in monotonically decreasing order;

vec: float *vec[1:n,1:n]*;

exit: the calculated eigenvectors corresponding to the eigenvalues in *val[1:n]* are delivered in the columns of *vec*.

Functions used: eqilbr, tfmreahes, bakreahes2, baklbr, reavalqri, reaveches, reascl.

```
int reaeig1(float **a, int n, float em[], float val[], float **vec)
{
    int *allocate_integer_vector(int, int);
    float *allocate_real_vector(int, int);
    float **allocate_real_matrix(int, int, int, int);
    void free_integer_vector(int *, int);
    void free_real_vector(float *, int);
    void free_real_matrix(float **, int, int, int);
    void tfmreahes(float **, int, float [], int []);
    void bakreahes2(float **, int, int, int, int [], float **);
    void eqilbr(float **, int, float [], float [], int []);
    void baklbr(int, int, int, float [], int [], float **);
    int reavalqri(float **, int, float [], float []);
    void reaveches(float **, int, float, float [], float []);
    void reascl(float **, int, int, int);
    int i,k,max,j,l,*ind,*ind0;
    float residu,r,machtol,*d,*v,**b;

    ind=allocate_integer_vector(1,n);
    ind0=allocate_integer_vector(1,n);
    d=allocate_real_vector(1,n);
    v=allocate_real_vector(1,n);
```

```
    b=allocate_real_matrix(1,n,1,n);

    residu=0.0;
    max=0;
    eqilbr(a,n,em,d,ind0);
    tfmreahes(a,n,em,ind);
    for (i=1; i<=n; i++)
        for (j=((i == 1) ? 1 : i-1); j<=n; j++) b[i][j]=a[i][j];
    k=reavalqri(b,n,em,val);
    for (i=k+1; i<=n; i++)
        for (j=i+1; j<=n; j++)
            if (val[j] > val[i]) {
                r=val[i];
                val[i]=val[j];
                val[j]=r;
            }
    machtol=em[0]*em[1];
    for (l=k+1; l<=n; l++) {
        if (l > 1)
            if (val[l-1]-val[l] < machtol) val[l]=val[l-1]-machtol;
        for (i=1; i<=n; i++)
            for (j=((i == 1) ? 1 : i-1); j<=n; j++) b[i][j]=a[i][j];
        reaveches(b,n,val[l],em,v);
        if (em[7] > residu) residu=em[7];
        if (em[9] > max) max=em[9];
        for (j=1; j<=n; j++) vec[j][l]=v[j];
    }
    em[7]=residu;
    em[9]=max;
    bakreahes2(a,n,k+1,n,ind,vec);
    baklbr(n,k+1,n,d,ind0,vec);
    reascl(vec,n,k+1,n);
    free_integer_vector(ind,1);
    free_integer_vector(ind0,1);
    free_real_vector(d,1);
    free_real_vector(v,1);
    free_real_matrix(b,1,n,1);
    return k;
}
```

C. reaeig3

Determines all eigenvalues λ_j (assumed to be real) and corresponding eigenvectors $u^{(j)}$ (j=1,...,n) of a real $n \times n$ matrix A by equilibration to the form $A'=PDAD^{-1}P^{-1}$ (calling *eqilbr*), transformation to similar real upper Hessenberg form H (calling *tfmreahes*), computation of the eigenvalues of H by QR iteration and direct determination of the eigenvectors of H (calling *reaqri*); if all eigenvalues of H have been determined, there follow back transformation from the eigenvectors of H to the eigenvectors of A' (calling *bakreahes2*) further back transformation from the eigenvectors of A' to those of A (calling *baklbr*) and finally scaling of the eigenvectors of A in such a way that the element of maximum size in each eigenvector is 1 (calling *reascl*). The procedure *reaeig3* should be used only if all eigenvalues are real.

Function Parameters:

int reaeig3 (*a,n,em,val,vec*)

reaeig3: given the value 0 provided that the process is completed within *em[4]* iterations; otherwise *reaeig3* is given the value *k*, of the number of eigenvalues not calculated;

a: float *a[1:n,1:n]*;
entry: the matrix whose eigenvalues and eigenvectors are to be calculated;
exit: the array elements are altered;

n: int;
entry: the order of the given matrix;
em: float *em[0:5]*;
entry:
em[0]: the machine precision;
em[2]: the relative tolerance used for the QR iteration (*em[2]* > *em[0]*);
em[4]: maximum allowed number of QR iterations (for example, *em[4]*=10*n*);
exit:
em[1]: the infinity norm of the equilibrated matrix;
em[3]: the maximum absolute value of the subdiagonal elements neglected;
em[5]: the number of QR iterations performed; if the iteration process is not completed within *em[4]* iterations then the value *em[4]*+1 is delivered and in this case only the last *n-k* elements of *val* are approximate eigenvalues of the given matrix and no useful eigenvectors are delivered, where *k* is delivered in *reaeig3*;
val: float *val[1:n]*;
exit: the eigenvalues of the given matrix are delivered;
vec: float *vec[1:n,1:n]*;
exit: the calculated eigenvectors corresponding to the eigenvalues in *val[1:n]* are delivered in the columns of *vec*.

Functions used: eqilbr, tfmreahes, bakreahes2, baklbr, reaqri, reascl.

```
int reaeig3(float **a, int n, float em[], float val[], float **vec)
{
    int *allocate_integer_vector(int, int);
    float *allocate_real_vector(int, int);
    void free_integer_vector(int *, int);
    void free_real_vector(float *, int);
    void tfmreahes(float **, int, float [], int []);
    void bakreahes2(float **, int, int, int, int [], float **);
    void eqilbr(float **, int, float [], float [], int []);
    void baklbr(int, int, int, float [], int [], float **);
    void reascl(float **, int, int, int);
    int reaqri(float **, int, float [], float [], float **);
    int i,*ind,*ind0;
    float *d;

    ind=allocate_integer_vector(1,n);
    ind0=allocate_integer_vector(1,n);
    d=allocate_real_vector(1,n);
    eqilbr(a,n,em,d,ind0);
    tfmreahes(a,n,em,ind);
    i=reaqri(a,n,em,val,vec);
    if (i == 0) {
        bakreahes2(a,n,1,n,ind,vec);
        baklbr(n,1,n,d,ind0,vec);
        reascl(vec,n,1,n);
    }
    free_integer_vector(ind,1);
    free_integer_vector(ind0,1);
    free_real_vector(d,1);
    return i;
}
```

D. comeigval

Determines the real and complex eigenvalues λ_i (i=1,...,*n*) of a real *n*×*n* matrix *A* by equilibration to the form $A'=PDAD^{-1}P^{-1}$ (calling *eqilbr*) transformation to similar real upper

Hessenberg form H (calling *tfmreahes*) and computation of the eigenvalues of H by double QR iteration (calling *comvalqri*).

Function Parameters:

int **comeigval** (*a,n,em,re,im*)

comeigval: given the value 0 provided that the process is completed within *em[4]* iterations; otherwise *comeigval* is given the value *k*, of the number of eigenvalues not calculated;

a: float *a[1:n,1:n]*;
entry: the matrix whose eigenvalues are to be calculated;
exit: the array elements are altered;

n: int;
entry: the order of the given matrix;

em: float *em[0:5]*;
entry:
em[0]: the machine precision;
em[2]: the relative tolerance used for the QR iteration (*em[2]* > *em[0]*);
em[4]: the maximum allowed number of iterations (for example, *em[4]*=10*n*);
exit:
em[1]: the infinity norm of the equilibrated matrix;
em[3]: the maximum absolute value of the subdiagonal elements neglected;
em[5]: the number of QR iterations performed; if the iteration process is not completed within *em[4]* iterations then the value *em[4]*+1 is delivered and in this case only the last *n-k* elements of *re* and *im* are approximate eigenvalues of the given matrix, where *k* is delivered in *comeigval*;

re,im: float *re[1:n], im[1:n]*;
exit: the real and imaginary parts of the calculated eigenvalues of the given matrix are delivered in *re, im[1:n]*, the members of each nonreal complex conjugate pair being consecutive.

Functions used: eqilbr, tfmreahes, comvalqri.

```
int comeigval(float **a, int n, float em[], float re[], float im[])
{
    int *allocate_integer_vector(int, int);
    float *allocate_real_vector(int, int);
    void free_integer_vector(int *, int);
    void free_real_vector(float *, int);
    void eqilbr(float **, int, float [], float [], int []);
    void tfmreahes(float **, int, float [], int []);
    int comvalqri(float **, int, float [], float [], float []);
    int i,*ind,*ind0;
    float *d;

    ind=allocate_integer_vector(1,n);
    ind0=allocate_integer_vector(1,n);
    d=allocate_real_vector(1,n);
    eqilbr(a,n,em,d,ind0);
    tfmreahes(a,n,em,ind);
    i=comvalqri(a,n,em,re,im);
    free_integer_vector(ind,1);
    free_integer_vector(ind0,1);
    free_real_vector(d,1);
    return i;
```

E. comeigl

Determines the real and complex eigenvalues λ_j and corresponding eigenvectors $u^{(j)}$ (j=1,...,n) of a real $n \times n$ matrix A. A is equilibrated to the form $A'=PDAD^{-1}P^{-1}$ (by means of a call of *eqilbr*) and transformed to similar real upper Hessenberg form H (by means of a call of *tfmreahes*). The eigenvalues of H are then computed by double QR iteration (by means of a call of *comvalqri*). The real eigenvectors and complex eigenvectors of H are determined either by direct use of an iterative scheme of the form

$$(H - \lambda_j I)y_k^{(j)} = x_k^{(j)}, \qquad x_{k+1}^{(j)} = y_k^{(j)} / \|y_k^{(j)}\|_E \tag{1}$$

or by delayed application of such a scheme. If $\min|\lambda_j-\lambda_i| \leq \varepsilon\|H\|$ $(j \neq i)$ for λ_i ranging over the previously determined eigenvalues, λ_j is replaced in (1) by μ_j, where $\min|\mu_j-\lambda_i|=\varepsilon\|H\|$, ε being the value of the machine precision supplied by the user. The inverse iteration scheme is terminated if (a) $\|(H-\lambda_j I)x_k^{(j)}\| \leq \|H\|\tau$, where τ is a relative tolerance prescribed by the user, or (b) $k=kmax+1$, where the integer value of *kmax*, the maximum permitted number of inverse iterations, is also prescribed by the user. The above inverse iteration is performed, when λ_j is real, by *reaveches*, and when λ_j is complex, by *comveches*. The eigenvectors $v^{(j)}$ of the equilibrated matrix A' are then obtained from those, $x^{(j)}$, of H by back transformation (by means of a call of *bakreahes2*) and the eigenvectors $w^{(j)}$ of the original matrix A are recovered from the $v^{(j)}$ by means of a call of *baklbr*. Finally, the $w^{(j)}$ are scaled to $u^{(j)}$ by imposing the condition that the largest element of $u^{(j)}$ is 1 (by means of a call of *comscl*).

Function Parameters:

int comeigl (*a,n,em,re,im,vec*)

comeigl: given the value 0 provided that the process is completed within *em[4]* iterations; otherwise *comeigl* is given the value *k*, of the number of eigenvalues and eigenvectors not calculated;

a: float *a[1:n,1:n]*;
- entry: the matrix whose eigenvalues and eigenvectors are to be calculated;
- exit: the array elements are altered;

n: int;
- entry: the order of the given matrix;

em: float *em[0:9]*;
- entry:
  - *em[0]*: the machine precision (the value of ε above);
  - *em[2]*: the relative tolerance used for the QR iteration (*em[2] > em[0]*);
  - *em[4]*: the maximum allowed number of iterations (for example, *em[4]*=10*n*);
  - *em[6]*: the tolerance used for the eigenvectors (value of τ above; *em[6] > em[2]*); for each eigenvector the inverse iteration ends if the Euclidean norm of the residue vector is smaller than *em[1]*em[6]*;
  - *em[8]*: the maximum allowed number of inverse iterations for the calculation of each eigenvector (value of *kmax* above; for example, *em[8]*=5);
- exit:
  - *em[1]*: the infinity norm of the equilibrated matrix;
  - *em[3]*: the maximum absolute value of the subdiagonal elements neglected;
  - *em[5]*: the number of QR iterations performed; if the iteration process is not

completed within *em[4]* iterations then the value *em[4]*+1 is delivered and in this case only the last *n-k* elements of *re, im* and columns of *vec* are approximate eigenvalues and eigenvectors of the given matrix, where *k* is delivered in *comeig1*;

em[7]: the maximum Euclidean norm of the residues of the calculated eigenvectors of the transformed matrix;

em[9]: the largest number of inverse iterations performed for the calculation of some eigenvector; if the Euclidean norm of the residue for one or more eigenvectors remains larger than *em[1]*em[6]*, then the value *em[8]*+1 is delivered; nevertheless the eigenvectors may then very well be useful, this should be judged from the value delivered in *em[7]* or from some other test;

re,im: float *re[1:n], im[1:n]*;

exit: the real and imaginary parts of the calculated eigenvalues of the given matrix are delivered in arrays *re[1:n]* and *im[1:n]*, the members of each nonreal complex conjugate pair being consecutive;

vec: float *vec[1:n,1:n]*;

exit: the calculated eigenvectors are delivered in the columns of *vec*; an eigenvector corresponding to a real eigenvalue given in array *re* is delivered in the corresponding column of array *vec*; the real and imaginary part of an eigenvector corresponding to the first member of a nonreal complex conjugate pair of eigenvalues given in the arrays *re, im* are delivered in the two consecutive columns of array *vec* corresponding to this pair (the eigenvectors corresponding to the second members of nonreal complex conjugate pairs are not delivered, since they are simply the complex conjugate of those corresponding to the first member of such pairs).

Functions used: eqilbr, tfmreahes, bakreahes2, baklbr, reaveches, comvalqri, comveches, comscl.

```
int comeig1(float **a, int n, float em[], float re[], float im[],
            float **vec)
{
    int *allocate_integer_vector(int, int);
    float *allocate_real_vector(int, int);
    float **allocate_real_matrix(int, int, int, int);
    void free_integer_vector(int *, int);
    void free_real_vector(float *, int);
    void free_real_matrix(float **, int, int, int);
    void eqilbr(float **, int, float [], float [], int []);
    void tfmreahes(float **, int, float [], int []);
    void bakreahes2(float **, int, int, int, int [], float **);
    void baklbr(int, int, int, float [], int [], float **);
    void reaveches(float **, int, float, float [], float []);
    void comscl(float **, int, int, int, float []);
    int comvalqri(float **, int, float [], float [], float []);
    void comveches(float **, int, float, float,
                        float [], float [], float []);
    int i,j,k,pj,itt,again,*ind,*ind0;
    float x,y,max,neps,**ab,*d,*u,*v,temp1,temp2;

    ind=allocate_integer_vector(1,n);
    ind0=allocate_integer_vector(1,n);
    d=allocate_real_vector(1,n);
    u=allocate_real_vector(1,n);
    v=allocate_real_vector(1,n);
    ab=allocate_real_matrix(1,n,1,n);

    eqilbr(a,n,em,d,ind0);
    tfmreahes(a,n,em,ind);
```

```
    for (i=1; i<=n; i++)
        for (j=((i == 1) ? 1 : i-1); j<=n; j++) ab[i][j]=a[i][j];
    k=comvalqri(ab,n,em,re,im);
    neps=em[0]*em[1];
    max=0.0;
    itt=0;
    for (i=k+1; i<=n; i++) {
        x=re[i];
        y=im[i];
        pj=0;
        again=1;
        do {
            for (j=k+1; j<=i-1; j++) {
                temp1=x-re[j];
                temp2=y-im[j];
                if (temp1*temp1+temp2*temp2 <= neps*neps) {
                    if (pj == j)
                        neps=em[2]*em[1];
                    else
                        pj=j;
                    x += 2.0*neps;
                    again = (!again);
                    break;
                }
            }
            again = (!again);
        } while (again);
        re[i]=x;
        for (i=1; i<=n; i++)
            for (j=((i == 1) ? 1 : i-1); j<=n; j++) ab[i][j]=a[i][j];
        if (y != 0.0) {
            comveches(ab,n,re[i],im[i],em,u,v);
            for (j=1; j<=n; j++) vec[j][i]=u[j];
            i++;
            re[i]=x;
        } else
            reaveches(ab,n,x,em,v);
        for (j=1; j<=n; j++) vec[j][i]=v[j];
        if (em[7] > max) max=em[7];
        if (itt < em[9]) itt=em[9];
    }
    em[7]=max;
    em[9]=itt;
    bakreahes2(a,n,k+1,n,ind,vec);
    baklbr(n,k+1,n,d,ind0,vec);
    comscl(vec,n,k+1,n,im);
    free_integer_vector(ind,1);
    free_integer_vector(ind0,1);
    free_real_vector(d,1);
    free_real_vector(u,1);
    free_real_vector(v,1);
    free_real_matrix(ab,1,n,1);
    return k;
}
```

3.13.8 Complex Hermitian matrices

A. eigvalhrm

Determines the n' largest eigenvalues λ_j of the $n \times n$ Hermitian matrix A. A is first reduced by a similarity transformation to real tridiagonal form T by calling *hshhrmtrival*. The n' largest eigenvalues λ_j of T are then determined by means of a call of *valsymtri*.

Function Parameters:

void eigvalhrm (*a,n,numval,val,em*)

a: float *a[1:n,1:n]*;

entry: the real part of the upper triangle of the Hermitian matrix must be given in the upper triangular part of *a* (the elements *a[i,j]*, $i \leq j$); the imaginary part of the strict lower triangle of the Hermitian matrix must be given in the strict lower part of *a* (the elements *a[i,j]*, $i>j$);

exit: the array elements are altered;

n: int; entry: the order of the given matrix;

numval: int;

entry: *eigvalhrm* calculates the largest *numval* eigenvalues of the Hermitian matrix (the value of *n'* above);

val: float *val[1:numval]*;

exit: in array *val* the largest *numval* eigenvalues are delivered in monotonically nonincreasing order;

em: float *em[0:3]*;

entry:

em[0]: the machine precision (value of ε above);

em[2]: the relative tolerance used for the eigenvalues (value of μ above); more precisely, the tolerance for each eigenvalue λ is $|\lambda|$*em[2]+em[1]*em[0]*;

exit:

em[1]: an estimate of a norm of the original matrix;

em[3]: the number of iterations performed.

Functions used: hshhrmtrival, valsymtri.

```
void eigvalhrm(float **a, int n, int numval, float val[], float em[])
{
	float *allocate_real_vector(int, int);
	void free_real_vector(float *, int);
	void hshhrmtrival(float **, int, float [], float [], float []);
	void valsymtri(float [], float [], int, int, int,
						float [], float []);
	float *d,*bb;

	d=allocate_real_vector(1,n);
	bb=allocate_real_vector(1,n-1);
	hshhrmtrival(a,n,d,bb,em);
	valsymtri(d,bb,n,1,numval,val,em);
	free_real_vector(d,1);
	free_real_vector(bb,1);
}
```

B. eighrm

Determines the n' largest eigenvalues λ_j (j=1,...,n') and corresponding eigenvectors $u^{(j)}$ (j=1,...,n') of the $n \times n$ Hermitian matrix *A*. *A* is first reduced to similar real tridiagonal form *T* by a call of *hshhrmtri*. The required eigenvalues of this tridiagonal matrix are then obtained by a call of *valsymtri*. The corresponding eigenvectors of *T* are determined by inverse iteration and (possibly) Gram-Schmidt orthogonalization by a call of *vecsymtri*; the required eigenvectors of *A* are then recovered from those of *T* by means of a call of *bakhrmtri*.

Function Parameters:

void eighrm (*a,n,numval,val,vecr,veci,em*)

a: float *a[1:n,1:n]*;

entry: the real part of the upper triangle of the Hermitian matrix must be given in the upper triangular part of *a* (the elements *a[i,j]*, $i \leq j$); the imaginary part of the strict lower triangle of the Hermitian matrix must be given in the strict lower part of *a* (the elements *a[i,j]*, $i>j$);
exit: the array elements are altered;

n: int; entry: the order of the given matrix;

numval: int;
entry: *eighrm* calculates the largest *numval* eigenvalues of the Hermitian matrix (the value of *n'* above);

val: float *val[1:numval]*;
exit: in array *val* the largest *numval* eigenvalues are delivered in monotonically nonincreasing order;

vecr,veci: float *vecr[1:n,1:numval]*, *veci[1:n,1:numval]*;
exit: the calculated eigenvectors; the complex eigenvector with real part *vecr[1:n,i]* and the imaginary part *veci[1:n,i]* corresponds to the eigenvalue *val[i]*, *i=1,...,numval*;

em: float *em[0:9]*;
entry:
em[0]: the machine precision;
em[2]: the relative tolerance used for the eigenvalues; more precisely, the tolerance for each eigenvalue λ is $|\lambda|$**em[2]*+*em[1]***em[0]*;
em[4]: the orthogonalization parameter (for example, *em[4]*=0.01);
em[6]: the tolerance for the eigenvectors;
em[8]: the maximum number of inverse iterations allowed for the calculation of each eigenvector;
exit:
em[1]: an estimate of a norm of the original matrix;
em[3]: the number of iterations performed;
em[5]: the number of eigenvectors involved in the last Gram-Schmidt orthogonalization;
em[7]: the maximum Euclidean norm of the residues of the calculated eigenvectors;
em[9]: the largest number of inverse iterations performed for the calculation of some eigenvector; if, however, for some calculated eigenvector, the Euclidean norm of the residues remains greater than *em[1]***em[6]*, then *em[9]*=*em[8]*+1.

Functions used: hshhrmtri, valsymtri, vecsymtri, bakhrmtri.

```
void eighrm(float **a, int n, int numval, float val[], float **vecr,
            float **veci, float em[])
{
   float *allocate_real_vector(int, int);
   void free_real_vector(float *, int);
   void hshhrmtri(float **, int, float [], float [], float [],
                     float [], float [], float []);
   void valsymtri(float [], float [], int, int, int,
                     float [], float []);
   void vecsymtri(float [], float [], int, int, int,
                     float [], float **, float []);
   void bakhrmtri(float **, int, int, int, float **,
                     float **, float [], float []);
   float *bb,*tr,*ti,*d,*b;
```

```
    bb=allocate_real_vector(1,n-1);
    tr=allocate_real_vector(1,n-1);
    ti=allocate_real_vector(1,n-1);
    d=allocate_real_vector(1,n);
    b=allocate_real_vector(1,n);
    hshhrmtri(a,n,d,b,bb,em,tr,ti);
    valsymtri(d,bb,n,1,numval,val,em);
    b[n]=0.0;
    vecsymtri(d,b,n,1,numval,val,vecr,em);
    bakhrmtri(a,n,1,numval,vecr,veci,tr,ti);
    free_real_vector(bb,1);
    free_real_vector(tr,1);
    free_real_vector(ti,1);
    free_real_vector(d,1);
    free_real_vector(b,1);
}
```

C. qrivalhrm

Determines all eigenvalues λ_j of the $n \times n$ Hermitian matrix A. A is first reduced by a similarity transformation to real tridiagonal form T by calling *hshhrmtrival*. The eigenvalues λ_j of T are then determined QR iteration using a call of *qrivalsymtri*.

Function Parameters:

int qrivalhrm *(a,n,val,em)*

qrivalhrm: given the value 0 provided the QR iteration is completed within *em[4]* iterations; otherwise, *qrivalhrm* is given the number of eigenvalues, *k*, not calculated and only the last *n-k* elements of *val* are approximate eigenvalues of the original Hermitian matrix;

a: float *a[1:n,1:n]*;
entry: the real part of the upper triangle of the Hermitian matrix must be given in the upper triangular part of *a* (the elements *a[i,j]*, $i \leq j$); the imaginary part of the strict lower triangle of the Hermitian matrix must be given in the strict lower part of *a* (the elements *a[i,j]*, $i > j$);
exit: the array elements are altered;

n: int;
entry: the order of the given matrix;

val: float *val[1:n]*;
exit: the calculated eigenvalues;

em: float *em[0:5]*;
entry:
em[0]: the machine precision;
em[2]: the relative tolerance used for the QR iteration;
em[4]: the maximum allowed number of iterations;
exit:
em[1]: an estimate of a norm of the original matrix;
em[3]: the maximum absolute value of the codiagonal elements neglected;
em[5]: number of iterations performed; *em[5]*=*em[4]*+1 when *qrivalhrm*≠0.

Functions used: hshhrmtrival, qrivalsymtri.

```
int qrivalhrm(float **a, int n, float val[], float em[])
{
    float *allocate_real_vector(int, int);
    void free_real_vector(float *, int);
```

```
    void hshhrmtrival(float **, int, float [], float [], float []);
    int qrivalsymtri(float [], float [], int, float []);
    int i;
    float *bb;

    bb=allocate_real_vector(1,n);
    hshhrmtrival(a,n,val,bb,em);
    bb[n]=0.0;
    i=qrivalsymtri(val,bb,n,em);
    free_real_vector(bb,1);
    return i;
}
```

D. qrihrm

Determines all eigenvalues λ_j and corresponding eigenvectors $u^{(j)}$ of the $n \times n$ Hermitian matrix A. A is first reduced to similar real tridiagonal form T by a call of *hshhrmtri*. The eigenvalues λ_j and corresponding eigenvectors $v^{(j)}$ of T are then determined QR iteration, using a call of *qrisymtri*. The eigenvectors $u^{(j)}$ of A are then obtained from the $v^{(j)}$ by means of a call of *bakhrmtri*.

Function Parameters:

int qrihrm (*a,n,val,vr,vi,em*)

qrihrm: *qrihrm*=0, provided the process is completed within *em[4]* iterations; otherwise, *qrihrm* is given the number of eigenvalues, *k*, not calculated and only the last *n-k* elements of *val* are approximate eigenvalues and the columns of the arrays *vr, vi[1:n,n-k:n]* are approximate eigenvectors of the original Hermitian matrix;

a: float *a[1:n,1:n]*;
- entry: the real part of the upper triangle of the Hermitian matrix must be given in the upper triangular part of *a* (the elements *a[i,j]*, $i \leq j$); the imaginary part of the strict lower triangle of the Hermitian matrix must be given in the strict lower part of *a* (the elements *a[i,j]*, $i > j$);
- exit: the array elements are altered;

n: int;
- entry: the order of the given matrix;

val: float *val[1:n]*;
- exit: the calculated eigenvalues;

vr,vi: float *vr[1:n,1:n]*, *vi[1:n,1:n]*;
- exit: the calculated eigenvectors; the complex eigenvector with real part *vr[1:n,i]* and the imaginary part *vi[1:n,i]* corresponds to the eigenvalue *val[i]*, $i=1,...,n$;

em: float *em[0:5]*;
- entry:
  - *em[0]*: the machine precision;
  - *em[2]*: the relative tolerance for the QR iteration;
  - *em[4]*: maximum allowed number of iterations (for example, *em[4]*=10*n*);
- exit:
  - *em[1]*: an estimate of a norm of the original matrix;
  - *em[3]*: the maximum absolute value of the codiagonal elements neglected;
  - *em[5]*: number of iterations performed; *em[5]*=*em[4]*+1 when *qrihrm*$\neq$0.

Functions used: hshhrmtri, qrisymtri, bakhrmtri.

```
int qrihrm(float **a, int n, float val[], float **vr, float **vi,
        float em[])
{
    float *allocate_real_vector(int, int);
    void free_real_vector(float *, int);
    void hshhrmtri(float **, int, float [], float [], float [],
                    float [], float [], float []);
    int qrisymtri(float **, int, float [], float [], float [], float []);
    void bakhrmtri(float **, int, int, int, float **,
                    float **, float [], float []);
    int i,j;
    float *b,*bb,*tr,*ti;

    b=allocate_real_vector(1,n);
    bb=allocate_real_vector(1,n);
    tr=allocate_real_vector(1,n-1);
    ti=allocate_real_vector(1,n-1);
    hshhrmtri(a,n,val,b,bb,em,tr,ti);
    for (i=1; i<=n; i++) {
        vr[i][i]=1.0;
        for (j=i+1; j<=n; j++) vr[i][j]=vr[j][i]=0.0;
    }
    b[n]=bb[n]=0.0;
    i=qrisymtri(vr,n,val,b,bb,em);
    bakhrmtri(a,n,i+1,n,vr,vi,tr,ti);
    free_real_vector(b,1);
    free_real_vector(bb,1);
    free_real_vector(tr,1);
    free_real_vector(ti,1);
    return i;
}
```

3.13.9 Complex upper-Hessenberg matrices

A. valqricom

Determines the eigenvalues λ_j (j=1,...,n) of an $n \times n$ complex upper Hessenberg matrix A ($A_{i,j}$=0 for i>j+1) with real subdiagonal elements $A_{j+1,j}$, by QR iteration. For further details see the documentation of the procedure *qricom*.

Function Parameters:

int valqricom (*a1,a2,b,n,em,val1,val2*)

valqricom: given the value 0 provided the process is computed within *em[4]* iterations; otherwise given the number, *k*, of eigenvalues not calculated and only the last *n-k* elements of the arrays *val1* and *val2* are approximate eigenvalues of the upper Hessenberg matrix;

a1,a2: float *a1[1:n,1:n], a2[1:n,1:n]*;
- entry: the real part and the imaginary part of the upper triangle of the upper Hessenberg matrix must be given in the corresponding parts of the arrays *a1* and *a2*;
- exit: the array elements in the upper triangle of *a1* and *a2* are altered;

b: float *b[1:n-1]*;
- entry: the subdiagonal of the upper Hessenberg matrix;
- exit: the elements of *b* are altered;

n: int;
- entry: the order of the given matrix;

em: float *em[0:5]*;
- entry:

em[0]: the machine precision;
em[1]: an estimate of the norm of the upper Hessenberg matrix (e.g. the sum of the infinity norms of the real and imaginary parts of the matrix);
em[2]: the relative tolerance for the QR iteration;
em[4]: the maximum allowed number of iterations (e.g. $10n$);
exit:
em[3]: the maximum absolute value of the subdiagonal elements neglected;
em[5]: the number of iterations performed; *em[5]*=*em[4]*+1 in the case *valqricom*≠0;

val1,val2: float *val1[1:n], val2[1:n]*;
exit: the real part and the imaginary part of the calculated eigenvalues are delivered in *val1* and *val2*, respectively.

Functions used: comkwd, rotcomrow, rotcomcol, comcolcst.

```
#include <math.h>

int valqricom(float **a1, float **a2, float b[], int n, float em[],
              float val1[], float val2[])
{
   void comcolcst(int, int, int, float **, float **, float, float);
   void rotcomcol(int, int, int, int, float **, float **,
                  float, float, float);
   void rotcomrow(int, int, int, int, float **, float **,
                  float, float, float);
   void comkwd(float, float, float, float,
               float *, float *, float *, float *);
   int nm1,i,i1,q,q1,max,count;
   float r,z1,z2,dd1,dd2,cc,g1,g2,k1,k2,hc,a1nn,a2nn,aij1,aij2,
         ai1i,kappa,nui,mui1,mui2,muim11,muim12,nuim1,tol;

   tol=em[1]*em[2];
   max=em[4];
   count=0;
   r=0.0;
   if (n > 1) hc=b[n-1];
   do {
      nm1=n-1;
      i=n;
      do {
         q=i;
         i--;
      } while ((i >= 1) ? (fabs(b[i]) > tol) : 0);
      if (q > 1)
         if (fabs(b[q-1]) > r) r=fabs(b[q-1]);
      if (q == n) {
         val1[n]=a1[n][n];
         val2[n]=a2[n][n];
         n=nm1;
         if (n > 1) hc=b[n-1];
      } else {
         dd1=a1[n][n];
         dd2=a2[n][n];
         cc=b[nm1];
         comkwd((a1[nm1][nm1]-dd1)/2.0, (a2[nm1][nm1]-dd2)/2.0,
               cc*a1[nm1][n],cc*a2[nm1][n],&g1,&g2,&k1,&k2);
         if (q == nm1) {
            val1[nm1]=g1+dd1;
            val2[nm1]=g2+dd2;
            val1[n]=k1+dd1;
            val2[n]=k2+dd2;
            n -= 2;
            if (n > 1) hc=b[n-1];
         } else {
            count++;
            if (count > max) break;
```

```
                    z1=k1+dd1;
                    z2=k2+dd2;
                    if (fabs(cc) > fabs(hc)) z1 += fabs(cc);
                    hc=cc/2.0;
                    i=q1=q+1;
                    aij1=a1[q][q]-z1;
                    aij2=a2[q][q]-z2;
                    ai1i=b[q];
                    kappa=sqrt(aij1*aij1+aij2*aij2+ai1i*ai1i);
                    mui1=aij1/kappa;
                    mui2=aij2/kappa;
                    nui=ai1i/kappa;
                    a1[q][q]=kappa;
                    a2[q][q]=0.0;
                    a1[q1][q1] -= z1;
                    a2[q1][q1] -= z2;
                    rotcomrow(q1,n,q,q1,a1,a2,mui1,mui2,nui);
                    rotcomcol(q,q,q,q1,a1,a2,mui1,-mui2,-nui);
                    a1[q][q] += z1;
                    a2[q][q] += z2;
                    for (i1=q1+1; i1<=n; i1++) {
                        aij1=a1[i][i];
                        aij2=a2[i][i];
                        ai1i=b[i];
                        kappa=sqrt(aij1*aij1+aij2*aij2+ai1i*ai1i);
                        muim11=mui1;
                        muim12=mui2;
                        nuim1=nui;
                        mui1=aij1/kappa;
                        mui2=aij2/kappa;
                        nui=ai1i/kappa;
                        a1[i1][i1] -= z1;
                        a2[i1][i1] -= z2;
                        rotcomrow(i1,n,i,i1,a1,a2,mui1,mui2,nui);
                        a1[i][i]=muim11*kappa;
                        a2[i][i] = -muim12*kappa;
                        b[i-1]=nuim1*kappa;
                        rotcomcol(q,i,i,i1,a1,a2,mui1,-mui2,-nui);
                        a1[i][i] += z1;
                        a2[i][i] += z2;
                        i=i1;
                    }
                    aij1=a1[n][n];
                    aij2=a2[n][n];
                    kappa=sqrt(aij1*aij1+aij2*aij2);
                    if ((kappa < tol) ? 1 : (aij2*aij2 <= em[0]*aij1*aij1)) {
                        b[nm1]=nui*aij1;
                        a1[n][n]=aij1*mui1+z1;
                        a2[n][n] = -aij1*mui2+z2;
                    } else {
                        b[nm1]=nui*kappa;
                        a1nn=mui1*kappa;
                        a2nn = -mui2*kappa;
                        mui1=aij1/kappa;
                        mui2=aij2/kappa;
                        comcolcst(q,nm1,n,a1,a2,mui1,mui2);
                        a1[n][n]=mui1*a1nn-mui2*a2nn+z1;
                        a2[n][n]=mui1*a2nn+mui2*a1nn+z2;
                    }
                }
            }
    } while (n > 0);
    em[3]=r;
    em[5]=count;
    return n;
}
```

B. qricom

Computes the eigenvalues λ_j and corresponding eigenvectors $v^{(j)}$ (j=1,...,n) of an $n \times n$

complex upper Hessenberg matrix A ($A_{i,j}$=0 for i>j+1) with real subdiagonal elements $A_{j+1,j}$. A is transformed into a complex upper triangular matrix U by means of the Francis QR iteration [Fr61, Wi65]. The diagonal elements of U are its eigenvalues; the eigenvectors of U are obtained by solving the associated upper triangular system, and the eigenvectors of A are recovered by back transformation.

Function Parameters:

int qricom (*a1,a2,b,n,em,val1,val2,vec1,vec2*)

qricom: given the value 0 provided the process is computed within *em[4]* iterations; otherwise given the number, *k*, of eigenvalues not calculated and only the last *n-k* elements of the arrays *val1* and *val2* are approximate eigenvalues of the upper Hessenberg matrix and no useful eigenvectors are delivered;

a1,a2: float *a1[1:n,1:n], a2[1:n,1:n]*;
- entry: the real part and the imaginary part of the upper triangle of the upper Hessenberg matrix must be given in the corresponding parts of the arrays *a1* and *a2*;
- exit: the array elements in the upper triangle of *a1* and *a2* are altered;

b: float *b[1:n-1]*;
- entry: the real subdiagonal of the upper Hessenberg matrix;
- exit: the elements of *b* are altered;

n: int;
- entry: the order of the given matrix;

em: float *em[0:5]*;
- entry:
  - *em[0]*: the machine precision;
  - *em[1]*: an estimate of the norm of the upper Hessenberg matrix (e.g. the sum of the infinity norms of the real and imaginary parts of the matrix);
  - *em[2]*: the relative tolerance for the QR iteration;
  - *em[4]*: the maximum allowed number of iterations (e.g. 10*n*);
- exit:
  - *em[3]*: the maximum absolute value of the subdiagonal elements neglected;
  - *em[5]*: the number of iterations performed; *em[5]=em[4]*+1 in the case *qricom*≠0;

val1,val2: float *val1[1:n], val2[1:n]*;
- exit: the real part and the imaginary part of the calculated eigenvalues are delivered in *val1* and *val2*, respectively.

vec1,vec2: float *vec1[1:n,1:n], vec2[1:n,1:n]*;
- exit: the eigenvectors of the upper Hessenberg matrix; the eigenvector with real part *vec1[1:n,j]* and imaginary part *vec2[1:n,j]* corresponds to the eigenvalue *val1[j]+val2[j]**i, j=1,...,*n*.

Functions used: comkwd, rotcomrow, rotcomcol, comcolcst, comrowcst, matvec, commatvec, comdiv.

```
#include <math.h>

int qricom(float **a1, float **a2, float b[], int n, float em[],
            float val1[], float val2[], float **vec1, float **vec2)
{
    float *allocate_real_vector(int, int);
    void free_real_vector(float *, int);
```

```
void comkwd(float, float, float, float,
                float *, float *, float *, float *);
void rotcomrow(int, int, int, int, float **, float **,
                float, float, float);
void rotcomcol(int, int, int, int, float **, float **,
                float, float, float);
void comcolcst(int, int, int, float **, float **, float, float);
void comrowcst(int, int, int, float **, float **, float, float);
float matvec(int, int, int, float **, float []);
void commatvec(int, int, int, float **, float **,
                float [], float [], float *, float *);
void comdiv(float, float, float, float, float *, float *);
int m,nm1,i,i1,j,q,q1,max,count;
float r,z1,z2,dd1,dd2,cc,p1,p2,t1,t2,delta1,delta2,mv1,mv2,h,h1,
        h2,g1,g2,k1,k2,hc,aij12,aij22,a1nn,a2nn,aij1,aij2,aii1,
        kappa,nui,mui1,mui2,muim11,muim12,nuim1,tol,machtol,
        *tf1,*tf2;

tf1=allocate_real_vector(1,n);
tf2=allocate_real_vector(1,n);
tol=em[1]*em[2];
machtol=em[0]*em[1];
max=em[4];
count=0;
r=0.0;
m=n;
if (n > 1) hc=b[n-1];
for (i=1; i<=n; i++) {
    vec1[i][i]=1.0;
    vec2[i][i]=0.0;
    for (j=i+1; j<=n; j++)
        vec1[i][j]=vec1[j][i]=vec2[i][j]=vec2[j][i]=0.0;
}
do {
    nm1=n-1;
    i=n;
    do {
        q=i;
        i--;
    } while ((i >= 1) ? (fabs(b[i]) > tol) : 0);
    if (q > 1)
        if (fabs(b[q-1]) > r) r=fabs(b[q-1]);
    if (q == n) {
        val1[n]=a1[n][n];
        val2[n]=a2[n][n];
        n=nm1;
        if (n > 1) hc=b[n-1];
    } else {
        dd1=a1[n][n];
        dd2=a2[n][n];
        cc=b[nm1];
        p1=(a1[nm1][nm1]-dd1)*0.5;
        p2=(a2[nm1][nm1]-dd2)*0.5;
        comkwd(p1,p2,cc*a1[nm1][n],cc*a2[nm1][n],&g1,&g2,&k1,&k2);
        if (q == nm1) {
            a1[n][n]=val1[n]=g1+dd1;
            a2[n][n]=val2[n]=g2+dd2;
            a1[q][q]=val1[q]=k1+dd1;
            a2[q][q]=val2[q]=k2+dd2;
            kappa=sqrt(k1*k1+k2*k2+cc*cc);
            nui=cc/kappa;
            mui1=k1/kappa;
            mui2=k2/kappa;
            aij1=a1[q][n];
            aij2=a2[q][n];
            h1=mui1*mui1-mui2*mui2;
            h2=2.0*mui1*mui2;
            h = -nui*2.0;
            a1[q][n]=h*(p1*mui1+p2*mui2)-nui*nui*cc+aij1*h1+aij2*h2;
            a2[q][n]=h*(p2*mui1-p1*mui2)+aij2*h1-aij1*h2;
            rotcomrow(q+2,m,q,n,a1,a2,mui1,mui2,nui);
            rotcomcol(1,q-1,q,n,a1,a2,mui1,-mui2,-nui);
            rotcomcol(1,m,q,n,vec1,vec2,mui1,-mui2,-nui);
            n -= 2;
```

```
            if (n > 1) hc=b[n-1];
            b[q]=0.0;
        } else {
            count++;
            if (count > max) {
                em[3]=r;
                em[5]=count;
                free_real_vector(tf1,1);
                free_real_vector(tf2,1);
                return n;
            }
            z1=k1+dd1;
            z2=k2+dd2;
            if (fabs(cc) > fabs(hc)) z1 += fabs(cc);
            hc=cc/2.0;
            q1=q+1;
            aij1=a1[q][q]-z1;
            aij2=a2[q][q]-z2;
            ai1i=b[q];
            kappa=sqrt(aij1*aij1+aij2*aij2+ai1i*ai1i);
            mui1=aij1/kappa;
            mui2=aij2/kappa;
            nui=ai1i/kappa;
            a1[q][q]=kappa;
            a2[q][q]=0.0;
            a1[q1][q1] -= z1;
            a2[q1][q1] -= z2;
            rotcomrow(q1,m,q,q1,a1,a2,mui1,mui2,nui);
            rotcomcol(1,q,q,q1,a1,a2,mui1,-mui2,-nui);
            a1[q][q] += z1;
            a2[q][q] += z2;
            rotcomcol(1,m,q,q1,vec1,vec2,mui1,-mui2,-nui);
            for (i=q1; i<=nm1; i++) {
                i1=i+1;
                aij1=a1[i][i];
                aij2=a2[i][i];
                ai1i=b[i];
                kappa=sqrt(aij1*aij1+aij2*aij2+ai1i*ai1i);
                muim11=mui1;
                muim12=mui2;
                nuim1=nui;
                mui1=aij1/kappa;
                mui2=aij2/kappa;
                nui=ai1i/kappa;
                a1[i1][i1] -= z1;
                a2[i1][i1] -= z2;
                rotcomrow(i1,m,i,i1,a1,a2,mui1,mui2,nui);
                a1[i][i]=muim11*kappa;
                a2[i][i] = -muim12*kappa;
                b[i-1]=nuim1*kappa;
                rotcomcol(1,i,i,i1,a1,a2,mui1,-mui2,-nui);
                a1[i][i] += z1;
                a2[i][i] += z2;
                rotcomcol(1,m,i,i1,vec1,vec2,mui1,-mui2,-nui);
            }
            aij1=a1[n][n];
            aij2=a2[n][n];
            aij12=aij1*aij1;
            aij22=aij2*aij2;
            kappa=sqrt(aij12+aij22);
            if ((kappa < tol) ? 1 : (aij22 <= em[0]*aij12)) {
                b[nm1]=nui*aij1;
                a1[n][n]=aij1*mui1+z1;
                a2[n][n] = -aij1*mui2+z2;
            } else {
                b[nm1]=nui*kappa;
                a1nn=mui1*kappa;
                a2nn = -mui2*kappa;
                mui1=aij1/kappa;
                mui2=aij2/kappa;
                comcolcst(1,nm1,n,a1,a2,mui1,mui2);
                comcolcst(1,nm1,n,vec1,vec2,mui1,mui2);
                comrowcst(n+1,m,n,a1,a2,mui1,-mui2);
                comcolcst(n,m,n,vec1,vec2,mui1,mui2);
```

```
                    a1[n][n]=mui1*a1nn-mui2*a2nn+z1;
                    a2[n][n]=mui1*a2nn+mui2*a1nn+z2;
                }
            }
        }
    } while (n > 0);
    for (j=m; j>=2; j--) {
        tf1[j]=1.0;
        tf2[j]=0.0;
        t1=a1[j][j];
        t2=a2[j][j];
        for (i=j-1; i>=1; i--) {
            delta1=t1-a1[i][i];
            delta2=t2-a2[i][i];
            commatvec(i+1,j,i,a1,a2,tf1,tf2,&mv1,&mv2);
            if (fabs(delta1) < machtol && fabs(delta2) < machtol) {
                tf1[i]=mv1/machtol;
                tf2[i]=mv2/machtol;
            } else
                comdiv(mv1,mv2,delta1,delta2,&tf1[i],&tf2[i]);
        }
        for (i=1; i<=m; i++)
            commatvec(1,j,i,vec1,vec2,tf1,tf2,&vec1[i][j],&vec2[i][j]);
    }
    em[3]=r;
    em[5]=count;
    free_real_vector(tf1,1);
    free_real_vector(tf2,1);
    return n;
}
```

3.13.10 Complex full matrices

A. eigvalcom

Computes the eigenvalues λ_j (j=1,...,n) of an n×n complex matrix A. A is first transformed to equilibrated form A' (by means of a call of *eqilbrcom*); A' is then transformed to complex upper Hessenberg form H with real subdiagonal elements (by means of a call of *hshcomhes*). The eigenvalues λ_j (j=1,...,n) of H are then determined by QR iteration (by means of a call of *valqricom*).

Function Parameters:

int eigvalcom (*ar,ai,n,em,valr,vali*)

eigvalcom: given the value 0 provided the process is computed within *em[4]* iterations; otherwise given the number, *k*, of eigenvalues not calculated and only the last *n-k* elements of the arrays *valr* and *vali* are approximate eigenvalues of the original matrix;

ar,ai: float *ar[1:n,1:n], ai[1:n,1:n]*;
- entry: the real part and the imaginary part of the matrix must be given in the arrays *ar* and *ai*, respectively;
- exit: the array elements of *ar* and *ai* are altered;

n: int;
- entry: the order of the given matrix;

em: float *em[0:7]*;
- entry:
- *em[0]*: the machine precision;
- *em[2]*: the relative tolerance for the QR iteration;
- *em[4]*: the maximum allowed number of QR iterations (e.g. 10*n*);

em[6]: the maximum allowed number of iterations for equilibrating the original matrix (e.g. *em[6]*=$n*n/2$);

exit:

em[1]: the Euclidean norm of the equilibrated matrix;

em[3]: the maximum absolute value of the subdiagonal elements neglected in the QR iteration;

em[5]: the number of QR iterations performed; *em[5]*=*em[4]*+1 in the case *eigvalcom*≠0;

em[7]: the number of iterations performed for equilibrating the original matrix;

valr,vali: float *valr[1:n], vali[1:n]*;

exit: the real part and the imaginary part of the calculated eigenvalues are delivered in *valr* and *vali*, respectively.

Functions used: eqilbrcom, comeucnrm, hshcomhes, valqricom.

```
int eigvalcom(float **ar, float **ai, int n, float em[],
              float valr[], float vali[])
{
    int *allocate_integer_vector(int, int);
    float *allocate_real_vector(int, int);
    void free_integer_vector(int *, int);
    void free_real_vector(float *, int);
    void hshcomhes(float **, float **, int, float [], float [],
                    float [], float [], float []);
    float comeucnrm(float **, float **, int, int);
    void eqilbrcom(float **, float **, int, float [], float [], int []);
    int valqricom(float **, float **, float [], int, float [],
                    float [], float []);
    int i,*ind;
    float *d,*b,*del,*tr,*ti;

    ind=allocate_integer_vector(1,n);
    d=allocate_real_vector(1,n);
    b=allocate_real_vector(1,n);
    del=allocate_real_vector(1,n);
    tr=allocate_real_vector(1,n);
    ti=allocate_real_vector(1,n);
    eqilbrcom(ar,ai,n,em,d,ind);
    em[1]=comeucnrm(ar,ai,n-1,n);
    hshcomhes(ar,ai,n,em,b,tr,ti,del);
    i=valqricom(ar,ai,b,n,em,valr,vali);
    free_integer_vector(ind,1);
    free_real_vector(d,1);
    free_real_vector(b,1);
    free_real_vector(del,1);
    free_real_vector(tr,1);
    free_real_vector(ti,1);
    return i;
}
```

B. eigcom

Computes the eigenvalues λ_j and eigenvectors $u^{(j)}$ (j=1,...,n) of an $n \times n$ complex matrix A. A is first transformed to equilibrated form A' (by means of a call of *eqilbrcom*); A' is then transformed to complex upper Hessenberg form H with real subdiagonal elements (by means of a call of *hshcomhes*). The eigenvalues λ_j and eigenvectors $v^{(j)}$ (j=1,...,n) of H are then determined by QR iteration (by means of a call of *qricom*), and the $u^{(j)}$ are then recovered from the $v^{(j)}$ by two successive back transformation (by means of calls of *bakcomhes* and *baklbrcom*).

Function Parameters:

int eigcom (*ar,ai,n,em,valr,vali,vr,vi*)

eigcom: given the value 0 provided the process is computed within *em[4]* iterations; otherwise given the number, *k*, of eigenvalues not calculated and only the last *n-k* elements of the arrays *valr* and *vali* are approximate eigenvalues of the original matrix and no useful eigenvectors are delivered;

ar,ai: float *ar[1:n,1:n], ai[1:n,1:n]*;
- entry: the real part and the imaginary part of the matrix must be given in the arrays *ar* and *ai*, respectively;
- exit: the array elements of *ar* and *ai* are altered;

n: int;
- entry: the order of the given matrix;

em: float *em[0:7]*;
- entry:
  - *em[0]*: the machine precision;
  - *em[2]*: the relative tolerance for the QR iteration;
  - *em[4]*: the maximum allowed number of QR iterations (e.g. 10*n*);
  - *em[6]*: the maximum allowed number of iterations for equilibrating the original matrix (e.g. *em[6]=n*n*/2);
- exit:
  - *em[1]*: the Euclidean norm of the equilibrated matrix;
  - *em[3]*: the maximum absolute value of the subdiagonal elements neglected in the QR iteration;
  - *em[5]*: the number of QR iterations performed; *em[5]=em[4]*+1 in the case *eigcom*≠0;
  - *em[7]*: the number of iterations performed for equilibrating the original matrix;

valr,vali: float *valr[1:n], vali[1:n]*;
- exit: the real part and the imaginary part of the calculated eigenvalues are delivered in *valr* and *vali*, respectively;

vr,vi: float *vr[1:n,1:n], vi[1:n,1:n]*;
- exit: the eigenvectors of the matrix; the normalized eigenvector with real part *vr[1:n,j]* and imaginary part *vi[1:n,j]* corresponds to the eigenvalue *valr[j]+vali[j]**i, j=1,...,*n*.

Functions used: eqilbrcom, comeucnrm, hshcomhes, qricom, bakcomhes, baklbrcom, sclcom.

```
int eigcom(float **ar, float **ai, int n, float em[], float valr[],
           float vali[], float **vr, float **vi)
{
   int *allocate_integer_vector(int, int);
   float *allocate_real_vector(int, int);
   void free_integer_vector(int *, int);
   void free_real_vector(float *, int);
   void eqilbrcom(float **, float **, int, float [], float [], int []);
   float comeucnrm(float **, float **, int, int);
   void hshcomhes(float **, float **, int, float [], float [],
                  float [], float [], float []);
   int qricom(float **, float **, float [], int, float [],
              float [], float [], float **, float **);
   void bakcomhes(float **, float **, float [], float [], float [],
                  float **, float **, int, int, int);
   void baklbrcom(int, int, int, float [], int [], float **, float **);
```

```
    void sclcom(float **, float **, int, int, int);
    int i,*ind;
    float *d,*b,*del,*tr,*ti;

    ind=allocate_integer_vector(1,n);
    d=allocate_real_vector(1,n);
    b=allocate_real_vector(1,n);
    del=allocate_real_vector(1,n);
    tr=allocate_real_vector(1,n);
    ti=allocate_real_vector(1,n);
    eqilbrcom(ar,ai,n,em,d,ind);
    em[1]=comeucnrm(ar,ai,n-1,n);
    hshcomhes(ar,ai,n,em,b,tr,ti,del);
    i=qricom(ar,ai,b,n,em,valr,vali,vr,vi);
    if (i == 0) {
        bakcomhes(ar,ai,tr,ti,del,vr,vi,n,1,n);
        baklbrcom(n,1,n,d,ind,vr,vi);
        sclcom(vr,vi,n,1,n);
    }
    free_integer_vector(ind,1);
    free_real_vector(d,1);
    free_real_vector(b,1);
    free_real_vector(del,1);
    free_real_vector(tr,1);
    free_real_vector(ti,1);
    return i;
}
```

3.14 The generalized eigenvalue problem

3.14.1 Real asymmetric matrices

A. qzival

Solves the generalized matrix eigenvalue problem $Ax=\lambda Bx$ by means of QZ iteration [MS71]. Given two $n{\times}n$ matrices A and B, *qzival* determines complex $\alpha^{(j)}$ and real $\beta^{(j)}$ such that $\beta^{(j)}A-\alpha^{(j)}B$ is singular (j=1,...,n).

QZ iteration may fail to converge in the determination of a sequence $\alpha^{(i)}$, $\beta^{(i)}$ (i=1,...,m). Such failure is signalled at return from *qzival* by the allocation of the value -1 to the array *iter[i]*, (i=1,...,m) and allocation of a nonnegative value, that of the number of iterations required in each case, to the array *iter[j]* (j=m+1,...,n) to signal success in the determination of the remaining pairs of numbers $\alpha^{(j)}$, $\beta^{(j)}$. In particular, if *iter[1]* contains the value 0 at exit, then the computations have been completely successful.

If QZ iteration is completely successful (i.e. m=0 above), A and B are both reduced by unitary transformations to quasi-upper triangular form U and upper triangular form V respectively: U has 1×1 or 2×2 blocks on the principal diagonal, but $U_{i,j}=0$ for i>j+1 and $U_{i+1,i}=0$ for those elements not belonging to the blocks; $V_{i,j}=0$ for i>j.

The sets $\alpha^{(j)}$, $\beta^{(j)}$ (j=m+1,...,n) may contain complex conjugate pairs in the sense that for some $k\geq m+1$, $\alpha^{(k)}/\beta^{(k)}$ = conjugate of $\{\alpha^{(k+1)}/\beta^{(k+1)}\}$; the components of such pairs are stored consecutively in the output arrays *alfr, alfi* and *beta*.

Function Parameters:

void qzival (*n,a,b,alfr,alfi,beta,iter,em*)

n: int;
entry: the number of rows and columns of the matrices *a* and *b*;

a: float *a[1:n,1:n]*;
entry: the given matrix;

exit: a quasi upper triangular matrix (value of U above);

b: float *b[1:n,1:n]*;
entry: the given matrix;
exit: an upper triangular matrix (value of V above);

alfr: float *alfr[1:n]*;
exit: the real part of $\alpha^{(j)}$ given in *alfr[j]*, j=*m*+1,...,*n*, in the above;

alfi: float *alfi[1:n]*;
exit: the imaginary part of $\alpha^{(j)}$ given in *alfi[j]*, j=*m*+1,...,*n*, in the above;

beta: float *beta[1:n]*;
exit: the value of $\beta^{(j)}$ given in *beta[j]*, j=*m*+1,...,*n*, in the above;

iter: int *iter[1:n]*;
exit: trouble indicator and iteration counter, see above; if *iter[1]*=0 then no trouble is signalized;

em: float *em[0:1]*;
entry:
em[0]: the smallest positive machine number;
em[1]: the relative precision of elements of *a* and *b*.

Functions used: elmcol, hshdecmul, hestgl2, hsh2col, hsh3col, hsh2row2, hsh3row2, chsh2, hshvecmat, hshvectam.

```
#include <math.h>

void qzival(int n, float **a, float **b, float alfr[], float alfi[],
            float beta[], int iter[], float em[])
{
    void elmcol(int, int, int, int, float **, float **, float);
    void hshdecmul(int, float **, float **, float);
    void hestgl2(int, float **, float **);
    void hsh2row2(int, int, int, int, float, float,
                  float **, float **);
    void hsh3row2(int, int, int, float, float, float,
                  float **, float **);
    void hsh2col(int, int, int, int, float, float, float **, float **);
    void hsh3col(int, int, int, int, float, float, float,
                float **, float **);
    void chsh2(float, float, float, float, float *, float *, float *);
    void hshvecmat(int, int, int, int, float, float [], float **);
    void hshvectam(int, int, int, int, float, float [], float **);
    int i,q,m,m1,q1,j,k,k1,k2,k3,km1,stationary,goon,l,out;
    float dwarf,eps,epsa,epsb,
          anorm,bnorm,ani,bni,constt,a10,a20,a30,b11,b22,b33,b44,a11,
          a12,a21,a22,a33,a34,a43,a44,b12,b34,old1,old2,
          an,bn,e,c,d,er,ei,a11r,a11i,a12r,a12i,a21r,a21i,a22r,a22i,
          cz,szr,szi,cq,sqr,sqi,ssr,ssi,tr,ti,bdr,bdi,r;

    dwarf=em[0];
    eps=em[1];
    hshdecmul(n,a,b,dwarf);
    hestgl2(n,a,b);
    anorm=bnorm=0.0;
    for (i=1; i<=n; i++) {
        bni=0.0;
        iter[i]=0;
        ani = (i > 1) ? fabs(a[i][i-1]) : 0.0;
        for (j=i; j<=n; j++) {
            ani += fabs(a[i][j]);
            bni += fabs(b[i][j]);
        }
        if (ani > anorm) anorm=ani;
        if (bni > bnorm) bnorm=bni;
    }
    if (anorm == 0.0) anorm=eps;
```

```
if (bnorm == 0.0) bnorm=eps;
epsa=eps*anorm;
epsb=eps*bnorm;
m=n;
out=0;
do {
    i=q=m;
    while ((i > 1) ? fabs(a[i][i-1]) > epsa : 0) {
        q=i-1;
        i--;
    }
    if (q > 1) a[q][q-1]=0.0;
    goon=1;
    while (goon) {
        if (q >= m-1) {
            m=q-1;
            goon=0;
        } else {
            if (fabs(b[q][q]) <= epsb) {
                b[q][q]=0.0;
                q1=q+1;
                hsh2col(q,q,n,q,a[q][q],a[q1][q],a,b);
                a[q1][q]=0.0;
                q=q1;
            } else {
                goon=0;
                m1=m-1;
                q1=q+1;
                constt=0.75;
                (iter[m])++;
                stationary = (iter[m] == 1) ? 1 :
                    (fabs(a[m][m-1]) >= constt*old1 &&
                    fabs(a[m-1][m-2]) >= constt*old2);
                if (iter[m] > 30 && stationary) {
                    for (i=1; i<=m; i++) iter[i] = -1;
                    out=1;
                    break;
                }
                if (iter[m] == 10 && stationary) {
                    a10=0.0;
                    a20=1.0;
                    a30=1.1605;
                } else {
                    b11=b[q][q];
                    b22 = (fabs(b[q1][q1]) < epsb) ? epsb : b[q1][q1];
                    b33 = (fabs(b[m1][m1]) < epsb) ? epsb : b[m1][m1];
                    b44 = (fabs(b[m][m]) < epsb) ? epsb : b[m][m];
                    a11=a[q][q]/b11;
                    a12=a[q][q1]/b22;
                    a21=a[q1][q]/b11;
                    a22=a[q1][q1]/b22;
                    a33=a[m1][m1]/b33;
                    a34=a[m1][m]/b44;
                    a43=a[m][m1]/b33;
                    a44=a[m][m]/b44;
                    b12=b[q][q1]/b22;
                    b34=b[m1][m]/b44;
                    a10=((a33-a11)*(a44-a11)-a34*a43+a43*b34*a11)/a21+
                            a12-a11*b12;
                    a20=(a22-a11-a21*b12)-(a33-a11)-(a44-a11)+a43*b34;
                    a30=a[q+2][q1]/b22;
                }
                old1=fabs(a[m][m-1]);
                old2=fabs(a[m-1][m-2]);
                for (k=q; k<=m1; k++) {
                    k1=k+1;
                    k2=k+2;
                    k3 = (k+3 > m) ? m : k+3;
                    km1 = (k-1 < q) ? q : k-1;
                    if (k != m1) {
                        if (k == q)
                            hsh3col(km1,km1,n,k,a10,a20,a30,a,b);
                        else {
                            hsh3col(km1,km1,n,k,a[k][km1],
```

```
                                        a[k1][km1],a[k2][km1],a,b);
                            a[k1][km1]=a[k2][km1]=0.0;
                        }
                        hsh3row2(1,k3,k,b[k2][k2],b[k2][k1],
                                        b[k2][k],a,b);
                        b[k2][k]=b[k2][k1]=0.0;
                    } else {
                        hsh2col(km1,km1,n,k,a[k][km1],a[k1][km1],a,b);
                        a[k1][km1]=0.0;
                    }
                    hsh2row2(1,k3,k3,k,b[k1][k1],b[k1][k],a,b);
                    b[k1][k]=0.0;
                }
            }
        }
    }   /* goon loop */
    if (out) break;
} while (m >= 3);

m=n;
do {
    if ((m > 1) ? (a[m][m-1] == 0) : 1) {
        alfr[m]=a[m][m];
        beta[m]=b[m][m];
        alfi[m]=0.0;
        m--;
    } else {
        l=m-1;
        if (fabs(b[l][l]) <= epsb) {
            b[l][l]=0.0;
            hsh2col(l,l,n,l,a[l][l],a[m][l],a,b);
            a[m][l]=b[m][l]=0.0;
            alfr[l]=a[l][l];
            alfr[m]=a[m][m];
            beta[l]=b[l][l];
            beta[m]=b[m][m];
            alfi[m]=alfi[l]=0.0;
        } else
            if (fabs(b[m][m]) <= epsb) {
                b[m][m]=0.0;
                hsh2row2(1,m,m,l,a[m][m],a[m][l],a,b);
                a[m][l]=b[m][l]=0.0;
                alfr[l]=a[l][l];
                alfr[m]=a[m][m];
                beta[l]=b[l][l];
                beta[m]=b[m][m];
                alfi[m]=alfi[l]=0.0;
            } else {
                an=fabs(a[l][l])+fabs(a[l][m])+fabs(a[m][l])+
                    fabs(a[m][m]);
                bn=fabs(b[l][l])+fabs(b[l][m])+fabs(b[m][m]);
                a11=a[l][l]/an;
                a12=a[l][m]/an;
                a21=a[m][l]/an;
                a22=a[m][m]/an;
                b11=b[l][l]/bn;
                b12=b[l][m]/bn;
                b22=b[m][m]/bn;
                e=a11/b11;
                c=((a22-e*b22)/b22-(a21*b12)/(b11*b22))/2.0;
                d=c*c+(a21*(a12-e*b12))/(b11*b22);
                if (d >= 0.0) {
                    e += ((c < 0.0) ? c-sqrt(d) : c+sqrt(d));
                    a11 -= e*b11;
                    a12 -= e*b12;
                    a22 -= e*b22;
                    if (fabs(a11)+fabs(a12) >= fabs(a21)+fabs(a22))
                        hsh2row2(1,m,m,l,a12,a11,a,b);
                    else
                        hsh2row2(1,m,m,l,a22,a21,a,b);
                    if (an >= fabs(e)*bn)
                        hsh2col(l,l,n,l,b[l][l],b[m][l],a,b);
                    else
                        hsh2col(l,l,n,l,a[l][l],a[m][l],a,b);
```

```
                        a[m][l]=b[m][l]=0.0;
                        alfr[l]=a[l][l];
                        alfr[m]=a[m][m];
                        beta[l]=b[l][l];
                        beta[m]=b[m][m];
                        alfi[m]=alfi[l]=0.0;
                    } else {
                        er=e+c;
                        ei=sqrt(-d);
                        a11r=a11-er*b11;
                        a11i=ei*b11;
                        a12r=a12-er*b12;
                        a12i=ei*b12;
                        a21r=a21;
                        a21i=0.0;
                        a22r=a22-er*b22;
                        a22i=ei*b22;
                        if (fabs(a11r)+fabs(a11i)+fabs(a12r)+fabs(a12i) >=
                                fabs(a21r)+fabs(a22r)+fabs(a22i))
                            chsh2(a12r,a12i,-a11r,-a11i,&cz,&szr,&szi);
                        else
                            chsh2(a22r,a22i,-a21r,-a21i,&cz,&szr,&szi);
                        if (an >= (fabs(er)+fabs(ei))*bn)
                            chsh2(cz*b11+szr*b12,szi*b12,szr*b22,szi*b22,
                                    &cq,&sqr,&sqi);
                        else
                            chsh2(cz*a11+szr*a12,szi*a12,cz*a21+szr*a22,
                                    szi*a22,&cq,&sqr,&sqi);
                        ssr=sqr*szr+sqi*szi;
                        ssi=sqr*szi-sqi*szr;
                        tr=cq*cz*a11+cq*szr*a12+sqr*cz*a21+ssr*a22;
                        ti=cq*szi*a12-sqi*cz*a21+ssi*a22;
                        bdr=cq*cz*b11+cq*szr*b12+ssr*b22;
                        bdi=cq*szi*b12+ssi*b22;
                        r=sqrt(bdr*bdr+bdi*bdi);
                        beta[l]=bn*r;
                        alfr[l]=an*(tr*bdr+ti*bdi)/r;
                        alfi[l]=an*(tr*bdi-ti*bdr)/r;
                        tr=ssr*a11-sqr*cz*a12-cq*szr*a21+cq*cz*a22;
                        ti = -ssi*a11-sqi*cz*a12+cq*szi*a21;
                        bdr=ssr*b11-sqr*cz*b12+cq*cz*b22;
                        bdi = -ssi*b11-sqi*cz*b12;
                        r=sqrt(bdr*bdr+bdi*bdi);
                        beta[m]=bn*r;
                        alfr[m]=an*(tr*bdr+ti*bdi)/r;
                        alfi[m]=an*(tr*bdi-ti*bdr)/r;
                    }
                }
            m -= 2;
        }
    } while (m > 0);
}
```

B. qzi

Solves the generalized matrix eigenvalue problem $Ax=\lambda Bx$ by means of QZ iteration [MS71]. The procedure *qzi* applies the same method as *qzival*. Given two $n\times n$ matrices A and B, *qzi* determines complex $\alpha^{(j)}$ and real $\beta^{(j)}$ such that $\beta^{(j)}A-\alpha^{(j)}B$ is singular and vectors $x^{(j)}$ such that $\beta^{(j)}Ax^{(j)}=\alpha^{(j)}Bx^{(j)}$ (j=1,...,n), the latter being normalized by the condition that max(|real($x_i^{(j)}$)|, |imag($x_i^{(j)}$)|)=1 ($1\le i\le n$) for each $x^{(j)}$ and either real($x_i^{(j)}$)=1 or imag($x_i^{(j)}$) for some $x_i^{(j)}$.

With regard to the determination of the $\alpha^{(j)}$, $\beta^{(j)}$, the remarks made in the documentation to *qzival* apply with equal force here. In particular, (a) QZ iteration may fail to converge in the determination of a sequence $\alpha^{(i)}$, $\beta^{(i)}$ (i=1,...,m); this failure is signalled by the insertion of -1 in the array *iter[i]*, (i=1,...,m) and for those $\alpha^{(j)}$, $\beta^{(j)}$ that are determined, the required number of QZ iterations required in each case is allocated to the array *iter[j]* (j=m+1,...,n); (b) a quasi-upper triangular matrix U and an upper triangular matrix V are produced and (c) the sets $\alpha^{(j)}$, $\beta^{(j)}$ (j=m+1,...,n) may contain complex conjugate pairs in the sense that for some $k \geq m+1$, $\alpha^{(k)}/\beta^{(k)}$ = conjugate of $\{\alpha^{(k+1)}/\beta^{(k+1)}\}$, and the components of such pairs are stored consecutively in the output arrays *alfr, alfi* and *beta*.

Function Parameters:

void qzi (*n,a,b,x,alfr,alfi,beta,iter,em*)

n: int;
  entry: the number of rows and columns of the matrices *a*, *b* and *x*;
a: float *a[1:n,1:n]*;
  entry: the given matrix;
  exit: a quasi upper triangular matrix (value of U above);
b: float *b[1:n,1:n]*;
  entry: the given matrix;
  exit: an upper triangular matrix (value of V above);
x: float *x[1:n,1:n]*;
  entry: the $n \times n$ unit matrix;
  exit: the matrix of eigenvectors (components of $x^{(j)}$ above); the eigenvectors are stored in *x* as follows: if *alfi[m]*=0 then *x[.,m]* is the *m*-th real eigenvector; otherwise, for each pair of consecutive columns *x[.,m]* and *x[.,m+1]* are the real and imaginary parts of the *m*-th complex eigenvector, *x[.,m]* and -*x[.,m+1]* are the real and imaginary parts of the (m+1)-st complex eigenvector; the eigenvectors are normalized such that the largest component is 1 or 1+0*i;
alfr: float *alfr[1:n]*;
  exit: the real part of $\alpha^{(j)}$ given in *alfr[j]*, j=m+1,...,n, in the above;
alfi: float *alfi[1:n]*;
  exit: the imaginary part of $\alpha^{(j)}$ given in *alfi[j]*, j=m+1,...,n, in the above;
beta: float *beta[1:n]*;
  exit: the value of $\beta^{(j)}$ given in *beta[j]*, j=m+1,...,n, in the above;
iter: int *iter[1:n]*;
  exit: trouble indicator and iteration counter, see *qzival*; if *iter[1]*=0 then no trouble is signalized;
em: float *em[0:1]*;
  entry:
  em[0]: the smallest positive machine number;
  em[1]: the relative precision of elements of *a* and *b*.

Functions used: matmat, hshdecmul, hestgl3, hsh2col, hsh2row3, hsh3row3, hsh3col, chsh2, comdiv.

```
#include <math.h>

void qzi(int n, float **a, float **b, float **x, float alfr[],
         float alfi[], float beta[], int iter[], float em[])
{
    float matmat(int, int, int, int, float **, float **);
    void hshdecmul(int, float **, float **, float);
    void hestgl3(int, float **, float **, float **);
    void hsh2row3(int, int, int, int, int, float, float,
                        float **, float **, float **);
    void hsh3row3(int, int, int, int, float, float, float,
                        float **, float **, float **);
    void hsh2col(int, int, int, int, float, float, float **, float **);
```

```
void hsh3col(int, int, int, int, float, float, float,
               float **, float **);
void chsh2(float, float, float, float, float *, float *, float *);
void comdiv(float, float, float, float, float *, float *);
int i,q,m,m1,q1,j,k,k1,k2,k3,km1,stationary,goon,l,mr,mi,l1,out;
float dwarf,eps,epsa,epsb,
        anorm,bnorm,ani,bni,constt,a10,a20,a30,b11,b22,b33,b44,a11,
        a12,a21,a22,a33,a34,a43,a44,b12,b34,old1,old2,
        an,bn,e,c,d,er,ei,a11r,a11i,a12r,a12i,a21r,a21i,a22r,a22i,
        cz,szr,szi,cq,sqr,sqi,ssr,ssi,tr,ti,bdr,bdi,r,
        betm,alfm,sl,sk,tkk,tkl,tlk,tll,almi,almr,slr,sli,skr,ski,
        dr,di,tkkr,tkki,tklr,tkli,tlkr,tlki,tllr,tlli,s;

dwarf=em[0];
eps=em[1];
hshdecmul(n,a,b,dwarf);
hestgl3(n,a,b,x);
anorm=bnorm=0.0;
for (i=1; i<=n; i++) {
    bni=0.0;
    iter[i]=0;
    ani = (i > 1) ? fabs(a[i][i-1]) : 0.0;
    for (j=i; j<=n; j++) {
        ani += fabs(a[i][j]);
        bni += fabs(b[i][j]);
    }
    if (ani > anorm) anorm=ani;
    if (bni > bnorm) bnorm=bni;
}
if (anorm == 0.0) anorm=eps;
if (bnorm == 0.0) bnorm=eps;
epsa=eps*anorm;
epsb=eps*bnorm;
m=n;
out=0;
do {
    i=q=m;
    while ((i > 1) ? fabs(a[i][i-1]) > epsa : 0) {
        q=i-1;
        i--;
    }
    if (q > 1) a[q][q-1]=0.0;
    goon=1;
    while (goon) {
        if (q >= m-1) {
            m=q-1;
            goon=0;
        } else {
            if (fabs(b[q][q]) <= epsb) {
                b[q][q]=0.0;
                q1=q+1;
                hsh2col(q,q,n,q,a[q][q],a[q1][q],a,b);
                a[q1][q]=0.0;
                q=q1;
            } else {
                goon=0;
                m1=m-1;
                q1=q+1;
                constt=0.75;
                (iter[m])++;
                stationary = (iter[m] == 1) ? 1 :
                    (fabs(a[m][m-1]) >= constt*old1 &&
                    fabs(a[m-1][m-2]) >= constt*old2);
                if (iter[m] > 30 && stationary) {
                    for (i=1; i<=m; i++) iter[i] = -1;
                    out=1;
                    break;
                }
                if (iter[m] == 10 && stationary) {
                    a10=0.0;
                    a20=1.0;
                    a30=1.1605;
                } else {
                    b11=b[q][q];
```

```
                b22 = (fabs(b[q1][q1]) < epsb) ? epsb : b[q1][q1];
                b33 = (fabs(b[m1][m1]) < epsb) ? epsb : b[m1][m1];
                b44 = (fabs(b[m][m]) < epsb) ? epsb : b[m][m];
                a11=a[q][q]/b11;
                a12=a[q][q1]/b22;
                a21=a[q1][q]/b11;
                a22=a[q1][q1]/b22;
                a33=a[m1][m1]/b33;
                a34=a[m1][m]/b44;
                a43=a[m][m1]/b33;
                a44=a[m][m]/b44;
                b12=b[q][q1]/b22;
                b34=b[m1][m]/b44;
                a10=((a33-a11)*(a44-a11)-a34*a43+a43*b34*a11)/a21+
                    a12-a11*b12;
                a20=(a22-a11-a21*b12)-(a33-a11)-(a44-a11)+a43*b34;
                a30=a[q+2][q1]/b22;
            }
            old1=fabs(a[m][m-1]);
            old2=fabs(a[m-1][m-2]);
            for (k=q; k<=m1; k++) {
                k1=k+1;
                k2=k+2;
                k3 = (k+3 > m) ? m : k+3;
                km1 = (k-1 < q) ? q : k-1;
                if (k != m1) {
                    if (k == q)
                        hsh3col(km1,km1,n,k,a10,a20,a30,a,b);
                    else {
                        hsh3col(km1,km1,n,k,a[k][km1],
                                  a[k1][km1],a[k2][km1],a,b);
                        a[k1][km1]=a[k2][km1]=0.0;
                    }
                    hsh3row3(1,k3,n,k,b[k2][k2],b[k2][k1],
                                b[k2][k],a,b,x);
                    b[k2][k]=b[k2][k1]=0.0;
                } else {
                    hsh2col(km1,km1,n,k,a[k][km1],a[k1][km1],a,b);
                    a[k1][km1]=0.0;
                }
                hsh2row3(1,k3,k3,n,k,b[k1][k1],b[k1][k],a,b,x);
                b[k1][k]=0.0;
            }
        }
      }
    }   /* goon loop */
    if (out) break;
} while (m >= 3);

m=n;
do {
    if ((m > 1) ? (a[m][m-1] == 0) : 1) {
        alfr[m]=a[m][m];
        beta[m]=b[m][m];
        alfi[m]=0.0;
        m--;
    } else {
        l=m-1;
        if (fabs(b[l][l]) <= epsb) {
            b[l][l]=0.0;
            hsh2col(l,l,n,l,a[l][l],a[m][l],a,b);
            a[m][l]=b[m][l]=0.0;
            alfr[l]=a[l][l];
            alfr[m]=a[m][m];
            beta[l]=b[l][l];
            beta[m]=b[m][m];
            alfi[m]=alfi[l]=0.0;
        } else
            if (fabs(b[m][m]) <= epsb) {
                b[m][m]=0.0;
                hsh2row3(1,m,m,n,l,a[m][m],a[m][l],a,b,x);
                a[m][l]=b[m][l]=0.0;
                alfr[l]=a[l][l];
                alfr[m]=a[m][m];
```

```
            beta[l]=b[l][l];
            beta[m]=b[m][m];
            alfi[m]=alfi[l]=0.0;
        } else {
            an=fabs(a[l][l])+fabs(a[l][m])+fabs(a[m][l])+
                fabs(a[m][m]);
            bn=fabs(b[l][l])+fabs(b[l][m])+fabs(b[m][m]);
            a11=a[l][l]/an;
            a12=a[l][m]/an;
            a21=a[m][l]/an;
            a22=a[m][m]/an;
            b11=b[l][l]/bn;
            b12=b[l][m]/bn;
            b22=b[m][m]/bn;
            e=a11/b11;
            c=((a22-e*b22)/b22-(a21*b12)/(b11*b22))/2.0;
            d=c*c+(a21*(a12-e*b12))/(b11*b22);
            if (d >= 0.0) {
                e += ((c < 0.0) ? c-sqrt(d) : c+sqrt(d));
                a11 -= e*b11;
                a12 -= e*b12;
                a22 -= e*b22;
                if (fabs(a11)+fabs(a12) >= fabs(a21)+fabs(a22))
                    hsh2row3(1,m,m,n,l,a12,a11,a,b,x);
                else
                    hsh2row3(1,m,m,n,l,a22,a21,a,b,x);
                if (an >= fabs(e)*bn)
                    hsh2col(l,l,n,l,b[l][l],b[m][l],a,b);
                else
                    hsh2col(l,l,n,l,a[l][l],a[m][l],a,b);
                a[m][l]=b[m][l]=0.0;
                alfr[l]=a[l][l];
                alfr[m]=a[m][m];
                beta[l]=b[l][l];
                beta[m]=b[m][m];
                alfi[m]=alfi[l]=0.0;
            } else {
                er=e+c;
                ei=sqrt(-d);
                a11r=a11-er*b11;
                a11i=ei*b11;
                a12r=a12-er*b12;
                a12i=ei*b12;
                a21r=a21;
                a21i=0.0;
                a22r=a22-er*b22;
                a22i=ei*b22;
                if (fabs(a11r)+fabs(a11i)+fabs(a12r)+fabs(a12i) >=
                        fabs(a21r)+fabs(a22r)+fabs(a22i))
                    chsh2(a12r,a12i,-a11r,-a11i,&cz,&szr,&szi);
                else
                    chsh2(a22r,a22i,-a21r,-a21i,&cz,&szr,&szi);
                if (an >= (fabs(er)+fabs(ei))*bn)
                    chsh2(cz*b11+szr*b12,szi*b12,szr*b22,szi*b22,
                            &cq,&sqr,&sqi);
                else
                    chsh2(cz*a11+szr*a12,szi*a12,cz*a21+szr*a22,
                            szi*a22,&cq,&sqr,&sqi);
                ssr=sqr*szr+sqi*szi;
                ssi=sqr*szi-sqi*szr;
                tr=cq*cz*a11+cq*szr*a12+sqr*cz*a21+ssr*a22;
                ti=cq*szi*a12-sqi*cz*a21+ssi*a22;
                bdr=cq*cz*b11+cq*szr*b12+ssr*b22;
                bdi=cq*szi*b12+ssi*b22;
                r=sqrt(bdr*bdr+bdi*bdi);
                beta[l]=bn*r;
                alfr[l]=an*(tr*bdr+ti*bdi)/r;
                alfi[l]=an*(tr*bdi-ti*bdr)/r;
                tr=ssr*a11-sqr*cz*a12-cq*szr*a21+cq*cz*a22;
                ti = -ssi*a11-sqi*cz*a12+cq*szi*a21;
                bdr=ssr*b11-sqr*cz*b12+cq*cz*b22;
                bdi = -ssi*b11-sqi*cz*b12;
                r=sqrt(bdr*bdr+bdi*bdi);
                beta[m]=bn*r;
```

```
                        alfr[m]=an*(tr*bdr+ti*bdi)/r;
                        alfi[m]=an*(tr*bdi-ti*bdr)/r;
                    }
                }
            m -= 2;
        }
    } while (m > 0);

    for (m=n; m>=1; m--)
        if (alfi[m] == 0.0) {
            alfm=alfr[m];
            betm=beta[m];
            b[m][m]=1.0;
            l1=m;
            for (l=m-1; l>=1; l--) {
                sl=0.0;
                for (j=l1; j<=m; j++)
                    sl += (betm*a[l][j]-alfm*b[l][j])*b[j][m];
                if ((l != 1) ? (betm*a[l][l-1] == 0.0) : 1) {
                    d=betm*a[l][l]-alfm*b[l][l];
                    if (d == 0.0) d=(epsa+epsb)/2.0;
                    b[l][m] = -sl/d;
                } else {
                    k=l-1;
                    sk=0.0;
                    for (j=l1; j<=m; j++)
                        sk += (betm*a[k][j]-alfm*b[k][j])*b[j][m];
                    tkk=betm*a[k][k]-alfm*b[k][k];
                    tkl=betm*a[k][l]-alfm*b[k][l];
                    tlk=betm*a[l][k];
                    tll=betm*a[l][l]-alfm*b[l][l];
                    d=tkk*tll-tkl*tlk;
                    if (d == 0.0) d=(epsa+epsb)/2.0;
                    b[l][m]=(tlk*sk-tkk*sl)/d;
                    b[k][m] = (fabs(tkk) >= fabs(tlk)) ?
                                -(sk+tkl*b[l][m])/tkk :
                                -(sl+tll*b[l][m])/tlk;
                    l--;
                }
                l1=l;
            }
        } else {
            almr=alfr[m-1];
            almi=alfi[m-1];
            betm=beta[m-1];
            mr=m-1;
            mi=m;
            b[m-1][mr]=almi*b[m][m]/(betm*a[m][m-1]);
            b[m-1][mi]=(betm*a[m][m]-almr*b[m][m])/(betm*a[m][m-1]);
            b[m][mr]=0.0;
            b[m][mi] = -1.0;
            l1=m-1;
            for (l=m-2; l>=1; l--) {
                slr=sli=0.0;
                for (j=l1; j<=m; j++) {
                    tr=betm*a[l][j]-almr*b[l][j];
                    ti = -almi*b[l][j];
                    slr += tr*b[j][mr]-ti*b[j][mi];
                    sli += tr*b[j][mi]+ti*b[j][mr];
                }
                if ((l != 1) ? (betm*a[l][l-1] == 0.0) : 1) {
                    dr=betm*a[l][l]-almr*b[l][l];
                    di = -almi*b[l][l];
                    comdiv(-slr,-sli,dr,di,&b[l][mr],&b[l][mi]);
                } else {
                    k=l-1;
                    skr=ski=0.0;
                    for (j=l1; j<=m; j++) {
                        tr=betm*a[k][j]-almr*b[k][j];
                        ti = -almi*b[k][j];
                        skr += tr*b[j][mr]-ti*b[j][mi];
                        ski += tr*b[j][mi]+ti*b[j][mr];
                    }
                    tkkr=betm*a[k][k]-almr*b[k][k];
```

```
                    tkki = -almi*b[k][k];
                    tklr=betm*a[k][l]-almr*b[k][l];
                    tkli = -almi*b[k][l];
                    tlkr=betm*a[l][k];
                    tlki=0.0;
                    tllr=betm*a[l][l]-almr*b[l][l];
                    tlli = -almi*b[l][l];
                    dr=tkkr*tllr-tkki*tlli-tklr*tlkr;
                    di=tkkr*tlli+tkki*tllr-tkli*tlkr;
                    if (dr == 0.0 && di == 0.0) dr=(epsa+epsb)/2.0;
                    comdiv(tlkr*skr-tkkr*slr+tkki*sli,
                           tlkr*ski-tkkr*sli-tkki*slr,
                           dr,di,&b[l][mr],&b[l][mi]);
                    if (fabs(tkkr)+fabs(tkki) >= fabs(tlkr))
                        comdiv(-skr-tklr*b[l][mr]+tkli*b[l][mi],
                               -ski-tklr*b[l][mi]-tkli*b[l][mr],
                               tkkr,tkki,&b[k][mr],&b[k][mi]);
                    else
                        comdiv(-slr-tllr*b[l][mr]+tlli*b[l][mi],
                               -sli-tllr*b[l][mi]-tlli*b[l][mr],
                               tlkr,tlki,&b[k][mr],&b[k][mi]);
                    l--;
                }
                l1=l;
            }
            m--;
        }
    for (m=n; m>=1; m--)
        for (k=1; k<=n; k++) x[k][m]=matmat(1,m,k,m,x,b);
    for (m=n; m>=1; m--) {
        s=0.0;
        if (alfi[m] == 0.0) {
            for (k=1; k<=n; k++) {
                r=fabs(x[k][m]);
                if (r >= s) {
                    s=r;
                    d=x[k][m];
                }
            }
            for (k=1; k<=n; k++) x[k][m] /= d;
        } else {
            for (k=1; k<=n; k++) {
                r=fabs(x[k][m-1])+fabs(x[k][m]);
                an=x[k][m-1]/r;
                bn=x[k][m]/r;
                r *= sqrt(an*an+bn*bn);
                if (r >= s) {
                    s=r;
                    dr=x[k][m-1];
                    di=x[k][m];
                }
            }
            for (k=1; k<=n; k++)
                comdiv(x[k][m-1],x[k][m],dr,di,&x[k][m-1],&x[k][m]);
            m--;
        }
    }
}
```

C. hshdecmul

Given an $n \times n$ matrix A, determines vectors $u^{(i)}$ (i=1,...,n'-1; $n' \leq n$) such that with

$$E = \prod_{i=1}^{n'-1} (I - 2u^{(i)T}u^{(i)}/u^{(i)}u^{(i)T}) ,$$

$A'=EA$ is of upper triangular form and, also given the $n \times n$ matrix B, forms $B'=EB$.

hshdecmul is used in *qzi* and *qzival*.

Function Parameters:

void hshdecmul (*n,a,b,dwarf*)

n: int;
entry: the order of the given matrices;
a: float *a[1:n,1:n]*;
entry: the given matrix;
exit: the transformed matrix (value of A' above);
b: float *b[1:n,1:n]*;
entry: the given matrix;
exit: the upper triangular matrix (value of B' above);
dwarf: float;
entry: the smallest positive machine number.

Functions used: tammat, hshvecmat.

```
#include <math.h>

void hshdecmul(int n, float **a, float **b, float dwarf)
{
	float *allocate_real_vector(int, int);
	void free_real_vector(float *, int);
	float tammat(int, int, int, int, float **, float **);
	void hshvecmat(int, int, int, int, float, float [], float **);
	int j,k,k1,n1;
	float r,t,c,*v;

	v=allocate_real_vector(1,n);
	k=1;
	n1=n+1;
	for (k1=2; k1<=n1; k1++) {
		r=tammat(k1,n,k,k,b,b);
		if (r > dwarf) {
			r = (b[k][k] < 0.0) ? -sqrt(r+b[k][k]*b[k][k]) :
			                       sqrt(r+b[k][k]*b[k][k]);
			t=b[k][k]+r;
			c = -t/r;
			b[k][k] = -r;
			v[k]=1.0;
			for (j=k1; j<=n; j++) v[j]=b[j][k]/t;
			hshvecmat(k,n,k1,n,c,v,b);
			hshvecmat(k,n,1,n,c,v,a);
		}
		k=k1;
	}
	free_real_vector(v,1);
}
```

D. hestgl3

Given an $n \times n$ matrix A and an $n \times n$ upper triangular matrix U, obtains vectors $u_1^{(i)}$, $u_2^{(i)}$ such that with

$$Q_1 = \prod_{i=1}^{n'-1} (I - 2u_1^{(i)}u_1^{(i)T}/u_1^{(i)T}u_1^{(i)}) , \qquad (n' \le n)$$

and Q_2 similarly defined, $Q_1AQ_2=H$ is an upper Hessenberg matrix and $Q_1UQ_2=U'$ is an upper triangular matrix and, also given an $n \times n$ matrix X, forms $X'=Q_1XQ_2$. *hestgl3* is used in *qzi*.

Function Parameters:

void hestgl3 (n,a,b,x)

n: int; entry: the order of the given matrices;

a: float $a[1:n,1:n]$;
entry: the given matrix;
exit: the upper Hessenberg matrix (value of H above);

b: float $b[1:n,1:n]$;
entry: the given upper triangular matrix (value of U above);
exit: the upper triangular matrix (value of U' above);

x: float $x[1:n,1:n]$;
entry: the given matrix (value of X above);
exit: the transformed matrix (value of X' above).

Functions used: hsh2col, hsh2row3.

```
void hestgl3(int n, float **a, float **b, float **x)
{
	float *allocate_real_vector(int, int);
	void free_real_vector(float *, int);
	void hsh2col(int, int, int, int, float, float, float **, float **);
	void hsh2row3(int, int, int, int, int, float, float,
					float **, float **, float **);
	int nm1,k,l,k1,l1;

	if (n > 2) {
		for (k=2; k<=n; k++)
			for (l=1; l<=k-1; l++) b[k][l]=0.0;
		nm1=n-1;
		k=1;
		for (k1=2; k1<=nm1; k1++) {
			l1=n;
			for (l=n-1; l>=k1; l--) {
				hsh2col(k,l,n,l,a[l][k],a[l1][k],a,b);
				a[l1][k]=0.0;
				hsh2row3(1,n,l1,n,l,b[l1][l1],b[l1][l],a,b,x);
				b[l1][l]=0.0;
				l1=l;
			}
			k=k1;
		}
	}
}
```

E. hestgl2

Given an $n \times n$ matrix A and an $n \times n$ upper triangular matrix U, obtains vectors $u_1^{(i)}$, $u_2^{(i)}$ such that with

$$Q_1 = \prod_{i=1}^{n'-1} (I - 2u_1^{(i)}u_1^{(i)T}/u_1^{(i)T}u_1^{(i)}) , \qquad (n' \leq n)$$

and Q_2 similarly defined, $Q_1AQ_2=H$ is an upper Hessenberg matrix and $Q_1UQ_2=U'$ is an upper triangular matrix. *hestgl2* is used in *qzival*.

Function Parameters:

void hestgl2 (*n,a,b*)

n: int;
  entry: the order of the given matrices;
a: float *a[1:n,1:n]*;
  entry: the given matrix;
  exit: the upper Hessenberg matrix (value of *H* above);
b: float *b[1:n,1:n]*;
  entry: the given upper triangular matrix (value of *U* above);
  exit: the upper triangular matrix (value of *U'* above).

Functions used: hsh2col, hsh2row2.

```
void hestgl2(int n, float **a, float **b)
{
    float *allocate_real_vector(int, int);
    void free_real_vector(float *, int);
    void hsh2col(int, int, int, int, float, float, float **, float **);
    void hsh2row2(int, int, int, int, float, float,
                        float **, float **);
    int nm1,k,l,k1,l1;

    if (n > 2) {
        for (k=2; k<=n; k++)
            for (l=1; l<=k-1; l++) b[k][l]=0.0;
        nm1=n-1;
        k=1;
        for (k1=2; k1<=nm1; k1++) {
            l1=n;
            for (l=n-1; l>=k1; l--) {
                hsh2col(k,l,n,l,a[l][k],a[l1][k],a,b);
                a[l1][k]=0.0;
                hsh2row2(1,n,l1,l,b[l1][l1],b[l1][l],a,b);
                b[l1][l]=0.0;
                l1=l;
            }
            k=k1;
        }
    }
}
```

F. hsh2col

(a) Given the values of two elements $M_{k,j}$ (k=i,i+1) belonging to a certain column of a rectangular matrix *M*, determines a vector *v* such that all rows except the *i*-th and (*i*+1)-th of *M* and $M'=EM$ agree, where $E=I-2vv^T/v^Tv$ and $M'_{i,j}=0$ and, (b) given the elements $A_{k,j}$ (k=i,i+1; j=la,...,u) of the rectangular matrix *A*, determines the corresponding elements of $A'=EA$ and, (c) given the elements $B_{k,j}$ (k=i,i+1; j=lb,...,u) of the rectangular matrix *B*, determines the corresponding elements of $B'=EB$. *hsh2col* is used in *qzival* and *qzi*.

Function Parameters:

void hsh2col (*la,lb,u,i,a1,a2,a,b*)

la: int;
  entry: the lower bound of the running column subscript of *a* (value of *la* above);
lb: int;
  entry: the lower bound of the running column subscript of *b* (value of *lb* above);
u: int;

entry: the upper bound of the running column subscript of a and b (value of u above);

i: int;
entry: the lower bound of the running row subscript of a and b (value of i above); i+1 is the upper bound;

$a1,a2$: float;
entry: $a1$ and $a2$ are the i-th and (i+1)-th component of the vector to be transformed, respectively (values of $M_{k,j}$ (k=i,i+1) above);

a: float $a[i:i+1,la:u]$;
entry: the given matrix (value of A above);
exit: the transformed matrix (value of A' above);

b: float $b[i:i+1,lb:u]$;
entry: the given matrix (value of B above);
exit: the transformed matrix (value of B' above).

Function used: hshvecmat.

```
#include <math.h>

void hsh2col(int la, int lb, int u, int i, float a1, float a2,
             float **a, float **b)
{
    float *allocate_real_vector(int, int);
    void free_real_vector(float *, int);
    void hshvecmat(int, int, int, int, float, float [], float **);
    float *v,d1,d2,s1,s2,r,d,c;

    if (a2 != 0.0) {
        v=allocate_real_vector(i,i+1);
        d1=fabs(a1);
        d2=fabs(a2);
        s1 = (a1 >= 0.0) ? 1.0 : -1.0;
        s2 = (a2 >= 0.0) ? 1.0 : -1.0;
        if (d2 <= d1) {
            r=d2/d1;
            d=sqrt(1.0+r*r);
            c = -1.0-1.0/d;
            v[i+1]=s1*s2*r/(1.0+d);
        } else {
            r=d1/d2;
            d=sqrt(1.0+r*r);
            c = -1.0-r/d;
            v[i+1]=s1*s2/(r+d);
        }
        v[i]=1.0;
        hshvecmat(i,i+1,la,u,c,v,a);
        hshvecmat(i,i+1,lb,u,c,v,b);
        free_real_vector(v,i);
    }
}
```

G. hsh3col

(a) Given the values of three elements $M_{k,j}$ (k=i,i+1,i+2) belonging to a certain column of a rectangular matrix M, determines a vector v such that all rows except the i-th, (i+1)-th and (i+2)-th of M and $M'=EM$ agree, where

$$E = I - 2vv^T/v^Tv \quad \text{and} \quad M'_{i,j} = M'_{i,j+1} = 0$$

and, (b) given the elements $A_{k,j}$ (k=i,i+1,i+2; j=la,...,u) of the rectangular matrix A, determines the corresponding elements of $A'=EA$ and, (c) given the elements $B_{k,j}$

(k=i,i+1,i+2; j=lb,...,u) of the rectangular matrix B, determines the corresponding elements of $B'=EB$. *hsh3col* is used in *qzival* and *qzi*.

Function Parameters:

void hsh3col (*la,lb,u,i,a1,a2,a3,a,b*)

la: int;
entry: the lower bound of the running column subscript of a (value of la above);

lb: int;
entry: the lower bound of the running column subscript of b (value of lb above);

u: int;
entry: the upper bound of the running column subscript of a and b (value of u above);

i: int;
entry: the lower bound of the running row subscript of a and b (value of i above);
i+2 is the upper bound;

a1,a2,a3: float;
entry: $a1$, $a2$ and $a3$ are the i-th, (i+1)-th and (i+2)-th component of the vector to be transformed, respectively (values of $M_{k,j}$ (k=i,i+1,i+2) above);

a: float *a[i:i+2,la:u]*;
entry: the given matrix (value of A above);
exit: the transformed matrix (value of A' above);

b: float *b[i:i+2,lb:u]*;
entry: the given matrix (value of B above);
exit: the transformed matrix (value of B' above).

Function used: hshvecmat.

```
#include <math.h>

void hsh3col(int la, int lb, int u, int i, float a1, float a2,
             float a3, float **a, float **b)
{
    float *allocate_real_vector(int, int);
    void free_real_vector(float *, int);
    void hshvecmat(int, int, int, int, float, float [], float **);
    float c,*v,d1,d2,d3,s1,s2,s3,r1,r2,r3,d;

    if (a2 != 0.0 || a3 != 0.0) {
        v=allocate_real_vector(i,i+2);
        d1=fabs(a1);
        d2=fabs(a2);
        d3=fabs(a3);
        s1 = (a1 >= 0.0) ? 1.0 : -1.0;
        s2 = (a2 >= 0.0) ? 1.0 : -1.0;
        s3 = (a3 >= 0.0) ? 1.0 : -1.0;
        if (d1 >= d2 && d1 >= d3) {
            r2=d2/d1;
            r3=d3/d1;
            d=sqrt(1.0+r2*r2+r3*r3);
            c = -1.0-(1.0/d);
            d=1.0/(1.0+d);
            v[i+1]=s1*s2*r2*d;
            v[i+2]=s1*s3*r3*d;
        } else if (d2 >= d1 && d2 >= d3) {
            r1=d1/d2;
```

```
            r3=d3/d2;
            d=sqrt(1.0+r1*r1+r3*r3);
            c = -1.0-(s1*r1/d);
            d=1.0/(r1+d);
            v[i+1]=s1*s2*d;
            v[i+2]=s1*s3*r3*d;
        } else {
            r1=d1/d3;
            r2=d2/d3;
            d=sqrt(1.0+r1*r1+r2*r2);
            c = -1.0-(s1*r1/d);
            d=1.0/(r1+d);
            v[i+1]=s1*s2*r2*d;
            v[i+2]=s1*s3*d;
        }
        v[i]=1.0;
        hshvecmat(i,i+2,la,u,c,v,a);
        hshvecmat(i,i+2,lb,u,c,v,b);
        free_real_vector(v,i);
    }
}
```

H. hsh2row3

(a) Given the values of two elements $M_{i,k}$ (k=*j*,*j*+1) belonging to a certain row of a rectangular matrix *M*, determines a vector *v* such that all columns except the *j*-th, (*j*+1)-th of *M* and *M'*=*ME* agree, where $E=I-2vv^T/v^Tv$ and $M'_{i,j}=0$ and, (b) given the elements $A_{i,k}$ (i=*l*,...,*ua*; k=*j*,*j*+1) of the rectangular matrix *A*, determines the corresponding elements of *A'*=*AE* and, (c) given the elements $B_{i,k}$ (i=*l*,...,*ub*; k=*j*,*j*+1) of the rectangular matrix *B*, determines the corresponding elements of *B'*=*BE* and, (d) given the elements $X_{i,k}$ (i=*l*,...,*ux*; k=*j*,*j*+1) of the rectangular matrix *X*, determines the corresponding elements of *X'*=*XE*. *hsh2row3* is used in *qzi*.

Function Parameters:

void hsh2row3 (*l*,*ua*,*ub*,*ux*,*j*,*a1*,*a2*,*a*,*b*,*x*)

l: int;
entry: the lower bound of the running row subscript of *a*, *b* and *x* (value of *l* above);

ua: int;
entry: the upper bound of the running row subscript of *a* (value of *ua* above);

ub: int;
entry: the upper bound of the running row subscript of *b* (value of *ub* above);

ux: int;
entry: the upper bound of the running row subscript of *x* (value of *ux* above);

j: int;
entry: the lower bound of the running column subscript of *a*, *b* and *x* (value of *j* above);
j+1 is the upper bound;

a1,*a2*: float;
entry: *a1* and *a2* are the *j*-th and (*j*+1)-th component of the vector to be transformed, respectively (values of $M_{i,k}$ (k=*j*,*j*+1) above);

a: float *a[l:ua,j:j+1]*;
entry: the given matrix (value of *A* above);
exit: the transformed matrix (value of *A'* above);

b: float *b[l:ub,j:j+1]*;
entry: the given matrix (value of *B* above);

exit: the transformed matrix (value of B' above);

x: float $x[l:ux,j:j+1]$;
entry: the given matrix (value of X above);
exit: the transformed matrix (value of X' above).

Function used: hshvectam.

```
#include <math.h>

void hsh2row3(int l, int ua, int ub, int ux, int j, float a1,
              float a2, float **a, float **b, float **x)
{
    float *allocate_real_vector(int, int);
    void free_real_vector(float *, int);
    void hshvectam(int, int, int, int, float, float [], float **);
    float *v,d1,d2,s1,s2,r,d,c;

    if (a2 != 0.0) {
        v=allocate_real_vector(j,j+1);
        d1=fabs(a1);
        d2=fabs(a2);
        s1 = (a1 >= 0.0) ? 1.0 : -1.0;
        s2 = (a2 >= 0.0) ? 1.0 : -1.0;
        if (d2 <= d1) {
            r=d2/d1;
            d=sqrt(1.0+r*r);
            c = -1.0-1.0/d;
            v[j]=s1*s2*r/(1.0+d);
        } else {
            r=d1/d2;
            d=sqrt(1.0+r*r);
            c = -1.0-r/d;
            v[j]=s1*s2/(r+d);
        }
        v[j+1]=1.0;
        hshvectam(l,ua,j,j+1,c,v,a);
        hshvectam(l,ub,j,j+1,c,v,b);
        hshvectam(l,ux,j,j+1,c,v,x);
        free_real_vector(v,j);
    }
}
```

I. hsh2row2

(a) Given the values of two elements $M_{i,k}$ (k=j,j+1) belonging to a certain row of a rectangular matrix M, determines a vector v such that all columns except the j-th, (j+1)-th of M and $M'=ME$ agree, where $E=I-2vv^T/v^Tv$ and $M'_{i,j}=0$ and, (b) given the elements $A_{i,k}$ (i=l,...,ua; k=j,j+1) of the rectangular matrix A, determines the corresponding elements of $A'=AE$ and, (c) given the elements $B_{i,k}$ (i=l,...,ub; k=j,j+1) of the rectangular matrix B, determines the corresponding elements of $B'=BE$. *hsh2row2* is used in *qzival*.

Function Parameters:

void hsh2row2 (*l,ua,ub,j,a1,a2,a,b*)

l: int;
entry: the lower bound of the running row subscript of a and b (value of l above);

ua: int;
entry: the upper bound of the running row subscript of a (value of ua above);

ub: int;
entry: the upper bound of the running row subscript of b (value of ub above);

j: int;

entry: the lower bound of the running column subscript of *a* and *b* (value of *j* above); *j*+1 is the upper bound;

a1,a2: float;
entry: *a1* and *a2* are the *j*-th and (*j*+1)-th component of the vector to be transformed, respectively (values of $M_{i,k}$ (k=*j*,*j*+1) above);

a: float *a[la:ua,j:j+1]*;
entry: the given matrix (value of *A* above);
exit: the transformed matrix (value of *A'* above);

b: float *b[lb:ub,j:j+1]*;
entry: the given matrix (value of *B* above);
exit: the transformed matrix (value of *B'* above).

Function used: hshvectam.

```
#include <math.h>

void hsh2row2(int l, int ua, int ub, int j, float a1, float a2,
                    float **a, float **b)
{
    float *allocate_real_vector(int, int);
    void free_real_vector(float *, int);
    void hshvectam(int, int, int, int, float, float [], float **);
    float *v,d1,d2,s1,s2,r,d,c;

    if (a2 != 0.0) {
        v=allocate_real_vector(j,j+1);
        d1=fabs(a1);
        d2=fabs(a2);
        s1 = (a1 >= 0.0) ? 1.0 : -1.0;
        s2 = (a2 >= 0.0) ? 1.0 : -1.0;
        if (d2 <= d1) {
            r=d2/d1;
            d=sqrt(1.0+r*r);
            c = -1.0-1.0/d;
            v[j]=s1*s2*r/(1.0+d);
        } else {
            r=d1/d2;
            d=sqrt(1.0+r*r);
            c = -1.0-r/d;
            v[j]=s1*s2/(r+d);
        }
        v[j+1]=1.0;
        hshvectam(l,ua,j,j+1,c,v,a);
        hshvectam(l,ub,j,j+1,c,v,b);
        free_real_vector(v,j);
    }
}
```

J. hsh3row3

(a) Given the values of three elements $M_{i,k}$ (k=*j*,*j*+1,*j*+2) belonging to a certain row of a rectangular matrix *M*, determines a vector *v* such that all columns except the *j*-th, (*j*+1)-th and (*j*+2)-th of *M* and *M'*=*ME* agree, where $E=I-2vv^T/v^Tv$ and $M'_{i,j}=M'_{i+1,j}=0$ and, (b) given the elements $A_{i,k}$ (i=*l*,...,*u*; k=*j*,*j*+1,*j*+2) of the rectangular matrix *A*, determines the corresponding elements of *A'*=*AE* and, (c) given the elements $B_{i,k}$ (i=*l*,...,*u*; k=*j*,*j*+1,*j*+2) of the rectangular matrix *B*, determines the corresponding elements of *B'*=*BE* and, (c) given the elements $X_{i,k}$ (i=*l*,...,*ux*; k=*j*,*j*+1,*j*+2) of the rectangular matrix *X*, determines the corresponding elements of *X'*=*XE*. *hsh3row3* is used in *qzi*.

Function Parameters:

void hsh3row3 (*l,u,ux,j,a1,a2,a3,a,b,x*)

l: int;
entry: the lower bound of the running row subscript of *a*, *b* and *x* (value of *l* above);

u: int;
entry: the upper bound of the running row subscript of *a* and *b* (value of *u* above);

ux: int;
entry: the upper bound of the running row subscript of *x* (value of *ux* above);

j: int;
entry: the lower bound of the running column subscript of *a*, *b* and *x* (value of *j* above); *j*+2 is the upper bound;

a1,a2,a3: float;
entry: *a1, a2* and *a3* are the *j*-th, (*j*+1)-th and (*j*+2)-th component of the vector to be transformed, respectively (values of $M_{i,k}$ (k=*j*,*j*+1,*j*+2) above);

a: float *a[l:u,j:j+2]*;
entry: the given matrix (value of *A* above);
exit: the transformed matrix (value of *A'* above);

b: float *b[l:u,j:j+2]*;
entry: the given matrix (value of *B* above);
exit: the transformed matrix (value of *B'* above);

x: float *x[l:ux,j:j+2]*;
entry: the given matrix (value of *X* above);
exit: the transformed matrix (value of *X'* above).

Function used: hshvectam.

```
#include <math.h>

void hsh3row3(int l, int u, int ux, int j, float a1, float a2,
                float a3, float **a, float **b, float **x)
{
    float *allocate_real_vector(int, int);
    void free_real_vector(float *, int);
    void hshvectam(int, int, int, int, float, float [], float **);
    float *v,c,d1,d2,d3,s1,s2,s3,r1,r2,r3,d;

    if (a2 != 0.0 || a3 != 0.0) {
        v=allocate_real_vector(j,j+2);
        d1=fabs(a1);
        d2=fabs(a2);
        d3=fabs(a3);
        s1 = (a1 >= 0.0) ? 1.0 : -1.0;
        s2 = (a2 >= 0.0) ? 1.0 : -1.0;
        s3 = (a3 >= 0.0) ? 1.0 : -1.0;
        if (d1 >= d2 && d1 >= d3) {
            r2=d2/d1;
            r3=d3/d1;
            d=sqrt(1.0+r2*r2+r3*r3);
            c = -1.0-(1.0/d);
            d=1.0/(1.0+d);
            v[j+1]=s1*s2*r2*d;
            v[j]=s1*s3*r3*d;
        } else if (d2 >= d1 && d2 >= d3) {
            r1=d1/d2;
            r3=d3/d2;
            d=sqrt(1.0+r1*r1+r3*r3);
```

```
            c = -1.0-(s1*r1/d);
            d=1.0/(r1+d);
            v[j+1]=s1*s2*d;
            v[j]=s1*s3*r3*d;
        } else {
            r1=d1/d3;
            r2=d2/d3;
            d=sqrt(1.0+r1*r1+r2*r2);
            c = -1.0-(s1*r1/d);
            d=1.0/(r1+d);
            v[j+1]=s1*s2*r2*d;
            v[j]=s1*s3*d;
        }
        v[j+2]=1.0;
        hshvectam(l,u,j,j+2,c,v,a);
        hshvectam(l,u,j,j+2,c,v,b);
        hshvectam(l,ux,j,j+2,c,v,x);
        free_real_vector(v,j);
    }
}
```

K. hsh3row2

(a) Given the values of three elements $M_{i,k}$ (k=j,j+1,j+2) belonging to a certain row of a rectangular matrix M, determines a vector v such that all columns except the j-th, (j+1)-th and (j+2)-th of M and $M'=ME$ agree, where $E=I-2vv^T/v^Tv$ and $M'_{i,j}=M'_{i+1,j}=0$ and, (b) given the elements $A_{i,k}$ (i=l,...,u; k=j,j+1,j+2) of the rectangular matrix A, determines the corresponding elements of $A'=AE$ and, (c) given the elements $B_{i,k}$ (i=l,...,u; k=j,j+1,j+2) of the rectangular matrix B, determines the corresponding elements of $B'=BE$. *hsh3row2* is used in *qzival*.

Function Parameters:

void hsh3row2 (*l,u,j,a1,a2,a3,a,b*)

l: int;
  entry: the lower bound of the running row subscript of *a* and *b* (value of *l* above);

u: int;
  entry: the upper bound of the running row subscript of *a* and *b* (value of *u* above);

j: int;
  entry: the lower bound of the running column subscript of *a* and *b* (value of *j* above);
  j+2 is the upper bound;

a1,a2,a3: float;
  entry: *a1*, *a2* and *a3* are the *j*-th, (*j*+1)-th and (*j*+2)-th component of the vector to be transformed, respectively (values of $M_{i,k}$ (k=j,j+1,j+2) above);

a: float *a[l:u,j:j+2]*;
  entry: the given matrix (value of A above);
  exit: the transformed matrix (value of A' above);

b: float *b[l:u,j:j+2]*;
  entry: the given matrix (value of B above);
  exit: the transformed matrix (value of B' above).

Function used: hshvectam.

```
#include <math.h>

void hsh3row2(int l, int u, int j, float a1, float a2,
              float a3, float **a, float **b)
{
    float *allocate_real_vector(int, int);
    void free_real_vector(float *, int);
    void hshvectam(int, int, int, int, float, float [], float **);
    float *v,c,d1,d2,d3,s1,s2,s3,r1,r2,r3,d;

    if (a2 != 0.0 || a3 != 0.0) {
        v=allocate_real_vector(j,j+2);
        d1=fabs(a1);
        d2=fabs(a2);
        d3=fabs(a3);
        s1 = (a1 >= 0.0) ? 1.0 : -1.0;
        s2 = (a2 >= 0.0) ? 1.0 : -1.0;
        s3 = (a3 >= 0.0) ? 1.0 : -1.0;
        if (d1 >= d2 && d1 >= d3) {
            r2=d2/d1;
            r3=d3/d1;
            d=sqrt(1.0+r2*r2+r3*r3);
            c = -1.0-(1.0/d);
            d=1.0/(1.0+d);
            v[j+1]=s1*s2*r2*d;
            v[j]=s1*s3*r3*d;
        } else if (d2 >= d1 && d2 >= d3) {
            r1=d1/d2;
            r3=d3/d2;
            d=sqrt(1.0+r1*r1+r3*r3);
            c = -1.0-(s1*r1/d);
            d=1.0/(r1+d);
            v[j+1]=s1*s2*d;
            v[j]=s1*s3*r3*d;
        } else {
            r1=d1/d3;
            r2=d2/d3;
            d=sqrt(1.0+r1*r1+r2*r2);
            c = -1.0-(s1*r1/d);
            d=1.0/(r1+d);
            v[j+1]=s1*s2*r2*d;
            v[j]=s1*s3*d;
        }
        v[j+2]=1.0;
        hshvectam(l,u,j,j+2,c,v,a);
        hshvectam(l,u,j,j+2,c,v,b);
        free_real_vector(v,j);
    }
}
```

3.15 Singular values

3.15.1 Real bidiagonal matrices

A. qrisngvalbid

Computes, by use of a variant of the QR algorithm the singular values of a bidiagonal $n \times n$ matrix A, i.e. the elements $d_1,...,d_n$ of the diagonal matrix D for which $A=UDV^T$, where $U^TU=V^TV=I$ ($n \times n$ unit matrix).

Function Parameters:

int qrisngvalbid (*d,b,n,em*)

qrisngvalbid: given the number of singular values not found, i.e. a number not equal to zero if the number of iterations exceeds *em[4]*;

d: float *d[1:n]*;
entry: the diagonal of the bidiagonal matrix;
exit: the singular values;

b: float *b[1:n]*;
entry: the super diagonal of the bidiagonal matrix in *b[1:n-1]*;

n: int;
entry: the length of *b* and *d*;

em: float *em[1:7]*;
entry:
em[1]: the infinity norm of the matrix;
em[2]: the relative precision in the singular values;
em[4]: the maximal number of iterations to be performed;
em[6]: the minimal non-neglectable singular value;
exit:
em[3]: the maximal neglected superdiagonal element;
em[5]: the number of iterations performed;
em[7]: the numerical rank of the matrix; i.e. the number of singular values greater than or equal to *em[6]*.

Method: The method is described in detail in [WiR71]. *qrisngvalbid* is a rewriting of part of the procedure SVD published there.

```
#include <math.h>

int qrisngvalbid(float d[], float b[], int n, float em[])
{
    int n1,k,k1,i,i1,count,max,rnk;
    float tol,bmax,z,x,y,g,h,f,c,s,min;

    tol=em[2]*em[1];
    count=0;
    bmax=0.0;
    max=em[4];
    min=em[6];
    rnk=n;
    do {
        k=n;
        n1=n-1;
        while (1) {
            k--;
            if (k <= 0) break;
            if (fabs(b[k]) >= tol) {
                if (fabs(d[k]) < tol) {
                    c=0.0;
                    s=1.0;
                    for (i=k; i<=n1; i++) {
                        f=s*b[i];
                        b[i] *= c;
                        i1=i+1;
                        if (fabs(f) < tol) break;
                        g=d[i1];
                        d[i1]=h=sqrt(f*f+g*g);
                        c=g/h;
                        s = -f/h;
                    }
                    break;
                }
            } else {
                if (fabs(b[k]) > bmax) bmax=fabs(b[k]);
                break;
            }
        }
```

```
            if (k == n1) {
                if (d[n] < 0.0) d[n] = -d[n];
                if (d[n] <= min) rnk--;
                n=n1;
            } else {
                count++;
                if (count > max) break;
                k1=k+1;
                z=d[n];
                x=d[k1];
                y=d[n1];
                g = (n1 == 1) ? 0.0 : b[n1-1];
                h=b[n1];
                f=((y-z)*(y+z)+(g-h)*(g+h))/(2.0*h*y);
                g=sqrt(f*f+1.0);
                f=((x-z)*(x+z)+h*(y/((f < 0.0) ? f-g : f+g)-h))/x;
                c=s=1.0;
                for (i=k1+1; i<=n; i++) {
                    i1=i-1;
                    g=b[i1];
                    y=d[i];
                    h=s*g;
                    g *= c;
                    z=sqrt(f*f+h*h);
                    c=f/z;
                    s=h/z;
                    if (i1 != k1) b[i1-1]=z;
                    f=x*c+g*s;
                    g=g*c-x*s;
                    h=y*s;
                    y *= c;
                    d[i1]=z=sqrt(f*f+h*h);
                    c=f/z;
                    s=h/z;
                    f=c*g+s*y;
                    x=c*y-s*g;
                }
                b[n1]=f;
                d[n]=x;
            }
        } while (n > 0);
        em[3]=bmax;
        em[5]=count;
        em[7]=rnk;
        return n;
}
```

B. qrisngvaldecbid

Computes by use of a variant of the QR algorithm the singular value decomposition of an $m \times n$ matrix A ($m \geq n$), i.e. the $m \times n$ matrix U, the $n \times n$ diagonal matrix D, and the $n \times n$ matrix V for which $A=UDV^T$, where $U^TU=V^TV=I$ ($n \times n$ unit matrix). It is assumed that A has been reduced to bidiagonal form by preliminary pre- and post-multiplication by matrices of the form

$$I - 2uu^T/u^Tu,$$

by use of *hshreabid*.

Function Parameters:

int qrisngvaldecbid (*d,b,m,n,u,v,em*)

qrisngvaldecbid: given the number of singular values not found, i.e. a number not equal to zero if the number of iterations exceeds *em[4]*;

d: float *d[1:n]*;
  entry: the diagonal of the bidiagonal matrix;
  exit: the singular values;

b: float *b[1:n]*;
  entry: the super diagonal of the bidiagonal matrix in *b[1:n-1]*;
m: int;
  entry: the number of rows of the matrix *u*;
n: int;
  entry: the length of *b* and *d*, the number of columns of *u* and the number of columns and rows of *v*;
u: float *u[1:m,1:n]*;
  entry: the premultiplying matrix as produced by *pretfmmat*;
  exit: the premultiplying matrix *U* of the singular value decomposition UDV^T;
v: float *v[1:n,1:n]*;
  entry: the transpose of the postmultiplying matrix as produced by *psttfmmat*;
  exit: the transpose of the postmultiplying matrix *V* of the singular value decomposition;
em: float *em[1:7]*;
  entry:
  em[1]: the infinity norm of the matrix;
  em[2]: the relative precision in the singular values;
  em[4]: the maximal number of iterations to be performed;
  em[6]: the minimal non-neglectable singular value;
  exit:
  em[3]: the maximal neglected superdiagonal element;
  em[5]: the number of iterations performed;
  em[7]: the numerical rank of the matrix; i.e. the number of singular values greater than or equal to *em[6]*.

Function used: rotcol.

Method: The method is described in detail in [WiR71]. *qrisngvaldecbid* is a rewriting of part of the procedure SVD published there.

```
#include <math.h>

int qrisngvaldecbid(float d[], float b[], int m, int n, float **u,
                    float **v, float em[])
{
    void rotcol(int, int, int, int, float **, float, float);
    int n0,n1,k,k1,i,i1,count,max,rnk;
    float tol,bmax,z,x,y,g,h,f,c,s,min;

    tol=em[2]*em[1];
    count=0;
    bmax=0.0;
    max=em[4];
    min=em[6];
    rnk=n0=n;
    do {
        k=n;
        n1=n-1;
        while (1) {
            k--;
            if (k <= 0) break;
            if (fabs(b[k]) >= tol) {
                if (fabs(d[k]) < tol) {
                    c=0.0;
                    s=1.0;
                    for (i=k; i<=n1; i++) {
                        f=s*b[i];
```

```
                        b[i] *= c;
                        i1=i+1;
                        if (fabs(f) < tol) break;
                        g=d[i1];
                        d[i1]=h=sqrt(f*f+g*g);
                        c=g/h;
                        s = -f/h;
                        rotcol(1,m,k,i1,u,c,s);
                    }
                    break;
                }
            } else {
                if (fabs(b[k]) > bmax) bmax=fabs(b[k]);
                break;
            }
        }
        if (k == n1) {
            if (d[n] < 0.0) {
                d[n] = -d[n];
                for (i=1; i<=n0; i++) v[i][n] = -v[i][n];
            }
            if (d[n] <= min) rnk--;
            n=n1;
        } else {
            count++;
            if (count > max) break;
            k1=k+1;
            z=d[n];
            x=d[k1];
            y=d[n1];
            g = (n1 == 1) ? 0.0 : b[n1-1];
            h=b[n1];
            f=((y-z)*(y+z)+(g-h)*(g+h))/(2.0*h*y);
            g=sqrt(f*f+1.0);
            f=((x-z)*(x+z)+h*(y/((f < 0.0) ? f-g : f+g)-h))/x;
            c=s=1.0;
            for (i=k1+1; i<=n; i++) {
                i1=i-1;
                g=b[i1];
                y=d[i];
                h=s*g;
                g *= c;
                z=sqrt(f*f+h*h);
                c=f/z;
                s=h/z;
                if (i1 != k1) b[i1-1]=z;
                f=x*c+g*s;
                g=g*c-x*s;
                h=y*s;
                y *= c;
                rotcol(1,n0,i1,i,v,c,s);
                d[i1]=z=sqrt(f*f+h*h);
                c=f/z;
                s=h/z;
                f=c*g+s*y;
                x=c*y-s*g;
                rotcol(1,m,i1,i,u,c,s);
            }
            b[n1]=f;
            d[n]=x;
        }
    } while (n > 0);
    em[3]=bmax;
    em[5]=count;
    em[7]=rnk;
    return n;
}
```

3.15.2 Real full matrices

A. qrisngval

Computes the singular values of a given matrix. The matrix is first transformed to bidiagonal form by calling *hshreabid*, and then the singular values are calculated by *qrisngvalbid*.

Function Parameters:

int qrisngval (*a,m,n,val,em*)

qrisngval: given the number of singular values not found, i.e. a number not equal to zero if the number of iterations exceeds *em[4]*;

a: float *a[1:m,1:n]*;
- entry: the input matrix;
- exit: data concerning the transformation to bidiagonal form;

m: int;
- entry: the number of rows of the matrix *a*;

n: int;
- entry: the number of columns of *a*, *n* should satisfy $n \leq m$;

val: float *val[1:n]*;
- exit: the singular values;

em: float *em[0:7]*;
- entry:
  - *em[0]*: the machine precision;
  - *em[2]*: the relative precision in the singular values;
  - *em[4]*: the maximal number of iterations to be performed;
  - *em[6]*: the minimal non-neglectable singular value;
- exit:
  - *em[1]*: the infinity norm of the matrix;
  - *em[3]*: the maximal neglected superdiagonal element;
  - *em[5]*: the number of iterations performed;
  - *em[7]*: the numerical rank of the matrix; i.e. the number of singular values greater than or equal to *em[6]*.

Functions used: hshreabid, qrisngvalbid.

```
int qrisngval(float **a, int m, int n, float val[], float em[])
{
    float *allocate_real_vector(int, int);
    void free_real_vector(float *, int);
    void hshreabid(float **, int, int, float [], float [], float []);
    int qrisngvalbid(float [], float [], int, float []);
    int i;
    float *b;

    b=allocate_real_vector(1,n);
    hshreabid(a,m,n,val,b,em);
    i=qrisngvalbid(val,b,n,em);
    free_real_vector(b,1);
    return i;
}
```

B. qrisngvaldec

Calculates the singular value decomposition $A=UDV^T$ of a given matrix A. The matrix is first transformed to bidiagonal form by calling *hshreabid*, the two transforming matrices are calculated by calling *psttfmmat* and *pretfmmat*, and finally the singular value decomposition is calculated by *qrisngvaldecbid*.

Function Parameters:

int qrisngvaldec (*a,m,n,val,v,em*)

qrisngvaldec: given the number of singular values not found, i.e. a number not equal to zero if the number of iterations exceeds *em[4]*;

a: float *a[1:m,1:n]*;
- entry: the given matrix;
- exit: the matrix U in the singular value decomposition UDV^T;

m: int;
- entry: the number of rows of the matrix *a*;

n: int;
- entry: the number of columns of *a*, *n* should satisfy $n \leq m$;

val: float *val[1:n]*;
- exit: the singular values;

v: float *v[1:n,1:n]*;
- exit: the transpose of matrix V in the singular value decomposition;

em: float *em[0:7]*;
- entry:
  - *em[0]*: the machine precision;
  - *em[2]*: the relative precision in the singular values;
  - *em[4]*: the maximal number of iterations to be performed;
  - *em[6]*: the minimal non-neglectable singular value;
- exit:
  - *em[1]*: the infinity norm of the matrix;
  - *em[3]*: the maximal neglected superdiagonal element;
  - *em[5]*: the number of iterations performed;
  - *em[7]*: the numerical rank of the matrix; i.e. the number of singular values greater than or equal to *em[6]*.

Functions used: hshreabid, psttfmmat, pretfmmat, qrisngvaldecbid.

```
int qrisngvaldec(float **a, int m, int n, float val[],
                 float **v, float em[])
{
   float *allocate_real_vector(int, int);
   void free_real_vector(float *, int);
   void hshreabid(float **, int, int, float [], float [], float []);
   void psttfmmat(float **, int, float **, float []);
   void pretfmmat(float **, int, int, float []);
   int qrisngvaldecbid(float [], float [], int, int, float **,
                       float **, float []);
   int i;
   float *b;

   b=allocate_real_vector(1,n);
   hshreabid(a,m,n,val,b,em);
   psttfmmat(a,n,v,b);
   pretfmmat(a,m,n,val);
   i=qrisngvaldecbid(val,b,m,n,a,v,em);
   free_real_vector(b,1);
   return i;
}
```

3.16 Zeros of polynomials

3.16.1 Zeros of general real polynomials

A. zerpol

Attempts to determine, by use of Laguerre's method [Dek66] and composite deflation [Ad67, PeW71, Rey77], the zeros of the polynomial

$$P(z) = \sum_{i=0}^{n} a_i z^i,$$

where the a_i are real and $a_n \neq 0$.

For real zeros, Laguerre's method makes use of the recursion

$$z_{k+1} = z_k - \frac{nf_k}{f_k' \pm ((n-1)^2(f_k')^2 - n(n-1)f_k f_k'')}.$$

Again for real zeros, the deflation used may by described as follows: with α a derived approximation to a zero of $P(z)$, and r determined from the condition that $|a_i\alpha^i|$ takes its maximum value for $0 \leq i \leq n$ when $i=r$, p_i and q_i are obtained by use of the recursions

$$p_{n-1} = a_n, \quad p_{i-1} = a_i + \alpha p_i \qquad (i=n-1,...,r+1)$$
$$q_0 = -a_0/\alpha, \quad q_i = (a_i - q_{i-1})/\alpha \qquad (i=1,...,r-1)$$

(with obvious modifications if $r=0$ or $r=n$), the polynomial

$$\sum_{i=0}^{r-1} q_i z^i + \sum_{i=r}^{n-1} p_i z^i$$

is formed; its roots are those of $P(z)$ with α omitted. If all zeros of $P(z)$ are real, the above deflation may be continued.

Function Parameters:

int zerpol (*n,a,em,re,im,d*)

zerpol: given the number, *k*, of zeros not found;

n: int;
entry: the degree of the polynomial;

a: float *a[0:n]*;
entry: the coefficients of the polynomial (values of a_i above);

em: float *em[0:4]*;
entry:
em[0]: the machine precision;
em[1]: the maximal number of iterations allowed for each zero (e.g. 40);
exit:
em[2]: fail indication;
0 successful call;

1 upon entry degree $n \le 0$;
2 upon entry leading coefficient $a[n]$=0;
3 number of iterations exceeded $em[1]$;

em[3]: number of new starts in the last iteration; if upon exit, *em[2]*=3 and *em[3]*<5 then it may be useful to start again with a higher value of *em[1]*;

em[4]: total number of iterations performed;

re,im: float *re[1:n], im[1:n]*;
exit: the real and imaginary parts of the zeros of the polynomial; the members of each nonreal complex conjugate pair are consecutive;

d: float *d[0:n]*;
exit: if the call is unsuccessful and only *n-k* zeros have been found, then *d[0:k]* contains the coefficients of the (deflated) polynomial; furthermore the zeros found are delivered in *re[k+1:n], im[k+1:n]*, whereas the remaining parts of *re* and *im* contain no information.

Functions used: comabs, comsqrt.

```
#include <float.h>
#include <math.h>

int zerpol(int n, float a[], float em[], float re[],
            float im[], float d[])
{
    float *allocate_real_vector(int, int);
    void free_real_vector(float *, int);
    float comabs(float, float);
    void comsqrt(float, float, float *, float *);
    int zerpolfunction(int, float [], float [], float,
                        float, float, int *, float *);
    int i,totit,it,fail,start,up,max,giex,itmax,control,ih,m,split;
    float x,y,newf,oldf,maxrad,ae,tol,h1,h2,ln2,f[6],tries[11],
            h,side,s1re,s1im,s2re,s2im,dx,dy,h3,h4,h5,h6,*b;

    totit=it=fail=up=start=0;
    ln2=log(2.0);
    newf=FLT_MAX;
    ae=FLT_MIN;
    giex=log(newf)/ln2-40.0;
    tol=em[0];
    itmax=em[1];
    for (i=0; i<=n; i++) d[i]=a[n-i];
    if (n <= 0)
        fail=1;
    else
        if (d[0] == 0.0) fail=2;
    if (fail > 0) {
        em[2]=fail;
        em[3]=start;
        em[4]=totit;
        for (i=(n-1)/2; i>=0; i--) {
            tol=d[i];
            d[i]=d[n-i];
            d[n-i]=tol;
        }
        return n;
    }
    while (d[n] == 0.0 && n > 0) {
        re[n]=im[n]=0.0;
        n--;
    }
    x=y=0.0;
    while (n > 2) {
        /* control */
        if (it > itmax) {
            totit += it;
```

```
        fail=3;
        em[2]=fail;
        em[3]=start;
        em[4]=totit;
        for (i=(n-1)/2; i>=0; i--) {
            tol=d[i];
            d[i]=d[n-i];
            d[n-i]=tol;
        }
        return n;
    } else
        if (it == 0) {
            maxrad=0.0;
            max=(giex-log(fabs(d[0]))/ln2)/n;
            for (i=1; i<=n; i++) {
                h1 = (d[i] == 0.0) ? 0.0 : exp(log(fabs(d[i]/d[0]))/i);
                if (h1 > maxrad) maxrad=h1;
            }
            for (i=1; i<=n-1; i++)
                if (d[i] != 0.0) {
                    ih=(giex-log(fabs(d[i]))/ln2)/(n-i);
                    if (ih < max) max=ih;
                }
            max=max*ln2/log(n);
            side = -d[1]/d[0];
            side = (fabs(side) < tol) ? 0.0 :
                        ((side > 0.0) ? 1.0 : -1.0);
            if (side == 0.0) {
                tries[7]=tries[2]=maxrad;
                tries[9] = -maxrad;
                tries[6]=tries[4]=tries[3]=maxrad/sqrt(2.0);
                tries[5] = -tries[3];
                tries[10]=tries[8]=tries[1]=0.0;
            } else {
                tries[8]=tries[4]=maxrad/sqrt(2.0);
                tries[1]=side*maxrad;
                tries[3]=tries[4]*side;
                tries[6]=maxrad;
                tries[7] = -tries[3];
                tries[9] = -tries[1];
                tries[2]=tries[5]=tries[10]=0.0;
            }
            if (comabs(x,y) > 2.0*maxrad) x=y=0.0;
            control=0;
        } else {
            if (it > 1 && newf >= oldf) {
                up++;
                if (up == 5 && start < 5) {
                    start++;
                    up=0;
                    x=tries[2*start-1];
                    y=tries[2*start];
                    control=0;
                } else
                    control=1;
            } else
                control=1;
        }   /* end of control */
    if (control) {
        /* laguerre */
        if (fabs(f[0]) > fabs(f[1])) {
            h1=f[0];
            h6=f[1]/h1;
            h2=f[2]+h6*f[3];
            h3=f[3]-h6*f[2];
            h4=f[4]+h6*f[5];
            h5=f[5]-h6*f[4];
            h6=h6*f[1]+h1;
        } else {
            h1=f[1];
            h6=f[0]/h1;
            h2=h6*f[2]+f[3];
            h3=h6*f[3]-f[2];
            h4=h6*f[4]+f[5];
```

```
                h5=h6*f[5]-f[4];
                h6=h6*f[0]+f[1];
            }
            s1re=h2/h6;
            s1im=h3/h6;
            h2=s1re*s1re-s1im*s1im;
            h3=2.0*s1re*s1im;
            s2re=h2-h4/h6;
            s2im=h3-h5/h6;
            h1=s2re*s2re+s2im*s2im;
            h1 = (h1 != 0.0) ? (s2re*h2+s2im*h3)/h1 : 1.0;
            m = (h1 > n-1) ? ((n > 1) ? n-1 : 1) : ((h1 > 1.0) ? h1 : 1);
            h1=(float)(n-m)/(float) m;
            comsqrt(h1*(n*s2re-h2),h1*(n*s2im-h3),&h2,&h3);
            if (s1re*h2+s1im*h3 < 0.0) {
                h2 = -h2;
                h3 = -h3;
            }
            h2 += s1re;
            h3 += s1im;
            h1=h2*h2+h3*h3;
            if (h1 == 0.0) {
                dx = -n;
                dy=n;
            } else {
                dx = -n*h2/h1;
                dy=n*h3/h1;
            }
            h1=fabs(x)*tol+ae;
            h2=fabs(y)*tol+ae;
            if (fabs(dx) < h1 && fabs(dy) < h2) {
                dx = (dx == 0.0) ? h1 : ((dx > 0.0) ? h1 : -h1);
                dy = (dy == 0.0) ? h2 : ((dy > 0.0) ? h2 : -h2);
            }
            x += dx;
            y += dy;
            if (comabs(x,y) > 2.0*maxrad) {
                h1 = (fabs(x) > fabs(y)) ? fabs(x) : fabs(y);
                h2=log(h1)/ln2+1.0-max;
                if (h2 > 0.0) {
                    h2=pow(2.0,h2);
                    x /= h2;
                    y /= h2;
                }
            }   /* end of laguerre */
        }
        oldf=newf;
        if (zerpolfunction(n,d,f,x,y,tol,&it,&newf)) {
            if (y != 0.0 && fabs(y) < 0.1) {
                h=y;
                y=0.0;
                if (!zerpolfunction(n,d,f,x,y,tol,&it,&newf)) y=h;
            }
            re[n]=x;
            im[n]=y;
            if (y != 0.0) {
                re[n-1]=x;
                im[n-1] = -y;
            }
            /* deflation */
            if (x == 0.0 && y == 0.0)
                n--;
            else {
                b=allocate_real_vector(0,n-1);
                if (y == 0.0) {
                    n--;
                    b[n] = -d[n+1]/x;
                    for (i=1; i<=n; i++) b[n-i]=(b[n-i+1]-d[n-i+1])/x;
                    for (i=1; i<=n; i++) d[i] += d[i-1]*x;
                } else {
                    h1 = -2.0*x;
                    h2=x*x+y*y;
                    n -= 2;
                    b[n]=d[n+2]/h2;
```

```
                    b[n-1]=(d[n+1]-h1*b[n])/h2;
                    for (i=2; i<=n; i++)
                        b[n-i]=(d[n-i+2]-h1*b[n-i+1]-b[n-i+2])/h2;
                    d[1] -= h1*d[0];
                    for (i=2; i<=n; i++) d[i] -= h1*d[i-1]+h2*d[i-2];
                }
                split=n;
                h2=fabs(d[n]-b[n])/(fabs(d[n])+fabs(b[n]));
                for (i=n-1; i>=0; i--) {
                    h1=fabs(d[i])+fabs(b[i]);
                    if (h1 > tol) {
                        h1=fabs(d[i]-b[i])/h1;
                        if (h1 < h2) {
                            h2=h1;
                            split=i;
                        }
                    }
                }
                for (i=split+1; i<=n; i++) d[i]=b[i];
                d[split]=(d[split]+b[split])/2.0;
                free_real_vector(b,0);
            }   /* end of deflation */
            totit += it;
            up=start=it=0;
        }
    }
    if (n == 1) {
        re[1] = -d[1]/d[0];
        im[1]=0.0;
    } else {
        h1 = -0.5*d[1]/d[0];
        h2=h1*h1-d[2]/d[0];
        if (h2 >= 0.0) {
            re[2] = (h1 < 0.0) ? h1-sqrt(h2) : h1+sqrt(h2);
            re[1]=d[2]/(d[0]*re[2]);
            im[2]=im[1]=0.0;
        } else {
            re[2]=re[1]=h1;
            im[2]=sqrt(-h2);
            im[1] = -im[2];
        }
    }
    em[2]=fail;
    em[3]=start;
    em[4]=totit;
    return 0;
}

int zerpolfunction(int n, float d[], float f[], float x,
                        float y, float tol, int *it, float *newf)
{
    /* this function is used internally by ZERPOL */

    int k,m1,m2;
    float p,q,qsqrt,f01,f02,f03,f11,f12,f13,f21,f22,f23,stop;

    (*it)++;
    p=2.0*x;
    q = -(x*x+y*y);
    qsqrt=sqrt(-q);
    f01=f11=f21=d[0];
    f02=f12=f22=0.0;
    m1=n-4;
    m2=n-2;
    stop=fabs(f01)*0.8;
    for (k=1; k<=m1; k++) {
        f03=f02;
        f02=f01;
        f01=d[k]+p*f02+q*f03;
        f13=f12;
        f12=f11;
        f11=f01+p*f12+q*f13;
        f23=f22;
        f22=f21;
```

```
            f21=f11+p*f22+q*f23;
            stop=qsqrt*stop+fabs(f01);
        }
        if (m1 < 0) m1=0;
        for (k=m1+1; k<=m2; k++) {
            f03=f02;
            f02=f01;
            f01=d[k]+p*f02+q*f03;
            f13=f12;
            f12=f11;
            f11=f01+p*f12+q*f13;
            stop=qsqrt*stop+fabs(f01);
        }
        if (n == 3) f21=0.0;
        f03=f02;
        f02=f01;
        f01=d[n-1]+p*f02+q*f03;
        f[0]=d[n]+x*f01+q*f02;
        f[1]=y*f01;
        f[2]=f01-2.0*f12*y*y;
        f[3]=2.0*y*(-x*f12+f11);
        f[4]=2.0*(-x*f12+f11)-8.0*y*y*(-x*f22+f21);
        f[5]=y*(6.0*f12-8.0*y*y*f22);
        stop=qsqrt*(qsqrt*stop+fabs(f01))+fabs(f[0]);
        *newf=f02=comabs(f[0],f[1]);
        return (f02 < (2.0*fabs(x*f01)-8.0*(fabs(f[0])+fabs(f01)*qsqrt)+
                    10.0*stop)*tol*pow(1.0+tol,4*n+3.0));
}
```

B. bounds

Calculates upper bounds for the absolute error in given approximated zeros of a polynomial with real coefficients.

Given approximations α_i (i=1,...,n) to the zeros of a polynomial

$$P(z) = \sum_{i=0}^{n} a_i z^i$$

with real coefficients, determines centers γ_l and radii r_l of disjoint discs D_l together with positive integers m_l such that m_l zeros of $P(z)$ are contained in D_l (l=1,...,n'; m_1+...+$m_{n'}$=n).

The results upon which the method used is based may be described as follows. Let

$$R(z) = a_n \prod_{i=1}^{n} (z - \alpha_i) \quad \text{and} \quad Q(z) = R(z) - P(z) = \sum_{k=1}^{n-1} \varepsilon_k z^k .$$

Let m of the α_i, namely $\alpha_{k(i)}$ (i=1,...,m) lie near to each other; let

$$\prod_{i=1}^{n} .^{(k)} \quad \text{denote a product} \quad \prod_{i=1}^{n}$$

from which the terms with suffix $k(i)$ (i=1,...,m) have been removed; set

$$\gamma = (\sum_{i=1}^{m} \alpha_{k(i)})/m$$

and $\beta_i = \alpha_i - \gamma$ (i=1,...,n). For z on the circle C with center γ and suitable radius r,

$$|R(z)| \geq |a_n| \prod_{i=1}^{m} (r - |\beta_{k(i)}|) \prod_{i=1}^{n}{}^{.(k)} (|\beta_i| - r)$$

$$|Q(z)| \leq \sum_{k=0}^{n-1} |\varepsilon_k| (|\gamma| + r)^k .$$

If $|Q(z)| < |R(z)|$ for all z on C, then $R(z)$ and $R(z)-Q(z)=P(z)$ have the same number of zeros in C, i.e. if

$$\frac{\sum_{k=0}^{m-1} |\varepsilon_k| (|\gamma| + r)^k}{|a_n| \prod_{i=1}^{n}{}^{.(k)} (|\beta_i| - r)} < \prod_{i=1}^{m} (r - |\beta_{k(i)}|)$$

C contains m zeros of $P(z)$. A suitable value of r may be estimated by iteration: r_1 is determined from the condition

$$(r_1 - d)^m = \frac{\sum_{k=0}^{n-1} \varepsilon_k |\gamma|^k}{|a_n| \prod_{i=1}^{n}{}^{.(k)} |\beta_i|}$$

where $d = \max |\beta_{k(i)}|$ ($1 \leq i \leq m$), and r from the condition that $(r-d)^m = 1.1(r_1-d)^m$. For a more detailed description see [PeW71, Rey77].

Function Parameters:

void bounds (*n,a,re,im,rele,abse,recentre,imcentre,bound*)

n: int;
  entry: the degree of the polynomial;

a: float *a[0:n]*;
  entry: the coefficients of the polynomial of which *re[j]+im[j]**i are the approximated zeros (values of a_i above);

re,im: float *re[1:n], im[1:n]*;
  entry: real and imaginary parts of approximated zeros of a polynomial such that the members of each nonreal complex conjugate pair are consecutive (values of α_i above);
  exit: a permutation of the input data;

rele: float;
  entry: relative error in the non-vanishing coefficients *a[j]* of the given polynomial;

abse: float;
  entry: absolute error in the vanishing coefficients *a[j]* of the given polynomial; if there are no vanishing coefficients, *abse* should be zero;

recentre,imcentre: float *recentre[1:n], imcentre[1:n]*;
  exit: real and imaginary parts of the centers of disks in which some number of zeros of the polynomial given by *a* are situated; the number of identical centers denotes the number of zeros in that disk;

bound: float *bound[1:n]*;
  exit: radius of the disks whose centers are given correspondingly in *recentre* and *imcentre*.

```
#include <float.h>
#include <math.h>

void bounds(int n, float a[], float re[], float im[], float rele,
            float abse, float recentre[], float imcentre[],
            float bound[])
{
   float *allocate_real_vector(int, int);
   void free_real_vector(float *, int);
   void kcluster(int, int, int, float [], float [], float [],
                 float [], float [], float [], float []);
   int i,j,k,index1,index2,goon,place,clustin;
   float h,min,recent,imcent,xk,yk,zk,corr,*rc,*c,*rce,*clust,
         boundin,*wa1,*wa2,temp1,temp2;

   rc=allocate_real_vector(0,n);
   c=allocate_real_vector(0,n);
   rce=allocate_real_vector(0,n);
   clust=allocate_real_vector(1,n);
   rc[0]=c[0]=a[n];
   rce[0]=fabs(c[0]);
```

```
k=0;
for (i=1; i<=n; i++) {
    rc[i]=rce[i]=0.0;
    c[i]=a[n-i];
}
while (k < n) {
    k++;
    xk=re[k];
    yk=im[k];
    zk=xk*xk+yk*yk;
    for (j=k; j>=1; j--) rce[j] += rce[j-1]*sqrt(zk);
    if (yk == 0.0)
        for (j=k; j>=1; j--) rc[j] -= xk*rc[j-1];
    else {
        k++;
        if (k <= n && xk == re[k] && yk == -im[k]) {
            xk = -2.0*xk;
            for (j=k; j>=1; j--) rce[j] += rce[j-1]*sqrt(zk);
            for (j=k; j>=2; j--) rc[j] += xk*rc[j-1]+zk*rc[j-2];
            rc[1] += xk*rc[0];
        }
    }
}
rc[0]=rce[0];
corr=1.06*FLT_MIN;
for (i=1; i<=n-1; i++)
    rc[i]=fabs(rc[i]-c[i])+rce[i]*corr*(n+i-2)+
            rele*fabs(c[i])+abse;
rc[n]=fabs(rc[n]-c[n])+rce[n]*corr*(n-1)+rele*fabs(c[n])+abse;
for (i=1; i<=n; i++)
    kcluster(1,i,n,rc,re,im,recentre,imcentre,bound,clust);
goon=1;
while (goon) {
    index1=index2=0;
    min=FLT_MAX;
    i=n-clust[n]+1;
    while (i >= 2) {
        j=i;
        recent=recentre[i];
        imcent=imcentre[i];
        while (j >= 2) {
            j -= clust[j-1];
            temp1=recent-recentre[j];
            temp2=imcent-imcentre[j];
            h=sqrt(temp1*temp1+temp2*temp2);
            if (h < bound[i]+bound[j] && h <= min) {
                index1=j;
                index2=i;
                min=h;
            }
        }
        i -= clust[i-1];
    }
    if (index1 == 0)
        goon=0;
    else {
        if (imcentre[index1] == 0.0) {
            if (imcentre[index2] != 0.0) clust[index2] *= 2.0;
        }
        else
            if (imcentre[index2] == 0.0) clust[index1] *= 2.0;
        k=index1+clust[index1];
        if (k != index2) {
            /* shift */
            wa1=allocate_real_vector(1,clust[index2]);
            wa2=allocate_real_vector(1,clust[index2]);
            clustin=clust[index2];
            boundin=bound[index2];
            imcent=imcentre[index2];
            recent=recentre[index2];
            for (j=1; j<=clustin; j++) {
                place=index2+j-1;
                wa1[j]=re[place];
                wa2[j]=im[place];
```

```
                }
                for (j=index2-1; j>=k; j--) {
                    place=j+clustin;
                    re[place]=re[j];
                    im[place]=im[j];
                    clust[place]=clust[j];
                    bound[place]=bound[j];
                    recentre[place]=recentre[j];
                    imcentre[place]=imcentre[j];
                }
                for (j=k+clustin-1; j>=k; j--) {
                    place=j+1-k;
                    re[j]=wa1[place];
                    im[j]=wa2[place];
                    bound[j]=boundin;
                    clust[j]=clustin;
                    recentre[j]=recent;
                    imcentre[j]=imcent;
                }
                free_real_vector(wa1,1);
                free_real_vector(wa2,1);
            }   /* end of shift */
            k=clust[index1]+clust[k];
            kcluster(k,index1,n,rc,re,im,recentre,imcentre,
                        bound,clust);
        }
    }
    free_real_vector(rc,0);
    free_real_vector(c,0);
    free_real_vector(rce,0);
    free_real_vector(clust,1);
}

void kcluster(int k, int m, int n, float rc[], float re[],
                float im[], float recentre[], float imcentre[],
                float bound[], float clust[])
{
    /* this function is used internally by BOUNDS */

    int i,stop,l,nonzero;
    float recent,imcent,d,prod,rad,gr,r,*dist,s,h1,h2,temp1,temp2;

    dist=allocate_real_vector(m,m+k-1);
    recent=re[m];
    imcent=im[m];
    stop=m+k-1;
    l = (imcent == 0.0) ? 0 : ((imcent > 0.0) ? 1 : -1);
    nonzero = (l != 0);
    for (i=m+1; i<=stop; i++) {
        recent += re[i];
        if (nonzero) {
            nonzero=(l == ((im[i] == 0.0) ? 0 :
                        ((im[i]>0.0) ? 1 : -1)));
            imcent += im[i];
        }
    }
    recent /= k;
    imcent = (nonzero ? imcent/k : 0.0);
    d=0.0;
    rad=0.0;
    for (i=m; i<=stop; i++) {
        recentre[i]=recent;
        imcentre[i]=imcent;
        temp1=re[i]-recent;
        temp2=im[i]-imcent;
        dist[i]=sqrt(temp1*temp1+temp2*temp2);
        if (d < dist[i]) d=dist[i];
    }
    s=sqrt(recent*recent+imcent*imcent);
    h1=rc[1];
    h2=rc[0];
    for (i=2; i<=n; i++) h1=h1*s+rc[i];
    for (i=1; i<=m-1; i++) {
        temp1=re[i]-recent;
```

```
            temp2=im[i]-imcent;
            h2 *= fabs(sqrt(temp1*temp1+temp2*temp2));
        }
        for (i=m+k; i<=n; i++) {
            temp1=re[i]-recent;
            temp2=im[i]-imcent;
            h2 *= fabs(sqrt(temp1*temp1+temp2*temp2));
        }
        gr=fabs((h1 == 0.0) ? 0.0 : ((h2 == 0.0) ? 10.0 : h1/h2));
        if (gr > 0.0)
            do {
                r=rad;
                rad=d+exp(log(1.1*gr)/k);
                if (rad == r) rad *= exp(log(1.1)/k);
                s=sqrt(recent*recent+imcent*imcent)+rad;
                h1=rc[1];
                h2=rc[0];
                for (i=2; i<=n; i++) h1=h1*s+rc[i];
                for (i=1; i<=m-1; i++) {
                    temp1=re[i]-recent;
                    temp2=im[i]-imcent;
                    h2 *= fabs(sqrt(temp1*temp1+temp2*temp2)-rad);
                }
                for (i=m+k; i<=n; i++) {
                    temp1=re[i]-recent;
                    temp2=im[i]-imcent;
                    h2 *= fabs(sqrt(temp1*temp1+temp2*temp2)-rad);
                }
                gr=(h1 == 0.0) ? 0.0 : ((h2 == 0.0) ? -10.0 : h1/h2);
                prod=1.0;
                for (i=m; i<=stop; i++) prod *= (rad-dist[i]);
            } while (prod <= gr);
        for (i=m; i<=stop; i++) {
            bound[i]=rad;
            clust[i]=k;
        }
        free_real_vector(dist,m);
}
```

3.16.2 Zeros of orthogonal polynomials

A. allzerortpol

Calculates all zeros of an orthogonal polynomial given by the coefficients of their recurrence relation. *allzerortpol* determines the roots of the polynomial $p_n(x)$, where

$$p_0(x) = 1, \quad p_1(x) = x - b_0$$
$$p_{k+1}(x) = (x-b_k)p_k(x) - c_k p_{k-1}(x) \qquad (k=1,...,n-1)$$

the b_k and c_k are being given ($c_k>0$, $k=1,...,n-1$).

The roots of $p_n(x)$ are the eigenvalues of the $n \times n$ tridiagonal matrix T for which

$$T_{i,i}=b_{i-1} \ (i=1,...,n), \quad T_{i,i+1}=T_{i+1,i}=c_i^{1/2} \ (i=1,...,n-1).$$

These eigenvalues are obtained by QR iteration, using a call of *qrivalsymtri*.

Function Parameters:

void allzerortpol (*n,b,c,zer,em*)

n: int;
  entry: the degree of the orthogonal polynomial of which the zeros are to be calculated;

b,c: float *b[0:n-1]*, *c[0:n-1]*;
  entry: the elements *b[i]* and *c[i]*, *i=0,1,...,n-1*, contain the coefficients of the recurrence relation $p_{i+1}(x)=(x-b[i])*p_i(x)-c[i]*p_{i-1}(x)$, $i=0,1,...,n-1$, assuming *c[0]*=0;

exit: the contents of the arrays b and c are not altered;

zer: float *zer[1:n]*;
exit: the zeros of the n-th degree orthogonal polynomial;

em: float *em[0:5]*;
entry:
em[0]: the machine precision;
em[2]: the relative tolerance of the zeros;
em[4]: the maximal allowed number of iterations (e.g. 5*n);
exit:
em[1]: the value of max(|*b[0]*|+1, *c[i]*+|*b[i]*|+1 ($i=1,...,n-2$), *c[n-1]*+|*b[n-1]*|);
em[3]: the maximum absolute value of the codiagonal elements neglected;
em[5]: the number of iterations performed.

Functions used: qrivalsymtri, dupvec.

References: see the procedure *selzerortpol*.

```
#include <math.h>

void allzerortpol(int n, float b[], float c[], float zer[], float em[])
{
    float *allocate_real_vector(int, int);
    void free_real_vector(float *, int);
    int qrivalsymtri(float [], float [], int, float []);
    void dupvec(int, int, int, float [], float []);
    int i;
    float nrm,*bb;

    bb=allocate_real_vector(1,n);
    nrm=fabs(b[0]);
    for (i=1; i<=n-2; i++)
        if (c[i]+fabs(b[i]) > nrm) nrm=c[i]+fabs(b[i]);
    if (n > 1)
        nrm = (nrm+1 >= c[n-1]+fabs(b[n-1])) ? nrm+1.0 :
                (c[n-1]+fabs(b[n-1]));
    em[1]=nrm;
    for (i=n; i>=1; i--) zer[i]=b[i-1];
    dupvec(1,n-1,0,bb,c);
    qrivalsymtri(zer,bb,n,em);
    free_real_vector(bb,1);
}
```

B. lupzerortpol

Calculates a number of adjacent upper or lower zeros of an orthogonal polynomial given by the coefficients of their recurrence relation. *lupzerortpol* determines either the m smaller roots or the m larger roots of the polynomial $p_n(x)$, where

$$p_0(x) = 1, \quad p_1(x) = x - b_0$$

$$p_{k+1}(x) = (x-b_k)p_k(x) - c_k p_{k-1}(x) \qquad (k=1,...,n-1)$$

the b_k and c_k are being given ($c_k>0$, $k=1,...,n-1$).

The roots of $p_n(x)$ are the eigenvalues of the $n \times n$ tridiagonal matrix T for which

$$T_{i,i}=b_{i-1} \; (i=1,...,n), \quad T_{i,i+1}=T_{i+1,i}=c_i^{1/2} \; (i=1,...,n-1).$$

The m smaller eigenvalues of this matrix are determined by use of a rational variant of the QR algorithm. The m larger roots of $p_n(x)$ may be obtained by supplying $-b_k$ in place of b_k in the above ($k=0,...,n-1$) so that the roots of $p_n(-x)$ are found, and reversing the signs attached to the values of the roots determined in this way.

Function Parameters:

void lupzerortpol (n,m,b,c,zer,em)

n: int;
 entry: the degree of the orthogonal polynomial of which the zeros are to be calculated;

m: int;
 entry: the number of zeros to be calculated;

b,c: float $b[0:n-1]$, $c[0:n-1]$;
 entry: the elements $b[i]$ and $c[i]$, $i=0,1,...,n-1$, contain the coefficients of the recurrence relation $p_{i+1}(x)=(x-b[i])*p_i(x)-c[i]*p_{i-1}(x)$, $i=0,1,...,n-1$, assuming $c[0]=0$;
 exit: the contents of the arrays b and c are altered;

zer: float $zer[1:m]$;
 exit: the m lowest zeros are delivered; if the array $b[0:n-1]$ contained the opposite values of the corresponding recurrence coefficients then the opposite values of the m upper zeros are delivered; in either case $zer[i] < zer[i+1]$, $i=1,...,m-1$;

em: float $em[0:6]$;
 entry:
 $em[0]$: the machine precision;
 $em[2]$: the relative tolerance of the zeros;
 $em[4]$: the maximal allowed number of iterations (e.g. $15*m$);
 $em[6]$: if all zeros are known to be positive then 1 else 0;
 exit:
 $em[1]$: the value of $\max(|b[0]|+1,\ c[i]+|b[i]|+1\ (i=1,...,n-2),\ c[n-1]+|b[n-1]|)$;
 $em[3]$: the maximum absolute value of the theoretical errors of the zeros;
 $em[5]$: the number of iterations performed.

Functions used: dupvec, infnrmvec.

Reference: see the procedure *selzerortpol*.

```
#include <math.h>

void lupzerortpol(int n, int m, float b[], float c[], float zer[],
                  float em[])
{
    float infnrmvec(int, int, int *, float []);
    void dupvec(int, int, int, float [], float []);
    int i,posdef,j,k,t,converge;
    float nrm,dlam,eps,delta,e,ep,err,p,q,qp,r,s,tot;

    nrm=fabs(b[0]);
    for (i=1; i<=n-2; i++)
        if (c[i]+fabs(b[i]) > nrm) nrm=c[i]+fabs(b[i]);
    if (n > 1)
        nrm = (nrm+1 >= c[n-1]+fabs(b[n-1])) ? nrm+1.0 :
                  (c[n-1]+fabs(b[n-1]));
    em[1]=nrm;
    for (i=n; i>=1; i--) b[i]=b[i-1];
    for (i=n; i>=2; i--) c[i]=c[i-1];
    posdef = (em[6] == 1.0);
    dlam=em[2];
    eps=em[0];
    c[1]=err=q=s=0.0;
    tot=b[1];
    for (i=n; i>=1; i--) {
        p=q;
        q=sqrt(c[i]);
```

```
            e=b[i]-p-q;
            if (e < tot) tot=e;
        }
        if (posdef && (tot < 0.0))
            tot=0.0;
        else
            for(i=1; i<=n; i++) b[i] -= tot;
        t=0;
        for (k=1; k<=m; k++) {
            converge=0;
            /* next qr transformation */
            do {
                t++;
                tot += s;
                delta=b[n]-s;
                i=n;
                e=fabs(eps*tot);
                if (dlam < e) dlam=e;
                if (delta <= dlam) {
                    converge=1;
                    break;
                }
                e=c[n]/delta;
                qp=delta+e;
                p=1.0;
                for (i=n-1; i>=k; i--) {
                    q=b[i]-s-e;
                    r=q/qp;
                    p=p*r+1.0;
                    ep=e*r;
                    b[i+1]=qp+ep;
                    delta=q-ep;
                    if (delta <= dlam) {
                        converge=1;
                        break;
                    }
                    e=c[i]/q;
                    qp=delta+e;
                    c[i+1]=qp*ep;
                }
                if (converge) break;
                b[k]=qp;
                s=qp/p;
            } while (tot+s > tot);      /* end of qr transformation */
            if (!converge) {
                /* irregular end of iteration,
                    deflate minimum diagonal element */
                s=0.0;
                i=k;
                delta=qp;
                for (j=k+1; j<=n; j++)
                    if (b[j] < delta) {
                        i=j;
                        delta=b[j];
                    }
            }
            /* convergence */
            if (i < n) c[i+1]=c[i]*e/qp;
            for (j=i-1; j>=k; j--) {
                b[j+1]=b[j]-s;
                c[j+1]=c[j];
            }
            b[k]=tot;
            c[k] = err += fabs(delta);
        }
        em[5]=t;
        em[3]=infnrmvec(1,m,&t,c);
        dupvec(1,m,0,zer,b);
}
```

C. selzerortpol

Calculates a number of adjacent zeros of an orthogonal polynomial given by the coefficients of their recurrence relation [GolW69, L57, St72]. *selzerortpol* determines the subset λ_k ($k=n1,n1+1,...,n2$) of the total set of roots λ_k ($k=1,...,n$) (arranged in descending order) of the polynomial $p_n(x)$, where

$$p_0(x) = 1, \quad p_1(x) = x - b_0$$
$$p_{k+1}(x) = (x-b_k)p_k(x) - c_k p_{k-1}(x) \qquad (k=1,...,n-1)$$

the b_k and c_k are being given ($c_k>0$, $k=1,...,n-1$).

The roots of $p_n(x)$ are the eigenvalues of the $n\times n$ tridiagonal matrix T for which

$$T_{i,i}=b_{i-1} \; (i=1,...,n), \quad T_{i,i+1}=T_{i+1,i}=c_i^{1/2} \; (i=1,...,n-1).$$

These eigenvalues are obtained by means of a call of *valsymtri*.

It is efficient to use *allzerortpol* if more than 50 percent of extreme zeros or more than 25 percent of selected zeros are wanted.

Function Parameters:

void selzerortpol (*n,n1,n2,b,c,zer,em*)

n,n1,n2: int;
- entry: the degree of the orthogonal polynomial of which the *n1*-th up to and including *n2*-th zeros are to be calculated (*zer[n1]* ≥ *zer[n2]*);

b,c: float *b[0:n-1], c[0:n-1]*;
- entry: the elements *b[i]* and *c[i]*, *i=0,1,...,n-1*, contain the coefficients of the recurrence relation $p_{i+1}(x)=(x-b[i])*p_i(x)-c[i]*p_{i-1}(x)$, $i=0,1,...,n-1$, assuming *c[0]*=0;
- exit: the contents of the arrays *b* and *c* are not altered;

zer: float *zer[n1:n2]*;
- exit: the *n2-n1+1* calculated zeros in decreasing order;

em: float *em[0:5]*;
- entry:
  - *em[0]*: the machine precision;
  - *em[2]*: the relative tolerance of the zeros;
- exit:
  - *em[1]*: the value of max(|*b[0]*|+1, *c[i]*+|*b[i]*|+1 (*i=1,...,n-2*), *c[n-1]*+|*b[n-1]*|);
  - *em[5]*: the number of iterations performed.

Function used: valsymtri.

```
#include <math.h>

void selzerortpol(int n, int n1, int n2, float b[], float c[],
                  float zer[], float em[])
{
    float *allocate_real_vector(int, int);
    void free_real_vector(float *, int);
    void valsymtri(float [], float [], int, int, int,
                   float [], float []);
    int i;
    float nrm,*d;

    d=allocate_real_vector(1,n);
    nrm=fabs(b[0]);
    for (i=n-2; i>=1; i--)
        if (c[i]+fabs(b[i]) > nrm) nrm=c[i]+fabs(b[i]);
    if (n > 1)
        nrm = (nrm+1 >= c[n-1]+fabs(b[n-1])) ? nrm+1.0 :
                (c[n-1]+fabs(b[n-1]));
    em[1]=nrm;
```

```
    for (i=n; i>=1; i--) d[i]=b[i-1];
    valsymtri(d,c,n,n1,n2,zer,em);
    em[5]=em[3];
    free_real_vector(d,1);
}
```

D. alljaczer

Calculates all zeros of the n-th Jacobi polynomial $P_n^{(\alpha,\beta)}(x)$, see [AbS65]. The Jacobi polynomials satisfy the recursion

$$P_0^{(\alpha,\beta)}(x) = 1, \qquad P_1^{(\alpha,\beta)}(x) = ½(\alpha+\beta+2)x+½(\alpha-\beta),$$

$$2(k+1)(k+\alpha+\beta+1)(2k+\alpha+\beta)\ P_{k+1}^{(\alpha,\beta)}(x) =$$
$$(2k+\alpha+\beta+1)((2k+\alpha+\beta)(2k+\alpha+\beta+2)x+\alpha^2-\beta^2)\ P_k^{(\alpha,\beta)}(x) -$$
$$2k(k+\alpha)(k+\beta)(2k+\alpha+\beta+2)\ P_{k-1}^{(\alpha,\beta)}(x) \qquad (k=2,3,...)$$

and the coefficient of x^k in $P_k^{(\alpha,\beta)}(x)$ is

$$c_k = 2^{-k}\binom{2k+\alpha+\beta}{k}.$$

The polynomials $p_k(x) = P_k^{(\alpha,\beta)}(x)/c_k$ satisfy the recursion

$$p_0(x) = 1, \qquad p_1(x) = x - \frac{\beta-\alpha}{\alpha+\beta+2},$$

$$p_{k+1}(x) = (x - \frac{\beta^2-\alpha^2}{(2k+\alpha+\beta)(2k+\alpha+\beta+2)})p_k(x) -$$
$$\frac{4k(k+\alpha)(k+\beta)(k+\alpha+\beta)}{(2k+\alpha+\beta)^2(2k+\alpha+\beta-1)(2k+\alpha+\beta+1)}p_{k-1}(x).$$

The roots of $p_n(x)$ (i.e. those of $P_n^{(\alpha,\beta)}(x)$) may be obtained by a call of *allzerortpol*. However, for the special case in which α=ß,

$$P_{2m}^{(\alpha,\alpha)}(x) = c_m P_m^{(\alpha,-1/2)}(2x^2-1)$$
$$P_{2m-1}^{(\alpha,\alpha)}(x) = d_m x P_m^{(\alpha,1/2)}(2x^2-1)$$

where c_m and d_m are independent of x. Thus, in this special case, the determination of the roots of $P_n^{(\alpha,\beta)}(x)$ may slightly be simplified.

Function Parameters:

void alljaczer (*n,alfa,beta,zer*)

n: int;
entry: the upper bound of the array *zer*, $n \geq 1$;

alfa,beta: float;
entry: the parameters of the Jacobi polynomial (values of α and ß above);
alfa, beta > -1;

zer: float *zer[1:n]*;
exit: the zeros of the n-th Jacobi polynomial with parameters *alfa* and *beta*.

Function used: allzerortpol.

```
#include <float.h>
#include <math.h>
```

```
void alljaczer(int n, float alfa, float beta, float zer[])
{
    float *allocate_real_vector(int, int);
    void free_real_vector(float *, int);
    void allzerortpol(int, float [], float [], float [], float []);
    int i,m;
    float sum,min,*a,*b,em[6],gamma,zeri;

    if (alfa == beta) {
        a=allocate_real_vector(0,n/2);
        b=allocate_real_vector(0,n/2);
        m=n/2;
        if (n != 2*m) {
            gamma=0.5;
            zer[m+1]=0.0;
        } else
            gamma = -0.5;
        min=0.25-alfa*alfa;
        sum=alfa+gamma+2.0;
        a[0]=(gamma-alfa)/sum;
        a[1]=min/sum/(sum+2.0);
        b[1]=4.0*(1.0+alfa)*(1.0+gamma)/sum/sum/(sum+1.0);
        for (i=2; i<=m-1; i++) {
            sum=i+i+alfa+gamma;
            a[i]=min/sum/(sum+2.0);
            sum *= sum;
            b[i]=4.0*i*(i+alfa+gamma)*(i+alfa)*(i+gamma)/sum/(sum-1.0);
        }
        em[0]=FLT_MIN;
        em[2]=FLT_EPSILON;
        em[4]=6*m;
        allzerortpol(m,a,b,zer,em);
        for (i=1; i<=m; i++) {
            zer[i] = zeri = -sqrt((1.0+zer[i])/2.0);
            zer[n+1-i] = -zeri;
        }
        free_real_vector(a,0);
        free_real_vector(b,0);
    } else {
        a=allocate_real_vector(0,n);
        b=allocate_real_vector(0,n);
        min=(beta-alfa)*(beta+alfa);
        sum=alfa+beta+2.0;
        b[0]=0.0;
        a[0]=(beta-alfa)/sum;
        a[1]=min/sum/(sum+2.0);
        b[1]=4.0*(1.0+alfa)*(1.0+beta)/sum/sum/(sum+1.0);
        for (i=2; i<=n-1; i++) {
            sum=i+i+alfa+beta;
            a[i]=min/sum/(sum+2.0);
            sum *= sum;
            b[i]=4.0*i*(i+alfa+beta)*(i+alfa)*(i+beta)/(sum-1.0)/sum;
        }
        em[0]=FLT_MIN;
        em[2]=FLT_EPSILON;
        em[4]=6*n;
        allzerortpol(n,a,b,zer,em);
        free_real_vector(a,0);
        free_real_vector(b,0);
    }
}
```

E. alllagzer

Calculates all zeros of the n-th Laguerre polynomial $L_n^{(\alpha)}(x)$, see [AbS65]. The Laguerre polynomials satisfy the recursion

$$L_0^{(\alpha)}(x) = 1, \qquad L_1^{(\alpha)}(x) = \alpha+1-x,$$

$$(k+1)\, L_{k+1}^{(\alpha)}(x) = (2k+\alpha+1-x)\, L_k^{(\alpha)}(x) - (k+\alpha)\, L_{k-1}^{(\alpha)}(x)$$

and the coefficient of x^k in $L_k^{(\alpha)}(x)$ is $c_k=(-1)^k/k!$. The polynomials $p_k(x) = L_k^{(\alpha)}(x)/c_k$ satisfy the recursion

$$p_0(x) = 1, \qquad p_1(x) = x-\alpha-1,$$
$$p_{k+1}(x) = (x-2k-\alpha-1)\ p_k(x) - k(k+\alpha)\ p_{k-1}(x) \qquad (k=2,3,...).$$

The roots of $p_n(x)$ (i.e. those of $L_n^{(\alpha)}(x)$) are obtained by means of a call of *allzerortpol*.

Function Parameters:

void alllagzer (*n,alfa,zer*)

n: int;
entry: the upper bound of the array *zer*, $n \geq 1$;
alfa: float;
entry: the parameter of the Laguerre polynomial (value of α above); *alfa* > -1;
zer: float *zer[1:n]*;
exit: the zeros of the *n*-th Laguerre polynomial with parameters *alfa*.

Function used: allzerortpol.

```
#include <float.h>

void alllagzer(int n, float alfa, float zer[])
{
    float *allocate_real_vector(int, int);
    void free_real_vector(float *, int);
    void allzerortpol(int, float [], float [], float [], float []);
    int i;
    float *a,*b,em[6];

    a=allocate_real_vector(0,n);
    b=allocate_real_vector(0,n);
    b[0]=0.0;
    a[n-1]=n+n+alfa-1.0;
    for (i=1; i<=n-1; i++) {
        a[i-1]=i+i+alfa-1.0;
        b[i]=i*(i+alfa);
    }
    em[0]=FLT_MIN;
    em[2]=FLT_EPSILON;
    em[4]=6*n;
    allzerortpol(n,a,b,zer,em);
    free_real_vector(a,0);
    free_real_vector(b,0);
}
```

3.16.3 Zeros of complex polynomials

comkwd

Determines the roots *g* and *k* of the quadratic equation z^2-$2pz$-$q=0$ with complex *p* and *q*.

Function Parameters:

void comkwd (*pr,pi,qr,qi,gr,gi,kr,ki*)

pr,pi,qr,qi: float;
entry: *pr, qr* are the real parts and *pi, qi* are the imaginary parts of the coefficients of the quadratic equation:
$z^2 - 2(pr+pi*i)z - (qr+qi*i) = 0;$
gr,gi,kr,ki: float *;
exit: the real parts and the imaginary parts of the dinomial are delivered in *gr*,

kr and *gi, ki*, respectively; moreover, the modulus of *gr*+*gi**i is greater than or equal to the modulus of *kr*+*ki**i.

Functions used: commul, comdiv, comsqrt.

```
#include <math.h>

void comkwd(float pr, float pi, float qr, float qi,
            float *gr, float *gi, float *kr, float *ki)
{
    void commul(float, float, float, float, float *, float *);
    void comdiv(float, float, float, float, float *, float *);
    void comsqrt(float, float, float *, float *);
    float hr,hi;

    if (qr == 0.0 && qi == 0.0) {
        *kr = *ki = 0.0;
        *gr = pr*2.0;
        *gi = pi*2.0;
        return;
    }
    if (pr == 0.0 && pi == 0.0) {
        comsqrt(qr,qi,gr,gi);
        *kr = -(*gr);
        *ki = -(*gi);
        return;
    }
    if (fabs(pr) > 1.0 || fabs(pi) > 1.0) {
        comdiv(qr,qi,pr,pi,&hr,&hi);
        comdiv(hr,hi,pr,pi,&hr,&hi);
        comsqrt(1.0+hr,hi,&hr,&hi);
        commul(pr,pi,hr+1.0,hi,gr,gi);
    } else {
        comsqrt(qr+(pr+pi)*(pr-pi),qi+pr*pi*2.0,&hr,&hi);
        if (pr*hr+pi*hi > 0.0) {
            *gr = pr+hr;
            *gi = pi+hi;
        } else {
            *gr = pr-hr;
            *gi = pi-hi;
        }
    }
    comdiv(-qr,-qi,*gr,*gi,kr,ki);
}
```

4. Analytic Evaluations

4.1 Evaluation of an infinite series

A. euler

Applies the Euler transformation to the series

$$\sum_{i=1}^{\infty} a_i \, .$$

The course of the computations is determined by two parameters, both prescribed by the user: *eps*, a real tolerance; *tim*, a positive integer specifying the number of consecutive transformed sums whose agreement to within the tolerance *eps* is taken to imply that the last of them is an acceptable approximation to the Euler sum of the original series.

A set of numbers $M_{i,j}$, a sequence of integers $J(i)$, and sequences of partial sums S_i and terms t_i for which $S_{i+1}=S_i+t_{i+1}$ are computed as follows. Initially $M_{1,1}=a_1$, $J(1)=0$, $S_1=\frac{1}{2}a_1$. For $i \geq 1$, and with $M_{i+1,1}=a_{i+1}$, the numbers $M_{i+1,j+1} = \frac{1}{2}(M_{i,j}+M_{i+1,j})$ $(j=1,...,J(i))$ are computed. If $|M_{i+1,J(i)+1}| < |M_{i,J(i)}|$ then $J(i+1)=J(i)+1$ and $t_{i+1}=\frac{1}{2}M_{i+1,J(i)+1}$; otherwise $J(i+1)=J(i)$ and $t_{i+1}=M_{i+1,J(i)+1}$. If $|t_{i+1}|<eps$ for $i=I,...,I+tim-1$ for some $I>0$, the process terminated.

Function Parameters:

float euler (*ai,eps,tim*)

euler: delivers the computed sum of the infinite series;

ai: float *(*ai)(i)*; *int i*;
this function is the *i*-th term of the series, $i \geq 0$;

eps: float;
entry: summation criterion, see *tim* below;

tim: int;
entry: the summation is continued until *tim* successive terms of the transformed series are in absolute value less than *eps*.

```
#include <math.h>

float euler(float (*ai)(int), float eps, int tim)
{
    int i,k,n,t;
    float mn,mp,ds,sum,m[16];

    n=t=i=0;
    m[0]=(*ai)(i);
    sum=m[0]/2.0;
    do {
        i++;
        mn=(*ai)(i);
        for (k=0; k<=n; k++) {
            mp=(mn+m[k])/2.0;
            m[k]=mn;
            mn=mp;
        }
        if (fabs(mn) < fabs(m[n]) && n < 15) {
```

```
            ds=mn/2.0;
            n++;
            m[n]=mn;
        } else
            ds=mn;
        sum += ds;
        t = (fabs(ds) < eps) ? t+1 : 0;
    } while (t < tim);
    return sum;
}
```

B. sumposseries

Performs the summation of a convergent series with positive monotonically decreasing terms using the Van Wijngaarden transformation [Dan69, Vnw65] of the series to an alternating series.

sumposseries estimates the sum of a series of real terms

$$\sum_{i=1}^{\infty} v(i) \ . \tag{1}$$

by use of the transformation

$$\sum_{i=1}^{\infty} v(i) = \sum_{j=1}^{\infty} (-1)^{j-1} \sum_{k=1}^{\infty} 2^{k-1} v(2^{k-1}j) \ . \tag{2}$$

The above transformation may be derived from the identity

$$(1-x)^{-1} = \prod_{k=1}^{\infty} (1 + x^{2^{k-1}}). \qquad (|x|<1)$$

Differentiating throughout, and dividing the resulting relationship by the original version

$$x(1-x)^{-1} = \sum_{k=1}^{\infty} \frac{2^{k-1} x^{2^{k-1}}}{1+x^{2^{k-1}}}$$

$$= \sum_{j=1}^{\infty} (-1)^{j-1} \sum_{k=1}^{\infty} 2^{k-1} x^{2^{k-1}j} \ .$$

Replacing x by the displacement operator D ($Dv(i) = v(i+1)$) and applying both sides of the derived relationship to $v(0)$, the transformation (2) is obtained.

Euler's transformation may be applied to the series on the right hand side of relationship (2): denoting the Euler sum of a series by $E\{...\}$, relationship (2) yields

$$\sum_{i=1}^{\infty} v(i) = E\{\sum_{j=1}^{\infty} (-1)^{j-1} v'(j)\}$$

where

$$v'(j) = \sum_{k=1}^{\infty} 2^{k-1} v(2^{k-1} j) . \tag{3}$$

Each series (3) involves a subsequent of the terms of the series (1), and may converge far more rapidly than the latter. For example, when $v(i) = i^{-\alpha}$ $(\alpha>1)$,

$$\sum_{i=1}^{\infty} v(i) - \sum_{i=1}^{n} v(i) = O(n^{1-\alpha})$$

but

$$v'(j) - \sum_{k=1}^{n} 2^{k-1} v(2^{k-1} j) = O(2^{(1-\alpha)n}) .$$

(e.g. when $\alpha=2$, 2^{-n} decreases to zero more rapidly than n^{-1}.) Again when $v(i)=x^i$ $(|x|<1)$ the two remainder terms are $O(x^n)$ and $O(x^l)$ (where $l=2^n$) respectively. The auxiliary sums (3) with even values of j may be obtained from values of sums previously determined:

$$v'(2j) = \tfrac{1}{2}(v'(j) - v(j)) . \tag{4}$$

Thus the sum of the component series (3) may often be evaluated directly or indirectly by direct summation; the sum of the alternating series of sums on the right hand side of relationship (2) may then be obtained by use of Euler's transformation. In this way the numerical sum of a large number of terms of the series (1) is effectively approximated by a weighted sum which involves relatively few of them.

It may occur that the auxiliary series (3) do not converge with sufficient rapidity. In such a case, application of the transformation (2) may be repeated:

$$v'(j) = E\{\sum_{l=1}^{\infty} (-1)^{l-1} v''(j,l)\} \tag{5}$$

where

$$v''(j,l) = \sum_{m=1}^{\infty} 2^{m-1} 2^{2^{m-1} l-1} v(2^{2^{m-1} l-1} j) .$$

Again, for fixed j,

$$v''(j,2l) = \tfrac{1}{2}(v''(j,l) - 2^{l-1}v(2^{l-1}j))$$

and relationship (3) still holds. For odd j, the terms $v''(j,l)$ are obtained, directly or indirectly, by summation; they are inserted into an Euler process to yield terms $v'(j)$ given by formula (5); the $v'(j)$ with even j are derived by use of relationship (4); the $v'(j)$ are inserted into a further Euler process to yield the sum (1).

Clearly the above process of recursion may be extended: letting i be an integer vector with components $i(1)$, $i(2)$,... and $v^{(j)}(i)$ be a function of the first j components of i,

$$\sum_{i=1}^{\infty} v(i) = E\{\sum_{i(1)=1}^{\infty} (-1)^{i(1)-1}v^{(1)}(i)\}$$

where

$$v^{(j)}(i) = E\{\sum_{i(j+1)=1}^{\infty} (-1)^{i(j+1)-1}v^{j+1}(i)\} \qquad (j=1,...,k-1)$$

and $v^{(k)}(i)$ is obtained by direct summation:

$$v^{(k)}(i) = \sum_{i(k+1)=1}^{\infty} f^{(k+1)}(i)v(s^{(k+1)}(i)) \tag{6}$$

where $f^{(k+1)}(i)$, $s^{(k+1)}(i)$ are integer functions of the first $k+1$ components if i, and may be derived by use of the following recursion:

$f(1) = 1, \qquad s(1) = 2^{i(k+1)-1}$

$f(j+1) = f(j)s(j)$

$s(j+1) = 2^{s(j)i(k-j+1)-1} \qquad (j=1,...,k-1)$

$f^{(k+1)}(i) = f(k)s(k), \qquad s^{(k+1)}(i) = i(1)s(k).$

For fixed $i(1)$, $i(2)$,..., $i(j-1)$, $v^{(j)}(i)$ with even values of $i(j)$ may be determined without summation or transformation: writing $v^{(j)}(i)$ as $v^{(j)}(i(1),i(2),...,i(j))$, etc.

$$v^{(j)}(i(1),i(2),...,i(j-1),2i(j)) = \tfrac{1}{2}(v^{(j)}(i(1),i(2),...,i(j-1),i(j)) - f^{(j+1)}(i(1),i(2),...,i(j),1)v(s^{(j+1)}(i(1),i(2),...,i(j),1))) \tag{7}$$

for $j=1,...,k$.

For many series

$$\sum_{i=1}^{\infty} u(i)$$

of real terms with consistent sign

$$\left|\sum_{i=1}^{\infty} u(i)\right| \le R(n) = \left|\frac{2u(n)}{1-u(n+1)/u(n)}\right| .$$

If $R(n) \leq \varepsilon$ for $n=m,m+1,...,m+tim-1$, ε being a real tolerance, m and tim being positive integers, it may reasonably be inferred that the partial sum

$$\sum_{i=1}^{n-1} u(i)$$

represents the sum of the series in question to the stated tolerance. This test is used to establish the level k at which the recursion functions during the computation. With integer values of m and tim supplied by the user, the above test is first applied to the series (1) itself, and if the test is passed, no transformation is carried out. If the test fails, then with $i(1)=1$, the test is applied to the series (6) with $k=1$; if this test fails, it is applied to the series (6) with $k=2$ and $i(1)=i(2)=1$, and so on; acceptance of the test determines the value of k for which $v^{(k)}(i)$ is to be determined, directly or indirectly, by summation.

With $i(1)=i(2)=...=i(k)=1$, $v^{(k)}(i)$ is determined by summation, and inserted into the Euler process to determine $v^{(k-1)}(i)$. With $i(1)=i(2)=...=i(k-1)=1$, $i(k)=2$, $v^{(k)}(i)$ is determined by use of relationship (7) with $j=k$, and $-v^{(k)}(i)$ is inserted into the above Euler process. With $i(1)=i(2)=...=i(k-1)=1$, $i(k)=3$, $v^{(k)}(i)$ is determined by summation, and so on. When the Euler process produces a transformed sum (namely $v^{(k-1)}(1,1,...,1)$) it is inserted into the penultimate Euler process. $v^{(k-1)}(i)$ with $i(1)=i(2)=...=i(k-1)$, $i(k-1)=2$ is determined by use of relationship (7) with $j=k-1$. The rate of convergence of the series (6) with k replaced by k-1 is now tested as described above; if the test is passed, the $v^{(k-1)}(i)$ are determined by direct summation; otherwise k is increased to a level at which convergence is sufficiently rapid, and the process described above is continued; testing rate of convergence takes place after the termination of each Euler process. Finally the Euler process involving the terms $v^{(1)}(i)$ terminates, and the transformed sum of the series (1) is obtained.

Function Parameters:

float sumposseries (*ai,maxaddup,maxzero,maxrecurs,machexp,tim*)

sumposseries: delivers the computed sum of the infinite series;

ai: float *(*ai)(i)*; float *i*;
this function is the *i*-th term of the series, $i \geq 1$;

maxaddup: int;
entry: upper limit for the number of straightforward additions (value of *m* above);

maxzero: float;
entry: tolerance in the Euler summation, see *tim* below; *maxzero* is also used as a tolerance for *maxaddup* straightforward additions;

maxrecurs: int;
entry: upper limit for the recursion depth of the Van Wijngaarden transformation;

machexp: int;
entry: in order to avoid overflow and evaluation of those terms which can be neglected, *machexp* has to be the largest admissible value for which terms with index $k*(2^{machexp})$ can be computed (k is small); otherwise overflow might occur in computing a value for the parameter *i* (in function *ai*), which can be an usually high power of 2;

tim: int;
entry: tolerance in the Euler summation; the summation is continued until *tim* successive terms of the transformed series are in absolute value less than *maxzero*.

```
#include <math.h>

float sumposseries(float (*ai)(float), int maxaddup, float maxzero,
                   int maxrecurs, int machexp, int tim)
{
    float sumposseriessumup(int, float (*)(float), int, float, int,
                int, int, int, int, int, int, int);
    int recurs,vl,vl2,vl4;

    recurs=0;
    vl=1000;
    vl2=2*vl;
    vl4=2*vl2;
    return sumposseriessumup(0,ai,maxaddup,maxzero,maxrecurs,
                   machexp,tim,recurs,vl,vl2,vl4,0);
}

float sumposseriessumup(int bjk, float (*ai)(float),
                int maxaddup, float maxzero, int maxrecurs,
                int machexp, int tim, int recurs, int vl,
                int vl2, int vl4, int jj)
{
    /* this function is internally used by SUMPOSSERIES */

    float *allocate_real_vector(int, int);
    void free_real_vector(float *, int);
    float sumposseriesbjk(int, float, float, float (*)(float));
    int j,transform,j2,jodd,k,n,t;
    float i,sum,nextterm,*v,mn,mp,ds,esum,m[16],temp,vj;

    i=maxaddup+1;
    j=1;
    transform=0;
    while (1) {
        temp = (bjk) ? sumposseriesbjk(jj,i,machexp,ai) : (*ai)(i);
        if (temp <= maxzero) {
            if (j >= tim) break;
            j++;
            i++;
        } else {
            if (recurs != maxrecurs) transform=1;
            break;
        }
    }
    if (!transform) {
        sum=i=0.0;
        j=0;
        do {
            i++;
            nextterm = (bjk) ?
                sumposseriesbjk(jj,i,machexp,ai) : (*ai)(i);
            j = (nextterm <= maxzero) ? j+1 : 0;
            sum += nextterm;
        } while (j < tim);
        return sum;
    }
    /* transform series */
    v=allocate_real_vector(1,vl);
    j2=0;
    jodd=1;
    /* euler */
    n=t=j=0;
    jj=j+1;
    if (jodd) {
        jodd=0;
        recurs++;
        temp=vj=sumposseriessumup(1,ai,maxaddup,maxzero,
                   maxrecurs,machexp,tim,recurs,vl,vl2,vl4,jj);
        recurs--;
        if (jj <= vl)
            v[jj]=temp;
        else
            if (jj <= vl2) v[jj-vl]=temp;
    } else {
```

```
        jodd=1;
        if (jj > vl4) {
            recurs++;
            vj = -sumposseriessumup(1,ai,maxaddup,maxzero,
                        maxrecurs,machexp,tim,recurs,vl,vl2,vl4,jj);
            recurs--;
        } else {
            j2++;
            i=j2;
            if (jj > vl2) {
                temp = (bjk) ?
                            sumposseriesbjk(jj,i,machexp,ai) : (*ai)(i);
                vj = -(v[j2-vl]-temp)/2.0;
            }
            else {
                temp = (bjk) ?
                            sumposseriesbjk(jj,i,machexp,ai) : (*ai)(i);
                temp=v[(jj <= vl) ? jj : jj-vl]=(v[j2]-temp)/2.0;
                vj = -temp;
            }
        }
    }
    m[0]=vj;
    esum=m[0]/2.0;
    do {
        j++;
        jj=j+1;
        if (jodd) {
            jodd=0;
            recurs++;
            temp=vj=sumposseriessumup(1,ai,maxaddup,maxzero,
                            maxrecurs,machexp,tim,recurs,vl,vl2,vl4,jj);
            recurs--;
            if (jj <= vl)
                v[jj]=temp;
            else
                if (jj <= vl2) v[jj-vl]=temp;
        } else {
            jodd=1;
            if (jj > vl4) {
                recurs++;
                vj = -sumposseriessumup(1,ai,maxaddup,maxzero,
                            maxrecurs,machexp,tim,recurs,vl,vl2,vl4,jj);
                recurs--;
            } else {
                j2++;
                i=j2;
                if (jj > vl2) {
                    temp = (bjk) ?
                                sumposseriesbjk(jj,i,machexp,ai) : (*ai)(i);
                    vj = -(v[j2-vl]-temp)/2.0;
                } else {
                    temp = (bjk) ?
                                sumposseriesbjk(jj,i,machexp,ai) : (*ai)(i);
                    temp=v[(jj <= vl) ? jj : jj-vl]=(v[j2]-temp)/2.0;
                    vj = -temp;
                }
            }
        }
        mn=vj;
        for (k=0; k<=n; k++) {
            mp=(mn+m[k])/2.0;
            m[k]=mn;
            mn=mp;
        }
        if (fabs(mn) < fabs(m[n]) && n < 15) {
            ds=mn/2.0;
            n++;
            m[n]=mn;
        } else
            ds=mn;
        esum += ds;
        t = (fabs(ds) < maxzero) ? t+1 : 0;
    } while (t < tim);
```

```
    free_real_vector(v,1);
    return esum;
}

float sumposseriesbjk(int j, float i, float machexp,
                      float (*ai)(float))
{
    /* this function is internally used by SUMPOSSERIESSUMUP */

    float coeff;

    if (i > machexp) return 0.0;
    coeff=pow(2.0,i-1.0);
    return coeff*(*ai)(j*coeff);
}
```

4.2 Quadrature

4.2.1 One-dimensional quadrature

A. qadrat

Evaluates the integral

$$I(f;a,b) = \int_a^b f(x)dx$$

where f is real valued, a and b are finite ($b<a$ is permitted).

qadrat functions recursively [RoF72]. With $f_i=f(a+ih)$, where $h=(b-a)/32$, two sums of the form

$$U = u_1(f_2+f_{30}) + \sum_{i=2}^{5} u_i(f_{4i-8}+f_{40-4i}) + u_6 f_{16}$$

$$V = v_0(f_1+f_{31}) + v_1(f_2+f_{30}) + \sum_{i=2}^{5} v_i(f_{4i-8}+f_{40-4i}) + v_6 f_{16}$$

are computed (U and V are such that Uh and Vh are equal to $I(f:a,b)$ if f is a polynomial of degree ≤ 16). If either $|h|<hmin=|b-a|\eta_a$, where η_a is an absolute tolerance prescribed by the user, or $|U-V|<|V|\eta_r+\eta_a$, where η_r is a relative tolerance also prescribed by the user, then Vh is accepted as an approximation to $I(f;a,b)$. If neither of these conditions is satisfied, a further sum of the form

$$W = w_0(f_1+f_{31} + w_1(f_2+f_{30}) + \sum_{i=2}^{5} w_i(f_{4i-8}+f_{40-4i}) + w_6 f_{16} - w_7(f_6+f_{26})$$

is computed (again Wh is equal to $I(f;a,b)$ if f is a polynomial of degree ≤ 16). If now $|W-V|<|W|\eta_r+\eta_a$, then Wh is accepted as an approximation to $I(f;a,b)$. If this further condition is not satisfied, $I(f;a,b)$ is then expressed as

$$\int_a^{(a+b)/2} f(x)dx - \int_b^{(a+b)/2} f(x)dx \quad (1)$$

and the two integrals are each treated in the above manner (with *hmin* left unchanged). The first is approximated by linear sums involving the function values

$$f_i' = f(a+ih') \text{ where } h' = (b-a)/64.$$

Since $f_{2i}' = f_i$ $(i=0,1,2,4,6,8,12,16)$, this set of values may immediately be used in an attempt to evaluate the first of the integrals (1). The second integral is approximated by linear sums involving the function values $f_i' = f(b+ih')$, where $h' = (a-b)/64$. Since $f_{2i}' = f_{32-i}$ $(i=0,1,2,4,6,8,12,16)$.

Function Parameters:

float qadrat (*x,a,b,fx,e*)

qadrat: delivers the computed value of the definite integral from *a* to *b* of the function *f(x)*;

x: float *;
integration variable;

a,b: float;
entry: (*a,b*) denotes the interval of integration; *b*<*a* is allowed;

fx: float *(*fx)(x)*; *float x*;
fx denotes the integrand *f(x)*;

e: float *e[1:3]*;
entry:
e[1]: the relative accuracy required (value of η_r above);
e[2]: the absolute accuracy required (value of η_a above);
exit:
e[3]: the number of elementary integrations with $h < |b-a|*e[1]$.

```
#include <math.h>

float qadrat(float *x, float a, float b, float (*fx)(float),
             float e[])
{
   float lint(float *, float (*)(float), float [], float, float,
              float, float, float, float, float, float, float,
              float, float, float, float, float);
   float f0,f2,f3,f5,f6,f7,f9,f14,hmin,hmax,re,ae,result;

   hmax=(b-a)/16.0;
   if (hmax == 0.0) return 0.0;
   re=e[1];
   ae=2.0*e[2]/fabs(b-a);
   e[3]=0.0;
   hmin=fabs(b-a)*re;
   *x=a;
   f0=(*fx)(*x);
   *x=a+hmax;
   f2=(*fx)(*x);
   *x=a+2.0*hmax;
   f3=(*fx)(*x);
   *x=a+4.0*hmax;
   f5=(*fx)(*x);
   *x=a+6.0*hmax;
   f6=(*fx)(*x);
   *x=a+8.0*hmax;
   f7=(*fx)(*x);
   *x=b-4.0*hmax;
   f9=(*fx)(*x);
   *x=b;
   f14=(*fx)(*x);
   result = lint(x,fx,e,a,b,f0,f2,f3,f5,f6,f7,f9,f14,
                 hmin,hmax,re,ae)*16.0;
   return result;
}
```

```
float lint(float *x, float (*fx)(float), float e[], float x0,
               float xn, float f0, float f2, float f3, float f5,
               float f6, float f7, float f9, float f14,
               float hmin, float hmax, float re, float ae)
{
    /* this function is internally used by QADRAT */

    float v,w,h,xm,f1,f4,f8,f10,f11,f12,f13;

    xm=(x0+xn)/2.0;
    h=(xn-x0)/32.0;
    *x=xm+4.0*h;
    f8=(*fx)(*x);
    *x=xn-4.0*h;
    f11=(*fx)(*x);
    *x=xn-2.0*h;
    f12=(*fx)(*x);
    v=0.330580178199226*f7+0.173485115707338*(f6+f8)+
        0.321105426559972*(f5+f9)+0.135007708341042*(f3+f11)+
        0.165714514228223*(f2+f12)+0.393971460638127e-1*(f0+f14);
    *x=x0+h;
    f1=(*fx)(*x);
    *x=xn-h;
    f13=(*fx)(*x);
    w=0.260652434656970*f7+0.239063286684765*(f6+f8)+
        0.263062635477467*(f5+f9)+0.218681931383057*(f3+f11)+
        0.275789764664284e-1*(f2+f12)+0.105575010053846*(f1+f13)+
        0.157119426059518e-1*(f0+f14);
    if (fabs(h) < hmin) e[3] += 1.0;
    if (fabs(v-w) < fabs(w)*re+ae || fabs(h) < hmin)
        return h*w;
    else {
        *x=x0+6.0*h;
        f4=(*fx)(*x);
        *x=xn-6.0*h;
        f10=(*fx)(*x);
        v=0.245673430093324*f7+0.255786258286921*(f6+f8)+
            0.228526063690406*(f5+f9)+0.500557131525460e-1*(f4+f10)+
            0.177946487736780*(f3+f11)+0.584014599347449e-1*(f2+f12)+
            0.874830942871331e-1*(f1+f13)+
            0.189642078648079e-1*(f0+f14);
        return ((fabs(v-w) < fabs(v)*re+ae) ? h*v :
                    (lint(x,fx,e,x0,xm,f0,f1,f2,f3,f4,f5,f6,f7,
                          hmin,hmax,re,ae) -
                    lint(x,fx,e,xn,xm,f14,f13,f12,f11,f10,f9,f8,f7,
                          hmin,hmax,re,ae)));
    }
}
```

B. integral

Evaluates the integral

$$\int_{\alpha}^{\beta} f(x)\,dx \tag{1}$$

where f is real valued, $-\infty < \alpha < \infty$, and $-\infty \leq \beta \leq \infty$ (when $\beta < \infty$, $\beta < \alpha$ is permitted). This is done by means of Simpson's rule with Richardson correction [DekR71, RoF72]. If the fourth derivative is too large (and thus the correction term), the total interval is split into two equal parts and the integration process is invoked recursively. This is done in such a way that the total amount of Richardson corrections is slightly smaller than or equal to *e[1]** |integral of the function from a to b|+*e[2]*. For the integration of a definite integral over a finite interval, the use of *qadrat* is recommended, especially when high accuracy is required.

integral may be called a number of times to evaluate the integral of f over a sequence of intervals of the form $[\alpha_0,\alpha_1]$, $[\alpha_1,\alpha_2]$,..., $[\alpha_{n-1},\alpha_n]$ and, when used in this way, may be directed to accumulate the total integral over the range $[\alpha_0,\alpha_n]$.

The values of α and β in expression (1) are determined partly by the values of a and b occurring in the parameter list of *integral*. If *ua* is nonzero then $\alpha=a$; otherwise $\alpha=a'$. If *ub* is nonzero then $\beta=b$; if *ub* is zero, then $\beta=\infty$ if $b>a$, and $\beta=-\infty$ if $b<a$. Immediately before return from a successful call, *integral* sets $a'=b$ if *ub* is nonzero, and in this way the integrals over the above sequence of intervals may be evaluated without adjustment of a.

If *ub* is zero (so that $\beta=\infty$ or $\beta=-\infty$ above) the integral (1) is expressed as

$$\int_\alpha^b f(x)\,dx - \int_1^0 f(b-1+z^{-1})\,z^{-2}dz \tag{2}$$

if $\beta=\infty$, and as

$$\int_\alpha^b f(x)\,dx - \int_1^0 f(b+1+z^{-1})\,z^{-2}dz \tag{3}$$

if $\beta=-\infty$. *integral* evaluates the two relevant integrals over finite ranges, and forms their difference.

For integration over a finite subrange, use is made of the formula

$$\int_\gamma^\delta f(x)dx = I_1 - \frac{I_2-I_1}{15} + O(h^7) \tag{4}$$

where $h = (\delta-\gamma)/4$, and with $f_i = f(\gamma+ih)$

$I_1 = (h/3)(f_0 + 4f_1 + 2f_2 + 4f_3 + f_4)$

$I_2 = (h/3)(2f_0 + 8f_2 + 2f_4)$.

Acceptance of the value of the expression $I_1 - (I_2-I_1)/15$ as an approximation to the value of the intermediate integral (2) is decided by the values of two tolerances η_a and η_r provided by the user, from which a further tolerance τ_a is derived. If in expression (1) β is finite (i.e. *ub* is nonzero) then $\tau_a = 180/(\beta-\alpha)$. If $|\beta|$ is infinite, so that the pair (2) or (3) of integrals is to be evaluated, $\tau_a = 90\eta_a/|b-\alpha|$ for the first integral of the pair, and $\tau_a = 90\eta_a$ for the second. A minimum stepsize *hmin* is also determined. If in expression (1) β is finite, then $hmin = |\beta-\alpha|\eta_r$. If $|\beta|$ is infinite so that the pair (2) or (3) of integrals is to be evaluated, $hmin = |b-\alpha|\eta_r$ for the first integral, and $hmin = \eta_r$ for the second.

integral functions recursively. Formula (4) is first examined with γ and δ set equal to the endpoints of the relevant integral over a finite range being evaluated. The sums

$v = 15(f_0 + f_4 + 2f_2 + 4(f_1 + f_3))$

$t = f_0 + 6f_2 + f_4 - 4(f_1 + f_3)$

(so that $v = 180I_1/h$, $t = 180(I_2-I_1)/h$) are formed. If $|t| < |v|\eta_r + \tau_a$, or $|\delta-\gamma| < hmin$, where η_r, τ_a and *hmin* are as defined above, then $I_1 - (I_2-I_1)/15$ is accepted as an approximation to the value of the integral (4); otherwise, the formula

$$\int_\gamma^\delta f(x)dx = \int_\gamma^{(\gamma+\delta)/2} f(x)dx + \int_{(\gamma+\delta)/2}^\delta f(x)dx$$

is used, each constituent of the right hand side being treated as above. The computed approximation to the first constituent integral involves function values $f_i' = f(\gamma+ih')$ where

$h' = (\delta-\gamma)/8$, so that $f_{2i}' = f_i$ $(i=0,1,2)$; those even order f_i' have already been evaluated, and may be used directly. The computed approximation to the second constituent integral involves function values $f_i' = f(½(\gamma+\delta)+ih')$ so that $f_{2i}' = f_{i+2}$ $(i=0,1,2)$.

Function Parameters:

float integral (*a,b,fx,e,ua,ub*)

integral: delivers the computed value of the definite integral of the function from *a* to *b*; after successive calls of the procedure, the integral over the total interval is delivered, i.e. the value of *a* in the last call with *ua* equals nonzero is the starting point of the integral;

a,b: float;
entry: (*a,b*) denotes the interval of integration; *b*<*a* is allowed;

fx: float *(*fx)(x)*; *float x*;
fx denotes the integrand *f(x)*;

e: float *e[1:6]*;
entry:
e[1]: the relative accuracy required (value of η_r above);
e[2]: the absolute accuracy required (value of η_a above);
exit:
e[3]: the number of skipped integration steps;
e[4]: the value of the integral from *a* to *b*;
e[5]: if *ub* is zero then *e[5]*=0 else *e[5]*=*b*;
e[6]: if *ub* is zero then *e[5]*=0 else *e[5]*=*f(b)*;

ua: int;
entry: determines the starting point of the integration; if *ua* is zero then the starting point is *e[5]* otherwise the starting point is *a*;

ub: int;
entry: determines the final point of the integration;
if *ub* is nonzero then the final point is *b*; otherwise if *b* > *a* then the final point is ∞ else it is -∞.

```
#include <math.h>

float integral(float a, float b, float (*fx)(float), float e[],
               int ua, int ub)
{
   float integralqad(int, float (*)(float), float [], float *,
               float *, float *, float *, float *, float *, float,
               float, float);
   float x0,x1,x2,f0,f1,f2,re,ae,b1,x;

   re=e[1];
   if (ub)
      ae=e[2]*180.0/fabs(b-a);
   else
      ae=e[2]*90.0/fabs(b-a);
   if (ua) {
      e[3]=e[4]=0.0;
      x=x0=a;
      f0=(*fx)(x);
   } else {
      x=x0=a=e[5];
      f0=e[6];
   }
   e[5]=x=x2=b;
   e[6]=f2=(*fx)(x);
   e[4] += integralqad(0,fx,e,&x0,&x1,&x2,&f0,&f1,&f2,re,ae,b1);
   if (!ub) {
```

```
        if (a < b) {
            b1=b-1.0;
            x0=1.0;
        } else {
            b1=b+1.0;
            x0 = -1.0;
        }
        f0=e[6];
        e[5]=x2=0.0;
        e[6]=f2=0.0;
        ae=e[2]*90.0;
        e[4] -= integralqad(1,fx,e,&x0,&x1,&x2,&f0,&f1,&f2,re,ae,b1);
    }
    return e[4];
}

float integralqad(int transf, float (*fx)(float), float e[],
            float *x0, float *x1, float *x2, float *f0, float *f1,
            float *f2, float re, float ae, float b1)
{
    /* this function is internally used by INTEGRAL */

    void integralint(int, float (*)(float), float [], float *,
            float *, float *, float *, float *, float *,
            float *, float, float, float, float);
    float sum,hmin,x,z;

    hmin=fabs((*x0)-(*x2))*re;
    x=(*x1)=((*x0)+(*x2))*0.5;
    if (transf) {
        z=1.0/x;
        x=z+b1;
        (*f1)=(*fx)(x)*z*z;
    } else
        (*f1)=(*fx)(x);
    sum=0.0;
    integralint(transf,fx,e,x0,x1,x2,f0,f1,f2,&sum,re,ae,b1,hmin);
    return sum/180.0;
}

void integralint(int transf, float (*fx)(float), float e[],
            float *x0, float *x1, float *x2, float *f0, float *f1,
            float *f2, float *sum, float re, float ae, float b1,
            float hmin)
{
    /* this function is internally used by INTEGRALQAD of INTEGRAL */

    int anew;
    float x3,x4,f3,f4,h,x,z,v,t;

    x4=(*x2);
    (*x2)=(*x1);
    f4=(*f2);
    (*f2)=(*f1);
    anew=1;
    while (anew) {
        anew=0;
        x=(*x1)=((*x0)+(*x2))*0.5;
        if (transf) {
            z=1.0/x;
            x=z+b1;
            (*f1)=(*fx)(x)*z*z;
        } else
            (*f1)=(*fx)(x);
        x=x3=((*x2)+x4)*0.5;
        if (transf) {
            z=1.0/x;
            x=z+b1;
            f3=(*fx)(x)*z*z;
        } else
            f3=(*fx)(x);
        h=x4-(*x0);
        v=(4.0*((*f1)+f3)+2.0*(*f2)+(*f0)+f4)*15.0;
        t=6.0*(*f2)-4.0*((*f1)+f3)+(*f0)+f4;
```

```
        if (fabs(t) < fabs(v)*re+ae)
            (*sum) += (v-t)*h;
        else if (fabs(h) < hmin)
            e[3] += 1.0;
        else {
            integralint(transf,fx,e,x0,x1,x2,f0,f1,f2,sum,
                            re,ae,b1,hmin);
            *x2=x3;
            *f2=f3;
            anew=1;
        }
        if (!anew) {
            *x0=x4;
            *f0=f4;
        }
    }
}
```

4.2.2 Multidimensional quadrature

tricub

Evaluates the double integral

$$I(g;\Omega) = \iint_{\Omega} g(x,y)dxdy$$

where Ω is a triangle with vertices (x_i,y_i), (x_j,y_j) and (x_k,y_k), and g is real valued. A nested sequence of cubature rules of order 2, 3, 4 and 5 is applied [He73]. If the difference between the result with the 4-th degree rule and the result with the 5-th degree rule is too large, then the triangle is divided into four congruent triangles. This process is applied recursively in order to obtain an adaptive cubature algorithm.

The theory of the method adopted may be presented as follows. The area of Ω is

$$\Delta = \tfrac{1}{2}\left|x_iy_j - x_jy_i + x_ky_i - x_iy_k + x_jy_k - x_ky_j\right|.$$

Denote the argument pair (x_i,y_i) by z_i, and the other pairs similarly. Set

$$z_{j,k} = \tfrac{1}{2}(z_j + z_k)$$

and define $z_{k,i}$, $z_{i,j}$ similarly ($z_{j,k}$ is the midpoint of the line joining z_j and z_k). Set

$$z_c = (z_i + z_j + z_k)/3$$

(z_c is the barycenter is Ω). Set $z_i' = \tfrac{1}{2}(z_i + z_c)$ and define z_j', z_k' similarly. Set

$$z_{i;j,k} = \tfrac{1}{2}(z_i + z_{j,k})$$

and define $z_{k;i,j}$, $z_{j;k,i}$ similarly. Denote values of g at the above points by corresponding suffices, so that $g_i = g(z_i) = g(x_i,y_i)$, $g_{j,k} = g(z_{j,k}) = g(\tfrac{1}{2}(x_j+x_k),\tfrac{1}{2}(y_j+y_k))$ and so on. Set

$$A = g_i + g_j + g_k, \qquad B = g_{j,k} + g_{k,i} + g_{i,j}$$
$$C = g_i' + g_j' + g_k', \qquad D = g_{i;j,k} + g_{k;i,j} + g_{j;k,i}$$

and

$$S_2 = 5A + 45g_c, \qquad S_3 = 3A + 27g_c + 8B$$
$$S_4 = A + 9g_c + 4B + 12C$$
$$S_5 = (51A + 2187g_c + 276B + 972C - 768D)/63$$

and

$$I_j = S_j\Delta/60 .$$

If g is a polynomial in x and y of degree $\leq j$, then $I_j = I(g;\Omega)$ exactly.

The user is asked to provide values of tolerances η_r,η_a. *tricub* determines from these further tolerances: $\tau_r = 30\eta_r$, $\tau_r' = \Delta\eta_r$ and $\tau_a = 30\eta_a/\Delta$.

tricub functions recursively. S_2 and S_3 are evaluated, and if either $\Delta<\tau_r$' or $|S_3-S_2|<\tau_r$

then I_3 is accepted as an approximation to $I(g;\Omega)$. If neither of these conditions is satisfied, S_4 is evaluated and if $|S_4-S_3|<\tau_r$, I_4 is accepted as an approximation to $I(g;\Omega)$. If this condition is not satisfied, S_5 is evaluated and if $|S_5-S_4|<\tau_r$, I_5 is accepted as an approximation to $I(g;\Omega)$. If this condition is not satisfied Ω is subdivided into four congruent triangles whose vertices are $z_{i,j}$, $z_{j,k}$, $z_{k,i}$; z_i, $z_{i,j}$, $z_{i,k}$; z_j, $z_{j,k}$, $z_{j,i}$; z_k, $z_{k,j}$, $z_{k,i}$ and the integrals over these four triangles are individually treated as above. The barycenter of the first triangle is z_c (i.e. that of Ω). The barycenters of the remaining three triangles are z_i', z_j' *and* z_k' respectively. $z_{i;j,k}$, $z_{j;k,i}$ and $z_{k;i,j}$ are the midpoints of the sides of the first triangle. The values of g at the points just mentioned (and at z_i, z_j, z_k) are required in the treatment of the integrals over the four triangles. Further function values at the midpoints of the lines joining z_i and $z_{i,j}$, z_i and $z_{i,k}$, z_j and $z_{j,k}$, z_j and $z_{j,i}$, z_k and $z_{k,j}$, and z_k and $z_{k,i}$ are required.

Function Parameters:

float tricub (*xi,yi,xj,yj,xk,yk,g,re,ae*)

tricub: delivers the computed value of the definite integral of the function *g(x,y)* over the triangular domain with vertices (*xi,yi*), (*xj,yj*) and (*xk,yk*);

xi,yi: float;
entry: the coordinates of the first vertex of the triangular domain of integration;

xj,yj: float;
entry: the coordinates of the second vertex of the triangular domain of integration;

xk,yk: float;
entry: the coordinates of the third vertex of the triangular domain of integration; the algorithm is symmetric in the vertices; this implies that the result of the procedure (on all counts) is invariant for any permutation of the vertices;

g: float *(*g)(x,y)*; *float x, float y*;
g denotes the integrand *g(x,y)*;

re: float;
entry: the required relative error (value of η_r above);

ae: float;
entry: the required absolute error (value of η_a above); one should take for *ae* and *re* values which are greater than the absolute and relative error in the computation of the integrand *g*.

```
#include <math.h>

float tricub(float xi, float yi, float xj, float yj, float xk,
        float yk, float (*g)(float,float), float re, float ae)
{
    float tricubint(float, float, float, float,
            float, float, float, float, float,
            float, float, float, float, float,
            float, float, float, float, float,
            float, float, float (*)(float, float),
            float, float, float *, float);
    float surf,surfmin,xz,yz,gi,gj,gk;

    surf=0.5*fabs(xj*yk-xk*yj+xi*yj-xj*yi+xk*yi-xi*yk);
    surfmin=surf*re;
    re *= 30.0;
    ae=30.0*ae/surf;
    xz=(xi+xj+xk)/3.0;
    yz=(yi+yj+yk)/3.0;
    gi=(*g)(xi,yi);
    gj=(*g)(xj,yj);
    gk=(*g)(xk,yk);
```

```
    xi *= 0.5;
    yi *= 0.5;
    xj *= 0.5;
    yj *= 0.5;
    xk *= 0.5;
    yk *= 0.5;
    return tricubint(xi,yi,gi,xj,yj,gj,xk,yk,gk,
                     xj+xk,yj+yk,(*g)(xj+xk,yj+yk),
                     xk+xi,yk+yi,(*g)(xk+xi,yk+yi),
                     xi+xj,yi+yj,(*g)(xi+xj,yi+yj),
                     0.5*xz,0.5*yz,(*g)(xz,yz),
                     g,re,ae,&surf,surfmin)/60.0;
}

float tricubint(float ax1, float ay1, float af1, float ax2,
            float ay2, float af2, float ax3, float ay3, float af3,
            float bx1, float by1, float bf1, float bx2, float by2,
            float bf2, float bx3, float by3, float bf3, float px,
            float py, float pf, float (*g)(float, float),
            float re, float ae, float *surf, float surfmin)
{
    /* this function is internally used by TRICUB */

    float e,i3,i4,i5,a,b,c,sx1,sy1,sx2,sy2,sx3,sy3,cx1,cy1,cf1,
          cx2,cy2,cf2,cx3,cy3,cf3,dx1,dy1,df1,dx2,dy2,df2,dx3,dy3,
          df3,result;

    a=af1+af2+af3;
    b=bf1+bf2+bf3;
    i3=3.0*a+27.0*pf+8.0*b;
    e=fabs(i3)*re+ae;
    if (*surf < surfmin || fabs(5.0*a+45.0*pf-i3) < e)
       return i3*(*surf);
    else {
       cx1=ax1+px;
       cy1=ay1+py;
       cf1=(*g)(cx1,cy1);
       cx2=ax2+px;
       cy2=ay2+py;
       cf2=(*g)(cx2,cy2);
       cx3=ax3+px;
       cy3=ay3+py;
       cf3=(*g)(cx3,cy3);
       c=cf1+cf2+cf3;
       i4=a+9.0*pf+4.0*b+12.0*c;
       if (fabs(i3-i4) < e)
          return i4*(*surf);
       else {
          sx1=0.5*bx1;
          sy1=0.5*by1;
          dx1=ax1+sx1;
          dy1=ay1+sy1;
          df1=(*g)(dx1,dy1);
          sx2=0.5*bx2;
          sy2=0.5*by2;
          dx2=ax2+sx2;
          dy2=ay2+sy2;
          df2=(*g)(dx2,dy2);
          sx3=0.5*bx3;
          sy3=0.5*by3;
          dx3=ax3+sx3;
          dy3=ay3+sy3;
          df3=(*g)(dx3,dy3);
          i5=(51.0*a+2187.0*pf+276.0*b+972.0*c-
                768.0*(df1+df2+df3))/63.0;
          if (fabs(i4-i5) < e)
             return i5*(*surf);
          else {
             (*surf) *= 0.25;
             result=tricubint(sx1,sy1,bf1,sx2,sy2,bf2,sx3,sy3,bf3,
                      dx1,dy1,df1,dx2,dy2,df2,dx3,dy3,df3,px,py,pf,
                      g,re,ae,surf,surfmin)+
                    tricubint(ax1,ay1,af1,sx3,sy3,bf3,sx2,sy2,bf2,
                      dx1,dy1,df1,ax1+sx2,ay1+sy2,
```

```
                        (*g)(ax1+sx2,ay1+sy2),ax1+sx3,ay1+sy3,
                        (*g)(ax1+sx3,ay1+sy3),0.5*cx1,0.5*cy1,cf1,
                        g,re,ae,surf,surfmin)+
                    tricubint(ax2,ay2,af2,sx3,sy3,bf3,sx1,sy1,bf1,
                        dx2,dy2,df2,ax2+sx1,ay2+sy1,
                        (*g)(ax2+sx1,ay2+sy1),ax2+sx3,ay2+sy3,
                        (*g)(ax2+sx3,ay2+sy3),0.5*cx2,0.5*cy2,cf2,
                        g,re,ae,surf,surfmin)+
                    tricubint(ax3,ay3,af3,sx1,sy1,bf1,sx2,sy2,bf2,
                        dx3,dy3,df3,ax3+sx2,ay3+sy2,
                        (*g)(ax3+sx2,ay3+sy2),ax3+sx1,ay3+sy1,
                        (*g)(ax3+sx1,ay3+sy1),0.5*cx3,0.5*cy3,cf3,
                        g,re,ae,surf,surfmin);
            (*surf) *= 4.0;
            return result;
         }
      }
   }
}
```

4.2.3 Gaussian quadrature - General weights

A. reccof

Calculates from a given weight function on [-1,1] the recurrence coefficients of the corresponding orthogonal polynomials [G68a, G68b].

Given $w(x)$, defined for $-1 \le x \le 1$, and n, derives the coefficients b_k and c_k occurring in the recursion $p_0(x) = 1, \quad p_1(x) = x - b_0$

$$p_{k+1}(x) = (x-b_k)p_k(x) - c_k p_{k-1}(x) \qquad (k=1,...,n) \tag{1}$$

satisfied by the polynomials

$$p_k(x) = x^k + \sum_{l=0}^{k-1} p_{k,l} x^l$$

defined by the conditions

$$\int_{-1}^{1} p_i(x)\, p_j(x)\, w(x)\, dx = \begin{cases} 1 & (i=j) \\ 0 & (i \ne j) \end{cases} .$$

The formulae

$$L_k = \int_{-1}^{1} p_k(x)^2 w(x) dx \tag{2}$$

$$b_k = \left(\int_{-1}^{1} x p_k(x)^2 w(x) dx\right)/L_k \tag{3}$$

are used with $k=0$ to determine b_0 (and hence $p_1(x)$) and then with $c_k = L_k/L_{k-1}$ and (1) with $k=1,...,n-1$ to determine the b_k and c_k. The integrals (2) and (3) are evaluated by use of the formula

$$\int_{-1}^{1} f(x)\, dx = \frac{\pi}{m} \sum_{j=1}^{m} \sin(\theta_j)\, f(\cos(\theta_j)) \ .$$

The user must supply the value of *m*.

When *w* is symmetric (*w(-x)=w(x)*), the above formulae may be simplified (for example, by b_{2k} are zero). For this reason the user is asked to allocate a value to the variable *sym* upon call of *reccof*: nonzero if *w* is symmetric, zero otherwise.

Function Parameters:

void reccof (*n,m,x,wx,b,c,l,sym*)

n: int;
entry: upper bound for the indices of the arrays *b, c, l* ($n \geq 0$);

m: int;
entry: the number of points used in the Gauss-Chebyshev quadrature rule for calculating the approximation of the integral representations of *b[k], c[k]*;

x: float *;
entry: the weight function variable;

wx: float *(*wx)(x); float x*;
the weight function (*w(x)* above);

b,c,l: float *b[0:n], c[0:n], l[0:n]*;
exit: the approximate recurrence coefficients for $p_{k+1}(x)=(x-b[k])*p_k(x)-c[k]*p_{k-1}(x)$, $k=0,1,2,...,n$, and the approximate square lengths
l[k] = integral from -1 to 1 ($w(x)*p_k(x)*p_k(x)$) *dx*;

sym: int;
entry: if *sym* is nonzero then the weight function on [-1,1] is even, otherwise the weight function on [-1,1] is not even.

Function used: ortpol.

```
#include <math.h>

void reccof(int n, int m, float *x, float (*wx)(float), float b[],
            float c[], float l[], int sym)
{
    float ortpol(int, float, float [], float []);
    int i,j,up;
    float r,s,pim,h,hh,arg,sa,temp;

    pim=4.0*atan(1.0)/m;
    if (sym) {
        for (j=0; j<=n; j++) {
            r=b[j]=0.0;
            up=m/2;
            for (i=1; i<=up; i++) {
                arg=(i-0.5)*pim;
                *x=cos(arg);
                temp=ortpol(j,*x,b,c);
                r += sin(arg)*(*wx)(*x)*temp*temp;
            }
            if (up*2 == m)
                l[j]=2.0*r*pim;
            else {
                *x=0.0;
                temp=ortpol(j,0.0,b,c);
                l[j]=(2.0*r+(*wx)(*x)*temp*temp)*pim;
            }
```

```
            c[j] = (j == 0) ? 0.0 : l[j]/l[j-1];
        }
    } else
        for (j=0; j<=n; j++) {
            r=s=0.0;
            up=m/2;
            for (i=1; i<=up; i++) {
                arg=(i-0.5)*pim;
                sa=sin(arg);
                *x=cos(arg);
                temp=ortpol(j,*x,b,c);
                h=(*wx)(*x)*temp*temp;
                *x = -(*x);
                temp=ortpol(j,*x,b,c);
                hh=(*wx)(*x)*temp*temp;
                r += (h+hh)*sa;
                s += (hh-h)*(*x)*sa;
            }
            b[j]=s*pim;
            if (up*2 == m)
                l[j]=r*pim;
            else {
                *x=0.0;
                temp=ortpol(j,0.0,b,c);
                l[j]=(r+(*wx)(*x)*temp*temp)*pim;
            }
            c[j] = (j == 0) ? 0.0 : l[j]/l[j-1];
        }
}
```

B. gsswts

Calculates from the recurrence coefficients of orthogonal polynomials the Gaussian weights of the corresponding weight function [G70].
Given the coefficients b_k and c_k occurring in the recursion

$$p_0(x) = 1, \qquad p_1(x) = x - b_0$$
$$p_{k+1}(x) = (x-b_k)p_k(x) - c_k p_{k-1}(x) \qquad (k=1,...,n-1)$$

satisfied by the polynomials

$$p_k(x) = x^k + \sum_{l=0}^{k-1} p_{k,l} x^l$$

defined by the conditions

$$\int_\alpha^\beta p_i(x) p_j(x) w(x) dx = \begin{cases} 1 & (i=j) \\ 0 & (i \neq y) \end{cases}$$

and the n real roots $z_1,...,z_n$ of p_n, derives the weights $w_1,...,w_n$ which render the quadrature formula

$$\sum_{i=1}^{n} w_i f(z_i) = \int_\alpha^\beta w(x) f(x) dx \Big/ \int_\alpha^\beta w(x) dx$$

exact for all polynomials of degree < $2n$. The w_i are given by

$$w_i = \left(1 + \sum_{j=1}^{n-1} p_j(z_i)^2 / c_j\right)^{-1}.$$

Function Parameters:

void gsswts (n,zer,b,c,w)

n: int;
entry: the number of weights to be computed; upper bound for arrays z and w $(n \geq 1)$;

zer: float $zer[1:n]$;
entry: the zeros of the n-th degree orthogonal polynomial;

b,c: float $b[0:n-1]$, $c[1:n-1]$;
entry: the recurrence coefficients;

w: float $w[1:n]$;
exit: the Gaussian weights divided by the integral over the weight function.

Function used: allortpol.

```
void gsswts(int n, float zer[], float b[], float c[], float w[])
{
    float *allocate_real_vector(int, int);
    void free_real_vector(float *, int);
    void allortpol(int, float, float [], float [], float []);
    int j,k;
    float s,*p;

    p=allocate_real_vector(0,n-1);
    for (j=1; j<=n; j++) {
        allortpol(n-1,zer[j],b,c,p);
        s=0.0;
        for (k=n-1; k>=1; k--) s=(s+p[k]*p[k])/c[k];
        w[j]=1.0/(1.0+s);
    }
    free_real_vector(p,0);
}
```

C. gsswtssym

Calculates from the recurrence coefficients of orthogonal polynomials the Gaussian weights of the corresponding weight function.
Given the coefficients c_k occurring in the recursion

$$p_0(x) = 1, \quad p_1(x) = x$$
$$p_{k+1}(x) = xp_k(x) - c_k p_{k-1}(x) \quad (k=1,...,n-1)$$

satisfied by the polynomials

$$p_k(x) = x^k + \sum_{l=0}^{k-1} p_{k,l} x^l$$

defined by the conditions

$$\int_{-\infty}^{\alpha} p_i(x) p_j(x) w(x) dx = \begin{cases} 1 & (i=j) \\ 0 & (i \neq y) \end{cases}$$

where w is symmetric ($w(-x)=w(x)$), and $n'=n/2$ nonpositive roots $z_1,...,z_{n'}$ ($z_i<z_{i+1}$) of p_n, derives the weights $w_1,...,w_{n''}$ ($n''=(n+1)/2$) which with $w_{n''+1},...,w_n$ further defined by $w_j=w_{n+1-j}$ ($j=1,...,n''-1$) render the quadrature formula

$$\sum_{i=1}^{n} w_i f(z_i) = \int_\alpha^\beta w(x)f(x)dx \;/\; \int_\alpha^\beta w(x)dx$$

exact for all polynomials of degree < $2n$. (For the roots of p_n, $z_j=-z_{n+1-j}$ ($j=1,...,n'-1$) also; in providing n' negative roots, one provides them all). The w_i above are given by

$$w_i = \left(1 + \sum_{j=1}^{n-1} p_j(z_i)^2/c_j\right)^{-1}.$$

Function Parameters:

void gsswtssym (*n,zer,c,w*)

n: int;
entry: the weights of an *n*-points Gauss rule are to be computed (because of symmetry only $(n+1)/2$ of the values are delivered);

zer: float *zer[1:n/2]*;
entry: the negative zeros of the *n*-th degree orthogonal polynomial (*zer[i]*<*zer[i+1]*, $i=1,2,...,n/2-1$; if *n* is odd then 0 is also a zero);

c: float *c[1:n-1]*;
entry: the recurrence coefficients;

w: float *w[1:(n+1)/2]*;
exit: part of the Gaussian weights divided by the integral of the weight function (note that *w[n+1-k]=w[k]*, $k=1,2,...,(n+1)/2$).

Function used: allortpolsym.

Method: see the procedure *gsswts*, this procedure is supplied for storage economical reasons.

```
void gsswtssym(int n, float zer[], float c[], float w[])
{
    float *allocate_real_vector(int, int);
    void free_real_vector(float *, int);
    void allortpolsym(int, float, float [], float []);
    int i,twoi,low,up;
    float s,*p;

    p=allocate_real_vector(0,n-1);
    low=1;
    up=n;
    while (low < up) {
        allortpolsym(n-1,zer[low],c,p);
        s=p[n-1]*p[n-1];
        for (i=n-1; i>=1; i--) s=s/c[i]+p[i-1]*p[i-1];
        w[low]=1.0/s;
        low++;
        up--;
    }
    if (low == up) {
        s=1.0;
```

```
        for (twoi=n-1; twoi>=2; twoi -= 2) s=s*c[twoi-1]/c[twoi]+1.0;
        w[low]=1.0/s;
    }
    free_real_vector(p,0);
}
```

4.2.4 Gaussian quadrature - Special weights

A. gssjacwghts

Computes the n roots, x_i, of the n-th Jacobi polynomial $P_n^{(\alpha,\beta)}$ and corresponding Gauss-Christoffel numbers w_i needed for the n-point Gauss-Jacobi quadrature over [-1,1] with weight function $(1-x)^\alpha(1+x)^\beta$. The quadrature formula

$$\int_{-1}^{1}(1-x)^\alpha(1+x)^\beta f(x)dx = \sum_{i=1}^{n} w_i f(x_i)$$

is exact if f is a polynomial of degree $\leq 2n$-1.

The x_i are first determined by means of a call of *alljaczer*, and the w_i then derived by use of the formula

$$w_i = \left(\sum_{j=0}^{n-1} P_j^{(\alpha,\beta)}(x_i)^2 \right)^{-1} .$$

Function Parameters:

void gssjacwghts (*n,alfa,beta,x,w*)

n: int;
entry: the upper bound of the arrays x and w ($n \geq 1$);

alfa,beta: float;
entry: the parameters of the weight function for the Jacobi polynomials, *alfa, beta* > -1;

x: float *x[1:n]*;
exit: *x[i]* is the i-th zero of the n-th Jacobi polynomial;

w: float *w[1:n]*;
exit: *w[i]* is the Gauss-Christoffel number associated with the i-th zero of the n-th Jacobi polynomial.

Functions used: gamma, alljaczer.

Method: See [AbS65, St72].

```
#include <math.h>

void gssjacwghts(int n, float alfa, float beta, float x[], float w[])
{
    float *allocate_real_vector(int, int);
    void free_real_vector(float *, int);
    float gamma(float);
    void alljaczer(int, float, float, float []);
    int i,j,m;
```

```
    float r0,r1,r2,s,h0,alfa2,xi,*a,*b,min,sum,alfabeta,temp;

    if (alfa == beta) {
        b=allocate_real_vector(1,n-1);
        alljaczer(n,alfa,alfa,x);
        alfa2=2.0*alfa;
        temp=gamma(1.0+alfa);
        h0=pow(2.0,alfa2+1.0)*temp*temp/gamma(alfa2+2.0);
        b[1]=1.0/sqrt(3.0+alfa2);
        m=n-n/2;
        for (i=2; i<=n-1; i++)
            b[i]=sqrt(i*(i+alfa2)/(4.0*(i+alfa)*(i+alfa)-1.0));
        for (i=1; i<=m; i++) {
            xi=fabs(x[i]);
            r0=1.0;
            r1=xi/b[1];
            s=1.0+r1*r1;
            for (j=2; j<=n-1; j++) {
                r2=(xi*r1-b[j-1]*r0)/b[j];
                r0=r1;
                r1=r2;
                s += r2*r2;
            }
            w[i]=w[n+1-i]=h0/s;
        }
        free_real_vector(b,1);
    } else {
        a=allocate_real_vector(0,n);
        b=allocate_real_vector(0,n);
        alfabeta=alfa+beta;
        min=(beta-alfa)*alfabeta;
        b[0]=0.0;
        sum=alfabeta+2.0;
        a[0]=(beta-alfa)/sum;
        a[1]=min/sum/(sum+2.0);
        b[1]=2.0*sqrt((1.0+alfa)*(1.0+beta)/(sum+1.0))/sum;
        for (i=2; i<=n-1; i++) {
            sum=i+i+alfabeta;
            a[i]=min/sum/(sum+2.0);
            b[i]=(2.0/sum)*sqrt(i*(sum-i)*(i+alfa)*(i+beta)/(sum*sum-1.0));
        }
        h0=pow(2.0,alfabeta+1.0)*gamma(1.0+alfa)*gamma(1.0+beta)/
                    gamma(2.0+alfabeta);
        alljaczer(n,alfa,beta,x);
        for (i=1; i<=n; i++) {
            xi=x[i];
            r0=1.0;
            r1=(xi-a[0])/b[1];
            sum=1.0+r1*r1;
            for (j=2; j<=n-1; j++) {
                r2=((xi-a[j-1])*r1-b[j-1]*r0)/b[j];
                sum += r2*r2;
                r0=r1;
                r1=r2;
            }
            w[i]=h0/sum;
        }
        free_real_vector(a,0);
        free_real_vector(b,0);
    }
}
```

B. gsslagwghts

Computes the n roots, x_i, of the n-th Laguerre polynomial $L_n^{(\alpha)}$ and corresponding Gauss-Christoffel numbers w_i needed for the n-point Gauss-Laguerre quadrature of $f(x)$ over $(0,\infty]$ with respect to the weight function $x^\alpha e^{-x}$. The quadrature formula

$$\int_0^\infty x^\alpha e^{-x} f(x)dx = \sum_{i=1}^{n} w_i f(x_i)$$

is exact if f is a polynomial of degree $\leq 2n-1$.

The x_i are first determined by means of a call of *alllagzer*, and the w_i then derived by use of the formula

$$w_i = \left(\sum_{j=0}^{n-1} L_j^{(\alpha)}(x_i)^2 \right)^{-1} .$$

Function Parameters:

void gsslagwghts (*n,alfa,x,w*)

n: int;
entry: the upper bound of the arrays *x* and *w* ($n \geq 1$);

alfa: float;
entry: the parameter of the weight function for the Laguerre polynomials, *alfa* > -1;

x: float *x[1:n]*;
exit: *x[i]* is the *i*-th zero of the *n*-th Laguerre polynomial;

w: float *w[1:n]*;
exit: *w[i]* is the Gaussian weight associated with the *i*-th zero of the *n*-th Laguerre polynomial.

Functions used: gamma, alllagzer.

Method: The zeros and weights are computed in the same way as in the procedure *gssjacwghts*.

```
#include <math.h>

void gsslagwghts(int n, float alfa, float x[], float w[])
{
    float *allocate_real_vector(int, int);
    void free_real_vector(float *, int);
    void alllagzer(int, float, float []);
    float gamma(float);
    int i,j;
    float h0,s,r0,r1,r2,xi,*a,*b;

    a=allocate_real_vector(0,n);
    b=allocate_real_vector(0,n);
    a[0]=1.0+alfa;
    a[1]=3.0+alfa;
    b[1]=sqrt(a[0]);
    for (i=2; i<=n-1; i++) {
        a[i]=i+i+alfa+1.0;
        b[i]=sqrt(i*(i+alfa));
    }
    alllagzer(n,alfa,x);
    h0=gamma(1.0+alfa);
    for (i=1; i<=n; i++) {
        xi=x[i];
        r0=1.0;
        r1=(xi-a[0])/b[1];
        s=1.0+r1*r1;
        for (j=2; j<=n-1; j++) {
            r2=((xi-a[j-1])*r1-b[j-1]*r0)/b[j];
            r0=r1;
```

```
            r1=r2;
            s += r2*r2;
        }
        w[i]=h0/s;
    }
    free_real_vector(a,0);
    free_real_vector(b,0);
}
```

4.3 Numerical differentiation

4.3.1 Calculation with difference formulas

A. jacobnnf

Calculates the Jacobian matrix of an *n*-dimensional function of *n* variables using forward differences. *jacobnnf* computes first order difference quotient approximations

$$J_{i,j} = (f_i(x_1,...,x_{j-1},x_j+\delta_j,x_{j+1},...,x_n)-f_i(x_1,...,x_{j-1},x_j,x_{j+1},...,x_n))/\delta_i \quad (i,j=1,...,n)$$

to the partial derivatives $J_{i,j} = \partial f_i(x)/\partial x_j$ of the components of the function $f(x)$ ($f,x \in R^n$).

Function Parameters:

void jacobnnf (*n,x,f,jac,di,funct*)

n: int;
  entry: the number of independent variables and the dimension of the function;
x: float *x[1:n]*;
  entry: the point at which the Jacobian has to be calculated;
f: float *f[1:n]*;
  entry: the values of the function components at the point given in array *x*;
jac: float *jac[1:n,1:n]*;
  exit: the Jacobian matrix in such a way that the partial derivative of *f[i]* with respect to *x[j]* is given in *jac[i,j]*, *i,j=1,...,n*;
di: float *(*di)(i), int i*;
  entry: the partial derivatives to *x[i]* are approximated by forward differences, using an increment in the *i*-th variable equal to the value of *di*, *i=1,...,n*;
funct: void *(*funct)(n,x,f)*;
  entry: the meaning of the parameters of the function *funct* is as follows:
    n: the number of independent variables of the function *f*;
    x: the independent variables are given in *x[1:n]*;
    f: after a call of *funct* the function components should be given in *f[1:n]*.

```
void jacobnnf(int n, float x[], float f[], float **jac,
                float (*di)(int), void (*funct)(int, float[], float[]))
{
    float *allocate_real_vector(int, int);
    void free_real_vector(float *, int);
    int i,j;
    float step,aid,*f1;

    f1=allocate_real_vector(1,n);
    for (i=1; i<=n; i++) {
        step=(*di)(i);
        aid=x[i];
        x[i]=aid+step;
        step=1.0/step;
        (*funct)(n,x,f1);
        for (j=1; j<=n; j++) jac[j][i]=(f1[j]-f[j])*step;
```

```
        x[i]=aid;
    }
    free_real_vector(f1,1);
}
```

B. jacobnmf

Calculates the Jacobian matrix of an *n*-dimensional function of *m* variables using forward differences. *jacobnmf* computes first order difference quotient approximations

$$J_{i,j} = (f_i(x_1,...,x_{j-1},x_j+\delta_j,x_{j+1},...,x_m) - f_i(x_1,...,x_{j-1},x_j,x_{j+1},...,x_m))/\delta_i \qquad (i,j=1,...,n;j=1,...m)$$

to the partial derivatives $J_{i,j} = \partial f_i(x)/\partial x_j$ of the components of the function $f(x)$ $(f \epsilon R^n, x \epsilon R^m)$.

Function Parameters:

void jacobnmf (*n,m,x,f,jac,di,funct*)

n: int;
  entry: the number of function components;
m: int;
  entry: the number of independent variables;
x: float *x[1:m]*;
  entry: the point at which the Jacobian has to be calculated;
f: float *f[1:n]*;
  entry: the values of the function components at the point given in array *x*;
jac: float *jac[1:n,1:m]*;
  exit: the Jacobian matrix in such a way that the partial derivative of *f[i]* with respect to *x[j]* is given in *jac[i,j]*, *i=1,...,n, j=1,...,m*;
di: float *(*di)(i), int i*;
  entry: the partial derivatives to *x[i]* are approximated by forward differences, using an increment in the *i*-th variable equal to the value of *di*, *i=1,...,m*;
funct: void *(*funct)(n,m,x,f)*;
  entry: the meaning of the parameters of the function *funct* is as follows:
    n: the number of function components;
    m: the number of independent variables of the function *f*;
    x: the independent variables are given in *x[1:m]*;
    f: after a call of *funct* the function components should be given in *f[1:n]*.

```
void jacobnmf(int n, int m, float x[], float f[], float **jac,
              float (*di)(int), void (*funct)(int, int, float[], float[]))
{
    float *allocate_real_vector(int, int);
    void free_real_vector(float *, int);
    int i,j;
    float step,aid,*f1;

    f1=allocate_real_vector(1,n);
    for (i=1; i<=m; i++) {
        step=(*di)(i);
        aid=x[i];
        x[i]=aid+step;
        step=1.0/step;
        (*funct)(n,m,x,f1);
        for (j=1; j<=n; j++) jac[j][i]=(f1[j]-f[j])*step;
        x[i]=aid;
    }
    free_real_vector(f1,1);
}
```

C. jacobnbndf

Calculates the Jacobian matrix of an n-dimensional function of n variables, if this Jacobian is known to be a band matrix and have to be stored rowwise in a one-dimensional array. *jacobnbndf* computes first order difference quotient approximations

$$J_{i,j} = (f_i(x_1,...,x_{j-1},x_j+\delta_i,x_{j+1},...,x_n)-f_i(x_1,...,x_{j-1},x_j,x_{j+1},...,x_n))/\delta i$$

for $i=1,...n$; $\max(1,i-lw)\leq j\leq\min(n,i+rw)$ to the partial derivatives $J_{i,j} = \partial f_i(x)/\partial x_j$ of the components of the function $f(x)$ ($f,x\epsilon R^n$).

Function Parameters:

void jacobnbndf (*n,lw,rw,x,f,jac,di,funct*)

n: int;
    entry: the number of independent variables and the dimension of the function;
lw: int;
    entry: the number of codiagonals to the left of the main diagonal of the Jacobian matrix, which is known to be a band matrix;
rw: int;
    entry: the number of codiagonals to the right of the main diagonal of the Jacobian matrix;
x: float *x[1:n]*;
    entry: the point at which the Jacobian has to be calculated;
f: float *f[1:n]*;
    entry: the values of the function components at the point given in array *x*;
jac: float *jac[1:(lw+rw)*(n-1)+n]*;
    exit: the Jacobian matrix in such a way that the (*i,j*)-th element of the Jacobian, i.e. the partial derivative of *f[i]* to *x[j]* is given in *jac[(lw+rw)*(i-1)+j]*, *i=1,...,n*, *j*=max(1,*i-lw*),...,min(*n,i+rw*);
di: float *(*di)(i), int i*;
    entry: the partial derivatives to *x[i]* are approximated with forward differences, using an increment to the *i*-th variable that equals the value of *di*, *i=1,...,n*;
funct: void *(*funct)(n,l,u,x,f)*;
    entry: the meaning of the parameters of the function *funct* is as follows:
        n: the number of function components;
        l,u: int; the lower and upper bound of the function component subscript;
        x: the independent variables are given in *x[1:n]*;
        f: after a call of *funct* the function components *f[i]*, *i=l,...,u*, should be given in *f[l:u]*.

```
void jacobnbndf(int n, int lw, int rw, float x[], float f[],
                    float jac[], float (*di)(int),
                    int (*funct)(int, int, int, float[], float[]))
{
    float *allocate_real_vector(int, int);
    void free_real_vector(float *, int);
    int i,j,k,l,u,t,b,ll;
    float aid,stepi,*f1;

    l=1;
    u=lw+1;
    t=rw+1;
    b=lw+rw;
    for (i=1; i<=n; i++) {
        ll=l;
```

```
        f1=allocate_real_vector(ll,u);
        stepi=(*di)(i);
        aid=x[i];
        x[i]=aid+stepi;
        (*funct)(n,l,u,x,f1);
        x[i]=aid;
        k = i+((i <= t) ? 0 : i-t)*b;
        for (j=l; j<=u; j++) {
            jac[k]=(f1[j]-f[j])/stepi;
            k += b;
        }
        if (i >= t) l++;
        if (u < n) u++;
        free_real_vector(f1,ll);
    }
}
```

5. Analytic Problems

5.1 Non-linear equations

5.1.1 Single equation - No derivative available

A. zeroin

Given the values, x_0 and y_0 say, of the end points of an interval assumed to contain a zero of the function $f(x)$, *zeroin* attempts to find, by use of a combination of linear interpolation, extrapolation, and bisection, values, x_n and y_n say, of the end points of a smaller interval containing the zero. The method used is described in detail in [BusD75]. At successful exit $f(x_n)f(y_n) \leq 0$, $|f(x_n)| \leq |f(y_n)|$ and $|x_n - y_n| \leq 2*tol(x_n)$, where *tol* is a tolerance function prescribed by the user. The value of $tol(x)$ may be changed during the course of the computation or left constant, as the user requires. For example $tol(x)$ may be the function $|x|*re+ae$ (where *re*, the permissible relative error, should be greater than the machine precision, and *ae*, the permissible absolute error should not be set equal to zero if the prescribed interval contains the origin).

The algorithm used results in the construction of five sequences of real numbers x_i, a_i, b_i, c_i and d_i, and two integer sequences $j(i)$, $k(i)$ $(i=1,...,n)$ as follows. Initially, with $x_1=y_0$, if $|f(x_1)| \leq |f(x_0)|$ then $b_1=x_1$, $a_1=c_1=x_0$, otherwise $b_1=x_0$, $a_1=c_1=x_1$. Then, for $i=2,...,n$: (a) $j(i)$ is taken to be the largest integer in the range $1<j(i)<i$ for which $|b_{j(i)}-c_{j(i)}| \leq ½ |b_{j(i)-1}-c_{j(i)-1}|$ with $j(i)=1$ if this condition is never satisfied; (b) with

$$h(b,c) = b + sign(c-b)tol(b)$$
$$m(b,c) = (b+c)/2$$
$$l(b,a) = b - f(b)(b-a)/(f(b)-f(a)) \qquad (f(a) \neq f(b))$$
$$l(b,a) = \infty \qquad (f(a)=f(b) \neq 0)$$
$$l(b,a) = b \qquad (f(a)=f(b)=0)$$

and

$w(l,b,c) = l$ if l is between $h(b,c)$ and $m(b,c)$;

$w(l,b,c) = h(b,c)$ if $|l-b| \leq tol(b)$ and l does not lie outside the interval bounded by b and $m(b,c)$;

$w(l,b,c) = m(b,c)$ otherwise

and $r(b,a,d)$ determined by the condition that

$$(x-r(b,a,d))/(px+q) = f(x)$$

when $x=a,b,d$, x_i is determined by use of the formula

$$x_i = w(l(b_{i-1},a_{i-1}),\ b_{i-1},c_{i-1}) \qquad \text{if } j(i) \geq i-2$$
$$x_i = w(r(b_{i-1},a_{i-1},d_{i-1}),\ b_{i-1},c_{i-1}) \qquad \text{if } j(i) = i-3$$
$$x_i = m(b_{i-1},c_{i-1}) \qquad \text{otherwise}$$

(c) $k(i)$ is taken to be the largest integer in the range $0 \leq k(i)<i$ for which $f(x_{k(i)})f(x_i) \leq 0$, and if $|f(x_i)| \leq |f(x_{k(i)})|$, then $b_i=x_i$, $c_i=x_{k(i)}$, $a_i=b_{i-1}$, otherwise $b_i=x_{k(i)}$, $a_i=c_i=x_i$; also, if $b_i=x_i$ or $b_i=b_{i-1}$ then $d_i=a_{i-1}$, otherwise $d_i=b_{i-1}$; (d) n is the smallest integer for which $|b_n-c_n| \leq 2*tol(b_n)$, and then $x_n=b_n$, $y_n=c_n$.

The number of evaluations of *fx* and *tolx* is at most $4*\log_2(|x-y|)/tau$, where x and y are the argument values given upon entry, *tau* is the minimum of the tolerance function *tolx* on the initial interval. If upon entry x and y satisfy $f(x)*f(y)\leq 0$, then convergence is guaranteed and the asymptotic order of convergence is 1.618 for simple zeros.

Function Parameters:

int zeroin ($x,y,fx,tolx$)

zeroin: on exit *zeroin* is given a nonzero value when a sufficiently small subinterval of J containing a zero of the function $f(x)$ has been found; otherwise *zeroin* is given the value zero;

x: float *;
- entry: one endpoint of interval J in which a zero is searched for;
- exit: a value approximating the zero within the tolerance $2*tol(x)$ when *zeroin* has the value nonzero, and a presumably worthless argument value otherwise;

y: float *;
- entry: the other endpoint of interval J in which a zero is searched for; upon entry $x < y$ as well as $y < x$ is allowed;
- exit: the other straddling approximation of the zero, i.e. upon exit the values of y and x satisfy (a) $f(x)*f(y)\leq 0$, (b) $|x-y|\leq 2*tol(x)$ and (c) $|f(x)|\leq|f(y)|$ when *zeroin* has a nonzero value;

fx: float (*fx)(x);
- entry: defines function f as a function depending on the actual parameter for x;

tolx: float (*tolx)(x);
- entry: defines the tolerance function *tol* which may depend on the actual parameter for x; one should choose *tolx* positive and never smaller than the precision of the machine's arithmetic at x, i.e. in this arithmetic x+*tolx* and x-*tolx* should always yield values distinct from x; otherwise the procedure may get into a loop.

```
#include <float.h>
#include <math.h>

int zeroin(float *x, float *y, float (*fx)(float), float (*tolx)(float))
{
    int ext,extrapolate;
    float c,fc,b,fb,a,fa,d,fd,fdb,fda,w,mb,tol,m,p,q;

    b = *x;
    fb=(*fx)(*x);
    a = *x = *y;
    fa=(*fx)(*x);
    c=a;
    fc=fa;
    ext=0;
    extrapolate=1;
    while (extrapolate) {
        if (fabs(fc) < fabs(fb)) {
            if (c != a) {
                d=a;
                fd=fa;
            }
            a=b;
            fa=fb;
            b = *x =c;
            fb=fc;
            c=a;
            fc=fa;
        }
        tol=(*tolx)(*x);
        m=(c+b)*0.5;
        mb=m-b;
        if (fabs(mb) > tol) {
```

```
            if (ext > 2)
                w=mb;
            else {
                if (mb == 0.0)
                    tol=0.0;
                else
                    if (mb < 0.0) tol = -tol;
                p=(b-a)*fb;
                if (ext <= 1)
                    q=fa-fb;
                else {
                    fdb=(fd-fb)/(d-b);
                    fda=(fd-fa)/(d-a);
                    p *= fda;
                    q=fdb*fa-fda*fb;
                }
                if (p < 0.0) {
                    p = -p;
                    q = -q;
                }
                w=(p<FLT_MIN || p<=q*tol) ? tol : ((p<mb*q) ? p/q : mb);
            }
            d=a;
            fd=fa;
            a=b;
            fa=fb;
            *x = b += w;
            fb=(*fx)(*x);
            if ((fc >= 0.0) ? (fb >= 0.0) : (fb <= 0.0)) {
                c=a;
                fc=fa;
                ext=0;
            } else
                ext = (w == mb) ? 0 : ext+1;
        } else
            break;
    }
    *y = c;
    return ((fc >= 0.0) ? (fb <= 0.0) : (fb >= 0.0));
}
```

B. zeroinrat

Given the values, x_0 and y_0 say, of the end points of an interval assumed to contain a zero of the function $f(x)$, *zeroinrat* attempts to find, by use of a combination of linear interpolation, extrapolation, and bisection, values, x_n and y_n say, of the end points of a smaller interval containing the zero. The method used is described in detail in [BusD75]. At successful exit $f(x_n)f(y_n) \le 0$, $|f(x_n)| \le |f(y_n)|$ and $|x_n - y_n| \le 2*tol(x_n)$, where *tol* is a tolerance function prescribed by the user (see the documentation of *zeroin*).

The algorithm used by *zeroinrat* closely resembles that used by *zeroin*, except that x_i is now determined by use of the formula

$x_i = w(l(b_{i-1},a_{i-1}),\ b_{i-1},c_{i-1})$ if $j(i) \ge i-2$

$x_i = w(r(b_{i-1},a_{i-1},d_{i-1}),\ b_{i-1},c_{i-1})$ if $j(i) = i-3$

$x_i = w(2r(b_{i-1},a_{i-1},d_{i-1})-b_{i-1},b_{i-1},c_{i-1})$ if $i \ge 3$ and $j(i)=i-4$

$x_i = m(b_{i-1},c_{i-1})$ otherwise.

The number of evaluations of *fx* and *tolx* is at most $5*\log_2(|x-y|)/tau$, where x and y are the argument values given upon entry, *tau* is the minimum of the tolerance function *tolx* on the initial interval. If upon entry x and y satisfy $f(x)*f(y) \le 0$, then convergence is guaranteed and the asymptotic order of convergence is 1.839 for simple zeros.

zeroin is preferable for simple (i.e. cheaply to calculate) functions and/or when no high precision is required. *zeroinrat* is preferable for complicated (i.e. expensive) functions when a zero is required in rather high precision and also for functions having a pole near the zero.

When the analytic derivative of the function is easily obtained, then *zeroinder* should be taken into consideration.

Function Parameters:

int zeroinrat (*x,y,fx,tolx*)

zeroinrat: on exit *zeroinrat* is given a nonzero value when a sufficiently small subinterval of *J* containing a zero of the function *f(x)* has been found; otherwise *zeroinrat* is given the value zero;

x: float *;
  entry: one endpoint of interval *J* in which a zero is searched for;
  exit: a value approximating the zero within the tolerance 2**tol(x)* when *zeroinrat* has a nonzero value;

y: float *;
  entry: the other endpoint of interval *J* in which a zero is searched for; upon entry $x < y$ as well as $y < x$ is allowed;
  exit: the other straddling approximation of the zero, i.e. upon exit the values of *y* and *x* satisfy (a) $f(x)*f(y) \leq 0$, (b) $|x-y| \leq 2*tol(x)$ and (c) $|f(x)| \leq |f(y)|$ when *zeroinrat* has a nonzero value;

fx: float (**fx*)(*x*);
  entry: defines function *f* as a function depending on the actual parameter for *x*;

tolx: float (**tolx*)(*x*);
  entry: defines the tolerance function *tol* which may depend on the actual parameter for *x*; one should choose *tolx* positive and never smaller than the precision of the machine's arithmetic at *x*, i.e. in this arithmetic *x+tolx* and *x-tolx* should always yield values distinct from *x*; otherwise the procedure may get into a loop.

```
#include <float.h>
#include <math.h>

int zeroinrat(float *x, float *y, float (*fx)(float),
              float (*tolx)(float))
{
    int ext,first,extrapolate;
    float b,fb,a,fa,d,fd,c,fc,fdb,fda,w,mb,tol,m,p,q;

    b = *x;
    fb=(*fx)(*x);
    a = *x = *y;
    fa=(*fx)(*x);
    first=1;
    c=a;
    fc=fa;
    ext=0;
    extrapolate=1;
    while (extrapolate) {
        if (fabs(fc) < fabs(fb)) {
            if (c != a) {
                d=a;
                fd=fa;
            }
            a=b;
            fa=fb;
            b = *x =c;
            fb=fc;
            c=a;
            fc=fa;
        }
        tol=(*tolx)(*x);
        m=(c+b)*0.5;
```

```
		mb=m-b;
		if (fabs(mb) > tol) {
			if (ext > 3)
				w=mb;
			else {
				if (mb == 0.0)
					tol=0.0;
				else
					if (mb < 0.0) tol = -tol;
				p=(b-a)*fb;
				if (first) {
					q=fa-fb;
					first=0;
				} else {
					fdb=(fd-fb)/(d-b);
					fda=(fd-fa)/(d-a);
					p *= fda;
					q=fdb*fa-fda*fb;
				}
				if (p < 0.0) {
					p = -p;
					q = -q;
				}
				if (ext == 3) p *= 2.0;
				w=(p<FLT_MIN || p<=q*tol) ? tol : ((p<mb*q) ? p/q : mb);
			}
			d=a;
			fd=fa;
			a=b;
			fa=fb;
			*x = b += w;
			fb=(*fx)(*x);
			if ((fc >= 0.0) ? (fb >= 0.0) : (fb <= 0.0)) {
				c=a;
				fc=fa;
				ext=0;
			} else
				ext = (w == mb) ? 0 : ext+1;
		} else
			break;
	}
	*y = c;
	return ((fc >= 0.0) ? (fb <= 0.0) : (fb >= 0.0));
}
```

5.1.2 Single equation - Derivative available

zeroinder

Given the values, x_0 and y_0 say, of the end points of an interval assumed to contain a zero of the function $f(x)$, *zeroinder* attempts to find, by use of a combination of interpolation and extrapolation based upon the use of a fractional linear function, and of bisection, values, x_n and y_n say, of the end points of a smaller interval containing the zero. At successful exit $f(x_n)f(y_n) \leq 0$, $|f(x_n)| \leq |f(y_n)|$ and $|x_n - y_n| \leq 2*tol(x_n)$, where *tol* is a tolerance function prescribed by the user (see the documentation of *zeroin*).

The algorithm used by *zeroinder* is similar to that used by *zeroin*, with the difference that the estimate $l(b,a)$ of the zero of $f(x)$ obtained by linear interpolation between the two function values at $x=a$ and $x=b$ is replaced by an estimate obtained by interpolation based upon use of a fractional linear function $(x-u)/(vx+w)$ whose value agrees with that of $f(x)$ at $x=a$ and $x=b$, and whose derivative is also equal in value to that of $f(x)$ at one of these points.

zeroinder is to prefer to *zeroin* or *zeroinrat* if the derivative is (much) cheaper to evaluate than the function.

The number of evaluations of *fx*, *dfx* and *tolx* is at most $4*\log_2(|x-y|)/tau$, where x and y are the argument values given upon entry, *tau* is the minimum of the tolerance function *tolx* on the initial interval (i.e. *zeroinder* requires at most 4 times the number of steps required for bisection). If upon entry x and y satisfy $f(x)*f(y) \leq 0$, then convergence is guaranteed and the asymptotic order of convergence is 2.414 for a simple zero of f.

Function Parameters:

int zeroinder (*x,y,fx,dfx,tolx*)

zeroinder: on exit *zeroinder* is given a nonzero value when a sufficiently small subinterval of J containing a zero of the function $f(x)$ has been found; otherwise *zeroin* is given the value zero;

x: float *;
- entry: one endpoint of interval J in which a zero is searched for;
- exit: a value approximating the zero within the tolerance $2*tol(x)$ when *zeroinder* has a nonzero value;

y: float *;
- entry: the other endpoint of interval J in which a zero is searched for; upon entry $x < y$ as well as $y < x$ is allowed;
- exit: the other straddling approximation of the zero, i.e. upon exit the values of y and x satisfy (a) $f(x)*f(y) \leq 0$, (b) $|x-y| \leq 2*tol(x)$ and (c) $|f(x)| \leq |f(y)|$ when *zeroinder* has a nonzero value;

fx: float (**fx*)(*x*);
- entry: defines function f as a function depending on the actual parameter for x;

dfx: float (**dfx*)(*x*);
- entry: defines derivative *df* of f as a function depending on the actual parameter for x;

tolx: float (**tolx*)(*x*);
- entry: defines the tolerance function *tol* which may depend on the actual parameter for x; one should choose *tolx* positive and never smaller than the precision of the machine's arithmetic at x, i.e. in this arithmetic $x+tolx$ and $x-tolx$ should always yield values distinct from x; otherwise the procedure may get into a loop.

```
#include <float.h>
#include <math.h>

int zeroinder(float *x, float *y, float (*fx)(float),
              float (*dfx)(float), float (*tolx)(float))
{
    int ext,extrapolate;
    float b,fb,dfb,a,fa,dfa,c,fc,dfc,d,w,mb,tol,m,p,q;

    b = *x;
    fb=(*fx)(*x);
    dfb=(*dfx)(*x);
    a = *x = *y;
    fa=(*fx)(*x);
    dfa=(*dfx)(*x);
    c=a;
    fc=fa;
    dfc=dfa;
    ext=0;
    extrapolate=1;
    while (extrapolate) {
        if (fabs(fc) < fabs(fb)) {
            a=b;
            fa=fb;
```

```
            dfa=dfb;
            b = *x =c;
            fb=fc;
            dfb=dfc;
            c=a;
            fc=fa;
            dfc=dfa;
        }
        tol=(*tolx)(*x);
        m=(c+b)*0.5;
        mb=m-b;
        if (fabs(mb) > tol) {
            if (ext > 2)
                w=mb;
            else {
                if (mb == 0.0)
                    tol=0.0;
                else
                    if (mb < 0.0) tol = -tol;
                d = (ext == 2) ? dfa : (fb-fa)/(b-a);
                p=fb*d*(b-a);
                q=fa*dfb-fb*d;
                if (p < 0.0) {
                    p = -p;
                    q = -q;
                }
                w=(p<FLT_MIN || p<=q*tol) ? tol : ((p<mb*q) ? p/q : mb);
            }
            a=b;
            fa=fb;
            dfa=dfb;
            *x = b += w;
            fb=(*fx)(*x);
            dfb=(*dfx)(*x);
            if ((fc >= 0.0) ? (fb >= 0.0) : (fb <= 0.0)) {
                c=a;
                fc=fa;
                dfc=dfa;
                ext=0;
            } else
                ext = (w == mb) ? 0 : ext+1;
        } else
            break;
    }
    *y = c;
    return ((fc >= 0.0) ? (fb <= 0.0) : (fb >= 0.0));
}
```

5.1.3 System of equations - No Jacobian matrix

A. quanewbnd

Solves systems of non-linear equations of which the Jacobian is known to be a band matrix and an approximation of the Jacobian is assumed to be available at the initial guess. The method used is the same as given in [Bro71].

quanewbnd computes an approximation to the solution x of the system of equations $f(x)=0$ $(f,x\epsilon R^n)$ for which the components $J^{(k,j)}=\partial f^{(k)}(x)/\partial x^{(j)}$ of the associated Jacobian matrix are zero when $j>i+rw$ or $i>j+lw$, an approximation x_0 to x and an approximation J_0 to $\partial f(x_0)/\partial x_0$ being assumed to be available.

At the i-th stage of the iterative method used, δ_i is obtained by solving (*decsolbnd* is called for this purpose) the band matrix system of equations

$$J_i\delta_i = -f(x_i) \tag{1}$$

with $x_{i+1} = x_i + \delta_i$ and with $p_i^{(j)} = (0,...,0,\delta_i^{k(j)},...,\delta_i^{h(j)},0,...,0)$ where $h(j)$=max$(1,j-lw)$, $k(j)$=min$(n,j+rw)$, and with I_j being the matrix whose (j,j)-th element is

unity, all other elements being zero, and (the bars denoting Euclidean norms) with
$\Delta_i^{(j)} = I_j f(x_{i+1})(p_i^{(j)})^T/(p_i^{(j)},p_i^{(j)})$ if $\|p_i^{(j)}\| \geq \varepsilon^2 \|\delta_i\| \, \|p_i\|$, and 0 otherwise
(ε being the machine precision) J_{i+1} is obtained from the relationship

$$J_{i+1} = J_i + \sum_{j=1}^{n} \Delta_i^{(j)} ;$$

formula (1) is reapplied and the process continued.

Function Parameters:

void quanewbnd (*n,lw,rw,x,f,jac,funct,in,out*)

n: int;
entry: the number of independent variables; the number of equations should also be equal to *n*;

lw: int;
entry: the number of codiagonals to the left of the main diagonal of the Jacobian;

rw: int;
entry: the number of codiagonals to the right of the main diagonal of the Jacobian;

x: float *x[1:n]*;
entry: an initial estimate of the solution of the system that has to be solved;
exit: the calculated solution of the system;

f: float *f[1:n]*;
entry: the values of the function components at the initial guess;
exit: the values of the function components at the calculated solution;

jac: float *jac[1:(lw+rw)*(n-1)+n]*;
entry: an approximation of the Jacobian at the initial estimate of the solution; an approximation of the (*i,j*)-th element of the Jacobian is given in *jac[(lw+rw)*(i-1)+j]*, for *i*=1,...,*n* and *j*=max(1,*i-lw*),...,min(*n,i+rw*);
exit: an approximation to the Jacobian at the calculated solution;

funct: int *(*funct)(n,l,u,x,f)*;
entry: the meaning of the parameters of the function *funct* is as follows:
n: the number of independent variables of the function *f*;
l,u: int; the lower and upper bound of the function component subscript;
x: the independent variables are given in *x[1:n]*;
f: after a call of *funct* the function components *f[i]*, *i=l,...,u*, should be given in *f[l:u]*;
exit: if the value of *funct* is zero then the execution of *quanewbnd* will be terminated, while the value of *out[5]* is set equal to 2;

in: float *in[0:4]*;
entry:
in[0]: the machine precision;
in[1]: the relative precision asked for;
in[2]: the absolute precision asked for; if the value, delivered in *out[5]* equals zero then the last correction vector *d*, say, which is a measure for the error in the solution, satisfies the inequality $\|d\| \leq \|x\|$ * *in[1]+in[2]*, whereby *x* denotes the calculated solution, given in array *x* and $\|.\|$ denotes the Euclidean norm; however, we cannot guarantee that the true error in the solution satisfies this inequality, especially if the Jacobian is (nearly) singular at the solution;

in[3]: the maximum value of the norm of the residual vector allowed; if *out[5]*=0 then this residual vector *r*, say, satisfies: $\|r\| \leq in[3]$;
in[4]: the maximum number of function component evaluations allowed; *l-u+1* function component evaluations are counted for each call of *funct(n,l,u,x,f)*; if *out[5]*=1 then the process is terminated because the number of evaluations exceeded the value given in *in[4]*;

out: float *out[1:5]*;
exit:
out[1]: the Euclidean norm of the last step accepted;
out[2]: the Euclidean norm of the residual vector at the calculated solution;
out[3]: the number of function component evaluations performed;
out[4]: the number of iterations carried out;
out[5]: the integer value delivered in *out[5]* gives some information about the termination of the process;
out[5]=0: the process is terminated in a normal way; the last step and the norm of the residual vector satisfy the conditions (see *in[2]*, *in[3]*); if *out[5]*≠0 then the process is terminated prematurely;
out[5]=1: the number of function component evaluations exceeds the value given in *in[4]*;
out[5]=2: a call of *funct* delivered the value zero;
out[5]=3: the approximation to the Jacobian matrix turns out to be singular.

Functions used: mulvec, dupvec, vecvec, elmvec, decsolbnd.

```
#include <math.h>

void quanewbnd(int n, int lw, int rw,
               float x[], float f[], float jac[],
               int (*funct)(int, int, int, float[], float[]),
               float in[], float out[])
{
    float *allocate_real_vector(int, int);
    void free_real_vector(float *, int);
    float vecvec(int, int, int, float [], float []);
    void elmvec(int, int, int, float [], float [], float);
    void mulvec(int, int, int, float [], float [], float);
    void dupvec(int, int, int, float [], float []);
    void decsolbnd(float [], int, int, int, float [], float []);
    int l,it,fcnt,fmax,err,b,i,j,k,r,m;
    float macheps,reltol,abstol,tolres,nd,mz,res,*delta,mul,crit,
          *pp,*s,aux[6],*lu;

    delta=allocate_real_vector(1,n);
    nd=0.0;
    macheps=in[0];
    reltol=in[1];
    abstol=in[2];
    tolres=in[3];
    fmax=in[4];
    mz=macheps*macheps;
    it=fcnt=0;
    b=lw+rw;
    l=(n-1)*b+n;
    b++;
    res=sqrt(vecvec(1,n,0,f,f));
    err=0;
    while (1) {
        if (err != 0 || (res < tolres &&
            sqrt(nd) < sqrt(vecvec(1,n,0,x,x))*reltol+abstol)) break;
        it++;
        if (it != 1) {
```

```
            /* update jac */
            pp=allocate_real_vector(1,n);
            s=allocate_real_vector(1,n);
            crit=nd*mz;
            for (i=1; i<=n; i++) pp[i]=delta[i]*delta[i];
            r=k=1;
            m=rw+1;
            for (i=1; i<=n; i++) {
                mul=0.0;
                for (j=r; j<=m; j++) mul += pp[j];
                j=r-k;
                if (fabs(mul) > crit) elmvec(k,m-j,j,jac,delta,f[i]/mul);
                k += b;
                if (i > lw)
                    r++;
                else
                    k--;
                if (m < n) m++;
            }
            free_real_vector(pp,1);
            free_real_vector(s,1);
        }
        /* direction */
        lu=allocate_real_vector(1,l);
        aux[2]=macheps;
        mulvec(1,n,0,delta,f,-1.0);
        dupvec(1,l,0,lu,jac);
        decsolbnd(lu,n,lw,rw,aux,delta);
        free_real_vector(lu,1);
        if (aux[3] != n) {
            err=3;
            break;
        } else {
            elmvec(1,n,0,x,delta,1.0);
            nd=vecvec(1,n,0,delta,delta);
            /* evaluate */
            fcnt += n;
            if (!((*funct)(n,1,n,x,f))) {
                err=2;
                break;
            }
            if (fcnt > fmax) err=1;
            res=sqrt(vecvec(1,n,0,f,f));
        }
    }
    out[1]=sqrt(nd);
    out[2]=res;
    out[3]=fcnt;
    out[4]=it;
    out[5]=err;
    free_real_vector(delta,1);
}
```

B. quanewbnd1

Solves systems of non-linear equations of which the Jacobian is known to be a band matrix and an approximation to the Jacobian at the initial guess is calculated using forward differences.

Computes an approximation to the solution x of the system of equations $f(x)=0$ ($f,x \epsilon R^n$) for which the components $J^{(k,j)}=\partial f^{(k)}(x)/\partial x^{(j)}$ to the associated Jacobian matrix are zero when $j>i+rw$ or $i>j+lw$, an approximation x_0 to x being assumed to be available.

An approximation J_0 to $\partial f(x_0)/\partial x_0$, based upon the use of first order finite difference approximations using equal increments d for all components, is first obtained by means of a call of *jacobnbndf*, and *quanewbnd* is then called to obtain the desired approximation.

Function Parameters:

void quanewbnd1 (*n,lw,rw,x,f,funct,in,out*)

n: int;
entry: the number of independent variables; the number of equations should also be equal to *n*;

lw: int;
entry: the number of codiagonals to the left of the main diagonal of the Jacobian;

rw: int;
entry: the number of codiagonals to the right of the main diagonal of the Jacobian;

x: float *x[1:n]*;
entry: an initial estimate of the solution of the system that has to be solved;
exit: the calculated solution of the system;

f: float *f[1:n]*;
exit: the values of the function components at the calculated solution;

funct: int *(*funct)(n,l,u,x,f)*;
entry: the meaning of the parameters of the function *funct* is as follows:
- *n*: the number of independent variables of the function *f*;
- *l,u*: int; the lower and upper bound of the function component subscript;
- *x*: the independent variables are given in *x[1:n]*;
- *f*: after a call of *funct* the function components *f[i]*, *i=l,...,u*, should be given in *f[l:u]*;

exit: if the value of *funct* is zero then the execution of *quanewbnd1* will be terminated, while the value of *out[5]* is set equal to 2;

in: float *in[0:5]*;
entry:
- *in[0]*: the machine precision;
- *in[1]*: the relative precision asked for;
- *in[2]*: the absolute precision asked for; if the value delivered in *out[5]* equals zero then the last correction vector *d*, say, which is a measure for the error in the solution, satisfies the inequality $\|d\| \leq \|x\| * in[1] + in[2]$, whereby *x* denotes the calculated solution, given in array *x* and $\|.\|$ denotes the Euclidean norm; however, we cannot guarantee that the true error in the solution satisfies this inequality, especially if the Jacobian is (nearly) singular at the solution;
- *in[3]*: the maximum value of the norm of the residual vector allowed; if *out[5]*=0 then this residual vector *r*, say, satisfies: $\|r\| \leq in[3]$;
- *in[4]*: the maximum number of function component evaluations allowed; *l-u+1* function component evaluations are counted each call of *funct(n,l,u,x,f)*; if *out[5]*=1 then the process is terminated because the number of evaluations exceeded the value given in *in[4]*;
- *in[5]*: the Jacobian matrix at the initial guess is approximated using forward differences, with a fixed increment to each variable that equals the value given in *in[5]*;

out: float *out[1:5]*;
exit:
- *out[1]*: the Euclidean norm of the last step accepted;
- *out[2]*: the Euclidean norm of the residual vector at the calculated solution;
- *out[3]*: the number of function component evaluations performed;
- *out[4]*: the number of iterations carried out;
- *out[5]*: the integer value delivered in *out[5]* gives some information about the

termination of the process;

out[5]=0: the process is terminated in a normal way; the last step and the norm of the residual vector satisfy the conditions (see *in[2]*, *in[3]*); if *out[5]*≠0 then the process is terminated prematurely;

out[5]=1: the number of function component evaluations exceeds the value given in *in[4]*;

out[5]=2: a call of *funct* delivered the value zero;

out[5]=3: the approximation of the Jacobian matrix turns out to be singular.

Functions used: jacobnbndf, quanewbnd.

```
void quanewbnd1(int n, int lw, int rw, float x[], float f[],
                int (*funct)(int, int, int, float[], float[]),
                float in[], float out[])
{
   float *allocate_real_vector(int, int);
   void free_real_vector(float *, int);
   float quanewbnd1t(float, int);
   void quanewbnd(int, int, int, float [], float [], float [],
                  int (*)(int, int, int, float[], float[]),
                  float [], float []);
   void jacobnbndf(int, int, int, float [], float [],
                  float [], float (*)(int),
                  int (*)(int, int, int, float[], float[]));
   int k;
   float *jac;

   jac=allocate_real_vector(1,(lw+rw)*(n-1)+n);
   (*funct)(n,1,n,x,f);
   k=(lw+rw)*(n-1)+n*2-((lw-1)*lw+(rw-1)*rw)/2;
   in[4] -= k;
   quanewbnd1t(in[5], 1);
   jacobnbndf(n,lw,rw,x,f,jac,quanewbnd1s,funct);
   quanewbnd(n,lw,rw,x,f,jac,funct,in,out);
   in[4] += k;
   out[3] += k;
   free_real_vector(jac,1);
}

float quanewbnd1s(int i)
{
   /* this function is used internally by QUANEWBND1 */

   float quanewbnd1t(float, int);

   return (quanewbnd1t(0.0,0));
}

float quanewbnd1t(float x, int i)
{
   /* this function is used internally by QUANEWBND1 */

   static float y;

   y = (i ? x : y);
   return (y);
}
```

5.2 Unconstrained optimization

5.2.1 One variable - No derivative

minin

Determines a point $x\epsilon[a,b]$ at which the real valued function $f(x)$ assumes a minimum value. It is assumed that, for some point $\mu\epsilon(a,b)$ either (a) f is strictly monotonically decreasing on $[a,\mu)$ and strictly monotonically increasing on $[\mu,b]$ or (b) these two intervals may be replaced by $[a,\mu]$ and $(\mu,b]$ respectively. Use is made of function values alone.

The method used involves the determination of five sequences of points a_n, b_n, v_n, w_n, x_n and further auxiliary numbers, u_n, p_n and q_n (n=0,1,...). Initially

$$a_0 = a,\quad b_0 = b,\quad v_0 = w_0 = x_0 = a_0 + ½(3-\sqrt{5})(b_0-a_0).$$

At the n-th stage of the computations a local minimum is known to lie in $[a_n,b_n]$ and $x_n\epsilon[a_n,b_n]$. If $\max(x_n-a_n, b_n-x_n) \leq 2*tol(x_n)$, where $tol(x)$ is a tolerance function prescribed by the user, the computations terminate, x_n is such that $f(x_n) < f(x_r)$ either (a) for r=0,1,...,n-1 or (b) for r=0,1,...,n-2 with $f(x_n) = f(x_{n-1})$; $f(w_n) < f(x_r)$ for r=0,1,...,n-2 in case (a) and for r=0,1,...,n-3 in case (b). If the computations are not terminated, p_n, q_n are determined such that $x_n + p_n/q_n$ is a turning point of the parabola through $(v_n,f(v_n))$, $(w_n,f(w_n))$ and $(x_n,f(x_n))$:

$$p_n = \pm\ ((x_n-v_n)^2(f(x_n)-f(w_n))-(x_n-w_n)^2(f(x_n)-f(v_n)))$$

$$q_n = \pm 2((x_n-v_n)^2(f(x_n)-f(w_n))-(x_n-w_n)(f(w_n)-f(v_n))).$$

If either $|p_n/q_n|\leq tol(x_n)$ or q_n=0 or $x_n+p_n/q_n \notin (a_n,b_n)$ then

$$u_{n+1} = ½(\sqrt{5}-1)x_n+½(3-\sqrt{5})a_n \text{ if } x_n\geq ½(a_n+b_n)$$

$$u_{n+1} = ½(\sqrt{5}-1)x_n+½(3-\sqrt{5})b_n \text{ if } x_n< ½(a_n+b_n);$$

otherwise $u_{n+1} = x_n + p_n/q_n$. The points a_n,...,w_n are updated as follows: if $f(u_{n+1}) \leq f(x_n)$ then (a) $a_{n+1}=a_n$ and $b_{n+1}=x_n$ if $u_{n+1} < x_n$, while if $u_{n+1} \geq x_n$, $a_{n+1}=x_n$ and $b_{n+1}=b_n$, and also (b) $v_{n+1}=w_n$, $w_{n+1}=x_n$ and $x_{n+1}=u_{n+1}$; if however, $f(u_{n+1}) > f(x_n)$ then (a) $a_{n+1}=u_{n+1}$ and $b_{n+1}=b_n$ if $u_{n+1}<x_n$, while if $u_{n+1} \geq x_n$, $a_{n+1}=a_n$ and $b_{n+1}=u_{n+1}$ and also (b) $x_{n+1}=x_n$, $v_{n+1}=x_n$ and $w_{n+1}=u_{n+1}$ if $f(u_{n+1}) \leq f(w_n)$ or $w_n=x_n$ while, if neither of these conditions is satisfied and either $f(u_{n+1}) \leq f(v_n)$ or $v_n=x_n$ or $v_n=w_n$, then $x_{n+1}=x_n$, $v_{n+1}=u_{n+1}$, $w_{n+1}=w_n$.

The user should be aware of the fact that the choice of *tolx* may highly affect the behavior of the algorithm, although convergence to a point for which the given function is minimal on the interval is assured. The asymptotic behavior will usually be fine as long as the numerical function is strictly δ-unimodal on the given interval (see [Bre73]) and the tolerance function satisfies $tol(x)\geq\delta$, for all x in the given interval.

Function Parameters:

float minin (*x,a,b,fx,tolx*)

minin: delivers the calculated minimum value of the function, defined by *fx*, on the interval with endpoints *a* and *b*;

x: float *;
- entry: the actual parameter for *fx* and *tolx* depends on *x*;
- exit: the calculated approximation of the position of the minimum;

a,b: float *;
- entry: the endpoints of the interval on which a minimum is searched for;
- exit: the endpoints of the interval with length less than 4**tol(x)* such that *a*<*x*<*b*;

fx: float (**fx*)(*x*);
- entry: the function is given by the actual parameter *fx* which depends on *x*;

tolx: float (**tolx*)(*x*);
- entry: defines the tolerance function which may depend on *x*; a suitable tolerance function is: $|x|*re+ae$, where *re* is the relative precision desired and *ae* is an absolute precision which should not be chosen equal to zero.

```
#include <math.h>
```

```
float minin(float *x, float *a, float *b, float (*fx)(float),
            float (*tolx)(float))
{
    float z,c,d,e,m,p,q,r,tol,t,u,v,w,fu,fv,fw,fz;

    c=(3.0-sqrt(5.0))/2.0;
    if (*a > *b) {
        z = *a;
        *a = *b;
        *b=z;
    }
    w = *x = *a;
    fw=(*fx)(*x);
    z = *x = *b;
    fz=(*fx)(*x);
    if (fz > fw) {
        z=w;
        w = *x;
        v=fz;
        fz=fw;
        fw=v;
    }
    v=w;
    fv=fw;
    e=0.0;
    while (1) {
        m=(*a + *b)*0.5;
        tol=(*tolx)(*x);
        t=tol*2.0;
        if (fabs(z-m) <= t-(*b - *a)*0.5) break;
        p=q=r=0.0;
        if (fabs(e) > tol) {
            r=(z-w)*(fz-fv);
            q=(z-v)*(fz-fw);
            p=(z-v)*q-(z-w)*r;
            q=(q-r)*2.0;
            if (q > 0.0)
                p = -p;
            else
                q = -q;
            r=e;
            e=d;
        }
        if (fabs(p) < fabs(q*r*0.5) && p > (*a-z)*q && p < (*b-z)*q) {
            d=p/q;
            u=z+d;
            if (u-(*a) < t || (*b)-u < t) d = ((z < m) ? tol : -tol);
        } else {
            e = ((z < m) ? *b : *a) - z;
            d=c*e;
        }
        u = *x = z + ((fabs(d) >= tol) ? d : ((d>0.0) ? tol : -tol));
        fu=(*fx)(*x);
        if (fu <= fz) {
            if (u < z)
                *b=z;
            else
                *a=z;
            v=w;
            fv=fw;
            w=z;
            fw=fz;
            z=u;
            fz=fu;
        } else {
            if (u < z)
                *a=u;
            else
                *b=u;
            if (fu <= fw) {
                v=w;
                fv=fw;
                w=u;
                fw=fu;
```

```
            } else
                if (fu <= fv || v == w) {
                    v=u;
                    fv=fu;
                }
        }
    }
    *x=z;
    return fz;
}
```

5.2.3 One variable - Derivative available

mininder

Determines a point $x \in [a,b]$ at which the real valued function $f(x)$ of one variable assumes a minimum value, when the analytical derivative $Df(x)$ of the function is available.

mininder has almost the same structure as the procedure given in [Bre73]. However, cubic interpolation (see [Dav59]) is used instead of quadratic interpolation to approximate the minimum. The method used involves the determination of a sequence of intervals $[a_n,b_n]$ (n=0,1,...) known to contain the required minimum point, and a further sequence of points u_n (n=1,2,...). Initially $a_0=a$, $b_0=b$. u_{n+1} is the point at which g has a minimum value, where g is the osculating cubic for which $g(x)=f(x)$ and $Dg(x)=Df(x)$ when $x=a_n$ and $x=b_n$. With $Df(a_n) \leq 0$, $Df(b_n)>0$, $a_{n+1}=u_{n+1}$ and $b_{n+1}=b_n$ if $Df(u_n) \leq 0$, while $a_{n+1}=a_n$, $b_{n+1}=u_{n+1}$ if $Df(u_n) > 0$. Termination occurs when $b_{n+1}-a_{n+1} \leq 3*tol(u_{n+1})$, where $tol(x)$ is a tolerance function prescribed by the user.

The user should be aware of the fact that the choice of *tolx* may highly affect the behavior of the algorithm, although convergence to a point for which the given function is minimal on the interval is assured. The asymptotic behavior will usually be fine as long as the numerical function is strictly δ-unimodal on the given interval (see [Bre73]) and the tolerance function satisfies $tol(x) \geq \delta$, for all x in the given interval. Let the value of *dfx* at the begin and end point of the initial interval be denoted by *dfa* and *dfb*, respectively, then, finding a global minimum is only guaranteed if the function is convex and *dfa* ≤ 0 and *dfb* ≥ 0. If these conditions are not satisfied then a local minimum or a minimum at one of the end points might be found.

Function Parameters:

float mininder (*x,y,fx,dfx,tolx*)

mininder: delivers the calculated minimum value of the function, defined by *fx*, on the interval with endpoints *a* and *b*;

x: float *;
- entry: one of the end points of the interval on which the function has to be minimized;
- exit: the calculated approximation of the position of the minimum;

y: float *;
- entry: the other end point of the interval on which the function has to be minimized;
- exit: a value such that $|x-y| \leq 3*tol(x)$;

fx: float (*fx)(x);
- entry: the function is given by the actual parameter *fx* which depends on *x*;

dfx: float (*dfx)(x);
- entry: the derivative of the function is given by the actual parameter *dfx* which

depends on *x*; *fx* and *dfx* are evaluated successively for a certain value of *x*;

tolx: float (**tolx*)(*x*);

entry: defines the tolerance function which may depend on *x*; a suitable tolerance function is: $|x|*re+ae$, where *re* is the relative precision desired and *ae* is an absolute precision which should not be chosen equal to zero.

```
#include <math.h>

float mininder(float *x, float *y, float (*fx)(float),
               float (*dfx)(float), float (*tolx)(float))
{
    int sgn;
    float a,b,c,fa,fb,fu,dfa,dfb,dfu,e,d,tol,ba,z,p,q,s;

    if (*x <= *y) {
        a = *x;
        fa=(*fx)(*x);
        dfa=(*dfx)(*x);
        b = *x = *y;
        fb=(*fx)(*x);
        dfb=(*dfx)(*x);
    } else {
        b = *x;
        fb=(*fx)(*x);
        dfb=(*dfx)(*x);
        a = *x = *y;
        fa=(*fx)(*x);
        dfa=(*dfx)(*x);
    }
    c=(3.0-sqrt(5.0))/2.0;
    d=b-a;
    e=d*2.0;
    z=e*2.0;
    while (1) {
        ba=b-a;
        tol=(*tolx)(*x);
        if (ba < tol*3.0) break;
        if (fabs(dfa) <= fabs(dfb)) {
            *x=a;
            sgn=1;
        } else {
            *x=b;
            sgn = -1;
        }
        if (dfa <= 0.0 && dfb >= 0.0) {
            z=(fa-fb)*3.0/ba+dfa+dfb;
            s=sqrt(z*z-dfa*dfb);
            p = (sgn == 1) ? dfa-s-z : dfb+s-z;
            p *= ba;
            q=dfb-dfa+s*2.0;
            z=e;
            e=d;
            d = (fabs(p) <= fabs(q)*tol) ? tol*sgn : -p/q;
        } else
            d=ba;
        if (fabs(d) >= fabs(z*0.5) || fabs(d) > ba*0.5) {
            e=ba;
            d=c*ba*sgn;
        }
        *x += d;
        fu=(*fx)(*x);
        dfu=(*dfx)(*x);
        if (dfu >= 0.0 || (fu >= fa && dfa <= 0.0)) {
            b = *x;
            fb=fu;
            dfb=dfu;
        } else {
            a = *x;
            fa=fu;
            dfa=dfu;
        }
```

```
    }
    if (fa <= fb) {
        *x=a;
        *y=b;
        return fa;
    } else {
        *x=b;
        *y=a;
        return fb;
    }
}
```

5.2.4 More variables - Auxiliary procedures

A. linemin

Determines the real value α_{min} (>0) of α for which $f(\alpha) = F(x_0+\alpha d)$ attains a minimum, where $F(x)$ is a real valued function of $x \epsilon R^n$; $x_0 \epsilon R^n$ is fixed, and $d \epsilon R^n$ is a fixed direction. It is assumed that the partial derivatives $g_i(x) = \partial F(x)/\partial x_i$ $(i=1,...,n)$ may easily be computed; we then have $f'(\alpha) = \partial f(\alpha)/\partial\alpha = d^T g(x_0+\alpha d)$.

The method employed utilises an iterative process, based upon cubic interpolation (see [Dav59]), for locating a minimum of the real-valued function f. The process involves an initial approximation α_0, three real sequences u_k, v_k and y_k, a fixed real parameter $\mu \epsilon (0,½)$, and relative and absolute tolerances, ε_r and ε_a respectively, used to terminate the process. Initially $y_0=u_0=0$, $v_0=\alpha_0$. At step k:

(a) if $f'(\alpha_k) \geq 0$, compute $y = v_k - (v_k-u_k)(f'(v_k)+w-z)/(f'(v_k)-f'(u_k)+2w)$ where $z = 3(f(u_k)-f(v_k))/(v_k-u_k) + f'(u_k) + f'(v_k)$ and $w = (z^2 - f'(u_k)f'(v_k))^{1/2}$; set $\varepsilon_k = \|x_0 + y_k d\| \varepsilon_v + \varepsilon_a$; y_{k+1} is determined as follows: (i) if $(y-u_k) < \varepsilon_k$ then $y_{k+1} = u_k+\varepsilon_k$, else (ii) if $(v_k-y)<\varepsilon_k$ then $y_{k+1}=v_k-\varepsilon_k$, else (iii) $y_{k+1}=y$; u_{k+1} and v_{k+1} are determined as follows: (iv) if $f'(y_{k+1}) \geq 0$ then $v_{k+1}=y_{k+1}$ and $u_{k+1}=u_k$; else (v) $v_{k+1}=v_k$ and $u_{k+1}=y_{k+1}$.

(b) if $f'(v_k) < 0$ then (i) if $(f(v_k)-f(0))/v_k f'(0) > \mu$ (now $0 < v_k < \varepsilon_{min}$) set $u_{k+1}=v_k$ and $y_{k+1}=v_{k+1}=2v_k$; else (ii) (now $u_k < \varepsilon_{min} < v_k$) set $y_{k+1} = ½(u_k+v_k)$ and if $f'(v_{k+1}) \geq 0$ or $(f(y_{k+1})-f(0))/y_{k+1}f'(0) > \mu$ then $v_{k+1}=y_{k+1}$ and $u_{k+1}=u_k$, else $v_{k+1}=v_k$ and $u_{k+1}=y_{k+1}$.

The above process is terminated if either

I. $½|v_k - u_k| < \|x_0 + y_{k+1}d\| \varepsilon_r + \varepsilon_a$,

II. $\mu \leq (f(y_k)-f(0))/y_k f'(0) \leq 1 - \mu$, or

III. the maximal permitted number of simultaneous evaluations of $f(\varepsilon)$ and $f'(\varepsilon)$ is exceeded.

In cases I and II, y_{k+1} is an acceptable approximation to ε_{min}.

The direction vector d and the initial values, x_0, $f(0)=F(x_0)$, and $f'(0)=d^T g(0)$ must be supplied by the user. The user must also prescribe the value of the variable *strongsearch*: if this is nonzero then the stopping criteria I and III alone are used; if it is zero then all criteria are used. The stopping criterion used when the value of *strongsearch* is zero is described in [Fle63] and [GolP67]. A detailed description of this procedure is given in [Bus72b].

Function Parameters:

void linemin (*n,x,d,nd,alfa,g,funct,f0,f1,df0,df1,evlmax,strongsearch,in*)

n: int;
entry: the number of variables of the given function f;

x: float *x[1:n]*;

entry: a vector x_0, such that *f* is decreasing in x_0, in the direction given by *d*;
exit: the calculated approximation of the vector for which *f* is minimal on the line defined by: x_0 + *alfa*d*, (*alfa*>0);

d: float *d[1:n]*;
entry: the direction of the line on which *f* has to be minimized;

nd: float;
entry: the Euclidean norm of the vector given in *d[1:n]*;

alfa: float *;
the independent variable that defines the position on the line on which *f* has to be minimized; this line is defined by x_0 + *alfa*d*, (*alfa*>0);
entry: an estimate *alfa0* of the value for which *h(alfa)=F(*x_0*+alfa*d)* is minimal;
exit: the calculated approximation *alfam* of the value for which *h(alfa)* is minimal;

g: float *g[1:n]*;
exit: the gradient of *f* at the calculated approximation of the minimum;

funct: float (**funct*)(*n,x,g*);
entry: a call of *funct* should effectuate:
1. *funct=f(x)*;
2. the value of *g[i]*, (*i=1,...,n*), becomes the value of the *i*-th component of the gradient of *f* at *x*;

f0: float;
entry: the value of *h(0)*, (see *alfa*);

f1: float *;
entry: the value of *h(alfa0)*, (see *alfa*);
exit: the value of *h(alfam)*, (see *alfa*);

df0: float;
entry: the value of the derivative of *h* at *alfa*=0;

df1: float *;
entry: the value of the derivative of *h* at *alfa*=*alfa0*;
exit: the value of the derivative of *h* at *alfa*=*alfam*;

evlmax: int *;
entry: the maximum allowed number of calls of *funct*;
exit: the number of times *funct* has been called;

strongsearch: int:
entry: if the value of *strongsearch* is nonzero then the process makes use of two stopping criteria:
A: the number of times *funct* has been called exceeds the given value of *evlmax*;
B: an interval is found with length less than two times the prescribed precision, on which a minimum is expected;
if the value of *strongsearch* is zero then the process makes also use of a third stopping criterion:
C: $\mu \le$ *(h(alfak)-h(alfa0))/(alfak*df0)* $\le 1 - \mu$, whereby *alfak* is the current iterate and μ a prescribed constant;

in: float *in[1:3]*;
entry:
in[1]: relative precision, ε_r, necessary for the stopping criterion B (see *strongsearch*);
in[2]: absolute precision, ε_a, necessary for the stopping criterion B (see *strongsearch*);
the prescribed precision, ε, at *alfa=alfak* is given by:

$\varepsilon = \|x_0+alfa*d\| \varepsilon_r+\varepsilon_a$, where $\|.\|$ denotes the Euclidean norm;

in[3]: the parameter μ necessary for stopping criterion C; this parameter must satisfy: 0<μ<0.5; in practice, a choice of μ=0.0001 is advised.

Functions used: vecvec, elmvec, dupvec.

```
#include <math.h>

void linemin(int n, float x[], float d[], float nd, float *alfa,
             float g[], float (*funct)(int, float[], float[]),
             float f0, float *f1, float df0, float *df1, int *evlmax,
             int strongsearch, float in[])
{
   float *allocate_real_vector(int, int);
   void free_real_vector(float *, int);
   float vecvec(int, int, int, float [], float []);
   void elmvec(int, int, int, float [], float [], float);
   void dupvec(int, int, int, float [], float []);
   int evl,notinint;
   float f,oldf,df,olddf,mu,alfa0,q,w,y,z,reltol,abstol,eps,aid,*x0;

   x0=allocate_real_vector(1,n);
   reltol=in[1];
   abstol=in[2];
   mu=in[3];
   evl=0;
   alfa0=0.0;
   oldf=f0;
   olddf=df0;
   y = *alfa;
   notinint=1;
   dupvec(1,n,0,x0,x);
   eps=(sqrt(vecvec(1,n,0,x,x))*reltol+abstol)/nd;
   q=((*f1)-f0)/((*alfa)*df0);
   while (1) {
      if (notinint) notinint = ((*df1) < 0.0 && q > mu);
      aid = *alfa;
      if (*df1 >= 0.0) {
         /* cubic interpolation */
         z=3.0*(oldf-(*f1))/(*alfa)+olddf+(*df1);
         w=sqrt(z*z-olddf*(*df1));
         *alfa = (*alfa)*(1.0-((*df1)+w-z)/((*df1)-olddf+w*2.0));
         if (*alfa < eps)
            *alfa=eps;
         else
            if (aid-(*alfa) < eps) *alfa=aid-eps;
      } else
         if (notinint) {
            alfa0 = *alfa = y;
            olddf = *df1;
            oldf = *f1;
         } else
            *alfa *= 0.5;
      y = (*alfa)+alfa0;
      dupvec(1,n,0,x,x0);
      elmvec(1,n,0,x,d,y);
      eps=(sqrt(vecvec(1,n,0,x,x))*reltol+abstol)/nd;
      f=(*funct)(n,x,g);
      evl++;
      df=vecvec(1,n,0,d,g);
      q=(f-f0)/(y*df0);
      if (!(((notinint || strongsearch) ? 1 : (q < mu || q > 1.0-mu))
            && (evl < *evlmax))) break;
      if (notinint || df > 0.0 || q < mu) {
         *df1=df;
         *f1=f;
      } else {
         alfa0=y;
         *alfa=aid-(*alfa);
         olddf=df;
         oldf=f;
```

```
        }
        if (*alfa <= eps*2.0) break;
    }
    *alfa=y;
    *evlmax=evl;
    *df1=df;
    *f1=f;
    free_real_vector(x0,1);
}
```

B. rnk1upd

Determines the upper triangular part of the symmetric $n \times n$ matrix g by use of the formula $g = h + cvv^T$. h being a given symmetric matrix whose upper triangular part is available, c being a real constant, and $v \epsilon R^n$ a vector.

Function Parameters:

void rnk1upd (h,n,v,c)

h: float $h[1:n*(n+1)/2]$;
- entry: the upper triangle (stored columnwise, i.e. $a_{ij}=h[(j-1)*j/2+i]$, $1 \leq i \leq j \leq n$) of the symmetric matrix that has to be updated;
- exit: the upper triangle (stored columnwise) of the updated matrix;

n: int;
- entry: the order of the symmetric matrix whose upper triangle is stored columnwise in the one-dimensional array h;

v: float $v[1:n]$;
- entry: the given matrix is updated (another matrix is added to it) with a symmetric matrix u, of rank one, defined by: $u_{i,j}=c*v[i]*v[j]$;

c: float; see v.

Function used: elmvec.

```
void rnk1upd(float h[], int n, float v[], float c)
{
    void elmvec(int, int, int, float [], float [], float);
    int j,k;

    k=0;
    j=1;
    do {
        k++;
        elmvec(j,j+k-1,1-j,h,v,v[k]*c);
        j += k;
    } while (k < n);
}
```

C. davupd

Determines the upper triangular part of the symmetric $n \times n$ matrix g by use of the formula

$$g = h + c_1 vv^T - c_2 ww^T.$$

h being a given symmetric matrix whose upper triangular part is available, c_1 and c_2 being real constants, and $v,w \epsilon R^n$ vectors.

Function Parameters:

void davupd $(h,n,v,w,c1,c2)$

h: float *h[1:n*(n+1)/2]*;
 entry: the upper triangle (stored columnwise, i.e. a_{ij}=*h[(j-1)*j/2+i]*, $1 \le i \le j \le n$) of the symmetric matrix that has to be updated;
 exit: the upper triangle (stored columnwise) of the updated matrix;

n: int;
 entry: the order of the symmetric matrix whose upper triangle is stored columnwise in the one-dimensional array *h*;

v,w: float *v[1:n], w[1:n]*;
 entry: the given matrix is updated with a symmetric matrix *u*, of rank two, defined by: $u_{i,j}$=*c1*v[i]*v[j]-c2*w[i]*w[j]*;

c1,c2: float; see *v* and *w* above.

```
void davupd(float h[], int n, float v[], float w[], float c1, float c2)
{
    int i,j,k;
    float vk,wk;

    k=0;
    j=1;
    do {
        k++;
        vk=v[k]*c1;
        wk=w[k]*c2;
        for (i=0; i<=k-1; i++) h[i+j] += v[i+1]*vk-w[i+1]*wk;
        j += k;
    } while (k < n);
}
```

D. fleupd

Determines the upper triangular part of the symmetric $n \times n$ matrix g by use of the formula

$$g = h - c_1 vw^T - c_2 wv^T + (1 + (c_1/c_2)c_1 vv^T).$$

h being a given symmetric matrix whose upper triangular part is available, c_1 and $c_2(\neq 0)$ being real constants, and $v,w \in R^n$ vectors.

Function Parameters:

void fleupd $(h,n,v,w,c1,c2)$

h: float *h[1:n(n+1)/2]*;
 entry: the upper triangle (stored columnwise, i.e. a_{ij}=*h[(j-1)j/2+i]*, $1 \le i \le j \le n$) of the symmetric matrix that has to be updated;
 exit: the upper triangle (stored columnwise) of the updated matrix;

n: int;
 entry: the order of the symmetric matrix whose upper triangle is stored columnwise in the one-dimensional array *h*;

v,w: float *v[1:n], w[1:n]*;
 entry: the given matrix is updated with a symmetric matrix *u*, of rank two, defined by: $u_{i,j}$=*c2*v[i]*v[j]-c1*(v[i]*w[j]+w[i]*v[j])*;

c1,c2: float; see *v* and *w* above.

```
void fleupd(float h[], int n, float v[], float w[], float c1, float c2)
{
    int i,j,k;
```

```
    float vk,wk;

    k=0;
    j=1;
    do {
        k++;
        vk = -w[k]*c1+v[k]*c2;
        wk=v[k]*c1;
        for (i=0; i<=k-1; i++) h[i+j] += v[i+1]*vk-w[i+1]*wk;
        j += k;
    } while (k < n);
}
```

5.2.5 More variables - No derivatives

praxis

Determines a point x at which the function $f(x)$ is a minimum, where $x=(x_1,...,x_n)$, by use of an iterative algorithm due to Brent [Bre73]. An initial approximation to x is to be provided.

Function Parameters:

void praxis (*n,x,funct,in,out*)

n: int;
entry: the number of variables of the function to be minimized;

x: float *x[1:n]*;
entry: an approximation of the position of the minimum;
exit: the calculated position of the minimum;

funct: float *(*funct)(n,x)*;
funct should deliver the value of the function to be minimized, at the point given by *x[1:n]*; the meaning of the parameters of the function *funct* is as follows:
n: the number of variables;
x: the values of the variables for which the function has to be evaluated;

in: float *in[0:9]*;
entry:
in[0]: the machine precision;
in[1]: the relative tolerance for the stepvector (relative to the current estimates of the variables), see *in[2]*;
in[2]: the absolute tolerance for the stepvector (relative to the current estimates of the variables); the process is terminated when in *in[8]*+1 successive iteration steps the Euclidean norm of the step vector is less than $(in[1]*\|x\|+in[2])*0.5$; *in[1]* should be chosen in agreement with the precision in which the function is calculated; usually *in[1]* should be chosen such that $in[1] \geq \sqrt{(in[0])}$; *in[0]* should be chosen different from zero;
in[3],in[4]: *in[3]* and *in[4]* are neither used nor changed;
in[5]: the maximum number of function evaluations allowed (i.e. calls of *funct*);
in[6]: the maximum step size; *in[6]* should be equal to the maximum expected distance between the guess and the minimum; if *in[6]* is too small or too large then the initial rate of convergence will be slow;
in[7]: the maximum scaling factor; the value of *in[7]* may be used to obtain automatic scaling of the variables; however, this scaling is worthwhile but may be unreliable; therefore, the user should try to scale his problem as well as

possible and set *in[7]*=1; in either case, *in[7]* should not be chosen greater than 10;

in[8]: the process terminates if no substantial improvement of the values of the variables is obtained in *in[8]+1* successive iteration steps (see *in[1], in[2]*); *in[8]*=4 is very cautious; usually, *in[8]*=1 is satisfactory;

in[9]: if the problem is known to be ill-conditioned (see [Bre73]) then the value of *in[9]* should be negative, otherwise *in[9]*≥0;

out: float *out[1:6]*;

exit:

out[1]: this value gives information about the termination of the process;

out[1]=0: normal termination;

out[1]=1: the process is broken off because at the end of an iteration step the number of calls of *funct* exceeded the value given in *in[5]*;

out[1]=2: the process is broken off because the condition of the problem is too bad;

out[2]: the calculated minimum of the function;

out[3]: the value of the function at the initial guess;

out[4]: the number of function evaluations needed to obtain this result;

out[5]: the number of line searches (see [Bre73]);

out[6]: the step size in the last iteration step.

Functions used: inivec, inimat, dupvec, dupmat, dupcolvec, mulrow, mulcol, vecvec, tammat, mattam, ichrowcol, elmveccol, qrisngvaldec.

```
#include <stdlib.h>
#include <math.h>
#include <float.h>

void praxis(int n, float x[], float (*funct)(int, float[]),
            float in[], float out[])
{
    float *allocate_real_vector(int, int);
    float **allocate_real_matrix(int, int, int, int);
    void free_real_vector(float *, int);
    void free_real_matrix(float **, int, int, int);
    void inivec(int, int, float [], float);
    void inimat(int, int, int, int, float **, float);
    void dupvec(int, int, int, float [], float []);
    void dupmat(int, int, int, int, float **, float **);
    void dupcolvec(int, int, int, float **, float []);
    void mulrow(int, int, int, int, float **, float **, float);
    void mulcol(int, int, int, int, float **, float **, float);
    float vecvec(int, int, int, float [], float []);
    float tammat(int, int, int, int, float **, float **);
    float mattam(int, int, int, int, float **, float **);
    void ichrowcol(int, int, int, int, float **);
    void elmveccol(int, int, int, float [], float **, float);
    int qrisngvaldec(float **, int, int, float [], float **, float []);
    void praxismin(int, int, float *, float *, float *, int, int,
            float [], float **, float *, float *, float *, float,   float,
            float [], float [], int *, int *, float *, float, float,
            float, float, float, float, float, float,
            float (*)(int, float[]));
    int illc,i,j,k,k2,nl,maxf,nf,kl,kt,ktm,emergency;
    float s,sl,dn,dmin,fx,f1,lds,ldt,sf,df,qf1,qd0,qd1,qa,qb,qc,m2,m4,
            small,vsmall,large,vlarge,scbd,ldfac,t2,macheps,reltol,
            abstol,h,**v,*d,*y,*z,*q0,*q1,**a,em[8],l;

    d=allocate_real_vector(1,n);
    y=allocate_real_vector(1,n);
    z=allocate_real_vector(1,n);
```

```
q0=allocate_real_vector(1,n);
q1=allocate_real_vector(1,n);
v=allocate_real_matrix(1,n,1,n);
a=allocate_real_matrix(1,n,1,n);

macheps=in[0];
reltol=in[1];
abstol=in[2];
maxf=in[5];
h=in[6];
scbd=in[7];
ktm=in[8];
illc = in[9] < 0.0;
small=macheps*macheps;
vsmall=small*small;
large=1.0/small;
vlarge=1.0/vsmall;
m2=reltol;
m4=sqrt(m2);
srand(1);
ldfac = (illc ? 0.1 : 0.01);
kt=nl=0;
nf=1;
out[3]=qf1=fx=(*funct)(n,x);
abstol=t2=small+fabs(abstol);
dmin=small;
if (h < abstol*100.0) h=abstol*100;
ldt=h;
inimat(1,n,1,n,v,0.0);
for (i=1; i<=n; i++) v[i][i]=1.0;
d[1]=qd0=qd1=0.0;
dupvec(1,n,0,q1,x);
inivec(1,n,q0,0.0);
emergency=0;

while (1) {
    sf=d[1];
    d[1]=s=0.0;
    praxismin(1,2,&(d[1]),&s,&fx,0,
              n,x,v,&qa,&qb,&qc,qd0,qd1,q0,q1,&nf,
              &nl,&fx,m2,m4,dmin,ldt,reltol,abstol,small,h,funct);
    if (s <= 0.0) mulcol(1,n,1,1,v,v,-1.0);
    if (sf <= 0.9*d[1] || 0.9*sf >= d[1]) inivec(2,n,d,0.0);
    for (k=2; k<=n; k++) {
        dupvec(1,n,0,y,x);
        sf=fx;
        illc = (illc || kt > 0);
        while (1) {
            kl=k;
            df=0.0;
            if (illc) {
                /* random stop to get off resulting valley */
                for (i=1; i<=n; i++) {
                    s=z[i]=(0.1*ldt+t2*pow(10.0,kt))*
                              (rand()/(float)RAND_MAX-0.5);
                    elmveccol(1,n,i,x,v,s);
                }
                fx=(*funct)(n,x);
                nf++;
            }
            for (k2=k; k2<=n; k2++) {
                sl=fx;
                s=0.0;
                praxismin(k2,2,&(d[k2]),&s,&fx,0,
                    n,x,v,&qa,&qb,&qc,qd0,qd1,q0,q1,&nf,
                    &nl,&fx,m2,m4,dmin,ldt,reltol,abstol,small,h,funct);
                s = illc ? d[k2]*(s+z[k2])*(s+z[k2]) : sl-fx;
                if (df < s) {
                    df=s;
                    kl=k2;
                }
            }
            if (!illc && df < fabs(100.0*macheps*fx))
                illc=1;
```

```
            else
                break;
        }
        for (k2=1; k2<=k-1; k2++) {
            s=0.0;
            praxismin(k2,2,&(d[k2]),&s,&fx,0,
                n,x,v,&qa,&qb,&qc,qd0,qd1,q0,q1,&nf,
                &nl,&fx,m2,m4,dmin,ldt,reltol,abstol,small,h,funct);
        }
        f1=fx;
        fx=sf;
        lds=0.0;
        for (i=1; i<=n; i++) {
            sl=x[i];
            x[i]=y[i];
            y[i] = sl -= y[i];
            lds += sl*sl;
        }
        lds=sqrt(lds);
        if (lds > small) {
            for (i=kl-1; i>=k; i--) {
                for (j=1; j<=n; j++) v[j][i+1]=v[j][i];
                d[i+1]=d[i];
            }
            d[k]=0.0;
            dupcolvec(1,n,k,v,y);
            mulcol(1,n,k,k,v,v,1.0/lds);
            praxismin(k,4,&(d[k]),&lds,&f1,1,
                n,x,v,&qa,&qb,&qc,qd0,qd1,q0,q1,&nf,
                &nl,&fx,m2,m4,dmin,ldt,reltol,abstol,small,h,funct);
            if (lds <= 0.0) {
                lds = -lds;
                mulcol(1,n,k,k,v,v,-1.0);
            }
        }
        ldt *= ldfac;
        if (ldt < lds) ldt=lds;
        t2=m2*sqrt(vecvec(1,n,0,x,x))+abstol;
        kt = (ldt > 0.5*t2) ? 0 : kt+1;
        if (kt > ktm) {
            out[1]=0.0;
            emergency=1;
        }
    }
    if (emergency) break;
    /* quad */
    s=fx;
    fx=qf1;
    qf1=s;
    qd1=0.0;
    for (i=1; i<=n; i++) {
        s=x[i];
        x[i]=l=q1[i];
        q1[i]=s;
        qd1 += (s-l)*(s-l);
    }
    l=qd1=sqrt(qd1);
    s=0.0;
    if ((qd0*qd1 > FLT_MIN) && (nl >=3*n*n)) {
        praxismin(0,2,&s,&l,&qf1,1,
                n,x,v,&qa,&qb,&qc,qd0,qd1,q0,q1,&nf,
                &nl,&fx,m2,m4,dmin,ldt,reltol,abstol,small,h,funct);
        qa=l*(l-qd1)/(qd0*(qd0+qd1));
        qb=(l+qd0)*(qd1-l)/(qd0*qd1);
        qc=l*(l+qd0)/(qd1*(qd0+qd1));
    } else {
        fx=qf1;
        qa=qb=0.0;
        qc=1.0;
    }
    qd0=qd1;
    for (i=1; i<=n; i++) {
        s=q0[i];
        q0[i]=x[i];
```

```
        x[i]=qa*s+qb*x[i]+qc*q1[i];
    }
    /* end of quad */
    dn=0.0;
    for (i=1; i<=n; i++) {
        d[i]=1.0/sqrt(d[i]);
        if (dn < d[i]) dn=d[i];
    }
    for (j=1; j<=n; j++) {
        s=d[j]/dn;
        mulcol(1,n,j,j,v,v,s);
    }
    if (scbd > 1.0) {
        s=vlarge;
        for (i=1; i<=n; i++) {
            sl=z[i]=sqrt(mattam(1,n,i,i,v,v));
            if (sl < m4) z[i]=m4;
            if (s > sl) s=sl;
        }
        for (i=1; i<=n; i++) {
            sl=s/z[i];
            z[i]=1.0/sl;
            if (z[i] > scbd) {
                sl=1.0/scbd;
                z[i]=scbd;
            }
            mulrow(1,n,i,i,v,v,sl);
        }
    }
    for (i=1; i<=n; i++) ichrowcol(i+1,n,i,i,v);
    em[0]=em[2]=macheps;
    em[4]=10*n;
    em[6]=vsmall;
    dupmat(1,n,1,n,a,v);
    if (qrisngvaldec(a,n,n,d,v,em) != 0) {
        out[1]=2.0;
        emergency=1;
    }
    if (emergency) break;
    if (scbd > 1.0) {
        for (i=1; i<=n; i++) mulrow(1,n,i,i,v,v,z[i]);
        for (i=1; i<=n; i++) {
            s=sqrt(tammat(1,n,i,i,v,v));
            d[i] *= s;
            s=1.0/s;
            mulcol(1,n,i,i,v,v,s);
        }
    }
    for (i=1; i<=n; i++) {
        s=dn*d[i];
        d[i] = (s > large) ? vsmall :
                    ((s < small) ? vlarge : 1.0/(s*s));
    }
    /* sort */
    for (i=1; i<=n-1; i++) {
        k=i;
        s=d[i];
        for (j=i+1; j<=n; j++)
            if (d[j] > s) {
                k=j;
                s=d[j];
            }
        if (k > i) {
            d[k]=d[i];
            d[i]=s;
            for (j=1; j<=n; j++) {
                s=v[j][i];
                v[j][i]=v[j][k];
                v[j][k]=s;
            }
        }
    }
    /* end of sort */
    dmin=d[n];
```

```
        if (dmin < small) dmin=small;
        illc = (m2*d[1]) > dmin;
        if (nf >= maxf) {
            out[1]=1.0;
            break;
        }
    }
    out[2]=fx;
    out[4]=nf;
    out[5]=nl;
    out[6]=ldt;
    free_real_vector(d,1);
    free_real_vector(y,1);
    free_real_vector(z,1);
    free_real_vector(q0,1);
    free_real_vector(q1,1);
    free_real_matrix(v,1,n,1);
    free_real_matrix(a,1,n,1);
}

void praxismin(int j, int nits, float *d2, float *x1, float *f1,
            int fk, int n, float x[], float **v, float *qa, float *qb,
            float *qc, float qd0, float qd1, float q0[], float q1[],
            int *nf, int *nl, float *fx, float m2, float m4,
            float dmin, float ldt, float reltol, float abstol,
            float small, float h, float (*funct)(int, float[]))
{
    /* this function is internally used by PRAXIS */

    float praxisflin(float, int, int, float [], float **, float *,
                    float *, float *, float, float, float [],
                    float [], int *,  float (*)(int, float[]));
    int k,dz,loop;
    float x2,xm,f0,f2,fm,d1,t2,s,sf1,sx1;

    sf1 = *f1;
    sx1 = *x1;
    k=0;
    xm=0.0;
    f0 = fm = *fx;
    dz = *d2 < reltol;
    s=sqrt(vecvec(1,n,0,x,x));
    t2=m4*sqrt(fabs(*fx)/(dz ? dmin : *d2)+s*ldt)+m2*ldt;
    s=s*m4+abstol;
    if (dz && (t2 > s)) t2=s;
    if (t2 < small) t2=small;
    if (t2 > 0.01*h) t2=0.01*h;
    if (fk && (*f1 <= fm)) {
        xm = *x1;
        fm = *f1;
    }
    if (!fk || (fabs(*x1) < t2)) {
        *x1 = (*x1 > 0.0) ? t2 : -t2;
        *f1=praxisflin(*x1,j,n,x,v,qa,qb,qc,qd0,qd1,q0,q1,nf,funct);
    }
    if (*f1 <= fm) {
        xm = *x1;
        fm = *f1;
    }
    loop=1;
    while (loop) {
        if (dz) {
            /* evaluate praxisflin at another point and
               estimate the second derivative */
            x2 = (f0 < *f1) ? -(*x1) : (*x1)*2.0;
            f2=praxisflin(x2,j,n,x,v,qa,qb,qc,qd0,qd1,q0,q1,nf,funct);
            if (f2 <= fm) {
                xm=x2;
                fm=f2;
            }
            *d2=(x2*((*f1)-f0)-(*x1)*(f2-f0))/((*x1)*x2*((*x1)-x2));
        }
        /* estimate first derivative at 0 */
        d1=((*f1)-f0)/(*x1)-(*x1)*(*d2);
```

```
        dz=1;
        x2 = (*d2 <= small) ? ((d1 < 0.0) ? h : -h) : -0.5*d1/(*d2);
        if (fabs(x2) > h) x2 = (x2 > 0.0) ? h : -h;
        while (1) {
            f2=praxisflin(x2,j,n,x,v,qa,qb,qc,qd0,qd1,q0,q1,nf,funct);
            if (k < nits && f2 > f0) {
                k++;
                if (f0 < *f1 && (*x1)*x2 > 0.0) break;
                x2=0.5*x2;
            } else {
                loop=0;
                break;
            }
        }
    }
    (*nl)++;
    if (f2 > fm)
        x2=xm;
    else
        fm=f2;
    *d2 = (fabs(x2*(x2-(*x1))) > small) ?
              ((x2*((*f1)-f0)-(*x1)*(fm-f0))/((*x1)*x2*((*x1)-x2))) :
              ((k > 0) ? 0.0 : *d2);
    if (*d2 <= small) *d2=small;
    *x1=x2;
    *fx=fm;
    if (sf1 < *fx) {
        *fx=sf1;
        *x1=sx1;
    }
    if (j > 0) elmveccol(1,n,j,x,v,*x1);
}

float praxisflin(float l, int j, int n, float x[], float **v,
            float *qa, float *qb, float *qc, float qd0, float qd1,
            float q0[], float q1[], int *nf,
            float (*funct)(int, float[]))
{
    /* this function is internally used by PRAXISMIN */

    int i;
    float *t,result;

    t=allocate_real_vector(1,n);
    if (j > 0)
        for (i=1; i<=n; i++) t[i]=x[i]+l*v[i][j];
    else {
        /* search along parabolic space curve */
        *qa=l*(l-qd1)/(qd0*(qd0+qd1));
        *qb=(l+qd0)*(qd1-l)/(qd0*qd1);
        *qc=l*(l+qd0)/(qd1*(qd0+qd1));
        for (i=1; i<=n; i++) t[i]=(*qa)*q0[i]+(*qb)*x[i]+(*qc)*q1[i];
    }
    (*nf)++;
    result=(*funct)(n,t);
    free_real_vector(t,1);
    return result;
}
```

5.2.6 More variables - Gradient available

A. rnk1min

Determines a vector $x \in R^n$ for which the real valued function $F(x)$ attains a minimum. It is assumed that the partial derivatives $g_i(x)=\partial F(x)/\partial x_i$ $(i=1,...,n)$ may easily be computed. *rnk1min* is suitable, in particular, for use in connection with problems for which the $n \times n$ Hessian matrix $G(x)$, whose components are $G_{i,j}(x)=\partial^2 F(x)/\partial x_i \partial x_j$ $(i,j=1,...,n)$, is almost singular at the minimum.

With $H(x)=G(x)^{-1}$, and the initial vector $x^{(0)}$ prescribed, the sequence of vectors $x^{(k)}$ produced by use of the Newton scheme $x^{(k+1)} = x^{(k)} - H(x^{(k)})g(x^{(k)})^T$ (k=0,1,...) under certain conditions converges quadratically. Use of this scheme requires the evaluation and inversion of a Hessian matrix at each stage, and in order to avoid this *rnk1min* determines a sequence of vectors $x^{(k)}$ by use of a scheme of the form $x^{(k+1)} = x^{(k)} - \alpha^{(k)}H^{(k)}g(x^{(k)})$ in which the metric $H^{(k)}$ is an approximation to $H(x^{(k)})$, and is corrected by use of a simple updating formula of the form $H^{(k+1)} = H^{(k)} + C^{(k)}$ where $C^{(k)}$ is of rank one (or possibly rank two), and the $\alpha^{(k)}$ are suitably determined real numbers.

The user is asked to provide at outset the initial approximation vector $x^{(0)}$ and values of the machine precision ε, the required relative and absolute precisions, ε_r and ε_a respectively, in the determination of the minimum, a descent parameter μ in the range $0<\mu<1/2$ (often μ=0.0001 is suitable), an upper bound ε_g for $\|g(x)\|$ (here and in the sequel all norms are Euclidean) and a lower bound F_{min} for $F(x)$, both holding for all x in a domain containing the minimum and all $x^{(k)}$ produced, either a rough estimate $c>0$ of $\|G(x^{(0)})\|$ (1.0 is often suitable) or approximate values of the elements of the upper triangular part of the inverse Hessian matrix $H(x^{(0)})$, the maximum permitted number of simultaneous evaluations of $F(x)$ and $g(x)$ for the required values of x, and an orthogonality parameter ß for which $(\varepsilon/\varepsilon_r)^{1/2}/n \leq ß < 1.0$ (often 0.01 is suitable).

The algorithm used consists first of an initialization, and then in determinations of sequences of directions $d^{(k)}$, virtual steplengths $\alpha^{(k)}$, steps $\delta^{(k)}$, approximations $x^{(k)}$ to the minimum, increments in derivatives $\xi^{(k)}$, and metrics $H^{(k)}$. Initially, either $H^{(0)}=cI$ (I being the $n{\times}n$ unit matrix) or $H^{(0)}$ is the approximation to the inverse Hessian matrix, whose elements have been supplied by the user. Then, for k=0,1,... steps I to VI as follows are carried out.

I. If

$$g(x^{(k)})^T H^{(k)} g(x^{(k)}) > 0 \tag{1}$$

then set $d^{(k)} = -H^{(k)}g(x^{(k)})$. If condition (1) is violated, and the new direction $d^{(k)}$ were to be taken as just given then $(d^{(k)})^T g(x^{(k)}) \geq 0$, i.e. the derivative of $F(x)$ in the direction $d^{(k)}$ at the point $x^{(k)}$ would be decreasing. To avoid this, $H^{(k)}$ is decomposed in the form $H^{(k)} = U\Lambda U^T$, where U is a matrix of eigenvectors, and Λ an $n{\times}n$ diagonal matrix of eigenvalues λ_i ($i=1,...,n$); defining $|\Lambda|$ to be the diagonal matrix with diagonal elements $|\lambda_i|$ (i=1,...,n), and $K^{(k)}$ by $K^{(k)} = U|\Lambda|U^T$ (so that $K^{(k)}$ is positive definite) we take $d^{(k)} = -K^{(k)}g(x^{(k)})$ when condition (1) is violated.

II. If k=0, set $\theta^{(k)} = \min(1,\ 2(F_{min}-F(x^{(0)}))/(-d^{(0)})^T g(x^{(0)}))$, and determine the smallest integer $r \geq 0$ for which either

$$(d^{(0)})^T g(x^{(0)}+2^r\theta^{(0)}d^{(0)}) \geq 0$$

or

$$(F(x^{(0)}-2^r\theta^{(0)}d^{(0)})-F(x^{(0)}))/2^r\theta^{(0)}(d^{(0)})^T g(x^{(0)}) < \mu$$

and set $\theta^{(0)} = 2^r\theta^{(0)}$ for this r. $F(x)$ now decreases and then increases upon the line $x=x^{(0)}+\alpha d^{(0)}$ ($\alpha>0$). An approximation to the value of α specifying the point at which F is minimum upon this line is determined by cubic interpolation (see the theory of *linemin* and [Dav59]) or (if $(d^{(0)})^T g(x^{(0)}+\theta^{(0)}d^{(0)}) < 0$ by bisection. For the value of α so determined

$$\mu \leq (F(x^{(k)}+\alpha d^{(k)}) - F(x^{(k)}))/\alpha(d^{(k)})^T g(x^{(k)}) \leq 1-\mu. \tag{2}$$

If $k \neq 0$, set $\theta^{(k)} = \|\delta^{(k-1)}\|/\|d^{(k)}\|$ if $k<n$, and $\theta^{(k)}$=1 otherwise. If

$$(F(x^{(k)}+\theta^{(k)}d^{(k)}) - F(x^{(k)}))/\theta^{(k)}(d^{(k)})^T g(x^{(k)}) \geq \mu \tag{3}$$

set $\alpha^{(k)} = \theta^{(k)}$. If condition (3) is violated, the distance from $x=x^{(k)}$ to the minimum of $F(x)$ upon the line $x^{(k)}+\alpha d^{(k)}$ ($\alpha>0$) is overestimated by setting $\alpha=\theta^{(k)}$, and a value of α

in the range $0<\alpha<\theta^{(k)}$ is determined by cubic interpolation or bisection as above, such that condition (2) is satisfied for this α, for which we set $\alpha^{(k)}=\alpha$.

III. Set $\delta^{(k)} = \alpha^{(k)}d^{(k)}$ and $x^{(k+1)} = x^{(k)} + \delta^{(k)}$.

IV. If condition (1) is satisfied, set $\xi^{(k)} = -\alpha^{(k)}g(x^{(k)})$. Otherwise set $\xi^{(k)} = -\alpha^{(k)}U\Delta U^T d^{(k)}$ where, with U and Δ being determined as in stage I, Δ is the $n\times n$ diagonal matrix with diagonal elements sign (λ_i) $(i=1,...,n)$.

V. If, with $\gamma^{(k)} = g(x^{(k+1)}) - g(x^{(k)})$,
$$|(\gamma^{(k)}-\xi^{(k)})^T\delta^{(k)}| > \beta\|\gamma^{(k)}-\xi^{(k)}\|\,\|\delta^{(k)}\| \tag{4}$$
then $H^{(k+1)}$ is determined by use of the rank-one updating formula
$$H^{(k+1)} = H^{(k)} + c^{(k)}u^{(k)}(u^{(k)})^T \tag{5}$$
where $u^{(k)} = \delta^{(k)} - H^{(k)}\gamma^{(k)}$, $c^{(k)} = 1/((\gamma^{(k)})^T u^{(k)})$.
If condition (4) is violated, but
$$c^{(k)}(\gamma^{(k)})^T\delta^{(k)} > 0 \tag{6}$$
then the rank-two updating formula
$$H^{(k+1)} = H^{(k)} - c_1^{(k)}v^{(k)}(w^{(k)})^T - c_2^{(k)}w^{(k)}(v^{(k)})^T + (1 + (c_1^{(k)}/c_2^{(k)})c_1^{(k)}v^{(k)}(v^{(k)})^T \tag{7}$$
with
$$c_1^{(k)} = 1/(\delta^{(k)})^T\gamma^{(k)},\quad c_2^{(k)} = -1/(\gamma^{(k)})^T H^{(k)}\gamma^{(k)},\quad v_{(k)}=\delta^{(k)},\quad w^{(k)}=H^{(k)}\gamma^{(k)} \tag{8}$$
is used. If both condition (4) and (6) are violated, the rank-two updating formula
$$H^{(k+1)} = H^{(k)} + c_1^{(k)}v^{(k)}(v^{(k)})^T + c_2^{(k)}w^{(k)}w^{(k)} \tag{9}$$
with $c_1^{(k)},...,w^{(k)}$ as defined by formulae (8) is used.

VI. If $\|H^{(k+1)}g(x^{(k+1)})\| \le \varepsilon_r\|x^{(k+1)}\| + \varepsilon_a$ and $\|g(x^{(k+1)})\| \le \varepsilon_g$ and $g(x^{(k+1)})^T H^{(k+1)}g(x^{(k+1)}) \ge 0$ and $k \ge n$ then $x^{(k+1)}$ is accepted as an approximation to the minimum. If k is equal to the number specified by the maximum number of function evaluations then the process is terminated (and $x^{(k+1)}$ is then not an acceptable approximation).

The eigen-value/vector determination possibly required at stage I (in most practical cases this requirement is of infrequent occurrence) is carried out by a call of *eigsym1*. The line minimization required at stage II is carried out by *linemin*. The updating formulae (5,7,9) possibly required at stage V are implemented by *rnk1upd*, *fleupd* and *davupd* respectively. A detailed description of the algorithm and some results about its convergence is given in [Bus72b].

Function Parameters:

float rnk1min (*n,x,g,h,funct,in,out*)

rnk1min: delivers the calculated least value of the given function;

n: int;
entry: the number of variables of the function to be minimized;

x: float *x[1:n]*;
entry: an approximation of a minimum of the function;
exit: the calculated minimum of the function;

g: float *g[1:n]*;
exit: the gradient of the function at the calculated minimum;

h: float *h[1:n(n+1)/2]*;
the upper triangle of an approximation of the inverse Hessian is stored columnwise in *h* (i.e. the (*i,j*)-th element = *h[(j-1)j/2+i]*, $1\le i\le j\le n$); if *in[6]*>0 initializing of *h* will be done automatically and the initial approximation of the inverse Hessian will equal the unit matrix multiplied with the value of *in[6]*; if *in[6]*<0 then no initializing of *h* will be done and the user should give in *h* an approximation of the inverse Hessian, at the starting point; the upper triangle of an approximation of the inverse Hessian at the

calculated minimum is delivered in *h*;

funct: float *(*funct)(n,x,g)*;

a call of *funct* must have the following results:

1. *funct* becomes the value of the function to be minimized at the point *x*;
2. the value of *g[i]*, (i=1,...,n), becomes the value of the i-th component of the gradient of the function at *x*;

in: float *in[0:9]*;

entry:

in[0]: the machine precision;

in[1]: the relative tolerance for the solution; this tolerance should not be chosen smaller than *in[0]*;

in[2]: the absolute tolerance for the solution;

in[3]: a parameter used for controlling line minimization (see [Fle63, GolP67]); usually a suitable value is 0.0001;

in[4]: the absolute tolerance for the Euclidean norm of the gradient at the solution;

in[5]: a lower bound for the function value;

in[6]: this parameter controls the initialization of the approximation of the inverse Hessian (metric), see *h*; usually the choice *in[6]*=1 will give good results;

in[7]: the maximum allowed number of calls of *funct*;

in[8]: a parameter used for controlling the updating of the metric; it is used to avoid unboundedness of the metric (see [Po70]); the value of *in[8]* should satisfy: $\sqrt{(in[0]/in[1])}/n < in[8] < 1$; usually a suitable value will be 0.01;

out: float *out[0:4]*;

exit:

out[0]: the Euclidean norm of the product of the metric and the gradient at the calculated minimum;

out[1]: the Euclidean norm of the gradient at the calculated minimum;

out[2]: the number of calls of *funct* necessary to attain this result;

out[3]: the number of iterations in which a line search was necessary;

out[4]: the number of iterations in which a direction had to be calculated with the method given in [Gr67]; in such an iteration a calculation of the eigenvalues and eigenvectors of the metric is necessary.

Functions used: vecvec, matvec, tamvec, elmvec, symmatvec, inivec, inisymd, mulvec, dupvec, eigsym1, linemin, rnk1upd, davupd, fleupd.

```
#include <math.h>

float rnk1min(int n, float x[], float g[], float h[],
              float (*funct)(int, float[], float[]),
              float in[], float out[])
{
   float *allocate_real_vector(int, int);
   float **allocate_real_matrix(int, int, int, int);
   void free_real_vector(float *, int);
   void free_real_matrix(float **, int, int, int);
   float vecvec(int, int, int, float [], float []);
   float matvec(int, int, int, float **, float []);
   float tamvec(int, int, int, float **, float []);
   void elmvec(int, int, int, float [], float [], float);
   float symmatvec(int, int, int, float [], float []);
   void inivec(int, int, float [], float);
   void inisymd(int, int, int, float [], float);
```

```
void mulvec(int, int, int, float [], float [], float);
void dupvec(int, int, int, float [], float []);
void eigsym1(float [], int, int, float [], float **, float []);
void linemin(int, float [], float [], float, float *, float [],
             float (*)(int, float[], float[]), float, float *,
             float, float *, int *, int, float []);
void rnk1upd(float [], int, float [], float);
void davupd(float [], int, float [], float [], float, float);
void fleupd(float [], int, float [], float [], float, float);
int i,it,n2,cntl,cnte,evl,evlmax,ok;
float f,f0,fmin,mu,dg,dg0,ghg,gs,nrmdelta,alfa,macheps,reltol,
      abstol,eps,tolg,orth,aid,*v,*delta,*gamma,*s,*p,**vec,*th,
      em[10],temp1,temp2;

v=allocate_real_vector(1,n);
delta=allocate_real_vector(1,n);
gamma=allocate_real_vector(1,n);
s=allocate_real_vector(1,n);
p=allocate_real_vector(1,n);
vec=allocate_real_matrix(1,n,1,n);

macheps=in[0];
reltol=in[1];
abstol=in[2];
mu=in[3];
tolg=in[4];
fmin=in[5];
it=0;
alfa=in[6];
evlmax=in[7];
orth=in[8];
n2=(n*(n+1))/2;
cntl=cnte=0;
if (alfa > 0.0) {
    inivec(1,n2,h,0.0);
    inisymd(1,n,0,h,alfa);
}
f=(*funct)(n,x,g);
evl=1;
dg=sqrt(vecvec(1,n,0,g,g));
for (i=1; i<=n; i++) delta[i] = -symmatvec(1,n,i,h,g);
nrmdelta=sqrt(vecvec(1,n,0,delta,delta));
dg0=vecvec(1,n,0,delta,g);
ok = dg0 < 0.0;
eps=sqrt(vecvec(1,n,0,x,x))*reltol+abstol;
it++;
while ((nrmdelta > eps || dg > tolg || !ok) && (evl < evlmax)) {
    if (!ok) {
        /* calculating greenstadts direction */
        th=allocate_real_vector(1,n2);
        em[0]=macheps;
        em[2]=aid=sqrt(macheps*reltol);
        em[4]=orth;
        em[6]=aid*n;
        em[8]=5.0;
        cnte++;
        dupvec(1,n2,0,th,h);
        eigsym1(th,n,n,v,vec,em);
        for (i=1; i<=n; i++) {
            aid = -tamvec(1,n,i,vec,g);
            s[i]=aid*fabs(v[i]);
            v[i]=((v[i] == 0.0) ? 0.0 : ((v[i] > 0.0) ? aid : -aid));
        }
        for (i=1; i<=n; i++) {
            delta[i]=matvec(1,n,i,vec,s);
            p[i]=matvec(1,n,i,vec,v);
        }
        dg0=vecvec(1,n,0,delta,g);
        nrmdelta=sqrt(vecvec(1,n,0,delta,delta));
        free_real_vector(th,1);
    }
    dupvec(1,n,0,s,x);
    dupvec(1,n,0,v,g);
    if (it > n)
```

```
            alfa=1.0;
        else {
            if (it != 1)
                alfa /= nrmdelta;
            else {
                alfa=2.0*(fmin-f)/dg0;
                if (alfa > 1.0) alfa=1.0;
            }
        }
        elmvec(1,n,0,x,delta,alfa);
        f0=f;
        f=(*funct)(n,x,g);
        evl++;
        dg=vecvec(1,n,0,delta,g);
        if (it == 1 || f0-f < -mu*dg0*alfa) {
            /* line minimization */
            i=evlmax-evl;
            cntl++;
            linemin(n,s,delta,nrmdelta,&alfa,g,funct,f0,&f,
                        dg0,&dg,&i,0,in);
            evl += i;
            dupvec(1,n,0,x,s);
        }
        dupvec(1,n,0,gamma,g);
        elmvec(1,n,0,gamma,v,-1.0);
        if (!ok) mulvec(1,n,0,v,p,-1.0);
        dg -= dg0;
        if (alfa != 1.0) {
            mulvec(1,n,0,delta,delta,alfa);
            mulvec(1,n,0,v,v,alfa);
            nrmdelta *= alfa;
            dg *= alfa;
        }
        dupvec(1,n,0,p,gamma);
        elmvec(1,n,0,p,v,1.0);
        for (i=1; i<=n; i++) v[i]=symmatvec(1,n,i,h,gamma);
        dupvec(1,n,0,s,delta);
        elmvec(1,n,0,s,v,-1.0);
        gs=vecvec(1,n,0,gamma,s);
        ghg=vecvec(1,n,0,v,gamma);
        aid=dg/gs;
        temp1=vecvec(1,n,0,delta,p);
        temp2=orth*nrmdelta;
        if (temp1*temp1 > vecvec(1,n,0,p,p)*temp2*temp2)
            rnk1upd(h,n,s,1.0/gs);
        else
            if (aid >= 0.0)
                fleupd(h,n,delta,v,1.0/dg,(1.0+ghg/dg)/dg);
            else
                davupd(h,n,delta,v,1.0/dg,1.0/ghg);
        for (i=1; i<=n; i++) delta[i] = -symmatvec(1,n,i,h,g);
        alfa=nrmdelta;
        nrmdelta=sqrt(vecvec(1,n,0,delta,delta));
        eps=sqrt(vecvec(1,n,0,x,x))*reltol+abstol;
        dg=sqrt(vecvec(1,n,0,g,g));
        dg0=vecvec(1,n,0,delta,g);
        ok = dg0 <= 0.0;
        it++;
    }
    out[0]=nrmdelta;
    out[1]=dg;
    out[2]=evl;
    out[3]=cntl;
    out[4]=cnte;
    free_real_vector(v,1);
    free_real_vector(delta,1);
    free_real_vector(gamma,1);
    free_real_vector(s,1);
    free_real_vector(p,1);
    free_real_matrix(vec,1,n,1);
    return f;
}
```

B. flemin

Determines a vector $x \epsilon R^n$ for which the real valued function $F(x)$ attains a minimum by means of the variable metric algorithm given in [Fle63], except for some details (see [Bus72b]). It is assumed that the partial derivatives $g_i(x)=\partial F(x)/\partial x_i$ $(i=1,...,n)$ may easily be computed. *flemin* is suitable, in particular, for use in connection with problems for which n is relatively large and the computation of $F(x)$ and $g(x)$ is relatively cheap.

The theory underlying the operation of *flemin* is similar to that underlying the operation of *rnk1min*, and the algorithms employed are similar in many respects.

The user is asked to provide at outset the initial approximation vector $x^{(0)}$ and the same parameters ε_r, ε_a, μ, ε_g, F_{min}, (possibly) c, and a number specifying the maximum permitted number of simultaneous evaluations of $F(x)$ and $g(x)$, together with (possibly) approximations to the elements in the upper triangular part of $H(x^{(0)})$, as for *rnk1min*.

The initialization for the algorithm employed is the same as that for *rnk1min*. The stages are as follows.

I. Simply set $d^{(k)} = -H^{(k)}g(x^{(k)})$.

II and III. As for *rnk1min*.

IV. If

$$(\delta^{(k)})^T\gamma^{(k)} < 0 \tag{1}$$

then set $H^{(k+1)} = H^{(k)}$. If condition (1) is violated and

$$(\delta^{(k)})^T\gamma^{(k)} \geq (\gamma^{(k)})^T H^{(k)}\gamma^{(k)} \tag{2}$$

then the rank-two updating formulae (7,8) of *rnk1min* are used. If both condition (1,2) are violated, the rank-two updating formulae (8.9) of *rnk1min* are used.

V. If $\|H^{(k+1)}g(x^{(k+1)})\| \leq \varepsilon_r\|x^{(k+1)}\| + \varepsilon_a$ and $\|g(x^{(k+1)})\| \leq \varepsilon_g$ and $k \geq n$ then $x^{(k+1)}$ is accepted as an approximation to the minimum. If k is equal to the number specified by the maximum number of function evaluations then the process is terminated (and $x^{(k+1)}$ is then not an acceptable approximation).

The above algorithm differs from one given by Fletcher (see [Fle63]), (which may fail when $F(x)$ is not approximately a quadratic function in the components of x) only in the determination of $\theta^{(k)}$ at stage II (the method adopted by *flemin* to a large extent overcomes the difficulty mentioned).

Function Parameters:

float flemin (*n,x,g,h,funct,in,out*)

flemin: delivers the calculated least value of the given function;

n: int;
entry: the number of variables of the function to be minimized;

x: float *x[1:n]*;
entry: an approximation of a minimum of the function;
exit: the calculated minimum of the function;

g: float *g[1:n]*;
exit: the gradient of the function at the calculated minimum;

h: float *h[1:n(n+1)/2]*;
the upper triangle of an approximation of the inverse Hessian is stored columnwise in h (i.e. the (i,j)-th element = $h[(j-1)j/2+i]$, $1 \leq i \leq j \leq n$); if $in[6]>0$ initializing of h will be done automatically and the initial approximation of the inverse Hessian will equal the unit matrix multiplied with the value of $in[6]$; if $in[6]<0$ then no initializing of h will be done and the user should give in h an approximation of the inverse Hessian, at the

starting point; the upper triangle of an approximation of the inverse Hessian at the calculated minimum is delivered in *h*;

funct: float *(*funct)(n,x,g)*;
a call of *funct* must have the following results:
1. *funct* becomes the value of the function to be minimized at the point *x*;
2. the value of *g[i]*, (*i*=1,...,*n*), becomes the value of the *i*-th component of the gradient of the function at *x*;

in: float *in[1:7]*;
entry:
in[1]: the relative tolerance for the solution;
in[2]: the absolute tolerance for the solution;
in[3]: a parameter used for controlling line minimization (see *rnk1min*); usually a suitable value is 0.0001;
in[4]: the absolute tolerance for the Euclidean norm of the gradient at the solution;
in[5]: a lower bound for the function value;
in[6]: this parameter controls the initialization of the approximation of the inverse Hessian (metric), see *h*; usually the choice *in[6]*=1 will give good results;
in[7]: the maximum allowed number of calls of *funct*;

out: float *out[0:4]*;
exit:
out[0]: the Euclidean norm of the product of the metric and the gradient at the calculated minimum;
out[1]: the Euclidean norm of the gradient at the calculated minimum;
out[2]: the number of calls of *funct*, necessary to attain this result;
out[3]: the number of iterations in which a line search was necessary;
out[4]: if *out[4]*=-1 then the process is broken off because no down hill direction could be calculated; the precision asked for may not be attained and is possibly chosen too high; normally *out[4]*=0.

Functions used: vecvec, elmvec, symmatvec, inivec, inisymd, mulvec, dupvec, linemin, davupd, fleupd.

```
#include <math.h>

float flemin(int n, float x[], float g[], float h[],
            float (*funct)(int, float[], float[]),
            float in[], float out[])
{
    float *allocate_real_vector(int, int);
    void free_real_vector(float *, int);
    float vecvec(int, int, int, float [], float []);
    void elmvec(int, int, int, float [], float [], float);
    float symmatvec(int, int, int, float [], float []);
    void inivec(int, int, float [], float);
    void inisymd(int, int, int, float [], float);
    void mulvec(int, int, int, float [], float [], float);
    void dupvec(int, int, int, float [], float []);
    void linemin(int, float [], float [], float, float *, float [],
                float (*)(int, float[], float[]), float, float *,
                float, float *, int *, int, float []);
    void davupd(float [], int, float [], float [], float, float);
    void fleupd(float [], int, float [], float [], float, float);
    int i,it,cntl,evl,evlmax;
    float f,f0,fmin,mu,dg,dg0,nrmdelta,alfa,reltol,abstol,eps,tolg,
            aid,*v,*delta,*s;

    v=allocate_real_vector(1,n);
```

```
delta=allocate_real_vector(1,n);
s=allocate_real_vector(1,n);

reltol=in[1];
abstol=in[2];
mu=in[3];
tolg=in[4];
fmin=in[5];
alfa=in[6];
evlmax=in[7];
out[4]=0.0;
it=0;
f=(*funct)(n,x,g);
evl=1;
cntl=0;
if (alfa > 0.0) {
    inivec(1,(n*(n+1))/2,h,0.0);
    inisymd(1,n,0,h,alfa);
}
for (i=1; i<=n; i++) delta[i] = -symmatvec(1,n,i,h,g);
dg=sqrt(vecvec(1,n,0,g,g));
nrmdelta=sqrt(vecvec(1,n,0,delta,delta));
eps=sqrt(vecvec(1,n,0,x,x))*reltol+abstol;
dg0=vecvec(1,n,0,delta,g);
it++;
while ((nrmdelta > eps || dg > tolg) && (evl < evlmax)) {
    dupvec(1,n,0,s,x);
    dupvec(1,n,0,v,g);
    if (it >= n)
        alfa=1.0;
    else {
        if (it != 1)
            alfa /= nrmdelta;
        else {
            alfa=2.0*(fmin-f)/dg0;
            if (alfa > 1.0) alfa=1.0;
        }
    }
    elmvec(1,n,0,x,delta,alfa);
    f0=f;
    f=(*funct)(n,x,g);
    evl++;
    dg=vecvec(1,n,0,delta,g);
    if (it == 1 || f0-f < -mu*dg0*alfa) {
        /* line minimization */
        i=evlmax-evl;
        cntl++;
        linemin(n,s,delta,nrmdelta,&alfa,g,funct,f0,&f,
                dg0,&dg,&i,0,in);
        evl += i;
        dupvec(1,n,0,x,s);
    }
    if (alfa != 1.0) mulvec(1,n,0,delta,delta,alfa);
    mulvec(1,n,0,v,v,-1.0);
    elmvec(1,n,0,v,g,1.0);
    for (i=1; i<=n; i++) s[i]=symmatvec(1,n,i,h,v);
    aid=vecvec(1,n,0,v,s);
    dg=(dg-dg0)*alfa;
    if (dg > 0.0)
        if (dg >= aid)
            fleupd(h,n,delta,s,1.0/dg,(1.0+aid/dg)/dg);
        else
            davupd(h,n,delta,s,1.0/dg,1.0/aid);
    for (i=1; i<=n; i++) delta[i] = -symmatvec(1,n,i,h,g);
    alfa *= nrmdelta;
    nrmdelta=sqrt(vecvec(1,n,0,delta,delta));
    eps=sqrt(vecvec(1,n,0,x,x))*reltol+abstol;
    dg=sqrt(vecvec(1,n,0,g,g));
    dg0=vecvec(1,n,0,delta,g);
    if (dg0 > 0.0) {
        out[4] = -1.0;
        break;
    }
    it++;
```

```
	}
	out[0]=nrmdelta;
	out[1]=dg;
	out[2]=evl;
	out[3]=cntl;
	free_real_vector(v,1);
	free_real_vector(delta,1);
	free_real_vector(s,1);
	return f;
}
```

5.3 Overdetermined nonlinear systems

5.3.1 Least squares - With Jacobian matrix

A. marquardt

Calculates the least squares solution of an overdetermined system of nonlinear equations with Marquardt's method [BusDK75, M63].

marquardt computes the values of the components $p_1,...,p_n$ of $p \epsilon R^n$ such that

$$\Phi(p) = \sum_{i=1}^{m} [f(t_i,p) - d_i]^2$$

is a minimum at p, where $f(t,p)$ is a real-valued function, and d_i $(i=1,...,m)$ are given data readings $(m \geq n)$ by use of an improved version of the Levenberg-Marquardt algorithm (the p_j offers a parameter set of best fit of the function $f(t,p)$ to the data at the points t_i).

An initial approximation $p^{(0)}$ to p, two tolerances ε_{re} and ε_a, and a real number ξ expressing the ratio between the gradient and Gauss-Newton directions must be specified (ξ may be taken to be 0.01 for well-conditioned problem; in any case ξ must satisfy the inequality:
machine precision $< \xi \leq$ 1/machine precision).

At the k-th stage of the algorithm, the $m \times n$ Jacobian matrix $J(p^{(k)})$, whose components are given by $J_{i,j}(p) = \partial f(t_i,p)/\partial p_j$ $(i=1,...,m;\ j=1,...,n)$ with $p=p^{(k)}$, is decomposed in the form $J(p^{(k)})=U^{(k)}D^{(k)}V^{(k)}$ where $U^{(k)}$ is an orthogonal $m \times m$ matrix, $D^{(k)}$ is an $m \times n$ matrix whose principal $n \times n$ submatrix is the diagonal matrix $D^{(k)}$ of singular values of $J(p^{(k)})$ ($D^{(k)}$ having zeros elsewhere), and $V^{(k)}$ is an $n \times n$ orthogonal matrix.

The scalar $\rho^{(p)} = \xi \| D^{(k)} \|^2$ is determined. The further number $\lambda'^{(k)}$ is determined by setting

$$\lambda'^{(0)} = \lambda^{(-1)} = \rho^{(0)} \text{ if } k=0 \text{ and,}$$

if $k \geq 1$, $\lambda'^{(k)} = w\lambda^{(k-1)}$ if $\lambda^{(k-1)} \leq \lambda^{(k-2)}$ and $\lambda'^{(k)} \leq \lambda^{(k-1)}$ if $\lambda^{(k-1)} > \lambda^{(k-2)}$ (w is a real number in the range $0 < w < 1$; *marquardt* uses the value 0.5). If $h(\lambda'^{(k)}) \geq \mu$ (μ being a real number in the range $0 < \mu \leq 0.5$; *marquardt* uses the value 0.5) where, with $s^{(k)}(\lambda)$ defined by

$$(J(p^{(k)})^T J(p^{(k)}) + \lambda I)s^{(k)}(\lambda) = -J(p^{(k)})^T f(p_{(k)}),$$

$$h^{(k)}(\lambda) = (\Phi(p^{(k)}) - \Phi(p^{(k)}+s^{(k)}(\lambda))) \,/\, -s_{(k)}(\lambda)^T J(p^{(k)})^T f(p^{(k)}),$$

the components of $f(p^{(k)})$ being $f(t_i,p^{(k)})-d_i x$ $(i=1...,m)$ then $\lambda^{(k)}$ is set equal to $\lambda'^{(k)}$. If, however, $h(\lambda'^{(k)}) < \mu$, $\lambda^{(k)}$ is taken to be $\nu_r \max(\lambda'^{(k)},\rho^{(k)})$ (ν being a real number in the range $1<\nu<\infty$; *marquardt* uses the value 10) where r is the smallest nonnegative integer for which the relationship $h(\nu^r \max(\lambda'^{(k)},\rho^{(k)})) \geq \mu$, is satisfied, $p^{(k+1)}$ is then determined by $p^{(k+1)}=p^{(k)}+s^{(k)}(\lambda^{(k)})$.

The algorithm is terminated (with $p^{(k)}$ accepted as an approximation to p) when $\Phi(p^{(k-1)}) - \Phi(p^{(k)}) \le \varepsilon_r\Phi(p^{(k)}) + \varepsilon_a$, or $\Phi(p^{(k)}) \le \varepsilon_a$.

Function Parameters:

void marquardt (*m,n,par,g,v,funct,jacobian,in,out*)

m: int;
entry: the number of equations;

n: int;
entry: the number of unknown variables; n should satisfy $n \le m$;

par: float *par[1:n]*;
entry: an approximation to a least squares solution of the system;
exit: the calculated least squares solution;

g: float *g[1:m]*;
exit: the residual vector at the calculated solution;

v: float *v[1:n,1:n]*;
exit: the inverse of the matrix J^TJ where J denotes the transpose of the matrix of partial derivatives $\partial g[i]/\partial par[j]$ (i=1,...,m; j=1,...,n); i.e. *v[i,j]* contains the (i,j)-th element of $(J(p)^TJ(p))^{-1}$ (i,j=1,...,n) where p is the value of $p^{(k)}$ in the above at termination;
note that the standard deviation of σ_j associated with the computed estimate of p_j (j=1,...,n) and the correlation $\rho_{i,j}$ between the estimates of p_i and p_j (i=1,...,n; j=i+1,...,n) may easily be extracted from the contents of the array v by use of the formula $\sigma = \sqrt{(v[j,j]/m)}$ (j=1,...,n), $\rho_{i,j} = v[i,j]/\sqrt{(v[i,i]*v[j,j])}$ (i=1,...,n; j=i+1,...,n);

funct: int *(*funct)(m,n,par,g)*;
entry: *m,n,par*;
m, n have the same meaning as in the procedure *marquardt*;
array *par[1:n]* contains the current values of the unknowns and should not be altered;
exit: upon completion of a call of *funct*, the array *g[1:m]* should contain the residual vector obtained with the current values of the unknowns; e.g. in curve fitting problems, *g[i]* = theoretical value *f(x[i], par)* - observed value *y[i]*;
after a successful call of *funct*, the function should deliver the value nonzero; however, if *funct* delivers the value zero then it is assumed that the current estimates of the unknowns lie outside a feasible region and the process is terminated (see *out[1]*); hence, proper programming of *funct* makes it possible to avoid calculation of a residual vector with values of the unknown variables which make no sense or which even may cause overflow in the computation;

jacobian: void *(*jacobian)(m,n,par,g,jac)*;
entry: *m,n,par,g*;
m, n, par: see *funct*;
g contains the residual vector obtained with the current values of the unknowns and should not be altered;
exit: float *jac[1:m,1:n]*;
upon completion of a call of *jacobian*, the array *jac* should contain the partial derivatives $\partial g[i]/\partial par[j]$, obtained with the current values of the unknown variables given in *par[1:n]*;
it is a prerequisite for the proper operation of the procedure *marquardt* that the precision of the elements of the matrix *jac* is at least the precision defined by *in[3]* and

in[4];

in: float *in[0:6]*;

entry:

in[0]: the machine precision;

in[1], in[2]: these are not used by *marquardt*;

in[3]: the relative tolerance for the difference between the Euclidean norm of the ultimate and penultimate residual vector (value of ε_{re} above); see *in[4]*;

in[4]: the absolute tolerance for the difference between the Euclidean norm of the ultimate and penultimate residual vector (value of ε_a above);

the process is terminated if the improvement of the sum of squares is less than *in[3]**(sum of squares)+*in[4]***in[4]*; these tolerances should be chosen greater than the corresponding errors of the calculated residual vector;

in[5]: the maximum number of calls of *funct* allowed;

in[6]: a starting value used for the relation between the gradient and the Gauss-Newton direction (value of ξ above); if the problem is well conditioned then a suitable value for *in[6]* will be 0.01; if the problem is ill conditioned then *in[6]* should be greater, but the value of *in[6]* should satisfy:

$in[0] < in[6] \leq 1/in[0]$;

out: float *out[1:7]*;

exit:

out[1]: this value gives information about the termination of the process;

out[1]=0: normal termination;

out[1]=1: the process has been broken off because the number of calls of *funct* exceeded the number given in *in[5]*;

out[1]=2: the process has been broken off because a call of *funct* delivered the value zero;

out[1]=3: *funct* became zero when called with the initial estimates of *par[1:n]*; the iteration process was not started and so *v[1:n,1:n]* cannot be used;

out[1]=4: the process has been broken off because the precision asked for cannot be attained; this precision is possibly chosen too high, relative to the precision in which the residual vector is calculated (see *in[3]*);

out[2]: the Euclidean norm of the residual vector calculated with values of the unknowns delivered;

out[3]: the Euclidean norm of the residual vector calculated with the initial values of the unknown variables;

out[4]: the number of calls of *funct* necessary to attain the calculated result;

out[5]: the total number of iterations performed; note that in each iteration one evaluation of the Jacobian matrix had to be made;

out[6]: the improvement vector in the last iteration step;

out[7]: the condition number of J^TJ, i.e. the ratio of its largest to smallest eigenvalues.

Functions used: mulcol, dupvec, vecvec, matvec, tamvec, mattam, qrisngvaldec.

```
#include <math.h>

void marquardt(int m, int n, float par[], float g[], float **v,
               int (*funct)(int, int, float[], float[]),
               void (*jacobian)(int, int, float[], float[], float **),
               float in[], float out[])
{
    float *allocate_real_vector(int, int);
    float **allocate_real_matrix(int, int, int, int);
```

```
void free_real_vector(float *, int);
void free_real_matrix(float **, int, int, int);
void mulcol(int, int, int, int, float **, float **, float);
void dupvec(int, int, int, float [], float []);
float vecvec(int, int, int, float [], float []);
float matvec(int, int, int, float **, float []);
float tamvec(int, int, int, float **, float []);
float mattam(int, int, int, int, float **, float **);
int qrisngvaldec(float **, int, int, float [], float **, float []);
int maxfe,fe,it,i,j,err,emergency;
float vv,ww,w,mu,res,fpar,fparpres,lambda,lambdamin,p,pw,reltolres,
      abstolres,em[8],*val,*b,*bb,*parpres,**jac,temp;

val=allocate_real_vector(1,n);
b=allocate_real_vector(1,n);
bb=allocate_real_vector(1,n);
parpres=allocate_real_vector(1,n);
jac=allocate_real_matrix(1,m,1,n);

vv=10.0;
w=0.5;
mu=0.01;
ww = (in[6] < 1.0e-7) ? 1.0e-8 : 1.0e-1*in[6];
em[0]=em[2]=em[6]=in[0];
em[4]=10*n;
reltolres=in[3];
abstolres=in[4]*in[4];
maxfe=in[5];
err=0;
fe=it=1;
p=fpar=res=0.0;
pw = -log(ww*in[0])/2.30;
if (!(*funct)(m,n,par,g)) {
   err=3;
   out[4]=fe;
   out[5]=it-1;
   out[1]=err;
   free_real_vector(val,1);
   free_real_vector(b,1);
   free_real_vector(bb,1);
   free_real_vector(parpres,1);
   free_real_matrix(jac,1,m,1);
   return;
}
fpar=vecvec(1,m,0,g,g);
out[3]=sqrt(fpar);
emergency=0;
it=1;
do {
   (*jacobian)(m,n,par,g,jac);
   i=qrisngvaldec(jac,m,n,val,v,em);
   if (it == 1)
      lambda=in[6]*vecvec(1,n,0,val,val);
   else
      if (p == 0.0) lambda *= w;
   for (i=1; i<=n; i++) b[i]=val[i]*tamvec(1,m,i,jac,g);
   while (1) {
      for (i=1; i<=n; i++) bb[i]=b[i]/(val[i]*val[i]+lambda);
      for (i=1; i<=n; i++) parpres[i]=par[i]-matvec(1,n,i,v,bb);
      fe++;
      if (fe >= maxfe)
         err=1;
      else
         if (!(*funct)(m,n,parpres,g)) err=2;
      if (err != 0) {
         emergency=1;
         break;
      }
      fparpres=vecvec(1,m,0,g,g);
      res=fpar-fparpres;
      if (res < mu*vecvec(1,n,0,b,bb)) {
         p += 1.0;
         lambda *= vv;
         if (p == 1.0) {
```

```
                lambdamin=ww*vecvec(1,n,0,val,val);
                if (lambda < lambdamin) lambda=lambdamin;
            }
            if (p >= pw) {
                err=4;
                emergency=1;
                break;
            }
        } else {
            dupvec(1,n,0,par,parpres);
            fpar=fparpres;
            break;
        }
    }
    if (emergency) break;
    it++;
} while (fpar > abstolres && res > reltolres*fpar+abstolres);
for (i=1; i<=n; i++) mulcol(1,n,i,i,jac,v,1.0/(val[i]+in[0]));
for (i=1; i<=n; i++)
    for (j=1; j<=i; j++) v[i][j]=v[j][i]=mattam(1,n,i,j,jac,jac);
lambda=lambdamin=val[1];
for (i=2; i<=n; i++)
    if (val[i] > lambda)
        lambda=val[i];
    else
        if (val[i] < lambdamin) lambdamin=val[i];
temp=lambda/(lambdamin+in[0]);
out[7]=temp*temp;
out[2]=sqrt(fpar);
out[6]=sqrt(res+fpar)-out[2];
out[4]=fe;
out[5]=it-1;
out[1]=err;
free_real_vector(val,1);
free_real_vector(b,1);
free_real_vector(bb,1);
free_real_vector(parpres,1);
free_real_matrix(jac,1,m,1);
}
```

B. gssnewton

Calculates the least squares solution of an overdetermined system of nonlinear equations with the Gauss-Newton method [BusDK75, Har61, Sp67].

gssnewton computes the values of the components $p_1,...,p_n$ of $p \in R^n$ such that

$$\Phi(p) = \sum_{i=1}^{m} [f(t_i,p) - d_i]^2$$

is a minimum at p, where $f(t,p)$ is a real-valued function, and d_i $(i=1,...,m)$ are given data readings $(m \geq n)$ by use of an improved version of the Gauss-Newton algorithm (the p_j offer a parameter set of best fit of the function $f(t,p)$ to the data at the points t_i).

An initial approximation $p^{(0)}$ to p, three tolerances ε_{re}, ε_a and ε_0 must be given. At the k-th stage of the algorithm, the $m \times n$ Jacobian matrix $J(p^{(k)})$, whose components are given by $J_{i,j}(p) = \partial f(t_i,p)/\partial p_j$ $(i=1,...,m;\ j=1,...,n)$ with $p=p^{(k)}$, is decomposed in the form $J(p^{(k)})=Q^{(k)}R^{(k)}$ where $Q^{(k)}$ is an $m \times n$ orthogonal matrix, and $R^{(k)}$ is an $m \times n$ upper triangular matrix. The direction vector $d^{(k)} \in R^n$ is determined by solving the $n \times n$ upper triangular system derived from the first n equations of the system $R^{(k)}d^{(k)} = -(Q^{(k)})^T f(p^{(k)})$ where the components of $f(p^{(k)}) \in R^m$ are $f(t_i,p^{(k)})-d_i$ $(i=1,...,m)$.

The scalar $\alpha^{(p)}$ is determined (a) by setting $\alpha^{(k)}=1$ if $\Phi(p^{(k)}+d^{(k)}) \leq \Phi(p^{(k)})$ or, if this condition is violated, (b) by setting $\alpha^{(k)}=2^{-r}$, where r is the smallest nonnegative integer for which

$$\Phi(p^{(k)} + 2^{-r}d^{(k)}) < \Phi(p^{(k)}),$$
$$\Phi(p^{(k)} + 2^{-r}d^{(k)}) < \Phi(p^{(k)} + 2^{-(r+1)}d^{(k)})$$

(these inequalities are inspected for r replaced by r' (r'=0,1,...,r)) $p^{(k+1)}$ is then $p^{(k)} + \alpha^{(k)}d^{(k)}$.

The algorithm is terminated (with $p^{(k)}$ accepted as an approximation to p) when (c) $\Phi(p^{(k)})\le\varepsilon_0^2$ or (d) $\|p^{(k+1)}-p^{(k)}\| \le \|p^{(k+1)}\|\varepsilon_{re} + \varepsilon_a$, and the standard deviations of the elements of p are then calculated.

Function Parameters:

void gssnewton (*m,n,par,g,v,funct,jacobian,in,out*)

m: int;
  entry: the number of equations;

n: int;
  entry: the number of unknowns in the *m* equations; ($n \le m$);

par: float *par[1:n]*;
  entry: an approximation to a least squares solution of the system;
  exit: the calculated least squares solution;

rv: float *rv[1:m]*;
  exit: the residual vector of the system at the calculated solution;

jjinv: float *jjinv[1:n,1:n]*;
  exit: the inverse of the matrix J^TJ where J denotes the transpose of the Jacobian matrix at the solution;

funct: int *(*funct)(m,n,par,rv)*;
  entry: *m,n,par*;
    m, n have the same meaning as in the procedure *gssnewton*;
    array *par[1:n]* contains the current values of the unknowns and should not be altered;
  exit: upon completion of a call of *funct*, the array *rv[1:m]* should contain the residual vector obtained with the current values of the unknowns;
    the programmer of *funct* may decide that some current estimates of the unknowns lie outside a feasible region; in this case *funct* should deliver the value zero and the process is terminated (see *out[1]*), otherwise *funct* should deliver the value nonzero;

jacobian: void *(*jacobian)(m,n,par,rv,jac)*;
  the parameters of *jacobian* are:
  m, n: see *gssnewton*;
  par: entry: current estimate of the unknowns, these values should not be changed;
  rv: entry: the residual vector of the system of equations corresponding to the vector of unknowns as given in *par*; on exit the values are not changed;
  jac: exit: float *jac[1:m,1:n]*;
    a call of the procedure *jacobian* should deliver the Jacobian matrix evaluated at the current estimates of the unknown variables given in *par* in such a way, that the partial derivative $\partial rv[i]/\partial par[j]$ is delivered in *jac[i,j]*, i=1,...,m, j=1,...,n;

in: float *in[0:7]*;
  entry:
  in[0]: the machine precision;
  in[1]: the relative tolerance for the step vector (relative to the vector of current

estimates in *par*) (value of ε_{re} above); see *in[2]*; *in[1]* should not be chosen smaller than *in[0]*;

in[2]: the absolute tolerance for the step vector (relative to the vector of current estimates in *par*) (value of ε_a above);

the process is terminated if in some iteration (but not the first) the Euclidean norm of the calculated Newton step is less than *in[1]** $\|par\|$+*in[2]*;

in[3]: not used by *gssnewton*;

in[4]: the absolute tolerance for the Euclidean norm of the residual vector (value of ε_0 above); the process is terminated when this norm is less than *in[4]*;

in[5]: the maximum allowed number of function evaluations (i.e. calls of *funct*);

in[6]: the maximum allowed number of halvings of a calculated Newton step vector, a suitable value is 15;

in[7]: the maximum allowed number of successive *in[6]* times halved step vector, suitable values are 1 and 2;

out: float *out[1:9]*;

exit:

the process was terminated because *out[1]* =

1: the norm of the residual vector is small with respect to *in[4]*;

2: the calculated Newton step is sufficiently small (see *in[1]*, *in[2]*);

3: the calculated step was completely damped (halved) in *in[7]* successive iterations;

4: *out[4]* exceeds *in[5]*, the maximum allowed number of calls of *funct*;

5: the Jacobian was not full-rank (see *out[8]*);

6: *funct* delivered zero at a new vector of estimates of the unknowns;

7: *funct* delivered zero in a call from *jacobian*;

out[2]: the Euclidean norm of the last residual vector;

out[3]: the Euclidean norm of the initial residual vector;

out[4]: the total number of calls of *funct*; *out[4]* will be less than *in[5]+in[6]*;

out[5]: the total number of iterations;

out[6]: the Euclidean norm of the last step vector;

out[7]: iteration number of the last iteration in which the Newton step was halved;

out[8],out[9]: rank and maximum column norm of the Jacobian matrix in the last iteration, as delivered by *lsqortdec* in *aux[3]* and *aux[5]*.

Functions used: dupvec, vecvec, elmvec, lsqortdec, lsqsol, lsqinv.

```
#include <math.h>

void gssnewton(int m, int n, float par[], float rv[], float **jjinv,
               int (*funct)(int, int, float[], float[]),
               void (*jacobian)(int, int, float[], float[], float **),
               float in[], float out[])
{
   int *allocate_integer_vector(int, int);
   float *allocate_real_vector(int, int);
   float **allocate_real_matrix(int, int, int, int);
   void free_integer_vector(int *, int);
   void free_real_vector(float *, int);
   void free_real_matrix(float **, int, int, int);
   float vecvec(int, int, int, float [], float []);
   void dupvec(int, int, int, float [], float []);
   void elmvec(int, int, int, float [], float [], float);
   void lsqortdec(float **, int, int, float [], float [], int []);
   void lsqsol(float **, int, int, float [], int [], float []);
   void lsqinv(float **, int, float [], int []);
   int i,j,inr,mit,text,it,itmax,inrmax,tim,feval,fevalmax,conv,
         testthf,dampingon,*ci,fail;
```

```
float rho,res1,res2,rn,reltolpar,abstolpar,abstolres,stap,normx,
      **jac,*pr,*aid,*sol,*fu2,aux[6];

ci=allocate_integer_vector(1,n);
pr=allocate_real_vector(1,n);
aid=allocate_real_vector(1,n);
sol=allocate_real_vector(1,n);
fu2=allocate_real_vector(1,m);
jac=allocate_real_matrix(1,m+1,1,n);

itmax=fevalmax=in[5];
aux[2]=n*in[0];
tim=in[7];
reltolpar=in[1]*in[1];
abstolpar=in[2]*in[2];
abstolres=in[4]*in[4];
inrmax=in[6];
dupvec(1,n,0,pr,par);
if (m < n)
   for (i=1; i<=n; i++) jac[m+1][i]=0.0;
text=4;
mit=0;
testthf=1;
res2=stap=out[5]=out[6]=out[7]=0.0;
(*funct)(m,n,par,fu2);
rn=vecvec(1,m,0,fu2,fu2);
out[3]=sqrt(rn);
feval=1;
dampingon=0;
fail=0;
it=1;
do {
   out[5]=it;
   (*jacobian)(m,n,par,fu2,jac);
   if (!testthf) {
      text=7;
      fail=1;
      break;
   }
   lsqortdec(jac,m,n,aux,aid,ci);
   if (aux[3] != n) {
      text=5;
      fail=1;
      break;
   }
   lsqsol(jac,m,n,aid,ci,fu2);
   dupvec(1,n,0,sol,fu2);
   stap=vecvec(1,n,0,sol,sol);
   rho=2.0;
   normx=vecvec(1,n,0,par,par);
   if (stap > reltolpar*normx+abstolpar || it == 1 && stap > 0.0) {
      inr=0;
      do {
         rho /= 2.0;
         if (inr > 0) {
            res1=res2;
            dupvec(1,m,0,rv,fu2);
            dampingon = inr > 1;
         }
         for (i=1; i<=n; i++) pr[i]=par[i]-sol[i]*rho;
         feval++;
         if (!(*funct)(m,n,pr,fu2)) {
            text=6;
            fail=1;
            break;
         }
         res2=vecvec(1,m,0,fu2,fu2);
         conv = inr >= inrmax;
         inr++;
      } while ((inr == 1) ? (dampingon || res2 >= rn) :
                 (!conv && (rn <= res1 || res2 < res1)));
      if (fail) break;
      if (conv) {
         mit++;
```

```
                if (mit < tim) conv=0;
            } else
                mit=0;
            if (inr > 1) {
                rho *= 2.0;
                elmvec(1,n,0,par,sol,-rho);
                rn=res1;
                if (inr > 2) out[7]=it;
            } else {
                dupvec(1,n,0,par,pr);
                rn=res2;
                dupvec(1,m,0,rv,fu2);
            }
            if (rn <= abstolres) {
                text=1;
                itmax=it;
            } else
                if (conv && inrmax > 0) {
                    text=3;
                    itmax=it;
                } else
                    dupvec(1,m,0,fu2,rv);
        } else {
            text=2;
            rho=1.0;
            itmax=it;
        }
        it++;
    } while (it <= itmax && feval < fevalmax);
    if (!fail) {
        lsqinv(jac,n,aid,ci);
        for (i=1; i<=n; i++) {
            jjinv[i][i]=jac[i][i];
            for (j=i+1; j<=n; j++) jjinv[i][j]=jjinv[j][i]=jac[i][j];
        }
    }
    out[6]=sqrt(stap)*rho;
    out[2]=sqrt(rn);
    out[4]=feval;
    out[1]=text;
    out[8]=aux[3];
    out[9]=aux[5];
    free_integer_vector(ci,1);
    free_real_vector(pr,1);
    free_real_vector(aid,1);
    free_real_vector(sol,1);
    free_real_vector(fu2,1);
    free_real_matrix(jac,1,m+1,1);
}
```

5.4 Differential equations - Initial value problems

5.4.1 First order - No derivatives right hand side

A. rk1

Solves an initial value problem for a single first order ordinary differential equation $dy/dx = f(x,y)$ by means of a 5-th order Runge-Kutta method [see Z64]. The equation is assumed to be non-stiff.

rk1 is based on an explicit 5-th order Runge-Kutta method and is provided with step length and error control. The error control is based on the last term of the Taylor series which is taken into account. A step is rejected if the absolute value of this last term is greater than $(|fxy|*e[1]+e[2])*|h|/k$, where $k = |b$-(if *fi* is nonzero then *a* else $d[3])|$ denotes the length of the integration interval, otherwise a step is accepted. *rk1* uses as its minimal absolute step length $hmin = e[1]*k+e[2]$. If a step of length $|h| = hmin$ is rejected then the step is skipped.

Function Parameters:

void rk1 (*x,a,b,y,ya,fxy,e,d,fi*)

x: float *;
  entry: the independent variable;
  exit: upon completion of a call, *x* is equal to *b*;
a: float;
  entry: the initial value of *x*;
b: float;
  entry: a value parameter, giving the end value of *x*;
y: float *;
  entry: the dependent variable;
ya: float;
  entry: the value of *y* at *x=a*;
fxy: float (**fxy*)(*x,y*);
  entry: the function giving the value of *dy/dx*;
e: float *e[1:2]*;
  entry:
  e[1]: a relative tolerance;
  e[2]: an absolute tolerance;
d: float *d[1:4]*;
  exit:
  d[1]: number of steps skipped;
  d[2]: equals the step length;
  d[3]: equals *b*;
  d[4]: equals *y(b)*;
fi: int;
  entry: if *fi* is nonzero then *rk1* integrates from *x=a* to *x=b* with initial value *y(a)=ya* and trial step *b-a*; if *fi* is zero then *rk1* integrates from *x=d[3]* to *x=b* with initial value *y(d[3])=d[4]* and step length *h=d[2]*sign(b-d[3])*, while *a* and *ya* are ignored.

```
#include <math.h>

void rk1(float *x, float a, float b, float *y, float ya,
        float (*fxy)(float, float), float e[], float d[], int fi)
{
    int last,first,reject,test,ta,tb;
    float e1,e2,xl,yl,h,ind,hmin,absh,k0,k1,k2,k3,k4,k5,discr,tol,mu,
            mu1,fh,hl;

    if (fi) {
        d[3]=a;
        d[4]=ya;
    }
    d[1]=0.0;
    xl=d[3];
    yl=d[4];
    if (fi) d[2]=b-d[3];
    absh=h=fabs(d[2]);
    if (b-xl < 0.0) h = -h;
    ind=fabs(b-xl);
    hmin=ind*e[1]+e[2];
    e1=e[1]/ind;
```

```
    e2=e[2]/ind;
    first=1;
    test=1;
    if (fi) {
        last=1;
        test=0;
    }
    while (1) {
        if (test) {
            absh=fabs(h);
            if (absh < hmin) {
                h = (h > 0.0) ? hmin : -hmin;
                absh=hmin;
            }
            ta=(h >= b-xl);
            tb=(h >= 0.0);
            if ((ta && tb) || (!(ta || tb))) {
                d[2]=h;
                last=1;
                h=b-xl;
                absh=fabs(h);
            } else
                last=0;
        }
        test=1;
        *x=xl;
        *y=yl;
        k0=(*fxy)(*x,*y)*h;
        *x=xl+h/4.5;
        *y=yl+k0/4.5;
        k1=(*fxy)(*x,*y)*h;
        *x=xl+h/3.0;
        *y=yl+(k0+k1*3.0)/12.0;
        k2=(*fxy)(*x,*y)*h;
        *x=xl+h*0.5;
        *y=yl+(k0+k2*3.0)/8.0;
        k3=(*fxy)(*x,*y)*h;
        *x=xl+h*0.8;
        *y=yl+(k0*53.0-k1*135.0+k2*126.0+k3*56.0)/125.0;
        k4=(*fxy)(*x,*y)*h;
        *x = (last ? b : xl+h);
        *y=yl+(k0*133.0-k1*378.0+k2*276.0+k3*112.0+k4*25.0)/168.0;
        k5=(*fxy)(*x,*y)*h;
        discr=fabs(k0*21.0-k2*162.0+k3*224.0-k4*125.0+k5*42.0)/14.0;
        tol=fabs(k0)*e1+absh*e2;
        reject = discr > tol;
        mu=tol/(tol+discr)+0.45;
        if (reject) {
            if (absh <= hmin) {
                d[1] += 1.0;
                *y=yl;
                first=1;
                if (b == *x) break;
                xl = *x;
                yl = *y;
            } else
                h *= mu;
        } else {
            if (first) {
                first=0;
                hl=h;
                h *= mu;
            } else {
                fh=mu*h/hl+mu-mu1;
                hl=h;
                h *= fh;
            }
            mu1=mu;
            *y=yl+(-k0*63.0+k1*189.0-k2*36.0-k3*112.0+k4*50.0)/28.0;
            k5=(*fxy)(*x,*y)*hl;
            *y=yl+(k0*35.0+k2*162.0+k4*125.0+k5*14.0)/336.0;
            if (b == *x) break;
            xl = *x;
            yl = *y;
```

```
        }
    }
    if (!last) d[2]=h;
    d[3] = *x;
    d[4] = *y;
}
```

B. rke

Solves an initial value problem for a system of first order ordinary differential equations $dy/dx = f(x,y)$, from $x = x0$ to $x = xe$ where $y(x0) = y0$, by means of a 5-th order Runge-Kutta method. The system is assumed to be non-stiff.

The method upon which *rke* is based, is a member of a class of 5-th order Runge-Kutta formulas presented in [Eng69]. Automatic stepsize control is implemented in a way proposed in [Z64]. For further information see [Be74].

Function Parameters:

void rke (*x,xe,n,y,der,data,fi,out*)

x: float *;
entry: the independent variable; the initial value *x0*;

xe: float *;
entry: the final value of *x*;

n: int;
entry: the number of equations of the system;

y: float *y[1:n]*;
entry: the dependent variable; the initial values at $x = x0$;
exit: the values of the solution at $x = xe$;

der: void (**der*)(*n,t,v*);
this procedure performs an evaluation of the right hand side of the system with dependent variables *v[1:n]* and independent variable *t*; upon completion of *der* the right hand side should be overwritten on *v[1:n]*;

data: float *data[1:6]*;
in array *data* one should give:
data[1]: the relative tolerance;
data[2]: the absolute tolerance;
after each step *data[3:6]* contains:
data[3]: the steplength used for the last step;
data[4]: the number of integration steps performed;
data[5]: the number of integration steps rejected;
data[6]: the number of integration steps skipped;
if upon completion of *rke* *data[6]* > 0, then results should be considered most critically;

fi: int;
entry: if *fi* is nonzero then the integration starts at *x0* with a trial step *xe - x0*; if *fi* is zero then the integration is continued with a step length
*data[3] * sign(xe - x0)*;

out: void (**out*)(*n,x,xe,y,data*);
after each integration step performed, *out* can be used to obtain information from the solution process, e.g. the values of *x*, *y[1:n]*, and *data[3:6]*; *out* can also be used to update *data*, but *x* and *xe* remain unchanged.

```
#include <math.h>

void rke(float *x, float *xe, int n, float y[],
            void (*der)(int, float, float[]), float data[], int fi,
            void (*out)(int, float, float, float [], float []))
{
    float *allocate_real_vector(int, int);
    void free_real_vector(float *, int);
    int j,last,first,reject,test,ta,tb;
    float xt,h,hmin,ind,hl,ht,absh,fhm,discr,tol,mu,mu1,fh,e1,e2,
            *k0,*k1,*k2,*k3,*k4;

    k0=allocate_real_vector(1,n);
    k1=allocate_real_vector(1,n);
    k2=allocate_real_vector(1,n);
    k3=allocate_real_vector(1,n);
    k4=allocate_real_vector(1,n);
    if (fi) {
        data[3]=(*xe)-(*x);
        data[4]=data[5]=data[6]=0.0;
    }
    absh=h=fabs(data[3]);
    if (*xe < *x) h = -h;
    ind=fabs((*xe)-(*x));
    hmin=ind*data[1]+data[2];
    e1=12.0*data[1]/ind;
    e2=12.0*data[2]/ind;
    first=1;
    reject=0;
    test=1;
    if (fi) {
        last=1;
        test=0;
    }
    while (1) {
        if (test) {
            absh=fabs(h);
            if (absh < hmin) {
                h = (*xe == *x) ? 0.0 : ((*xe > *x) ? hmin : -hmin);
                absh=hmin;
            }
            ta=(h >= (*xe)-(*x));
            tb=(h >= 0.0);
            if ((ta && tb) || (!(ta || tb))) {
                last=1;
                h=(*xe)-(*x);
                absh=fabs(h);
            } else
                last=0;
        }
        test=1;
        if (!reject) {
            for (j=1; j<=n; j++) k0[j]=y[j];
            (*der)(n,*x,k0);
        }
        ht=0.184262134833347*h;
        xt = *x+ht;
        for (j=1; j<=n; j++) k1[j]=k0[j]*ht+y[j];
        (*der)(n,xt,k1);
        ht=0.690983005625053e-1*h;
        xt=4.0*ht+(*x);
        for (j=1; j<=n; j++) k2[j]=(3.0*k1[j]+k0[j])*ht+y[j];
        (*der)(n,xt,k2);
        xt=0.5*h+(*x);
        ht=0.1875*h;
        for (j=1; j<=n; j++)
            k3[j]=((1.74535599249993*k2[j]-k1[j])*2.23606797749979+
                        k0[j])*ht+y[j];
        (*der)(n,xt,k3);
        xt=0.723606797749979*h+(*x);
        ht=0.4*h;
        for (j=1; j<=n; j++)
            k4[j]=(((0.517595468166681*k0[j]-k1[j])*0.927050983124840+
                        k2[j])*1.46352549156242+k3[j])*ht+y[j];
```

```
        (*der)(n,xt,k4);
        xt = (last ? *xe : *x+h);
        ht=2.0*h;
        for (j=1; j<=n; j++)
            k1[j]=((((2.0*k4[j]+k2[j])*0.412022659166595+k1[j])*
                        2.23606797749979-k0[j])*0.375-k3[j])*ht+y[j];
        (*der)(n,xt,k1);
        reject=0;
        fhm=0.0;
        for (j=1; j<=n; j++) {
            discr=fabs((1.6*k3[j]-k2[j]-k4[j])*5.0+k0[j]+k1[j]);
            tol=fabs(k0[j])*e1+e2;
            reject = (discr > tol || reject);
            fh=discr/tol;
            if (fh > fhm) fhm=fh;
        }
        mu=1.0/(1.0+fhm)+0.45;
        if (reject) {
            data[5] += 1.0;
            if (absh <= hmin) {
                data[6] += 1.0;
                hl=h;
                reject=0;
                first=1;
                data[3]=hl;
                data[4] += 1.0;
                *x=xt;
                (*out)(n,*x,*xe,y,data);
                if (*x == *xe) break;
            } else
                h *= mu;
        } else {
            if (first) {
                first=0;
                hl=h;
                h *= mu;
            } else {
                fh=mu*h/hl+mu-mu1;
                hl=h;
                h *= fh;
            }
            mu1=mu;
            ht=hl/12.0;
            for (j=1; j<=n; j++)
                y[j]=((k2[j]+k4[j])*5.0+k0[j]+k1[j])*ht+y[j];
            data[3]=hl;
            data[4] += 1.0;
            *x=xt;
            (*out)(n,*x,*xe,y,data);
            if (*x == *xe) break;
        }
    }
    free_real_vector(k0,1);
    free_real_vector(k1,1);
    free_real_vector(k2,1);
    free_real_vector(k3,1);
    free_real_vector(k4,1);
}
```

C. rk4a

Solves an initial value problem for a single first order ordinary differential equation $dy/dx = f(x,y)$, where $f(x,y)$ may become large, e.g. in the neighbourhood of a singularity, by means of a 5-th order Runge-Kutta method [see Z64]. The equation is assumed to be non-stiff.

rk4a integrates the given differential equation as it stands if $|f(x,y)| \leq 1$. If, however, $|f(x,y)| > 1$ then y is selected as the variable of integration, and the equation actually solved is $dx/dy = 1/f(x,y)$. The procedure is provided with step size and error control. *rk4a* integrates the differential equation from $x = a$ ($y(a)$ being given), and the direction of integration (i.e. x increasing or decreasing, or y increasing or decreasing) specified at outset, until to within stated tolerances a zero of the function $b(x,y)$ is encountered.

To explain the use of the function $b(x,y)$, we remark that if it is desired to continue integration from $x = a$ to $x = c$, then $b(x,y)$ is simply taken to be $x-c$; if it is desired to continue integration until $y(x)$ and a given function $g(x)$ have equal values (so that the graphs of y and g intersect), then $b(x,y)$ is taken to be $y(x) - g(x)$; if it is believed but not known that the graphs of $y(x)$ and $g(x)$ intersect in the range $x = a$ to $x = c$, $b(x,y)$ can be taken to be
$(x-c)(y(x)-g(x))$ and by subsequent inspection of x, y and g upon exit from *rk4a*, it can be decided whether the calculations were terminated because a point of intersection had been found or because the end of the range had been reached. The value of $b(x,y)$ for the current values of x and y upon call of *rk4a* is not inspected: it is thus possible of progress systematically from one zero of b to the next.

Function Parameters:

void rk4a (*x,xa,b,y,ya,fxy,e,d,fi,xdir,pos*)

x: float *;
entry: the independent variable;
exit: upon completion of a call, x is equal to the most recent value of the independent variable;

xa: float;
entry: the initial value of x;

b: float (**b*)(*x,y*);
the equation $b=0$ fulfilled within the tolerances *e[4]* and *e[5]* specifies the end of the integration interval, at the end of each integration step b is evaluated and is tested for change of sign (it delivers the value of $b(x,y)$ described above);

y: float *;
entry: the dependent variable;

ya: float;
entry: the value of y at $x = xa$;

fxy: float (**fxy*)(*x,y*);
fxy gives the value of dy/dx;

e: float *e[0:5]*;
entry:
e[0], *e[2]*: relative tolerances for x and y respectively;
e[1], *e[3]*: absolute tolerances for x and y respectively;
e[4], *e[5]*: tolerances used in the determination of the zero of b;

d: float *d[0:4]*;
After completion of each step we have:
if $d[0]>0$ then x is the integration variable;
if $d[0]<0$ then y is the integration variable;
d[1] is the number of steps skipped;
d[2] is the step size;
d[3] is equal to the last value of x;
d[4] is equal to the last value of y;

fi: int;
entry: if *fi* is nonzero then the integration is started with initial values

$x = xa$, $y = ya$; if *fi* is zero then the integration is started with $x = d[3]$, $y = d[4]$;

xdir,pos: int;

entry: if *fi* is nonzero then the integration starts in such a way that
if *pos* is nonzero and *xdir* is nonzero then x increases,
if *pos* is nonzero and *xdir* is zero then y increases,
if *pos* is zero and *xdir* is nonzero then x decreases,
if *pos* is zero and *xdir* is zero then y decreases.

```
#include <float.h>
#include <math.h>

void rk4a(float *x, float xa, float (*b)(float, float),
          float *y, float ya, float (*fxy)(float, float),
          float e[], float d[], int fi, int xdir, int pos)
{
   void rk4arkstep(float *, float, float, float *, float, float,
            float (*)(float, float), int, int, float *, float *,
            float *, float *, float *,   float *, float *, float);
   int i,iv,first,fir,rej,t,next,ext,extrapolate;
   float k0,k1,k2,k3,k4,k5,fhm,absh,discr,s,xl,cond0,s1,cond1,
         yl,hmin,h,zl,tol,hl,mu,mu1,e1[3],fzero,
         c,fc,bb,fb,a,fa,dd,fd,fdb,fda,w,mb,m,p,q;

   if (fi) {
      d[3]=xa;
      d[4]=ya;
      d[0]=1.0;
   }
   d[1]=0.0;
   *x=xl=d[3];
   *y=yl=d[4];
   iv = d[0] > 0.0;
   first=fir=1;
   hmin=e[0]+e[1];
   h=e[2]+e[3];
   if (h < hmin) hmin=h;
   while (1) {
      zl=(*fxy)(*x,*y);
      if (fabs(zl) <= 1.0) {
         if (!iv) {
            d[2] = h /= zl;
            d[0]=1.0;
            iv=first=1;
         }
         if (fir) {
            t=(((iv && xdir) || (!(iv || xdir))) ? h : h*zl) < 0.0;
            if (fi ? ((t && pos) || (!(t || pos))) : (h*d[2] < 0))
                     h = -h;
         }
         i=1;
      } else {
         if (iv) {
            if (!fir) d[2] = h *= zl;
            d[0] = -1.0;
            iv=0;
            first=1;
         }
         if (fir) {
            h=e[0]+e[1];
            t=(((iv && xdir) || (!(iv || xdir))) ? h : h*zl) < 0.0;
            if (fi ? ((t && pos) || (!(t || pos))) : (h*d[2] < 0))
                     h = -h;
         }
         i=1;
      }
      next=0;
      while (1) {
         absh=fabs(h);
```

```
            if (absh < hmin) {
                h = (h == 0.0) ? 0.0 : ((h > 0.0) ? hmin : -hmin);
                absh=hmin;
            }
            if (iv) {
                rk4arkstep(x,xl,h,y,yl,zl,fxy,i,0,
                        &k0,&k1,&k2,&k3,&k4,&k5,&discr,mu);
                tol=e[2]*fabs(k0)+e[3]*absh;
            } else {
                rk4arkstep(y,yl,h,x,xl,1.0/zl,fxy,i,1,
                        &k0,&k1,&k2,&k3,&k4,&k5,&discr,mu);
                tol=e[0]*fabs(k0)+e[1]*absh;
            }
            rej = discr > tol;
            mu=tol/(tol+discr)+0.45;
            if (!rej) break;
            if (absh <= hmin) {
                if (iv) {
                    *x=xl+h;
                    *y=yl+k0;
                } else {
                    *x=xl+k0;
                    *y=yl+h;
                }
                d[1] += 1.0;
                first=1;
                next=1;
                break;
            }
            h *= mu;
            i=0;
        }
        if (!next) {
            if (first) {
                first=fir;
                hl=h;
                h *= mu;
            } else {
                fhm=mu*h/hl+mu-mu1;
                hl=h;
                h *= fhm;
            }
            if (iv)
                rk4arkstep(x,xl,hl,y,yl,zl,fxy,2,0,
                        &k0,&k1,&k2,&k3,&k4,&k5,&discr,mu);
            else
                rk4arkstep(y,yl,hl,x,xl,zl,fxy,2,1,
                        &k0,&k1,&k2,&k3,&k4,&k5,&discr,mu);
            mu1=mu;
        }
        if (fir) {
            fir=0;
            cond0=(*b)(*x,*y);
            if (!(fi || rej)) h=d[2];
        } else {
            d[2]=h;
            cond1=(*b)(*x,*y);
            if (cond0*cond1 <= 0.0) break;
            cond0=cond1;
        }
        d[3]=xl=(*x);
        d[4]=yl=(*y);
    }
    e1[1]=e[4];
    e1[2]=e[5];
    s1 = iv ? *x : *y;
    s = iv ? xl : yl;
    /* find zero */
    bb=s;
    if (iv) {
        if (s == xl)
            fzero=cond0;
        else
            if (s == s1)
```

```
                fzero=cond1;
            else {
                rk4arkstep(x,xl,s-xl,y,yl,zl,fxy,3,0,
                        &k0,&k1,&k2,&k3,&k4,&k5,&discr,mu);
                fzero=(*b)(*x,*y);
            }
    } else {
        if (s == yl)
            fzero=cond0;
        else
            if (s == s1)
                fzero=cond1;
            else {
                rk4arkstep(y,yl,s-yl,x,xl,zl,fxy,3,1,
                        &k0,&k1,&k2,&k3,&k4,&k5,&discr,mu);
                fzero=(*b)(*x,*y);
            }
    }
    fb=fzero;
    a=s=s1;
    if (iv) {
        if (s == xl)
            fzero=cond0;
        else
            if (s == s1)
                fzero=cond1;
            else {
                rk4arkstep(x,xl,s-xl,y,yl,zl,fxy,3,0,
                        &k0,&k1,&k2,&k3,&k4,&k5,&discr,mu);
                fzero=(*b)(*x,*y);
            }
    } else {
        if (s == yl)
            fzero=cond0;
        else
            if (s == s1)
                fzero=cond1;
            else {
                rk4arkstep(y,yl,s-yl,x,xl,zl,fxy,3,1,
                        &k0,&k1,&k2,&k3,&k4,&k5,&discr,mu);
                fzero=(*b)(*x,*y);
            }
    }
    fa=fzero;
    c=a;
    fc=fa;
    ext=0;
    extrapolate=1;
    while (extrapolate) {
        if (fabs(fc) < fabs(fb)) {
            if (c != a) {
                dd=a;
                fd=fa;
            }
            a=bb;
            fa=fb;
            bb=s=c;
            fb=fc;
            c=a;
            fc=fa;
        }
        tol=fabs(e1[1]*s)+fabs(e1[2]);
        m=(c+bb)*0.5;
        mb=m-bb;
        if (fabs(mb) > tol) {
            if (ext > 2)
                w=mb;
            else {
                if (mb == 0.0)
                    tol=0.0;
                else
                    if (mb < 0.0) tol = -tol;
                p=(bb-a)*fb;
                if (ext <= 1)
```

```
                q=fa-fb;
            else {
                fdb=(fd-fb)/(dd-bb);
                fda=(fd-fa)/(dd-a);
                p *= fda;
                q=fdb*fa-fda*fb;
            }
            if (p < 0.0) {
                p = -p;
                q = -q;
            }
            w=(p<FLT_MIN || p<=q*tol) ? tol : ((p<mb*q) ? p/q : mb);
        }
        dd=a;
        fd=fa;
        a=bb;
        fa=fb;
        s = bb += w;
        if (iv) {
            if (s == xl)
                fzero=cond0;
            else
                if (s == s1)
                    fzero=cond1;
                else {
                    rk4arkstep(x,xl,s-xl,y,yl,zl,fxy,3,0,
                            &k0,&k1,&k2,&k3,&k4,&k5,&discr,mu);
                    fzero=(*b)(*x,*y);
                }
        } else {
            if (s == yl)
                fzero=cond0;
            else
                if (s == s1)
                    fzero=cond1;
                else {
                    rk4arkstep(y,yl,s-yl,x,xl,zl,fxy,3,1,
                            &k0,&k1,&k2,&k3,&k4,&k5,&discr,mu);
                    fzero=(*b)(*x,*y);
                }
        }
        fb=fzero;
        if ((fc >= 0.0) ? (fb >= 0.0) : (fb <= 0.0)) {
            c=a;
            fc=fa;
            ext=0;
        } else
            ext = (w == mb) ? 0 : ext+1;
    } else
        break;
    }
    /* end of finding zero */
    s1 = iv ? *x : *y;
    if (iv)
        rk4arkstep(x,xl,s-xl,y,yl,zl,fxy,3,0,
                &k0,&k1,&k2,&k3,&k4,&k5,&discr,mu);
    else
        rk4arkstep(y,yl,s-yl,x,xl,zl,fxy,3,1,
                &k0,&k1,&k2,&k3,&k4,&k5,&discr,mu);
    d[3]=(*x);
    d[4]=(*y);
}

void rk4arkstep(float *x, float xl, float h, float *y, float yl,
            float zl, float (*fxy)(float, float), int d, int invf,
            float *k0, float *k1, float *k2, float *k3, float *k4,
            float *k5, float *discr, float mu)
{
    /* this function is internally used by RK4A */

    if (d != 2) {
        if (d == 3) {
            *x=xl;
            *y=yl;
```

```
            *k0=(invf ? (1.0/(*fxy)(*x,*y)) : (*fxy)(*x,*y))*h;
        } else
            if (d == 1)
                *k0=zl*h;
            else
                *k0 *= mu;
        *x=xl+h/4.5;
        *y=yl+(*k0)/4.5;
        *k1=(invf ? (1.0/(*fxy)(*x,*y)) : (*fxy)(*x,*y))*h;
        *x=xl+h/3.0;
        *y=yl+((*k0)+(*k1)*3.0)/12.0;
        *k2=(invf ? (1.0/(*fxy)(*x,*y)) : (*fxy)(*x,*y))*h;
        *x=xl+h*0.5;
        *y=yl+((*k0)+(*k2)*3.0)/8.0;
        *k3=(invf ? (1.0/(*fxy)(*x,*y)) : (*fxy)(*x,*y))*h;
        *x=xl+h*0.8;
        *y=yl+((*k0)*53.0-(*k1)*135.0+(*k2)*126.0+(*k3)*56.0)/125.0;
        *k4=(invf ? (1.0/(*fxy)(*x,*y)) : (*fxy)(*x,*y))*h;
        if (d <= 1) {
            *x=xl+h;
            *y=yl+((*k0)*133.0-(*k1)*378.0+(*k2)*276.0+
                    (*k3)*112.0+(*k4)*25.0)/168.0;
            *k5=(invf ? (1.0/(*fxy)(*x,*y)) : (*fxy)(*x,*y))*h;
            *discr=fabs((*k0)*21.0-(*k2)*162.0+(*k3)*224.0-
                        (*k4)*125.0+(*k5)*42.0)/14.0;
            return;
        }
    }
    *x=xl+h;
    *y=yl+(-(*k0)*63.0+(*k1)*189.0-(*k2)*36.0-
            (*k3)*112.0+(*k4)*50.0)/28.0;
    *k5 = (invf ? (1.0/(*fxy)(*x,*y)) : (*fxy)(*x,*y))*h;
    *y=yl+((*k0)*35.0+(*k2)*162.0+(*k4)*125.0+(*k5)*14.0)/336.0;
}
```

D. rk4na

Solves an initial value problem for a system of first order ordinary differential equations $dx_j(x_0)/dx_0 = f_j(x)$, $(j=1,...,n)$ of which the derivative components are supposed to become large, e.g. in the neighbourhood of a singularities, by means of a 5-th order Runge-Kutta method [Z64]. The system is assumed to be non-stiff.

rk4na integrates the given system of differential equations from $x_0 = a$ ($x_j(a)$, $j=1,...,n$ being given) and the direction of integration (i.e. with l specified, $0 \leq l \leq n$, x_l increasing or decreasing) prescribed at outset until, to within stated tolerances, a zero of the function $b(x)$ is encountered. ($b(x)$ is, of course, effectively a function of the single variable x_0).

The role of the function b is similar to that played in the implementation of *rk4a*; now, for example, we may take $b(x)$ to be $x_1 - x_2$, to determine the point at which $x_1(x_0)$ and $x_2(x_0)$ become equal.

At each integration step, the quantities $|dx_j(x_0)/dx_0|$ $(j=0,1,...,n)$ are inspected, and the variable of integration is taken to be $x_{j'}$, where j' is that value of j corresponding to the maximum of these quantities. The system of equations actually solved has the form

$$dx_j/dx_{j'} = f_j(x)/f_{j'}(x) \qquad (j=0,1,...,n;\ j \neq j').$$

Function Parameters:

void rk4na (*x,xa,b,fxj,e,d,fi,n,l,pos*)

x: float *x[0:n]*;
  entry: *x[0]* is the independent variable, *x[1],...,x[n]* are the dependent variables;
  exit: the solution at *b*=0;

xa: float *xa[0:n]*;
  entry: the initial values of *x[0],...,x[n]*;

b: float (**b*)(*n,x*);
b depends on *x[0],...,x[n]*; if the equation *b=0* is satisfied within a certain tolerance (see parameter *e*), the integration is terminated; *b* is evaluated and tested for change of sign at the end of each step;

fxj: float (**fxj*)(*n,j,x*);
fxj depends on *x[0],..,x[n]* and *j*, defining the right hand side of the differential equation; at each call it delivers: *dx[j]/dx[0]*;

e: float *e[0:2n+3]*;
entry: *e[2j]* and *e[2j+1]*, $0 \le j \le n$, are the relative and the absolute tolerance, respectively, associated with *x[j]*; *e[2n+2]* and *e[2n+3]* are the relative and absolute tolerance used in the determination of the zero of *b*;

d: float *d[0:n+3]*;
After completion of each step we have:
d[0] is the number of steps skipped;
d[2] is the step length;
d[j+3] is the last value of *x[j]*, *j*=0,...,*n*;

fi: int;
entry: if *fi* is nonzero then the integration is started with initial condition *x[j]* = *xa[j]*;
if *fi* is zero then the integration is continued with *x[j]* = *d[j+3]*;

n: int;
entry: the number of equations;

l: int;
entry: an integer to be supplied by the user, $0 \le l \le n$ (see *pos*);

pos: int;
entry: if *fi* is nonzero then the integration starts in such a way that *x[l]* increases if *pos* is nonzero and *x[l]* decreases if *pos* is zero;
if *fi* is zero then *pos* is of no significance.

```
#include <float.h>
#include <math.h>

void rk4na(float x[], float xa[], float (*b)(int, float[]),
            float (*fxj)(int, int, float[]), float e[], float d[],
            int fi, int n, int l, int pos)
{
    float *allocate_real_vector(int, int);
    float **allocate_real_matrix(int, int, int, int);
    void free_real_vector(float *, int);
    void free_real_matrix(float **, int, int, int);
    void rk4narkstep(float, int, int, int, float,
                float (*)(int, int, float[]), float [], float [],
                float [], float [], float **);
    int j,i,iv,iv0,fir,first,rej,change,t,next,ext,extrapolate;
    float h,cond0,cond1,fhm,absh,tol,fh,max,x0,x1,s,hmin,hl,mu,mu1,
            p,fzero,*xl,*discr,*y,**k,e1[3],
            c,fc,bb,fb,a,fa,dd,fd,fdb,fda,w,mb,m,q;

    xl=allocate_real_vector(0,n);
    discr=allocate_real_vector(0,n);
    y=allocate_real_vector(0,n);
    k=allocate_real_matrix(0,5,0,n);
    if (fi) {
        for (i=0; i<=n; i++) d[i+3]=xa[i];
        d[0]=d[2]=0.0;
    }
    d[1]=0.0;
    for (i=0; i<=n; i++) x[i]=xl[i]=d[i+3];
    iv=d[0];
    h=d[2];
```

```
    first=fir=1;
    y[0]=1.0;
    next=0;
    change=1;
    while (1) {
        if (!change) {
            while (1) {
                absh=fabs(h);
                if (absh < hmin) {
                    h = (h > 0.0) ? hmin : -hmin;
                    absh=fabs(h);
                }
                rk4narkstep(h,i,n,iv,mu,fxj,x,xl,y,discr,k);
                rej=0;
                fhm=0.0;
                for (i=0; i<=n; i++)
                    if (i != iv) {
                        tol=e[2*i]*fabs(k[0][i])+e[2*i+1]*absh;
                        rej=(tol < discr[i] || rej);
                        fh=discr[i]/tol;
                        if (fh > fhm) fhm=fh;
                    }
                mu=1.0/(1.0+fhm)+0.45;
                if (!rej) break;
                if (absh <= hmin) {
                    for (i=0; i<=n; i++)
                        if (i != iv)
                            x[i]=xl[i]+k[0][i];
                        else
                            x[i]=xl[i]+h;
                    d[1] += 1.0;
                    first=1;
                    next=1;
                    break;
                }
                h *= mu;
                i=0;
            }
            if (!next) {
                if (first) {
                    first=fir;
                    hl=h;
                    h *= mu;
                } else {
                    fh=mu*h/hl+mu-mu1;
                    hl=h;
                    h *= fh;
                }
                rk4narkstep(hl,2,n,iv,mu,fxj,x,xl,y,discr,k);
                mu1=mu;
            }
            next=0;
            if (fir) {
                fir=0;
                cond0=(*b)(n,x);
                if (!(fi || rej)) h=d[2];
            } else {
                d[2]=h;
                cond1=(*b)(n,x);
                if (cond0*cond1 <= 0.0) break;
                cond0=cond1;
            }
            for (i=0; i<=n; i++) d[i+3]=xl[i]=x[i];
        }
        change=0;
        iv0=iv;
        for (j=1; j<=n; j++) y[j]=(*fxj)(n,j,x);
        max=fabs(y[iv]);
        for (i=0; i<=n; i++)
            if (fabs(y[i]) > max) {
                max=fabs(y[i]);
                iv=i;
            }
        if (iv0 != iv) {
```

```
                first=1;
                d[0]=iv;
                d[2]=h=y[iv]/y[iv0]*h;
            }
            x0=xl[iv];
            if (fir) {
                hmin=e[0]+e[1];
                for (i=1; i<=n; i++) {
                    h=e[2*i]+e[2*i+1];
                    if (h < hmin) hmin=h;
                }
                h=e[2*iv]+e[2*iv+1];
                t=y[1]/y[iv]*h < 0.0;
                if ((fi && ((t && pos) || !(t || pos))) ||
                        (!fi && d[2]*h < 0.0)) h = -h;
            }
            i=1;
        }
        e1[1]=e[2*n+2];
        e1[2]=e[2*n+3];
        x1=x[iv];
        s=x0;
        /* find zero */
        bb=s;
        if (s == x0)
            fzero=cond0;
        else
            if (s == x1)
                fzero=cond1;
            else {
                rk4narkstep(s-xl[iv],3,n,iv,mu,fxj,x,xl,y,discr,k);
                fzero=(*b)(n,x);
            }
        fb=fzero;
        a=s=x1;
        if (s == x0)
            fzero=cond0;
        else
            if (s == x1)
                fzero=cond1;
            else {
                rk4narkstep(s-xl[iv],3,n,iv,mu,fxj,x,xl,y,discr,k);
                fzero=(*b)(n,x);
            }
        fa=fzero;
        c=a;
        fc=fa;
        ext=0;
        extrapolate=1;
        while (extrapolate) {
            if (fabs(fc) < fabs(fb)) {
                if (c != a) {
                    dd=a;
                    fd=fa;
                }
                a=bb;
                fa=fb;
                bb=s=c;
                fb=fc;
                c=a;
                fc=fa;
            }
            tol=fabs(e1[1]*s)+fabs(e1[2]);
            m=(c+bb)*0.5;
            mb=m-bb;
            if (fabs(mb) > tol) {
                if (ext > 2)
                    w=mb;
                else {
                    if (mb == 0.0)
                        tol=0.0;
                    else
                        if (mb < 0.0) tol = -tol;
                    p=(bb-a)*fb;
```

```
                if (ext <= 1)
                    q=fa-fb;
                else {
                    fdb=(fd-fb)/(dd-bb);
                    fda=(fd-fa)/(dd-a);
                    p *= fda;
                    q=fdb*fa-fda*fb;
                }
                if (p < 0.0) {
                    p = -p;
                    q = -q;
                }
                w=(p<FLT_MIN || p<=q*tol) ? tol : ((p<mb*q) ? p/q : mb);
            }
            dd=a;
            fd=fa;
            a=bb;
            fa=fb;
            s = bb += w;
            if (s == x0)
                fzero=cond0;
            else
                if (s == x1)
                    fzero=cond1;
                else {
                    rk4narkstep(s-xl[iv],3,n,iv,mu,fxj,x,xl,y,discr,k);
                    fzero=(*b)(n,x);
                }
            fb=fzero;
            if ((fc >= 0.0) ? (fb >= 0.0) : (fb <= 0.0)) {
                c=a;
                fc=fa;
                ext=0;
            } else
                ext = (w == mb) ? 0 : ext+1;
        } else
            break;
    }
    /* end of finding zero */
    x0=s;
    x1=x[iv];
    rk4narkstep(x0-xl[iv],3,n,iv,mu,fxj,x,xl,y,discr,k);
    for (i=0; i<=n; i++) d[i+3]=x[i];
    free_real_vector(xl,0);
    free_real_vector(discr,0);
    free_real_vector(y,0);
    free_real_matrix(k,0,5,0);
}

void rk4narkstep(float h, int d, int n, int iv, float mu,
                float (*fxj)(int, int, float[]), float x[], float xl[],
                float y[], float discr[], float **k)
{
    /* this function is internally used by RK4NA */

    int i,j;
    float p;

    if (d != 2) {
        if (d == 3) {
            for (i=0; i<=n; i++) x[i]=xl[i];
            for (j=1; j<=n; j++) y[j]=(*fxj)(n,j,x);
            p=h/y[iv];
            for (i=0; i<=n; i++)
                if (i != iv) k[0][i]=y[i]*p;
        } else
            if (d == 1) {
                p=h/y[iv];
                for (i=0; i<=n; i++)
                    if (i != iv) k[0][i]=p*y[i];
            } else
                for (i=0; i<=n; i++)
                    if (i != iv) k[0][i] *= mu;
        for (i=0; i<=n; i++)
```

```
            x[i]=xl[i]+((i == iv) ? h : k[0][i])/4.5;
        for (j=1; j<=n; j++) y[j]=(*fxj)(n,j,x);
        p=h/y[iv];
        for (i=0; i<=n; i++)
            if (i != iv) k[1][i]=y[i]*p;
        for (i=0; i<=n; i++)
            x[i]=xl[i]+
                    ((i == iv) ? h*4.0 : (k[0][i]+k[1][i]*3.0))/12.0;
        for (j=1; j<=n; j++) y[j]=(*fxj)(n,j,x);
        p=h/y[iv];
        for (i=0; i<=n; i++)
            if (i != iv) k[2][i]=y[i]*p;
        for (i=0; i<=n; i++)
            x[i]=xl[i]+
                    ((i == iv) ? h*0.5 : (k[0][i]+k[2][i]*3.0)/8.0);
        for (j=1; j<=n; j++) y[j]=(*fxj)(n,j,x);
        p=h/y[iv];
        for (i=0; i<=n; i++)
            if (i != iv) k[3][i]=y[i]*p;
        for (i=0; i<=n; i++)
            x[i]=xl[i]+((i == iv) ? h*0.8 : (k[0][i]*53.0-
                k[1][i]*135.0+k[2][i]*126.0+k[3][i]*56.0)/125.0);
        for (j=1; j<=n; j++) y[j]=(*fxj)(n,j,x);
        p=h/y[iv];
        for (i=0; i<=n; i++)
            if (i != iv) k[4][i]=y[i]*p;
        if (d <= 1) {
            for (i=0; i<=n; i++)
                x[i]=xl[i]+((i == iv) ? h : (k[0][i]*133.0-
                    k[1][i]*378.0+k[2][i]*276.0+k[3][i]*112.0+
                    k[4][i]*25.0)/168.0);
            for (j=1; j<=n; j++) y[j]=(*fxj)(n,j,x);
            p=h/y[iv];
            for (i=0; i<=n; i++)
                if (i != iv) k[5][i]=y[i]*p;
            for (i=0; i<=n; i++)
                if (i != iv)
                    discr[i]=fabs(k[0][i]*21.0-k[2][i]*162.0+
                        k[3][i]*224.0-k[4][i]*125.0+k[5][i]*42.0)/14.0;
            return;
        }
    }
    for (i=0; i<=n; i++)
        x[i]=xl[i]+((i == iv) ? h :
            (-k[0][i]*63.0+k[1][i]*189.0-k[2][i]*36.0-k[3][i]*112.0+
                k[4][i]*50.0)/28.0);
    for (j=1; j<=n; j++) y[j]=(*fxj)(n,j,x);
    p=h/y[iv];
    for (i=0; i<=n; i++)
        if (i != iv) k[5][i]=y[i]*p;
    for (i=0; i<=n; i++)
        if (i != iv)
            x[i]=xl[i]+(k[0][i]*35.0+k[2][i]*162.0+k[4][i]*125.0+
                            k[5][i]*14.0)/336.0;
}
```

E. rk5na

Solves an initial value problem for a system of first order ordinary differential equations

$$dx_j/dx_0 = f_j(x)/f_0(x), \quad (j=1,2,...,n)$$

of which the derivative components are supposed to become large, e.g. in the neighbourhood of a singularities, by means of a 5-th order Runge-Kutta method [Z64]. The system is assumed to be non-stiff.

rk5na integrates the given system of differential equations from $x_0 = a$ ($x_j(a)$, j=1,...,n, being given) and the direction of integration (i.e. with l specified, $0 \le l \le n$, x_l increasing or decreasing) prescribed at outset until, to within stated tolerances, a zero of the function $b(x)$ is encountered.

The arc length s is used as the variable of integration. Since $ds^2 = dx_0^2 + ... + dx_n^2$, it follows that $ds = (dx_0/f_0(x))a$, where

$$a = \sqrt{\sum_{j=0}^{n} f_j(x)^2} \ .$$

The system of equations actually solved is

$$dx_j/dx_0 = f_j(x)/a \quad (j=0,1,...,n).$$

Thus, points at which $f_0(x) = 0$ and dx_j/dx_0 may become infinite cause no special trouble.

The role of the function b is similar to that played by the same function in the implementation of *rk4na*.

Function Parameters:

void rk5na (*x,xa,b,fxj,e,d,fi,n,l,pos*)

x: float *x[0:n]*;
entry: the dependent variables;

xa: float *xa[0:n]*;
entry: the initial values of *x[0],...,x[n]*;

b: float (*b)(*n,x*);
b depends on *x[0],...,x[n]*; if, within some tolerance, *b=0* then the integration is terminated; see parameter *e*;

fxj: float (*fxj)(*n,j,x*);
fxj depends on *x[0],..,x[n]* and *j*, giving the value of *dx[j]/dx[0]*;

e: float *e[0:2n+3]*;
entry: *e[2j]* and *e[2j+1]*, $0 \leq j \leq n$, are the relative and the absolute tolerance, respectively, associated with *x[j]*; while *e[2n+2]* and *e[2n+3]* are the relative and absolute tolerance used in the determination of the zero of *b*;

d: float *d[1:n+3]*;
After completion of each step we have:
|*d[1]*| is the arc length;
d[2] is the step length;
d[j+3] is the latest value of *x[j]*, *j*=0,...,*n*;

fi: int;
entry: if *fi* is nonzero then the integration is started with initial conditions *x[j]=xa[j]*, *j*=0,...,*n*; if *fi* is zero then the integration is continued with *x[j]=d[j+3]*;

n: int;
entry: the number of equations;

l: int;
entry: an integer to be supplied by the user, $1 \leq l \leq n$ (see *pos*);

pos: int;
entry: if *fi* is nonzero then the integration starts in such a way that *x[l]* increases if *pos* is nonzero and *x[l]* decreases if *pos* is zero;
if *fi* is zero then *pos* is ignored.

```
#include <float.h>
#include <math.h>

void rk5na(float x[], float xa[], float (*b)(int, float[]),
            float (*fxj)(int, int, float[]), float e[], float d[],
            int fi, int n, int l, int pos)
```

```
{
    float *allocate_real_vector(int, int);
    float **allocate_real_matrix(int, int, int, int);
    void free_real_vector(float *, int);
    void free_real_matrix(float **, int, int, int);
    void rk5narkstep(float, int, int, float,
                float (*)(int, int, float[]), float [], float [],
                float [], float [], float **);
    int j,i,first,fir,rej,t,ext,extrapolate;
    float fhm,s,s0,cond0,s1,cond1,h,absh,tol,fh,hl,mu,mu1,
            fzero,*y,*xl,*discr,**k,e1[3],
            c,fc,bb,fb,a,fa,dd,fd,fdb,fda,w,mb,m,p,q;

    y=allocate_real_vector(0,n);
    xl=allocate_real_vector(0,n);
    discr=allocate_real_vector(0,n);
    k=allocate_real_matrix(0,5,0,n);
    if (fi) {
        for (i=0; i<=n; i++) d[i+3]=xa[i];
        d[1]=d[2]=0.0;
    }
    for (i=0; i<=n; i++) x[i]=xl[i]=d[i+3];
    s=d[1];
    first=fir=1;
    h=e[0]+e[1];
    for (i=1; i<=n; i++) {
        absh=e[2*i]+e[2*i+1];
        if (h > absh) h=absh;
    }
    if (fi) {
        j=1;
        t=(*fxj)(n,j,x)*h < 0.0;
        if ((t && pos) || !(t || pos)) h = -h;
    } else
        if (d[2]*h < 0.0) h = -h;
    i=0;
    while (1) {
        rk5narkstep(h,i,n,mu,fxj,x,xl,y,discr,k);
        rej=0;
        fhm=0.0;
        absh=fabs(h);
        for (i=0; i<=n; i++) {
            tol=e[2*i]*fabs(k[0][i])+e[2*i+1]*absh;
            rej=(tol < discr[i] || rej);
            fh=discr[i]/tol;
            if (fh > fhm) fhm=fh;
        }
        mu=1.0/(1.0+fhm)+0.45;
        if (rej) {
            h *= mu;
            i=1;
        } else {
            if (first) {
                first=fir;
                hl=h;
                h *= mu;
            } else {
                fh=mu*h/hl+mu-mu1;
                hl=h;
                h *= fh;
            }
            rk5narkstep(hl,2,n,mu,fxj,x,xl,y,discr,k);
            mu1=mu;
            s += hl;
            if (fir) {
                cond0=(*b)(n,x);
                fir=0;
                if (!fi) h=d[2];
            } else {
                d[2]=h;
                cond1=(*b)(n,x);
                if (cond0*cond1 <= 0.0) break;
                cond0=cond1;
            }
```

```
            for (i=0; i<=n; i++) d[i+3]=xl[i]=x[i];
            d[1]=s0=s;
            i=0;
        }
    }
    e1[1]=e[2*n+2];
    e1[2]=e[2*n+3];
    s1=s;
    s=s0;
    /* find zero */
    bb=s;
    if (s == s0)
        fzero=cond0;
    else
        if (s == s1)
            fzero=cond1;
        else {
            rk5narkstep(s-s0,3,n,mu,fxj,x,xl,y,discr,k);
            fzero=(*b)(n,x);
        }
    fb=fzero;
    a=s=s1;
    if (s == s0)
        fzero=cond0;
    else
        if (s == s1)
            fzero=cond1;
        else {
            rk5narkstep(s-s0,3,n,mu,fxj,x,xl,y,discr,k);
            fzero=(*b)(n,x);
        }
    fa=fzero;
    c=a;
    fc=fa;
    ext=0;
    extrapolate=1;
    while (extrapolate) {
        if (fabs(fc) < fabs(fb)) {
            if (c != a) {
                dd=a;
                fd=fa;
            }
            a=bb;
            fa=fb;
            bb=s=c;
            fb=fc;
            c=a;
            fc=fa;
        }
        tol=fabs(e1[1]*s)+fabs(e1[2]);
        m=(c+bb)*0.5;
        mb=m-bb;
        if (fabs(mb) > tol) {
            if (ext > 2)
                w=mb;
            else {
                if (mb == 0.0)
                    tol=0.0;
                else
                    if (mb < 0.0) tol = -tol;
                p=(bb-a)*fb;
                if (ext <= 1)
                    q=fa-fb;
                else {
                    fdb=(fd-fb)/(dd-bb);
                    fda=(fd-fa)/(dd-a);
                    p *= fda;
                    q=fdb*fa-fda*fb;
                }
                if (p < 0.0) {
                    p = -p;
                    q = -q;
                }
                w=(p<FLT_MIN || p<=q*tol) ? tol : ((p<mb*q) ? p/q : mb);
```

```
            }
            dd=a;
            fd=fa;
            a=bb;
            fa=fb;
            s = bb += w;
            if (s == s0)
                fzero=cond0;
            else
                if (s == s1)
                    fzero=cond1;
                else {
                    rk5narkstep(s-s0,3,n,mu,fxj,x,xl,y,discr,k);
                    fzero=(*b)(n,x);
                }
            fb=fzero;
            if ((fc >= 0.0) ? (fb >= 0.0) : (fb <= 0.0)) {
                c=a;
                fc=fa;
                ext=0;
            } else
                ext = (w == mb) ? 0 : ext+1;
        } else
            break;
    }
    /* end of finding zero */
    rk5narkstep(s-s0,3,n,mu,fxj,x,xl,y,discr,k);
    for (i=0; i<=n; i++) d[i+3]=x[i];
    d[1]=s;
    free_real_vector(y,0);
    free_real_vector(xl,0);
    free_real_vector(discr,0);
    free_real_matrix(k,0,5,0);
}

void rk5narkstep(float h, int d, int n, float mu,
                float (*fxj)(int, int, float[]), float x[], float xl[],
                float y[], float discr[], float **k)
{
    /* this function is internally used by RK5NA */

    int i,j;
    float p,s;

    if (d != 2) {
        if (d == 1)
            for (i=0; i<=n; i++) k[0][i] *= mu;
        else {
            for (i=0; i<=n; i++) x[i]=xl[i];
            for (j=0; j<=n; j++) y[j]=(*fxj)(n,j,x);
            s=0.0;
            for (j=0; j<=n; j++) s += y[j]*y[j];
            p=h/sqrt(s);
            for (i=0; i<=n; i++) k[0][i]=y[i]*p;
        }
        for (i=0; i<=n; i++)   x[i]=xl[i]+k[0][i]/4.5;
        for (j=0; j<=n; j++) y[j]=(*fxj)(n,j,x);
        s=0.0;
        for (j=0; j<=n; j++) s += y[j]*y[j];
        p=h/sqrt(s);
        for (i=0; i<=n; i++) k[1][i]=y[i]*p;
        for (i=0; i<=n; i++)   x[i]=xl[i]+(k[0][i]+k[1][i]*3.0)/12.0;
        for (j=0; j<=n; j++) y[j]=(*fxj)(n,j,x);
        s=0.0;
        for (j=0; j<=n; j++) s += y[j]*y[j];
        p=h/sqrt(s);
        for (i=0; i<=n; i++) k[2][i]=y[i]*p;
        for (i=0; i<=n; i++)   x[i]=xl[i]+(k[0][i]+k[2][i]*3.0)/8.0;
        for (j=0; j<=n; j++) y[j]=(*fxj)(n,j,x);
        s=0.0;
        for (j=0; j<=n; j++) s += y[j]*y[j];
        p=h/sqrt(s);
        for (i=0; i<=n; i++) k[3][i]=y[i]*p;
        for (i=0; i<=n; i++)
```

```
          x[i]=xl[i]+(k[0][i]*53.0-k[1][i]*135.0+k[2][i]*126.0+
                              k[3][i]*56.0)/125.0;
      for (j=0; j<=n; j++) y[j]=(*fxj)(n,j,x);
      s=0.0;
      for (j=0; j<=n; j++) s += y[j]*y[j];
      p=h/sqrt(s);
      for (i=0; i<=n; i++) k[4][i]=y[i]*p;
      if (d <= 1) {
          for (i=0; i<=n; i++)
              x[i]=xl[i]+(k[0][i]*133.0-k[1][i]*378.0+k[2][i]*276.0+
                              k[3][i]*112.0+k[4][i]*25.0)/168.0;
          for (j=0; j<=n; j++) y[j]=(*fxj)(n,j,x);
          s=0.0;
          for (j=0; j<=n; j++) s += y[j]*y[j];
          p=h/sqrt(s);
          for (i=0; i<=n; i++) k[5][i]=y[i]*p;
          for (i=0; i<=n; i++)
              discr[i]=fabs(k[0][i]*21.0-k[2][i]*162.0+
                        k[3][i]*224.0-k[4][i]*125.0+k[5][i]*42.0)/14.0;
          return;
      }
  }
  for (i=0; i<=n; i++)
      x[i]=xl[i]+(-k[0][i]*63.0+k[1][i]*189.0-k[2][i]*36.0-
                        k[3][i]*112.0+k[4][i]*50.0)/28.0;
  for (j=0; j<=n; j++) y[j]=(*fxj)(n,j,x);
  s=0.0;
  for (j=0; j<=n; j++) s += y[j]*y[j];
  p=h/sqrt(s);
  for (i=0; i<=n; i++) k[5][i]=y[i]*p;
  for (i=0; i<=n; i++)
      x[i]=xl[i]+(k[0][i]*35.0+k[2][i]*162.0+k[4][i]*125.0+
                        k[5][i]*14.0)/336.0;
}
```

F. multistep

Solves an initial value problem for a system of first order ordinary differential equations

$$dy_i(x)/dx = f_i(x,y(x)) \qquad (i=1,2,...,n)$$

from $x = x_0$ to $x = x_{end}$ ($x_0 < x_{end}$), $y(x_0)$ being given, based on two linear multistep methods. For stiff problems it uses the backward differentiation method, and for non-stiff problems the Adams-Bashforth-Moulton method [He71].

During the implementation of a typical stage of the variant of the Adams-Moulton method employed, (a) approximations $f_i^{(p+\nu)}$ to $f_i(x,y(x))$ at $x = x_{p+\nu}$ ($\nu=0,...,k$; $i=1,...,n$) are available, together with (b) approximations $y_i^{(p+k,\nu)}$ to $d^\nu y_i(x)/dx^\nu$ at $x = x_{p+k}$ ($\nu=0,...,k$; $i=1,...,n$), and (c) approximations to the derivatives $d^\nu y_i(x)/dx^\nu$ ($\nu=0,...,k$; $i=1,...,n$) at $x=x_{p+k+1}$ are determined by prediction and iterated correction. If the order k is preserved then the $f_i^{(p)}$ are discarded, and the process is repeated.

At a typical stage in the implementation of the variant of the Curtiss-Hirschfelder method employed, approximations $y_i^{(p+\nu)}$ ($\nu=0,...,k$; $i=1,...,n$) to $y_i(x)$ at $x = x_{p+\nu}$ ($\nu=0,...,k$; $i=1,...,n$) are available, together with the derivative approximations (b) above, and the derivative approximations (c) above are also determined by prediction and iterated correction.

The above process of iterated correction may, if necessary, be accelerated by use of the Jacobian matrix whose elements are given by $\partial f_i(x,y(x))/\partial y_j(x)$ ($i,j=1,...,n$).

The Adam-Moulton method is a relatively rapid process for high-accuracy computation of the solution of systems of differential equations with slowly varying dependent variables. The Curtiss-Hirschfelder method is suitable for the solution of stiff systems, whose solution vector contains one or more rapidly varying components. *multistep* uses a steplength and order control mechanism due to Gear (the steplength is $x_{p+k+1} - x_{p+k}$ above; k is the order).

If so requested, *multistep* begins by attempting an unaccelerated Adams-Moulton method; if this is too time-consuming, acceleration using a Jacobian matrix is embarked upon; if this also fails, *multistep* switches to Gear's implementation of the Curtiss-Hirschfelder method, (after the determination of a Jacobian matrix at the commencement of one step, the same matrix is used to accelerate correction over as many subsequent steps as possible). However, it is possible to request *multistep* to use Gear's method at outset. Tactical considerations concerning steplength and order are reversible; strategic decisions concerning the method are not; once *multistep* has embarked upon Gear's method, it continues with it for the duration of the call.

Function Parameters:

int multistep (*x,xend,y,hmin,hmax,yamx,eps,first,save,deriv,available,jacobian,stiff,n,out*)

multistep: if difficulties are encountered during the integration (i.e. *save[-1]* $\neq$ 0 or *save[-2]* $\neq$ 0) then *multistep* is set to zero, otherwise *multistep* is set to nonzero;

x: float *;
- entry: the independent variable; the initial value *x0*;
- exit: the final value *xend*;

xend: float;
- entry: the final value of *x* (*xend* $\geq$ *x*);

y: float *y[1:6n]*;
- entry: the dependent variable; *y[1:n]* are the initial values of the solution of the system of differential equations at *x* = *x0*;
- exit: *y[1:n]* are the final values of the solution at *x* = *xend*;

hmin,hmax: float;
- entry: the minimum and maximum steplength allowed, respectively;

ymax: float *ymax[1:n]*;
- entry: the absolute local error bound divided by *eps*;
- exit: *ymax[i]* gives the maximal value of the entry value of *ymax[i]* and the values of |*y[i]*| during integration;

eps: float;
- entry: the relative local error bound;

first: int *;

if *first* is nonzero then the procedure starts the integration with a first order Adams method and a steplength equal to *hmin*, upon completion of a call, *first* is set to zero; if *first* is zero then the procedure continues integration;

save: float *save[-38:6n]*;

in this array the procedure stores information which can be used in a continuing call with *first* = zero; also the following messages are delivered:
- *save[0]*=0: an Adams method has been used;
- *save[0]*=1: the procedure switched to Gear's method;
- *save[-1]*=0: no error message;
- *save[-1]*=1: with the *hmin* specified the procedure cannot handle the nonlinearity (decrease *hmin*!);
- *save[-2]*: number of times that the requested local error bound was exceeded;
- *save[-3]*: if *save[-2]* is nonzero then *save[-3]* gives an estimate of the maximal local error bound, otherwise *save[-3]*=0;

deriv: void (**deriv*)(*df,n,x,y*); float *df[]*, int *n*, float *x*, float *y[]*;

this procedure should deliver *dy[i]/dx* in *df[i]*;

available: int (**available*)(*n,x,y,jacobian*);
if an analytic expression of the Jacobian matrix is not available the *available* is set to zero, otherwise *available* is set to nonzero and the evaluation of this procedure must have the following side-effect: the entries of the Jacobian matrix *d(dy[i]/dx)/dy[j]* are delivered in the array elements *jacobian[i,j]*;

jacobian: float *jacobian[1:n,1:n]*;
at each evaluation of the user supplied function *available* with the result *available* equals nonzero, the Jacobian matrix has to be assigned to this array;

stiff: int;
entry: if *stiff* equals nonzero then the procedure skips an attempt to solve the problem with Adams-Bashforth or Adams-Moulton methods, directly using Gear's method;

n: int;
entry: the number of equations;

out: void (**out*)(*h,k,n,x,y*); float *h*, int *k*, int *n*, float *x*, float *y[]*;
at the end of each accepted step of integration process this procedure is called, the last steplength used, *h*, and the order of the method, *k*, are delivered;
at each call of the procedure *out*, the current values of the independent variable *x* and of the solution *y[i](x)* are available for use; moreover, in the neighbourhood of the current value of *x*, any value of *y[i](x')* can be computed by means of the following interpolation formula

$$y[i](x') = \sum_{j=0}^{k} y[i+j*n]\left(\frac{x'-x}{h}\right)^j .$$

Functions used: matvec, dec, sol.

```
#include <float.h>
#include <math.h>

int multistep(float *x, float xend, float y[], float hmin, float hmax,
              float ymax[], float eps, int *first, float save[],
              void (*deriv)(float [], int, float, float[]),
              int (*available)(int, float, float [], float **),
              float **jacobian, int stiff, int n,
              void (*out)(float, int, int, float, float []))
{
    int *allocate_integer_vector(int, int);
    float *allocate_real_vector(int, int);
    float **allocate_real_matrix(int, int, int, int);
    void free_integer_vector(int *, int);
    void free_real_vector(float *, int);
    void free_real_matrix(float **, int, int, int);
    float matvec(int, int, int, float **, float []);
    void dec(float **, int, float [], int []);
    void sol(float **, int, int [], float []);
    void multistepreset(float [], float [], float *, float *,
           float *, float *, int *, float, float,
           float, float, int, int, int);
    void multisteporder(float [], float [], float *, float *,
              float *, float *, float *, int *, float, int, int);
    void multistepstep(int *, float *, float, float, float,
                 float [], float,  float [], float [], float [],
                 int, int, int, int);
    void multistepjacobian(int, float, float [], float,
```

```
        float [], float [], float [], float **,
        void (*)(float [], int, float, float[]),
        int (*)(int, float, float [], float **),
        int *, int *, int *);
static float adams1[35] = {1.0, 1.0, 144.0, 4.0, 0.0, 0.5, 1.0,
    0.5, 576.0, 144.0, 1.0, 5.0/12.0, 1.0, 0.75, 1.0/6.0, 1436.0,
    576.0, 4.0, 0.375, 1.0, 11.0/12.0, 1.0/3.0, 1.0/24.0, 2844.0,
    1436.0, 1.0, 251.0/720.0, 1.0, 25.0/24.0, 35.0/72.0, 5.0/48.0,
    1.0/120.0, 0.0, 2844.0, 0.1};
static float adams2[35] = {1.0, 1.0, 9.0, 4.0, 0.0, 2.0/3.0, 1.0,
    1.0/3.0, 36.0, 20.25, 1.0, 6.0/11.0, 1.0, 6.0/11.0, 1.0/11.0,
    84.028, 53.778, 0.25, 0.48, 1.0, 0.7, 0.2, 0.02, 156.25, 108.51,
    0.027778, 120.0/274.0, 1.0, 225.0/274.0, 85.0/274.0, 15.0/274.0,
    1.0/274.0, 0.0, 187.69, 0.0047361};
static int adams,withjacobian,m,same,kold;
static float xold,hold,a0,tolup,tol,toldwn,tolconv;
int evaluate,evaluated,decompose,decomposed,conv,i,j,l,k,knew,
        fails,*p;
float h,ch,chnew,error,dfi,c,a[6],*delta,*lastdelta,*df,**jac,
        aux[4],*fixy,*fixdy,*dy,ss,aa;

p=allocate_integer_vector(1,n);
delta=allocate_real_vector(1,n);
lastdelta=allocate_real_vector(1,n);
df=allocate_real_vector(1,n);
fixy=allocate_real_vector(1,n);
fixdy=allocate_real_vector(1,n);
dy=allocate_real_vector(1,n);
jac=allocate_real_matrix(1,n,1,n);

if (*first) {
    *first = 0;
    m=n;
    save[-1]=save[-2]=save[-3]=0.0;
    (*out)(0.0,0,n,*x,y);
    adams=(!stiff);
    withjacobian=(!adams);
    if (withjacobian)
        multistepjacobian(n,*x,y,eps,fixy,fixdy,dy,jacobian,deriv,
                        available,&evaluate,&decompose,&evaluated);
    if (adams)
        for (j=0; j<=34; j++) save[j-38]=adams1[j];
    else
        for (j=0; j<=34; j++) save[j-38]=adams2[j];
    NewStart:
    k=1;
    same=2;
    multisteporder(a,save,&tolup,&tol,&toldwn,&tolconv,
                        &a0,&decompose,eps,k,n);
    (*deriv)(df,n,*x,y);
    if (!withjacobian)
        h=hmin;
    else {
        ss=FLT_MIN;
        for (i=1; i<=n; i++) {
            aa=matvec(1,n,i,jacobian,df)/ymax[i];
            ss += aa*aa;
        }
        h=sqrt(2.0*eps/sqrt(ss));
    }
    if (h > hmax)
        h=hmax;
    else
        if (h < hmin) h=hmin;
    xold=(*x);
    hold=h;
    kold=k;
    ch=1.0;
    for (i=1; i<=n; i++) {
        save[i]=y[i];
        save[m+i]=y[m+i]=df[i]*h;
    }
    (*out)(0.0,0,n,*x,y);
} else {
```

```
        withjacobian=(!adams);
        ch=1.0;
        k=kold;
        multistepreset(y,save,x,&ch,&c,&h,&decomposed,hmin,hmax,
                       hold,xold,m,k,n);
        multisteporder(a,save,&tolup,&tol,&toldwn,&tolconv,
                       &a0,&decompose,eps,k,n);
        decompose=withjacobian;
    }
    fails=0;
    while (*x < xend) {
        if ((*x)+h <= xend)
            *x += h;
        else {
            h=xend-(*x);
            *x = xend;
            ch=h/hold;
            c=1.0;
            for (j=m; j<=k*m; j+=m) {
                c *= ch;
                for (i=j+1; i<=j+n; i++) y[i] *= c;
            }
            same = ((same < 3) ? 3 : same+1);
        }
        /* prediction */
        for (l=1; l<=n; l++) {
            for (i=l; i<=(k-1)*m+l; i+=m)
                for (j=(k-1)*m+l; j>=i; j-=m) y[j] += y[j+m];
            delta[l]=0.0;
        }
        evaluated=0;
        /* correction and estimation local error */
        for (l=1; l<=3; l++) {
            (*deriv)(df,n,*x,y);
            for (i=1; i<=n; i++) df[i]=df[i]*h-y[m+i];
            if (withjacobian) {
                if (evaluate)
                    multistepjacobian(n,*x,y,eps,fixy,fixdy,dy,jacobian,
                        deriv,available,&evaluate,&decompose,&evaluated);
                if (decompose) {
                    /* decompose jacobian */
                    decompose=0;
                    decomposed=1;
                    c = -a0*h;
                    for (j=1; j<=n; j++) {
                        for (i=1; i<=n; i++) jac[i][j]=jacobian[i][j]*c;
                        jac[j][j] += 1.0;
                    }
                    aux[2]=FLT_EPSILON;
                    dec(jac,n,aux,p);
                }
                sol(jac,n,p,df);
            }
            conv=1;
            for (i=1; i<=n; i++) {
                dfi=df[i];
                y[i] += a0*dfi;
                y[m+i] += dfi;
                delta[i] += dfi;
                conv=(conv && (fabs(dfi) < tolconv*ymax[i]));
            }
            if (conv) {
                ss=FLT_MIN;
                for (i=1; i<=n; i++) {
                    aa=delta[i]/ymax[i];
                    ss += aa*aa;
                }
                error=ss;
                break;
            }
        }
        /* acceptance or rejection */
        if (!conv) {
            if (!withjacobian) {
```

```
            evaluate=withjacobian=((same >= k) || (h < 1.1*hmin));
            if (!withjacobian) ch /= 4.0;
        }
        else if (!decomposed) decompose=1;
        else if (!evaluated) evaluate=1;
        else if (h > 1.1*hmin) ch /= 4.0;
        else if (adams) goto TryCurtiss;
        else {
            save[-1]=1.0;
            break;
        }
        multistepreset(y,save,x,&ch,&c,&h,&decomposed,hmin,hmax,
                        hold,xold,m,k,n);
    } else
        if (error > tol) {
            fails++;
            if (h > 1.1*hmin) {
                if (fails > 2) {
                    if (adams) {
                        adams=0;
                        for (j=0; j<=34; j++) save[j-38]=adams2[j];
                    }
                    kold=0;
                    multistepreset(y,save,x,&ch,&c,&h,&decomposed,
                                    hmin,hmax,hold,xold,m,k,n);
                    goto NewStart;
                } else {
                    multistepstep(&knew,&chnew,tolup,tol,toldwn,delta,
                                error,lastdelta,y,ymax,fails,m,k,n);
                    if (knew != k) {
                        k=knew;
                        multisteporder(a,save,&tolup,&tol,&toldwn,
                                    &tolconv,&a0,&decompose,eps,k,n);
                    }
                    ch *= chnew;
                    multistepreset(y,save,x,&ch,&c,&h,&decomposed,
                                    hmin,hmax,hold,xold,m,k,n);
                }
            } else {
                if (adams) {
                    TryCurtiss:
                    adams=0;
                    for (j=0; j<=34; j++) save[j-38]=adams2[j];
                } else
                    if (k == 1) {
                        /* violate eps criterion */
                        c=eps*sqrt(error/tol);
                        if (c > save[-3]) save[-3]=c;
                        save[-2] += 1.0;
                        same=4;
                        goto ErrorTestOk;
                    }
                k=kold=1;
                multistepreset(y,save,x,&ch,&c,&h,&decomposed,
                                hmin,hmax,hold,xold,m,k,n);
                multisteporder(a,save,&tolup,&tol,&toldwn,
                                &tolconv,&a0,&decompose,eps,k,n);
                same=2;
            }
        } else {
            ErrorTestOk:
            fails=0;
            for (i=1; i<=n; i++) {
                c=delta[i];
                for (l=2; l<=k; l++) y[l*m+i] += a[l]*c;
                if (fabs(y[i]) > ymax[i]) ymax[i]=fabs(y[i]);
            }
            same--;
            if (same == 1)
                for (i=1; i<=n; i++) lastdelta[i]=delta[i];
            else
                if (same == 0) {
                    multistepstep(&knew,&chnew,tolup,tol,toldwn,delta,
                                error,lastdelta,y,ymax,fails,m,k,n);
```

```
                    if (chnew > 1.1) {
                        decomposed=0;
                        if (k != knew) {
                            if (knew > k)
                                for (i=1; i<=n; i++)
                                    y[knew*m+i]=delta[i]*a[k]/knew;
                            k=knew;
                            multisteporder(a,save,&tolup,&tol,&toldwn,
                                        &tolconv,&a0,&decompose,eps,k,n);
                        }
                        same=k+1;
                        if (chnew*h > hmax) chnew=hmax/h;
                        h *= chnew;
                        c=1.0;
                        for (j=m; j<=k*m; j+=m) {
                            c *= chnew;
                            for (i=j+1; i<=j+n; i++) y[i] *= c;
                        }
                    } else
                        same=10;
                }
            if (*x != xend) {
                xold=(*x);
                hold=h;
                kold=k;
                ch=1.0;
                for (i=k*m+n; i>=1; i--) save[i]=y[i];
                (*out)(h,k,n,*x,y);
            }
        }
    }
    save[0]=(adams ? 0.0 : 1.0);
    free_integer_vector(p,1);
    free_real_vector(delta,1);
    free_real_vector(lastdelta,1);
    free_real_vector(df,1);
    free_real_vector(fixy,1);
    free_real_vector(fixdy,1);
    free_real_vector(dy,1);
    free_real_matrix(jac,1,n,1);
    return ((save[-1] == 0.0) && (save[-2] == 0.0));
}

void multistepreset(float y[], float save[], float *x, float *ch,
            float *c, float *h, int *decomposed, float hmin, float hmax,
            float hold, float xold, int m, int k, int n)
{
    /* this function is internally used by MULTISTEP */

    int i,j;

    if (*ch < hmin/hold)
        *ch = hmin/hold;
    else
        if (*ch > hmax/hold) *ch = hmax/hold;
    *x = xold;
    *h = hold*(*ch);
    *c = 1.0;
    for (j=0; j<=k*m; j+=m) {
        for (i=1; i<=n; i++) y[j+i]=save[j+i]*(*c);
        (*c) *= (*ch);
    }
    *decomposed = 0;
}

void multisteporder(float a[], float save[], float *tolup,
            float *tol, float *toldwn, float *tolconv, float *a0,
            int *decompose, float eps, int k, int n)
{
    /* this function is internally used by MULTISTEP */

    int i,j;
    float c;
```

```
    c=eps*eps;
    j=(k-1)*(k+8)/2-38;
    for (i=0; i<=k; i++) a[i]=save[i+j];
    *tolup = c*save[j+k+1];
    *tol = c*save[j+k+2];
    *toldwn = c*save[j+k+3];
    *tolconv = eps/(2*n*(k+2));
    *a0 = a[0];
    *decompose = 1;
}

void multistepstep(int *knew, float *chnew, float tolup, float tol,
                   float toldwn, float delta[], float error,
                   float lastdelta[], float y[], float ymax[],
                   int fails, int m, int k, int n)
{
    /* this function is internally used by MULTISTEP */

    int i;
    float a1,a2,a3,aa,ss;

    if (k <= 1)
        a1=0.0;
    else {
        ss=FLT_MIN;
        for (i=1; i<=n; i++) {
            aa=y[k*m+i]/ymax[i];
            ss += aa*aa;
        }
        a1=0.75*pow(toldwn/ss,0.5/k);
    }
    a2=0.80*pow(tol/error,0.5/(k+1));
    if (k >= 5 || fails != 0)
        a3=0.0;
    else {
        ss=FLT_MIN;
        for (i=1; i<=n; i++) {
            aa=(delta[i]-lastdelta[i])/ymax[i];
            ss += aa*aa;
        }
        a3=0.70*pow(tolup/ss,0.5/(k+2));
    }
    if (a1 > a2 && a1 > a3) {
        *knew = k-1;
        *chnew = a1;
    } else
        if (a2 > a3) {
            *knew = k;
            *chnew = a2;
        } else {
            *knew = k+1;
            *chnew = a3;
        }
}

void multistepjacobian(int n, float x, float y[], float eps,
            float fixy[], float fixdy[], float dy[], float **jacobian,
            void (*deriv)(float [], int, float, float[]),
            int (*available)(int, float, float [], float **),
            int *evaluate, int *decompose, int *evaluated)
{
    /* this function is internally used by MULTISTEP */

    int i,j;
    float d;

    *evaluate = 0;
    *decompose = *evaluated = 1;
    if (!(*available)(n,x,y,jacobian)) {
        for (i=1; i<=n; i++) fixy[i]=y[i];
        (*deriv)(fixdy,n,x,y);
        for (j=1; j<=n; j++) {
            d=((eps > fabs(fixy[j])) ? eps*eps : eps*fabs(fixy[j]));
            y[j] += d;
```

```
                (*deriv)(dy,n,x,y);
                for (i=1; i<=n; i++) jacobian[i][j]=(dy[i]-fixdy[i])/d;
                y[j]=fixy[j];
            }
        }
    }
```

G. diffsys

Integrates the system of first order ordinary differential equations

$$dy_i(x)/dx = f_i(x,y)$$

from $x = x0$ to $x = xe$ by means of a high order extrapolation method based on the modified midpoint rule. The method is suitable for high accuracy problems and is not suited for stiff equations.

diffsys is a slight modification of the algorithm in [BulS65]. By this modification integration from *x0* until *xe* can be performed by one call of *diffsys*. A number of integration steps are taken, starting with the initial step *h0*, in each integration step a number of solutions are computed by means of the modified midpoint rule. Extrapolation is used to improve these solutions, until the required accuracy is met. An integration step is rejected if the accuracy requirements are not fulfilled after nine extrapolation steps. In these cases the integration step is rejected, and integration is tried again with the integration step halved.

The algorithm is for each step a variable order method (the highest order is 14), and uses a variable number of function evaluations, depending on the order (minimum is 3, maximum is 217). The algorithm is less sensitive to too small values of the initial stepsize than the original algorithm (see [Fox72, HEFS72]). However, bad guesses require still some more computations.

Function Parameters:

void diffsys (*x,xe,n,y,derivative,aeta,reta,s,h0,output*)

x: float *;
  entry: the independent variable; the initial value *x0*;
  exit: the final value *xe*;
xe: float;
  entry: the final value of *x* ($xe \geq x$);
n: int;
  entry: the number of equations;
y: float *y[1:n]*;
  entry: the dependent variable; *y[1:n]* are the initial values of the solution of the system of differential equations at $x = x0$;
  exit: *y[1:n]* are the final values of the solution at $x = xe$;
derivative: void (**derivative*)(*n,x,y,dy*);
  this procedure should deliver the right hand side of the *i*-th differential equation at the point (*x,y*) as *dy[i]*, *i*=1,...,*n*;
aeta: float;
  entry: required absolute precision in the integration process;
reta: float;
  entry: required relative precision in the integration process;
s: float *s[1:n]*;
  the array *s* is used to control the accuracy of the computed values of *y*;
  entry: it is advisable to set *s[i]*=0, *i*=1,...,*n*;

exit: the maximum value of $|y[i]|$, encountered during integration, if this value exceeds the value of *s[i]* on entry;

h0: float;
entry: the initial step to be taken;

output: void (**output*)(*n,x,xe,y,s*);
this procedure is called at the end of each integration step, the user can ask for output of parameters *x, xe, y* and *s*.

```
#include <float.h>
#include <math.h>

void diffsys(float *x, float xe, int n, float y[],
            void (*derivative)(int, float, float [], float []),
            float aeta, float reta, float s[], float h0,
            void (*output)(int, float, float, float [], float []))
{
    float *allocate_real_vector(int, int);
    float **allocate_real_matrix(int, int, int, int);
    void free_real_vector(float *, int);
    void free_real_matrix(float **, int, int, int);
    int i,j,k,kk,jj,l,m,r,sr,konv,b0,bh,last,next;
    float a,b,b1,c,g,h,u,v,ta,fc,*ya,*yl,*ym,*dy,*dz,**dt,d[7],
         **yg,**yh;

    ya=allocate_real_vector(1,n);
    yl=allocate_real_vector(1,n);
    ym=allocate_real_vector(1,n);
    dy=allocate_real_vector(1,n);
    dz=allocate_real_vector(1,n);
    dt=allocate_real_matrix(1,n,0,6);
    yg=allocate_real_matrix(0,7,1,n);
    yh=allocate_real_matrix(0,7,1,n);

    last=0;
    h=h0;
    do {
        next=0;
        if (h*1.1 >= xe-(*x)) {
            last=1;
            h0=h;
            h=xe-(*x)+FLT_EPSILON;
        }
        (*derivative)(n,*x,y,dz);
        bh=0;
        for (i=1; i<=n; i++) ya[i]=y[i];
        while (1) {
            a=h+(*x);
            fc=1.5;
            b0=0;
            m=1;
            r=2;
            sr=3;
            jj = -1;
            for (j=0; j<=9; j++) {
                if (b0) {
                    d[1]=16.0/9.0;
                    d[3]=64.0/9.0;
                    d[5]=256.0/9.0;
                } else {
                    d[1]=9.0/4.0;
                    d[3]=9.0;
                    d[5]=36.0;
                }
                konv=1;
                if (j > 6) {
                    l=6;
                    d[6]=64.0;
                    fc *= 0.6;
                } else {
                    l=j;
```

```
                d[l]=m*m;
            }
            m *= 2;
            g=h/m;
            b=g*2.0;
            if (bh && j < 8)
                for (i=1; i<=n; i++) {
                    ym[i]=yh[j][i];
                    yl[i]=yg[j][i];
                }
            else {
                kk=(m-2)/2;
                m--;
                for (i=1; i<=n; i++) {
                    yl[i]=ya[i];
                    ym[i]=ya[i]+g*dz[i];
                }
                for (k=1; k<=m; k++) {
                    (*derivative)(n,(*x)+k*g,ym,dy);
                    for (i=1; i<=n; i++) {
                        u=yl[i]+b*dy[i];
                        yl[i]=ym[i];
                        ym[i]=u;
                        u=fabs(u);
                        if (u > s[i]) s[i]=u;
                    }
                    if (k == kk && k != 2) {
                        jj++;
                        for (i=1; i<=n; i++) {
                            yh[jj][i]=ym[i];
                            yg[jj][i]=yl[i];
                        }
                    }
                }
            }
            (*derivative)(n,a,ym,dy);
            for (i=1; i<=n; i++) {
                v=dt[i][0];
                ta=c=dt[i][0]=(ym[i]+yl[i]+g*dy[i])/2.0;
                for (k=1; k<=l; k++) {
                    b1=d[k]*v;
                    b=b1-c;
                    u=v;
                    if (b != 0.0) {
                        b=(c-v)/b;
                        u=c*b;
                        c=b1*b;
                    }
                    v=dt[i][k];
                    dt[i][k]=u;
                    ta += u;
                }
                if (fabs(y[i]-ta) > reta*s[i]+aeta) konv=0;
                y[i]=ta;
            }
            if (konv) {
                next=1;
                break;
            }
            d[2]=4.0;
            d[4]=16.0;
            b0 = !b0;
            m=r;
            r=sr;
            sr=m*2;
        }
        if (next) break;
        bh = !bh;
        last=0;
        h /= 2.0;
    }
    h *= fc;
    (*x)=a;
    (*output)(n,*x,xe,y,s);
```

```
	} while (!last);
	free_real_vector(ya,1);
	free_real_vector(yl,1);
	free_real_vector(ym,1);
	free_real_vector(dy,1);
	free_real_vector(dz,1);
	free_real_matrix(dt,1,n,0);
	free_real_matrix(yg,0,7,1);
	free_real_matrix(yh,0,7,1);
}
```

H. ark

Solves an initial value problem for a system of first order ordinary differential equations which is obtained from semi-discretization of an initial boundary value problem for a parabolic or hyperbolic equation, based on stabilized, explicit Runge-Kutta methods of low order [Be72, Vh71, VhBDS71, VhK71]. Automatic stepsize control is provided but step rejection has been excluded in order to save storage. Because of its limited storage requirements and adaptive stability facilities the method is well suited for the solution of initial boundary value problems for partial differential equations.

ark solves the system of differential equations

$$Du_i(t) = f_i(t,u(t)) \qquad (i=m0,m0+1,...,m)$$

where $D \equiv d/dt$, with $u(t_0)$ given, from $t = t0$ to $t = te$. The theory upon which the operation of *ark* is based is, in its early stages, the same as that supporting *efrk*. However, *ark* operates with a fixed stability polynomial

$$p_n(z) = 1 + \beta_1 z + ... + \beta_n z^n$$

all of whose coefficients the user supplies at call of *ark*; furthermore, it is assumed that those eigenvalues of $J(t)$ (the Jacobian matrix whose elements are given by

$$J_{i,j}(t) = \partial f_{i+m0-1}(t,u(t))/\partial u_{j+m-1}(t) \qquad (i,j=1,...,m-m0+1)$$

which lie in the closed left half-plane Real(δ) ≤ 0 are either all pure real or pure imaginary for $t0 \leq t \leq te$.

With $\sigma(J)$ the spectral radius of $J(t)$ with respect to its eigenvalues in the closed left half-plane, stability polynomials $p_n(z)$ have been constructed such that the associated Runge-Kutta scheme (2) of *efrk* is stable if the stepsize (in t) h satisfies the inequality $h \leq \beta(n)/\sigma(J)$; $\beta(n)$ is a constant associated with the polynomial $p_n(z)$ in question, and in each case $p_n(z)$ has been so determined that $\beta(n)$ assumes its maximum permissible value. Examples of such polynomials $p_n(z)$ and associated constants $\beta(n)$, together with the types of equations (i.e. those for which the eigenvalues in the closed left half-plane Real(δ) ≤ 0 of the associated Jacobian matrix $J(t)$ are either all pure real or all pure imaginary) to which they refer, and the order r of exactness (see the documentation to *efrk*) of the polynomial in question are given in the following table.

n	β_1	β_2	β_3	β_4	β_5	$\beta(n)$	type	r
3	1	1/2	1/4			2	imaginary	2
4	1	1/2	1/6	1/24		$2\sqrt{2}$	imaginary	3
4	1	5/32	1/128	1/8192		32	real	1
4	1	1/2	0.078	0.0036		12	real	2
4	1	1/2	1/6	0.0185		6	real	3
5	1	1/2	3/16	1/32	1.128	4	imaginary	2

Function Parameters:

void ark (*t,te,m0,m,u,derivative,data,out*)

t: float *;
the independent variable *t* of the systems of ordinary differential equations
du/dt = *f(t,u)*, *u=u0* at *t=t0*;
entry: the initial value *t0*;
exit: the final value *te*;
te: float *;
entry: the final value of *t* (*te* ≥ *t*);
m0,m: int *;
indices of the first and last equation of the system;
u: float *u[m0:m]*;
entry: the initial values of the solution of the system of differential equations at *t=t0*;
exit: the values of the solution at *t* = *te*;
derivative: void (**derivative*)(*m0,m,t,v*);
this procedure performs an evaluation of the right hand side of the system with dependent variables *v[m0:m]* and independent variable *t*, upon completion of *derivative*, the right hand side should be overwritten on *v[m0:m]*;
data: float *data[1:10+data[1]]*;
in array *data* one should give:
data[1]: the number of evaluations of *f(t,u)* per integration step (*data[1]*≥*data[2]*);
data[2]: the order of accuracy of the method (*data[2]*≤3) (value of *r* above);
data[3]: stability bound (value of *ß(n)* above);
data[4]: the spectral radius of the Jacobian matrix with respect to those eigenvalues which are located in the non-positive half plane (value of σ(*J(x)*) above);
data[5]: the minimal stepsize;
data[6]: the absolute tolerance;
data[7]: the relative tolerance;
if both *data[6]* and *dat[7]* are negative then the integration is performed with a constant step *data[5]*;
data[8]: *data[8]* should be zero if *ark* is called for a first time, for continued integration *data[8]* should not be changed;
data[11],...,data[10+data[1]]: polynomial coefficients (the values of $ß_i$ above in *data[i+10]*, *i*=1,...,*data[1]*);
after each step the following by-products are delivered:
data[8]: the number of integration steps performed;
data[9]: an estimate of the local error last made;
data[10]: information messages:
data[10]=0: no difficulty;
data[10]=1: minimal steplength exceeds the steplength prescribed by stability theory, i.e. *data[5]*>*data[3]/data[4]*; termination of *ark*; decrease minimal steplength;
if necessary, *data[i]*, *i*=4,...,7, can be updated (after each step by means of procedure *out*);
out: void (**out*)(*m0,m,t,te,u,data*);
after each integration step performed information can be obtained or updated by this procedure.

Functions used: inivec, mulvec, dupvec, vecvec, elmvec, decsol.

```
#include <float.h>
#include <math.h>

void ark(float *t, float *te, int *m0, int *m, float u[],
        void (*derivative)(int *, int *, float *, float[]),
        float data[],
        void (*out)(int *, int *, float *, float *, float [], float []))
{
    float *allocate_real_vector(int, int);
    float **allocate_real_matrix(int, int, int, int);
    void free_real_vector(float *, int);
    void free_real_matrix(float **, int, int, int);
    void inivec(int, int, float [], float);
    void mulvec(int, int, int, float [], float [], float);
    void dupvec(int, int, int, float [], float []);
    float vecvec(int, int, int, float [], float []);
    void elmvec(int, int, int, float [], float [], float);
    void decsol(float **, int, float [], float []);
    float arkmui(int, int, int, float []);
    float arklabda(int, int, int, int, float []);
    static float th1[8] = {1.0, 0.5, 1.0/6.0, 1.0/3.0, 1.0/24.0,
        1.0/12.0, 0.125, 0.25};
    static float ec0,ec1,ec2,tau0,tau1,tau2,taus,t2;
    int p,n,q,start,step1,last,i,j,k,l,n1,m00;
    float thetanm1,tau,betan,qinv,eta,*mu,*lambda,*thetha,*ro,*r,
            **alfa,th[9],aux[4],s,ss,theta0,tauacc,taustab,
            aa,bb,cc,ec,mt,lt;

    n=data[1];
    m00=(*m0);
    mu=allocate_real_vector(1,n);
    lambda=allocate_real_vector(1,n);
    thetha=allocate_real_vector(0,n);
    ro=allocate_real_vector(m00,*m);
    r=allocate_real_vector(m00,*m);
    alfa=allocate_real_matrix(1,8,1,n+1);

    p=data[2];
    ec1=ec2=0.0;
    betan=data[3];
    thetanm1 = (p == 3) ? 0.75 : 1.0;
    theta0=1.0-thetanm1;
    s=1.0;
    for (j=n-1; j>=1; j--) {
        s = -s*theta0+data[n+10-j];
        mu[j]=data[n+11-j]/s;
        lambda[j]=mu[j]-theta0;
    }
    for (i=1; i<=8; i++)
        for (j=0; j<=n; j++)
            if (i == 1) alfa[i][j+1]=1.0;
            else if (j == 0) alfa[i][j+1]=0.0;
            else if (i == 2 || i == 4 || i == 8)
                    alfa[i][j+1]=pow(arkmui(j,n,p,lambda),(i+2)/3);
            else if ((i == 3 || i == 6) && j > 1) {
                s=0.0;
                for (l=1; l<=j-1; l++)
                    s += arklabda(j,l,n,p,lambda)*
                                pow(arkmui(l,n,p,lambda),i/3);
                alfa[i][j+1]=s;
            }
            else if (i == 5 && j > 2) {
                s=0.0;
                for (l=2; l<=j-1; l++) {
                    ss=0.0;
                    for (k=1; k<=l-1; k++)
                        ss += arklabda(l,k,n,p,lambda)*
                                    arkmui(k,n,p,lambda);
                    s += arklabda(j,l,n,p,lambda)*ss;
                }
                alfa[i][j+1]=s;
            }
            else if (i == 7 && j > 1) {
                s=0.0;
```

```
                for (l=1; l<=j-1; l++)
                    s += arklabda(j,l,n,p,lambda)*arkmui(l,n,p,lambda);
                alfa[i][j+1]=s*arkmui(j,n,p,lambda);
            }
            else alfa[i][j+1]=0.0;
    n1 = ((n < 4) ? n+1 : ((n < 7) ? 4 : 8));
    for (i=1; i<=8; i++) th[i]=th1[i-1];
    if (p == 3 && n < 7) th[1]=th[2]=0.0;
    aux[2]=FLT_EPSILON;
    decsol(alfa,n1,aux,th);
    inivec(0,n,thetha,0.0);
    dupvec(0,n1-1,1,thetha,th);
    if (!(p == 3 && n < 7)) {
        thetha[0] -= theta0;
        thetha[n-1] -= thetanm1;
        q=p+1;
    } else
        q=3;
    qinv=1.0/q;
    start=(data[8] == 0.0);
    data[10]=0.0;
    last=0;
    dupvec(*m0,*m,0,r,u);
    (*derivative)(m0,m,t,r);
    do {
        /* stepsize */
        eta=sqrt(vecvec(*m0,*m,0,u,u))*data[7]+data[6];
        if (eta > 0.0) {
            if (start) {
                if (data[8] == 0) {
                    tauacc=data[5];
                    step1=1;
                } else
                    if (step1) {
                        tauacc=pow(eta/ec2,qinv);
                        if (tauacc > 10.0*tau2)
                            tauacc=10.0*tau2;
                        else
                            step1=0;
                    } else {
                        bb=(ec2-ec1)/tau1;
                        cc = -bb*t2+ec2;
                        ec=bb*(*t)+cc;
                        tauacc = (ec < 0.0) ? tau2 : pow(eta/ec,qinv);
                        start=0;
                    }
            } else {
                aa=((ec0-ec1)/tau0+(ec2-ec1)/tau1)/(tau1+tau0);
                bb=(ec2-ec1)/tau1-(2.0*t2-tau1)*aa;
                cc = -(aa*t2+bb)*t2+ec2;
                ec=(aa*(*t)+bb)*(*t)+cc;
                tauacc = ((ec < 0.0) ? taus : pow(eta/ec,qinv));
                if (tauacc > 2.0*taus) tauacc=2.0*taus;
                if (tauacc < taus/2.0) tauacc=taus/2.0;
            }
        } else
            tauacc=data[5];
        if (tauacc < data[5]) tauacc=data[5];
        taustab=betan/data[4];
        if (taustab < data[5]) {
            data[10]=1.0;
            break;
        }
        tau = ((tauacc > taustab) ? taustab : tauacc);
        taus=tau;
        if (tau >= (*te)-(*t)) {
            tau=(*te)-(*t);
            last=1;
        }
        tau0=tau1;
        tau1=tau2;
        tau2=tau;
        /* difference scheme */
        mulvec(*m0,*m,0,ro,r,thetha[0]);
```

```
        if (p == 3) elmvec(*m0,*m,0,u,r,0.25*tau);
        for (i=1; i<=n-1; i++) {
            mt=mu[i]*tau;
            lt=lambda[i]*tau;
            for (j=(*m0); j<=(*m); j++) r[j]=lt*r[j]+u[j];
            s=(*t)+mt;
            (*derivative)(m0,m,&s,r);
            if (thetha[i] != 0.0) elmvec(*m0,*m,0,ro,r,thetha[i]);
            if (i == n) {
                data[9]=sqrt(vecvec(*m0,*m,0,ro,ro))*tau;
                ec0=ec1;
                ec1=ec2;
                ec2=data[9]/pow(tau,q);
            }
        }
        elmvec(*m0,*m,0,u,r,thetanm1*tau);
        dupvec(*m0,*m,0,r,u);
        s=(*t)+tau;
        (*derivative)(m0,m,&s,r);
        if (thetha[n] != 0.0) elmvec(*m0,*m,0,ro,r,thetha[n]);
        data[9]=sqrt(vecvec(*m0,*m,0,ro,ro))*tau;
        ec0=ec1;
        ec1=ec2;
        ec2=data[9]/pow(tau,q);
        t2=(*t);
        if (last) {
            last=0;
            (*t)=(*te);
        } else
            (*t) += tau;
        data[8] += 1.0;
        (*out)(m0,m,t,te,u,data);
    } while ((*t) != (*te));
    free_real_vector(mu,1);
    free_real_vector(lambda,1);
    free_real_vector(thetha,0);
    free_real_vector(ro,m00);
    free_real_vector(r,m00);
    free_real_matrix(alfa,1,8,1);
}

float arkmui(int i, int n, int p, float lambda[])
{
    /* this function is internally used by ARK */

    return ((i==n) ? 1.0 : ((i<1 || i>n) ? 0.0 :
            ((p<3) ? lambda[i] : ((p==3) ? lambda[i]+0.25 : 0.0))));
}

float arklabda(int i, int j, int n, int p, float lambda[])
{
    /* this function is internally used by ARK */

    float arkmui(int, int, int, float []);

    return ((p<3) ? ((j==i-1) ? arkmui(i,n,p,lambda) : 0.0) :
              ((p==3) ? ((i==n) ? ((j==0) ? 0.25 :
              ((j==n-1) ? 0.75 : 0.0)) :
              ((j==0) ? ((i==1) ? arkmui(1,n,p,lambda) : 0.25) :
              ((j==i-1) ? lambda[i] : 0.0))) : 0.0));
}
```

I. efrk

Solves an initial value problem for a system of first order ordinary differential equations by means of an exponentially fitted explicit Runge-Kutta method of first, second or third order. *efrk* is a special purpose procedure for stiff equations with a known, clustered eigenvalue spectrum; automatic error control is not provided. A detailed description of the method is given in [De72], see also [BeDHKW73, BeDHV74, DekHH72].

efrk solves the system of differential equations

$$Du_i(t) = f_i(t,u(t)) \qquad i=m0,m0+1,...,m \tag{1}$$

where $D \equiv d/dt$, with $u(t_0)$ given, from $t = t0$ to $t = te$. Consider the N-th order Runge-Kutta scheme

$$u^{(n+1)} = u^{(n)} + \sum_{j=0}^{N-1} \theta_j k^{(j)} \tag{2}$$

$$k^{(j)} = h_n u(t_n + \mu_j h_n,\ u^{(n)} + \sum_{l=0}^{j-1} \lambda_{j,l} k^{(l)}) \qquad j=0,1,...,N-1$$

where $h_n = t_{n+1}-t_n$. For the purposes of theoretical investigation, we work in a space whose coordinates are numbered $m0,m0+1,...,m$, and it is assumed that, f being a suitable differentiable function, $u^{(n+1)}$ given by formula (2) may be expressed as the sum of a power series in the variable h_n, with coefficients involving the derivatives of $u^{(n)}$:

$$u^{(n+1)} = u^{(n)} + \beta_1(Du^{(n)})h_n + \beta_2(D^2u^{(n)})h_n^2 + \beta_3 J^{(n)}D^2u^{(n)}h_n^3 + \tfrac{1}{2}\beta_{3,1}(D^3u^{(n)}-J^{(n)}D^2u^{(n)})h_n^3 + ... \tag{3}$$

where $J^{(n)}$ is the $(m-m0+1)\times(m-m0+1)$ Jacobian matrix whose elements are $\partial f_i(t_n,u^{(n)}(t))/\partial u_j^{(n)}(t)$ $(i,j=m0,m0+1,...,m)$. Expanding each of the $k^{(j)}$ in (2) in ascending powers of h_n, it is found that the θ_j, μ_j and $\lambda_{i,j}$ of (2) and the β_j and $\beta_{i,j}$ of (3) are connected by the system of relationships

$$\beta_1 = \sum_{j=0}^{N-1} \theta_j, \qquad \beta_2 = \sum_{j=1}^{N-1} \theta_j \mu_j$$

$$\beta_3 = \sum_{j=2}^{N-1} \theta_j \sum_{l=1}^{j-1} \lambda_{j,l}\mu_l, \qquad \beta_{3,1} = \sum_{j=1}^{N-1} \theta_j \mu_j^2$$

$$\beta_4 = \sum_{j=3}^{N-1} \theta_j \sum_{l=2}^{j-1} \lambda_{j,l} \sum_{i=1}^{l-1} \lambda_{l,i}\mu_i, \qquad \beta_{4,1} = \sum_{j=2}^{N-1} \theta_j \sum_{l=1}^{j-1} \lambda_{j,l}\mu_l^2$$

$$\beta_{4,2} = \sum_{j=2}^{N-1} \theta_j \mu_j \sum_{l=1}^{j-1} \lambda_{j,l}\mu_l, \qquad \beta_{4,3} = \sum_{j=1}^{N-1} \theta_j \mu_j^2, \quad ...$$

By imposing appropriate conditions upon the β_i and $\beta_{i,j}$, the θ_j, μ_j and $\lambda_{i,j}$ are determined, and the Runge-Kutta scheme (2) is established.

One type of condition concerns the extent of agreement between the series (3) and the Taylor series expansion of $u(t_n+h_n)$ in ascending powers of h_n, given that u satisfies equation (1). If such agreement is to persist up to and including the terms in h_n^r (the scheme (2) is then said to be r-th order exact) the β_i and $\beta_{i,j}$ must have the following values

r	β_1	β_2	β_3	$\beta_{3,1}$	β_4	$\beta_{4,1}$	$\beta_{4,2}$	$\beta_{4,3}$
1	1							
2	1	1/2						
3	1	1/2	1/6	1/3				
4	1	1/2	1/6	1/3	1/24	1/12	1/8	1/4

A further type of condition concerns stability. Let $u'^{(n+1)}$ be the vector produced by means of the scheme (2) from the vector $u'^{(n)}$, and $u''^{(n+1)}$ similarly from $u''^{(n)}$, where $e^{(n)} = y'^{(n)} - y''^{(n)}$ is small. Then $y'^{(n+1)} - y''^{(n+1)} = p_N(h_n J^{(n)})e^{(n)} + O(e^{(n)})^2$
as $e^{(n)} \rightarrow 0$, where $p_N(z) = 1 + \beta_1 z + \beta_2 z^2 + ... + \beta_N z^N$.
The scheme (2) is stable if $|p_N(h_n\delta)| \leq 1$ for all eigenvalues δ of $J^{(n)}$ in the closed left half-plane Real(δ) ≤ 0.

If all of the above δ are very small, a relatively large stepsize h_n may be taken without violating the above stability condition. For the treatment of such differential equations, it is appropriate to choose the β_i and $\beta_{i,j}$, and by implication the θ_j, μ_j and $\lambda_{i,j}$ in (2) so as to ensure r-th order exactness for the highest possible value of r. If the above δ are not small (this phenomenon accompanies stiff equations and many systems arising from the numerical solution of partial differential equations) it is wiser to sacrifice a certain degree of exactness in order to obtain stability. One works with a stability polynomial whose first few coefficients β_i are fixed (thus ensuring a certain degree of exactness) and determines the remaining coefficients β_i in such a way that the stability condition $|p_N(h_n\delta)| \leq 1$ is satisfied for the largest possible stepsize h_n.

The problem of determining the coefficients β_i which satisfy the above criteria has been successfully resolved, by a process of exponential fitting, in two cases. The two cluster case in which, during an integration step, the above δ are to be found in two circles, one with center at the origin, and the other with center z_l upon the negative real axis, and the three-cluster case in which three circles at the origin and z_l and the conjugate of z_l in the left-half plane, are concerned.

For certain systems of differential equations of the form (1), the Jacobian matrix $J^{(n)}$ (and hence its distribution of eigenvalues) does not vary greatly over the range of integration; it is then possible to prescribe at outset a center z_l and the diameter of a circle with center at z_l (or two centers z_l and conjugate(z_l), and equal diameters of two circles) for which, over the entire range of integration, the above δ remain within the circle in question. For other examples, new estimates of z_l and the diameter must be determined at the commencement of each integration step, and a new Runge-Kutta scheme must be constructed (*efrk* carries this construction out). *efrk* caters for both cases.

Equation (1) may refer to a system in which the initial and final indices (*mo* and *m*) are fixed over the range of integration. It is, however, possible that these indices should vary during the course of the computation. (An example in which this occurs (it concerns the numerical solution of a partial differential equation) is given to illustrate the use of the procedure *ark*.) In such a case, the procedure which evaluates the derivatives (1) must also modify *m0* and *m* in a suitable manner.

When supplied with appropriate data:

(i) the values of the fixed first r coefficients β_i, i=1,...,r, of the stability polynomial $p_N(z)$ (these may be extracted from the above table);

(ii) the degree of exponential fitting: if z_l is negative real, l, the number supplied, is the degree of exponential fitting; if imag(z_l)≠0, l is twice this degree (and must therefore be even; in both cases $r+l$ is the total number of evaluations of the set of f_i in (1));

(iii) the value of z_l in the form $z_l = \sigma e^{i\theta}$, where z_l is the center of a circular region containing a cluster of eigenvalues δ of the Jacobian matrix $J^{(m)}$ as described above during the first integration step (this is step below);

(iv) the diameter of the above circular region;

(v) the value of an integer variable *thirdorder*: the value of *thirdorder* is nonzero if the user requires a third order exact polynomial $p_N(z)$ to be used (in this case, the supplied value of r must be ≥ 3 and the appropriate coefficients β_i, i=0,1,2,3, must be supplied), and *thirdorder* is zero otherwise;

(vi) an upper bound *tol* for the rounding errors in the computations of one Runge-Kutta step;

(vii) an upper bound step for the stepsize (the stepsize actually used during the computations may be reduced by stability constraints);

(viii) *t0* and *te*, the initial and final values of *t*;

(ix) *m0* and *m*, the indices of the first and last equations in the set (2) holding when *t=t0*;

(x) the initial values $u_i(t0)$ ($i=m0,m0+1,...,m$);

efrk constructs an appropriate Runge-Kutta scheme of the form (3) at the commencement of the first integration step and integrates the system (1), either retaining the constructed scheme throughout, or constructing a new scheme if the estimates of z_l and the diameter supplied during the course of the computation change sufficiently.

The user is also asked to provide a function *output* which is called by *efrk* at the conclusion of each integration step. *output* is intended partly as a monitoring device, and may also be used to change the values of σ and ϕ in the representation $z_l = \sigma e^{j\phi}$ and of the diameter of the circle with center z_l enclosing a cluster of eigenvalues during the computation.

Function Parameters:

void efrk (*t,te,m0,m,u,sigma,phi,diameter,derivative,k,step,r,l,beta,thirdorder,tol,output*)

t: float *;
the independent variable *t* of the systems of ordinary differential equations
$du/dt = f(t,u)$;
entry: the initial value *t0*;
exit: the final value *te*;

te: float;
entry: the final value of *t* ($te \geq t$);

m0,m: int;
indices of the first and last equation of the system;

u: float *u[m0:m]*; the dependent variable;
entry: the initial values of the solution of the system of differential equations at *t=t0*;
exit: the values of the solution at *t = te*;

sigma: float *;
entry: the modulus of the point at which exponential fitting is desired (the value of σ above); for example an approximation of the center of the left hand cluster;
exit: value of σ for the last step;

phi: float *;
entry: the argument of the center of the left hand cluster (the value of ϕ above); in the case of two complex conjugated clusters, the argument of the center in the second quadrant should be taken;
exit: value of ϕ for the last step;

diameter: float *;

entry: the diameter of the left hand cluster of eigenvalues of the Jacobian matrix of the system of differential equations; in case of non-linear equations *diameter* should have such a value that the variation of the eigenvalues in this cluster in the period (*t*, *t+step*) is less than half the diameter;

exit: value of the diameter for the last step;

derivative: void (**derivative*)(*m0,m,t,u*);

this procedure should deliver the value of *f(t,u)* in the point (*t,u*) in the array *u*;

k: int *;

counts the number of integration steps taken;

entry: an (arbitrarily) chosen value of *k0*, e.g. *k0* = 0;

exit: *k0* + the number of integration steps performed;

step: float *;

the stepsize chosen will be at most equal to *step*; this stepsize may be reduced by stability constraints, imposed by a positive diameter, or by considerations of internal stability;

r: float;

entry: the degree of exactness of the Runge-Kutta scheme to be employed (value of *r* above); *r+l* is the number of evaluations of *f(t,u)* on which the Runge-Kutta scheme is based; for *r*=1,2,≥3, first, second and third order accuracy may be obtained by an appropriate choice of the array *beta*;

l: float;

entry: if *phi*=4*$\tan^{-1}(1)$ then *l* is the order of the exponential fitting, otherwise *l* is twice the order of the exponential fitting; note that *l* should be even in the latter case;

beta: float *beta[0:r+l]*;

entry: the elements *beta[i]*, *i*=0,...,*r*, should have the value of the *r*+1 first coefficients of the stability polynomial;

thirdorder: int;

entry: if third order accuracy is desired then *thirdorder* should have the value nonzero, in combination with appropriate choices of *r* ($r \geq 3$) and the array *beta* (*beta[i]*=1/*i*!, *i*=0,1,2,3); in all other cases *thirdorder* must have the value zero;

tol: float;

entry: an upper bound for the rounding errors in the computations in one Runge-Kutta step; in some cases (e.g. large values of *sigma* and *r*) *tol* will cause a decrease of the stepsize;

output: void (**output*)(*m0,m,t,te,u,sigma,phi,diameter,k,step,r,l*);

this procedure is called at the end of each integration step; the user can ask for output of some parameters, for example *t, k, u, r, l* and compute new values for *sigma, phi, diameter* and *step*.

Functions used: elmvec, dec, sol.

```
#include <float.h>
#include <math.h>

void efrk(float *t, float te, int m0, int m, float u[],
        float *sigma, float *phi, float *diameter,
        void (*derivative)(int, int, float, float[]),
        int *k, float *step, float r, float l, float beta[],
        int thirdorder, float tol,
        void (*output)(int, int, float, float,
```

```
                  float [], float *, float *, float *,
                  int, float *, int, int))
{
    int *allocate_integer_vector(int, int);
    float *allocate_real_vector(int, int);
    float **allocate_real_matrix(int, int, int, int);
    void free_integer_vector(int *, int);
    void free_real_vector(float *, int);
    void free_real_matrix(float **, int, int, int);
    void elmvec(int, int, int, float [], float [], float);
    void sol(float **, int, int [], float []);
    void dec(float **, int, float [], int []);
    int n,first,last,complex,change,*p,i,j,c1,c3;
    float theta0,thetanm1,h,b,b0,phi0,phil,cosphi,sinphi,eps,
          betar,*mu,*labda,*pt,*fac,*betac,*rl,**a,aux[4],
          dd,hstab,hstabint,c,temp,c2,e,b1,zi,cosiphi,siniphi,
          cosphil,*d,bb,mt,lt,tht;

    p=allocate_integer_vector(1,l);
    mu=allocate_real_vector(0,r+l-1);
    labda=allocate_real_vector(0,r+l-1);
    pt=allocate_real_vector(0,r);
    fac=allocate_real_vector(0,l-1);
    betac=allocate_real_vector(0,l-1);
    rl=allocate_real_vector(m0,m);
    d=allocate_real_vector(1,l);
    a=allocate_real_matrix(1,l,1,l);

    n=r+l;
    first=1;
    b0 = -1.0;
    betar=pow(beta[r],1.0/r);
    last=0;
    eps=FLT_EPSILON;
    phi0=phil=4.0*atan(1.0);
    do {
        /* stepsize */
        h=(*step);
        dd=fabs((*sigma)*sin(*phi));
        complex=(((l/2)*2 == l) && 2.0*dd > *diameter);
        if (*diameter > 0.0) {
            temp=(*sigma)*(*sigma)/((*diameter)*((*diameter)*0.25+dd));
            hstab=pow(temp,l*0.5/r)/betar/(*sigma);
        } else
            hstab=h;
        dd = (thirdorder ? pow(2.0*tol/eps/beta[r],1.0/(n-1))*
                pow(4.0,(l-1.0)/(n-1.0)) : pow(tol/eps,(1.0/r)/betar));
        hstabint=fabs(dd/(*sigma));
        if (h > hstab) h=hstab;
        if (h > hstabint) h=hstabint;
        if ((*t)+h > te*(1.0-(*k)*eps)) {
            last=1;
            h=te-(*t);
        }
        b=h*(*sigma);
        dd=(*diameter)*0.1*h;
        dd *= dd;
        if (h < (*t)*eps) break;
        change=((b0 == -1.0) ||
                ((b-b0)*(b-b0)+b*b0*((*phi)-phi0)*((*phi)-phi0) > dd));
        if (change) {
            /* coefficient */
            b0=b;
            phi0=(*phi);
            if (b >= 1.0) {
                if (complex) {
                    /* solution of complex equations */
                    if (phi0 != phil) {
                        /* elements of matrix */
                        phil=phi0;
                        cosphi=cos(phil);
                        sinphi=sin(phil);
                        cosiphi=1.0;
                        siniphi=0.0;
```

```
            for (i=0; i<=l-1; i++) {
                c1=r+1+i;
                c2=1.0;
                for (j=l-1; j>=1; j-=2) {
                    a[j][l-i]=c2*cosiphi;
                    a[j+1][l-i]=c2*siniphi;
                    c2 *= c1;
                    c1--;
                }
                cosphil=cosiphi*cosphi-siniphi*sinphi;
                siniphi=cosiphi*sinphi+siniphi*cosphi;
                cosiphi=cosphil;
            }
            aux[2]=0.0;
            dec(a,l,aux,p);
        }
        /* right hand side */
        e=exp(b*cosphi);
        b1=b*sinphi-(r+1)*phil;
        cosiphi=e*cos(b1);
        siniphi=e*sin(b1);
        b1=1.0/b;
        zi=pow(b1,r);
        for (j=l; j>=2; j-=2) {
            d[j]=zi*siniphi;
            d[j-1]=zi*cosiphi;
            cosphil=cosiphi*cosphi-siniphi*sinphi;
            siniphi=cosiphi*sinphi+siniphi*cosphi;
            cosiphi=cosphil;
            zi *= b;
        }
        cosiphi=zi=1.0;
        siniphi=0.0;
        for (i=r; i>=0; i--) {
            c1=i;
            c2=beta[i];
            c3=((2*i > l-2) ? 2 : l-2*i);
            cosphil=cosiphi*cosphi-siniphi*sinphi;
            siniphi=cosiphi*sinphi+siniphi*cosphi;
            cosiphi=cosphil;
            for (j=l; j>=c3; j-=2) {
                d[j] += zi*c2*siniphi;
                d[j-1] -= zi*c2*cosiphi;
                c2 *=c1;
                c1--;
            }
            zi *= b1;
        }
        sol(a,l,p,d);
        for (i=1; i<=l; i++) beta[r+i]=d[l+1-i]*b1;
    } else {
        /* form beta */
        if (first) {
            /* form constants */
            first=0;
            fac[0]=1.0;
            for (i=1; i<=l-1; i++) fac[i]=i*fac[i-1];
            pt[r]=l*fac[l-1];
            for (i=1; i<=r; i++) pt[r-i]=pt[r-i+1]*(l+i)/i;
        }
        if (l == 1) {
            c=1.0-exp(-b);
            for (j=1; j<=r; j++) c=beta[j]-c/b;
            beta[r+1]=c/b;
        } else
            if (b > 40.0)
                for (i=r+1; i<=r+l; i++) {
                    c=0.0;
                    for (j=0; j<=r; j++)
                        c=beta[j]*pt[j]/(i-j)-c/b;
                    beta[i]=c/b/fac[l+r-i]/fac[i-r-1];
                }
            else {
                dd=c=exp(-b);
```

```
                        betac[l-1]=dd/fac[l-1];
                        for (i=1; i<=l-1; i++) {
                            c=b*c/i;
                            dd += c;
                            betac[l-1-i]=dd/fac[l-1-i];
                        }
                        bb=1.0;
                        for (i=r+1; i<=r+l; i++) {
                            c=0.0;
                            for (j=0; j<=r; j++)
                                c=(beta[j]-((j < l) ? betac[j] : 0.0))*
                                        pt[j]/(i-j)-c/b;
                            beta[i]=c/b/fac[l+r-i]/fac[i-r-1]+
                                        ((i < l) ? bb*betac[i] : 0.0);
                            bb *= b;
                        }
                    }
                }
            }
            labda[0]=mu[0]=0.0;
            if (thirdorder) {
                theta0=0.25;
                thetanm1=0.75;
                if (b < 1.0) {
                    c=mu[n-1]=2.0/3.0;
                    labda[n-1]=5.0/12.0;
                    for (j=n-2; j>=1; j--) {
                        c=mu[j]=c/(c-0.25)/(n-j+1);
                        labda[j]=c-0.25;
                    }
                } else {
                    c=mu[n-1]=beta[2]*4.0/3.0;
                    labda[n-1]=c-0.25;
                    for (j=n-2; j>=1; j--) {
                        c=mu[j]=c/(c-0.25)*beta[n-j+1]/beta[n-j]/
                                    ((j < l) ? b : 1.0);
                        labda[j]=c-0.25;
                    }
                }
            } else {
                theta0=0.0;
                thetanm1=1.0;
                if (b < 1.0)
                    for (j=n-1; j>=1; j--) mu[j]=labda[j]=1.0/(n-j+1);
                else {
                    labda[n-1]=mu[n-1]=beta[2];
                    for (j=n-2; j>=1; j--)
                        mu[j]=labda[j]=beta[n-j+1]/beta[n-j]/
                                        ((j < l) ? b : 1.0);
                }
            }
        }
        (*k)++;
        /* difference scheme */
        i = -1;
        do {
            i++;
            mt=mu[i]*h;
            lt=labda[i]*h;
            for (j=m0; j<=m; j++) rl[j]=u[j]+lt*rl[j];
            (*derivative)(m0,m,(*t)+mt,rl);
            if (i == 0 || i == n-1) {
                tht=((i == 0) ? theta0*h : thetanm1*h);
                elmvec(m0,m,0,u,rl,tht);
            }
        } while (i < n-1);
        (*t) += h;
        (*output)(m0,m,*t,te,u,sigma,phi,diameter,*k,step,r,l);
    } while (!last);
    free_integer_vector(p,1);
    free_real_vector(mu,0);
    free_real_vector(labda,0);
    free_real_vector(pt,0);
    free_real_vector(fac,0);
```

```
    free_real_vector(betac,0);
    free_real_vector(rl,m0);
    free_real_vector(d,1);
    free_real_matrix(a,1,l,1);
}
```

5.4.2 First Order - Jacobian matrix available

A. efsirk

Solves an initial value problem, given as an autonomous system of first order differential equations

$$Dy(x) = f(y) \tag{1}$$

where $D = d/dx$ and $f, y, \epsilon\ R^m$, with $y(x_0)$ prescribed, over the interval $x0 \leq x \leq xe$ by means of an exponentially fitted, A-stable, semi-implicit Runge-Kutta method of third order [BeDHKW73, BeDHV74, Vh72]. This procedure is suitable for the integration of stiff equations. The algorithm uses for each step two function evaluations and if the input parameter *linear* is zero one evaluation of the Jacobian matrix. The stepsize is not determined by the accuracy of the numerical solution, but by the amount by which the given differential equation differs from a linear equation. The procedure does not reject integration steps.

The method used has the form

$$y_{n+1} = y_n + (1/4)h_n f_n(y_n) + (3/4)h_n f(y_n + \Lambda_{1,0}^{(n)}(h_n J_n)h_n f(y_n)) \tag{2}$$

where $x^{n+1} = x_n + h_n$, and with $y_n \approx y(x_n)$, $J_n = \partial f(y_n)/\partial y_n$,

$$\Lambda_{1,0}^{(n)}(z) = \frac{2}{3}\frac{1 - \frac{1}{6}(1+3\alpha_1^{(n)})z}{1 - \frac{1}{2}(1+\alpha_1^{(n)})z + \frac{1}{12}(1+3\alpha_1^{(n)})z^2}.$$

The stability function for the above method is

$$R(z) = 1 + z + \frac{3}{4}z^2\Lambda_{1,0}^{(n)}(z) = \frac{1 + \frac{1}{2}(1-\alpha_1^{(n)})z + \frac{1}{12}(1-3\alpha_1^{(n)})z^2}{1 - \frac{1}{2}(1+\alpha_1^{(n)})z + \frac{1}{12}(1+3\alpha_1^{(n)})z^2}$$

and the parameter $\alpha_1^{(n)}$ in $\Lambda_{1,0}^{(n)}(z)$ is determined by use of the relationship

$$R_n(h_n\delta_n) = e^{h_n\delta_n},$$

where δ_n is the eigenvalue of J_n in the left half plane with largest absolute value, so that

$$\alpha_1^{(n)} = \frac{1}{3z_n}\frac{z_n^2 + 6z_n + 12 - e^{z_n}(z_n^2 - 6z_n + 12)}{e^{z_n}(z_n - 2) + z_n + 2}$$

where $z_n = h_n\delta_n$.

The right hand side vector in equation (1) is computed by means of a user supplied function *derivative*. The Jacobian matrix J_n is computed by another user supplied function *jacobian*. If the variable *linear* in the parameter list is given the value nonzero upon call then *jacobian* is called upon when $n=0$ in formula (2) and not otherwise; if *linear* has the value zero then *jacobian* is called at the commencement of each step of the integration method. When *linear* has the value zero upon call, the stepsize h_n is determined as follows by use of a second method alternative to (2) for the numerical solution of equations (1). This method produces approximations y'_n to $y(x_n)$ and has the form

$$y'_{n+1} = y_n + \nu_0^{(n)}h_nf(y_n) + \nu_1^{(n)}h_nf(y_n + \Lambda_{1,0}(h_nJ_n)h_nf(y_n)) + \nu_2^{(n)}\Lambda_{1,0}(h_nJ_n)h_nf(y_n) + \nu_3^{(n)}h_nf(y_{n+1}). \qquad (3)$$

When

$$\nu_3^{(n)} = \frac{1}{2}\frac{1 + 3\alpha_1^{(n)}}{5 - 3\alpha_1^{(n)}}, \qquad \nu_2^{(n)} = \frac{3}{4}(1 + 2\nu_3^{(n)}),$$

$$\nu_1^{(n)} = \frac{1}{2} - 2\nu_3^{(n)}, \qquad \nu_0^{(n)} = 0$$

the stability function of the method (3), namely

$$R'(z) = 1 + \nu_0^{(n)}z + \nu_1^{(n)}z(1 + \Lambda_{1,0}(z)) + \nu_2^{(n)}\Lambda_{1,0}(z)z + \nu_3^{(n)}zR(z)$$

is $R(z)$ above. If y_{n+1} *and* y'_{n+1} were to be determined by use of the schemes (2) and (3) respectively with step length h'_n in both cases, then, since the method (3) is second order exact, $\|y'_{n+1} - y_{n+1}\| \approx c_nh'^3_n$, where c_n is independent of the variable h'_n. If $c_n \approx c_{n-1}$, an optimal steplength h'_n may be determined by use of the relationships

$$h'_n = (tol/c_n)^{1/3} \approx (tol/c_{n-1})^{1/3} \approx h_{n-1}(tol/\|y'_n - y_n\|)^{1/3}$$

or, approximately,

$$h'_n = ((4/3)tol \;/\; (tol + \|y'_n - y_n\|) + 1/3)\; h_{n-1} \qquad (4)$$

where *tol* is an acceptable local tolerance (taken by *efsirk* to be $tol = \eta_a + \eta_r\|y_n\|$, where η_a and η_r are absolute and relative tolerances prescribed by the user). Thus y'_n is determined by use of formula (4). With two limits *hmin* and *hmax* prescribed by the user, h_n is taken to be h'_n if $hmin \le h'_n \le hmax$, to be *hmin* if $h'_n < hmin$, and to be *hmax* if $h'_n > hmax$.

When *linear* has the value nonzero upon call, a constant value $h_n = h_{max}$ is used in formula (2), and the reference values y'_n are not determined.

Function Parameters:

void efsirk (*x,xe,m,y,delta,derivative,jacobian,j,n,aeta,reta,hmin,hmax,linear,output*)

x: float *;
- entry: the independent variable; the initial value *x0*;
- exit: the final value *xe*;

xe: float;
- entry: the final value of *x*;

m: int;
- entry: the number of differential equations;

y: float *y[1:m]*;
the dependent variable; during the integration process the computed solution at the point *x* is associated with the array *y*;
- entry: the initial values of the solution of the system;
- exit: the final values of the solution;

delta: float *;
delta denotes the real part of the point at which exponential fitting is desired; alternatives:
delta= (an estimate of) the real part of the, in absolute value, largest eigenvalue of the Jacobian matrix of the system;
delta< -10^{14}, in order to obtain asymptotic stability;
delta= 0, in order to obtain a higher order of accuracy in case of linear or almost linear equations;

derivative: void (**derivative*)(*m,a,delta*);
in *efsirk* when *derivative* is called *a[i]* contains the values of *y[i]*;
upon completion of a call of *derivative*, the array *a* should contain the values of *f(y)*;
note that the variable *x* should not be used in *derivative* because the differential equation is supposed to be autonomous;

jacobian: void (**jacobian*)(*m,j,y,delta*);
in *efsirk* when *jacobian* is called the array *y* contains the values of the dependent variable;
upon completion of a call of *jacobian*, the array *j* should contain the values of the Jacobian matrix of *f(y)*;

j: float *j[1:m,1:m]*;
j is an auxiliary array which is used in the procedure *jacobian*;

n: int *;
counts the integration steps;

aeta,reta: float;
entry: required absolute and relative local accuracy;

hmin,hmax: float;
entry: minimal and maximal stepsize by which the integration is performed;

linear: int;
entry: if *linear* is nonzero then the procedure *jacobian* will only be called if $n = 1$; the integration will then be performed with a stepsize *hmax*; the corresponding reduction of computing time can be exploited in case of linear or almost linear equations;

output: void (**output*)(*x,xe,m,y,delta,j,n*);
the user can ask for output of parameters.

Functions used: vecvec, matvec, matmat, gsselm, solelm.

```
#include <float.h>
#include <math.h>

void efsirk(float *x, float xe, int m, float y[],
            float *delta, void (*derivative)(int, float[], float *),
            void (*jacobian)(int, float **, float [], float *),
            float **j, int *n, float aeta, float reta, float hmin,
            float hmax, int linear,
            void (*output)(float, float, int, float [],
                           float, float **, int))
{
    int *allocate_integer_vector(int, int);
    float *allocate_real_vector(int, int);
    float **allocate_real_matrix(int, int, int, int);
    void free_integer_vector(int *, int);
    void free_real_vector(float *, int);
    void free_real_matrix(float **, int, int, int);
    float vecvec(int, int, int, float [], float []);
    float matmat(int, int, int, int, float **, float **);
```

```
float matvec(int, int, int, float **, float []);
void gsselm(float **, int, float [], int [], int []);
void solelm(float **, int, int [], int [], float []);
int k,l,lin,*ri,*ci;
float step,h,mu0,mu1,mu2,theta0,theta1,nu1,nu2,nu3,yk,fk,c1,c2,
      d,*f,*k0,*labda,**j1,aux[8],discr,eta,s,z1,z2,e,alpha1,a,b;

ri=allocate_integer_vector(1,m);
ci=allocate_integer_vector(1,m);
f=allocate_real_vector(1,m);
k0=allocate_real_vector(1,m);
labda=allocate_real_vector(1,m);
j1=allocate_real_matrix(1,m,1,m);

aux[2]=FLT_EPSILON;
aux[4]=8.0;
for (k=1; k<=m; k++) f[k]=y[k];
*n = 0;
(*output)(*x,xe,m,y,*delta,j,*n);
step=0.0;
do {
    (*n)++;
    /* difference scheme */
    (*derivative)(m,f,delta);
    /* step size */
    if (linear)
        s=h=hmax;
    else
        if (*n == 1 || hmin == hmax)
            s=h=hmin;
        else {
            eta=aeta+reta*sqrt(vecvec(1,m,0,y,y));
            c1=nu3*step;
            for (k=1; k<=m; k++) labda[k] += c1*f[k]-y[k];
            discr=sqrt(vecvec(1,m,0,labda,labda));
            s=h=(eta/(0.75*(eta+discr))+0.33)*h;
            if (h < hmin)
                s=h=hmin;
            else
                if (h > hmax) s=h=hmax;
        }
    if ((*x)+s > xe) s=xe-(*x);
    lin=((step == s) && linear);
    step=s;
    if (!linear || *n == 1) (*jacobian)(m,j,y,delta);
    if (!lin) {
        /* coefficient */
        z1=step*(*delta);
        if (*n == 1) z2=z1+z1;
        if (fabs(z2-z1) > 1.0e-6*fabs(z1) || z2 > -1.0) {
            a=z1*z1+12.0;
            b=6.0*z1;
            if (fabs(z1) < 0.1)
                alpha1=(z1*z1/140.0-1.0)*z1/30.0;
            else if (z1 < 1.0e-14)
                alpha1=1.0/3.0;
            else if (z1 < -33.0)
                alpha1=(a+b)/(3.0*z1*(2.0+z1));
            else {
                e=((z1 < 230.0) ? exp(z1) : FLT_MAX);
                alpha1=((a-b)*e-a-b)/(((2.0-z1)*e-2.0-z1)*3.0*z1);
            }
            mu2=(1.0/3.0+alpha1)*0.25;
            mu1 = -(1.0+alpha1)*0.5;
            mu0=(6.0*mu1+2.0)/9.0;
            theta0=0.25;
            theta1=0.75;
            a=3.0*alpha1;
            nu3=(1.0+a)/(5.0-a)*0.5;
            a=nu3+nu3;
            nu1=0.5-a;
            nu2=(1.0+a)*0.75;
            z2=z1;
        }
```

```
            c1=step*mu1;
            d=step*step*mu2;
            for (k=1; k<=m; k++) {
                for (l=1; l<=m; l++)
                    j1[k][l]=d*matmat(1,m,k,l,j,j)+c1*j[k][l];
                j1[k][k] += 1.0;
            }
            gsselm(j1,m,aux,ri,ci);
        }
        c1=step*step*mu0;
        d=step*2.0/3.0;
        for (k=1; k<=m; k++) {
            k0[k]=fk=f[k];
            labda[k]=d*fk+c1*matvec(1,m,k,j,f);
        }
        solelm(j1,m,ri,ci,labda);
        for (k=1; k<=m; k++) f[k]=y[k]+labda[k];
        (*derivative)(m,f,delta);
        c1=theta0*step;
        c2=theta1*step;
        d=nu1*step;
        for (k=1; k<=m; k++) {
            yk=y[k];
            fk=f[k];
            labda[k]=yk+d*fk+nu2*labda[k];
            y[k]=f[k]=yk+c1*k0[k]+c2*fk;
        }
        (*x) += step;
        (*output)(*x,xe,m,y,*delta,j,*n);
    } while (*x < xe);
    free_integer_vector(ri,1);
    free_integer_vector(ci,1);
    free_real_vector(f,1);
    free_real_vector(k0,1);
    free_real_vector(labda,1);
    free_real_matrix(j1,1,m,1);
}
```

B. eferk

Solves an initial value problem, given as an autonomous system of first order differential equations

$$Dy(x) = f(y) \tag{1}$$

where $D \equiv d/dx$ and $f, y, \in \mathbb{R}^m$, with $y(x_0)$ prescribed, over the interval $x0 \le x \le xe$ by means of an exponentially fitted, semi-implicit Runge-Kutta method of third order [BeDHKW73, BeDHV74, DekHH72, Vh72]. This procedure is suitable for the integration of stiff differential equations. The algorithm uses for each step two function evaluations and if the input parameter *linear* is zero one evaluation of the Jacobian matrix. The stepsize is determined by an estimation of the local truncation error based on the residual function. Integration steps are not rejected.

The method used is one of a class having the form

$$y_{n+1} = y_n + \theta_0(z)h_n f_n(y_n) + \theta_1(z)h_n f(y_n + \Lambda_{1,0}(z)h_n f(y_n))$$

where, with $x^{n+1} = x_n + h_n$, $y_n \approx y(x_n)$ and $z = h_n J_n$ where $J_n = \partial f(y_n)/\partial y_n$, θ_0, θ_1 and $\Lambda_{1,0}$ being polynomials. Such a method is first order exact if $\theta_0(0) + \theta_1(0) = 1$, second order exact if also $\theta_0'(0) + \theta'_1(0) + \theta_1(0)\Lambda_{1,0}(0) = 1/2$, and third order consistent if also

$$\theta_0''(0) + \theta_1''(0) + 2\theta_1(0)\Lambda_{1,0}'(0) + 2\theta_1'(0)\Lambda_{1,0}(0) = 1/3$$

and

$$\theta_1(0)\Lambda_{1,0}^2(0) = 1/3.$$

The choice $\theta_0(z) = 1/4$, $\theta_1(z) = 3/4$ is made and the method used by *eferk* has the form

$$y_{n+1} = y_n + \frac{1}{4}h_n f(y_n) +$$
$$\frac{3}{4}h_n f(y_n + \frac{4}{3}(\frac{1}{2} + \frac{1}{6}h_n J_n + \sum_{j=4}^{m} \beta_j (h_n J_n)^{j-2}) h_n f(y_n)) \ . \tag{2}$$

The stability function for this method is

$$R_m(z) = \sum_{j=0}^{p} \frac{z^j}{j!} \sum_{j=p+1}^{m} \beta_j z^j$$

with $p = 3$. The β_j may be determined by exponential fitting of order r_j at the s points z_j (j=1,...,s), so that

$$\frac{d^k}{dz_j^k} R_m(z_j) = e^{z_j} \qquad (k=0,\dots,r_j; j=1,\dots,s)$$

where

$$\sum_{j=1}^{s} (r_j + 1) = m - p \ .$$

Approximately

$$R_m(z) \approx \left(\sum_{j=0}^{p} \frac{z^j}{j!} \right) \prod_{j=1}^{s} \left(\frac{z - z_j}{z_j} \right)^{r_j} .$$

That part of the stability region (i.e. the domain over which $|R_m(z)| \leq 1$) in the neighbourhood of z_j is a small circle with z_j as center and radius ρ_j where

$$(\rho_j)^{r_j} = p! z_j^{(r_j - p)} \prod_{\substack{i=1 \\ (i \neq j)}}^{s} \left| \frac{z_i}{z_i - z_j} \right|^{r_i} .$$

If exponential fitting takes place at one point z_0 alone

$$\rho_0 = |z_0| \ |p!z_0^{-p}|^{1/(m-p)}.$$

If fitting takes place at two complex conjugate points z_0 and conjugate(z_0), with arg(Λ_0) = ϕ

$$\rho_0 = |z_0| \ |(p!z_0^{-p})/2\sin\ \phi|^{2/(m-p)}.$$

Since these stability regions are small, it is necessary to determine z_0 precisely. Furthermore the regions prescribe a maximal step length h_{stab} for stability: if $d\delta_n$ is the distance between the smallest eigenvalue of J_n and the point of fitting δ_n then, when $p = 3$,

$$h_{stab} = \frac{1}{\delta_n} 6^{\frac{1}{3}} \left|\frac{\delta_n}{d\delta_n}\right|^{\frac{m}{3}-1}.$$

The step-choice strategy is based upon the following considerations. If $y_{n+1}(h)$ is the value of the solution to equation (1) at $x = x_n + h$ with initial values $y_{n+1}(0) = y_n$ produced by use of the scheme (2), and $\xi_n(h) = dy_{n+1}(h)/dh - f(y_{n+1}(h))$ then $h_n\|\xi_n(h)\|$ is a good estimate of the local error $\|y(x_n+h) - y_{n+1}(h)\|$, and $\rho_n'(h) = h|R(hJ_n)f(y_n) - f(y_{n+1})|$ is a good estimate of $h_n\|\xi_n(h)\|$. With h_n, $tol = \eta_a + \eta_r\|y_n\|$, (where η_a and η_r are absolute and relative tolerances prescribed by the user) and $\rho_n'(h_n)$ available, the quantity

$$h_{n+1}' = \left(\frac{4\ tol}{(3+j)\,tol + \rho_n'(h_n)} + \frac{1}{3-j}\right) h_n$$

with j=0 for nonlinear equations, and j=1 for approximately linear equations, is determined. With two limits *hmin* and *hmax* prescribed by the user, the steplength h_{n+1} is taken to be h_{n+1}' if $hmin \le h_{n+1}' \le hmax$; if h_{n+1}' lies outside this range, h_{n+1} is *hmax* if $h_{n+1}' > hmax$ and *hmin* if $h_{n+1}' < hmin$.

It can occur that the system of equations (1) is derived from the system $Dy(x)=g(x,y)$ with $y(x_0) = y_0$, where y, $g \in R^{m-1}$, simply by taking the first m-1 components of y in (1) to be those of y, and the m-th component of y to be x. In this case, the m-th component of f in (1) is always 1, and certain simplifications in the formulae can be exploited. If it is known that equations (1) have been derived in the above manner then the variable *aut* in the parameter list of *eferk* should be given the value zero upon call, otherwise nonzero.

Function Parameters:

void eferk
(*x,xe,m,y,sigma,phi,derivative,j,jacobian,k,l,aut,aeta,reta,hmin,hmax,linear,output*)

x: float *;
- entry: the independent variable; the initial value *x0*;
- exit: the final value *xe*;

xe: float;
- entry: the final value of *x* (*xe* ≥ *x*);

m: int;

entry: the number of equations;

y: float *y[1:m]*;
the dependent variable;
entry: the initial values of the system of differential equations: *y[i]* at *x=x0*;
exit: the final values of the solution: *y[i]* at *x=xe*;

sigma: float *;
the modulus of the point at which exponential fitting is desired, for example the largest negative eigenvalue of the Jacobian matrix of the system of differential equations;

phi: float;
the argument of the complex point at which exponential fitting is desired;

derivative: void (**derivative*)(*m,y*);
this procedure should deliver the right hand side of the *i*-th differential equation at the point (*y*) as *y[i]*;

j: float *j[1:m,1:m]*;
the Jacobian matrix of the system; array *j* should be updated in the procedure *jacobian*;

jacobian: void (**jacobian*)(*m,j,y,sigma*);
in this procedure the Jacobian at the point (*y*) has to be assigned to the array *j*;

k: int *;
counts the integration steps taken;

l: int;
entry: if *phi*=4*$\tan^{-1}(1)$ then *l* is the order of the exponential fitting, else *l* is twice the order of the exponential fitting;

aut: int;
entry: if the system has been written in autonomous form by adding the equation *dy[m]/dx = 1* to the system then *aut* may have the value zero, else *aut* should have the value nonzero;

aeta: float;
entry: required absolute precision in the integration process, *aeta* has to be positive;

reta: float;
entry: required relative precision in the integration process, *reta* has to be positive;

hmin: float;
entry: the steplength chosen will be at least equal to *hmin*;

hmax: float;
entry: the steplength chosen will be at most equal to *hmax*;

linear: int;
entry: the procedure *jacobian* is called only if *linear* is zero or *k* = 0; so if the system is linear then *linear* may have the value nonzero;

output: void (**output*)(*x,xe,m,y,j,k*);
this procedure is called at the end of each integration step; the user can ask for output of some parameters, for example, *x, y, j, k*.

Functions used: vecvec, matvec, dec, sol.

```
#include <math.h>

void eferk(float *x, float xe, int m, float y[],
           float *sigma, float phi,
           void (*derivative)(int, float[]),
           float **j,
           void (*jacobian)(int, float **, float [], float *),
           int *k, int l, int aut, float aeta, float reta, float hmin,
           float hmax, int linear,
```

```
          void (*output)(float, float, int, float [], float **, int))
{
    int *allocate_integer_vector(int, int);
    float *allocate_real_vector(int, int);
    float **allocate_real_matrix(int, int, int, int);
    void free_integer_vector(int *, int);
    void free_real_vector(float *, int);
    void free_real_matrix(float **, int, int, int);
    float vecvec(int, int, int, float [], float []);
    float matvec(int, int, int, float **, float []);
    void dec(float **, int, float [], int []);
    void sol(float **, int, int [], float []);
    int m1,i,change,last,*p,c1,q;
    float h,b,b0,phi0,cosphi,sinphi,eta,discr,fac,pi,*beta,*betha,
          *betac,*k0,*d,*d1,*d2,**a,aux[4],s,cos2phi,sina,e,zi,
          c2,cosiphi,siniphi,cosphil,*dd,emin1,b1,b2,a0,a1,a2,a3,
          c,ddd,betai,bethai;

    p=allocate_integer_vector(1,l);
    beta=allocate_real_vector(0,l);
    betha=allocate_real_vector(0,l);
    betac=allocate_real_vector(0,l+3);
    k0=allocate_real_vector(1,m);
    d=allocate_real_vector(1,m);
    d1=allocate_real_vector(1,m);
    d2=allocate_real_vector(1,m);
    dd=allocate_real_vector(1,l);
    a=allocate_real_matrix(1,l,1,l);

    b0 = phi0 = -1.0;
    pi=4.0*atan(1.0);
    betac[l]=betac[l+1]=betac[l+2]=betac[l+3]=0.0;
    beta[0]=1.0/6.0;
    betha[0]=0.5;
    fac=1.0;
    for (i=2; i<=l-1; i++) fac *= i;
    m1=(aut ? m : m-1);
    *k = 0;
    last=0;
    do {
        for (i=1; i<=m; i++) d[i]=y[i];
        (*derivative)(m,d);
        if (!linear || *k == 0) (*jacobian)(m,j,y,sigma);
        /* step size */
        eta=aeta+reta*sqrt(vecvec(1,m1,0,y,y));
        if (*k == 0) {
            discr=sqrt(vecvec(1,m1,0,d,d));
            h=eta/discr;
        } else {
            s=0.0;
            for (i=1; i<=m1; i++) {
                ddd=d[i]-d2[i];
                s += ddd*ddd;
            }
            discr=h*sqrt(s)/eta;
            h *= (linear ? 4.0/(4.0+discr)+0.5 :
                           4.0/(3.0+discr)+1.0/3.0);
        }
        if (h < hmin) h=hmin;
        if (h > hmax) h=hmax;
        b=fabs(h*(*sigma));
        change=(fabs(1.0-b/b0) > 0.05 || phi != phi0);
        if (1.1*h >= xe-(*x)) {
            change=last=1;
            h=xe-(*x);
        }
        if (!change) h=h*b0/b;
        if (change) {
            /* coefficient */
            b0=b=fabs(h*(*sigma));
            if (b >= 0.1) {
                if (phi != pi && l == 2 || fabs(phi-pi) > 0.01) {
                    /* solution of complex equations */
                    if (l == 2) {
```

```
        phi0=phi;
        cosphi=cos(phi0);
        sinphi=sin(phi0);
        e=exp(b*cosphi);
        zi=b*sinphi-3.0*phi0;
        sina=((fabs(sinphi) < 1.0e-6) ? -e*(b+3.0) :
                e*sin(zi)/sinphi);
        cos2phi=2.0*cosphi*cosphi-1.0;
        betha[2]=(0.5+(2.0*cosphi+(1.0+2.0*cos2phi+
                        sina)/b)/b)/b/b;
        sina=((fabs(sinphi) < 1.0e-6) ? e*(b+4.0) :
                sina*cosphi-e*cos(zi));
        betha[1] = -(cosphi+(1.0+2.0*cos2phi+
                        (4.0*cosphi*cos2phi+sina)/b)/b)/b;
        beta[1]=betha[2]+2.0*cosphi*(betha[1]-1.0/6.0)/b;
        beta[2]=(1.0/6.0-betha[1])/b/b;
    } else {
        if (phi0 != phi) {
            /* elements of matrix */
            phi0=phi;
            cosphi=cos(phi0);
            sinphi=sin(phi0);
            cosiphi=1.0;
            siniphi=0.0;
            for (i=0; i<=l-1; i++) {
                c1=4+i;
                c2=1.0;
                for (q=l-1; q>=1; q-=2) {
                    a[q][l-i]=c2*cosiphi;
                    a[q+1][l-i]=c2*siniphi;
                    c2 *= c1;
                    c1--;
                }
                cosphil=cosiphi*cosphi-siniphi*sinphi;
                siniphi=cosiphi*sinphi+siniphi*cosphi;
                cosiphi=cosphil;
            }
            aux[2]=0.0;
            dec(a,l,aux,p);
        }
        /* right hand side */
        e=exp(b*cosphi);
        zi=b*sinphi-4.0*phi0;
        cosiphi=e*cos(zi);
        siniphi=e*sin(zi);
        zi=1.0/b/b/b;
        for (q=l; q>=2; q-=2) {
            dd[q]=zi*siniphi;
            dd[q-1]=zi*cosiphi;
            cosphil=cosiphi*cosphi-siniphi*sinphi;
            siniphi=cosiphi*sinphi+siniphi*cosphi;
            cosiphi=cosphil;
            zi *= b;
        }
        siniphi=2.0*sinphi*cosphi;
        cosiphi=2.0*cosphi*cosphi-1.0;
        cosphil=cosphi*(2.0*cosiphi-1.0);
        dd[l] += sinphi*(1.0/6.0+(cosphi+(1.0+2.0*
                    cosiphi*(1.0+2.0*cosphi/b))/b)/b);
        dd[l-1] -= cosphi/6.0+(0.5*cosiphi+(cosphil+
                        (2.0*cosiphi*cosiphi-1.0)/b)/b)/b;
        dd[l-2] += sinphi*(0.5+(2.0*cosphi+
                        (2.0*cosiphi+1.0)/b)/b);
        dd[l-3] -= 0.5*cosphi-(cosiphi+cosphil/b)/b;
        if (l >= 5) {
            dd[l-4] += sinphi+siniphi/b;
            dd[l-5] -= cosphi+cosiphi/b;
            if (l >= 7) {
                dd[l-6] += sinphi;
                dd[l-7] -= cosphi;
            }
        }
        sol(a,l,p,dd);
        zi=1.0/b;
```

```
                    for (i=1; i<=l; i++) {
                        beta[i]=dd[l+1-i]*zi;
                        betha[i]=(i+3)*beta[i];
                        zi /= b;
                    }
                }
            } else {
                /* form beta */
                if (l == 1) {
                    betha[1]=(0.5-(1.0-(1.0-exp(-b))/b)/b)/b;
                    beta[1]=(1.0/6.0-betha[1])/b;
                } else if (l == 2) {
                    e=exp(-b);
                    emin1=e-1.0;
                    betha[1]=(1.0-(3.0+e+4.0*emin1/b)/b)/b;
                    betha[2]=(0.5-(2.0+e+3.0*emin1/b)/b)/b/b;
                    beta[2]=(1.0/6.0-betha[1])/b/b;
                    beta[1]=(1.0/3.0-(1.5-(4.0+e+5.0*emin1/b)/b)/b)/b;
                } else {
                    betac[l-1]=c=ddd=exp(-b)/fac;
                    for (i=l-1; i>=1; i--) {
                        c=i*b*c/(l-i);
                        betac[i-1]=ddd=ddd*i+c;
                    }
                    b2=0.5-betac[2];
                    b1=(1.0-betac[1])*(l+1)/b;
                    b0=(1.0-betac[0])*(l+2)*(l+1)*0.5/b/b;
                    a3=1.0/6.0-betac[3];
                    a2=b2*(l+1)/b;
                    a1=b1*(l+2)*0.5/b;
                    a0=b0*(l+3)/3.0/b;
                    ddd=l/b;
                    for (i=1; i<=l; i++) {
                        beta[i]=(a3/i-a2/(i+1)+a1/(i+2)-a0/(i+3))*ddd+
                                    betac[i+3];
                        betha[i]=(b2/i-b1/(i+1)+b0/(i+2))*ddd+
                                    betac[i+2];
                        ddd=ddd*(l-i)/i/b;
                    }
                }
            }
        } else
            for (i=1; i<=l; i++) {
                betha[i]=beta[i-1];
                beta[i]=beta[i-1]/(i+3);
            }
    }
    (*output)(*x,xe,m,y,j,*k);
    /* difference scheme */
    if (m1 < m) {
        d2[m]=1.0;
        k0[m]=y[m]+2.0*h/3.0;
        y[m] += 0.25*h;
    }
    for (q=1; q<=m1; q++) {
        k0[q]=y[q]+2.0*h/3.0*d[q];
        y[q] += 0.25*h*d[q];
        d1[q]=h*matvec(1,m,q,j,d);
        d2[q]=d1[q]+d[q];
    }
    for (i=0; i<=l; i++) {
        betai=4.0*beta[i]/3.0;
        bethai=betha[i];
        for (q=1; q<=m1; q++) d[q]=h*d1[q];
        for (q=1; q<=m1; q++) {
            k0[q] += betai*d[q];
            d1[q]=matvec(1,m1,q,j,d);
            d2[q] += bethai*d1[q];
        }
    }
    (*derivative)(m,k0);
    for (q=1; q<=m; q++) y[q] += 0.75*h*k0[q];
    (*k)++;
    (*x) += h;
```

```
    } while (!last);
    (*output)(*x,xe,m,y,j,*k);
    free_integer_vector(p,1);
    free_real_vector(beta,0);
    free_real_vector(betha,0);
    free_real_vector(betac,0);
    free_real_vector(k0,1);
    free_real_vector(d,1);
    free_real_vector(d1,1);
    free_real_vector(d2,1);
    free_real_vector(dd,1);
    free_real_matrix(a,1,1,1);
}
```

C. liniger1vs

Solves an initial value problem, given as an autonomous system of first order differential equations

$$Dy(x) = f(y) \qquad (1)$$

where $D \equiv d/dx$ and $f,y, \in R^m$, with $y(x_0)$ prescribed, over the interval $x0 \leq x \leq xe$ by means of an implicit, first order accurate, exponentially fitted one-step method. Automatic stepsize control is provided. This procedure is suitable for the integration of stiff differential equations. The algorithm is based on [LiW70]. The stepsize strategy requires many extra array operations. The user may avoid this extra work by giving the parameters *aeta* and *reta* a negative value and prescribing a stepsize *hmax*.

The method used has the form

$$y_{n+1} = y_n + h\,(\mu_n f_n + (1 - \mu_n) f_{n+1}) \qquad (2)$$

where, with $x^n = x_0 + nh$ (n=0,1,...), $y_n \approx y(x_n)$ and $f_n = f(y_n)$.

Equation (1) may be written as

$$Dy = f_n + J_n\,(y - y_n) + \dots$$

J_n being the Jacobian matrix $\partial f(y_n)/\partial y_n$. The linearized version of (1) is thus $Dy' = f(y')$ where $f(y')=f_n-J_n y_n+J_n y'$. The exact solution of the linearized equation with $y'(x_n)=y_n$ is

$$y'(x) = y_n - (x - x_n)\, e_2\,(J_n(x - x_n))f_n$$

where $e_2(z) = (e^z-1)/z$ so that $y_{n+1}' = y_n - he_2(hJ_n)f_n$. With y_n'' a computed value of y_n (i.e. of y_n') the scheme (2) yields

$$(I - (1-\mu_n)hJ_n)y_{n+1}'' = (I + \mu_n hJ_n)y_n'' + h(f_n-J_n y_n).$$

Thus

$$y_{n+1}'' - y_{n+1}' = R_n(hJ_n)(y_n'' - y_n') + h(R_n^{(2)}(hJ_n) - e_2(hJ_n))f_n \qquad (3)$$

where R_n is the stability function

$$R_n(z) = (1 + \mu_n z) \,/\, (1 - (1-\mu_n)z)$$

and $R_n^{(2)}(z) = (R_n(z) - 1)/z$. If all eigenvalues of J_n not in the neighbourhood of the origin are clustered about the single point δ_n, and μ_n is determined by the condition

$$R_n(h\delta_n) = e^{h\delta_n}, \qquad \text{i.e.} \quad \mu_n = (1 - e^{-h\delta_n})^{-1} - (h\delta_n)^{-1}$$

the second term on the right hand side of equation (3) is negligible, and

$$\| y_{n+1}'' - y_{n+1}' \| \leq \| R_n(hJ_n) \| \; \| y_n'' - y_n' \|.$$

When $\delta_n \in (-\infty,0)$, $\mu_n \in [0,½]$ and $|R_n(z)| < 1$ for $z \in (-\infty,0)$. Thus, with such δ_n and μ_n determined by exponential fitting in this way, the scheme (2) is stable.

y_{n+1} in formula (2) is determined by setting $y_{n+1}^{(0)} = y_n$ and using the Newton scheme

$$[I - (1-\mu_n)hJ_{n+1}^{(l)}]\,\Delta_{n+1}^{(l)} = y_n - y_{n+1}^{(l)} + \mu_n hf_n + (1-\mu_n)hf_{n+1}^{(l)} \quad (l=0,1,\dots) \qquad (4)$$

where $\Delta_{n+1}^{(l)} = y_{n+1}^{(l+1)} - y_{n+1}^{(l)}$, and $J_{n+1}^{(l)}$ is an approximation to the Jacobian matrix $Df(y_{n+1}^{(l)})/Dy_{n+1}^{(l)}$. This scheme is continued until either $\|\Delta_{n+1}^{(l)}\| \leq \eta_a + \|y_{n+1}^{(l)}\|\eta_r$, where η_a and η_r are absolute and relative tolerances prescribed by the user, or $l = itmax$, where *itmax* is a similarly prescribed integer.

The Jacobian matrix $J_{n+1}^{(l)}$ occurring in formula (4) is updated at every iteration and every integration step.

Function Parameters:

void **liniger1vs**
(*x,xe,m,y,sigma,derivative,j,jacobian,itmax,hmin,hmax,aeta,reta,info,output*)

x: float *;
entry: the independent variable; the initial value *x0*;
exit: the final value *xe*;

xe: float;
entry: the final value of *x* ($xe \geq x$);

m: int;
entry: the number of equations;

y: float *y[1:m]*;
the dependent variable;
entry: the initial values of the system of differential equations: *y[i]* at *x=x0*;
exit: the final values of the solution: *y[i]* at *x=xe*;

sigma: float *;
the modulus of the point at which exponential fitting is desired, for example the largest negative eigenvalue of the Jacobian matrix of the system of differential equations;

derivative: void (**derivative*)(*m,y,sigma*);
this procedure should deliver the right hand side of the *i*-th differential equation at the point (*y*) as *y[i]*;

j: float *j[1:m,1:m]*;
the Jacobian matrix of the system; array *j* should be updated in the procedure *jacobian*;

jacobian: void (**jacobian*)(*m,j,y,sigma*);
in this procedure (an approximation to) the Jacobian has to be assigned to the array *j*;

itmax: int;
entry: an upper bound for the number of iterations in Newton's process, used to solve the implicit equations;

hmin: float;
entry: minimal stepsize by which the integration is performed;

hmax: float;
entry: maximal stepsize by which the integration is performed;

aeta: float;
entry: required absolute precision in the integration process;

reta: float;
entry: required relative precision in the integration process;
if both *aeta* and *reta* have negative values then integration will be performed with a stepsize equal to *hmax*, which may be varied by the user; in this case the absolute values of *aeta* and *reta* will control the Newton iteration;

info: float *info[1:9]*;
during integration and upon exit this array contains the following information:
info[1]: number of integration steps taken;

info[2]: number of derivative evaluations used;
info[3]: number of Jacobian evaluations used;
info[4]: number of integration steps equal to *hmin* taken;
info[5]: number of integration steps equal to *hmax* taken;
info[6]: maximal number of integrations taken in the Newton process;
info[7]: local error tolerance;
info[8]: estimated local error;
info[9]: maximum value of the estimated local error;

output: void (**output*)(*x,xe,m,y,sigma,j,info*);
this procedure is called at the end of each integration step; the user can ask for output of some parameters, for example, *x, y, j, info.*

Functions used: inivec, mulvec, mulrow, dupvec, matvec, elmvec, vecvec, dec, sol.

```
#include <math.h>

void liniger1vs(float *x, float xe, int m, float y[], float *sigma,
          void (*derivative)(int, float[], float *), float **j,
          void (*jacobian)(int, float **, float [], float *),
          int itmax, float hmin, float hmax, float aeta, float reta,
          float info[],
          void (*output)(float, float, int, float [], float,
                              float **, float []))
{
    int *allocate_integer_vector(int, int);
    float *allocate_real_vector(int, int);
    float **allocate_real_matrix(int, int, int, int);
    void free_integer_vector(int *, int);
    void free_real_vector(float *, int);
    void free_real_matrix(float **, int, int, int);
    void inivec(int, int, float [], float);
    void mulvec(int, int, int, float [], float [], float);
    void mulrow(int, int, int, int, float **, float **, float);
    void dupvec(int, int, int, float [], float []);
    float vecvec(int, int, int, float [], float []);
    float matvec(int, int, int, float **, float []);
    void elmvec(int, int, int, float [], float [], float);
    void dec(float **, int, float [], int []);
    void sol(float **, int, int [], float []);
    void liniger1vscoef(int, float **, float **, float [], int [],
              float, float, float *, float *, float *, float *);
    int i,st,lastjac,last,first,evaljac,evalcoef,*pi,k,q;
    float h,hnew,mu,mu1,beta,p,e,e1,eta,eta1,discr,*dy,*yl,*yr,*f,
          **a,aux[4],hl,b;

    pi=allocate_integer_vector(1,m);
    dy=allocate_real_vector(1,m);
    yl=allocate_real_vector(1,m);
    yr=allocate_real_vector(1,m);
    f=allocate_real_vector(1,m);
    a=allocate_real_matrix(1,m,1,m);

    first=evaljac=1;
    last=evalcoef=0;
    inivec(1,9,info,0.0);
    eta=reta*sqrt(vecvec(1,m,0,y,y))+aeta;
    eta1=eta/sqrt(fabs(reta));
    dupvec(1,m,0,f,y);
    (*derivative)(m,f,sigma);
    info[2]=1.0;
    st=1;
    do {
        /* step size */
        if (eta < 0.0) {
            hl=h;
            h=hnew=hmax;
```

```
        info[5] += 1.0;
        if (1.1*hnew > xe-(*x)) {
            last=1;
            h=hnew=xe-(*x);
        }
        evalcoef=(h != hl);
    } else if (first) {
        h=hnew=hmin;
        first=0;
        info[4] += 1.0;
    } else {
        b=discr/eta;
        hl=h;
        if (b < 0.01) b=0.01;
        hnew = (b > 0.0) ? h*pow(b,-1.0/p) : hmax;
        if (hnew < hmin) {
            hnew=hmin;
            info[4] += 1.0;
        } else
            if (hnew > hmax) {
                hnew=hmax;
                info[5] += 1.0;
            }
        if (1.1*hnew >= xe-(*x)) {
            last=1;
            h=hnew=xe-(*x);
        } else
            if (fabs(h/hnew-1.0) > 0.1) h=hnew;
        evalcoef=(h !=hl);
    }
    info[1] += 1.0;
    if (evaljac) {
        (*jacobian)(m,j,y,sigma);
        info[3] += 1.0;
        h=hnew;
        liniger1vscoef(m,a,j,aux,pi,h,*sigma,&mu,&mu1,&beta,&p);
        evaljac=0;
        lastjac=st;
    } else
        if (evalcoef)
            liniger1vscoef(m,a,j,aux,pi,h,*sigma,&mu,&mu1,&beta,&p);
    i=1;
    do {
        /* iteration */
        if (reta < 0.0) {
            if (i == 1) {
                mulvec(1,m,0,dy,f,h);
                for (k=1; k<=m; k++) yl[k]=y[k]+mu*dy[k];
                sol(a,m,pi,dy);
                e=1.0;
            } else {
                for (k=1; k<=m; k++) dy[k]=yl[k]-y[k]+mu1*f[k];
                if (e*sqrt(vecvec(1,m,0,y,y)) > e1*e1) {
                    evaljac=(i >= 3);
                    if (i > 3) {
                        info[3] += 1.0;
                        (*jacobian)(m,j,y,sigma);
                        for (q=1; q<=m; q++) {
                            mulrow(1,m,q,q,a,j,-mu1);
                            a[q][q] += 1.0;
                        }
                        aux[2]=0.0;
                        dec(a,m,aux,pi);
                    }
                }
                sol(a,m,pi,dy);
            }
            e1=e;
            e=sqrt(vecvec(1,m,0,dy,dy));
            elmvec(1,m,0,y,dy,1.0);
            eta=sqrt(vecvec(1,m,0,y,y))*reta+aeta;
            discr=0.0;
            dupvec(1,m,0,f,y);
            (*derivative)(m,f,sigma);
```

```
                info[2] += 1.0;
            } else {
                if (i == 1) {
                    /* linearity */
                    for (k=1; k<=m; k++) dy[k]=y[k]-mu1*f[k];
                    sol(a,m,pi,dy);
                    elmvec(1,m,0,dy,y,-1.0);
                    e=sqrt(vecvec(1,m,0,dy,dy));
                    if (e*(st-lastjac) > eta) {
                        (*jacobian)(m,j,y,sigma);
                        lastjac=st;
                        info[3] += 1.0;
                        h=hnew;
                        liniger1vscoef(m,a,j,aux,pi,h,*sigma,&mu,&mu1,
                                        &beta,&p);
                        /* linearity */
                        for (k=1; k<=m; k++) dy[k]=y[k]-mu1*f[k];
                        sol(a,m,pi,dy);
                        elmvec(1,m,0,dy,y,-1.0);
                        e=sqrt(vecvec(1,m,0,dy,dy));
                    }
                    evaljac=(e*(st+1-lastjac) > eta);
                    mulvec(1,m,0,dy,f,h);
                    for (k=1; k<=m; k++) yl[k]=y[k]+mu*dy[k];
                    sol(a,m,pi,dy);
                    for (k=1; k<=m; k++) yr[k]=h*beta*matvec(1,m,k,j,dy);
                    sol(a,m,pi,yr);
                    elmvec(1,m,0,yr,dy,1.0);
                } else {
                    for (k=1; k<=m; k++) dy[k]=yl[k]-y[k]+mu1*f[k];
                    if (e > eta1 && discr > eta1) {
                        info[3] += 1.0;
                        (*jacobian)(m,j,y,sigma);
                        for (q=1; q<=m; q++) {
                            mulrow(1,m,q,q,a,j,-mu1);
                            a[q][q] += 1.0;
                        }
                        aux[2]=0.0;
                        dec(a,m,aux,pi);
                    }
                    sol(a,m,pi,dy);
                    e=sqrt(vecvec(1,m,0,dy,dy));
                }
                elmvec(1,m,0,y,dy,1.0);
                eta=sqrt(vecvec(1,m,0,y,y))*reta+aeta;
                eta1=eta/sqrt(reta);
                dupvec(1,m,0,f,y);
                (*derivative)(m,f,sigma);
                info[2] += 1.0;
                for (k=1; k<=m; k++) dy[k]=yr[k]-h*f[k];
                discr=sqrt(vecvec(1,m,0,dy,dy))/2.0;
            }
            if (i > info[6]) info[6]=i;
            i++;
        } while (e > fabs(eta) && discr > 1.3*eta && i <= itmax);
        info[7]=eta;
        info[8]=discr;
        (*x) += h;
        if (discr > info[9]) info[9]=discr;
        (*output)(*x,xe,m,y,*sigma,j,info);
        st++;
    } while (!last);
    free_integer_vector(pi,1);
    free_real_vector(dy,1);
    free_real_vector(yl,1);
    free_real_vector(yr,1);
    free_real_vector(f,1);
    free_real_matrix(a,1,m,1);
}

void liniger1vscoef(int m, float **a, float **j, float aux[],
                int pi[], float h, float sigma, float *mu,
                float *mu1, float *beta, float *p)
{
```

```
	/* this function is internally used by LINIGER1VS */

	int q;
	float b,e;

	b=fabs(h*sigma);
	if (b > 40.0) {
		*mu = 1.0/b;
		*beta = 1.0;
		*p = 2.0+2.0/(b-2.0);
	} else if (b < 0.04) {
		e=b*b/30.0;
		*p = 3.0-e;
		*mu = 0.5-b/12.0*(1.0-e/2.0);
		*beta = 0.5+b/6.0*(1.0-e);
	} else {
		e=exp(b)-1.0;
		*mu = 1.0/b-1.0/e;
		*beta = (1.0-b/e)*(1.0+1.0/e);
		*p = ((*beta)-(*mu))/(0.5-(*mu));
	}
	*mul = h*(1.0-(*mu));
	for (q=1; q<=m; q++) {
		mulrow(1,m,q,q,a,j,-(*mul));
		a[q][q] += 1.0;
	}
	aux[2]=0.0;
	dec(a,m,aux,pi);
}
```

D. liniger2

Solves an initial value problem, given as an autonomous system of first order differential equations

$$Dy(x) = f(y) \tag{1}$$

where $D \equiv d/dx$ and $f, y, \in R^m$, with $y(x_0)$ prescribed, over the interval $x0 \leq x \leq xe$ by means of an implicit, second (or possibly third) order accurate, exponentially fitted one-step method [see BeDHKW73, DekHH72, LiW70]. No automatic stepsize control is provided. This procedure is suitable for the integration of stiff differential equations.

The method used has the form

$$y_{n+1} = y_n + (1/2)h[(1-a_n)f_n + (1+a_n)f_{n+1}] + (1/4)h^2[(b_n-a_n)J_nf_n - (b_n+a_n)J_{n+1}f_{n+1}] \tag{2}$$

where, with $x^n = x_0 + nh$ $(n=0,1,...)$, $y_n \approx y(x_n)$, $f_n = f(y_n)$ and $J_n = \partial f(y_n)/\partial y_n$.
An analysis similar to that given in the documentation of *liniger1vs* reveals that the stability function for the above method has the form

$$R(z) = [1 + (1/2)(1-a_n)z + (1/4)(b_n-a_n)z^2] \,/\, [1 - (1/2)(1+a_n)z + (1/4)(b_n+a_n)z^2].$$

Exponential fitting is now possible at two points: with distinct nonzero z_0 and z_1,

$$R(z_0) = e^{z_0} \qquad \text{and} \qquad R(z_1) = e^{z_1}$$

if $a_n = 2[g(z_1)-g(z_0)]/[z_1g(z_0)-z_0g(z_1)]$, $b_n = 2(z_1-z_0)/[z_1g(z_0)-z_0g(z_1)]$
where $g(z) = z^2(1-e^z)/[e^z(2-z)-(2+z)]$.

It may be shown that the method (2) is stable in three cases. In the first (a) the eigenvalues of J_n not in the neighbourhood of the origin are clustered about two points $\delta_0^{(n)}$ and $\delta_1^{(n)}$ on the negative real axis; a_n and b_n are then obtained as above by setting $z_0=h\delta_0^{(n)}$, $z_1=h\delta_1^{(n)}$;
(b) the above eigenvalues are clustered about two complex conjugate points $\delta_0^{(n)}=\sigma e^{i\phi}$, $\delta_1^{(n)}=\sigma e^{-i\phi}$, in the left half plane $\pi/2 < \phi < 3\pi/2$; a_n and b_n are then determined as in (a); and

(c) the above eigenvalues are clustered about a single point $\delta_0^{(n)}$ on the negative real axis; a_n and b_n are then obtained by limiting relationships of the above form with $z_0 = h\delta_0^{(n)}$ and $z_1=0$.

In case (a) and (b) above, the method (2) is second order accurate; in case (c) it is third order accurate.

y_{n+1} in formula (2) is determined by setting $y_{n+1}^{(n)} = y_n$ and using the modified Newton scheme

$$[4I - 2h(1+a_n)J_{n+1}^{(l)} + h^2(b_n+a_n)(J_{n+1}^{(l)})^2]\,\Delta y_{n+1}^{(l)} =$$
$$h[2(1+a_n)I - h(b_n+a_n)J_{n+1}^{(l)}]f_{n+1}^{(l)} + 4y_n + 2h(1-a_n)f_n + h^2(b_n-a_n)J_nf_n - 4y_{n+1}^{(l)}$$

where $\Delta_{n+1}^{(l)} = y_{n+1}^{(l+1)} - y_{n+1}^{(l)}$ and $J_{n+1}^{(l)}$ is an approximation to the Jacobian matrix $\partial f(y_{n+1}^{(l)}/\partial y_{n+1}^{(l)}$. This scheme is continued until conditions analogous to those given in the documentation to *liniger1vs* are satisfied.

Function Parameters:

void liniger2 (*x,xe,m,y,sigma1,sigma2,f,evaluate,j,jacobian,k,itmax,step,aeta,reta,output*)

x: float *;
- entry: the independent variable; the initial value *x0*;
- exit: the final value *xe*;

xe: float;
- entry: the final value of *x* (*xe* ≥ *x*);

m: int;
- entry: the number of equations;

y: float *y[1:m]*;
the dependent variable;
- entry: the initial values of the system of differential equations: *y[i]* at *x=x0*;
- exit: the final values of the solution: *y[i]* at *x=xe*;

sigma1: float *;
the modulus of the point at which exponential fitting is desired, this point may be complex or real and negative;

sigma2: float *;
sigma2 may define three different types of exponential fitting; fitting in two complex conjugated points, fitting in two real negative points, or fitting in one point combined with third order accuracy;
if third order is desired then *sigma2* should have the value 0;
if fitting in a second negative point is desired then *sigma2* should have the value of the modulus of this point;
if fitting in two complex conjugated points is desired then *sigma2* should be minus the value of the argument of the point in the second quadrant
(thus a value between -π and -π/2);

f: float (**f*)(*m,y,i,sigma1,sigma2*); int *i*;
this procedure should deliver the right hand side of the *i*-th differential equation as *f*;

derivative: void (**derivative*)(*m,y,sigma*);
this procedure should deliver the right hand side of the *i*-th differential equation at the point (*y*) as *y[i]*;

evaluate: int (**evaluate*)(*itnum*); int *itnum*;
evaluate should be assigned the value nonzero if it is desired that the Jacobian of the system is updated in the *itnum*-th iteration step of the Newton process;

j: float *j[1:m,1:m]*;

the Jacobian matrix of the system; array *j* should be updated in the procedure *jacobian*;

jacobian: void (**jacobian*)(*m,j,y,sigma1,sigma2*);
in this procedure the Jacobian has to be assigned to the array *j*; or an approximation of the Jacobian, if only second order accuracy is required;

k: int *;
counts the number of integration steps taken;

itmax: int;
entry: an upper bound for the number of iterations in Newton's process, used to solve the implicit equations;

step: float;
entry: the length of the integration step, to be prescribed by the user;

aeta: float;
entry: required absolute precision in the Newton process, used to solve the implicit equations;

reta: float;
entry: required relative precision in the Newton process, used to solve the implicit equations;

output: void (**output*)(*x,xe,m,y,sigma1,sigma2,j,k*);
this procedure is called at the end of each integration step; the user can ask for output of some parameters, for example, *x, y, j, k*.

Functions used: vecvec, matvec, matmat, dec, sol.

```
#include <math.h>

void liniger2(float *x, float xe, int m, float y[], float *sigma1,
            float *sigma2,
            float (*f)(int, float[], int, float *, float *),
            int (*evaluate)(int), float **j,
            void (*jacobian)(int, float **, float [], float *, float *),
            int *k, int itmax, float step, float aeta, float reta,
            void (*output)(float, float, int, float [], float, float,
                                float **, int))
{
    int *allocate_integer_vector(int, int);
    float *allocate_real_vector(int, int);
    float **allocate_real_matrix(int, int, int, int);
    void free_integer_vector(int *, int);
    void free_real_vector(float *, int);
    void free_real_matrix(float **, int, int, int);
    float vecvec(int, int, int, float [], float []);
    float matvec(int, int, int, float **, float []);
    float matmat(int, int, int, int, float **, float **);
    void dec(float **, int, float [], int []);
    void sol(float **, int, int [], float []);
    void liniger2coef(int, float **, float **, float [], int [],
            float, float, float, float *, float *, float *,
            float *, float *);
    int i,last,*pi,itnum;
    float h,hl,c0,c1,c2,c3,c4,*dy,*yl,*fl,**a,aux[4],
            jfl,eta,discr;

    pi=allocate_integer_vector(1,m);
    dy=allocate_real_vector(1,m);
    yl=allocate_real_vector(1,m);
    fl=allocate_real_vector(1,m);
    a=allocate_real_matrix(1,m,1,m);

    last=0;
    *k = 0;
    hl=0.0;
    do {
```

```
        (*k)++;
        /* step size */
        h=step;
        if (1.1*h >= xe-(*x)) {
            last=1;
            h=xe-(*x);
            (*x)=xe;
        } else
            (*x) += h;
        /* newton iteration */
        itnum=0;
        while (1) {
            itnum++;
            if ((*evaluate)(itnum)) {
                (*jacobian)(m,j,y,sigma1,sigma2);
                liniger2coef(m,j,a,aux,pi,h,*sigma1,*sigma2,
                                &c0,&c1,&c2,&c3,&c4);
            } else
                if (itnum == 1 && h != hl)
                    liniger2coef(m,j,a,aux,pi,h,*sigma1,*sigma2,
                                    &c0,&c1,&c2,&c3,&c4);
            for (i=1; i<=m; i++) fl[i]=(*f)(m,y,i,sigma1,sigma2);
            if (itnum == 1)
                for (i=1; i<=m; i++) {
                    jfl=matvec(1,m,i,j,fl);
                    dy[i]=h*(fl[i]-c4*jfl);
                    yl[i]=y[i]+c2*fl[i]+c3*jfl;
                }
            else
                for (i=1; i<=m; i++)
                    dy[i]=yl[i]-y[i]+c1*fl[i]-c0*matvec(1,m,i,j,fl);
            sol(a,m,pi,dy);
            for (i=1; i<=m; i++) y[i] += dy[i];
            if (itnum >= itmax) break;
            eta=sqrt(vecvec(1,m,0,y,y))*reta+aeta;
            discr=sqrt(vecvec(1,m,0,dy,dy));
            if (eta >= discr) break;
        }
        hl=h;
        (*output)(*x,xe,m,y,*sigma1,*sigma2,j,*k);
    } while (!last);
    free_integer_vector(pi,1);
    free_real_vector(dy,1);
    free_real_vector(yl,1);
    free_real_vector(fl,1);
    free_real_matrix(a,1,m,1);
}

void liniger2coef(int m, float **j, float **a, float aux[], int pi[],
            float h, float sigma1, float sigma2, float *c0, float *c1,
            float *c2, float *c3, float *c4)
{
    /* this function is internally used by LINIGER2 */

    int i,k,out,doublefit;
    float b1,b2,r,r1,r2,ex,zeta,eta,sinl,cosl,sinh,cosh,d,p,q;

    out=0;
    doublefit=0;
    b1=h*sigma1;
    b2=h*sigma2;
    if (b1 < 0.1) {
        p=0.0;
        q=1.0/3.0;
        out=1;
    }
    if (!out) {
        if (b2 < 0.0) {
            /* complex */
            eta=fabs(b1*sin(sigma2));
            zeta=fabs(b1*cos(sigma2));
            if (eta < b1*b1*1.0e-6) {
                b1=b2=zeta;
                doublefit=1;
```

```
            }
            if (!doublefit)
                if (zeta > 40.0) {
                    p=1.0-4.0*zeta/b1/b1;
                    q=4.0*(1.0-zeta)/b1/b1+1.0;
                } else {
                    ex=exp(zeta);
                    sinl=sin(eta);
                    cosl=cos(eta);
                    sinh=0.5*(ex-1.0/ex);
                    cosh=0.5*(ex+1.0/ex);
                    d=eta*(cosh-cosl)-0.5*b1*b1*sinl;
                    p=(zeta*sinl+eta*sinh-
                                4.0*zeta*eta/b1/b1*(cosh-cosl))/d;
                    q=eta*((cosh-cosl-zeta*sinh-eta*sinl)*
                                4.0/b1/b1+cosh+cosl)/d;
                }
        } else if (b1 < 1.0 || b2 < 0.1) {
            /* third order */
            q=1.0/3.0;
            if (b1 > 40.0)
                r=b1/(b1-2.0);
            else {
                ex=exp(-b1);
                r=b1*(1.0-ex)/(b1-2.0+(b1+2.0)*ex);
            }
            p=r/3.0-2.0/b1;
        } else if (fabs(b1-b2) < b1*b1*1.0e-6)
            doublefit=1;
        else {
            if (b1 > 40.0)
                r=b1/(b1-2.0);
            else {
                ex=exp(-b1);
                r=b1*(1.0-ex)/(b1-2.0+(b1+2.0)*ex);
            }
            r1=r*b1;
            if (b2 > 40.0)
                r=b2/(b2-2.0);
            else {
                ex=exp(-b2);
                r=b2*(1.0-ex)/(b2-2.0+(b2+2.0)*ex);
            }
            r2=r*b2;
            d=b2*r1-b1*r2;
            p=2.0*(r2-r1)/d;
            q=2.0*(b2-b1)/d;
        }
        if (doublefit) {
            b1=0.5*(b1+b2);
            if (b1 > 40.0)
                r=b1/(b1-2.0);
            else {
                ex=exp(-b1);
                r=b1*(1.0-ex)/(b1-2.0+(b1+2.0)*ex);
            }
            r1=r;
            if (b1 > 40.0) ex=0.0;
            r2=b1/(1.0-ex);
            r2=1.0-ex*r2*r2;
            q=1.0/(r1*r1*r2);
            p=r1*q-2.0/b1;
        }
    }
    *c0 = 0.25*h*h*(p+q);
    *c1 = 0.5*h*(1.0+p);
    *c2 = h-(*c1);
    *c3 = 0.25*h*h*(q-p);
    *c4 = 0.5*h*p;
    for (i=1; i<=m; i++) {
        for (k=1; k<=m; k++)
            a[i][k]=(*c0)*matmat(1,m,i,k,j,j)-(*c1)*j[i][k];
        a[i][i] += 1.0;
    }
```

```
    aux[2]=0.0;
    dec(a,m,aux,pi);
}
```

E. gms

Solves an initial value problem, given as an autonomous system of first order differential equations

$$Dy(x) = f(y) \tag{1}$$

where D denotes differentiation with respect to the indicated argument and $y, f \in \mathrm{R}^r$, with $y(x_0)$ prescribed, over the interval $x0 \leq x \leq xe$ by means of a third order, exponentially fitted, generalized multistep method with automatic stepsize control [VhV74, Vw74]. This procedure is suitable for the integration of stiff differential equations.

The procedure *gms* describes an implementation of a third order three-step generalized linear multistep method with quasi-zero parasitic roots and quasi-adaptive stability function. The procedure supplies the additional starting values and performs a stepsize control which is based on the non-linearity of the differential equation. By this control the Jacobian matrix is incidentally evaluated. It is possible to eliminate the stepsize control. Then, one has to give the number of integration steps per Jacobian evaluation. For linear equations the stepsize control is automatically eliminated, while the procedure performs one evaluation of the Jacobian. However, in this case the three-step scheme is reduced to a one-step scheme. The procedure uses one function evaluation per integration step and it does not reject integration steps. Each change in the steplength or each reevaluation of the Jacobian costs one LU-decomposition. It is possible to fit exponentially, this fitting is equivalent to fitting in the sense of Liniger and Willoughby, only when the Jacobian matrix is evaluated at each integration step. When the system to be integrated is non-linear and the Jacobian matrix is not evaluated at each integration step, it is recommended to fit at infinity (input parameter $delta \leq -10^{15}$).

The method used is one of a class having the form

$$y_{n+1} = R(h_n J_n^*) + h_n \sum_{l=1}^{k} B_l(h_n J_n^*)[f(y_{n+1-l}) - J_n^* y_{n+1-l}] \tag{2}$$

where, with $x^{n+l} = x_n + h_n$, $y_n \approx y(x_n)$, J_n^* is an approximation to the Jacobian matrix $J_n = \partial f(y_n)/\partial y_n$, R and the B_l being rational functions.

Considerations of stability demand that $\| B_l(h_n J_n^*) \|$ $(l=1,...,k)$ be as small as possible: the stabilizing condition $B_l(z) \to 0$ as Real$(z) \to -\infty$ $(l=1,...,k)$ is imposed.

The method (2) is k-th order accurate if, with

$$q_l = (x_{n-l} - x_n)/h_n \quad (l=0,1,...,k-1)$$

and

$$C_0(z) = R(z) - \sum_{l=1}^{k} z\, B_l(z) ,$$

$$C_j(z) = \frac{1}{j!}\sum_{l=1}^{k} (j - zq_{l-1})q_{l-1}^{j-1}B_l(z) ,$$

$$C_j^{(i)} = \lim D^i C_j(z) \quad (z \to 0)$$

the conditions

$$C_j^{(0)} = (1/j!) \quad (j=1,...,k) \qquad C_i^{(j-l)} = 0 \quad (j=1,...,k;\ i=0,...,j\text{-}1)$$
$$\lim D^i R(z) = 1 \quad (z \to 0) \qquad (i=0,...,k) \tag{3}$$

are satisfied or, alternatively, with

$$\sum_{l=1}^{k} q_{l-1}^{j-1} B_l(z) = G_j(z) \qquad (j=1,...,k)$$

$$G_1(z) = [R(z)-1+O(z^{k+1})]/z, \quad G_{j+1}(z) = [jG_j(z)-1-O(z^{k+1-j})]/z \qquad (j=1,...,k\text{-}1) \tag{4}$$

as $z \to 0$. The one-step scheme

$$y_{n+1} = y_n + (J^*)^{-1}[R(h_n J_n^*) - I]f(y_n)$$

is second order accurate if $J_n^* = J_n$ exactly, and (3) is satisfied for some $k > 1$.

The local truncation error of the scheme (2) at $x = x_n$ is

$$\sum_{j=1}^{k}\left[\left(\frac{1}{(k+j)!} - c_{k+j}^{(0)}\right) D^{k+j}y(x_n) - \sum_{i=0}^{k+j-1} \frac{c_i^{(k-i+j)}}{(k-i+j)!} J_n^{*k-i+j} D^i y(x_n)\right] h_n^{k+j} .$$

The choice of $B_l(z)$ resulting from taking the 0 terms in the expressions (4) for the G_j to be zero minimizes the local truncation error, and in this case

$$c_j(z) = 1/j! \quad (j=1,...,k\text{-}1), \qquad c_n(z) = 1/k! + O(z)$$

and $c_j^{(i)} = 0$ $(j=0,1,...,k\text{-}1;\ i=1,2,...)$, $c_k^{(i)} = 0$ $(i=2,3,...)$. The local truncation error then becomes

$$\sum_{j=1}^{k}\left[\left(\frac{1}{(k+j)!} - c_{k+j}^{(0)}\right) D^{k+j}y(x_n) - \sum_{i=1}^{j} c_{k+j-i}^{(i)} J_n^{*i}\, D^{k+j-i}y(x_n)\right] h_n^{k+j} .$$

The stability function R used by *gms* is

$$R(z) = \frac{1 + \frac{1}{2}(1 - \alpha^{(n)})z + \frac{1}{12}(1 - 3\alpha^{(n)})z^2}{1 - \frac{1}{2}(1 + \alpha^{(n)})z + \frac{1}{12}(1 + 3\alpha^{(n)})z^2} .$$

The parameter $\alpha^{(n)}$ is determined by exponential fitting at the point $z = h_n\delta_n$ where δ_n is the eigenvalue of greatest absolute value on the negative real axis of J_n^*:

$$R(h_n\delta_n) = e^{h_n\delta_n} \quad \text{if} \quad \alpha^{(n)} = \frac{1}{3z_n}\,\frac{e^{z_n}(z_n^2-6z_n+12) - (z_n^2+6z_n+12)}{e^{z_n}(2-z_n) - (2+z_n)}$$

where $z_n = h_n\delta_n$. If $\alpha^{(n)} = 0$, R is fourth order consistent; otherwise third.

Taking the denominator of $R(z)$ to be $Q(z)$, the G_j have, when k=3, the form $G_j(z)=(g_{1,j}+g_{2,j}z)/Q(z)$ where $g_{1,j} = 1/j$, $g_{2,1} = -\alpha^{(n)}/2$, $g_{2,2} = g_{2,3} = -(1 + 3\alpha^{(n)})/12$, and then

$$B_l(z) = \sum_{j=1}^{3} b_{l,j} G_j(z) \qquad (l=1,2,3)$$

where $b_{1,1} = 1$, $b_{2,1} = b_{3,1} = 0$, $b_{1,2} = -(q_1+q_2)/q_1q_2$, $b_{2,2} = -q_2/(q_1^2-q_1q_2)$, $b_{3,2} = -q_1/(q_2^2-q_1q_2)$, $b_{1,3} = 1/q_1q_2$, $b_{2,3} = 1/(q_1^2-q_1q_2)$, $b_{3,3} = 1/(q_2^2-q_1q_2)$.

The three step, third order scheme of the class (2) with minimized local truncation error then becomes

$$Q(h_nJ_n^*)y_{n+1} = y_n + \sum_{l=1}^{k}\sum_{j=1}^{k} h_n b_{l,j} d_{1,j} f(y_{n+1-l}) +$$

$$h_n^2 J_n^{*2}\left[\sum_{l=1}^{k}\sum_{j=1}^{k} [h_n b_{l,j} d_{2,j} f(y_{n+1-l}) - b_{l,j} d_{1,j} y_{n+1-l}] + \frac{1-\alpha^{(n)}}{2} y_n\right] + \tag{5}$$

$$h_n^2 J_n^{*2}\left[\sum_{l=1}^{k}\sum_{j=1}^{k} (-b_{l,j} d_{2,j} y_{n+1-l}) + \frac{1-3\alpha^{(n)}}{12} y_n\right]$$

with k=3. When k=2, the $b_{l,j}$ are given by $b_{1,1} = 1$, $b_{2,1} = 0$, $b_{1,2} = 1 - 1/q_1$, $b_{2,2} = 1/q_1$.

The user is asked to provide an initial stepsize h_0. $J_0^* = J_0$ is then evaluated exactly, and the one-step scheme is used to determine y_1. The two-step scheme (resulting from the substitution k=2 above) is used to determine y_2, and thereafter the three-step scheme (5) is used.

The user is also asked to provide two limits *hmin* and *hmax* for the minimal and maximal stepsizes respectively. If these are equal (their particular values are then ignored) the computations are continued using the scheme (5) with constant stepsize h_0. In this case the user must also provide an integer value *nsjev*; the approximation J_n^* is then updated every *nsjev* steps.

The user is also asked to allocate a value to the parameter *linear* upon call. If this value is nonzero then a constant stepsize h and a constant Jacobian $J_n^* = J_0$ are used throughout the computation.

If the value allocated to *linear* is zero then the following method of estimating a stepsize is used. If y'_{n+1} were to be determined by use of a two step scheme (with k=2 above) and y_{n+1} were to be computed by use of the three-step scheme (5) with stepsize h_n' in both cases, then $\|y_{n+1} - y_{n+1}'\| \approx c_n h_n'^3$ where c_n is independent of h_n'. Thus with a permitted local tolerance tol, and assuming $c_{n-1} = c_n$, the maximal value of h_n' is given by $h_n' = (tol/c_n)^{1/3} \approx (tol/c_{n-1})^{1/3} = h_{n-1}(tol/\|y_n-y_n'\|)^{1/3} \approx h_{n-1}[((4/3)tol)/(tol + \|y_n-y_n'\|) + 1/3]$ where y_n' has been computed as above (from y_{n-1}) by means of a two-step method with stepsize h_{n-1}. The bound $tol = \eta_a + \eta_r\|y_n\|$ is used, where η_a and η_r are absolute and relative tolerances prescribed by the user. A new stepsize is chosen (i.e. h_n is taken to be h_n' rather than h_{n-1}) if $|(h_{n-1}-h_n')/h_{n-1}| > 0.1$. A new J_n^* is determined if (i) h_n is to be changed and a new J_n^* was not determined at the preceding step and (ii) $10|h_{n-1}-h_n'| < 0.1|h_{n-1}|$. (Naturally h_n is taken to be *hmin* or *hmax* as is appropriate, if h_n' lies outside the range $hmin \leq h_n' \leq hmax$).

Formula (5) represents a system of linear equations. The matrix $Q(h_n, J_n^*)$ is decomposed in LU form and y_{n+1} is obtained from this decomposition. The parameters *jev* and *lu* record the current number of evaluations of J_n^* and of LU decompositions (since h_n may be changed while J_n^* remains unchanged, these two numbers may differ).

Function Parameters:

void gms

(*x,xe,r,y,h,hmin,hmax,delta,derivative,jacobian,aeta,reta,n,jev,lu,nsjev,linear,out*)

x: float *;
- entry: the independent variable; the initial value *x0*;
- exit: the final value *xe*;

xe: float;
entry: the final value of x ($xe \geq x$);
r: int;
entry: the number of differential equations;
y: float *y[1:r]*;
the dependent variable;
entry: the initial value of y;
exit: the solution y at the point x after each integration step;
h: float;
entry: the steplength when the integration has to be performed without the stepsize mechanism, otherwise the initial steplength (see *hmin* and *hmax* below);
hmin,hmax: float;
entry: minimal and maximal steplength by which the integration is allowed to be performed; by putting *hmin=hmax* the stepsize mechanism is eliminated; in this case the given values for *hmin* and *hmax* are irrelevant, while the integration is performed with the steplength given by *h*;
delta: float *;
entry:
the real part of the point at which exponential fitting is desired; alternatives:
delta= (an estimate of) the real part of the largest eigenvalue in modulus of the Jacobian matrix of the system;
delta≤ -10^{15}, in order to obtain asymptotic stability;
delta= 0, in order to obtain a higher order of accuracy in case of linear equations;
derivative: void (**derivative*)(*r,y,delta*);
this procedure should replace the i-th component of the solution y by the i-th component of the derivative $f(y)$, i=1,...,r;
jacobian: void (**jacobian*)(*r,j,y,delta*);
in *gms* when this procedure is called, the array y contains the values of the dependent variable; upon completion of a call of *jacobian* the array j should contain the values of the Jacobian matrix of $f(y)$;
aeta,reta: float;
entry: measure of the absolute and relative local accuracy required;
these values are irrelevant when the integration is performed without the stepsize mechanism;
n: int *;
exit: the number of integration steps;
jev: int *;
exit: the number of Jacobian evaluations;
lu: int *;
exit: the number of LU-decompositions;
nsjev: int;
entry: number of integration steps per Jacobian evaluation;
the value of *nsjev* is relevant only when the integration is performed without the stepsize mechanism and the system to be solved is non-linear;
linear: int;
entry: *linear* equals nonzero when the system to be integrated is linear, otherwise zero;
if *linear* is nonzero the stepsize mechanism is automatically eliminated;
out: void (**out*)(*x,xe,r,y,delta,n,jev,lu*);

this procedure is called after each integration step; the user can ask for output of some parameters, for example, *x, y, delta, n, jev, lu.*

Functions used: vecvec, matvec, matmat, elmrow, elmvec, dupvec, gsselm, solelm, colcst, mulvec.

```
#include <float.h>
#include <math.h>

void gms(float *x, float xe, int r, float y[], float h,
        float hmin, float hmax, float *delta,
        void (*derivative)(int, float [], float *),
        void (*jacobian)(int, float **, float [], float *),
        float aeta, float reta, int *n, int *jev, int *lu,
        int nsjev, int linear,
        void (*out)(float, float, int, float [], float,
                    int, int, int))
{
    int *allocate_integer_vector(int, int);
    float *allocate_real_vector(int, int);
    float **allocate_real_matrix(int, int, int, int);
    void free_integer_vector(int *, int);
    void free_real_vector(float *, int);
    void free_real_matrix(float **, int, int, int);
    float vecvec(int, int, int, float [], float []);
    float matvec(int, int, int, float **, float []);
    float matmat(int, int, int, int, float **, float **);
    void elmrow(int, int, int, int, float **, float **, float);
    void elmvec(int, int, int, float [], float [], float);
    void dupvec(int, int, int, float [], float []);
    void gsselm(float **, int, float [], int [], int []);
    void solelm(float **, int, int [], int [], float []);
    void colcst(int, int, int, float **, float);
    void mulvec(int, int, int, float [], float [], float);
    void gmsopconstruct(int *, int *, int, float **,
            float **, float **, float [], float [], int [],
            int [], int *, int *, int *, float *, float *,
            float, float, float *, float *,
            void (*)(int, float **, float [], float *));
    void gmscoefficient(float *, float *, float,
            int, int, float *, float *, float,
            float, float **, float **, int);
    void gmsdiffscheme(int, int, int, float [],
            float [], int *, int *, float [], float,
            float **, float **, float, float [], float **,
            float **, float **, int [], int [], float *,
            void (*)(int, float [], float *));
    int k,l,nsjev1,count,count1,kchange,update,change,reeval,
        strategy,*ri,*ci;
    float a,alfa,s1,s2,x0,xl0,xl1,eta,h0,h1,q1,q2,
        discr,aux[10],**bd1,**bd2,*y1,*y0,**hjac,**h2jac2,
        **rqz,*yl,*fl;

    ri=allocate_integer_vector(1,r);
    ci=allocate_integer_vector(1,r);
    y1=allocate_real_vector(1,r);
    y0=allocate_real_vector(1,r);
    yl=allocate_real_vector(1,3*r);
    fl=allocate_real_vector(1,3*r);
    bd1=allocate_real_matrix(1,3,1,3);
    bd2=allocate_real_matrix(1,3,1,3);
    hjac=allocate_real_matrix(1,r,1,r);
    h2jac2=allocate_real_matrix(1,r,1,r);
    rqz=allocate_real_matrix(1,r,1,r);

    /* initialization */
    (*lu)=(*jev)=(*n)=nsjev1=kchange=0;
    x0=(*x);
    discr=0.0;
    k=1;
    h1=h0=h;
```

```
count = -2;
aux[2]=FLT_EPSILON;
aux[4]=8.0;
dupvec(1,r,0,y1,y);
reeval=change=1;
strategy=((hmin != hmax) && !linear);
q1 = -1.0;
q2 = -2.0;
count1=0;
xl0=xl1=0.0;
(*out)(*x,xe,r,y,*delta,*n,*jev,*lu);
(*x) += h1;
/* operator construction */
gmsopconstruct(&reeval,&update,r,hjac,h2jac2,rqz,y,aux,ri,ci,
      lu,jev,&nsjev1,delta,&alfa,h1,h0,&s1,&s2,jacobian);
bd1[1][1]=1.0;
bd2[1][1] = -alfa*0.5;
if (!linear)
    gmscoefficient(&xl1,&xl0,x0,change,*n,&q1,&q2,h1,
                        alfa,bd1,bd2,strategy);
while (1) {
    gmsdiffscheme(k,count,r,fl,yl,n,&nsjev1,y0,alfa,bd1,
          bd2,h1,y,hjac,h2jac2,rqz,ri,ci,delta,derivative);
    if (strategy) count++;
    if (count == 1) {
        /* test accuracy */
        k=2;
        dupvec(1,r,0,y1,y);
        gmsdiffscheme(k,count,r,fl,yl,n,&nsjev1,y0,alfa,bd1,
              bd2,h1,y,hjac,h2jac2,rqz,ri,ci,delta,derivative);
        k=3;
        eta=aeta+reta*sqrt(vecvec(1,r,0,y1,y1));
        elmvec(1,r,0,y,y1,-1.0);
        discr=sqrt(vecvec(1,r,0,y,y));
        dupvec(1,r,0,y,y1);
    }
    (*out)(*x,xe,r,y,*delta,*n,*jev,*lu);
    if ((*x) >= xe) break;
    /* step size */
    x0=(*x);
    h0=h1;
    if ((*n <= 2) && !linear) (k)++;
    if (count == 1) {
        a=eta/(0.75*(eta+discr))+0.33;
        h1 = (a <= 0.9 || a >= 1.1) ? a*h0 : h0;
        count=0;
        reeval=(a <= 0.9 && nsjev1 != 1);
        count1 = (a >= 1.0 || reeval) ? 0 : count1+1;
        if (count1 == 10) {
            count1=0;
            reeval=1;
            h1=a*h0;
        }
    } else {
        h1=h;
        reeval=((nsjev == nsjev1) && !strategy && !linear);
    }
    if (strategy)
        h1 = (h1 > hmax) ? hmax : ((h1 < hmin) ? hmin : h1);
    (*x) += h1;
    if ((*x) >= xe) {
        h1=xe-x0;
        (*x)=xe;
    }
    if ((*n <= 2) && !linear) reeval=1;
    if (h1 != h0) {
        update=1;
        kchange=3;
    }
    if (reeval) update=1;
    change=((kchange > 0) && !linear);
    kchange--;
    if (update)
        /* operator construction */
```

```
        gmsopconstruct(&reeval,&update,r,hjac,h2jac2,rqz,y,aux,
            ri,ci,lu,jev,&nsjev1,delta,&alfa,h1,h0,&s1,&s2,
            jacobian);
    if (!linear)
        gmscoefficient(&xl1,&xl0,x0,change,*n,&q1,&q2,h1,
                        alfa,bd1,bd2,strategy);
    /* next integration step */
    for (l=2; l>=1; l--) {
        dupvec(l*r+1,(l+1)*r,-r,yl,yl);
        dupvec(l*r+1,(l+1)*r,-r,fl,fl);
    }
    dupvec(1,r,0,yl,y);
  }
  free_integer_vector(ri,1);
  free_integer_vector(ci,1);
  free_real_vector(y1,1);
  free_real_vector(y0,1);
  free_real_vector(yl,1);
  free_real_vector(fl,1);
  free_real_matrix(bd1,1,3,1);
  free_real_matrix(bd2,1,3,1);
  free_real_matrix(hjac,1,r,1);
  free_real_matrix(h2jac2,1,r,1);
  free_real_matrix(rqz,1,r,1);
}

void gmsopconstruct(int *reeval, int *update, int r,
        float **hjac, float **h2jac2, float **rqz, float y[],
        float aux[], int ri[], int ci[], int *lu, int *jev,
        int *nsjev1, float *delta, float *alfa, float h1,
        float h0, float *s1, float *s2,
        void (*jacobian)(int, float **, float [], float *))
{
  /* this function is internally used by GMS */

  int i,j;
  float a,a1,z1,e,q;

  if (*reeval) {
    (*jacobian)(r,hjac,y,delta);
    (*jev)++;
    *nsjev1 = 0;
    if (*delta <= 1.0e-15)
        (*alfa)=1.0/3.0;
    else {
        z1=h1*(*delta);
        a=z1*z1+12.0;
        a1=6.0*z1;
        if (fabs(z1) < 0.1)
            (*alfa)=(z1*z1/140.0-1.0)*z1/30.0;
        else if (z1 < -33.0)
            (*alfa)=(a+a1)/(3.0*z1*(2.0+z1));
        else {
            e=exp(z1);
            (*alfa)=((a-a1)*e-a-a1)/(((2.0-z1)*e-2.0-z1)*z1*3.0);
        }
    }
    (*s1) = -(1.0+(*alfa))*0.5;
    (*s2)=((*alfa)*3.0+1.0)/12.0;
  }
  a=h1/h0;
  a1=a*a;
  if (*reeval) a=h1;
  if (a != 1.0)
    for (j=1; j<=r; j++) colcst(1,r,j,hjac,a);
  for (i=1; i<=r; i++) {
    for (j=1; j<=r; j++) {
        q=h2jac2[i][j]=(*reeval ? matmat(1,r,i,j,hjac,hjac) :
                                    h2jac2[i][j]*a1);
        rqz[i][j]=(*s2)*q;
    }
    rqz[i][i] += 1.0;
    elmrow(1,r,i,i,rqz,hjac,*s1);
  }
```

```
    gsselm(rqz,r,aux,ri,ci);
    (*lu)++;
    (*reeval)=(*update)=0;
}

void gmscoefficient(float *xl1, float *xl0, float x0,
        int change, int n, float *q1, float *q2, float h1,
        float alfa, float **bd1, float **bd2, int strategy)
{
    /* this function is internally used by GMS */

    float a,q12,q22,q1q2,xl2;

    xl2=(*xl1);
    (*xl1)=(*xl0);
    (*xl0)=x0;
    if (change) {
        if (n > 2) {
            (*q1)=((*xl1)-(*xl0))/h1;
            (*q2)=(xl2-(*xl0))/h1;
        }
        q12=(*q1)*(*q1);
        q22=(*q2)*(*q2);
        q1q2=(*q1)*(*q2);
        a = -(3.0*alfa+1.0)/12.0;
        bd1[1][3]=1.0+(1.0/3.0-((*q1)+(*q2))*0.5)/q1q2;
        bd1[2][3]=(1.0/3.0-(*q2)*0.5)/(q12-q1q2);
        bd1[3][3]=(1.0/3.0-(*q1)*0.5)/(q22-q1q2);
        bd2[1][3] = -alfa*0.5+a*(1.0-(*q1)-(*q2))/q1q2;
        bd2[2][3]=a*(1.0-(*q2))/(q12-q1q2);
        bd2[3][3]=a*(1.0-(*q1))/(q22-q1q2);
        if (strategy || n <= 2) {
            bd1[2][2]=1.0/(2.0*(*q1));
            bd1[1][2]=1.0-bd1[2][2];
            bd2[2][2] = -(3.0*alfa+1.0)/(12.0*(*q1));
            bd2[1][2] = -bd2[2][2]-alfa*0.5;
        }
    }
}

void gmsdiffscheme(int k, int count, int r, float fl[],
            float yl[], int *n, int *nsjev1, float y0[],
            float alfa, float **bd1, float **bd2, float h1,
            float y[], float **hjac, float **h2jac2, float **rqz,
            int ri[], int ci[], float *delta,
            void (*derivative)(int, float [], float *))
{
    /* this function is internally used by GMS */

    int i,l;

    if (count != 1) {
        dupvec(1,r,0,fl,yl);
        (*derivative)(r,fl,delta);
        (*n)++;
        (*nsjev1)++;
    }
    mulvec(1,r,0,y0,yl,(1.0-alfa)/2.0-bd1[1][k]);
    for (l=2; l<=k; l++) elmvec(1,r,r*(l-1),y0,yl,-bd1[l][k]);
    for (l=1; l<=k; l++) elmvec(1,r,r*(l-1),y0,fl,h1*bd2[l][k]);
    for (i=1; i<=r; i++) y[i]=matvec(1,r,i,hjac,y0);
    mulvec(1,r,0,y0,yl,(1.0-3.0*alfa)/12.0-bd2[1][k]);
    for (l=2; l<=k; l++) elmvec(1,r,r*(l-1),y0,yl,-bd2[l][k]);
    for (i=1; i<=r; i++) y[i] += matvec(1,r,i,h2jac2,y0);
    dupvec(1,r,0,y0,yl);
    for (l=1; l<=k; l++) elmvec(1,r,r*(l-1),y0,fl,h1*bd1[l][k]);
    elmvec(1,r,0,y,y0,1.0);
    solelm(rqz,r,ri,ci,y);
}
```

F. impex

Solves an initial value problem, given as an autonomous system of first order differential equations

$$dy_i(t)/dt = f_i(t,y(t)) \qquad (i=1,...,n)$$

with $y(t_0)$ prescribed, over the interval $t0 \le t \le tend$ by means of an implicit midpoint rule with smoothing and extrapolation. Automatic stepsize control is provided. This procedure is suitable for the integration of stiff differential equations.

The integration method [see BeDHV74, Li73] is based on the computation of two independent solutions with stepsizes *h* and *h*/2 by the implicit midpoint rule. Passive smoothing and passive extrapolation are performed to obtain stability and high accuracy. The algorithm uses for each step at least three function evaluations, and on change of stepsize or at slow convergence in the iteration process an approximation of the Jacobian matrix (computed by divided differences or explicitly specified by the user). If the computed local error exceeds tolerance, the last step is rejected. Moreover, two global errors are computed.

Function Parameters:

void impex (*n,t0,tend,y0,deriv,available,h0,hmax,presch,eps,weights,update,fail,control*)

n: int;
  entry: the number of equations;

t0: float;
  entry: the initial value of the independent variable;

tend: float;
  entry: the final value of the independent variable;

y0: float *y0[1:n]*;
  entry: the initial values of the system of differential equations: *y0[i]* at *t=t0*;

deriv: void (**deriv*)(*t,y,f,n*);
  this procedure should deliver the values of *f(t,y)* in the array *f[1:n]*;

available: int (**available*)(*t,y,a,n*);
  if an analytic expression of the Jacobian matrix at the point (*t,y*) is not available then this procedure should deliver the value zero; otherwise this procedure should deliver the value nonzero, and the Jacobian matrix should be assigned to the array *a[1:n,1:n]*;

h0: float;
  entry: the initial stepsize;

hmax: float;
  entry: maximal stepsize by which the integration is performed;

presch: int;
  entry: indicator for choice of stepsize;
    the stepsize is automatically controlled if *presch* equals zero; otherwise the stepsize has to be prescribed, either only initially or also by the procedure *control*;

eps: float;
  entry: bound for the estimate of the local error;

weights: float *weights[1:n]*;
  entry: the initial weights for the computation of the weighted Euclidean norm of the errors; note that the choice *weights[i]*=1 implies an estimation of the absolute error, whereas *weights[i]=y[i]* defines a relative error;

update: void (**update*)(*weights,y2,n*);
  this procedure may change the arrays *weights*, according to the value of an approximation to *y(t)*, given in the array *y2[1:n]*;

fail: int *;
  exit: if the procedure fails to solve the system of equations, *fail* will have the value

nonzero upon exit; this may occur upon divergence of the iteration process used in the midpoint rule, while integration is performed with a user defined prescribed stepsize;

control: void (**control*)(*tprint,t,h,hnew,y,error,n,tend*);

control is called on entry to *impex* and further as soon as the inequality $tprint \le t$ holds; during a call of *control*, printing of results and change of stepsize (if *presch* has the value nonzero) is then possible; the meaning of the parameters of *control* is:

tprint: float *;

entry: the value of the independent variable at which a call of *control* is desired;

exit: a new value ($tprint > t$) at which a call of *control* is desired;

t: float;

the actual value of the independent variable, up to which integration has been performed;

h: float;

halve the actual stepsize;

hnew: float;

the new stepsize; if *presch* is nonzero then the user may prescribe a new stepsize by changing *hnew*;

y: float *y[1:5,1:n]*;

the value of the dependent variable and its first four divided differences at the point *t* are given in this array;

error: float *error[1:3]*;

the elements of this array contain the following errors:

error[1]: the local error;

error[2]: the global error of second order in *h*;

error[3]: the global error of fourth order in *h*;

n: int;

the number of equations;

tend: float;

the final value of the independent variable.

Functions used: inivec, inimat, mulvec, mulrow, dupvec, duprowvec, dupmat, vecvec, matvec, matmat, elmvec, elmrow, dec, sol.

```
#include <float.h>
#include <math.h>

void impex(int n, float t0, float tend, float y0[],
      void (*deriv)(float, float [], float [], int),
      int (*available)(float, float [], float **, int),
      float h0, float hmax, int presch, float eps,
      float weights[],
      void (*update)(float [], float [], int), int *fail,
      void (*control)(float *, float, float, float, float **,
                  float [], int, float))
{
   int *allocate_integer_vector(int, int);
   float *allocate_real_vector(int, int);
   float **allocate_real_matrix(int, int, int, int);
   void free_integer_vector(int *, int);
   void free_real_vector(float *, int);
   void free_real_matrix(float **, int, int, int);
   void inivec(int, int, float [], float);
   void inimat(int, int, int, int, float **, float);
   void mulvec(int, int, int, float [], float [], float);
   void mulrow(int, int, int, int, float **, float **, float);
```

```
void dupvec(int, int, int, float [], float []);
void duprowvec(int, int, int, float **, float []);
void dupmat(int, int, int, int, float **, float **);
float vecvec(int, int, int, float [], float []);
float matvec(int, int, int, float **, float []);
float matmat(int, int, int, int, float **, float **);
void elmvec(int, int, int, float [], float [], float);
void elmrow(int, int, int, int, float **, float **, float);
void dec(float **, int, float [], int []);
void sol(float **, int, int [], float []);
int impexrecomp(float **, float, float, float [], int [],
                 int, int (*)(float, float [], float **, int),
                 void (*)(float, float [], float [], int));
int impexlargestep(int, float [], float, float *,
                 float *, float *, float [], float [], float [],
                 float, float, float [], float [], float [],
                 float [], float [], float [], int [], int [],
                 float [], float **, float **, float,
                 void (*)(float, float [], float [], int),
                 int (*)(float, float [], float **, int));
void impexbackdiff(int, float [], float [], float [],
                   float [], float [], float [], float [],
                   float [], float **, float **);
int i,k,eci,*ps1,*ps2,start,two,halv;
float t,t1,t2,t3,tp,h,h2,hnew,alf,lq,*y,*z,*s1,*s2,*s3,*u1,
      *u3,*w1,*w2,*w3,*ehr,**r,**rf,**a1,**a2,err[4],
      alf1,c0,c1,c2,c3,**kof,e[5],d[5],b0,b1,b2,b3,w,sl1,sn,lr;

ps1=allocate_integer_vector(1,n);
ps2=allocate_integer_vector(1,n);
y=allocate_real_vector(1,n);
z=allocate_real_vector(1,n);
s1=allocate_real_vector(1,n);
s2=allocate_real_vector(1,n);
s3=allocate_real_vector(1,n);
u1=allocate_real_vector(1,n);
u3=allocate_real_vector(1,n);
w1=allocate_real_vector(1,n);
w2=allocate_real_vector(1,n);
w3=allocate_real_vector(1,n);
ehr=allocate_real_vector(1,n);
r=allocate_real_matrix(1,5,1,n);
rf=allocate_real_matrix(1,5,1,n);
a1=allocate_real_matrix(1,n,1,n);
a2=allocate_real_matrix(1,n,1,n);
kof=allocate_real_matrix(2,4,2,4);

if (presch)
   h=h0;
else {
   if (h0 > hmax)
      h=hmax;
   else
      h=h0;
   if (h > (tend-t0)/4.0) h=(tend-t0)/4.0;
}
hnew=h;
alf=0.0;
t=tp=t0;
inivec(1,3,err,0.0);
inivec(1,n,ehr,0.0);
duprowvec(1,n,1,r,y0);
(*control)(&tp,t,h,hnew,r,err,n,tend);
Init:
/* initialization */
h2=hnew;
h=h2/2.0;
dupvec(1,n,0,s1,y0);
dupvec(1,n,0,s2,y0);
dupvec(1,n,0,s3,y0);
dupvec(1,n,0,w1,y0);
duprowvec(1,n,1,r,y0);
inivec(1,n,u1,0.0);
inivec(1,n,w2,0.0);
```

```
inimat(2,5,1,n,r,0.0);
inimat(1,5,1,n,rf,0.0);
t=t1=t0;
t2=t0-2.0*h-1.0e-6;
t3=2.0*t2+1.0;
if (impexrecomp(a1,h,t,s1,ps1,n,available,deriv)) goto Miss;
if (impexrecomp(a2,h2,t,w1,ps2,n,available,deriv)) goto Miss;
start=1;
for (eci=0; eci<=3; eci++) {
    /* one large step */
    if (impexlargestep(n,y,t,&t1,&t2,&t3,s1,s2,s3,h,h2,
                       z,u1,u3,w1,w2,w3,ps1,ps2,weights,
                       a1,a2,eps,deriv,available)) goto Miss;
    t += h2;
    if (eci > 0) {
        /* backward differences */
        impexbackdiff(n,u1,u3,w1,w2,w3,s1,s2,s3,r,rf);
        (*update)(weights,s2,n);
    }
}
eci=4;
Mstp:
if (hnew != h2) {
    eci=1;
    /* change of information */
    c1=hnew/h2;
    c2=c1*c1;
    c3=c2*c1;
    kof[2][2]=c1;
    kof[2][3]=(c1-c2)/2.0;
    kof[2][4]=c3/6.0-c2/2.0+c1/3.0;
    kof[3][3]=c2;
    kof[3][4]=c2-c3;
    kof[4][4]=c3;
    for (i=1; i<=n; i++) u1[i]=r[2][i]+r[3][i]/2.0+r[4][i]/3.0;
    alf1=matvec(1,n,1,rf,u1)/vecvec(1,n,0,u1,u1);
    alf=(alf+alf1)*c1;
    for (i=1; i<=n; i++) {
        e[1]=rf[1][i]-alf1*u1[i];
        e[2]=rf[2][i]-alf1*2.0*r[3][i];
        e[3]=rf[3][i]-alf1*4.0*r[4][i];
        e[4]=rf[4][i];
        d[1]=r[1][i];
        rf[1][i] = e[1] *= c2;
        for (k=2; k<=4; k++) {
            r[k][i]=d[k]=matmat(k,4,k,i,kof,r);
            rf[k][i]=e[k]=c2*matvec(k,4,k,kof,e);
        }
        s1[i]=d[1]+e[1];
        w1[i]=d[1]+4.0*e[1];
        s2[i]=s1[i]-(d[2]+e[2]/2.0);
        s3[i]=s2[i]-(d[2]+e[2])+(d[3]+e[3]/2.0);
    }
    t3=t-hnew;
    t2=t-hnew/2.0;
    t1=t;
    h2=hnew;
    h=h2/2.0;
    err[1]=0.0;
    if (halv) {
        for (i=1; i<=n; i++) ps2[i]=ps1[i];
        dupmat(1,n,1,n,a2,a1);
    }
    if (two) {
        for (i=1; i<=n; i++) ps1[i]=ps2[i];
        dupmat(1,n,1,n,a1,a2);
    } else
        if (impexrecomp(a1,hnew/2.0,t,s1,ps1,
                        n,available,deriv)) goto Miss;
    if (!halv)
        if (impexrecomp(a2,hnew,t,w1,ps2,
                        n,available,deriv)) goto Miss;
    /* one large step */
    if (impexlargestep(n,y,t,&t1,&t2,&t3,s1,s2,s3,h,h2,
```

```
                            z,u1,u3,w1,w2,w3,ps1,ps2,weights,
                            a1,a2,eps,deriv,available)) goto Miss;
        t += h2;
        eci=2;
    }
    /* one large step */
    if (impexlargestep(n,y,t,&t1,&t2,&t3,s1,s2,s3,h,h2,
                            z,u1,u3,w1,w2,w3,ps1,ps2,weights,
                            a1,a2,eps,deriv,available)) goto Miss;
    /* backward differences */
    impexbackdiff(n,u1,u3,w1,w2,w3,s1,s2,s3,r,rf);
    (*update)(weights,s2,n);
    /* error estimates */
    c0=c1=c2=c3=0.0;
    for (i=1; i<=n; i++) {
        w=weights[i]*weights[i];
        b0=rf[4][i]/36.0;
        c0 += b0*b0*w;
        lr=fabs(b0);
        b1=rf[1][i]+alf*r[2][i];
        c1 += b1*b1*w;
        b2=rf[3][i];
        c2 += b2*b2*w;
        sl1=fabs(rf[1][i]-rf[2][i]);
        sn = (sl1 < 1.0e-10) ? 1.0 : fabs(rf[1][i]-r[4][i]/6.0)/sl1;
        if (sn > 1.0) sn=1.0;
        if (start) {
            sn *= sn*sn*sn;
            lr *= 4.0;
        }
        ehr[i]=b3=sn*ehr[i]+lr;
        c3 += b3*b3*w;
    }
    b0=err[1];
    err[1]=b1=sqrt(c0);
    err[2]=sqrt(c1);
    err[3]=sqrt(c3)+sqrt(c2)/2.0;
    lq=eps/((b0 < b1) ? b1 : b0);
    if (b0 < b1 && lq >= 80.0) lq=10.0;
    if (eci < 4 && lq > 80.0) lq=20.0;
    halv=two=0;
    if (!presch) {
        if (lq < 1.0) {
            /* reject */
            if (start) {
                hnew=pow(lq,1.0/5.0)*h/2.0;
                goto Init;
            } else {
                for (k=1; k<=4; k++) elmrow(1,n,k,k+1,r,r,-1.0);
                for (k=1; k<=3; k++) elmrow(1,n,k,k+1,r,r,-1.0);
                for (k=1; k<=4; k++) elmrow(1,n,k,k+1,rf,rf,-1.0);
                t -= h2;
                halv=1;
                hnew=h;
                goto Mstp;
            }
        } else {
            /* step size */
            if (lq < 2.0) {
                halv=1;
                hnew=h;
            } else {
                if (lq > 80.0)
                    hnew=((lq > 5120.0) ? pow(lq/5.0,1.0/5.0) : 2.0)*h2;
                if (hnew > hmax) hnew=hmax;
                if (tend > t && tend-t < hnew) hnew=tend-t;
                two=(hnew == 2.0*h2);
            }
        }
    }
    Tryck:
    if (tp <= t) (*control)(&tp,t,h,hnew,r,err,n,tend);
    if (start) start=0;
    if (hnew == h2) t += h2;
```

```
    eci++;
    if (t < tend+h2)
        goto Mstp;
    else
        goto End;
    Miss:
    *fail = presch;
    if (!(*fail)) {
        if (eci > 1) t -= h2;
        halv=two=0;
        hnew=h2/2.0;
        if (start)
            goto Init;
        else
            goto Tryck;
    }
    End:
    free_integer_vector(ps1,1);
    free_integer_vector(ps2,1);
    free_real_vector(y,1);
    free_real_vector(z,1);
    free_real_vector(s1,1);
    free_real_vector(s2,1);
    free_real_vector(s3,1);
    free_real_vector(u1,1);
    free_real_vector(u3,1);
    free_real_vector(w1,1);
    free_real_vector(w2,1);
    free_real_vector(w3,1);
    free_real_vector(ehr,1);
    free_real_matrix(r,1,5,1);
    free_real_matrix(rf,1,5,1);
    free_real_matrix(a1,1,n,1);
    free_real_matrix(a2,1,n,1);
    free_real_matrix(kof,2,4,2);
}

int impexrecomp(float **a, float h, float t, float y[], int ps[],
        int n, int (*available)(float, float [], float **, int),
        void (*deriv)(float, float [], float [], int))
{
    /* this function is internally used by IMPEX */

    int i,j;
    float sl,aux[4],ss,*f1,*f2;

    sl=h/2.0;
    if (!(*available)(t,y,a,n)) {
        f1=allocate_real_vector(1,n);
        f2=allocate_real_vector(1,n);
        (*deriv)(t,y,f1,n);
        for (i=1; i<=n; i++) {
            ss=1.0e-6*y[i];
            if (fabs(ss) < 1.0e-6) ss=1.0e-6;
            y[i] += ss;
            (*deriv)(t,y,f2,n);
            for (j=1; j<=n; j++) a[j][i]=(f2[j]-f1[j])/ss;
            y[i] -= ss;
        }
        free_real_vector(f1,1);
        free_real_vector(f2,1);
    }
    for (i=1; i<=n; i++) {
        mulrow(1,n,i,i,a,a,-sl);
        a[i][i] += 1.0;
    }
    aux[2]=FLT_EPSILON;
    dec(a,n,aux,ps);
    if (aux[3] < n)
        return 1;
    else
        return 0;
}
```

```
int impexlargestep(int n, float y[], float t, float *t1,
        float *t2, float *t3, float s1[], float s2[], float s3[],
        float h, float h2, float z[], float u1[], float u3[],
        float w1[], float w2[], float w3[], int ps1[], int ps2[],
        float weights[], float **a1, float **a2, float eps,
        void (*deriv)(float, float [], float [], int),
        int (*available)(float, float [], float **, int))
{
    /* this function is internally used by IMPEX */

    int impexiterate(float [], float [], float **, float, float,
                     float [], int [], int, float,
                     void (*)(float, float [], float [], int),
                     int (*)(float, float [], float **, int));
    float a,b,c;

    a=(t+h-(*t1))/((*t1)-(*t2));
    b=(t+h-(*t2))/((*t1)-(*t3));
    c=(t+h-(*t1))/((*t2)-(*t3))*b;
    b *= a;
    a += 1.0+b;
    b=a+c-1.0;
    mulvec(1,n,0,z,s1,a);
    elmvec(1,n,0,z,s2,-b);
    elmvec(1,n,0,z,s3,c);
    if (impexiterate(z,s1,a1,h,t+h/2.0,weights,ps1,
                     n,eps,deriv,available)) return 1;
    dupvec(1,n,0,y,z);
    a=(t+h2-(*t1))/((*t1)-(*t2));
    b=(t+h2-(*t2))/((*t1)-(*t3));
    c=(t+h2-(*t1))/((*t2)-(*t3))*b;
    b *= a;
    a += 1.0+b;
    b=a+c-1.0;
    mulvec(1,n,0,z,s1,a);
    elmvec(1,n,0,z,s2,-b);
    elmvec(1,n,0,z,s3,c);
    if (impexiterate(z,y,a1,h,t+3.0*h/2.0,weights,ps1,
                     n,eps,deriv,available)) return 1;
    dupvec(1,n,0,u3,u1);
    dupvec(1,n,0,u1,y);
    dupvec(1,n,0,s3,s2);
    dupvec(1,n,0,s2,s1);
    dupvec(1,n,0,s1,z);
    elmvec(1,n,0,z,w1,1.0);
    elmvec(1,n,0,z,s2,-1.0);
    if (impexiterate(z,w1,a2,h2,t+h,weights,ps2,
                     n,eps,deriv,available)) return 1;
    (*t3)=(*t2);
    (*t2)=(*t1);
    (*t1)=t+h2;
    dupvec(1,n,0,w3,w2);
    dupvec(1,n,0,w2,w1);
    dupvec(1,n,0,w1,z);
    return 0;
}

int impexiterate(float z[], float y[], float **a, float h, float t,
                 float weights[], int ps[], int n, float eps,
                 void (*deriv)(float, float [], float [], int),
                 int (*available)(float, float [], float **, int))
{
    /* this function is internally used by IMPEXLARGESTEP (IMPEX) */

    int i,it,lit,fail;
    float max,max1,conv,*dz,*f1,temp;

    f1=allocate_real_vector(1,n);
    dz=allocate_real_vector(1,n);
    for (i=1; i<=n; i++) z[i]=(z[i]+y[i])/2.0;
    it=lit=1;
    conv=1.0;
    fail=0;
    while (1) {
```

```
        deriv(t,z,f1,n);
        for (i=1; i<=n; i++) f1[i]=dz[i]=z[i]-h*f1[i]/2.0-y[i];
        sol(a,n,ps,dz);
        elmvec(1,n,0,z,dz,-1.0);
        max=0.0;
        for (i=1; i<=n; i++) {
            temp=weights[i]*dz[i];
            max += temp*temp;
        }
        max=sqrt(max);
        if (max*conv < eps/10.0) break;
        it++;
        if (it != 2) {
            conv=max/max1;
            if (conv > 0.2) {
                if (lit == 0) {
                    fail=1;
                    break;
                }
                lit=0;
                conv=1.0;
                it=1;
                if (impexrecomp(a,h,t,z,ps,n,available,deriv)) {
                    fail=1;
                    break;
                }
            }
        }
        max1=max;
    }
    if (!fail)
        for (i=1; i<=n; i++) z[i]=2.0*z[i]-y[i];
    free_real_vector(f1,1);
    free_real_vector(dz,1);
    return fail;
}

void impexbackdiff(int n, float u1[], float u3[], float w1[],
                   float w2[], float w3[], float s1[], float s2[],
                   float s3[], float **r, float **rf)
{
    /* this function is internally used by IMPEX */

    int i,k;
    float b0,b1,b2,b3;

    for (i=1; i<=n; i++) {
        b1=(u1[i]+2.0*s2[i]+u3[i])/4.0;
        b2=(w1[i]+2.0*w2[i]+w3[i])/4.0;
        b3=(s3[i]+2.0*u3[i]+s2[i])/4.0;
        b2=(b2-b1)/3.0;
        b0=b1-b2;
        b2 -= (s1[i]-2.0*s2[i]+s3[i])/16.0;
        b1=2.0*b3-(b2+rf[1][i])-(b0+r[1][i])/2.0;
        b3=0.0;
        for (k=1; k<=4; k++) {
            b1 -= b3;
            b3=r[k][i];
            r[k][i]=b0;
            b0 -= b1;
        }
        r[5][i]=b0;
        for (k=1; k<=4; k++) {
            b3=rf[k][i];
            rf[k][i]=b2;
            b2 -= b3;
        }
        rf[5][i]=b2;
    }
}
```

5.4.3 First Order - Several derivatives available

A. modifiedtaylor

Solves an initial (boundary) value problem, given as a system of first order differential equations

$$Du^{(j)}(t) = h^{(j)}(u,t) \qquad (j=m0,m0+1,...,m) \tag{1}$$

where $D=d/dt$, with $u(t_0)$ prescribed, over the interval $t0 \leq t \leq te$, by means of a one-step Taylor method [Vh70a, VhBDS71, VhK71]. This procedure is suitable for the integration of large systems arising from partial differential equations, provided that higher order derivatives can be easily obtained.

The method used has the form

$$u_{k+1} = \sum_{i=0}^{n} \beta_i c_k^{(i)} \tau_k^i \tag{2}$$

where, with u_k being an approximation to $u(t_k)$, $c_k^{(i)} = D^i u(t)$ at $t=t_k$, and τ_k is the stepsize, so that $t_{k+1} = t_k + \tau_k$ (k=0,1,...). The method is p-th order consistent if $\beta_j = 1/j!$ (j=0,1,...,p).

In the following $\|.\|$ denotes either the maximum or Euclidean norm. Which of these is used during the execution of *modifiedtaylor* depends upon the input parameter *norm* upon call: 1 in the first case, ≠1 in the second.

The stepsize is determined by considerations of accuracy and stability. With regard to the former, it is assumed that the local error $\rho(\tau) = \|u(t_{k+1}) - u_{k+1}\|$ is approximated by the discrepancy $\|v(t_{k+1}) - u_{k+1}\|$, where $v(t)$ is the analytic solution to equation (1) with the initial values $v(t_k) = u_k$. For the p-th order exact method (2)

$$\rho_k'(\tau_k) = \sum_{i=p+1}^{n} \left|\frac{1}{i!} - \beta_i\right| \|c_k^{(i)}\| \, \tau_k^i \qquad (p < n) \tag{3}$$

$$\rho_k'(\tau_k) = \frac{1}{n!} \|c_k^{(n)}\| \, \tau_k^n \qquad (p = n)$$

are taken to be approximations to the discrepancy. It is further assumed that these estimates of the discrepancy may themselves be approximated by expressions of the form

$$\rho_k'(\tau_k) = C\tau_k^q \quad (k=1), \qquad \rho_k'(\tau_k) = (Bt_k + C)\tau_k^q \quad (k=2)$$

$$\rho_k'(\tau_k) = (At_k^2 + Bt_k + C)\tau_k^q \quad (k\geq 3) \tag{4}$$

where A, B, C are constants determined from fitting the given estimates at previous values of t_k, q being n if $p = n$ and p+1 otherwise. With the estimates (4) of the discrepancy in hand, the stepsize τacc_k obtained from considerations of accuracy is to be determined by use of a relationship of the form

$$\rho_k'(\tau_k) = \eta_k \tag{5}$$

where $\eta_k = \eta_{abs} + \eta_{re}\|u_k\|$, η_{abs} and η_{re} being absolute and relative tolerance provided by the user. *modifiedtaylor* takes

$$\tau acc_0 = \eta_0 / \|c_0^{(1)}\|$$

at the first step,

$$\tau acc_1 = \min(10\tau acc_0, (\eta_0/\|c_0^{(1)}\|)(\eta_1/\rho_0'(\tau_0))^{1/q})$$

at the second, and

$$\tau acc_k = \min(10\tau acc_{k-1}, (\eta_{k-1}/\| c_0^{(1)} \|)(\eta_k/\rho_{k-1}'(\tau_{k-1}))^{1/q})$$

at subsequent step until the value of the left hand expression in equation (5) exceeds that of the expression upon the right; then equation (5) is used.

The stepsize bound $\tau stab_k$ determined from considerations of stability is given by

$$tstab_k = \beta(n) / \sigma(J_k)$$

where $\beta(n)$ is the range constant of the stability polynomial

$$\sum_{i=0}^{n} \beta_i z^i$$

associated with the method (2), and $\sigma(J_k)$ is an upper bound for the spectral radius with respect to the eigenvalues in the left half-plane of the Jacobian matrix J_k whose elements are

$$J_k^{(i,j)} = \partial h^{(i-m0+1)}(u,t)/\partial u^{(j-m0+1)}(t) \qquad (i,j=1,2,...,m-m0+1)$$

as t ranges over $[t_k, t_k+\tau bd_k]$, τbd_k being a bound for τ_k.

The user should provide values of τmin, a lower bound for the stepsize, and $\alpha > 1$, a stepsize amplification factor. τmin is overridden by considerations of stability, but not of accuracy (thus $\max(\tau min, \tau acc_k)$ is one of the lower bounds for the stepsize). The stepsize is taken to be $\min(\max(\tau acc_0, \tau min), \tau stab_0)$ when $k=0$ and

$$\tau_k = \max(\tfrac{1}{2}\tau_{k-1}, \min(\max(\tau acc_k, \tau min), \tau stab_k, \alpha\tau_{k-1}))$$

thereafter. The upper bound τbd_k upon τ_k required in the determination of $\sigma(J_k)$ above, is τacc_0 when $k=0$ and $\alpha\tau_{k-1}$ otherwise.

The successive derivatives $D^i u^{(j)}(t)$ $(i=1,...,n;\ j=m0,...,m)$ are computed by use of a procedure *derivative* provided by the user.

Function Parameters:

void modifiedtaylor
(*t,te,m0,m,u,sigma,taumin,derivative,k,data,alfa,norm,aeta,reta,eta,rho,out*)

t: float *;
the independent variable;
entry: initial value *t0*;
exit: the final value *te*;

te: float;
entry: the final value of *t* ($te \geq t$);

m0,m: int;
entry: indices of the first and last equation of the system to be solved;

u: float *u[m0:m]*;
the dependent variable;
entry: the initial values of the solution of the system of differential equations at *t=t0*;
exit: the values of the solution at *t=te*;

sigma: float (**sigma*)(*t,m0,m*);
this procedure should deliver the spectral radius of the Jacobian matrix with respect to those eigenvalues which are located in the left half-plane;
if *sigma* tends to infinity, procedure *modifiedtaylor* terminates;

taumin: float;
entry: minimal step length by which the integration is performed (value of τmin above); however, actual stepsizes will always be within the interval

[min(*hmin*,*hstab*),*hstab*], where *hstab* (= *data[0]/sigma*) is the steplength prescribed by stability considerations;

derivative: void (**derivative*)(*t,m0,m,i,a*);
when this procedure is called, array *a[m0,m]* contains the components of the (*i*-1)-th derivative of *u* at the point *t*; upon completion of *derivative*, array *a* should contain the components of the *i*-th derivative of *u* at the point *t*;

k: int *;
exit: indicates the number of integration steps performed; on entry *k*=0;

data: float *data[-2:data[-2]]*;
entry:
data[-2]: the order of the highest derivative upon which the Taylor method is based (value of *n* above);
data[-1]: order of accuracy of the method (value of *p* above);
data[0]: stability parameter (value of $\beta(n)$ above);
data[1],...,data[data[-2]]: polynomial coefficients (values of β_j above, *j*=1,..,*n*);

alfa: float;
entry: growth factor for the integration step length (value of α above);

norm: int;
entry: if *norm*=1 then discrepancy and tolerance are estimated in the maximum norm, otherwise in the Euclidean norm;

aeta: float (**aeta*)(*t,m0,m*);
this procedure should return the absolute accuracy (value of η_{abs} above);

reta: float (**reta*)(*t,m0,m*);
this procedure should return the relative accuracy; (value of η_{re} above);
if both *aeta* and *reta* are negative, accuracy conditions will be ignored;

eta: float *;
exit: computed tolerance;

rho: float *;
exit: computed discrepancy;

out: void (**out*)(*t,te,m0,m,u,k,eta,rho*);
after each integration step the user can print some parameters such as *t, u, k, eta, rho*.

Function used: vecvec.

```
#include <math.h>

void modifiedtaylor(float *t, float te, int m0, int m, float u[],
           float (*sigma)(float, int, int), float taumin,
           void (*derivative)(float, int, int, int, float []),
           int *k, float data[], float alfa, int norm,
           float (*aeta)(float, int, int),
           float (*reta)(float, int, int),
           float *eta, float *rho,
           void (*out)(float, float, int, int, float [],
                        int, float, float))
{
   float *allocate_real_vector(int, int);
   void free_real_vector(float *, int);
   float vecvec(int, int, int, float [], float []);
   int i,n,p,q,start,step1,last,j;
   float ec0,ec1,ec2,tau0,tau1,tau2,taus,t2,t0,tau,taui,tauec,
           ecl,betan,gamma,*c,*beta,*betha,ifac,tauacc,taustab,
           aa,bb,cc,ec,s,x,b;

   n=data[-2];
   beta=allocate_real_vector(1,n);
```

```
betha=allocate_real_vector(1,n);
c=allocate_real_vector(m0,m);

i=0;
start=(*k == 0);
t0=(*t);
/* coefficient */
ifac=1.0;
gamma=0.5;
p=data[-1];
betan=data[0];
q = (p < n) ? p+1 : n;
for (j=1; j<=n; j++) {
    beta[j]=data[j];
    ifac /= j;
    betha[j]=ifac-beta[j];
}
if (p == n) betha[n]=ifac;
last=0;
do {
    /* step size */
    s=0.0;
    if (norm == 1)
        for (j=m0; j<=m; j++) {
            x=fabs(u[j]);
            if (x > s) s=x;
        }
    else
        s=sqrt(vecvec(m0,m,0,u,u));
    /* local error bound */
    *eta = (*aeta)(*t,m0,m)+(*reta)(*t,m0,m)*s;
    if (*eta > 0.0) {
        if (start) {
            if (*k == 0) {
                for (j=m0; j<=m; j++) c[j]=u[j];
                i=1;
                (*derivative)(*t,m0,m,i,c);
                s=0.0;
                if (norm == 1)
                    for (j=m0; j<=m; j++) {
                        x=fabs(c[j]);
                        if (x > s) s=x;
                    }
                else
                    s=sqrt(vecvec(m0,m,0,c,c));
                tauacc=(*eta)/s;
                step1=1;
            } else if (step1) {
                tauacc=pow((*eta)/(*rho),1.0/q)*tau2;
                if (tauacc > 10.0*tau2)
                    tauacc=10.0*tau2;
                else
                    step1=0;
            } else {
                bb=(ec2-ec1)/tau1;
                cc=ec2-bb*t2;
                ec=bb*(*t)+cc;
                tauacc = (ec < 0.0) ? tau2 : pow((*eta)/ec,1.0/q);
                start=0;
            }
        } else {
            aa=((ec0-ec1)/tau0+(ec2-ec1)/tau1)/(tau1+tau0);
            bb=(ec2-ec1)/tau1-aa*(2.0*t2-tau1);
            cc=ec2-t2*(bb+aa*t2);
            ec=cc+(*t)*(bb+(*t)*aa);
            tauacc = (ec < 0.0) ? taus : pow((*eta)/ec,1.0/q);
            if (tauacc > alfa*taus) tauacc=alfa*taus;
            if (tauacc < gamma*taus) tauacc=gamma*taus;
        }
    } else
        tauacc=te-(*t);
    if (tauacc < taumin) tauacc=taumin;
    taustab=betan/(*sigma)(*t,m0,m);
    if (taustab < 1.0e-12*((*t)-t0)) {
```

```
            (*out)(*t,te,m0,m,u,*k,*eta,*rho);
            break;
        }
        tau = (tauacc > taustab) ? taustab : tauacc;
        taus=tau;
        if (tau >= te-(*t)) {
            tau=te-(*t);
            last=1;
        }
        tau0=tau1;
        tau1=tau2;
        tau2=tau;
        (*k)++;
        i=0;
        /* difference scheme */
        for (j=m0; j<=m; j++) c[j]=u[j];
        taui=1.0;
        do {
            i++;
            (*derivative)(*t,m0,m,i,c);
            taui *= tau;
            b=beta[i]*taui;
            if (*eta > 0.0 && i >= p) {
                /* local error construction */
                if (i == p) {
                    ec1=0.0;
                    tauec=1.0;
                }
                if (i > p+1) tauec *= tau;
                s=0.0;
                if (norm == 1)
                    for (j=m0; j<=m; j++) {
                        x=fabs(c[j]);
                        if (x > s) s=x;
                    }
                else
                    s=sqrt(vecvec(m0,m,0,c,c));
                ecl += fabs(betha[i])*tauec*s;
                if (i == n) {
                    ec0=ec1;
                    ec1=ec2;
                    ec2=ecl;
                    *rho = ecl*pow(tau,q);
                }
            }
            for (j=m0; j<=m; j++) u[j] += b*c[j];
        } while (i < n);
        t2=(*t);
        if (last) {
            last=0;
            (*t)=te;
        } else
            (*t) += tau;
        (*out)(*t,te,m0,m,u,*k,*eta,*rho);
    } while (*t != te);
    free_real_vector(beta,1);
    free_real_vector(betha,1);
    free_real_vector(c,m0);
}
```

B. eft

Solves an initial value problem, given as a system of first order differential equations

$$Du^{(j)}(t) = g^{(j)}(u,t) \qquad (j=m0,m0+1,...,m) \tag{1}$$

where $D=d/dt$, with $u(t_0)$ prescribed, over the interval $t0 \le t \le te$, by means of a third order, first order consistent, exponentially fitted Taylor series method [see Vh70b, VhBDS71]. Automatic stepsize control is provided. This procedure is suitable for the integration of stiff differential equations, provided that higher order derivatives can be easily obtained.

The method used has the form

$$u_{k+1} = u_k + c_k^{(1)}h_k + \beta_2(h_k)c_k^{(2)}h_k^2 + \beta_3(h_k)c_k^{(3)}h_k^3 \qquad (2)$$

where, with u_k being an approximation to $u(t_k)$, $c_k^{(i)} \approx D^i u(t)$ at $t=t_k$, and h_k is the stepsize, so that $t_{k+1} = t_k + h_k$ (k=0,1,...).

It is supposed that over the range $t_k \leq t \leq t_k+hbd_k$, where hbd_k is an upper bound for h_k, those eigenvalues of the Jacobian matrix J_k, whose elements are

$$J_k^{(i,j)} = \partial g^{(i-m0+1)}(u,t) \,/\, \partial u^{(j-m0+1)}(t) \quad (i,j=1,2,...,m-m0+1),$$

lying in the left half-plane and not in the neighbourhood of the origin are clustered within two circles with centers at

$$\delta_k = \sigma_k e^{i\phi_k} \qquad \text{and} \qquad \overline{\delta_k}\,,$$

($0 < \sigma_k < \infty$, $\pi/2 < \phi_k < \pi$) of equal radii $d\delta_k$. The coefficients $\beta_2(h_k)$ and $\beta_3(h_k)$ are determined by imposing the condition that, with

$$P_3(z) = 1 + z + \beta_2(h_k)z^2 + \beta_3(h_k)z^3$$

$P_3(z) = e^z$ when $z = h_k\delta_k$, $z = h_k\mathrm{conj}(\delta_k)$; it then follows that

$$\beta_2(h_k) = -\frac{1}{b^2}\left[2b\cos\phi_k + 4\cos^2\phi_k - 1 + e^{b\cos\phi_k}\frac{\sin(b\sin\phi_k - 3\phi_k)}{\sin\phi_k}\right]$$

$$\beta_3(h_k) = \frac{1}{b^3}\left[b + 2\cos\phi_k + be^{b\cos\phi_k}\frac{\sin(b\sin\phi_k - 2\phi_k)}{\sin\phi_k}\right]$$

where $b = h_k\delta_k$. When b is small, $\beta_2(h_k) \approx 1/2 - b^2/24$, $\beta_3(h_k) \approx 1/6 + b\cos\phi_k/12$.

In the following, $\|.\|$ denotes either the maximum or Euclidean norm. Which of these is used during the execution of *eft* depends upon the value of the parameter *norm* upon call: 1 in the first case, $\neq 1$ in the second.

The coefficients $\beta_2(h_k)$, $\beta_3(h_k)$ in the scheme (2) depend upon h_k: the discrepancy $\rho'(u_k)$ between the local analytic solution of equation (1) and its approximation yielded by use of the scheme (2) is not accurately estimated by substitution of the local solution into formula (2) (this process yields useful results only when the coefficients of the stability polynomial of the numerical integration procedure being used are constant, as is the case with, for example, *modifiedtaylor*). The discrepancy is now to be estimated in terms of a residual function $\zeta(h)$ (i.e. a measure of the difference between the values of the expressions upon the two sides of equation (1)) resulting from use of the scheme (2) in which the steplength is taken to be an independent variable. Taking $u'(h)$ to be an approximation to $u(t_k+h)$ produced by use of formula (2),

$$u'(h) = u_k + c_k^{(1)} + \beta_2(h)c_k^{(2)}h_2 + \beta_3(h)c_k^{(3)}h^3$$

from which

$$Du'(h) = c_k^{(1)} + \beta'_2(h)c_k^{(2)} + \beta'_3(h)c_k^{(3)}h^2$$

where, with $b = \sigma_k h$,

$$\beta'_2(h) = -\frac{1}{b}\left[2\cos\phi_k + e^{b\cos\phi_k}\frac{\sin(b\sin\phi_k - 2\phi_k)}{\sin\phi_k}\right]$$
$$\beta'_3(h) = \frac{1}{b^2}\left[1 + e^{b\cos\phi_k}\frac{\sin(b\sin\phi_k - \phi_k)}{\sin\phi_k}\right]. \quad (3)$$

The residual function $\zeta_k(h)$ is defined by the formula

$$\zeta_k(h) = Du'(h) - g(u'(h), t_k+h) \quad (4)$$

so that

$$h\|\zeta_k(h)\| = \|g(u'(h), t_k+h)h - c_k^{(1)}h - \beta'_2(h)c_k^{(2)}h^2 - \beta'_3(h)c_k^{(3)}h^3\|. \quad (5)$$

If the discrepancy $\rho'_k(h)$ is defined to be

$$\rho'_k(h) = \int_0^h \|\zeta_k(h')\| dh',$$

then approximately $\rho'_k(h) = h\|\zeta_k(h)\|$.
An upper bound $hacc_k$ upon h_k derived from considerations of accuracy is thus given by

$$\rho'_k(hacc_k) = \eta_k = \eta_a + \eta_{re}\|u_k\| \quad (6)$$

where, η_a and η_{re} being prescribed absolute and relative tolerances, ρ'_k is as defined by formulae (3,4,5). The value of $\rho'_k(h)$ is predicted from previous determinations by use of the formula

$$\rho'(h) = (Ah + Bt + C)h^{Q(b)}$$

where with $b = \sigma h$ (σ being the spectral radius of the Jacobian matrix derived from equation (1) at the point t) and $Q(b) = 4 - 2b/3$ if $0 < b < 3/2$, $(30-2b)/9$ if $3/2 \le b < 6$ and 2 otherwise, the coefficients A, B and C are determined by three substitutions t_{k-3}, h_{k-3}; t_{k-2}, h_{k-2}; t_{k-1}, h_{k-1}. If h_{k-1} has the critical value $h_{cr} = h_{k-2}^2/h_{k-3}$, the value of A determined as above becomes infinite. For this reason h_k is, at each stage, excluded from the interval $[(1-\sqrt{\varepsilon})h_{cr}, (1+\sqrt{\varepsilon})h_{cr}]$, where ε is the machine precision. Also it may occur that A is negative; in this case the extrapolation formula $\rho'(h) = Ch^{Q(b)}$ is used. Except when A is negative, $hacc_k$ is determined from the equation

$$(A\, hacc_k + B\, t_k + C)\, hacc^{q(\sigma h_{k-1})} = \eta_k .$$

eft takes

$$hacc_0 = \frac{\eta_0}{\|c_0^{(1)}\|}$$

at the first step

$$hacc_1 = \min\left(10\, hacc_0, \frac{\eta_0}{\|c_0^{(1)}\|}\left(\frac{\eta_1}{\rho'_0(h_0)}\right)^{1/Q(\sigma_1 h_0)}\right)$$

at the second, and

$$hacc_k = \min\left(10\ hacc_{k-1}, \frac{\eta_{k-1}}{\|c_{k-1}^{(1)}\|}\left(\frac{\eta_{k-1}}{\rho'_{k-1}(h_{k-1})}\right)^{1/Q(\sigma_k h_{k-1})}\right)$$

at subsequent steps until the value of the left hand side in equation (6) (predicted as above) exceeds that of the expression upon the right; then equation (6) is used to determine $hacc_k$.

When $\sin\phi_k$ is not small, $|P_3(z)| \leq 1$ over the disc $|z - h\delta_k| \leq 1 / 2\sin\phi_k$; when $\sin\phi_k$ is small, $|P_3(z)| \leq 1$ when $|z + h\delta_k| \leq (h\sigma_k)^{1/2}$. Thus if the eigenvalues of the Jacobian matrix J_k (defined above) clustered about two points δ_k, conjugate(δ_k) lie within discs of radius $d\delta_k$, the method (2) is stable if $h_k < hstab_k$ where

$$hstab_k = \min\ [\sigma_k/(d\delta_k \sin\phi_k),\ 4\sigma_k^2/d\delta_k^2]\ /\ \sigma_k.$$

The user should provide values of *hmin*, a lower bound for the stepsize, and $\alpha > 1$, a stepsize amplification factor. *hmin* is overridden by considerations of stability but not of accuracy (thus $\max(hmin, hacc_k)$ is one of the lower bounds for the stepsize). The stepsize is taken to be $\min(\max(hacc_0, hmin), hstab_0)$ when k=0, and $h_k = \max(h_{k-1}/2, \min(\max(hacc_k, hmin), hstab_k, \alpha_k h_{k-1}))$ thereafter, where until formula (6) is used to determine $hacc_k$ (see above) α_k is the user's amplification factor α, and thereafter

$$\alpha_k = \min\left(\alpha, \left(\frac{\eta_k}{\rho'_k(h_k)}\right)^{1/Q(\sigma_k h_{k-1})}\right).$$

The upper bound hbd_k upon h_k required in the determination of $\sigma(J_k)$ above, is $hacc_0$ when k=0 and $\alpha_k h_{k-1}$ otherwise.

It may occur that the user wishes to integrate equation (1) over a succession of intervals $[t_0^{(l)}, te^{(l)}]$ (l=0,1,...) where $t_0^{(l+1)} = te^{(l)}$. In this case he should set k=0 prior to the first call of *eft* and leave k unchanged at subsequent calls. During these subsequent calls, *eft* continues normal operation (without special determinations of $hacc_0$, $hacc_1$, etc.). The current value of h_k at return from a call of *eft* is allocated by this procedure to *hstart*, for use at the start of the next call.

The successive derivatives $D^i u^{(j)}(t)$ (i=1,2,3; j=$m0$,...,m) are computed by use of a procedure *derivative* which the user must provide.

Function Parameters:

void eft

(*t,te,m0,m,u,sigma,phi,diameter,derivative,k,alfa,norm,aeta,reta,eta,rho,hmin,hstart,out*)

t: float *;
 the independent variable;
 entry: initial value *t0*;
 exit: the final value *te*;

te: float;
 entry: the final value of *t* ($te \geq t$);

m0,m: int;
 entry: indices of the first and last equation of the system to be solved;

u: float *u[m0:m]*;
 the dependent variable;
 entry: the initial values of the solution of the system of differential equations at *t=t0*;
 exit: the values of the solution at *t=te*;

sigma: float (**sigma*)(*t,m0,m*);
this procedure should deliver the modulus of the (complex) point at which exponential fitting is desired (value of σ_k above); for example an approximation to the modulus of the center of the left hand cluster;

phi: float;
entry: the argument of the (complex) point at which exponential fitting is desired (value of ϕ_k above); *phi* should have a value in the range [π/2,π];

diameter: float (**diameter*)(*t,m0,m*);
this procedure should deliver the diameter of the left hand cluster (value of $d\delta_k$ above);

derivative: void (**derivative*)(*t,m0,m,i,a*);
i assumes the values 1,2,3 and *a* is a one-dimensional array *a[m0:m]*;
when this procedure is called, array *a* contains the components of the (*i*-1)-th derivative of *u* at the point *t*; upon completion of *derivative*, array *a* should contain the components of the *i*-th derivative of *u* at the point *t*;

k: int *;
exit: indicates the number of integration steps performed; on entry *k*=0;

alfa: float;
entry: maximal growth factor for the integration step length (value of α above);

norm: int;
entry: if *norm*=1 then discrepancy and tolerance are estimated in the maximum norm, otherwise in the Euclidean norm;

aeta: float (**aeta*)(*t,m0,m*);
this procedure should return the absolute local accuracy (value of η_a above); *aeta* should be positive;

reta: float (**reta*)(*t,m0,m*);
this procedure should return the relative local accuracy; (value of η_{re} above); *reta* should be positive;

eta: float *;
exit: computed tolerance;

rho: float *;
exit: computed discrepancy;

hmin: float;
entry: minimal stepsize by which the integration is performed;
however, a smaller step will be taken if *hmin* exceeds the stepsize *hstab*, prescribed by the stability conditions;
if *hstab* becomes zero, the procedure terminates;

hstart: float *;
entry: the initial stepsize;
however, if *k* = 0 on entry then the value of *hstart* is not taken into consideration;
exit: a suggestion for the stepsize (current value of h_k above), if the integration should be continued for *t* > *te*;
hstart may be used in successive calls of the procedure, in order to obtain the solution in several points *te1*, *te2*, etc.;

out: void (**out*)(*t,te,m0,m,u,k,eta,rho*);
after each integration step the user can print some parameters such as *t, u, k, eta, rho*.

Functions used: inivec, dupvec, vecvec, elmvec.

```
#include <float.h>
#include <math.h>

void eft(float *t, float te, int m0, int m, float u[],
       float (*sigma)(float, int, int), float phi,
       float (*diameter)(float, int, int),
       void (*derivative)(float, int, int, int, float []),
       int *k, float alfa, int norm,
       float (*aeta)(float, int, int),
       float (*reta)(float, int, int),
       float *eta, float *rho, float hmin, float *hstart,
       void (*out)(float, float, int, int, float [],
                   int, float, float))
{
    float *allocate_real_vector(int, int);
    void free_real_vector(float *, int);
    void inivec(int, int, float [], float);
    void dupvec(int, int, int, float [], float []);
    float vecvec(int, int, int, float [], float []);
    void elmvec(int, int, int, float [], float [], float);
    int kl,last,start,j,i,ext,extrapolate;
    float q,ec0,ec1,ec2,h,hi,h0,h1,h2,betan,t2,sigmal,phil,*c,
           *ro,beta[4],betha[4],s,x,hacc,hstab,hcr,hmax,a,b,cc,
           b1,b2,bb,e,beta2,beta3,
           c0,fc,b0,fb,a0,fa,d0,fd,fdb,fda,w,mb,tol,mm,p0,q0;

    c=allocate_real_vector(m0,m);
    ro=allocate_real_vector(m0,m);

    start=1;
    last=0;
    dupvec(m0,m,0,c,u);
    (*derivative)(*t,m0,m,1,c);
    if (*k == 0) {
        /* local error bound */
        s=0.0;
        if (norm == 1)
            for (j=m0; j<=m; j++) {
                x=fabs(u[j]);
                if (x > s) s=x;
            }
        else
            s=sqrt(vecvec(m0,m,0,u,u));
        *eta = (*aeta)(*t,m0,m)+(*reta)(*t,m0,m)*s;
        s=0.0;
        if (norm == 1)
            for (j=m0; j<=m; j++) {
                x=fabs(c[j]);
                if (x > s) s=x;
            }
        else
            s=sqrt(vecvec(m0,m,0,c,c));
        *hstart = (*eta)/s;
    }
    do {
        /* difference scheme */
        hi=1.0;
        sigmal=(*sigma)(*t,m0,m);
        phil=phi;
        /* step size */
        if (!start) {
            /* local error bound */
            s=0.0;
            if (norm == 1)
                for (j=m0; j<=m; j++) {
                    x=fabs(u[j]);
                    if (x > s) s=x;
                }
            else
                s=sqrt(vecvec(m0,m,0,u,u));
            *eta = (*aeta)(*t,m0,m)+(*reta)(*t,m0,m)*s;
        }
        if (start) {
            h1=h2=hacc=(*hstart);
```

```
            ec2=ec1=1.0;
            kl=1;
            start=0;
        } else if (kl < 3) {
            hacc=pow((*eta)/(*rho),1.0/q)*h2;
            if (hacc > 10.0*h2)
                hacc=10.0*h2;
            else
                kl++;
        } else {
            a=(h0*(ec2-ec1)-h1*(ec1-ec0))/(h2*h0-h1*h1);
            h=h2*((*eta < *rho) ? pow((*eta)/(*rho),1.0/q) : alfa);
            if (a > 0.0) {
                b=(ec2-ec1-a*(h2-h1))/h1;
                cc=ec2-a*h2-b*t2;
                hacc=0.0;
                hmax=h;
                /* find zero */
                b0=hacc;
                fb=pow(hacc,q)*(a*hacc+b*(*t)+cc)-(*eta);
                a0=hacc=h;
                fa=pow(hacc,q)*(a*hacc+b*(*t)+cc)-(*eta);
                c0=a0;
                fc=fa;
                ext=0;
                extrapolate=1;
                while (extrapolate) {
                    if (fabs(fc) < fabs(fb)) {
                        if (c0 != a0) {
                            d0=a0;
                            fd=fa;
                        }
                        a0=b0;
                        fa=fb;
                        b0=hacc=c0;
                        fb=fc;
                        c0=a0;
                        fc=fa;
                    }
                    tol=1.0e-3*h2;
                    mm=(c0+b0)*0.5;
                    mb=mm-b0;
                    if (fabs(mb) > tol) {
                        if (ext > 2)
                            w=mb;
                        else {
                            if (mb == 0.0)
                                tol=0.0;
                            else
                                if (mb < 0.0) tol = -tol;
                            p0=(b0-a0)*fb;
                            if (ext <= 1)
                                q0=fa-fb;
                            else {
                                fdb=(fd-fb)/(d0-b0);
                                fda=(fd-fa)/(d0-a0);
                                p0 *= fda;
                                q0=fdb*fa-fda*fb;
                            }
                            if (p0 < 0.0) {
                                p0 = -p0;
                                q0 = -q0;
                            }
                            w=(p0<FLT_MIN || p0<=q0*tol) ? tol :
                                    ((p0<mb*q0) ? p0/q0 : mb);
                        }
                        d0=a0;
                        fd=fa;
                        a0=b0;
                        fa=fb;
                        hacc = b0 += w;
                        fb=pow(hacc,q)*(a*hacc+b*(*t)+cc)-(*eta);
                        if ((fc >= 0.0) ? (fb >= 0.0) : (fb <= 0.0)) {
                            c0=a0;
```

```
                    fc=fa;
                    ext=0;
                } else
                    ext = (w == mb) ? 0 : ext+1;
            } else
                break;
        }
        h=c0;
        if (!((fc >= 0.0) ? (fb <= 0.0) : (fb >= 0.0)))
            hacc=hmax;
    } else
        hacc=h;
    if (hacc < 0.5*h2) hacc=0.5*h2;
}
if (hacc < hmin) hacc=hmin;
h=hacc;
if (h*sigmal > 1.0) {
    a=fabs((*diameter)(*t,m0,m)/sigmal+FLT_EPSILON)/2.0;
    b=2.0*fabs(sin(phil));
    betan=((a > b) ? 1.0/a : 1.0/b)/a;
    hstab=fabs(betan/sigmal);
    if (hstab < 1.0e-14*(*t)) break;
    if (h > hstab) h=hstab;
}
hcr=h2*h2/h1;
if (kl > 2 && fabs(h-hcr) < FLT_EPSILON*hcr)
    h = (h < hcr) ? hcr*(1.0-FLT_EPSILON) :
                    hcr*(1.0+FLT_EPSILON);
if ((*t)+h > te) {
    last=1;
    *hstart = h;
    h=te-(*t);
}
h0=h1;
h1=h2;
h2=h;
/* coefficient */
b=h*sigmal;
b1=b*cos(phil);
bb=b*b;
if (fabs(b) < 1.0e-3) {
    beta2=0.5-bb/24.0;
    beta3=1.0/6.0+b1/12.0;
    betha[3]=0.5+b1/3.0;
} else if (b1 < -40.0) {
    beta2=(-2.0*b1-4.0*b1*b1/bb+1.0)/bb;
    beta3=(1.0+2.0*b1/bb)/bb;
    betha[3]=1.0/bb;
} else {
    e=exp(b1)/bb;
    b2=b*sin(phil);
    beta2=(-2.0*b1-4.0*b1*b1/bb+1.0)/bb;
    beta3=(1.0+2.0*b1/bb)/bb;
    if (fabs(b2/b) < 1.0e-5) {
        beta2 -= e*(b1-3.0);
        beta3 += e*(b1-2.0)/b1;
        betha[3]=1.0/bb+e*(b1-1.0);
    } else {
        beta2 -= e*sin(b2-3.0*phil)/b2*b;
        beta3 += e*sin(b2-2.0*phil)/b2;
        betha[3]=1.0/bb+e*sin(b2-phil)/b2*b;
    }
}
beta[1]=betha[1]=1.0;
beta[2]=beta2;
beta[3]=beta3;
betha[2]=1.0-bb*beta3;
b=fabs(b);
q = (b < 1.5) ? 4.0-2.0*b/3.0 :
            ((b < 6.0) ? (30.0-2.0*b)/9.0 : 2.0);
for (i=1; i<=3; i++) {
    hi *= h;
    if (i > 1) (*derivative)(*t,m0,m,i,c);
    /* local error construction */
```

```
            if (i == 1) inivec(m0,m,ro,0.0);
            if (i < 4) elmvec(m0,m,0,ro,c,betha[i]*hi);
            if (i == 4) {
                elmvec(m0,m,0,ro,c,-h);
                s=0.0;
                if (norm == 1)
                    for (j=m0; j<=m; j++) {
                        x=fabs(ro[j]);
                        if (x > s) s=x;
                    }
                else
                    s=sqrt(vecvec(m0,m,0,ro,ro));
                *rho=s;
                ec0=ec1;
                ec1=ec2;
                ec2=(*rho)/pow(h,q);
            }
            elmvec(m0,m,0,u,c,beta[i]*hi);
        }
        t2=(*t);
        (*k)++;
        if (last) {
            last=0;
            (*t)=te;
            start=1;
        } else
            (*t) += h;
        dupvec(m0,m,0,c,u);
        (*derivative)(*t,m0,m,1,c);
        /* local error construction */
        elmvec(m0,m,0,ro,c,-h);
        s=0.0;
        if (norm == 1)
            for (j=m0; j<=m; j++) {
                x=fabs(ro[j]);
                if (x > s) s=x;
            }
        else
            s=sqrt(vecvec(m0,m,0,ro,ro));
        *rho=s;
        ec0=ec1;
        ec1=ec2;
        ec2=(*rho)/pow(h,q);
        (*out)(*t,te,m0,m,u,*k,*eta,*rho);
    } while (*t != te);
    free_real_vector(ro,m0);
}
```

5.4.4 Second order - No derivatives right hand side

A. rk2

Solves an initial value problem for a single second order ordinary differential equation $d^2y/dx^2 = f(x,y,dy/dx)$, from $x=a$ to $x=b$, $y(a)$ and $dy(a)/da$ being given, by means of a 5-th order Runge-Kutta method with steplength and error control [Z64].

Function Parameters:

void rk2 (*x,a,b,y,ya,z,za,fxyz,e,d,fi*)

x: float *;
    the independent variable;

a: float;
    entry: the initial value of *x*;

b: float;
    entry: the end value of *x*, ($b \leq a$ is allowed);

y: float *;
the dependent variable;
exit: the value of *y(x)* at $x = b$;
ya: float;
entry: the initial value of *y* at *x=a*;
z: float *;
the derivative *dy/dx*;
exit: the value of *z(x)* at $x = b$;
za: float;
entry: the initial value of *dy/dx* at *x=a*;
fxyz: float (**fxyz*)(*x,y,z*);
the right hand side of the differential equation;
fxyz depends on *x, y, z*, giving the value of d^2y/dx^2;
e: float *e[1:4]*;
entry: *e[1]* and *e[3]* are used as relative, *e[2]* and *e[4]* are used as absolute tolerances for *y* and *dy/dx*, respectively;
d: float *d[1:5]*;
exit:
d[1]: the number of steps skipped;
d[2]: the last step length used;
d[3]: equal to *b*;
d[4]: equal to *y(b)*;
d[5]: equal to *dy/dx*, for $x = b$;
fi: int;
entry: if *fi* is nonzero then the integration starts at *x=a* with a trial step *b-a*; if *fi* is zero then the integration is continued with, as initial conditions, *x=d[3], y=d[4], z=d[5]*, and *a, ya* and *za* are ignored.

```
#include <math.h>

void rk2(float *x, float a, float b, float *y, float ya,
        float *z, float za, float (*fxyz)(float, float, float),
        float e[], float d[], int fi)
{
    int last,first,reject,test,ta,tb;
    float e1,e2,e3,e4,xl,yl,zl,h,ind,hmin,hl,absh,k0,k1,k2,k3,k4,
          k5,discry,discrz,toly,tolz,mu,mu1,fhy,fhz;

    if (fi) {
        d[3]=a;
        d[4]=ya;
        d[5]=za;
    }
    d[1]=0.0;
    xl=d[3];
    yl=d[4];
    zl=d[5];
    if (fi) d[2]=b-d[3];
    absh=h=fabs(d[2]);
    if (b-xl < 0.0) h = -h;
    ind=fabs(b-xl);
    hmin=ind*e[1]+e[2];
    hl=ind*e[3]+e[4];
    if (hl < hmin) hmin=hl;
    e1=e[1]/ind;
    e2=e[2]/ind;
    e3=e[3]/ind;
    e4=e[4]/ind;
    first=1;
```

```
    test=1;
    if (fi) {
        last=1;
        test=0;
    }
    while (1) {
        if (test) {
            absh=fabs(h);
            if (absh < hmin) {
                h = (h > 0.0) ? hmin : -hmin;
                absh=hmin;
            }
            ta=(h >= b-xl);
            tb=(h >= 0.0);
            if ((ta && tb) || (!(ta || tb))) {
                d[2]=h;
                last=1;
                h=b-xl;
                absh=fabs(h);
            } else
                last=0;
        }
        test=1;
        *x=xl;
        *y=yl;
        *z=zl;
        k0=(*fxyz)(*x,*y,*z)*h;
        *x=xl+h/4.5;
        *y=yl+(zl*18.0+k0*2.0)/81.0*h;
        *z=zl+k0/4.5;
        k1=(*fxyz)(*x,*y,*z)*h;
        *x=xl+h/3.0;
        *y=yl+(zl*6.0+k0)/18.0*h;
        *z=zl+(k0+k1*3.0)/12.0;
        k2=(*fxyz)(*x,*y,*z)*h;
        *x=xl+h*0.5;
        *y=yl+(zl*8.0+k0+k2)/16.0*h;
        *z=zl+(k0+k2*3.0)/8.0;
        k3=(*fxyz)(*x,*y,*z)*h;
        *x=xl+h*0.8;
        *y=yl+(zl*100.0+k0*12.0+k3*28.0)/125.0*h;
        *z=zl+(k0*53.0-k1*135.0+k2*126.0+k3*56.0)/125.0;
        k4=(*fxyz)(*x,*y,*z)*h;
        *x = (last ? b : xl+h);
        *y=yl+(zl*336.0+k0*21.0+k2*92.0+k4*55.0)/336.0*h;
        *z=zl+(k0*133.0-k1*378.0+k2*276.0+k3*112.0+k4*25.0)/168.0;
        k5=(*fxyz)(*x,*y,*z)*h;
        discry=fabs((-k0*21.0+k2*108.0-k3*112.0+k4*25.0)/56.0*h);
        discrz=fabs(k0*21.0-k2*162.0+k3*224.0-k4*125.0+k5*42.0)/14.0;
        toly=absh*(fabs(zl)*e1+e2);
        tolz=fabs(k0)*e3+absh*e4;
        reject=(discry > toly || discrz > tolz);
        fhy=discry/toly;
        fhz=discrz/tolz;
        if (fhz > fhy) fhy=fhz;
        mu=1.0/(1.0+fhy)+0.45;
        if (reject) {
            if (absh <= hmin) {
                d[1] += 1.0;
                *y=yl;
                *z=zl;
                first=1;
                if (b == *x) break;
                xl = *x;
                yl = *y;
                zl = *z;
            } else
                h *= mu;
        } else {
            if (first) {
                first=0;
                hl=h;
                h *= mu;
            } else {
```

```
                fhy=mu*h/hl+mu-mul;
                hl=h;
                h *= fhy;
            }
            mul=mu;
            *y=yl+(zl*56.0+k0*7.0+k2*36.0-k4*15.0)/56.0*hl;
            *z=zl+(-k0*63.0+k1*189.0-k2*36.0-k3*112.0+k4*50.0)/28.0;
            k5=(*fxyz)(*x,*y,*z)*hl;
            *y=yl+(zl*336.0+k0*35.0+k2*108.0+k4*25.0)/336.0*hl;
            *z=zl+(k0*35.0+k2*162.0+k4*125.0+k5*14.0)/336.0;
            if (b == *x) break;
            xl = *x;
            yl = *y;
            zl = *z;
        }
    }
    if (!last) d[2]=h;
    d[3] = *x;
    d[4] = *y;
    d[5] = *z;
}
```

B. rk2n

Solves an initial value problem for a system of second order ordinary differential equations $d^2y_j(x)/dx^2 = f_j(x,y,dy(x)/dx)$, $(j=1,2,...,n)$, from $x=a$ to $x=b$, $y(a)$ and $dy(a)/da$ being given, by means of a 5-th order Runge-Kutta method [Z64].

Upon completion of a call of *rk2n* we have: $x=d[3]=b$, $y[j]=d[j+3]$ the value of the dependent variables for $x=b$, $z[j]=d[n+j+3]$, the value of the derivatives of $y[j]$ at $x=b$, $j=1,...,n$. *rk2n* uses as its minimal absolute step length $hmin = \min(e[2*j-1]*k+e[2*j])$ with $1 \le j \le 2*n$ and $k = |b - (\text{if } fi \text{ is nonzero then } a \text{ else } d[3])|$.

If a step of length $|h| \le hmin$ is rejected then a step $\text{sign}(h)*hmin$ is skipped. A step is rejected if the absolute value of the computed discretization error is greater than

$$(|z[j]|*e[2*j-1]+e[2*j])*|h|/k$$

or if that term is greater than

$$(|fxyzj|*e[2*(j+n)-1]+e[2*(j+n)])*|h|/k,$$

for any value of j, $1 \le j \le n$, $(k=|b-a|)$.

Function Parameters:

void rk2n (*x,a,b,y,ya,z,za,fxyzj,e,d,fi,n*)

x: float *;
the independent variable;
exit: upon completion of a call of *rk2n*, it is equal to *b*;

a: float;
entry: the starting value of *x*;

b: float;
entry: the end value of *x*;

y: float *y[1:n]*;
the vector of dependent variables;
exit: the values of *y[j]* at $x = b$, $j=1,...,n$;

ya: float *ya[1:n]*;
entry: the initial values of *y[j]*, i.e. values at $x=a$;

z: float *z[1:n]*;
the first derivatives of the dependent variables;
exit: the value of $(d/dx)y[j]$ at $x = b$, $j=1,...,n$;

za: float *za[1:n]*;
entry: the initial values of *z[j]*, i.e. the values at *x=a*;

fxyzj: float (**fxyzj*)(*n,j,x,y,z*);
fxyzj depends on *x,j,y[j],z[j]*, (*j*=1,...,*n*), giving the value of $d^2y[j]/dx^2$;

e: float *e[1:4*n]*;
the element *e[2*j-1]* is a relative and *e[2*j]* an absolute tolerance associated with *y[j]*; *e[2*(n+j)-1]* is a relative and *e[2*(n+j)]* an absolute tolerance associated with *z[j]*;

d: float *d[1:2*n+3]*;
exit:
d[1]: the number of steps skipped;
d[2]: the last step length used;
d[3]: equal to *b*;
d[4],...,d[n+3]: equal to *y[1],...,y[n]* for *x* = *b*;
*d[n+4],...,d[2*n+3]*: equal to the derivatives *z[1],...,z[n]* for *x* = *b*;

fi: int;
entry: if *fi* is nonzero then the integration starts at *x=a* with a trial step *b-a*;
if *fi* is zero then the integration is continued with, as initial conditions, *x=d[3]*, *y[j]=d[j+3]*, *z[j]=d[n+3+j]*, and step length *h=d[2]**sign(*b-d[3]*), and *a*, *ya* and *za* are ignored;

n: int;
entry: the number of equations.

```
#include <math.h>

void rk2n(float *x, float a, float b, float y[], float ya[],
        float z[], float za[],
        float (*fxyzj)(int, int, float, float [], float []),
        float e[], float d[], int fi, int n)
{
    float *allocate_real_vector(int, int);
    void free_real_vector(float *, int);
    int j,jj,last,first,reject,test,ta,tb;
    float xl,h,ind,hmin,hl,absh,fhm,discry,discrz,toly,tolz,
            mu,mu1,fhy,fhz,*yl,*zl,*k0,*k1,*k2,*k3,*k4,*k5,*ee;

    yl=allocate_real_vector(1,n);
    zl=allocate_real_vector(1,n);
    k0=allocate_real_vector(1,n);
    k1=allocate_real_vector(1,n);
    k2=allocate_real_vector(1,n);
    k3=allocate_real_vector(1,n);
    k4=allocate_real_vector(1,n);
    k5=allocate_real_vector(1,n);
    ee=allocate_real_vector(1,4*n);
    if (fi) {
        d[3]=a;
        for (jj=1; jj<=n; jj++) {
            d[jj+3]=ya[jj];
            d[n+jj+3]=za[jj];
        }
    }
    d[1]=0.0;
    xl=d[3];
    for (jj=1; jj<=n; jj++) {
        yl[jj]=d[jj+3];
        zl[jj]=d[n+jj+3];
    }
    if (fi) d[2]=b-d[3];
    absh=h=fabs(d[2]);
    if (b-xl < 0.0) h = -h;
    ind=fabs(b-xl);
    hmin=ind*e[1]+e[2];
```

```
for (jj=2; jj<=2*n; jj++) {
    hl=ind*e[2*jj-1]+e[2*jj];
    if (hl < hmin) hmin=hl;
}
for (jj=1; jj<=4*n; jj++) ee[jj]=e[jj]/ind;
first=1;
test=1;
if (fi) {
    last=1;
    test=0;
}
while (1) {
    if (test) {
        absh=fabs(h);
        if (absh < hmin) {
            h = (h > 0.0) ? hmin : -hmin;
            absh=fabs(h);
        }
        ta=(h >= b-xl);
        tb=(h >= 0.0);
        if ((ta && tb) || (!(ta || tb))) {
            d[2]=h;
            last=1;
            h=b-xl;
            absh=fabs(h);
        } else
            last=0;
    }
    test=1;
    *x=xl;
    for (jj=1; jj<=n; jj++) {
        y[jj]=yl[jj];
        z[jj]=zl[jj];
    }
    for (j=1; j<=n; j++)  k0[j]=(*fxyzj)(n,j,*x,y,z)*h;
    *x=xl+h/4.5;
    for (jj=1; jj<=n; jj++) {
        y[jj]=yl[jj]+(zl[jj]*18.0+k0[jj]*2.0)/81.0*h;
        z[jj]=zl[jj]+k0[jj]/4.5;
    }
    for (j=1; j<=n; j++) k1[j]=(*fxyzj)(n,j,*x,y,z)*h;
    *x=xl+h/3.0;
    for (jj=1; jj<=n; jj++) {
        y[jj]=yl[jj]+(zl[jj]*6.0+k0[jj])/18.0*h;
        z[jj]=zl[jj]+(k0[jj]+k1[jj]*3.0)/12.0;
    }
    for (j=1; j<=n; j++) k2[j]=(*fxyzj)(n,j,*x,y,z)*h;
    *x=xl+h*0.5;
    for (jj=1; jj<=n; jj++) {
        y[jj]=yl[jj]+(zl[jj]*8.0+k0[jj]+k2[jj])/16.0*h;
        z[jj]=zl[jj]+(k0[jj]+k2[jj]*3.0)/8.0;
    }
    for (j=1; j<=n; j++) k3[j]=(*fxyzj)(n,j,*x,y,z)*h;
    *x=xl+h*0.8;
    for (jj=1; jj<=n; jj++) {
        y[jj]=yl[jj]+(zl[jj]*100.0+k0[jj]*12.0+
                  k3[jj]*28.0)/125.0*h;
        z[jj]=zl[jj]+(k0[jj]*53.0-k1[jj]*135.0+k2[jj]*126.0+
                  k3[jj]*56.0)/125.0;
    }
    for (j=1; j<=n; j++) k4[j]=(*fxyzj)(n,j,*x,y,z)*h;
    *x = (last ? b : xl+h);
    for (jj=1; jj<=n; jj++) {
        y[jj]=yl[jj]+(zl[jj]*336.0+k0[jj]*21.0+k2[jj]*92.0+
                  k4[jj]*55.0)/336.0*h;
        z[jj]=zl[jj]+(k0[jj]*133.0-k1[jj]*378.0+k2[jj]*276.0+
                  k3[jj]*112.0+k4[jj]*25.0)/168.0;
    }
    for (j=1; j<=n; j++) k5[j]=(*fxyzj)(n,j,*x,y,z)*h;
    reject=0;
    fhm=0.0;
    for (jj=1; jj<=n; jj++) {
        discry=fabs((-k0[jj]*21.0+k2[jj]*108.0-k3[jj]*112.0+
                  k4[jj]*25.0)/56.0*h);
```

```
            discrz=fabs(k0[jj]*21.0-k2[jj]*162.0+k3[jj]*224.0-
                        k4[jj]*125.0+k5[jj]*42.0)/14.0;
            toly=absh*(fabs(zl[jj])*ee[2*jj-1]+ee[2*jj]);
            tolz=fabs(k0[jj])*ee[2*(jj+n)-1]+absh*ee[2*(jj+n)];
            reject=((discry > toly) || (discrz > tolz) || reject);
            fhy=discry/toly;
            fhz=discrz/tolz;
            if (fhz > fhy) fhy=fhz;
            if (fhy > fhm) fhm=fhy;
        }
        mu=1.0/(1.0+fhm)+0.45;
        if (reject) {
            if (absh <= hmin) {
                d[1] += 1.0;
                for (jj=1; jj<=n; jj++) {
                    y[jj]=yl[jj];
                    z[jj]=zl[jj];
                }
                first=1;
                if (b == *x) break;
                xl = *x;
                for (jj=1; jj<=n; jj++) {
                    yl[jj] = y[jj];
                    zl[jj] = z[jj];
                }
            } else
                h *= mu;
        } else {
            if (first) {
                first=0;
                hl=h;
                h *= mu;
            } else {
                fhm=mu*h/hl+mu-mu1;
                hl=h;
                h *= fhm;
            }
            mu1=mu;
            for (jj=1; jj<=n; jj++) {
                y[jj]=yl[jj]+(zl[jj]*56.0+k0[jj]*7.0+k2[jj]*36.0-
                            k4[jj]*15.0)/56.0*hl;
                z[jj]=zl[jj]+(-k0[jj]*63.0+k1[jj]*189.0-k2[jj]*36.0-
                            k3[jj]*112.0+k4[jj]*50.0)/28.0;
            }
            for (j=1; j<=n; j++) k5[j]=(*fxyzj)(n,j,*x,y,z)*hl;
            for (jj=1; jj<=n; jj++) {
                y[jj]=yl[jj]+(zl[jj]*336.0+k0[jj]*35.0+k2[jj]*108.0+
                            k4[jj]*25.0)/336.0*hl;
                z[jj]=zl[jj]+(k0[jj]*35.0+k2[jj]*162.0+k4[jj]*125.0+
                            k5[jj]*14.0)/336.0;
            }
            if (b == *x) break;
            xl = *x;
            for (jj=1; jj<=n; jj++) {
                yl[jj] = y[jj];
                zl[jj] = z[jj];
            }
        }
    }
    if (!last) d[2]=h;
    d[3] = *x;
    for (jj=1; jj<=n; jj++) {
        d[jj+3]=y[jj];
        d[n+jj+3]=z[jj];
    }
    free_real_vector(yl,1);
    free_real_vector(zl,1);
    free_real_vector(k0,1);
    free_real_vector(k1,1);
    free_real_vector(k2,1);
    free_real_vector(k3,1);
    free_real_vector(k4,1);
    free_real_vector(k5,1);
    free_real_vector(ee,1);
```

}

C. rk3

Solves an initial value problem for a single second order ordinary differential equation without first derivative

$$d^2y/dx^2 = f(x,y),$$

from $x=a$ to $x=b$, $y(a)$ and $dy(a)/da$ being given, by means of a 5-th order Runge-Kutta method [Z64].

Upon completion of a call of *rk3* we have $x=d[3]=b$, $y=d[4]=y[b]$, $z=d[5]$, i.e. the value of dy/dx for $x=b$. *rk3* uses as its minimal absolute step length

$$hmin = \min (e[2*j-1]*k+e[2*j]) \text{ with } 1 \leq j \leq 2$$

and $k=|b$ - (if *fi* is nonzero then a else $d[3])|$. If a step of length $|h| \leq hmin$ is rejected then a step sign(h)*$hmin$ is skipped. A step is rejected if the absolute value of the last term taken into account is greater than $(|dy/dx|*e[1]+e[2])*|h|/k$ or if that term is greater than $(|fxy|*e[3]+e[4])*|h|/k$ where $k=|b-a|$.

Function Parameters:

void rk3 (*x,a,b,y,ya,z,za,fxy,e,d,fi*)

x: float *; the independent variable;
  exit: upon completion of a call of *rk3*, it is equal to *b*;
a: float;
  entry: the initial value of *x*;
b: float;
  entry: the end value of *x*, ($b \leq a$ is allowed);
y: float *; the dependent variable;
  exit: the value of *y(x)* at $x = b$;
ya: float;
  entry: the initial value of *y* at $x=a$;
z: float *;
  the derivative *dy/dx*;
  exit: the value of *dy/dx* at $x = b$;
za: float;
  entry: the initial value of *dy/dx* at $x=a$;
fxy: float (**fxy*)(*x,y*);
  fxy depends on *x* and *y*, giving the value of d_2y/dx_2;
e: float *e[1:4]*;
  entry: *e[1]* and *e[3]* are used as relative tolerances, *e[2]* and *e[4]* are used as absolute tolerances for *y* and *dy/dx*, respectively;
d: float *d[1:5]*;
  exit:
  d[1]: the number of steps skipped;
  d[2]: the last step length used;
  d[3]: equal to *b*;
  d[4]: equal to *y(b)*;
  d[5]: equal to *dy/dx*, for $x = b$;
fi: int;
  entry: if *fi* is nonzero then the integration starts at $x=a$ with a trial step *b-a*;

if *fi* is zero then the integration is continued with, as initial conditions, *x=d[3]*, *y=d[4]*, *z=d[5]*, and steplength *h=d[2]**sign(*b-d[3]*), *a, ya* and *za* are ignored.

```
#include <math.h>

void rk3(float *x, float a, float b, float *y, float ya,
        float *z, float za, float (*fxy)(float, float),
        float e[], float d[], int fi)
{
    int last,first,reject,test,ta,tb;
    float e1,e2,e3,e4,xl,yl,zl,h,ind,hmin,hl,absh,k0,k1,k2,k3,k4,
            k5,discry,discrz,toly,tolz,mu,mu1,fhy,fhz;

    if (fi) {
        d[3]=a;
        d[4]=ya;
        d[5]=za;
    }
    d[1]=0.0;
    xl=d[3];
    yl=d[4];
    zl=d[5];
    if (fi) d[2]=b-d[3];
    absh=h=fabs(d[2]);
    if (b-xl < 0.0) h = -h;
    ind=fabs(b-xl);
    hmin=ind*e[1]+e[2];
    hl=ind*e[3]+e[4];
    if (hl < hmin) hmin=hl;
    e1=e[1]/ind;
    e2=e[2]/ind;
    e3=e[3]/ind;
    e4=e[4]/ind;
    first=reject=1;
    test=1;
    if (fi) {
        last=1;
        test=0;
    }
    while (1) {
        if (test) {
            absh=fabs(h);
            if (absh < hmin) {
                h = (h > 0.0) ? hmin : -hmin;
                absh=hmin;
            }
            ta=(h >= b-xl);
            tb=(h >= 0.0);
            if ((ta && tb) || (!(ta || tb))) {
                d[2]=h;
                last=1;
                h=b-xl;
                absh=fabs(h);
            } else
                last=0;
        }
        test=1;
        if (reject) {
            *x=xl;
            *y=yl;
            k0=(*fxy)(*x,*y)*h;
        } else
            k0=k5*h/hl;
        *x=xl+0.276393202250021*h;
        *y=yl+(zl*0.276393202250021+k0*0.038196601125011)*h;
        k1=(*fxy)(*x,*y)*h;
        *x=xl+0.723606797749979*h;
        *y=yl+(zl*0.723606797749979+k1*0.261803398874989)*h;
        k2=(*fxy)(*x,*y)*h;
        *x=xl+h*0.5;
        *y=yl+(zl*0.5+k0*0.046875+k1*0.079824155839840-
                k2*0.001699155839840)*h;
```

```
            k4=(*fxy)(*x,*y)*h;
            *x = (last ? b : xl+h);
            *y=yl+(zl+k0*0.309016994374947+k2*0.190983005625053)*h;
            k3=(*fxy)(*x,*y)*h;
            *y=yl+(zl+k0*0.0833333333333333+k1*0.301502832395825+
                        k2*0.115163834270842)*h;
            k5=(*fxy)(*x,*y)*h;
            discry=fabs((-k0*0.5+k1*1.809016994374947+
                        k2*0.690983005625053-k4*2.0)*h);
            discrz=fabs((k0-k3)*2.0-(k1+k2)*10.0+k4*16.0+k5*4.0);
            toly=absh*(fabs(zl)*e1+e2);
            tolz=fabs(k0)*e3+absh*e4;
            reject=(discry > toly || discrz > tolz);
            fhy=discry/toly;
            fhz=discrz/tolz;
            if (fhz > fhy) fhy=fhz;
            mu=1.0/(1.0+fhy)+0.45;
            if (reject) {
                if (absh <= hmin) {
                    d[1] += 1.0;
                    *y=yl;
                    *z=zl;
                    first=1;
                    if (b == *x) break;
                    xl = *x;
                    yl = *y;
                    zl = *z;
                } else
                    h *= mu;
            } else {
                if (first) {
                    first=0;
                    hl=h;
                    h *= mu;
                } else {
                    fhy=mu*h/hl+mu-mu1;
                    hl=h;
                    h *= fhy;
                }
                mu1=mu;
                *z=zl+(k0+k3)*0.0833333333333333+
                        (k1+k2)*0.416666666666667;
                if (b == *x) break;
                xl = *x;
                yl = *y;
                zl = *z;
            }
        }
        if (!last) d[2]=h;
        d[3] = *x;
        d[4] = *y;
        d[5] = *z;
}
```

D. rk3n

Solves an initial value problem for a system of second order ordinary differential equations without first derivative

$$d^2y_j(x)/dx^2 = f_j(x,y), \qquad (j=1,2,...,n)$$

from $x=a$ to $x=b$, $y(a)$ and $dy(a)/da$ being given, by means of a 5-th order Runge-Kutta method [Z64].

Upon completion of a call of *rk3n* we have: $x=d[3]=b$, $y[j]=d[j+3]$ the value of the dependent variables for $x=b$, $z[j]=d[n+j+3]$, the value of the derivatives of $y[j]$ at $x=b$. *rk3n* uses as its minimal absolute step length

$$hmin = \min\ (e[2*j-1]*k+e[2*j]) \text{ with } 1 \le j \le 2n$$

and

$k = |b$ - (if *fi* is nonzero then a else $d[3])|$.

If a step of length $|h| \leq hmin$ is rejected then a step sign(h)**hmin* is skipped. A step is rejected if the absolute value of the last term taken into account is greater than

$$(|z[j]|*e[2*j-1]+e[2*j])*|h|/k$$

or if that term is greater than

$$(|fxyj|*e[2*(j+n)-1]+e[2*(j+n)])*|h|/k,$$

for any value of j, $1 \leq j \leq n$, ($k=|b-a|$).

Function Parameters:

void rk3n (*x,a,b,y,ya,z,za,fxyj,e,d,fi,n*)

x: float *;
the independent variable;
exit: upon completion of a call of *rk3n*, it is equal to *b*;

a: float;
entry: the starting value of *x*;

b: float;
entry: the end value of *x* ($b \leq a$ is allowed);

y: float *y[1:n]*;
the vector of dependent variables;
exit: the values of *y[j]* at $x = b$, j=1,...,n;

ya: float *ya[1:n]*;
entry: the initial values of *y[j]*, i.e. values at $x=a$;

z: float *z[1:n]*;
the first derivatives of the dependent variables, *z[j]* = *dy[j]/dx*;
exit: the value of *z[j](x)* at $x = b$, j=1,...,n;

za: float *za[1:n]*;
entry: the initial values of *z[j]*, i.e. the values at $x=a$;

fxyj: float (**fxyj*)(*n,j,x,y*);
fxyj depends on *x,y[1],y[2],...,y[n],j*, giving the value of $d_2y[j]/dx_2$;

e: float *e[1:4*n]*;
the element *e[2*j-1]* is a relative and *e[2*j]* is an absolute tolerance associated with *y[j]*; *e[2*(n+j)-1]* is a relative and *e[2*(n+j)]* is an absolute tolerance associated with *z[j]*;

d: float *d[1:2*n+3]*;
exit:
d[1]: the number of steps skipped;
d[2]: the last step length used;
d[3]: equals to *b*;
d[4],...,d[n+3]: equal to *y[1],...,y[n]* for $x = b$;
*d[n+4],...,d[2*n+3]*: equal to the derivatives *z[1],...,z[n]* for $x = b$;

fi: int;
entry: if *fi* is nonzero then the integration starts at $x=a$ with a trial step *b-a*;
if *fi* is zero then the integration is continued with, as initial conditions, $x=d[3]$, $y[j]=d[j+3]$, $z[j]=d[n+3+j]$, and step length $h=d[2]$*sign($b-d[3]$), and *a*, *ya* and *za* are ignored;

n: int;
entry: the number of equations.

```
#include <math.h>

void rk3n(float *x, float a, float b, float y[], float ya[],
        float z[], float za[],
        float (*fxyj)(int, int, float, float[]),
        float e[], float d[], int fi, int n)
{
    float *allocate_real_vector(int, int);
    void free_real_vector(float *, int);
    int j,jj,last,first,reject,test,ta,tb;
    float xl,h,hmin,ind,hl,absh,fhm,discry,discrz,toly,tolz,mu,
          mu1,fhy,fhz,*yl,*zl,*k0,*k1,*k2,*k3,*k4,*k5,*ee;

    yl=allocate_real_vector(1,n);
    zl=allocate_real_vector(1,n);
    k0=allocate_real_vector(1,n);
    k1=allocate_real_vector(1,n);
    k2=allocate_real_vector(1,n);
    k3=allocate_real_vector(1,n);
    k4=allocate_real_vector(1,n);
    k5=allocate_real_vector(1,n);
    ee=allocate_real_vector(1,4*n);

    if (fi) {
        d[3]=a;
        for (jj=1; jj<=n; jj++) {
            d[jj+3]=ya[jj];
            d[n+jj+3]=za[jj];
        }
    }
    d[1]=0.0;
    xl=d[3];
    for (jj=1; jj<=n; jj++) {
        yl[jj]=d[jj+3];
        zl[jj]=d[n+jj+3];
    }
    if (fi) d[2]=b-d[3];
    absh=h=fabs(d[2]);
    if (b-xl < 0.0) h = -h;
    ind=fabs(b-xl);
    hmin=ind*e[1]+e[2];
    for (jj=2; jj<=2*n; jj++) {
        hl=ind*e[2*jj-1]+e[2*jj];
        if (hl < hmin) hmin=hl;
    }
    for (jj=1; jj<=4*n; jj++) ee[jj]=e[jj]/ind;
    first=reject=1;
    test=1;
    if (fi) {
        last=1;
        test=0;
    }
    while (1) {
        if (test) {
            absh=fabs(h);
            if (absh < hmin) {
                h = (h > 0.0) ? hmin : -hmin;
                absh=hmin;
            }
            ta=(h >= b-xl);
            tb=(h >= 0.0);
            if ((ta && tb) || (!(ta || tb))) {
                d[2]=h;
                last=1;
                h=b-xl;
                absh=fabs(h);
            } else
                last=0;
        }
        test=1;
        if (reject) {
            *x=xl;
            for (jj=1; jj<=n; jj++) y[jj]=yl[jj];
            for (j=1; j<=n; j++) k0[j]=(*fxyj)(n,j,*x,y)*h;
```

```
    } else {
        fhy=h/hl;
        for (jj=1; jj<=n; jj++) k0[jj]=k5[jj]*fhy;
    }
    *x=xl+0.276393202250021*h;
    for (jj=1; jj<=n; jj++)
        y[jj]=yl[jj]+(zl[jj]*0.276393202250021+
                        k0[jj]*0.038196601125011)*h;
    for (j=1; j<=n; j++) k1[j]=(*fxyj)(n,j,*x,y)*h;
    *x=xl+0.723606797749979*h;
    for (jj=1; jj<=n; jj++)
        y[jj]=yl[jj]+(zl[jj]*0.723606797749979+
                        k1[jj]*0.261803398874989)*h;
    for (j=1; j<=n; j++) k2[j]=(*fxyj)(n,j,*x,y)*h;
    *x=xl+h*0.5;
    for (jj=1; jj<=n; jj++)
        y[jj]=yl[jj]+(zl[jj]*0.5+k0[jj]*0.046875+k1[jj]*
                0.079824155839840-k2[jj]*0.001699155839840)*h;
    for (j=1; j<=n; j++) k4[j]=(*fxyj)(n,j,*x,y)*h;
    *x = (last ? b : xl+h);
    for (jj=1; jj<=n; jj++)
        y[jj]=yl[jj]+(zl[jj]+k0[jj]*0.309016994374947+
                        k2[jj]*0.190983005625053)*h;
    for (j=1; j<=n; j++) k3[j]=(*fxyj)(n,j,*x,y)*h;
    for (jj=1; jj<=n; jj++)
        y[jj]=yl[jj]+(zl[jj]+k0[jj]*0.083333333333333+k1[jj]*
                0.301502832395825+k2[jj]*0.115163834270842)*h;
    for (j=1; j<=n; j++) k5[j]=(*fxyj)(n,j,*x,y)*h;
    reject=0;
    fhm=0.0;
    for (jj=1; jj<=n; jj++) {
        discry=fabs((-k0[jj]*0.5+k1[jj]*1.809016994374947+
                    k2[jj]*0.690983005625053-k4[jj]*2.0)*h);
        discrz=fabs((k0[jj]-k3[jj])*2.0-(k1[jj]+k2[jj])*10.0+
                    k4[jj]*16.0+k5[jj]*4.0);
        toly=absh*(fabs(zl[jj])*ee[2*jj-1]+ee[2*jj]);
        tolz=fabs(k0[jj])*ee[2*(jj+n)-1]+absh*ee[2*(jj+n)];
        reject=((discry > toly) || (discrz > tolz) || reject);
        fhy=discry/toly;
        fhz=discrz/tolz;
        if (fhz > fhy) fhy=fhz;
        if (fhy > fhm) fhm=fhy;
    }
    mu=1.0/(1.0+fhm)+0.45;
    if (reject) {
        if (absh <= hmin) {
            d[1] += 1.0;
            for (jj=1; jj<=n; jj++) {
                y[jj]=yl[jj];
                z[jj]=zl[jj];
            }
            first=1;
            if (b == *x) break;
            xl = *x;
            for (jj=1; jj<=n; jj++) {
                yl[jj]=y[jj];
                zl[jj]=z[jj];
            }
        } else
            h *= mu;
    } else {
        if (first) {
            first=0;
            hl=h;
            h *= mu;
        } else {
            fhy=mu*h/hl+mu-mu1;
            hl=h;
            h *= fhy;
        }
        mu1=mu;
        for (jj=1; jj<=n; jj++)
            z[jj]=zl[jj]+(k0[jj]+k3[jj])*0.083333333333333+
                        (k1[jj]+k2[jj])*0.416666666666667;
```

```
            if (b == *x) break;
            xl = *x;
            for (jj=1; jj<=n; jj++) {
                yl[jj]=y[jj];
                zl[jj]=z[jj];
            }
        }
    }
    if (!last) d[2]=h;
    d[3] = *x;
    for (jj=1; jj<=n; jj++) {
        d[jj+3]=y[jj];
        d[n+jj+3]=z[jj];
    }
    free_real_vector(yl,1);
    free_real_vector(zl,1);
    free_real_vector(k0,1);
    free_real_vector(k1,1);
    free_real_vector(k2,1);
    free_real_vector(k3,1);
    free_real_vector(k4,1);
    free_real_vector(k5,1);
    free_real_vector(ee,1);
}
```

5.4.5 Initial boundary value problem

arkmat

Solves an initial value problem, given as a system of first order (non-linear) differential equations

$$DU(t) = F(t,U(t)) \tag{1}$$

where $D=d/dt$ and U is an $n \times m$ matrix, with $U(t_0)$ prescribed, over the range $t0 \leq t \leq te$, by means of a stabilized Runge-Kutta method [Vh71]. This procedure is suitable for the integration of systems where the dependent variable and the right hand side are stored in a rectangular array instead of a vector. The integration stepsize used will depend on (a) the type of system to be solved (i.e. hyperbolic or parabolic); (b) the spectral radius of the Jacobian matrix of the system; (c) the indicated order of the particular Runge-Kutta method. *arkmat* is especially intended for systems of differential equations arising from initial boundary value problems in two dimensions, e.g. when the method of lines is applied to this kind of problem, the right hand side of the resulting system is much easier to describe in matrix than in vector form.

The method used is associated with one of four prescribed ninth degree stability polynomials (the choice being open to the user): the first has a degree of polynomial exactness equal to 9; the second and third have absolute values less than unity when their arguments lie in the range [-ß(9),0] for the maximal values of ß(9), and have degrees of polynomial exactness equal to 1 and 2 respectively (such polynomials are suitable for the numerical solution of parabolic partial differential equations, the eigenvalues of the Jacobian matrix derived from the right hand side of equation (1) (F and U being expanded to vector form) which lie in the closed left half-plane being real); the fourth polynomial has absolute value less than unity when its argument lies in the range i[-8,8] (it is suitable for the numerical solution of hyperbolic partial differential equations, the eigenvalues of the derived Jacobian matrix lying in the closed left half-plane being confined to the imaginary axis). The Runge-Kutta scheme employed has the form (with N=9 below)

$$F_{k,0} = F(t_k, U_k)$$

$$F_{k,i} = F(t + \lambda_i\tau,\ U_k + \lambda_k\tau F_{k,i-1}) \qquad (i=1,...,N-1) \tag{2}$$

$$U_{k+1} = U_k + F_{k,n-1}$$

where $U_k \approx U(t_k)$ and $t_{k+1} = t_k + \tau$ (k=0,1,...), the stepsize τ being prescribed. The coefficients λ_i are derived from the coefficients β_i of the stability polynomial

$$P(z) = \sum_{i=0}^{N} \beta_i z^i$$

chosen, by use of the relationships

$$\lambda_i = \beta_{N-i+1} / \beta_{N-i} \quad (i=1,...,N-1).$$

The choice (among the four described above) of stability polynomial is determined by the values allocated to the input parameters *type* and *order* upon call of *arkmat*. If *type*=1 the first polynomial is chosen, if *type*=2 and *order*=1 the second, if *type*=2 and *order*=2 the third, and if *type*=3, the fourth (if *type*≠2, the value of *order* is ignored). The fixed stepsize τ is determined partly by the range constant $\beta(n)$ of the stability polynomial associated with the method being used, and partly by the value of the input parameter *spr* (this will in general be a multiple of the spectral radius with respect to the eigenvalues in the closed left half-plane of the Jacobian matrix derived from equation (1) upon call of *arkmat*. If *type*=1, τ=4.3/*spr*; if *type*=2 and *order*=1, τ=156/*spr*; if *type*=2 and *order*=2, τ=64/*spr*, and if *type*=3, τ=8/*spr*.

Function Parameters:

void arkmat (*t,te,m,n,u,der,type,order,spr,out*)

t: float *;
the independent variable;
entry: initial value *t0*;
exit: the final value *te*;

te: float;
entry: the final value of *t*;

m: int;
entry: the number of columns of *u*;

n: int;
entry: the number of rows of *u*;

u: float *u[1:n,1:m]*;
entry: the initial values of the solution of the system of differential equations at *t=t0*;
exit: the values of the solution at *t=te*;

der: void (**der*)(*m,n,t,v,ftv*);
this procedure must be given by the user and performs an evaluation of the right hand side *F(t,v)* of the system; upon completion of *der*, the right hand side should be stored in the array *ftv[1:n,1:m]*;

type: int;
entry: the type of the system of differential equations to be solved; the user should supply one of the following values:
- *type*=1: if no specification of the type can be made;
- *type*=2: if the eigenvalues of the Jacobian matrix of the right hand side are negative real;
- *type*=3: if the eigenvalues of the Jacobian matrix of the right hand side are purely imaginary;

order: int *;

the order of the Runge-Kutta method used;
entry: for *type*=2 the user may choose *order*=1 or *order*=2; *order* should be 2 for the other types;
exit: if *order* is set to another value, it is assumed to be (if *type*=2 then 1 else 2);

spr: float *;
entry: the spectral radius of the Jacobian matrix of the right hand side, when the system is written in one dimensional form (ie. vector form);
the integration step will equal constant/*spr* (see above);
if necessary *spr* can be updated (after each step) by means of the procedure *out*;

out: void (**out*)(*t,te,m,n,u,type,order,spr*);
float *t*, float *te*, int *m*, int *n*, float *u[1:n,1:m]*, int *type*, int *order*, float **spr*;
after each integration step the user can print some parameters such as *t, u, type, order,* or possibly update *spr*.

Functions used: elmcol, dupmat.

```
#include <math.h>

void arkmat(float *t, float te, int m, int n, float **u,
      void (*der)(int, int, float, float **, float **),
      int type, int *order, float *spr,
      void (*out)(float, float, int, int, float **, int, int, float *))
{
   float **allocate_real_matrix(int, int, int, int);
   void free_real_matrix(float **, int, int, int);
   void elmcol(int, int, int, int, float **, float **, float);
   void dupmat(int, int, int, int, float **, float **);
   int sig,l,last,ta,tb,i;
   float tau,lambda[10],**uh,**du,mlt;
   static float lbd1[9]={1.0/9.0, 1.0/8.0, 1.0/7.0, 1.0/6.0,
                  1.0/5.0, 1.0/4.0, 1.0/3.0, 1.0/2.0, 4.3};
   static float lbd2[9]={0.1418519249e-2, 0.3404154076e-2,
               0.0063118569, 0.01082794375, 0.01842733851,
               0.03278507942, 0.0653627415, 0.1691078577, 156.0};
   static float lbd3[9]={0.3534355908e-2, 0.8532600867e-2,
               0.015956206, 0.02772229155, 0.04812587964,
               0.08848689452, 0.1863578961, 0.5, 64.0};
   static float lbd4[9]={1.0/8.0, 1.0/20.0, 5.0/32.0, 2.0/17.0,
               17.0/80.0, 5.0/22.0, 11.0/32.0, 1.0/2.0, 8.0};

   uh=allocate_real_matrix(1,n,1,m);
   du=allocate_real_matrix(1,n,1,m);

   /* initialize */
   if (type != 2 && type != 3) type=1;
   if (type != 2)
      *order = 2;
   else
      if (*order != 2) *order = 1;
   switch ((type == 1) ? 1 : type+(*order)-1) {
      case 1:  for (i=0; i<=8; i++) lambda[i+1]=lbd1[i]; break;
      case 2:  for (i=0; i<=8; i++) lambda[i+1]=lbd2[i]; break;
      case 3:  for (i=0; i<=8; i++) lambda[i+1]=lbd3[i]; break;
      case 4:  for (i=0; i<=8; i++) lambda[i+1]=lbd4[i]; break;
   }
   sig = ((te == *t) ? 0 : ((te > *t) ? 1 : -1));
   last=0;
   do {
      tau=((*spr == 0.0) ? fabs(te-(*t)) :
               fabs(lambda[9]/(*spr)))*sig;
      ta = (*t)+tau >= te;
      tb = tau >= 0.0;
      if ((ta && tb) || (!(ta || tb))) {
         tau=te-(*t);
         last=1;
```

```
        }
        /* difference scheme */
        (*der)(m,n,*t,u,du);
        for (i=1; i<=8; i++) {
            mlt=lambda[i]*tau;
            dupmat(1,n,1,m,uh,u);
            for (l=1; l<=m; l++) elmcol(1,n,l,l,uh,du,mlt);
            (*der)(m,n,(*t)+mlt,uh,du);
        }
        for (l=1; l<=m; l++) elmcol(1,n,l,l,u,du,tau);
        *t = (last ? te : (*t)+tau);
        (*out)(*t,te,m,n,u,type,*order,spr);
    } while (!last);
    free_real_matrix(uh,1,n,1);
    free_real_matrix(du,1,n,1);
}
```

5.5 Two point boundary value problems

5.5.1 Linear methods - Second order self adjoint

A. femlagsym

Solves a second order self-adjoint linear two point boundary value problem by means of Galerkin's method with continuous piecewise polynomials [BakH76, He75, StF73]. *femlagsym* computes approximations y_n $(n=0,...,N)$ to the values at the points $x = x_n$, where

$$-\infty < a = x_0 < x_1 < ... < x_N = b < \infty$$

of the real valued function $y(x)$ which satisfies the equation

$$-D(p(x)Dy(x)) + r(x)y(x) = f(x) \tag{1}$$

where $D=d/dx$, with boundary conditions

$$e_1y(a) + e_2Dy(a) = e_3, \qquad e_4y(b) + e_5Dy(b) = e_6 \qquad (e_1, e_4 \neq 0). \tag{2}$$

It is assumed that

(a) $p(x)> 0$ on the interval $[x_{n-1}, x_n]$, $n=1,...,N$;

(b) $p(x)$, $r(x)$ and $f(x)$ are required to be sufficiently smooth on $[x_0,x_N]$ except at the grid points where $p(x)$ should be at least continuous; in that case the order of accuracy (2, 4 or 6) is preserved;

(c) $r(x) \geq 0$ on $[x_0,x_N]$;
if, however, the problem has pure Dirichlet boundary conditions (i.e. $e[2]=e[5]=0$) this condition can be weakened to the requirement that $r(x) > -p0*(\pi/(x_N-x_0))^2$, where $p0$ is the minimum of $p(x)$ on $[x_0,x_N]$; one should note that the problem may be ill-conditioned when $r(x)$ is quite near that lower bound; for other negative values of $r(x)$ the existence of a solution remains an open question.

The theory upon which the method is based is explained with reference to the more general equation

$$-D(p(x)Dy(x)) + q(x)Dy(x) + r(x)y(x) = f(x) \tag{3}$$

of which equation (1) and the equations treated by other procedures (*femlag, femlagskew*) are special cases. $y(x)$ is approximated by a sum involving $Nk+1$ functions $\phi_{1,0}$ and $\phi_{m,j}$:

$$y(x) = a_0\phi_{1,0}(x) + \sum_{m=1}^{N}\sum_{j=1}^{k} a_{m,j}\phi_{m,j}(x) . \tag{4}$$

Inserting this expansion into equation (3), multiplying throughout by $\phi_{n,l}(x)$ and integrating by parts, there follows

$$\left[p(x)a_0\phi_{n,l}(x)D\phi_{1,0}(x) + p(x)\sum_{m=1}^{N}\sum_{j=1}^{k} a_{m,j}\phi_{n,l}(x)D\phi_{m,j}(x)\right]_a^b$$

$$+ a_0\int_a^b \{p(x)D\phi_{n,l}(x)D\phi_{1,0}(x) + q(x)\phi_{n,l}(x)D\phi_{1,0}(x) + r(x)\phi_{n,l}(x)\phi_{1,0}(x)\}\, dx \tag{5}$$

$$+ \sum_{m=1}^{N}\sum_{j=1}^{k} a_{m,j}\int_a^b \{p(x)D\phi_{n,l}(x)D\phi_{m,j}(x) + q(x)\phi_{n,l}(x)D\phi_{m,j}(x) + r(x)\phi_{n,l}(x)\phi_{m,j}(x)\}\, dx$$

$$= \int_a^b f(x)\,\phi_{n,l}(x)\,dx .$$

Let $W_\nu^{(b)}$ and $\xi_\nu^{(k)}$ be the weights and arguments of the k+1 Radau-Lobatto quadrature formula over [0,1], so that for suitable g

$$\int_0^1 g(x)\,dx = \sum_{\nu=0}^{k} W_\nu^{(k)} g(\xi_\nu^{(k)})$$

where $\xi_0^{(k)} = 0$, $\xi_k^{(k)} = 1$, $W_0^{(k)} = W_k^{(k)}$. Denote the interval $[x_{n-1}, x_n]$ by I_n and its length $x_n - x_{n-1}$ by Δ_n (n=1,..,N), and let $\Xi_{n,\nu}^{)k)} = x_{n-1} + \xi_\nu^{(k)}\Delta_n$ (n=1,...,N; ν=0,...,k) (these displaced Radau-Lobatto points lie in I_n, and

$$\Xi_{n,0}^{(k)} = x_{n-1},\ \Xi_{n,k}^{(k)} = x_n = \Xi_{n+1,0}^{(k)},\ \Xi_{n,\nu}^{(k)} - \Xi_{n,r}^{(k)} = (\xi_{n,\nu}^{(k)} - \xi_{n,r}^{(k)})\Delta_n.$$

The functions $\phi_{1,0}$ and $\phi_{n,l}$ (n=1,...,N; l=1,...,k-1; n=N, l=1,...,k) above are taken to be polynomials of degree k over the interval I_1 and I_n (n=1,...,N) respectively and zero elsewhere in $[a,b]$; $\phi_{n,k}$ (n=1,...,N-1) is a polynomial of degree k over I_n, another such polynomial over I_{n+1}, and is zero elsewhere in $[a,b]$. These polynomials are fixed by the conditions

$$\phi_0(\Xi_{1,0}^{(k)}) = \phi_0(a) = 1, \quad \phi_0(\Xi_{1,\nu}^{(k)}) = 0 \quad (\nu=1,\ldots,k)$$
$$\phi_{n,l}(\Xi_{n,l}^{(k)}) = 1, \quad \phi_{n,l}(\Xi_{n,\nu}^{(k)}) = 0 \quad (0 \le \nu \le k\ (\nu \ne l))$$

for n=1,...,N, l=1,...,k-1; n=N, l=1,...,k, and

$$\phi_{n,k}(\Xi_{n,k}^{(k)}) = 1$$
$$\phi_{n,k}(\Xi_{n,\nu}^{(k)}) = 0 \quad (\nu=0,\ldots,k-1), \qquad \phi_{n,k}(\Xi_{n+1,\nu}^{(k)}) = 0 \quad (\nu=1,\ldots,k)$$

for n=1,...,N-1. For suitable g

$$\int_a^b \phi_{n,l}(x)g(x)dx = \int_{I_n} \phi_{n,l}(x)g(x)dx = W_l^{(k)} g(\Xi_{n,l}^{(k)})\Delta_n$$

for n=1, l=0; n=1,...,N, l=1,...,k-1; n=N, l=1,...,k and

$$\int_a^b \phi_{n,k}(x)g(x)dx = \int_{I_n+I_{n+1}} \phi_{n,k}(x)g(x)dx = W_0^{(k)} g(\Xi_{n,k}^{(k)})[\Delta_n + \Delta_{n+1}]$$

for n=1,...,N-1. Setting

$$\lambda(\nu) = \prod_{r=0}^{k\ (\nu)} (\xi_\nu^{(k)} - \xi_r^{(k)}) \qquad (\nu=0,...,k)$$

$$\lambda'(\nu,\tau) = \left(\prod_{r=0}^{k\ (\tau,\nu)} (\xi_\nu^{(k)} - \xi_r^{(k)}) \right) / \lambda(\tau) \qquad (\nu,\tau=0,...,k)$$

where the superscripts (ν) and (τ,ν) imply that the terms corresponding to $r=\nu$ and $r=\tau$, $r=\nu$ are to be omitted, the polynomials $\phi_{1,0}$, $\phi_{n,l}$ are given explicitly by the formulae

$$\phi_{n,l}(x) = \left(\prod_{r=0}^{k\ (l)} (x - \Xi_{n,r}^{(k)}) \right) / (\lambda(l)\Delta_n^k) \qquad (x \in I_n;\ \ l < k)$$

$$\phi_{n,k}(x) = \left(\prod_{r=0}^{k\ (k)} (x - \Xi_{n,r}^{(k)}) \right) / (\lambda(k)\Delta_n^k) \qquad (x \in I_n)$$

$$\phi_{n,k}(x) = \left(\prod_{r=0}^{k\ (0)} (x - \Xi_{n+1,r}^{(k)}) \right) / (\lambda(0)\Delta_{n+1}^k) \qquad (x \in I_{n+1})\ .$$

Their derivatives at the displaced Radau-Lobatto points lying in the intervals over which they are not identically zero are given by

$$D\phi_{n,l}(\Xi_{n,\nu}^{(k)}) = \lambda'(\nu,l) / \Delta_n \qquad (l < k)$$

$$D\phi_{n,k}(\Xi_{n,\nu}^{(k)}) = \lambda'(\nu,k) / \Delta_n, \qquad D\phi_{n,k}(\Xi_{n+1,\nu}^{(k)}) = \lambda'(\nu,0) / \Delta_{n+1}.$$

Over intervals in which the $\phi_{n,l}$ are zero, the derivative of $\phi_{n,l}$ is, of course, also zero. The numbers $\lambda(\nu)$, $\lambda'(\nu,l)$ are problem independent; they may be, and have been, computed in advance.

The first of the boundary conditions (2) implies that

$$e_1 a_{1,0} + \sum_{j=1}^{k} e_2 \frac{\lambda'(0,j)}{\Delta_1} a_{i,j} = e_3 .$$

When n=1,...,N-1, l=1,...,k; n=N, l=1,...,k-1, the expression denoted by square brackets in formula (5) vanishes, since $\phi_{n,l}(a) = \phi_{n,l}(b) = 0$. With n=1, the following term is equal to

$$\left[\frac{1}{\Delta_1}\sum_{\nu=0}^{k} W_\nu^{(k)} p(\Xi_{1,\nu}^{(k)})\,\lambda'(\nu,l)\,\lambda'(\nu,0) + q(\Xi_{1,l}^{(k)})\,\lambda'(\nu,0)\right] a_{1,0} \tag{6}$$

for l=1,...,k; when $n > 1$, this term vanishes. For n=1,...,N, l=1,...,k-1 the next term is equal to

$$\sum_{j=1}^{k} \left(\frac{1}{\Delta_n} \sum_{\nu=0}^{k} W_\nu^{(k)} p(\Xi_{n,\nu}^{(k)})\,\lambda'(\nu,l)\,\lambda'(\nu,j) + W_l^{(k)} q(\Xi_{n,l}^{(k)})\lambda'(l,j) + W_l^{(k)} \delta_{l,j} r(\Xi_{n,l}^{(k)})\Delta_n \right) a_{n,j} \tag{7}$$

and when n=1,...,N-1, l=k it is equal to the sum of the above expression with l=k and

$$\sum_{j=1}^{k} \left[\frac{1}{\Delta_{n+1}} \sum_{\nu=0}^{k} W_\nu^{(k)} p(\Xi_{n+1,\nu}^{(k)})\,\lambda'(\nu,l)\,\lambda'(\nu,j) + W_l^{(k)} q(\Xi_{n+1,l}^{(k)})\,\lambda'(l,j) + W_l^{(k)} \delta_{l,j} r(\Xi_{n+1,l}^{(k)})\Delta_{n+1}\right] a_{n+1,j} .$$

The value of the expression upon the right hand side of equation (5) is $W_l^{(k)} f(\Xi_{n,l}^{(k)})\,\Delta_n$ for n=1, l=0,...,k, and n=1,...,N, l=1,...,k-1, and is equal to $W_0^{(k)} f(\Xi_{n,k}^{(k)})(\Delta_n + \Delta_{n+l})$ for n=1,...,N-1, l=k.

The second of the boundary conditions (2) implies that

$$e_4 a_{N,k} + \sum_{j=1}^{k} e_5 \frac{\lambda'(k,j)}{\Delta_n} a_{N,j} = e_6 .$$

In summary, the two boundary conditions (2) and Nk-1 equations derived from equation (3) with n=1,...,N-1, l=1,...,k and n=N, l=1,...,k-1 yield Nk+1 linear equations for the Nk+1 coefficients a_0 and $a_{m,j}$ (m=1,...,N, j=1,...,k). The first k-1 of these equations involve a_0 and $a_{i,j}$ (j=1,...,k) alone. By transferring the terms involving a_0 and $a_{1,k}$ to the right hand side, and premultiplying by the inverse of the matrix multiplying the vector ($a_{1,2}$,...,$a_{1,k-1}$), this vector may be isolated. Its components may be eliminated from the next equation which involves the two sets $a_{1,j}$ and $a_{2,j}$ (j=1,...,k). The following k-1 equations involve $a_{2,j}$ (j=1,...,k) alone. The vector ($a_{2,2}$,...,$a_{2,k-1}$) may again be isolated and its components eliminated from the equation from which $a_{1,2}$,...,$a_{1,k-1}$ have been eliminated, and also from the next equation which involves the two sets $a_{2,j}$, $a_{3,j}$ (j=1,...,k). The process may be continued, and a tridiagonal system of equations for a_0 and $a_{m,k}$ (m=1,...,N) is obtained. The process of elimination may be carried out as soon as each complete set of equations involving $a_{m,1},...,a_{m,k-1}$ has been constructed (computer storage space is thereby saved). When k=1, no elimination takes place; when k=2 and 3 (the other two values of k permitted by *femlagsym*) the matrix elimination process is particularly simple and is programmed independently (rather than by calls of matrix operation procedures).

The tridiagonal system of equations is solved by a method of Babuska [see Bab72]. Stripping some redundant sequences from the original exposition, the set of equations $Ay=f$, where

$$A_{i+1,i} = c_i,$$
$$A_{i,i+1} = b_i \quad (i=1,...,n-1),$$
$$A_{1,1} = \tau_1 - b_1,$$
$$A_{i,i} = \tau_i - c_{i-1} - b_i \quad (i=2,...,n-1),$$

$A_{n,n} = \tau_n - c_n$ may be solved by means of the construction of four sequences g_i, χ_i, g_i^* and χ_i^* as follows: with $\chi_1=\tau_1$, $g_1=f_1$

$$g_{i+1} = f_{i+1} - g_i c_i/(\chi_i-b_i), \quad \chi_{i+1} = \tau_{i+1} - \chi_i c_i/(\chi_i-b_i) \qquad (i=1,...,n-1)$$

and with χ_n^*, $g_n^* = f_n$

$$g_i^* = f_i - b_i g_{i+1}^*/(\chi_{i+1}^*-c_i), \quad \chi_i^* = \tau_i - b_i \chi_{i+1}^*/(\chi_{i+1}^*-c_i) \qquad (i=n-1,...,1)$$

and then

$$y_i = (g_i + g_i^* - f_i) / (\chi_i + \chi_i^* - \tau_i) \qquad (i=n,...,1).$$

The above process can be further economized (three one-dimensional arrays, with another to contain the y_i are required). The coefficients $a_{1,0}$, $a_{n,k}$ (n=1,...,N) obtained by means of the above process are, in order, the required approximations y_n (n=0,...,N) (since $\phi_{n,k}(x_m) = \delta_{m,n}$).

Function Parameters:

void femlagsym (*x,y,n,p,r,f,order,e*)

x: float *x[0:n]*;
entry: $a = x_0 < x_1 < ... < x_n = b$ is a partition of the interval $[a,b]$;

y: float *y[0:n]*;
exit: *y[i]*, i=0,1,...,n, is the approximate solution at *x[i]* of the differential equation (1) with boundary conditions (2);

n: int;
entry: the upper bound of the arrays *x* and *y* (value of *N* above), $n > 1$;

p: float (**p*)(*x*);
the procedure to compute *p(x)*, the coefficient of *Dy(x)* in equation (1);

r: float (**r*)(*x*);
the procedure to compute *r(x)*, the coefficient of *y(x)* in equation (1);

f: float (**f*)(*x*);
the procedure to compute *f(x)*, the right hand side of equation (1);

order: int;
entry: *order* denotes the order of accuracy required for the approximate solution of the differential equation; let h = max(*x[i]-x[i-1]*), then $|y[i] - y(x[i])| \le c*h^{order}$, i=0,...,n; *order* can be chosen equal to 2, 4 or 6 only;

e: float *e[1:6]*;
entry: *e[1]*,...,*e[6]* describe the boundary conditions (values of e_i, i=1,...,6, in (2)); *e[1]* and *e[4]* are not allowed to vanish both.

```
#include <math.h>

void femlagsym(float x[], float y[], int n, float (*p)(float),
        float (*r)(float), float (*f)(float), int order, float e[])
```

```
{
    float *allocate_real_vector(int, int);
    void free_real_vector(float *, int);
    int l,l1;
    float xl1,xl,h,a12,b1,b2,tau1,tau2,ch,tl,g,yl,pp,p1,p2,p3,p4,
            r1,r2,r3,r4,f1,f2,f3,f4,e1,e2,e3,e4,e5,e6,*t,*sub,*chi,*gi,
            h2,x2,h6,h15,b3,tau3,c12,c32,a13,a22,a23,x3,h12,h24,det,
            c13,c42,c43,a14,a24,a33,a34,b4,tau4,aux;

    t=allocate_real_vector(0,n-1);
    sub=allocate_real_vector(0,n-1);
    chi=allocate_real_vector(0,n-1);
    gi=allocate_real_vector(0,n-1);

    l=1;
    xl=x[0];
    e1=e[1];
    e2=e[2];
    e3=e[3];
    e4=e[4];
    e5=e[5];
    e6=e[6];
    while (l <= n) {
        l1=l-1;
        xl1=xl;
        xl=x[l];
        h=xl-xl1;
        if (order == 2) {
            /* element mat vec evaluation 1 */
            if (l == 1) {
                p2=(*p)(xl1);
                r2=(*r)(xl1);
                f2=(*f)(xl1);
            }
            p1=p2;
            p2=(*p)(xl);
            r1=r2;
            r2=(*r)(xl);
            f1=f2;
            f2=(*f)(xl);
            h2=h/2.0;
            b1=h2*f1;
            b2=h2*f2;
            tau1=h2*r1;
            tau2=h2*r2;
            a12 = -0.5*(p1+p2)/h;
        } else if (order == 4) {
            /* element mat vec evaluation 2 */
            if (l == 1) {
                p3=(*p)(xl1);
                r3=(*r)(xl1);
                f3=(*f)(xl1);
            }
            x2=(xl1+xl)/2.0;
            h6=h/6.0;
            h15=h/1.5;
            p1=p3;
            p2=(*p)(x2);
            p3=(*p)(xl);
            r1=r3;
            r2=(*r)(x2);
            r3=(*r)(xl);
            f1=f3;
            f2=(*f)(x2);
            f3=(*f)(xl);
            b1=h6*f1;
            b2=h15*f2;
            b3=h6*f3;
            tau1=h6*r1;
            tau2=h15*r2;
            tau3=h6*r3;
            a12 = -(2.0*p1+p3/1.5)/h;
            a13=(0.5*(p1+p3)-p2/1.5)/h;
            a22=(p1+p3)/h/0.375+tau2;
```

```
        a23 = -(p1/3.0+p3)*2.0/h;
        c12 = -a12/a22;
        c32 = -a23/a22;
        a12=a13+c32*a12;
        b1 += c12*b2;
        b2=b3+c32*b2;
        tau1 += c12*tau2;
        tau2=tau3+c32*tau2;
    } else {
        /* element mat vec evaluation 3 */
        if (l == 1) {
            p4=(*p)(xl1);
            r4=(*r)(xl1);
            f4=(*f)(xl1);
        }
        x2=xl1+0.27639320225*h;
        x3=xl-x2+xl1;
        h12=h/12.0;
        h24=h/2.4;
        p1=p4;
        p2=(*p)(x2);
        p3=(*p)(x3);
        p4=(*p)(xl);
        r1=r4;
        r2=(*r)(x2);
        r3=(*r)(x3);
        r4=(*r)(xl);
        f1=f4;
        f2=(*f)(x2);
        f3=(*f)(x3);
        f4=(*f)(xl);
        b1=h12*f1;
        b2=h24*f2;
        b3=h24*f3;
        b4=h12*f4;
        tau1=h12*r1;
        tau2=h24*r2;
        tau3=h24*r3;
        tau4=h12*r4;
        a12 = -(4.04508497187450*p1+0.57581917135425*p3+
                    0.25751416197911*p4)/h;
        a13=(1.5450849718747*p1-1.5075141619791*p2+
                    0.6741808286458*p4)/h;
        a14=((p2+p3)/2.4-(p1+p4)/2.0)/h;
        a22=(5.454237476562*p1+p3/0.48+0.79576252343762*p4)/h+tau2;
        a23 = -(p1+p4)/(h*0.48);
        a24=(0.67418082864575*p1-1.50751416197910*p3+
                1.54508497187470*p4)/h;
        a33=(0.7957625234376*p1+p2/0.48+5.454237476562*p4)/h+tau3;
        a34 = -(0.25751416197911*p1+0.57581917135418*p2+
                4.0450849718747*p4)/h;
        det=a22*a33-a23*a23;
        c12=(a13*a23-a12*a33)/det;
        c13=(a12*a23-a13*a22)/det;
        c42=(a23*a34-a24*a33)/det;
        c43=(a24*a23-a34*a22)/det;
        tau1 += c12*tau2+c13*tau3;
        tau2=tau4+c42*tau2+c43*tau3;
        a12=a14+c42*a12+c43*a13;
        b1 += c12*b2+c13*b3;
        b2=b4+c42*b2+c43*b3;
    }
    if (l == 1 || l == n) {
        /* boundary conditions */
        if (l == 1 && e2 == 0.0) {
            tau1=1.0;
            b1=e3/e1;
            b2 -= a12*b1;
            tau2 -= a12;
            a12=0.0;
        } else if (l == 1 && e2 != 0.0) {
            aux=p1/e2;
            tau1 -= aux*e1;
            b1 -= e3*aux;
```

```
            } else if (l == n && e5 == 0.0) {
                tau2=1.0;
                b2=e6/e4;
                b1 -= a12*b2;
                tau1 -= a12;
                a12=0.0;
            } else if (l == n && e5 != 0.0) {
                aux=p2/e5;
                tau2 += aux*e4;
                b2 += aux*e6;
            }
        }
        /* forward babushka */
        if (l == 1) {
            chi[0]=ch=tl=tau1;
            t[0]=tl;
            gi[0]=g=yl=b1;
            y[0]=yl;
            sub[0]=a12;
            pp=a12/(ch-a12);
            ch=tau2-ch*pp;
            g=b2-g*pp;
            tl=tau2;
            yl=b2;
        } else {
            chi[l1] = ch += tau1;
            gi[l1] = g += b1;
            sub[l1]=a12;
            pp=a12/(ch-a12);
            ch=tau2-ch*pp;
            g=b2-g*pp;
            t[l1]=tl+tau1;
            tl=tau2;
            y[l1]=yl+b1;
            yl=b2;
        }
        l++;
    }
    /* backward babushka */
    pp=yl;
    y[n]=g/ch;
    g=pp;
    ch=tl;
    l=n-1;
    while (l >= 0) {
        pp=sub[l];
        pp /= (ch-pp);
        tl=t[l];
        ch=tl-ch*pp;
        yl=y[l];
        g=yl-g*pp;
        y[l]=(gi[l]+g-yl)/(chi[l]+ch-tl);
        l--;
    }
    free_real_vector(t,0);
    free_real_vector(sub,0);
    free_real_vector(chi,0);
    free_real_vector(gi,0);
}
```

B. femlag

Solves a second order self-adjoint linear two point boundary value problem by means of Galerkin's method with continuous piecewise polynomials [Bab72, BakH76, He75, StF73]. *femlag* computes approximations y_n (n=0,...,N) to the values at the points $x = x_n$, where

$$-\infty < a = x_0 < x_1 < \ldots < x_N = b < \infty$$

of the real valued function $y(x)$ which satisfies the equation

$$-D^2y(x) + r(x)y(x) = f(x) \qquad (1)$$

where $D=d/dx$, with boundary conditions

$$e_1y(a) + e_2Dy(a) = e_3, \quad e_4y(b) + e_5Dy(b) = e_6 \quad (e_1, e_4 \neq 0). \qquad (2)$$

It is assumed that $r(x) \geq 0$ $(a \leq x \leq b)$; if $e_2 = e_5 = 0$, this condition may be relaxed to

$$r(x) > -(\pi/(b-a))^2 \ (a \leq x \leq b).$$

The general theory underlying the method used is that described in *femlagsym*. Now, however, (since $p(x)$ in equation (1) of *femlagsym* is unity) the term

$$\frac{1}{\Delta_n} \sum_{\nu=0}^{k} w_\nu^{(k)} p(\Xi_{n,\nu}^{(k)}) \lambda'(\nu,l) \lambda'(\nu,j)$$

in expression (7) of *femlagsym* becomes

$$\frac{1}{\Delta_n} \sum_{\nu=0}^{k} w_\nu^{(k)} \lambda'(\nu,l) \lambda'(\nu,j) \ .$$

The value of the sum in this expression is problem independent and may be determined in advance (a similar remark holds with respect to expression (6) in *femlagsym*); the computations may be simplified.

Function Parameters:

void femlag (*x,y,n,r,f,order,e*)

x: float *x[0:n]*;
 entry: $a = x_0 < x_1 < \ldots < x_n = b$ is a partition of the segment $[a,b]$;

y: float *y[0:n]*;
 exit: *y[i]*, *i*=0,1,...,*n*, is the approximate solution at *x[i]* of the differential equation (1) with boundary conditions (2);

n: int;
 entry: the upper bound of the arrays *x* and *y* (value of *N* above), $n > 1$;

r: float (**r*)(*x*);
 the procedure to compute *r(x)*, the coefficient of *y(x)* in equation (1);

f: float (**f*)(*x*);
 the procedure to compute *f(x)*, the right hand side of equation (1);

order: int;
 entry: *order* denotes the order of accuracy required for the approximate solution of the differential equation; let $h = \max(x[i]-x[i-1])$, then $|y[i] - y(x[i])| \leq c*h^{order}$, *i*=0,...,*n*; *order* can be chosen equal to 2, 4 or 6 only;

e: float *e[1:6]*;
 entry: *e[1]*,...,*e[6]* describe the boundary conditions (values of e_i, *i*=1,...,6, in (2)); neither *e[1]* nor *e[4]* is allowed to vanish.

```
#include <math.h>

void femlag(float x[], float y[], int n, float (*r)(float),
            float (*f)(float), int order, float e[])
{
   float *allocate_real_vector(int, int);
   void free_real_vector(float *, int);
```

```
int l,l1;
float xl1,xl,h,a12,b1,b2,tau1,tau2,ch,tl,g,yl,pp,e1,e2,e3,e4,e5,
      e6,*t,*sub,*chi,*gi,f2,r2,r1,f1,h2,r3,f3,x2,h6,h15,b3,tau3,
      c12,a13,a22,a23,r4,f4,x3,h12,h24,det,c13,c42,c43,a14,a24,
      a33,a34,b4,tau4;

t=allocate_real_vector(0,n-1);
sub=allocate_real_vector(0,n-1);
chi=allocate_real_vector(0,n-1);
gi=allocate_real_vector(0,n-1);

l=1;
xl=x[0];
e1=e[1];
e2=e[2];
e3=e[3];
e4=e[4];
e5=e[5];
e6=e[6];
while (l <= n) {
    l1=l-1;
    xl1=xl;
    xl=x[l];
    h=xl-xl1;
    if (order == 2) {
        /* element mat vec evaluation 1 */
        if (l == 1) {
            f2=(*f)(xl1);
            r2=(*r)(xl1);
        }
        a12 = -1.0/h;
        h2=h/2.0;
        r1=r2;
        r2=(*r)(xl);
        f1=f2;
        f2=(*f)(xl);
        b1=h2*f1;
        b2=h2*f2;
        tau1=h2*r1;
        tau2=h2*r2;
    } else if (order == 4) {
        /* element mat vec evaluation 2 */
        if (l == 1) {
            r3=(*r)(xl1);
            f3=(*f)(xl1);
        }
        x2=(xl1+xl)/2.0;
        h6=h/6.0;
        h15=h/1.5;
        r1=r3;
        r2=(*r)(x2);
        r3=(*r)(xl);
        f1=f3;
        f2=(*f)(x2);
        f3=(*f)(xl);
        b1=h6*f1;
        b2=h15*f2;
        b3=h6*f3;
        tau1=h6*r1;
        tau2=h15*r2;
        tau3=h6*r3;
        a12 = a23 = -8.0/h/3.0;
        a13 = -a12/8.0;
        a22 = -2.0*a12+tau2;
        c12 = -a12/a22;
        a12=a13+c12*a12;
        b2 *= c12;
        b1 += b2;
        b2 += b3;
        tau2 *= c12;
        tau1 += tau2;
        tau2=tau3+tau2;
    } else {
        /* element mat vec evaluation 3 */
```

```
        if (l == 1) {
            r4=(*r)(xl1);
            f4=(*f)(xl1);
        }
        x2=xl1+0.27639320225*h;
        x3=xl-x2+xl1;
        r1=r4;
        r2=(*r)(x2);
        r3=(*r)(x3);
        r4=(*r)(xl);
        f1=f4;
        f2=(*f)(x2);
        f3=(*f)(x3);
        f4=(*f)(xl);
        h12=h/12.0;
        h24=h/2.4;
        b1=h12*f1;
        b2=h24*f2;
        b3=h24*f3;
        b4=h12*f4;
        tau1=h12*r1;
        tau2=h24*r2;
        tau3=h24*r3;
        tau4=h12*r4;
        a12 = a34 = -4.8784183052078/h;
        a13=a24=0.7117516385412/h;
        a14 = -0.16666666666667/h;
        a23=25.0*a14;
        a22 = -2.0*a23+tau2;
        a33 = -2.0*a23+tau3;
        det=a22*a33-a23*a23;
        c12=(a13*a23-a12*a33)/det;
        c13=(a12*a23-a13*a22)/det;
        c42=(a23*a34-a24*a33)/det;
        c43=(a24*a23-a34*a22)/det;
        tau1 += c12*tau2+c13*tau3;
        tau2=tau4+c42*tau2+c43*tau3;
        a12=a14+c42*a12+c43*a13;
        b1 += c12*b2+c13*b3;
        b2=b4+c42*b2+c43*b3;
    }
    if (l == 1 || l == n) {
        /* boundary conditions */
        if (l == 1 && e2 == 0.0) {
            tau1=1.0;
            b1=e3/e1;
            b2 -= a12*b1;
            tau2 -= a12;
            a12=0.0;
        } else if (l == 1 && e2 != 0.0) {
            tau1 -= e1/e2;
            b1 -= e3/e2;
        } else if (l == n && e5 == 0.0) {
            tau2=1.0;
            b2=e6/e4;
            b1 -= a12*b2;
            tau1 -= a12;
            a12=0.0;
        } else if (l == n && e5 != 0.0) {
            tau2 += e4/e5;
            b2 += e6/e5;
        }
    }
    /* forward babushka */
    if (l == 1) {
        chi[0]=ch=tl=tau1;
        t[0]=tl;
        gi[0]=g=yl=b1;
        y[0]=yl;
        sub[0]=a12;
        pp=a12/(ch-a12);
        ch=tau2-ch*pp;
        g=b2-g*pp;
        tl=tau2;
```

```
                yl=b2;
            } else {
                chi[l1] = ch += tau1;
                gi[l1] = g += b1;
                sub[l1]=a12;
                pp=a12/(ch-a12);
                ch=tau2-ch*pp;
                g=b2-g*pp;
                t[l1]=tl+tau1;
                tl=tau2;
                y[l1]=yl+b1;
                yl=b2;
            }
            l++;
        }
        /* backward babushka */
        pp=yl;
        y[n]=g/ch;
        g=pp;
        ch=tl;
        l=n-1;
        while (l >= 0) {
            pp=sub[l];
            pp /= (ch-pp);
            tl=t[l];
            ch=tl-ch*pp;
            yl=y[l];
            g=yl-g*pp;
            y[l]=(gi[l]+g-yl)/(chi[l]+ch-tl);
            l--;
        }
        free_real_vector(t,0);
        free_real_vector(sub,0);
        free_real_vector(chi,0);
        free_real_vector(gi,0);
}
```

C. femlagspher

Solves a second order self-adjoint linear two point boundary value problem with spherical coordinates by means of Galerkin's method with continuous piecewise polynomials [Bab72, BakH76, He75, StF73]. *femlagspher* computes approximations y_n (n=0,...,N) to the values at the points $x = x_n$, where

$$-\infty < a = x_0 < x_1 < ... < x_N = b < \infty$$

of the real valued function $y(x)$ which satisfies the equation

$$-D(x^{nc}Dy(x))/x^{nc} + r(x)y(x) = f(x) \tag{1}$$

where $D=d/dx$, with boundary conditions

$$e_1y(a) + e_2Dy(a) = e_3, \quad e_4y(b) + e_5Dy(b) = e_6 \quad (e_1,\ e_4 \neq 0). \tag{2}$$

It is assumed that $r(x)$ and $f(x)$ are sufficiently smooth on $[x_0,x_N]$ except at the grid points; furthermore $r(x)$ should be nonnegative.

The solution is approximated by a function which is continuous on the closed interval $[x_0,x_N]$ and a polynomial of degree less than or equal to k on each segment $[x_{j-1},x_j]$ (j=1,...,N). This piecewise polynomial is entirely determined by the values it has at the knots x_j and on k-1 interior knots on each segment $[x_{j-1},x_j]$. These values are obtained by the solution of an *order*+1 diagonal linear system with a specially structured matrix. The entries of the matrix and the vector are inner products which are approximated by some piecewise k-point Gaussian quadrature. The evaluation of the matrix and the vector is done segment by segment: on each segment the contributions to the entries of the matrix and the vector are computed and embedded in the global matrix and vector. Since the function values on the interior points of each segment are not coupled with the function values outside that segment, the resulting linear system can be reduced to a tridiagonal system by means of static condensation. The final tridiagonal system, since it is of finite difference type, is solved by means of Babuska's method. For further details, see the documentation of *femlagsym*.

Function Parameters:

void femlagspher (*x,y,n,nc,r,f,order,e*)

x: float *x[0:n]*;
entry: $a = x_0 < x_1 < \ldots < x_n = b$ is a partition of the interval $[a,b]$;

y: float *y[0:n]*;
exit: *y[i]*, *i*=0,1,...,*n*, is the approximate solution at *x[i]* of the differential equation (1) with boundary conditions (2);

n: int;
entry: the upper bound of the arrays *x* and *y* (value of *N* above), $n > 1$;

nc: int;
entry: if *nc* = 0, Cartesian coordinates are used;
if *nc* = 1, polar coordinates are used;
if *nc* = 2, spherical coordinates are used;

r: float (**r*)(*x*);
the procedure to compute *r(x)*, the coefficient of *y(x)* in equation (1);

f: float (**f*)(*x*);
the procedure to compute *f(x)*, the right hand side of equation (1);

order: int;
entry: *order* denotes the order of accuracy required for the approximate solution of the differential equation; let $h = \max(x[i]-x[i-1])$, then
$|y[i] - y(x[i])| \le c*h^{order}$, *i*=0,...,*n*; *order* can be chosen equal to 2 or 4 only;

e: float *e[1:6]*;
entry: *e[1]*,...,*e[6]* describe the boundary conditions (values of e_i, *i*=1,...,6, in (2)); *e[1]* and *e[4]* are not allowed to vanish both.

```
#include <math.h>

void femlagspher(float x[], float y[], int n, int nc,
        float (*r)(float), float (*f)(float), int order, float e[])
{
    float *allocate_real_vector(int, int);
    void free_real_vector(float *, int);
    int l,l1;
    float xl1,xl,h,a12,b1,b2,tau1,tau2,ch,tl,g,yl,pp,tau3,b3,a13,a22,
          a23,c32,c12,e1,e2,e3,e4,e5,e6,*t,*sub,*chi,*gi,xm,vl,vr,wl,
          wr,pr,rm,fm,xl2,xlxr,xr2,xlm,xrm,vlm,vrm,wlm,wrm,flm,frm,rlm,
          rrm,pl1,pl2,pl3,pr1,pr2,pr3,ql1,ql2,ql3,rlmpl1,rlmpl2,rrmpr1,
          rrmpr2,vlmql1,vlmql2,vrmqr1,vrmqr2,qr1,qr2,qr3,a,a2,a3,a4,
          b,b4,p4h,p2,p3,p4,aux1,aux2,a5,a6,a7,a8,b5,b6,b7,b8,ab4,
          a2b3,a3b2,a4b,p5,p8,p8h,aux,plm,prm;

    t=allocate_real_vector(0,n-1);
    sub=allocate_real_vector(0,n-1);
    chi=allocate_real_vector(0,n-1);
    gi=allocate_real_vector(0,n-1);

    l=1;
    xl=x[0];
    e1=e[1];
    e2=e[2];
    e3=e[3];
    e4=e[4];
```

```
e5=e[5];
e6=e[6];
while (l <= n) {
    l1=l-1;
    xl1=xl;
    xl=x[l];
    h=xl-xl1;
    if (order == 2) {
        /* element mat vec evaluation 1 */
        if (nc == 0)
            vl=vr=0.5;
        else if (nc == 1) {
            vl=(xl1*2.0+xl)/6.0;
            vr=(xl1+xl*2.0)/6.0;
        } else {
            xl2=xl1*xl1/12.0;
            xlxr=xl1*xl/6.0;
            xr2=xl*xl/12.0;
            vl=3.0*xl2+xlxr+xr2;
            vr=3.0*xr2+xlxr+xl2;
        }
        wl=h*vl;
        wr=h*vr;
        pr=vr/(vl+vr);
        xm=xl1+h*pr;
        fm=(*f)(xm);
        rm=(*r)(xm);
        tau1=wl*rm;
        tau2=wr*rm;
        b1=wl*fm;
        b2=wr*fm;
        a12 = -(vl+vr)/h+h*(1.0-pr)*pr*rm;
    } else {
        /* element mat vec evaluation 2 */
        if (nc == 0) {
            xlm=xl1+h*0.2113248654052;
            xrm=xl1+xl-xlm;
            vlm=vrm=0.5;
            pl1=pr3=0.45534180126148;
            pl3=pr1 = -0.12200846792815;
            pl2=pr2=1.0-pl1-pl3;
            ql1 = -2.15470053837925;
            ql3 = -0.15470053837925;
            ql2 = -ql1-ql3;
            qr1 = -ql3;
            qr3 = -ql1;
            qr2 = -ql2;
        } else if (nc == 1) {
            a=xl1;
            a2=a*a;
            a3=a*a2;
            a4=a*a3;
            b=xl;
            b2=b*b;
            b3=b*b2;
            b4=b*b3;
            p2=10.0*(a2+4.0*a*b+b2);
            p3=6.0*(a3+4.0*(a2*b+a*b2)+b3);
            p4=sqrt(6.0*(a4+10.0*(a*b3+a3*b)+28.0*a2*b2+b4));
            p4h=p4*h;
            xlm=(p3-p4h)/p2;
            xrm=(p3+p4h)/p2;
            aux1=(a+b)/4.0;
            aux2=h*(a2+7.0*a*b+b2)/6.0/p4;
            vlm=aux1-aux2;
            vrm=aux1+aux2;
        } else {
            a=xl1;
            a2=a*a;
            a3=a*a2;
            a4=a*a3;
            a5=a*a4;
            a6=a*a5;
            a7=a*a6;
```

```
        a8=a*a7;
        b=xl;
        b2=b*b;
        b3=b*b2;
        b4=b*b3;
        b5=b*b4;
        b6=b*b5;
        b7=b*b6;
        b8=b*b7;
        ab4=a*b4;
        a2b3=a2*b3;
        a3b2=a3*b2;
        a4b=a4*b;
        p4=15.0*(a4+4.0*(a3*b+a*b3)+10.0*a2*b2+b4);
        p5=10.0*(a5+4.0*(a4b+ab4)+10.0*(a3b2+a2b3)+b5);
        p8=sqrt(10.0*(a8+10.0*(a7*b+a*b7)+55.0*(a2*b6+a6*b2)+
                164.0*(a5*b3+a3*b5)+290.0*a4*b4+b8));
        aux1=(a2+a*b+b2)/6.0;
        p8h=p8*h;
        aux2=(h*(a5+7.0*(a4b+ab4)+28.0*(a3b2+a2b3)+b5))/4.8/p8;
        xlm=(p5-p8h)/p4;
        xrm=(p5+p8h)/p4;
        vlm=aux1-aux2;
        vrm=aux1+aux2;
    }
    if (nc > 0) {
        plm=(xlm-xl1)/h;
        prm=(xrm-xl1)/h;
        aux=2.0*plm-1.0;
        pl1=aux*(plm-1.0);
        pl3=aux*plm;
        pl2=1.0-pl1-pl3;
        aux=2.0*prm-1.0;
        pr1=aux*(prm-1.0);
        pr3=aux*prm;
        pr2=1.0-pr1-pr3;
        aux=4.0*plm;
        ql1=aux-3.0;
        ql3=aux-1.0;
        ql2 = -ql1-ql3;
        aux=4.0*prm;
        qr1=aux-3.0;
        qr3=aux-1.0;
        qr2 = -qr1-qr3;
    }
    wlm=h*vlm;
    wrm=h*vrm;
    vlm /= h;
    vrm /= h;
    flm=(*f)(xlm)*wlm;
    frm=wrm*(*f)(xrm);
    rlm=(*r)(xlm)*wlm;
    rrm=wrm*(*r)(xrm);
    tau1=pl1*rlm+pr1*rrm;
    tau2=pl2*rlm+pr2*rrm;
    tau3=pl3*rlm+pr3*rrm;
    b1=pl1*flm+pr1*frm;
    b2=pl2*flm+pr2*frm;
    b3=pl3*flm+pr3*frm;
    vlmql1=ql1*vlm;
    vrmqr1=qr1*vrm;
    vlmql2=ql2*vlm;
    vrmqr2=qr2*vrm;
    rlmpl1=rlm*pl1;
    rrmpr1=rrm*pr1;
    rlmpl2=rlm*pl2;
    rrmpr2=rrm*pr2;
    a12=vlmql1*ql2+vrmqr1*qr2+rlmpl1*pl2+rrmpr1*pr2;
    a13=vlmql1*ql3+vrmqr1*qr3+rlmpl1*pl3+rrmpr1*pr3;
    a22=vlmql2*ql2+vrmqr2*qr2+rlmpl2*pl2+rrmpr2*pr2;
    a23=vlmql2*ql3+vrmqr2*qr3+rlmpl2*pl3+rrmpr2*pr3;
    c12 = -a12/a22;
    c32 = -a23/a22;
    a12=a13+c32*a12;
```

```
            b1 += c12*b2;
            b2=b3+c32*b2;
            tau1 += c12*tau2;
            tau2=tau3+c32*tau2;
        }
        if (l == 1 || l == n) {
            /* boundary conditions */
            if (l == 1 && e2 == 0.0) {
                tau1=1.0;
                b1=e3/e1;
                b2 -= a12*b1;
                tau2 -= a12;
                a12=0.0;
            } else if (l == 1 && e2 != 0.0) {
                aux=((nc == 0) ? 1.0 : pow(x[0],nc))/e2;
                b1 -= e3*aux;
                tau1 -= e1*aux;
            } else if (l == n && e5 == 0.0) {
                tau2=1.0;
                b2=e6/e4;
                b1 -= a12*b2;
                tau1 -= a12;
                a12=0.0;
            } else if (l == n && e5 != 0.0) {
                aux=((nc == 0) ? 1.0 : pow(x[n],nc))/e5;
                tau2 += aux*e4;
                b2 += aux*e6;
            }
        }
        /* forward babushka */
        if (l == 1) {
            chi[0]=ch=tl=tau1;
            t[0]=tl;
            gi[0]=g=yl=b1;
            y[0]=yl;
            sub[0]=a12;
            pp=a12/(ch-a12);
            ch=tau2-ch*pp;
            g=b2-g*pp;
            tl=tau2;
            yl=b2;
        } else {
            chi[l1] = ch += tau1;
            gi[l1] = g += b1;
            sub[l1]=a12;
            pp=a12/(ch-a12);
            ch=tau2-ch*pp;
            g=b2-g*pp;
            t[l1]=tl+tau1;
            tl=tau2;
            y[l1]=yl+b1;
            yl=b2;
        }
        l++;
    }
    /* backward babushka */
    pp=yl;
    y[n]=g/ch;
    g=pp;
    ch=tl;
    l=n-1;
    while (l >= 0) {
        pp=sub[l];
        pp /= (ch-pp);
        tl=t[l];
        ch=tl-ch*pp;
        yl=y[l];
        g=yl-g*pp;
        y[l]=(gi[l]+g-yl)/(chi[l]+ch-tl);
        l--;
    }
    free_real_vector(t,0);
    free_real_vector(sub,0);
    free_real_vector(chi,0);
```

```
    free_real_vector(gi,0);
}
```

5.5.2 Linear methods - Second order skew adjoint

femlagskew

Solves a second order skew-adjoint linear two point boundary value problem by means of Galerkin's method with continuous piecewise polynomials [Bab72, BakH76, He75, StF73]. *femlagskew* computes approximations y_n (n=0,...,N) to the values at the points $x = x_n$, where

$$-\infty < a = x_0 < x_1 < \ldots < x_N = b < \infty$$

of the real valued function $y(x)$ which satisfies the equation

$$-D^2y(x) + q(x)Dy(x) + r(x)y(x) = f(x) \tag{1}$$

where $D=d/dx$, with boundary conditions

$$e_1y(a) + e_2Dy(a) = e_3, \quad e_4y(b) + e_5Dy(b) = e_6 \quad (e_1, e_4 \neq 0). \tag{2}$$

It is assumed that

(a) $r(x) \geq 0$ ($a \leq x \leq b$), if $e_2 = e_5 = 0$ then this condition may be relaxed to $r(x) > -(\pi/(b-a))^2$ ($a \leq x \leq b$);

(b) $q(x)$ is not allowed to have very large values in some sense: the product $q(x)*(x_j-x_{j-1})$ should not be too large on the interval $[x_{j-1},x_j]$, otherwise the boundary value problem may degenerate to a singular perturbation or boundary layer problem, for which either special methods or a suitably chosen grid are needed;

(c) $q(x)$, $r(x)$ and $f(x)$ are required to be sufficiently differentiable on the domain of the boundary value problem; however, the derivatives are allowed to have discontinuities at the grid points, in which case the order of accuracy (2, 4 or 6) is preserved;

(d) if $q(x)$ and $r(x)$ satisfy the inequality $r(x) \geq Dq(x)/2$, the existence of a unique solution is guaranteed, otherwise this remains an open question.

The general theory underlying the method used is that described in *femlagsym*. Now a term $q(x)Dy(x)$ is being treated (it is absent from equation (1) in *femlagsym*), however, the simplification described in *femlag* is possible.

Function Parameters:

void femlagskew (*x,y,n,q,r,f,order,e*)

x: float *x[0:n]*;
entry: $a = x_0 < x_1 < \ldots < x_n = b$ is a partition of the segment $[a,b]$;

y: float *y[0:n]*;
exit: *y[i]*, *i*=0,1,...,*n*, is the approximate solution at *x[i]* of the differential equation (1) with boundary conditions (2);

n: int;
entry: the upper bound of the arrays *x* and *y* (value of *N* above), $n > 1$;

q: float (*q)(x);
the procedure to compute $q(x)$, the coefficient of $Dy(x)$ in equation (1);

r: float (*r)(x);
the procedure to compute $r(x)$, the coefficient of $y(x)$ in equation (1);

f: float (*f)(x);
the procedure to compute $f(x)$, the right hand side of equation (1);

order: int;

entry: *order* denotes the order of accuracy required for the approximate solution of the differential equation; let $h = \max(x[i]-x[i-1])$, then $|y[i] - y(x[i])| \le c*h^{order}$, i=0,...,n; *order* can be chosen equal to 2, 4 or 6 only;

e: float *e[1:6]*;

entry: *e[1]*,...,*e[6]* describe the boundary conditions (values of e_i, i=1,...,6, in (2)); neither *e[1]* nor *e[4]* is allowed to vanish.

```
#include <math.h>

void femlagskew(float x[], float y[], int n, float (*q)(float),
        float (*r)(float), float (*f)(float), int order, float e[])
{
    float *allocate_real_vector(int, int);
    void free_real_vector(float *, int);
    int l,l1;
    float xl1,xl,h,a12,a21,b1,b2,tau1,tau2,ch,tl,g,yl,pp,e1,e2,e3,e4,
          e5,e6,*t,*super,*sub,*chi,*gi,q2,r2,f2,q1,r1,f1,h2,s12,q3,
          r3,f3,s13,s22,x2,h6,h15,c12,c32,a13,a31,a22,a23,a32,b3,
          tau3,q4,r4,f4,s14,s23,x3,h12,h24,det,c13,c42,c43,a14,a24,
          a33,a34,a41,a42,a43,b4,tau4;

    t=allocate_real_vector(0,n-1);
    super=allocate_real_vector(0,n-1);
    sub=allocate_real_vector(0,n-1);
    chi=allocate_real_vector(0,n-1);
    gi=allocate_real_vector(0,n-1);

    l=1;
    xl=x[0];
    e1=e[1];
    e2=e[2];
    e3=e[3];
    e4=e[4];
    e5=e[5];
    e6=e[6];
    while (l <= n) {
        xl1=xl;
        l1=l-1;
        xl=x[l];
        h=xl-xl1;
        if (order == 2) {
            /* element mat vec evaluation 1 */
            if (l == 1) {
                q2=(*q)(xl1);
                r2=(*r)(xl1);
                f2=(*f)(xl1);
            }
            h2=h/2.0;
            s12 = -1.0/h;
            q1=q2;
            q2=(*q)(xl);
            r1=r2;
            r2=(*r)(xl);
            f1=f2;
            f2=(*f)(xl);
            b1=h2*f1;
            b2=h2*f2;
            tau1=h2*r1;
            tau2=h2*r2;
            a12=s12+q1/2.0;
            a21=s12-q2/2.0;
        } else if (order == 4) {
            /* element mat vec evaluation 2 */
            if (l == 1) {
                q3=(*q)(xl1);
                r3=(*r)(xl1);
                f3=(*f)(xl1);
            }
            x2=(xl1+xl)/2.0;
```

```
        h6=h/6.0;
        h15=h/1.5;
        q1=q3;
        q2=(*q)(x2);
        q3=(*q)(xl);
        r1=r3;
        r2=(*r)(x2);
        r3=(*r)(xl);
        f1=f3;
        f2=(*f)(x2);
        f3=(*f)(xl);
        b1=h6*f1;
        b2=h15*f2;
        b3=h6*f3;
        tau1=h6*r1;
        tau2=h15*r2;
        tau3=h6*r3;
        s12 = -1.0/h/0.375;
        s13 = -s12/8.0;
        s22 = -2.0*s12;
        a12=s12+q1/1.5;
        a13=s13-q1/6.0;
        a21=s12-q2/1.5;
        a23=s12+q2/1.5;
        a22=s22+tau2;
        a31=s13+q3/6.0;
        a32=s12-q3/1.5;
        c12 = -a12/a22;
        c32 = -a32/a22;
        a12=a13+c12*a23;
        a21=a31+c32*a21;
        b1 += c12*b2;
        b2=b3+c32*b2;
        tau1 += c12*tau2;
        tau2=tau3+c32*tau2;
    } else {
        /* element mat vec evaluation 3 */
        if (l == 1) {
            q4=(*q)(xl1);
            r4=(*r)(xl1);
            f4=(*f)(xl1);
        }
        x2=xl1+0.27639320225*h;
        x3=xl-x2+xl1;
        h12=h/12.0;
        h24=h/2.4;
        q1=q4;
        q2=(*q)(x2);
        q3=(*q)(x3);
        q4=(*q)(xl);
        r1=r4;
        r2=(*r)(x2);
        r3=(*r)(x3);
        r4=(*r)(xl);
        f1=f4;
        f2=(*f)(x2);
        f3=(*f)(x3);
        f4=(*f)(xl);
        s12 = -4.8784183052080/h;
        s13=0.7117516385414/h;
        s14 = -0.16666666666667/h;
        s23=25.0*s14;
        s22 = -2.0*s23;
        b1=h12*f1;
        b2=h24*f2;
        b3=h24*f3;
        b4=h12*f4;
        tau1=h12*r1;
        tau2=h24*r2;
        tau3=h24*r3;
        tau4=h12*r4;
        a12=s12+0.67418082864578*q1;
        a13=s13-0.25751416197912*q1;
        a14=s14+q1/12.0;
```

```
            a21=s12-0.67418082864578*q2;
            a22=s22+tau2;
            a23=s23+0.93169499062490*q2;
            a24=s13-0.25751416197912*q2;
            a31=s13+0.25751416197912*q3;
            a32=s23-0.93169499062490*q3;
            a33=s22+tau3;
            a34=s12+0.67418082864578*q3;
            a41=s14-q4/12.0;
            a42=s13+0.25751416197912*q4;
            a43=s12-0.67418082864578*q4;
            det=a22*a33-a23*a32;
            c12=(a13*a32-a12*a33)/det;
            c13=(a12*a23-a13*a22)/det;
            c42=(a32*a43-a42*a33)/det;
            c43=(a42*a23-a43*a22)/det;
            tau1 += c12*tau2+c13*tau3;
            tau2=tau4+c42*tau2+c43*tau3;
            a12=a14+c12*a24+c13*a34;
            a21=a41+c42*a21+c43*a31;
            b1 += c12*b2+c13*b3;
            b2=b4+c42*b2+c43*b3;
        }
        if (l == 1 || l == n) {
            /* boundary conditions */
            if (l == 1 && e2 == 0.0) {
                tau1=1.0;
                b1=e3/e1;
                a12=0.0;
            } else if (l == 1 && e2 != 0.0) {
                tau1 -= e1/e2;
                b1 -= e3/e2;
            } else if (l == n && e5 == 0.0) {
                tau2=1.0;
                a21=0.0;
                b2=e6/e4;
            } else if (l == n && e5 != 0.0) {
                tau2 += e4/e5;
                b2 += e6/e5;
            }
        }
        /* forward babushka */
        if (l == 1) {
            chi[0]=ch=tl=tau1;
            t[0]=tl;
            gi[0]=g=yl=b1;
            y[0]=yl;
            sub[0]=a21;
            super[0]=a12;
            pp=a21/(ch-a12);
            ch=tau2-ch*pp;
            g=b2-g*pp;
            tl=tau2;
            yl=b2;
        } else {
            chi[l1] = ch += tau1;
            gi[l1] = g += b1;
            sub[l1]=a21;
            super[l1]=a12;
            pp=a21/(ch-a12);
            ch=tau2-ch*pp;
            g=b2-g*pp;
            t[l1]=tl+tau1;
            tl=tau2;
            y[l1]=yl+b1;
            yl=b2;
        }
        l++;
    }
    /* backward babushka */
    pp=yl;
    y[n]=g/ch;
    g=pp;
    ch=tl;
```

```
    l=n-1;
    while (l >= 0) {
        pp=super[l]/(ch-sub[l]);
        tl=t[l];
        ch=tl-ch*pp;
        yl=y[l];
        g=yl-g*pp;
        y[l]=(gi[l]+g-yl)/(chi[l]+ch-tl);
        l--;
    }
    free_real_vector(t,0);
    free_real_vector(super,0);
    free_real_vector(sub,0);
    free_real_vector(chi,0);
    free_real_vector(gi,0);
}
```

5.5.3 Linear methods - Fourth order self adjoint

femhermsym

Solves a fourth order self-adjoint linear two point boundary value problem by means of Galerkin's method with continuous differentiable piecewise polynomial functions [BakH76, He75, StF73]. *femhermsym* computes approximations y_n and dy_n (n=0,...,N) to the values at the points $x = x_n$, where

$$-\infty < a = x_0 < x_1 < ... < x_N = b < \infty$$

of the real valued function $y(x)$ and its derivative $Dy(x)$, where y satisfies the equation

$$-D^2(p(x)D^2y(x)) - D(q(x)Dy(x)) + r(x)y(x) = f(x) \quad (1)$$

where $D=d/dx$, with boundary conditions

$$y(a) = e_1, \quad Dy(a) = e_2, \quad y(b) = e_3, \quad Dy(b) = e_4. \quad (2)$$

It is assumed that

(a) $p(x)$ should be positive on the interval $[x_0,x_N]$, and $q(x)$ and $r(x)$ should be nonnegative there;

(b) $p(x)$, $q(x)$, $r(x)$ and $f(x)$ are required to be sufficiently smooth on the interval $[x_0,x_N]$ except at the knots, where discontinuities of the derivatives are allowed; in that case the order of accuracy is preserved.

The solution is approximated by a function which is continuously differentiable on the closed interval $[x_0,x_N]$ and a polynomial of degree less than or equal to k (k = 1 + *order*/2) on each closed segment $[x_{j-1},x_j]$ (j=1,...,N). This function is entirely determined by the values of the zeroth and first derivative at the knots x_j and by the value it has at k-3 interior knots on each closed segment $[x_{j-1},x_j]$. The values of the function and its derivative at the knots are obtained by the solution of an *order*+1 diagonal linear system of (k-1)N-2 unknowns. The entries of the matrix and the vector are inner products which are approximated by piecewise k-point Lobatto quadrature. The evaluation of the matrix and the vector is performed segment by segment. If k>3 then the resulting linear system can be reduced to a pentadiagonal system by means of static condensation. This is possible because the function values at the interior knots on each segment $[x_{j-1},x_j]$ do not depend on function values outside that segment. The final pentadiagonal system, since the matrix is symmetric positive definite, is solved by means of Cholesky's decomposition method.

The theory upon which the method of solving the above boundary value problem is based on an extension of that described in *femlagsym*. The function $y(x)$ is now approximated not, as was the case for the method of *femlagsym* (see formula (4) there) by a function which is simply continuous and equal to a polynomial of degree k over the interval $[x_{n-1},x_n]$, but by a (necessarily continuous) function whose first derivative possesses these properties. Radau-Lobatto quadrature is used, systems of linear equations are derived, and coefficients concerning internal points of $[x_{n-1},x_n]$ are eliminated just as is described in *femlagsym*. Now, however, the resulting system of linear equations involving coefficients equal to the y_n and dy_n is pentadiagonal and is solved by Cholesky's decomposition method.

Function Parameters:

void femhermsym *(x,y,n,p,q,r,f,order,e)*

x: float *x[0:n]*;
entry: $a = x_0 < x_1 < ... < x_n = b$ is a partition of the segment $[a,b]$;

y: float *y[1:2n-2]*;
exit: *y[2i-1]* is an approximation to *y(x[i])*, *y[2i]* is an approximation to *dy(x[i])*, where *y(x)* is the solution of the equation (1) with boundary conditions (2);

n: int;
entry: the upper bound of the arrays *x* (value of *N* above), $n > 1$;

p: float (**p*)(*x*);
the procedure to compute *p(x)*, the coefficient of $D^2y(x)$ in equation (1); *p(x)* should be strictly positive;

q: float (**q*)(*x*);
the procedure to compute *q(x)*, the coefficient of *Dy(x)* in equation (1); *q(x)* should be nonnegative;

r: float (**r*)(*x*);
the procedure to compute *r(x)*, the coefficient of *y(x)* in equation (1); *r(x)* should be nonnegative;

f: float (**f*)(*x*);
the procedure to compute *f(x)*, the right hand side of equation (1);

order: int;
entry: *order* denotes the order of accuracy required for the approximate solution of the differential equation; let $h = \max(x[i]-x[i-1])$, then
$|y[2i-1] - y(x[i])| \le c1*h^{order}$,
$|y[2i] - Dy(x[i])| \le c2*h^{order}$, $i=1,...,n-1$; *order* can be chosen equal to 4, 6 or 8 only;

e: float *e[1:4]*;
entry: *e[1],...,e[4]* describe the boundary conditions (values of e_i, $i=1,...,4$, in (2)).

Function used: chldecsolbnd.

```
#include <float.h>
#include <math.h>

void femhermsym(float x[], float y[], int n, float (*p)(float),
        float (*q)(float), float (*r)(float), float (*f)(float),
        int order, float e[])
{
    float *allocate_real_vector(int, int);
    void free_real_vector(float *, int);
    void chldecsolbnd(float [], int, int, float [], float []);
    void femhermsymeval(int, int, float (*)(float),
            float (*)(float), float (*)(float), float (*)(float),
            float *, float *, float *, float *, float *,
            float *, float *, float *, float *, float *,
            float *, float *, float *, float *, float *, float);
    int l,n2,v,w;
    float *a,em[4],a11,a12,a13,a14,a22,a23,a24,a33,a34,a44,ya,yb,za,
            zb,b1,b2,b3,b4,d1,d2,e1,r1,r2,xl1,xl;
```

```
    a=allocate_real_vector(1,8*(n-1));
    l=1;
    w=v=0;
    n2=n+n-2;
    xl1=x[0];
    xl=x[1];
    ya=e[1];
    za=e[2];
    yb=e[3];
    zb=e[4];
    /* element matvec evaluation */
    femhermsymeval(order,l,p,q,r,f,&a11,&a12,&a13,&a14,&a22,
            &a23,&a24,&a33,&a34,&a44,&b1,&b2,&b3,&b4,&xl,xl1);
    em[2]=FLT_EPSILON;
    r1=b3-a13*ya-a23*za;
    d1=a33;
    d2=a44;
    r2=b4-a14*ya-a24*za;
    e1=a34;
    l++;
    while (l < n) {
        xl1=xl;
        xl=x[l];
        /* element matvec evaluation */
        femhermsymeval(order,l,p,q,r,f,&a11,&a12,&a13,&a14,&a22,
                &a23,&a24,&a33,&a34,&a44,&b1,&b2,&b3,&b4,&xl,xl1);
        a[w+1]=d1+a11;
        a[w+4]=e1+a12;
        a[w+7]=a13;
        a[w+10]=a14;
        a[w+5]=d2+a22;
        a[w+8]=a23;
        a[w+11]=a24;
        a[w+14]=0.0;
        y[v+1]=r1+b1;
        y[v+2]=r2+b2;
        r1=b3;
        r2=b4;
        v += 2;
        w += 8;
        d1=a33;
        d2=a44;
        e1=a34;
        l++;
    }
    l=n;
    xl1=xl;
    xl=x[l];
    /* element matvec evaluation */
    femhermsymeval(order,l,p,q,r,f,&a11,&a12,&a13,&a14,&a22,
            &a23,&a24,&a33,&a34,&a44,&b1,&b2,&b3,&b4,&xl,xl1);
    y[n2-1]=r1+b1-a13*yb-a14*zb;
    y[n2]=r2+b2-a23*yb-a24*zb;
    a[w+1]=d1+a11;
    a[w+4]=e1+a12;
    a[w+5]=d2+a22;
    chldecsolbnd(a,n2,3,em,y);
    free_real_vector(a,1);
}

void femhermsymeval(int order, int l, float (*p)(float),
            float (*q)(float), float (*r)(float), float (*f)(float),
            float *a11, float *a12, float *a13, float *a14, float *a22,
            float *a23, float *a24, float *a33, float *a34, float *a44,
            float *b1, float *b2, float *b3, float *b4, float *xl,
            float xl1)
{
    /* this function is internally used by FEMHERMSYM */

    static float p3,p4,p5,q3,q4,q5,r3,r4,r5,f3,f4,f5;

    if (order == 4) {
        float x2,h,h2,h3,p1,p2,q1,q2,r1,r2,f1,f2,b11,b12,b13,b14,b22,
              b23,b24,b33,b34,b44,s11,s12,s13,s14,s22,s23,s24,s33,s34,
```

```
            s44,m11,m12,m13,m14,m22,m23,m24,m33,m34,m44;
    h=(*xl)-xl1;
    h2=h*h;
    h3=h*h2;
    x2=(xl1+(*xl))/2.0;
    if (l == 1) {
        p3=(*p)(xl1);
        q3=(*q)(xl1);
        r3=(*r)(xl1);
        f3=(*f)(xl1);
    }
    /* element bending matrix */
    p1=p3;
    p2=(*p)(x2);
    p3=(*p)(*xl);
    b11=6.0*(p1+p3);
    b12=4.0*p1+2.0*p3;
    b13 = -b11;
    b14=b11-b12;
    b22=(4.0*p1+p2+p3)/1.5;
    b23 = -b12;
    b24=b12-b22;
    b33=b11;
    b34 = -b14;
    b44=b14-b24;
    /* element stiffness matrix */
    q1=q3;
    q2=(*q)(x2);
    q3=(*q)(*xl);
    s11=1.5*q2;
    s12=q2/4.0;
    s13 = -s11;
    s14=s12;
    s24=q2/24.0;
    s22=q1/6.0+s24;
    s23 = -s12;
    s33=s11;
    s34 = -s12;
    s44=s24+q3/6.0;
    /* element mass matrix */
    r1=r3;
    r2=(*r)(x2);
    r3=(*r)(*xl);
    m11=(r1+r2)/6.0;
    m12=r2/24.0;
    m13=r2/6.0;
    m14 = -m12;
    m22=r2/96.0;
    m23 = -m14;
    m24 = -m22;
    m33=(r2+r3)/6.0;
    m34=m14;
    m44=m22;
    /* element load vector */
    f1=f3;
    f2=(*f)(x2);
    f3=(*f)(*xl);
    *b1=h*(f1+2.0*f2)/6.0;
    *b3=h*(f3+2.0*f2)/6.0;
    *b2=h2*f2/12.0;
    *b4 = -(*b2);
    *a11=b11/h3+s11/h+m11*h;
    *a12=b12/h2+s12+m12*h2;
    *a13=b13/h3+s13/h+m13*h;
    *a14=b14/h2+s14+m14*h2;
    *a22=b22/h+s22*h+m22*h3;
    *a23=b23/h2+s23+m23*h2;
    *a24=b24/h+s24*h+m24*h3;
    *a34=b34/h2+s34+m34*h2;
    *a33=b33/h3+s33/h+m33*h;
    *a44=b44/h+s44*h+m44*h3;
} else if (order == 6) {
    float h,h2,h3,x2,x3,p1,p2,p3,q1,q2,q3,r1,r2,r3,f1,f2,f3,b11,b12,
            b13,b14,b15,b22,b23,b24,b25,b33,b34,b35,b44,b45,b55,s11,
```

```
        s12,s13,s14,s15,s22,s23,s24,s25,s33,s34,s35,s44,s45,s55,
        m11,m12,m13,m14,m15,m22,m23,m24,m25,m33,m34,m35,m44,m45,
        m55,a15,a25,a35,a45,a55,c1,c2,c3,c4,b5;
if (l == 1) {
    p4=(*p)(xl1);
    q4=(*q)(xl1);
    r4=(*r)(xl1);
    f4=(*f)(xl1);
}
h=(*xl)-xl1;
h2=h*h;
h3=h*h2;
x2=0.27639320225*h+xl1;
x3=xl1+(*xl)-x2;
/* element bending matrix */
p1=p4;
p2=(*p)(x2);
p3=(*p)(x3);
p4=(*p)(*xl);
b11=4.0333333333333e1*p1+1.1124913866738e-1*p2+
        1.4422084194664e1*p3+8.3333333333333e0*p4;
b12=1.4666666666667e1*p1-3.3191425091659e-1*p2+
        2.7985809175818e0*p3+1.6666666666667e0*p4;
b13=1.8333333333333e1*(p1+p4)+1.2666666666667e0*(p2+p3);
b15 = -(b11+b13);
b14 = -(b12+b13+b15/2.0);
b22=5.3333333333333e0*p1+9.9027346441674e-1*p2+
        5.4305986891624e-1*p3+3.3333333333333e-1*p4;
b23=6.6666666666667e0*p1-3.7791278464167e0*p2+
        2.4579451308295e-1*p3+3.6666666666667e0*p4;
b25 = -(b12+b23);
b24 = -(b22+b23+b25/2.0);
b33=8.3333333333333e0*p1+1.4422084194666e1*p2+
        1.1124913866726e-1*p3+4.0333333333333e1*p4;
b35 = -(b13+b33);
b34 = -(b23+b33+b35/2.0);
b45 = -(b14+b34);
b44 = -(b24+b34+b45/2.0);
b55 = -(b15+b35);
/* element stiffness matrix */
q1=q4;
q2=(*q)(x2);
q3=(*q)(x3);
q4=(*q)(*xl);
s11=2.8844168389330e0*q2+2.2249827733448e-2*q3;
s12=2.5671051872498e-1*q2+3.2894812749994e-3*q3;
s13=2.5333333333333e-1*(q2+q3);
s14 = -3.7453559925005e-2*q2-2.2546440074988e-2*q3;
s15 = -(s13+s11);
s22=8.3333333333333e-2*q1+2.2847006554164e-2*q2+
        4.8632677916445e-4*q3;
s23=2.2546440075002e-2*q2+3.7453559924873e-2*q3;
s24 = -3.3333333333333e-3*(q2+q3);
s25 = -(s12+s23);
s33=2.2249827733471e-2*q2+2.8844168389330e0*q3;
s34 = -3.2894812750127e-3*q2-2.5671051872496e-1*q3;
s35 = -(s13+s33);
s44=4.8632677916788e-4*q2+2.2847006554161e-2*q3+
        8.3333333333338e-2*q4;
s45 = -(s14+s34);
s55 = -(s15+s35);
/* element mass matrix */
r1=r4;
r2=(*r)(x2);
r3=(*r)(x3);
r4=(*r)(*xl);
m11=8.3333333333333e-2*r1+1.0129076086083e-1*r2+
        7.3759058058380e-3*r3;
m12=1.3296181273333e-2*r2+1.3704853933353e-3*r3;
m13 = -2.7333333333333e-2*(r2+r3);
m14=5.0786893258335e-3*r2+3.5879773408333e-3*r3;
m15=1.3147987115999e-1*r2-3.5479871159991e-2*r3;
m22=1.7453559925000e-3*r2+2.5464400750059e-4*r3;
m23 = -3.5879773408336e-3*r2-5.0786893258385e-3*r3;
```

```
      m24=6.6666666666667e-4*(r2+r3);
      m25=1.7259029213333e-2*r2-6.5923625466719e-3*r3;
      m33=7.3759058058380e-3*r2+1.0129076086083e-1*r3+
              8.3333333333333e-2*r4;
      m34 = -1.3704853933333e-3*r2-1.3296181273333e-2*r3;
      m35 = -3.5479871159992e-2*r2+1.3147987115999e-1*r3;
      m44=2.5464400750008e-4*r2+1.7453559924997e-3*r3;
      m45=6.5923625466656e-3*r2-1.7259029213330e-2*r3;
      m55=0.17066666666667e0*(r2+r3);
      /* element load vector */
      f1=f4;
      f2=(*f)(x2);
      f3=(*f)(x3);
      f4=(*f)(*xl);
      *b1=8.3333333333333e-2*f1+2.0543729868749e-1*f2-
              5.5437298687489e-2*f3;
      *b2=2.6967233145832e-2*f2-1.0300566479175e-2*f3;
      *b3 = -5.5437298687489e-2*f2+2.0543729868749e-1*f3+
              8.3333333333333e-2*f4;
      *b4=1.0300566479165e-2*f2-2.6967233145830e-2*f3;
      b5=2.6666666666667e-1*(f2+f3);
      *a11=h2*(h2*m11+s11)+b11;
      *a12=h2*(h2*m12+s12)+b12;
      *a13=h2*(h2*m13+s13)+b13;
      *a14=h2*(h2*m14+s14)+b14;
      a15=h2*(h2*m15+s15)+b15;
      *a22=h2*(h2*m22+s22)+b22;
      *a23=h2*(h2*m23+s23)+b23;
      *a24=h2*(h2*m24+s24)+b24;
      a25=h2*(h2*m25+s25)+b25;
      *a33=h2*(h2*m33+s33)+b33;
      *a34=h2*(h2*m34+s34)+b34;
      a35=h2*(h2*m35+s35)+b35;
      *a44=h2*(h2*m44+s44)+b44;
      a45=h2*(h2*m45+s45)+b45;
      a55=h2*(h2*m55+s55)+b55;
      /* static condensation */
      c1=a15/a55;
      c2=a25/a55;
      c3=a35/a55;
      c4=a45/a55;
      *b1=((*b1)-c1*b5)*h;
      *b2=((*b2)-c2*b5)*h2;
      *b3=((*b3)-c3*b5)*h;
      *b4=((*b4)-c4*b5)*h2;
      *a11=((*a11)-c1*a15)/h3;
      *a12=((*a12)-c1*a25)/h2;
      *a13=((*a13)-c1*a35)/h3;
      *a14=((*a14)-c1*a45)/h2;
      *a22=((*a22)-c2*a25)/h;
      *a23=((*a23)-c2*a35)/h2;
      *a24=((*a24)-c2*a45)/h;
      *a33=((*a33)-c3*a35)/h3;
      *a34=((*a34)-c3*a45)/h2;
      *a44=((*a44)-c4*a45)/h;
   } else {
      float x2,x3,x4,h,h2,h3,p1,p2,p3,p4,q1,q2,q3,q4,r1,r2,r3,r4,
              f1,f2,f3,f4,b11,b12,b13,b14,b15,b16,b22,b23,b24,b25,b26,
              b33,b34,b35,b36,b44,b45,b46,b55,b56,b66,s11,s12,s13,s14,
              s15,s16,s22,s23,s24,s25,s26,s33,s34,s35,s36,s44,s45,s46,
              s55,s56,s66,m11,m12,m13,m14,m15,m16,m22,m23,m24,m25,m26,
              m33,m34,m35,m36,m44,m45,m46,m55,m56,m66,c15,c16,c25,c26,
              c35,c36,c45,c46,b5,b6,a15,a16,a25,a26,a35,a36,a45,a46,
              a55,a56,a66,det;
      if (l == 1) {
         p5=(*p)(xl1);
         q5=(*q)(xl1);
         r5=(*r)(xl1);
         f5=(*f)(xl1);
      }
      h=(*xl)-xl1;
      h2=h*h;
      h3=h*h2;
      x2=xl1+h*0.172673164646;
```

```
x3=xl1+h/2.0;
x4=xl1+(*xl)-x2;
/* element bending matrix */
p1=p5;
p2=(*p)(x2);
p3=(*p)(x3);
p4=(*p)(x4);
p5=(*p)(*xl);
b11=105.8*p1+9.8*p5+7.3593121303513e-2*p2+
        2.2755555555556e1*p3+7.0565656088553e0*p4;
b12=27.6*p1+1.4*p5-3.41554824811e-1*p2+
        2.8444444444444e0*p3+1.0113960946522e0*p4;
b13 = -32.2*(p1+p5)-7.2063492063505e-1*(p2+p4)+
        2.2755555555556e1*p3;
b14=4.6*p1+8.4*p5+1.0328641222944e-1*p2-
        2.8444444444444e0*p3-3.3445562534992e0*p4;
b15 = -(b11+b13);
b16 = -(b12+b13+b14+b15/2.0);
b22=7.2*p1+0.2*p5+1.5851984028581e0*p2+
        3.5555555555556e-1*p3+1.4496032730059e-1*p4;
b23 = -8.4*p1-4.6*p5+3.3445562534992e0*p2+
        2.8444444444444e0*p3-1.0328641222944e-1*p4;
b24=1.2*(p1+p5)-4.7936507936508e-1*(p2+p4)-
        3.5555555555556e-1*p3;
b25 = -(b12+b23);
b26 = -(b22+b23+b24+b25/2.0);
b33=7.0565656088553e0*p2+2.2755555555556e1*p3+
        7.3593121303513e-2*p4+105.8*p5+9.8*p1;
b34 = -1.4*p1-27.6*p5-1.0113960946522e0*p2-
        2.8444444444444e0*p3+3.4155482481100e-1*p4;
b35 = -(b13+b33);
b36 = -(b23+b33+b34+b35/2.0);
b44=7.2*p5+p1/5.0+1.4496032730059e-1*p2+
        3.5555555555556e-1*p3+1.5851984028581e0*p4;
b45 = -(b14+b34);
b46 = -(b24+b34+b44+b45/2.0);
b55 = -(b15+b35);
b56 = -(b16+b36);
b66 = -(b26+b36+b46+b56/2.0);
/* element stiffness matrix */
q1=q5;
q2=(*q)(x2);
q3=(*q)(x3);
q4=(*q)(x4);
q5=(*q)(*xl);
s11=3.0242424037951e0*q2+3.1539909130065e-2*q4;
s12=1.2575525581744e-1*q2+4.1767169716742e-3*q4;
s13 = -3.0884353741496e-1*(q2+q4);
s14=4.0899041243062e-2*q2+1.2842455355577e-2*q4;
s15 = -(s13+s11);
s16=5.9254861177068e-1*q2+6.0512612719116e-2*q4;
s22=5.2292052865422e-3*q2+5.5310763862796e-4*q4+q1/20.0;
s23 = -1.2842455355577e-2*q2-4.0899041243062e-2*q4;
s24=1.7006802721088e-3*(q2+q4);
s25 = -(s12+s23);
s26=2.4639593097426e-2*q2+8.0134681270641e-3*q4;
s33=3.1539909130065e-2*q2+3.0242424037951e0*q4;
s34 = -4.1767169716742e-3*q2-1.2575525581744e-1*q4;
s35 = -(s13+s33);
s36 = -6.0512612719116e-2*q2-5.9254861177068e-1*q4;
s44=5.5310763862796e-4*q2+5.2292052865422e-3*q4+q5/20.0;
s45 = -(s14+s34);
s46=8.0134681270641e-3*q2+2.4639593097426e-2*q4;
s55 = -(s15+s35);
s56 = -(s16+s36);
s66=1.1609977324263e-1*(q2+q4)+3.5555555555556e-1*q3;
/* element mass matrix */
r1=r5;
r2=(*r)(x2);
r3=(*r)(x3);
r4=(*r)(x4);
r5=(*r)(*xl);
m11=9.7107020727310e-2*r2+1.5810259199180e-3*r4+r1/20.0;
m12=8.2354889460254e-3*r2+2.1932154960071e-4*r4;
```

```
m13=1.2390670553936e-2*(r2+r4);
m14 = -1.7188466249968e-3*r2-1.0508326752939e-3*r4;
m15=5.3089789712119e-2*r2+6.7741558661060e-3*r4;
m16 = -1.7377712856076e-2*r2+2.2173630018466e-3*r4;
m22=6.9843846173145e-4*r2+3.0424512029349e-5*r4;
m23=1.0508326752947e-3*r2+1.7188466249936e-3*r4;
m24 = -1.4577259475206e-4*(r2+r4);
m25=4.5024589679127e-3*r2+9.3971790283374e-4*r4;
m26 = -1.4737756452780e-3*r2+3.0759488725998e-4*r4;
m33=1.5810259199209e-3*r2+9.7107020727290e-2*r4+r5/20.0;
m34 = -2.1932154960131e-4*r2-8.2354889460354e-3*r4;
m35=6.7741558661123e-3*r2+5.3089789712112e-2*r4;
m36 = -2.2173630018492e-3*r2+1.7377712856071e-2*r4;
m44=3.0424512029457e-5*r2+6.9843846173158e-4*r4;
m45 = -9.3971790283542e-4*r2-4.5024589679131e-3*r4;
m46=3.0759488726060e-4*r2-1.4737756452778e-3*r4;
m55=2.9024943310657e-2*(r2+r4)+3.5555555555556e-1*r3;
m56=9.5006428402050e-3*(r4-r2);
m66=3.1098153547125e-3*(r2+r4);
/* element load vector */
f1=f5;
f2=(*f)(x2);
f3=(*f)(x3);
f4=(*f)(x4);
f5=(*f)(*x1);
*b1=1.6258748099336e-1*f2+2.0745852339969e-2*f4+f1/20.0;
*b2=1.3788780589233e-2*f2+2.8778860774335e-3*f4;
*b3=2.0745852339969e-2*f2+1.6258748099336e-1*f4+f5/20.0;
*b4 = -2.8778860774335e-3*f2-1.3788780589233e-2*f4;
b5=(f2+f4)/11.25+3.5555555555556e-1*f3;
b6=2.9095718698132e-2*(f4-f2);
*a11=h2*(h2*m11+s11)+b11;
*a12=h2*(h2*m12+s12)+b12;
*a13=h2*(h2*m13+s13)+b13;
*a14=h2*(h2*m14+s14)+b14;
a15=h2*(h2*m15+s15)+b15;
a16=h2*(h2*m16+s16)+b16;
*a22=h2*(h2*m22+s22)+b22;
*a23=h2*(h2*m23+s23)+b23;
*a24=h2*(h2*m24+s24)+b24;
a25=h2*(h2*m25+s25)+b25;
a26=h2*(h2*m26+s26)+b26;
*a33=h2*(h2*m33+s33)+b33;
*a34=h2*(h2*m34+s34)+b34;
a35=h2*(h2*m35+s35)+b35;
a36=h2*(h2*m36+s36)+b36;
*a44=h2*(h2*m44+s44)+b44;
a45=h2*(h2*m45+s45)+b45;
a46=h2*(h2*m46+s46)+b46;
a55=h2*(h2*m55+s55)+b55;
a56=h2*(h2*m56+s56)+b56;
a66=h2*(h2*m66+s66)+b66;
/* static condensation */
det = -a55*a66+a56*a56;
c15=(a15*a66-a16*a56)/det;
c16=(a16*a55-a15*a56)/det;
c25=(a25*a66-a26*a56)/det;
c26=(a26*a55-a25*a56)/det;
c35=(a35*a66-a36*a56)/det;
c36=(a36*a55-a35*a56)/det;
c45=(a45*a66-a46*a56)/det;
c46=(a46*a55-a45*a56)/det;
*a11=((*a11)+c15*a15+c16*a16)/h3;
*a12=((*a12)+c15*a25+c16*a26)/h2;
*a13=((*a13)+c15*a35+c16*a36)/h3;
*a14=((*a14)+c15*a45+c16*a46)/h2;
*a22=((*a22)+c25*a25+c26*a26)/h;
*a23=((*a23)+c25*a35+c26*a36)/h2;
*a24=((*a24)+c25*a45+c26*a46)/h;
*a33=((*a33)+c35*a35+c36*a36)/h3;
*a34=((*a34)+c35*a45+c36*a46)/h2;
*a44=((*a44)+c45*a45+c46*a46)/h;
*b1=((*b1)+c15*b5+c16*b6)*h;
*b2=((*b2)+c25*b5+c26*b6)*h2;
```

```
            *b3=((*b3)+c35*b5+c36*b6)*h;
            *b4=((*b4)+c45*b5+c46*b6)*h2;
        }
}
```

5.5.4 Non-linear methods

nonlinfemlagskew

Solves a nonlinear two point boundary value problem with spherical coordinates [Bab72, Bak77, BakH76, StF73]. *nonlinfemlagskew* solves the differential equation

$$D(x^{nc}Dy(x))/x^{nc} = f(x,y,Dy(x)), \qquad a < x < b, \tag{1}$$

where $D=d/dx$, with boundary conditions

$$e_1y(a) + e_2Dy(a) = e_3, \qquad e_4y(b) + e_5Dy(b) = e_6 \qquad (e_1,\ e_4 \neq 0). \tag{2}$$

The functions *f*, *df/dy* and *df/dz* are required to be sufficiently smooth in their variables on the interior of every segment $[x_i,x_{i+1}]$ (i=0,...,n-1).

Let *y[0](x)* be some initial approximation of *y(x)*; then the nonlinear problem is solved by successively solving

$$-D(x^{nc}Dg[k](x))/x^{nc} + fy(x,y[k](x),Dy[k](x))*g[k](x) \quad + fz(x,y[k](x),Dy[k](x))*Dg[k](x)$$
$$= D(x^{nc}Dy[k](x)/x^{nc} - f(x,y[k],Dy[k](x)), \qquad x_0 < x < x_n,$$

with boundary conditions

$$e_1g[k](x_0) + e_2Dg[k](x_0) = 0, \qquad e_4g[k](x_n) + e_5Dg[k](x_n) = 0$$

with Galerkin's method (see *femlagsym*) and putting

$$y[k+1](x) = y[k](x) + g[k](x), \qquad k=0,1,...$$

This is the so-called Newton-Kantorowitch method.

Function Parameters:

void nonlinfemlagskew (*x,y,n,f,fy,fz,nc,e*)

x: float *x[0:n]*;
entry: $a = x_0 < x_1 < ... < x_n = b$ is a partition of the segment $[a,b]$;

y: float *y[0:n]*;
entry: *y[i]*, i=0,1,...,n, is an initial approximate solution at *x[i]* of the differential equation (1) with boundary conditions (2);
exit: *y[i]*, i=0,1,...,n, is the Galerkin solution at *x[i]* of the differential equation (1) with boundary conditions (2);

n: int;
entry: the upper bound of the arrays *x* and *y*, $n > 1$;

f: float (*f)(*x,y,z*); float *x,y,z*;
the procedure to compute the right hand side of (1); (*f(x,y,z)* is the right hand side of equation (1));

fy: float (*fy)(*x,y,z*); float *x,y,z*;
the procedure to compute *fy(x,y,z)*, the derivative of *f* with respect to *y*;

fz: float (*fz)(*x,y,z*); float *x,y,z*;
the procedure to compute *fz(x,y,z)*, the derivative of *f* with respect to *z*;

nc: int;
entry: if *nc* = 0, Cartesian coordinates are used;
if *nc* = 1, polar coordinates are used;
if *nc* = 2, spherical coordinates are used;

e: float *e[1:6]*;

entry: *e[1],...,e[6]* describe the boundary conditions (values of e_i, i=1,...,6, in (2)); *e[1]* and *e[4]* are not allowed to vanish both.

Function used: dupvec.

```
#include <float.h>
#include <math.h>

void nonlinfemlagskew(float x[], float y[], int n,
        float (*f)(float, float, float),
        float (*fy)(float, float, float),
        float (*fz)(float, float, float), int nc, float e[])
{
    float *allocate_real_vector(int, int);
    void free_real_vector(float *, int);
    void dupvec(int, int, int, float [], float []);
    int l,l1,it;
    float xl1,xl,h,a12,a21,b1,b2,tau1,tau2,ch,tl,g,yl,pp,
          zl1,zl,e1,e2,e4,e5,eps,rho,*t,*super,*sub,*chi,*gi,*z,
          xm,vl,vr,wl,wr,pr,qm,rm,fm,xl12,xl1xl,xl2,zm,zaccm;

    t=allocate_real_vector(0,n-1);
    super=allocate_real_vector(0,n-1);
    sub=allocate_real_vector(0,n-1);
    chi=allocate_real_vector(0,n-1);
    gi=allocate_real_vector(0,n-1);
    z=allocate_real_vector(0,n);

    dupvec(0,n,0,z,y);
    e1=e[1];
    e2=e[2];
    e4=e[4];
    e5=e[5];
    it=1;
    do {
        l=1;
        xl=x[0];
        zl=z[0];
        while (l <= n) {
            xl1=xl;
            l1=l-1;
            xl=x[l];
            h=xl-xl1;
            zl1=zl;
            zl=z[l];
            /* element mat vec evaluation 1 */
            if (nc == 0)
                vl=vr=0.5;
            else if (nc == 1) {
                vl=(xl1*2.0+xl)/6.0;
                vr=(xl1+xl*2.0)/6.0;
            } else {
                xl12=xl1*xl1/12.0;
                xl1xl=xl1*xl/6.0;
                xl2=xl*xl/12.0;
                vl=3.0*xl12+xl1xl+xl2;
                vr=3.0*xl2+xl1xl+xl12;
            }
            wl=h*vl;
            wr=h*vr;
            pr=vr/(vl+vr);
            xm=xl1+h*pr;
            zm=pr*zl+(1.0-pr)*zl1;
            zaccm=(zl-zl1)/h;
            qm=(*fz)(xm,zm,zaccm);
            rm=(*fy)(xm,zm,zaccm);
            fm=(*f)(xm,zm,zaccm);
            tau1=wl*rm;
            tau2=wr*rm;
            b1=wl*fm-zaccm*(vl+vr);
            b2=wr*fm+zaccm*(vl+vr);
```

```
            a12 = -(vl+vr)/h+vl*qm+(1.0-pr)*pr*rm*(wl+wr);
            a21 = -(vl+vr)/h-vr*qm+(1.0-pr)*pr*rm*(wl+wr);
         if (l == 1 || l == n) {
                /* boundary conditions */
                if (l == 1 && e2 == 0.0) {
                    tau1=1.0;
                    b1=a12=0.0;
                } else if (l == 1 && e2 != 0.0) {
                    tau1 -= e1/e2;
                } else if (l == n && e5 == 0.0) {
                    tau2=1.0;
                    b2=a21=0.0;
                } else if (l == n && e5 != 0.0) {
                    tau2 += e4/e5;
                }
            }
            /* forward babushka */
            if (l == 1) {
                chi[0]=ch=tl=tau1;
                t[0]=tl;
                gi[0]=g=yl=b1;
                y[0]=yl;
                sub[0]=a21;
                super[0]=a12;
                pp=a21/(ch-a12);
                ch=tau2-ch*pp;
                g=b2-g*pp;
                tl=tau2;
                yl=b2;
            } else {
                chi[l1] = ch += tau1;
                gi[l1] = g += b1;
                sub[l1]=a21;
                super[l1]=a12;
                pp=a21/(ch-a12);
                ch=tau2-ch*pp;
                g=b2-g*pp;
                t[l1]=tl+tau1;
                tl=tau2;
                y[l1]=yl+b1;
                yl=b2;
            }
            l++;
        }
        /* backward babushka */
        pp=yl;
        y[n]=g/ch;
        g=pp;
        ch=tl;
        l=n-1;
        while (l >= 0) {
            pp=super[l]/(ch-sub[l]);
            tl=t[l];
            ch=tl-ch*pp;
            yl=y[l];
            g=yl-g*pp;
            y[l]=(gi[l]+g-yl)/(chi[l]+ch-tl);
            l--;
        }
        eps=0.0;
        rho=1.0;
        for (l=0; l<=n; l++) {
            rho += fabs(z[l]);
            eps += fabs(y[l]);
            z[l] -= y[l];
        }
        rho *= FLT_EPSILON;
        it++;
    } while (eps > rho);
    dupvec(0,n,0,y,z);
    free_real_vector(t,0);
    free_real_vector(super,0);
    free_real_vector(sub,0);
    free_real_vector(chi,0);
```

```
    free_real_vector(gi,0);
    free_real_vector(z,0);
}
```

5.6 Two-dimensional boundary value problems

5.6.1 Elliptic special linear systems

A. richardson

Solves a system of linear equations with a coefficient matrix having positive real eigenvalues by means of a non-stationary second order iterative method: Richardson's method [CoHHS73, Vh68]. Since Richardson's method is particularly suitable for solving a system of linear equations that is obtained by discretizing a two-dimensional elliptic boundary value problem, the procedure *richardson* is programmed in such a way that the solution vector is given as a two-dimensional array $u[j,l]$, $lj \leq j \leq uj$, $ll \leq l \leq ul$. The coefficient matrix is not stored, but each row corresponding to a pair (j,l) is generated when needed. *richardson* can also be used to determine the eigenvalue of the coefficient matrix corresponding to the dominant eigenfunction.

richardson either (a) determines an approximation u_n to the solution of the system of equations

$$Au = f \tag{1}$$

where A is an $N\times N$ matrix ($N=(uj-lj+1)*(ul-ll+1)$; lj, $ll \geq 1$) whose eigenvalues λ_i are such that $0 < \lambda_1 < \lambda_2 < ... < \lambda_N = b$, and u is a vector whose components $u^{(i)}$ are stored in the locations of a two-dimensional real array U by means of the mapping $u^{((j-lj)(ul-ll+1)+l-ll+1)}$ in location (j,l) ($j=lj,lj+1,...,uj$; $l=ll,ll+1,...,ul$) or (b) estimates the eigenvalue λ_1 of the matrix A just described.

In both cases a sequence of vectors u_k (k=1,...,n) is constructed from an initial member u_0 (if the parameter *inap* is given the value nonzero upon call of *richardson*, the components of u_0 are those stored in the locations of U prior to call; if *inap* is given the value zero, u_0 is the unit vector (1,1,...,1)) by means of the recursion

$$u_1 = \beta_0 u_0 - \omega_0 v_0$$

$$u_{k+1} = \beta_k u_k + (1 - \beta_k)u_{k-1} - \omega_k v_k \tag{2}$$

where

$$v_k = Au_k - f \tag{3}$$

and, with $0 < a < b$, $y_0 = (a + b)/(b - a)$

$$\beta_0 = 1, \quad \omega_0 = 2/(b+a), \quad \beta_k = 2y_0 T_k(y_0)/T_{k+1}(y_0), \quad \omega_k = 4T_k(y_0)/((b-a)T_{k+1}(y_0)),$$

$T_k(y) = \cos(k \arccos(y))$ being a Tscheyscheff polynomial. Individually the u_k, v_k satisfy the recursions

$$u_1 = (\beta_0 - \omega_0 A)u_0 + \omega_0 f$$

$$u_{k+1} = (\beta_k - \omega_k A)u_k + (1 - \beta_k)u_{k-1} + \omega_k f$$

$$v_1 = (\beta_0 - \omega_0 A)v_0$$

$$v_{k+1} = (\beta_k - \omega_k A)v_k - (1 - \beta_k)v_{k-1} .$$

The residue vectors v_k are then given explicitly by the formula

$$v_k = C_k(a,b;A)v_0$$

where

$$C_k(a,b;\lambda) = T_k((b+a-2\lambda)/(b-a))/T_k((b+a)/(b-a)) .$$

Of all k-th degree polynomials p_k with $p_k(0)=1$, $C_k(a,b;\lambda)$ is that for which $\max|p_k(\lambda)|$ $(a \le \lambda \le b)$ is a minimum.

If, in the above, $a < \lambda_1$, $\|v_k\|$ tends to zero as k increases, and u_k tends to the solution u of equation (1). The mean rate of convergence over k steps may be defined to be

$$-\ln(\|v_k\|/\|v_0\|)/k \qquad \text{(where ln is the natural logarithm)}$$

and the measure of this quantity computed during the implementation of *richardson* is

$$r_k = -(1/(2k))[\ln\{\|v_k\|_2/\|v_0\|_2\} + \ln\{\|v_k\|_\infty/\|v_0\|_\infty\}].$$

If, however, $\lambda_1 < a$, $\|v_k\|$ may not tend to zero, but an estimate of λ_1 may be extracted from the u_k and v_k. If, in the decomposition of v_0 in terms of the eigenvalues of e_j or A,

$$v_0 = \sum_{j=1}^{N} c_j e_j$$

$c_1 \neq 0$, then Λ_k, where

$$\mu_k = \|v_k\|/\|u_k-u_{k-1}\|, \qquad \Lambda_k = \mu_k(\surd(ab) - \mu_k)/((\surd a+\surd b)^2/4 - \mu_k)$$

tends to λ_1. The estimates used in the implementation of *richardson* are $\Lambda_k = (\Lambda_k^{(2)} + \Lambda_k^{(\infty)})/2$ where with $\tau = 2, \infty$ in both cases,

$$\mu_k^{(\tau)} = \|v_k\|_\tau/\|u_k-u_{k-1}\|_\tau, \qquad \Lambda_k^{(\tau)} = \mu_k^{(\tau)}(\surd(ab) - \mu_k^{(\tau)})/((\surd a+\surd b)^2/4 - \mu_k^{(\tau)}) .$$

In both cases it is required that b should be an upper bound for the eigenvalues λ_i of A: it is remarked that, denoting the elements of A by $A_{j,l}$, for all λ_i

$$\lambda_i \le \min\left(\max_j \sum_{l=1}^{N} |A_{j,l}|,\ \max_l \sum_{j=1}^{N} |A_{j,l}|\right).$$

Function Parameters:

void richardson (*u,lj,uj,ll,ul,inap,residual,a,b,n,discr,k,rateconv,domeigval,out*)

u: float *u[lj:uj,ll:ul]*;
after each iteration the approximate solution calculated by *richardson* is stored in *u*;
entry: if *inap* is chosen to be nonzero then an initial approximation of the solution, otherwise arbitrary;
exit: the final approximation of the solution;

lj,uj: int;
entry: lower and upper bound for the first subscript of *u*;

ll,ul: int;
entry: lower and upper bound for the second subscript of *u*;

inap: int;
entry: if the user wishes to introduce an initial approximation then *inap* should be chosen to be nonzero, choosing *inap* to be zero has the effect that all components of *u* are set equal to 1 before the first iteration is performed;

residual: void (**residual*)(*lj,uj,ll,ul,u*);
suppose that the system of equation at hand is $Au=f$, for any entry u the procedure *residual* should calculate the residual $Au-f$ in each point j,l, where $lj \le j \le uj$, $ll \le l \le ul$, and substitute these values in the array u;

a,b: float;

entry: if one wishes to find the solution of the boundary value problem then in *a* and *b* the user should give a lower and upper bound for the eigenvalues for which the corresponding eigenfunctions in eigenfunction expansion of the residual *Au-f*, with *u* equals the initial approximation, should be reduced;
if the dominant eigenvalue is to be found then one should choose *a* greater than this eigenvalue;

n: int *;
entry: gives the total number of iterations to be performed;

discr: float *discr[1:2]*;
exit: after each iteration *richardson* delivers in *discr[1]* the Euclidean norm of the residual (value of $\|v_n\|_2$ above), and in *discr[2]* the maximum norm of the residual (value of $\|v_n\|_\infty$ above);

k: int *;
exit: counts the number of iterations *richardson* is performing;

rateconv: float *;
exit: after each iteration the average rate of convergence is assigned to *rateconv* (value of r_n above);

domeigval: float *;
exit: after each iteration the dominant eigenvalue, if present, is assigned to *domeigval* (value of Λ_n above); if there is no dominant eigenvalue then the value of *domeigval* is meaningless; this manifests itself by showing no convergence to a fixed value;

out: void (**out*)(*u,lj,uj,ll,ul,n,discr,k,rateconv,domeigval*);
by this procedure one has access to the following quantities:
for 0≤*k*≤*n* the *k*-th iterand in *u*, the Euclidean and maximum norm of the *k*-th residual in *discr[1]* and *discr[2]*, respectively;
for 0<*k*≤*n* also the average rate of convergence and the approximation to the dominant eigenvalue, both with respect to the *k*-th iterand in *u*, in *rateconv* and *domeigval*, respectively;
moreover, *out* can be used to let *n* be dependent on the accuracy reached in approximating the dominant eigenvalue.

```
#include <math.h>

void richardson(float **u, int lj, int uj, int ll, int ul,
        int inap, void (*residual)(int, int, int, int, float **),
        float a, float b, int *n, float discr[], int *k,
        float *rateconv, float *domeigval,
        void (*out)(float **, int, int, int, int, int *, float [],
                 int, float, float))
{
   float **allocate_real_matrix(int, int, int, int);
   void free_real_matrix(float **, int, int, int);
   int j,l;
   float x,y,z,y0,c,d,alfa,omega,omega0,eigmax,eigeucl,euclres,maxres,
         rcmax,rceucl,maxres0,euclres0,**v,**res,auxres0,auxv,auxu,
         auxres,eucluv,maxuv;

   v=allocate_real_matrix(lj,uj,ll,ul);
   res=allocate_real_matrix(lj,uj,ll,ul);

   alfa=2.0;
   omega=4.0/(b+a);
   y0=(b+a)/(b-a);
   x=0.5*(b+a);
   y=(b-a)*(b-a)/16.0;
   z=4.0*y0*y0;
```

```
    c=a*b;
    c=sqrt(c);
    d=sqrt(a)+sqrt(b);
    d=d*d;
    if (!inap)
        for (j=lj; j<=uj; j++)
            for (l=ll; l<=ul; l++) u[j][l]=1.0;
    *k=0;
    for (j=lj; j<=uj; j++)
        for (l=ll; l<=ul; l++) res[j][l]=u[j][l];
    (*residual)(lj,uj,ll,ul,res);
    omega0=2.0/(b+a);
    maxres0=euclres0=0.0;
    for (j=lj; j<=uj; j++)
        for (l=ll; l<=ul; l++) {
            auxres0=res[j][l];
            v[j][l]=u[j][l]-omega0*auxres0;
            auxres0=fabs(auxres0);
            maxres0 = (maxres0 < auxres0) ? auxres0 : maxres0;
            euclres0 += auxres0*auxres0;
        }
    euclres0=sqrt(euclres0);
    discr[1]=euclres0;
    discr[2]=maxres0;
    (*out)(u,lj,uj,ll,ul,n,discr,*k,*rateconv,*domeigval);
    while (*k < *n) {
        (*k)++;
        /* calculate parameters alfa and omega for each iteration */
        alfa=z/(z-alfa);
        omega=1.0/(x-omega*y);
        /* iteration */
        eucluv=euclres=maxuv=maxres=0.0;
        for (j=lj; j<=uj; j++)
            for (l=ll; l<=ul; l++) res[j][l]=v[j][l];
        (*residual)(lj,uj,ll,ul,res);
        for (j=lj; j<=uj; j++)
            for (l=ll; l<=ul; l++) {
                auxv=u[j][l];
                auxu=v[j][l];
                auxres=res[j][l];
                auxv=alfa*auxu-omega*auxres+(1.0-alfa)*auxv;
                v[j][l]=auxv;
                u[j][l]=auxu;
                auxu=fabs(auxu-auxv);
                auxres=fabs(auxres);
                maxuv = (maxuv < auxu) ? auxu : maxuv;
                maxres = (maxres < auxres) ? auxres : maxres;
                eucluv += auxu*auxu;
                euclres += auxres*auxres;
            }
        eucluv=sqrt(eucluv);
        euclres=sqrt(euclres);
        discr[1]=euclres;
        discr[2]=maxres;
        maxuv=maxres/maxuv;
        eucluv=euclres/eucluv;
        eigmax=maxuv*(c-maxuv)/(0.25*d-maxuv);
        eigeucl=eucluv*(c-eucluv)/(0.25*d-eucluv);
        *domeigval=0.5*(eigmax+eigeucl);
        rceucl = -log(euclres/euclres0)/(*k);
        rcmax = -log(maxres/maxres0)/(*k);
        *rateconv=0.5*(rceucl+rcmax);
        (*out)(u,lj,uj,ll,ul,n,discr,*k,*rateconv,*domeigval);
    }
    free_real_matrix(v,lj,uj,ll);
    free_real_matrix(res,lj,uj,ll);
}
```

B. elimination

Determines an approximation u'_n to the solution of the system of equations

$$Au = f \tag{1}$$

where A is an $N\times N$ matrix $(N=(uj-lj+1)*(ul-ll+1);\ lj,\ ll \geq 1)$ whose eigenvalues λ_i are such that $0 < \lambda_1 < \lambda_2 < ... < \lambda_N = b$, and u is a vector whose components $u^{(i)}$ are stored in the locations of a two-dimensional real array U by means of the mapping $u^{((j-lj)(ul-ll+1)+l-ll+1)}$ in location (j,l) $(j=lj,lj+1,...,uj;\ l=ll,ll+1,...,ul)$.

elimination is to be used in conjunction with *richardson*, one of whose possible functions is precisely the solution of the problem described above. In the implementation of *richardson*, a sequence of vectors u_k and residual vectors v_k, where $v_k = Au_k - f$, is produced by use of recursion (2,3) of the documentation to *richardson* and, with $0 < a < b$,

$$v_k = C_k(a,b;A)v_0, \qquad u_k = u + A^{-1}C_k(a,b;A)v_0$$

where of all polynomials p_k with $p_k(0) = 1$, $p_k(\lambda) = C_k(a,b;\lambda)$ is that for which $\max|p_k^{(\lambda)}|$ $(a \leq \lambda \leq b)$ is a minimum; explicitly

$$C_k(a,b;\lambda) = T_k((b+a-2\lambda)/(b-a))/T_k((b+a)/(b-a)) .$$

Decomposing v_0 in terms of the eigenvectors e_i of A,

$$v_0 = \sum_{i=1}^{N} c_i e_i$$

$$v_k = \sum_{i=1}^{N} c_i C_k(a,n;\lambda_i) e_i , \qquad u_k = u + \sum_{i=1}^{N} \lambda_i^{-1} c_i C_k(a,b;\lambda_i) e_i .$$

If $a < \lambda_1$, then $u_k \rightarrow u$. An upper bound b for the λ_i may easily be given (see formula (4) of *richardson*). A lower bound for the λ_i may also be obtained from the remark that the λ_i lie in the adjunction of the discs

$$|\lambda - A^{(i,i)}| \leq \sum_{\substack{j=1 \\ (j\neq i)}}^{N} |A^{(i,j)}| \qquad (i=1,...,n). \tag{2}$$

In the above, however, it is required that the lower bound a should be positive, and this may not be true of the lower bound derived from the expressions (2).

In one of its modes of application (with $a > \lambda_1$) *richardson* yields an approximation to λ_1, and hence a lower bound a' for the λ_i. Applying *richardson* yet again, with a' in place of a and u_n used as the initial approximation u'_0 to produce a new sequence of approximations u'_k and residual vectors v'_k, the residuals

$$\begin{aligned} v'_{n'} &= \sum_{j=1}^{N} C_n(a,b;\lambda_j) C_{n'}(a',b;\lambda_j) e_j \\ u'_{n'} &= u + \sum_{j=1}^{N} C_n(a,b;\lambda_j) C_{n'}(a',b;\lambda_j) \lambda_j^{-1} e_j . \end{aligned} \tag{3}$$

$C_{n'}(a',b;\lambda)$ has the same properties with respect to $[a',b]$ as $C_n(a,b;\lambda)$ has with respect to

$[a,b]$: $u'_{n'} \rightarrow u$.

Assuming that $a < \lambda_2$, the terms involving j=2,3,... in formulae (3) are negligible:

$$v'_{n'} \approx C_n(a,b,;\lambda_l)C_{n'}(a',b;\lambda_l)e_l$$

$$u'_{n'} \approx u + C_n(a,b,;\lambda_l)C_{n'}(a',b;\lambda_l)\lambda_l^{-1}e_l \ . \qquad (4)$$

By taking

$$a^* = a^*(u') = \{2\lambda_l + b[\cos(\pi/(2n'))-1]\} \ / \ \{\cos(\pi/(2n'))+1\}$$

$C_{n'}(a^*,b;\lambda_l) = 0$, and the second term upon the right hand side of formula (4) may be eliminated.

The above process requires a total of $n+n'$ vector constructions. The mean rate of convergence may therefore be defined as

$R(n,n') = -\ln(\|v'_{n'}\| / \|v_0\|) \ / \ (n+n')$ (where ln is the natural logarithm).

The question now arises as to what value of n' maximizes this rate of convergence. Adopting the above assumption

$$R(n,n') = -[1/(n+n')]\{\ln(C_n(a,b;\lambda_l)) + \ln(C_{n'}(a'(n'),b;\lambda_l))\}.$$

Since $T_n\{(b-a)/(b+a)\} \approx \cosh\{2n\surd(a/b)\}$ where $a \ll b$ and n is large,

$\max|C_n(a,b;\lambda) \approx \cosh\{2n\surd(a/b)\}^{-1}$ $(a \leq \lambda \leq b)$ and $-\ln\{C_n(a,b;\lambda_l)\} \approx 2n\surd(a/b) - m(2)$.

Thus, approximately

$$R(n,n') = 2\surd(a/b) - \ln(2)/(n+n') - \{2\surd(a/b) - \ln|T_p[(b+a^*(n'))/(b-a^*(n'))]|\}/(n+n').$$

For a fixed total number of constructions (i.e. $n+n'$ constant), $R(n,n')$ is maximized when $n' \approx p$ and

$$2\surd(a/b) - (d/dp)\ln|T_p\{(b+a'(p))/(b-a'(p))\}| =$$

$$2\surd(a/b) + [\arccos\{w(p)\} - \{b\pi\sin(\pi/(2p))\}/\{2p(b-a)\surd(1-w(p)^2)\}]\tan\{p \arccos w(p)\} \qquad (5)$$

is zero, where $w(p) = \{b\cos(\pi/(2p)) + a\} \ / \ (b - a)$.

The smallest positive zero p' of the function (5) is found and n' is taken to be p'.

In summary *elimination* is to be used as follows: firstly, *richardson* is called with $a>\lambda_l$ and the procedure *out* required by *richardson* having a form suitable for the estimation of λ_l (e.g. increasing n by steps of unity until two successive estimates of λ_l agree sufficiently closely); *richardson* then allocates the derived estimate of λ_l to *domeigval*. With the parameters *u,lj,...,domeigval* untouched, but with the procedure *out* required by *elimination* now having a form suitable for the solution of equation (1) (e.g. inspecting *discr[1]* and *discr[2]* when $k=n+n'$, and increasing n' by unity, and repeating this operation should this be necessary) the user now calls *elimination*.

Function Parameters:

void elimination (*u,lj,uj,ll,ul,residual,a,b,n,discr,k,rateconv,domeigval,out*)

u: float *u[lj:uj,ll:ul]*;
after each iteration the approximate solution calculated by *elimination* is stored in *u*;
entry: an initial approximation of the solution which is obtained by use of *richardson*;
exit: the final approximation to the solution;

lj,uj: int;
entry: lower and upper bound for the first subscript of *u*;

ll,ul: int;
entry: lower and upper bound for the second subscript of *u*;

residual: void (**residual*)(*lj,uj,ll,ul,u*);
suppose that the system of equation at hand is $Au=f$; for any entry *u* the procedure *residual* should calculate the residual $Au-f$ in each point j,l, where $lj \leq j \leq uj$, $ll \leq l \leq ul$, and substitute these values in the array *u*;

a,b: float;
a and *b* should have the same values as in the preceding call of *richardson* (see the description of *richardson*);

n: int *;
exit: the number of iterations the procedure *elimination* needs to eliminate the eigenfunction belonging to the dominant eigenvalue is assigned to *n*;

discr: float *discr[1:2]*;
exit: after each iteration *elimination* delivers in *discr[1]* the Euclidean norm of the residual and in *discr[2]* the maximum norm of the residual;

k: int *;
exit: counts the number of iterations *elimination* is performing;

rateconv: float *;
exit: after each iteration the average rate of convergence is assigned to *rateconv*;

domeigval: float *;
before a call of *elimination* the value of the eigenvalue for which the corresponding eigenfunction has to be eliminated should be assigned to *domeigval*; if after application of *elimination* there is a new dominant eigenfunction, then *domeigval* will be equal to the corresponding eigenvalue; otherwise the value of *domeigval* becomes meaningless;

out: void (**out*)(*u,lj,uj,ll,ul,n,discr,k,rateconv,domeigval*);
by this procedure one has access to the following quantities:
for $0 \le k \le n$ the *k*-th iterand in *u*, the Euclidean and maximum norm of the *k*-th residual in *discr[1]* and *discr[2]*, respectively;
for $0 < k \le n$ also the average rate of convergence with respect to the *k*-th iterand in *u*, in *rateconv*;
for $k=n$, possibly the dominant eigenvalue of the coefficient matrix of the equation $Au=f$, in *domeigval*.

Function used: richardson.

```
#include <float.h>
#include <math.h>

void elimination(float **u, int lj, int uj, int ll, int ul,
        void (*residual)(int, int, int, int, float **),
        float a, float b, int *n, float discr[], int *k,
        float *rateconv, float *domeigval,
        void (*out)(float **, int, int, int, int, int *, float [],
                    int, float, float))
{
    float **allocate_real_matrix(int, int, int, int);
    void free_real_matrix(float **, int, int, int);
    void richardson(float **, int, int, int, int,
            int, void (*)(int, int, int, int, float **),
            float, float, int *, float [], int *, float *, float *,
            void (*)(float **, int, int, int, int, int *, float [],
                        int, float, float));
    float optpol(float, float, float, float);
    int ext,extrapolate;
    float pi,auxcos,c,d,
            cc,fc,bb,fb,aa,fa,dd,fd,fdb,fda,w,mb,tol,m,p,q;

    pi=3.14159265358979;
    c=1.0;
    if (optpol(c,a,b,*domeigval) < 0.0) {
        d=0.5*pi*sqrt(fabs(b/(*domeigval)));
        while (1) {
            d += d;
            /* finding zero */
            bb=c;
```

```
            fb=optpol(c,a,b,*domeigval);
            aa=c=d;
            fa=optpol(c,a,b,*domeigval);
            cc=aa;
            fc=fa;
            ext=0;
            extrapolate=1;
            while (extrapolate) {
                if (fabs(fc) < fabs(fb)) {
                    if (cc != aa) {
                        dd=aa;
                        fd=fa;
                    }
                    aa=bb;
                    fa=fb;
                    bb=c=cc;
                    fb=fc;
                    cc=aa;
                    fc=fa;
                }
                tol=c*1.0e-3;
                m=(cc+bb)*0.5;
                mb=m-bb;
                if (fabs(mb) > tol) {
                    if (ext > 2)
                        w=mb;
                    else {
                        if (mb == 0.0)
                            tol=0.0;
                        else
                            if (mb < 0.0) tol = -tol;
                        p=(bb-aa)*fb;
                        if (ext <= 1)
                            q=fa-fb;
                        else {
                            fdb=(fd-fb)/(dd-bb);
                            fda=(fd-fa)/(dd-aa);
                            p *= fda;
                            q=fdb*fa-fda*fb;
                        }
                        if (p < 0.0) {
                            p = -p;
                            q = -q;
                        }
                        w=(p<FLT_MIN || p<=q*tol) ? tol :
                                ((p<mb*q) ? p/q : mb);
                    }
                    dd=aa;
                    fd=fa;
                    aa=bb;
                    fa=fb;
                    c = bb += w;
                    fb=optpol(c,a,b,*domeigval);
                    if ((fc >= 0.0) ? (fb >= 0.0) : (fb <= 0.0)) {
                        cc=aa;
                        fc=fa;
                        ext=0;
                    } else
                        ext = (w == mb) ? 0 : ext+1;
                } else
                    break;
            }
            d=cc;
            if ((fc >= 0.0) ? (fb <= 0.0) : (fb >= 0.0)) {
                *n=floor(c+0.5);
                break;
            }
        }
    } else
        *n=1;
    auxcos=cos(0.5*pi/(*n));
    richardson(u,lj,uj,ll,ul,1,residual,
        (2.0*(*domeigval)+b*(auxcos-1.0))/(auxcos+1.0),b,n,discr,k,
        rateconv,domeigval,out);
```

```
}

float optpol(float x, float a, float b, float domeigval)
{
   /* this function is internally used by ELIMINATION */

   float pi,w,y;

   pi=3.14159265358979;
   w=(b*cos(0.5*pi/x)+domeigval)/(b-domeigval);
   if (w < -1.0) w = -1.0;
   if (fabs(w) <= 1.0) {
      y=acos(w);
      return 2.0*sqrt(a/b)+tan(x*y)*(y-b*pi*sin(0.5*pi/x)*
                0.5/(x*(b-domeigval)*sqrt(fabs(1.0-w*w))));
   } else {
      y=log(w+sqrt(fabs(w*w-1.0)));
      return 2.0*sqrt(a/b)-tanh(x*y)*(y+b*pi*sin(0.5*pi/x)*
                0.5/(x*(b-domeigval)*sqrt(fabs(w*w-1.0))));
   }
}
```

5.6 Parameter estimation in differential equations

5.6.1 Initial value problems

peide

Estimates unknown variables in a system of first order differential equations by using a given set of observed values [Vn75]. The unknown variables may appear nonlinearly both in the differential equations and its initial values.

peide determines the components $p_1, p_2, ..., p_m$ of $p \in R^m$ occurring in the system of differential equations

$$dy_j(t,p)/dt = f_j(t,y,p) \qquad (j=1,...,n) \tag{1}$$

($y_j(t_0,p)$, $j=1,...,n$, being prescribed) for which the sum of squares

$$\sum_{i=1}^{nobs} [y^{(i)}(t^{(i)},p) - y'^{(i)}(t^{(i)})]^2 \tag{2}$$

i.e. the square of the Euclidean norm of the residual vector whose components are

$$rv_i(p) = y^{(i)}(t^{(i)},p) - y'^{(i)}(t^{(i)}) \qquad (i=1,...,nobs) \tag{3}$$

is a minimum, where the $\{y^{(i)}(t^{(i)},p)\}$ form a selection of values of components of $y(t,p)$ taken at prescribed values of $t = t^{(i)}$, and $y'^{(i)}(t^{(i)})$ are observed values of these components. The above selection is specified by an array of integers $\{cobs[i]\}$ fixing the component suffices of the observed values, and an array of real numbers $\{tobs[i]\}$ fixing the values of t at which the observations have been made. Thus if observations of all components of $y(t,p)$ at the points $t=t'_1,...,t'_N$ are available, we have $cobs[1]=1$, $cobs[2]=2,...,$ $cobs[n]=n$, $cobs[n+1]=1$, $cobs[n+2]=2,...,$ $cobs[nN]=n$, $tobs[1]=tobs[2]=...=tobs[n]=t'_1$, $tobs[n+1]=...=tobs[2n]=t'_2,...,$ $tobs[nN]=t'_N$, with $nobs=nN$. If the first component of $y(t,p)$ alone has been observed at $t=t'_1,...,t'_N$, we have $cobs[1]=cobs[2]=...=cobs[N]=1$, $tobs[1]=t'_1,...,tobs[N]=t'_N$ and $nobs=N$. Adopting this scheme, it is possible that the distribution of components of $y(t,p)$ for which observations are available may vary from one argument t of observation to another.

The numerical integration of the system of equations (1) is carried out by Gear's method. The vector $p = p'$ yielding a minimum of the function (2) is located by use of the Levenberg-Marquardt algorithm: the user is asked to provide an initial approximation $p^{(0)}$ to p'; a sequence of vectors $\{p^{(k)}\}$ (k=0,1...) is produced for which a member, $p^{(k')}$, is an acceptable approximation to p'.

It may occur that it is difficult to prescribe an excellent initial approximation $p^{(0)}$ to p', and for this reason the user may instruct *peide* to carry out fitting, during the early stages of the computation, over a system of subranges of the total argument range $t = t^{(1)}$ to $t = t^{(nobs)}$. The user is then asked to prescribe a system of break-points (integers) $bp[i']$ (i'=1,2,...,nbp) which, in particular, specify a nondecreasing sequence of values of t, namely $T^{(i')} = t^{(bp[i'])}$ (i'=1,...,nbp), for which observed values of selected components of $y(t,p)$ are available. The original minimization problem is now extended so as to concern functions $y(t,T^{(i)},p'')$ defined for $T^{(i)} \le t < T^{(i+1)}$ respectively (i=0,1,...,nbp) where $T^{(0)} = t^{(1)}$ and $T^{(nbp+1)} = t^{(nobs)}$, $p'' \in R^{m+nbp}$ being an extension of p. Each of these functions satisfies a system of equations of the form (1) over its range of definition. The components of the initial values $y_j(t,T^{(0)},p''^{(0)})$ at $t=T^{(0)}=t^{(1)}$ are simply the initial values $y_j(t_0,p)$ (j=1,...,n) for the original problem. During the first traversal of the range $[t^{(1)},t^{(nobs)}]$ the initial values $\{y_j(t,T^{(i)},p''^{(0)})\}$ at $t = T^{(i)}$ for the integration of the system corresponding to (1) over the range $[T^{(i)},T^{(i+1)}]$ are $y_j(T^{(i')},T^{(i'-1)},p''^{(0)})$ (obtained from integration up to the end point of the preceding interval) for all j except that, j' say, corresponding to a component for which an observed value is available, when the initial value is simply taken to be the observed value. The components of $p'' \in R^{m+nbp}$ are given by $p''_i = p_i$ (i=1,...,n) and $p''_{m+i'} = y(T^{(i')},T^{(i')},p)$ (i=1,...,nbp). Setting $i''(i) = i'$ for all i for which $T^{(i')} \le t^{(i)} < T^{(i'+1)}$, the function now to be minimized is

$$\sum_{i=1}^{nobx} \{y^{(i)}(t^{(i)},T^{i''(i)},p'') - y'^{(i)}(t^{(i)})\}^2 +$$
$$M^2 \sum_{i'=1}^{nbp} \{y^{(bp[i'])}(T^{(i')},T^{(i'-1)},p'') - y^{(bp[i'])}(T^{(i')},T^{(i')},p'')\}2 \tag{4}$$

i.e. the square of the Euclidean norm of the residual vector whose components are

$$res_i(p'') = y^{(i)}(t^{(i)},T^{i''(i)},p'') - y'^{(i)}(t^{(i)}) \qquad (i=1,...,nobs)$$
$$res_{nobs+i'}(p'') = M\{y^{(bp[i'])}(T^{(i')},T^{(i'-1)},p'') - y^{(bp[i'])}(T^{(i')},T^{(i')},p'')\} \quad (i'=1,...,nbp) \tag{5}$$

where M is a weight factor. After each traversal of the range $[t_0, t^{(nobs)}]$ a new vector $p''^{(k)}$ is determined by use of the Levenberg-Marquardt algorithm.

During traversals of the range subsequent to the first, the vector p used in the systems of equations corresponding to (1) is made up of the first m components of p. The initial values $y_j(t,T^{(i')},p''^{(k)})$ at $t = T^{(i')}$ for the integration of the system corresponding to (1) over the range $[T^{(i')},T^{(i'+1)}]$ are again $y_j(T^{(i')},T^{(i'-1)},p''^{(k)})$ for all j except j' as above, when the new initial value is taken to be $p''_{m+i'}$. The iterations of the Levenberg-Marquardt algorithm are continued until the vectors $\{p''^{(k)}\}$ have approached a limit sufficiently closely. The second sum in expression (4) corresponds to a continuity requirement. After the above $p''^{(k)}$ have attained a limit sufficiently closely, the weight factor M is increased (i.e. the continuity requirements are made more stringent). A new terminating sequence of further vectors $\{p^{(k)}\}$ is derived, again M is increased, and so on. When, at a certain break-point $bp[i'']$,

$$|y^{(bp[i'''])}(T^{(i''')},T^{(i'''-1)},p''^{(k)}) - y^{(bp[i'''])}(T^{(i''')},T^{(i''')},p''^{(k)})|$$

becomes sufficiently small (i.e. the discontinuity at that break-point has virtually disappeared) the term corresponding to $i' = i'''$ is removed from expression (4), and the component with suffix $m+i'''$ is discarded from p'', the computations being continued with reduced sum and vector. This process is continued until all break-points have been discarded, and the original minimization problem is solved by a final use of the Levenberg-Marquardt algorithm. Naturally cases occur for which an excellent approximation $p^{(0)}$ to p'' is available, and in such a case (by setting *nbp*=0) the user may avoid the above excursion. For the sake of completeness, it is mentioned that certain of the above break points may correspond to equal values of t, (the components of $y(t,p)$ for which observed values are available at this value of t having differing suffices). In such a case in the above exposition certain of the intervals of the form $[T^{(i')},T^{(i'+1)}]$ reduce to a single point. Numerical integration of equations of the form (1) over such an interval is dispensed with, but the process of eliminating discontinuities as described is retained.

That the initial values $y_j(t_0,p)$ (j=1,...,n) for the equations (1) may depend upon p is also permitted (see the procedure *ystart* below).

Function Parameters:

void peide
(*n,m,nobs,nbp,par,res,bp,jtjinv,in,out,deriv,jacdfdy,jacdfdp,callystart,data,monitor*)

n: int;
entry: the number of differential equations;

m: int;
entry: the number of unknown variables;

nobs: int;
entry: the number of observations; *nobs* should satisfy *nobs* ≥ *m*;

nbp: int *;
entry: the number of break-points; if no break-points are used then set *nbp* = 0;
exit: with normal termination of the process *nbp* = 0;
otherwise, if the process has been broken off (see *out[1]*), the value of *nbp* is the number of break-points used before the process broke off;

par: float *par[1:m+nbp]*;
entry: *par[1:m]* should contain an initial approximation to the required parameter vector (values of $p_i^{(0)}$, $i=1,...,m$, above);
exit: *par[1:m]* contains the calculated parameter vector;
if *out[1]* > 0 and *nbp* > 0 then *par[m+1:m+nbp]* contains the values of the newly introduced parameters before the process broke off;

res: float *res[1:nobs+nbp]*;
exit: *res[1:nobs]* contains the residual vector at the calculated minimum (*res[i]* contains $y^{(i)}(t^{(i)},p') - y^{(i)}(t^{(i)})$, i=1,...,*nobs*, in the above);
if *out[1]* > 0 and *nbp* > 0 then *res[nobs+1:nobs+nbp]* contains the additional continuity requirements at the break-points before the process broke off (*res[nobs+i']* contains $y^{(bp[i'])}(T^{(i')},T^{(i'-1)},p''^{(k)}) - y^{(bp[i'])}(T^{(i')},T^{(i')},p''^{(k)})$, i'=1,...,*nbp*, in the above);

bp: int *bp[0:nbp]*;
entry: *bp[i]*, i=1,...,*nbp*, should correspond to the index of that time of observation which will be used as a break-point (1 ≤ *bp[i]* ≤ *nobs*); the break-points have to be ordered such that *bp[i]* ≤ *bp[j]* if $i \le j$;
exit: with normal termination of the process *bp[1:nbp]* contains no information; otherwise, if *out[1]* > 0 and *nbp* > 0 then *bp[i]*, i=1,...,*nbp*, contains the index

of that time of observation which was used as a break-point before the process broke off;

jtjinv: float *jtjinv[1:m,1:m]*;

exit: the inverse of the matrix *J'*J* where *J* denotes the matrix of partial derivatives *dres[i]/dpar[k]* (*i*=1,...,*nobs*; *k*=1,...,*m*) and *J'* denotes the transpose of *J*; this matrix can be used if additional information about the result is required; e.g. statistical data such as the covariance matrix, correlation matrix and confidence intervals can easily be calculated from *jtjinv* and *out[2]*;

in: float *in[0:6]*;

entry:

in[0]: the machine precision;

in[1]: the ratio: the minimal steplength for the integration of the differential equations divided by the distance between two neighboring observations; mostly, a suitable value is 10^{-4};

in[2]: the relative local error bound for the integration process; this value should satisfy *in[2]* ≤ *in[3]*; this parameter controls the accuracy of the numerical integration; mostly, a suitable value is *in[3]*/100;

in[3]: the relative tolerance for the difference between the Euclidean norm of the ultimate and penultimate residual vector; see *in[4]* below;

in[4]: the absolute tolerance for the difference between the Euclidean norm of the ultimate and penultimate residual vector;

the process is terminated if the improvement of the sum of squares is less than *in[3]* * (sum of squares) + *in[4]* * *in[4]*;

in[3] and *in[4]* should be chosen in accordance with the relative and absolute errors in the observations; note that the Euclidean norm of the residual vector is defined as the square root of the sum of squares;

in[5]: the maximum number of times that the integration of the differential equations is performed;

in[6]: a starting value used for the relation between the gradient and the Gauss-Newton direction; if the problem is well conditioned then a suitable value for *in[6]* will be 0.01; if the problem is ill conditioned then *in[6]* should be greater, but the value of *in[6]* should satisfy: *in[0]* < *in[6]* ≤ *1/in[0]*;

out: float *out[1:7]*;

exit:

out[1]: this value gives information about the termination of the process;

out[1]=0: normal termination;

if *out[1]* > 0 then the process has been broken off and this may occur because of the following reasons:

out[1]=1: the number of integrations performed exceeded the number given in *in[5]*;

out[1]=2 the differential equations are very nonlinear; during an integration the value of *in[1]* was decreased by a factor 10000 and it is advised to decrease *in[1]*, although this will increase computing time;

out[1]=3 a call of *deriv* delivered the value zero;

out[1]=4 a call of *jacdfdy* delivered the value zero;

out[1]=5 a call of *jacdfdp* delivered the value zero;

out[1]=6 the precision asked for cannot be attained; this precision is possibly chosen too high, relative to the precision in which the residual vector is calculated (see *in[3]*);

out[2]: the Euclidean norm of the residual vector calculated with values of the unknowns delivered;
out[3]: the Euclidean norm of the residual vector calculated with the initial values of the unknown variables;
out[4]: the number of integrations performed, needed to obtain the calculated result; if *out[4]* = 1 and *out[1]* > 0 then the matrix *jtjinv* cannot be used;
out[5]: the maximum number of times that the requested local error bound was exceeded in one integration; if it is a large number then it may be better to decrease the value of *in[1]*;
out[6]: the improvement of the Euclidean norm of the residual vector in the last integration step of the process of Marquardt;
out[7]: the condition number of *J'*J*, i.e. the ratio of its largest to smallest eigenvalues;

deriv: int (**deriv*)(*n,m,par,y,t,df*);
entry: *par[1:m]* contains the current values of the unknowns and should not be altered;
y[1:n] contains the solutions of the differential equations at time *t* and should not be altered;
exit: an array element *df[i]* (*i*=1,...,*n*) should contain the right hand side of the *i*-th differential equation;
after a successful call of *deriv*, the procedure should deliver the value nonzero; however, if *deriv* delivers the value zero then the process is terminated (see *out[1]*); hence, proper programming of *deriv* makes it possible to avoid calculation of the right hand side with values of the unknown variables which cause overflow in the computation;

jacdfdy: int (**jacdfdy*)(*n,m,par,y,t,fy*); float *fy[1:n,1:n]*;
entry: for parameters *par*, *y* and *t*, see *deriv* above;
exit: an array element *fy[i,j]* (*i,j*=1,...,*n*) should contain the partial derivative of the right hand side of the *i*-th differential equation with respect to *y[j]*, i.e. *df[i]/dy[j]*;
the integer value should be assigned to this procedure in the same way as is done for the value of *deriv*;

jacdfdp: int (**jacdfdp*)(*n,m,par,y,t,fp*); float *fp[1:n,1:m]*;
entry: for parameters *par*, *y* and *t*, see *deriv* above;
exit: an array element *fp[i,j]* should contain the partial derivative of the right hand side of the *i*-th differential equation with respect to *par[j]*, i.e. *df[i]/dpar[j]*;
the integer value should be assigned to this procedure in the same way as is done for the value of *deriv*;

callystart: void (**callystart*)(*n,m,par,y,ymax*);
entry: *par[1:m]* contains the current values of the unknown variables and should not be altered;
exit: *y[1:n]* should contain the initial values of the corresponding differential equations; the initial values may be functions of the unknown variables *par*; in that case, the initial values of *dy/dpar* also have to be supplied; note that *dy[i]/dpar[j]* corresponds with *y[5*n+j*n+i]* (*i*=1,...,*n*, *j*=1,...,*m*);
ymax[i], *i*=1,...,*n*, should contain a rough estimate to the maximal absolute value of *y[i]* over the integration interval;

data: void (**data*)(*nobs,tobs,obs,cobs*);

this procedure takes the data to fit into the procedure *peide*;
entry:
nobs has the same meaning as in *peide*;
exit:
tobs: float *tobs[0:nobs]*;
the array element *tobs[0]* should contain the time, corresponding to the initial values of *y* given in the procedure *callystart*; an array element *tobs[i]*, $1 \le i \le nobs$, should contain the *i*-th time of observation; the observations have to be ordered such that $tobs[i] \le tobs[j]$ if $i \le j$;
cobs: int *cobs[1:nobs]*;
an array element *cobs[i]* should contain the component of *y* observed at time *tobs[i]*; note that $1 \le cobs[i] \le n$;
obs: float *obs[1:nobs]*;
an array element *obs[i]* should contain the observed value of the component *cobs[i]* of *y* at the time *tobs[i]*;

monitor: void (**monitor*)(*post,ncol,nrow,par,res,weight,nis*);
int *post,ncol,nrow,weight,nis*;
this procedure can be used to obtain information about the course of the iteration process; if no intermediate results are desired then a dummy procedure satisfies; inside *peide*, the procedure *monitor* is called at two different places and this is denoted by the value of *post*;
post=1: *monitor* is called after an integration of the differential equations; at this place are available: the current values of the unknown variables *par[1:ncol]*, where $ncol=m+nbp$, the calculated residual vector *res[1:nrow]*, where $nrow=nobs+nbp$, and the value of *nis*, which is the number of integration steps performed during the solution of the last initial value problem;
post=2: *monitor* is called before a minimization of the Euclidean norm of the residual vector with the Levenberg-Marquardt algorithm is started; available are the current values of *par[1:ncol]* and the value of the *weight*, with which the continuity requirements at the break-points are added to the original least squares problem.

Functions used: inivec, inimat, mulvec, mulrow, dupvec, dupmat, vecvec, matvec, elmevc, sol, dec, mulcol, tamvec, mattam, qrisngvaldec.

```
#include <float.h>
#include <math.h>

void peide(int n, int m, int nobs, int *nbp, float par[],
        float res[], int bp[], float **jtjinv,
        float in[], float out[],
        int (*deriv)(int,int,float [],float [],float,float []),
        int (*jacdfdy)(int,int,float [],float [],float,float **),
        int (*jacdfdp)(int,int,float [],float [],float,float **),
        void (*callystart)(int,int,float [],float [],float[]),
        void (*data)(int,float [],float [],int[]),
        void (*monitor)(int,int,int,float [],float [],int,int))
{
    int *allocate_integer_vector(int, int);
    float *allocate_real_vector(int, int);
    float **allocate_real_matrix(int, int, int, int);
    void free_integer_vector(int *, int);
    void free_real_vector(float *, int);
    void free_real_matrix(float **, int, int, int);
    int peidefunct(int nrow, int ncol, float par[], float res[],
        int n, int m, int nobs, int *nbp, int first, int *sec,
```

```
       int *max, int *nis, float eps1, int weight, int bp[],
       float save[], float ymax[], float y[], float **yp,
       float **fy, float **fp, int cobs[], float tobs[],
       float obs[], float in[], float aux[], int clean,
       int (*deriv)(int,int,float [],float [],float,float []),
       int (*jacdfdy)(int,int,float [],float [],float,float **),
       int (*jacdfdp)(int,int,float [],float [],float,float **),
       void (*callystart)(int,int,float [],float [],float[]),
       void (*monitor)(int,int,int,float [],float [],int,int));
   void inivec(int, int, float [], float);
   void inimat(int, int, int, int, float **, float);
   void mulvec(int, int, int, float [], float [], float);
   void mulrow(int, int, int, int, float **, float **, float);
   void dupvec(int, int, int, float [], float []);
   void dupmat(int, int, int, int, float **, float **);
   float vecvec(int, int, int, float [], float []);
   float matvec(int, int, int, float **, float []);
   void elmvec(int, int, int, float [], float [], float);
   void sol(float **, int, int [], float []);
   void dec(float **, int, float [], int []);
   void mulcol(int, int, int, int, float **, float **, float);
   float tamvec(int, int, int, float **, float []);
   float mattam(int, int, int, int, float **, float **);
   int qrisngvaldec(float **,int,int,float [],float **,float []);
   int i,j,weight,ncol,nrow,away,max,nfe,nis,*cobs,
          first,sec,clean,nbpold,maxfe,fe,it,err,emergency;
   float eps1,res1,in3,in4,fac3,fac4,aux[4],*obs,*save,*tobs,
          **yp,*ymax,*y,**fy,**fp,w,**aid,temp,
          vv,ww,w2,mu,res2,fpar,fparpres,lambda,lambdamin,p,pw,
          reltolres,abstolres,em[8],*val,*b,*bb,*parpres,**jaco;
   static float save1[35]={1.0, 1.0, 9.0, 4.0, 0.0, 2.0/3.0, 1.0,
          1.0/3.0, 36.0, 20.25, 1.0, 6.0/11.0, 1.0, 6.0/11.0,
          1.0/11.0, 84.028, 53.778, 0.25, 0.48, 1.0, 0.7, 0.2,
          0.02, 156.25, 108.51, 0.027778, 120.0/274.0, 1.0,
          225.0/274.0, 85.0/274.0, 15.0/274.0, 1.0/274.0, 0.0,
          187.69, 0.0047361};

   nbpold=(*nbp);
   cobs=allocate_integer_vector(1,nobs);
   obs=allocate_real_vector(1,nobs);
   save=allocate_real_vector(-38,6*n);
   tobs=allocate_real_vector(0,nobs);
   ymax=allocate_real_vector(1,n);
   y=allocate_real_vector(1,6*n*(nbpold+m+1));
   yp=allocate_real_matrix(1,nbpold+nobs,1,nbpold+m);
   fy=allocate_real_matrix(1,n,1,n);
   fp=allocate_real_matrix(1,n,1,m+nbpold);
   aid=allocate_real_matrix(1,m+nbpold,1,m+nbpold);

   for (i=0; i<=34; i++) save[-38+i]=save1[i];
   (*data)(nobs,tobs,obs,cobs);
   weight=1;
   first=sec=0;
   clean=(*nbp > 0);
   aux[2]=FLT_EPSILON;
   eps1=1.0e10;
   out[1]=0.0;
   bp[0]=max=0;
   /* smooth integration without break-points */
   if (!peidefunct(nobs,m,par,res,
          n,m,nobs,nbp,first,&sec,&max,&nis,eps1,weight,bp,
          save,ymax,y,yp,fy,fp,cobs,tobs,obs,in,aux,clean,deriv,
          jacdfdy,jacdfdp,callystart,monitor)) goto Escape;
   res1=sqrt(vecvec(1,nobs,0,res,res));
   nfe=1;
   if (in[5] == 1.0) {
      out[1]=1.0;
      goto Escape;
   }
   if (clean) {
      first=1;
      clean=0;
      fac3=sqrt(sqrt(in[3]/res1));
      fac4=sqrt(sqrt(in[4]/res1));
```

```
        eps1=res1*fac4;
        if (!peidefunct(nobs,m,par,res,
                n,m,nobs,nbp,first,&sec,&max,&nis,eps1,weight,bp,
                save,ymax,y,yp,fy,fp,cobs,tobs,obs,in,aux,clean,deriv,
                jacdfdy,jacdfdp,callystart,monitor)) goto Escape;
        first=0;
    } else
        nfe=0;
    ncol=m+(*nbp);
    nrow=nobs+(*nbp);
    sec=1;
    in3=in[3];
    in4=in[4];
    in[3]=res1;
    weight=away=0;
    out[4]=out[5]=w=0.0;
    temp=sqrt(weight)+1.0;
    weight=temp*temp;
    while (weight != 16 && *nbp > 0) {
        if (away == 0 && w != 0.0) {
            /* if no break-points were omitted then one function
                function evaluation is saved */
            w=weight/w;
            for (i=nobs+1; i<=nrow; i++) {
                for (j=1; j<=ncol; j++) yp[i][j] *= w;
                res[i] *= w;
            }
            sec=1;
            nfe--;
        }
        in[3] *= fac3*weight;
        in[4]=eps1;
        (*monitor)(2,ncol,nrow,par,res,weight,nis);
        /* marquardt's method */
        val=allocate_real_vector(1,ncol);
        b=allocate_real_vector(1,ncol);
        bb=allocate_real_vector(1,ncol);
        parpres=allocate_real_vector(1,ncol);
        jaco=allocate_real_matrix(1,nrow,1,ncol);
        vv=10.0;
        w2=0.5;
        mu=0.01;
        ww = (in[6] < 1.0e-7) ? 1.0e-8 : 1.0e-1*in[6];
        em[0]=em[2]=em[6]=in[0];
        em[4]=10*ncol;
        reltolres=in[3];
        abstolres=in[4]*in[4];
        maxfe=in[5];
        err=0;
        fe=it=1;
        p=fpar=res2=0.0;
        pw = -log(ww*in[0])/2.30;
        if (!peidefunct(nrow,ncol,par,res,
                    n,m,nobs,nbp,first,&sec,&max,&nis,eps1,
                    weight,bp,save,ymax,y,yp,fy,fp,cobs,tobs,obs,
                    in,aux,clean,deriv,jacdfdy,jacdfdp,
                    callystart,monitor))
            err=3;
        else {
            fpar=vecvec(1,nrow,0,res,res);
            out[3]=sqrt(fpar);
            emergency=0;
            it=1;
            do {
                dupmat(1,nrow,1,ncol,jaco,yp);
                i=qrisngvaldec(jaco,nrow,ncol,val,aid,em);
                if (it == 1)
                    lambda=in[6]*vecvec(1,ncol,0,val,val);
                else
                    if (p == 0.0) lambda *= w2;
                for (i=1; i<=ncol; i++)
                    b[i]=val[i]*tamvec(1,nrow,i,jaco,res);
                while (1) {
                    for (i=1; i<=ncol; i++)
```

```
                    bb[i]=b[i]/(val[i]*val[i]+lambda);
                for (i=1; i<=ncol; i++)
                    parpres[i]=par[i]-matvec(1,ncol,i,aid,bb);
                fe++;
                if (fe >= maxfe)
                    err=1;
                else
                    if (!peidefunct(nrow,ncol,parpres,res,
                            n,m,nobs,nbp,first,&sec,&max,&nis,
                            eps1,weight,bp,save,ymax,y,yp,fy,fp,
                            cobs,tobs,obs,in,aux,clean,deriv,
                            jacdfdy,jacdfdp,callystart,monitor))
                        err=2;
                if (err != 0) {
                    emergency=1;
                    break;
                }
                fparpres=vecvec(1,nrow,0,res,res);
                res2=fpar-fparpres;
                if (res2 < mu*vecvec(1,ncol,0,b,bb)) {
                    p += 1.0;
                    lambda *= vv;
                    if (p == 1.0) {
                        lambdamin=ww*vecvec(1,ncol,0,val,val);
                        if (lambda < lambdamin) lambda=lambdamin;
                    }
                    if (p >= pw) {
                        err=4;
                        emergency=1;
                        break;
                    }
                } else {
                    dupvec(1,ncol,0,par,parpres);
                    fpar=fparpres;
                    break;
                }
            }
            if (emergency) break;
            it++;
        } while (fpar>abstolres &&
                    res2>reltolres*fpar+abstolres);
        for (i=1; i<=ncol; i++)
            mulcol(1,ncol,i,i,jaco,aid,1.0/(val[i]+in[0]));
        for (i=1; i<=ncol; i++)
            for (j=1; j<=i; j++)
                aid[i][j]=aid[j][i]=mattam(1,ncol,i,j,jaco,jaco);
        lambda=lambdamin=val[1];
        for (i=2; i<=ncol; i++)
            if (val[i] > lambda)
                lambda=val[i];
            else
                if (val[i] < lambdamin) lambdamin=val[i];
        temp=lambda/(lambdamin+in[0]);
        out[7]=temp*temp;
        out[2]=sqrt(fpar);
        out[6]=sqrt(res2+fpar)-out[2];
    }
    out[4]=fe;
    out[5]=it-1;
    out[1]=err;
    free_real_vector(val,1);
    free_real_vector(b,1);
    free_real_vector(bb,1);
    free_real_vector(parpres,1);
    free_real_matrix(jaco,1,nrow,1);
    if (out[1] > 0.0) goto Escape;
    /* the relative starting value of lambda is adjusted
       to the last value of lambda used */
    away=out[4]-out[5]-1.0;
    in[6] *= pow(5.0,away)*pow(2.0,away-out[5]);
    nfe += out[4];
    w=weight;
    temp=sqrt(weight)+1.0;
    eps1=temp*temp*in[4]*fac4;
```

```
        away=0;
        /* omit useless break-points */
        for (j=1; j<=(*nbp); j++)
            if (fabs(obs[bp[j]]+res[bp[j]]-par[j+m]) < eps1) {
                (*nbp)--;
                for (i=j; i<=(*nbp); i++) bp[i]=bp[i+1];
                dupvec(j+m,(*nbp)+m,1,par,par);
                j--;
                away++;
                bp[*nbp+1]=0;
            }
        ncol -= away;
        nrow -= away;
        temp=sqrt(weight)+1.0;
        weight=temp*temp;
    }
    in[3]=in3;
    in[4]=in4;
    *nbp=0;
    weight=1;
    (*monitor)(2,m,nobs,par,res,weight,nis);
    /* marquardt's method */
    val=allocate_real_vector(1,m);
    b=allocate_real_vector(1,m);
    bb=allocate_real_vector(1,m);
    parpres=allocate_real_vector(1,m);
    jaco=allocate_real_matrix(1,nobs,1,m);
    vv=10.0;
    w2=0.5;
    mu=0.01;
    ww = (in[6] < 1.0e-7) ? 1.0e-8 : 1.0e-1*in[6];
    em[0]=em[2]=em[6]=in[0];
    em[4]=10*m;
    reltolres=in[3];
    abstolres=in[4]*in[4];
    maxfe=in[5];
    err=0;
    fe=it=1;
    p=fpar=res2=0.0;
    pw = -log(ww*in[0])/2.30;
    if (!peidefunct(nobs,m,par,res,
                n,m,nobs,nbp,first,&sec,&max,&nis,eps1,weight,bp,
                save,ymax,y,yp,fy,fp,cobs,tobs,obs,in,aux,clean,
                deriv,jacdfdy,jacdfdp,callystart,monitor))
        err=3;
    else {
        fpar=vecvec(1,nobs,0,res,res);
        out[3]=sqrt(fpar);
        emergency=0;
        it=1;
        do {
            dupmat(1,nobs,1,m,jaco,yp);
            i=qrisngvaldec(jaco,nobs,m,val,jtjinv,em);
            if (it == 1)
                lambda=in[6]*vecvec(1,m,0,val,val);
            else
                if (p == 0.0) lambda *= w2;
            for (i=1; i<=m; i++)
                b[i]=val[i]*tamvec(1,nobs,i,jaco,res);
            while (1) {
                for (i=1; i<=m; i++)
                    bb[i]=b[i]/(val[i]*val[i]+lambda);
                for (i=1; i<=m; i++)
                    parpres[i]=par[i]-matvec(1,m,i,jtjinv,bb);
                fe++;
                if (fe >= maxfe)
                    err=1;
                else
                    if (!peidefunct(nobs,m,parpres,res,
                            n,m,nobs,nbp,first,&sec,&max,&nis,eps1,
                            weight,bp,save,ymax,y,yp,fy,fp,cobs,tobs,
                            obs,in,aux,clean,deriv,jacdfdy,jacdfdp,
                            callystart,monitor))
                        err=2;
```

```
                if (err != 0) {
                    emergency=1;
                    break;
                }
                fparpres=vecvec(1,nobs,0,res,res);
                res2=fpar-fparpres;
                if (res2 < mu*vecvec(1,m,0,b,bb)) {
                    p += 1.0;
                    lambda *= vv;
                    if (p == 1.0) {
                        lambdamin=ww*vecvec(1,m,0,val,val);
                        if (lambda < lambdamin) lambda=lambdamin;
                    }
                    if (p >= pw) {
                        err=4;
                        emergency=1;
                        break;
                    }
                } else {
                    dupvec(1,m,0,par,parpres);
                    fpar=fparpres;
                    break;
                }
            }
            if (emergency) break;
            it++;
        } while (fpar>abstolres && res2>reltolres*fpar+abstolres);
        for (i=1; i<=m; i++)
            mulcol(1,m,i,i,jaco,jtjinv,1.0/(val[i]+in[0]));
        for (i=1; i<=m; i++)
            for (j=1; j<=i; j++)
                jtjinv[i][j]=jtjinv[j][i]=mattam(1,m,i,j,jaco,jaco);
        lambda=lambdamin=val[1];
        for (i=2; i<=m; i++)
            if (val[i] > lambda)
                lambda=val[i];
            else
                if (val[i] < lambdamin) lambdamin=val[i];
        temp=lambda/(lambdamin+in[0]);
        out[7]=temp*temp;
        out[2]=sqrt(fpar);
        out[6]=sqrt(res2+fpar)-out[2];
    }
    out[4]=fe;
    out[5]=it-1;
    out[1]=err;
    free_real_vector(val,1);
    free_real_vector(b,1);
    free_real_vector(bb,1);
    free_real_vector(parpres,1);
    free_real_matrix(jaco,1,nobs,1);
    nfe += out[4];

    Escape:
    if (out[1] == 3.0)
        out[1]=2.0;
    else
        if (out[1] == 4.0) out[1]=6.0;
    if (save[-3] != 0.0) out[1]=save[-3];
    out[3]=res1;
    out[4]=nfe;
    out[5]=max;
    free_integer_vector(cobs,1);
    free_real_vector(obs,1);
    free_real_vector(save,-38);
    free_real_vector(tobs,0);
    free_real_vector(ymax,1);
    free_real_vector(y,1);
    free_real_matrix(yp,1,nbpold+nobs,1);
    free_real_matrix(fy,1,n,1);
    free_real_matrix(fp,1,n,1);
    free_real_matrix(aid,1,m+nbpold,1);
}
```

```
int peidefunct(int nrow, int ncol, float par[], float res[],
        int n, int m, int nobs, int *nbp, int first, int *sec,
        int *max, int *nis, float eps1, int weight, int bp[],
        float save[], float ymax[], float y[], float **yp,
        float **fy, float **fp, int cobs[], float tobs[],
        float obs[], float in[], float aux[], int clean,
        int (*deriv)(int,int,float [],float [],float,float []),
        int (*jacdfdy)(int,int,float [],float [],float,float **),
        int (*jacdfdp)(int,int,float [],float [],float,float **),
        void (*callystart)(int,int,float [],float [],float[]),
        void (*monitor)(int,int,int,float [],float [],int,int))
{
    /* this function is internally used by PEIDE */

    void peidereset(int, int, float, float, float, float, float [],
            float [], float *, float *, float *, int *);
    void peideorder(int, int, float, float [], float [],
            float *, float *, float *, float *, float *, int *);
    void peidestep(int, int, int, float, float, float, float,
            float [], float [], float [], float [], int *, float *);
    float peideinterpol(int, int, int, float, float []);
    int l,k,knew,fails,same,kpold,n6,nnpar,j5n,cobsii,*p,evaluate,
            evaluated,decompose,conv,extra,npar,i,j,jj,ii;
    float xold,hold,a0,tolup,tol,toldwn,tolconv,h,ch,chnew,error,
            dfi,tobsdif,a[6],*delta,*lastdelta,*df,*y0,**jacob,xend,
            hmax,hmin,eps,s,aa,x,t,c;

    p=allocate_integer_vector(1,n);
    delta=allocate_real_vector(1,n);
    lastdelta=allocate_real_vector(1,n);
    df=allocate_real_vector(1,n);
    y0=allocate_real_vector(1,n);
    jacob=allocate_real_matrix(1,n,1,n);

    if (*sec) {
        *sec=0;
        goto Finish;
    }
    xend=tobs[nobs];
    eps=in[2];
    npar=m;
    extra=(*nis)=0;
    ii=1;
    jj = (*nbp == 0) ? 0 : 1;
    n6=n*6;
    inivec(-3,-1,save,0.0);
    inivec(n6+1,(6+m)*n,y,0.0);
    inimat(1,nobs+(*nbp),1,m+(*nbp),yp,0.0);
    t=tobs[1];
    x=tobs[0];
    (*callystart)(n,m,par,y,ymax);
    hmax=tobs[1]-tobs[0];
    hmin=hmax*in[1];
    /* evaluate jacobian */
    evaluate=0;
    decompose=evaluated=1;
    if (!(*jacdfdy)(n,m,par,y,x,fy)) {
        save[-3]=4.0;
        goto Finish;
    }
    nnpar=n*npar;

    Newstart:
    k=1;
    kpold=0;
    same=2;
    peideorder(n,k,eps,a,save,&tol,&tolup,&toldwn,&tolconv,
                &a0,&decompose);
    if (!(*deriv)(n,m,par,y,x,df)) {
        save[-3]=3.0;
        goto Finish;
    }
    s=FLT_MIN;
    for (i=1; i<=n; i++) {
```

```
        aa=matvec(1,n,i,fy,df)/ymax[i];
        s += aa*aa;
    }
    h=sqrt(2.0*eps/sqrt(s));
    if (h > hmax)
        h=hmax;
    else
        if (h < hmin) h=hmin;
    xold=x;
    hold=h;
    ch=1.0;
    for (i=1; i<=n; i++) {
        save[i]=y[i];
        save[n+i]=y[n+i]=df[i]*h;
    }
    fails=0;
    while (x < xend) {
        if (x+h <= xend)
            x += h;
        else {
            h=xend-x;
            x=xend;
            ch=h/hold;
            c=1.0;
            for (j=n; j<=k*n; j += n) {
                c *= ch;
                for (i=j+1; i<=j+n; i++) y[i] *= c;
            }
            same = (same < 3) ? 3 : same+1;
        }
        /* prediction */
        for (l=1; l<=n; l++) {
            for (i=l; i<=(k-1)*n+l; i += n)
                for (j=(k-1)*n+l; j>=i; j -= n) y[j] += y[j+n];
            delta[l]=0.0;
        }
        evaluated=0;
        /* correction and estimation local error */
        for (l=1; l<=3; l++) {
            if (!(*deriv)(n,m,par,y,x,df)) {
                save[-3]=3;
                goto Finish;
            }
            for (i=1; i<=n; i++) df[i]=df[i]*h-y[n+i];
            if (evaluate) {
                /* evaluate jacobian */
                evaluate=0;
                decompose=evaluated=1;
                if (!(*jacdfdy)(n,m,par,y,x,fy)) {
                    save[-3]=4.0;
                    goto Finish;
                }
            }
            if (decompose) {
                /* decompose jacobian */
                decompose=0;
                c = -a0*h;
                for (j=1; j<=n; j++) {
                    for (i=1; i<=n; i++) jacob[i][j]=fy[i][j]*c;
                    jacob[j][j] += 1.0;
                }
                dec(jacob,n,aux,p);
            }
            sol(jacob,n,p,df);
            conv=1;
            for (i=1; i<=n; i++) {
                dfi=df[i];
                y[i] += a0*dfi;
                y[n+i] += dfi;
                delta[i] += dfi;
                conv=(conv && (fabs(dfi) < tolconv*ymax[i]));
            }
            if (conv) {
                s=FLT_MIN;
```

```
                for (i=1; i<=n; i++) {
                    aa=delta[i]/ymax[i];
                    s += aa*aa;
                }
                error=s;
                break;
            }
        }
        /* acceptance or rejection */
        if (!conv) {
            if (!evaluated)
                evaluate=1;
            else {
                ch /= 4.0;
                if (h < 4.0*hmin) {
                    save[-1] += 10.0;
                    hmin /= 10.0;
                    if (save[-1] > 40.0) goto Finish;
                }
            }
            peidereset(n,k,hmin,hmax,hold,xold,y,save,&ch,&x,
                        &h,&decompose);
        } else if (error > tol) {
            fails++;
            if (h > 1.1*hmin) {
                if (fails > 2) {
                    peidereset(n,k,hmin,hmax,hold,xold,y,save,&ch,&x,
                                &h,&decompose);
                    goto Newstart;
                } else {
                    /* calculate step and order */
                    peidestep(n,k,fails,tolup,toldwn,tol,error,delta,
                                lastdelta,y,ymax,&knew,&chnew);
                    if (knew != k) {
                        k=knew;
                        peideorder(n,k,eps,a,save,&tol,&tolup,
                                    &toldwn,&tolconv,&a0,&decompose);
                    }
                    ch *= chnew;
                    peidereset(n,k,hmin,hmax,hold,xold,y,save,&ch,&x,
                                &h,&decompose);
                }
            } else {
                if (k == 1) {
                    /* violate eps criterion */
                    save[-2] += 1.0;
                    same=4;
                    goto Errortestok;
                }
                k=1;
                peidereset(n,k,hmin,hmax,hold,xold,y,save,&ch,&x,
                            &h,&decompose);
                peideorder(n,k,eps,a,save,&tol,&tolup,
                            &toldwn,&tolconv,&a0,&decompose);
                same=2;
            }
        } else {
            Errortestok:
            fails=0;
            for (i=1; i<=n; i++) {
                c=delta[i];
                for (l=2; l<=k; l++) y[l*n+i] += a[l]*c;
                if (fabs(y[i]) > ymax[i]) ymax[i]=fabs(y[i]);
            }
            same--;
            if (same == 1)
                dupvec(1,n,0,lastdelta,delta);
            else if (same == 0) {
                /* calculate step and order */
                peidestep(n,k,fails,tolup,toldwn,tol,error,delta,
                            lastdelta,y,ymax,&knew,&chnew);
                if (chnew > 1.1) {
                    if (k != knew) {
                        if (knew > k)
```

```
                mulvec(knew*n+1,knew*n+n,-knew*n,y,delta,
                        a[k]/knew);
            k=knew;
            peideorder(n,k,eps,a,save,&tol,&tolup,
                        &toldwn,&tolconv,&a0,&decompose);
        }
        same=k+1;
        if (chnew*h > hmax) chnew=hmax/h;
        h *= chnew;
        c=1.0;
        for (j=n; j<=k*n; j += n) {
            c *= chnew;
            mulvec(j+1,j+n,0,y,y,c);
        }
        decompose=1;
    } else
        same=10;
}
(*nis)++;
/* start of an integration step of yp */
if (clean) {
    hold=h;
    xold=x;
    kpold=k;
    ch=1.0;
    dupvec(1,k*n+n,0,save,y);
} else {
    if (h != hold) {
        ch=h/hold;
        c=1.0;
        for (j=n6+nnpar; j<=kpold*nnpar+n6; j += nnpar) {
            c *= ch;
            for (i=j+1; i<=j+nnpar; i++) y[i] *= c;
        }
        hold=h;
    }
    if (k > kpold)
        inivec(n6+k*nnpar+1,n6+k*nnpar+nnpar,y,0.0);
    xold=x;
    kpold=k;
    ch=1.0;
    dupvec(1,k*n+n,0,save,y);
    /* evaluate jacobian */
    evaluate=0;
    decompose=evaluated=1;
    if (!(*jacdfdy)(n,m,par,y,x,fy)) {
        save[-3]=4.0;
        goto Finish;
    }
    /* decompose jacobian */
    decompose=0;
    c = -a0*h;
    for (j=1; j<=n; j++) {
        for (i=1; i<=n; i++) jacob[i][j]=fy[i][j]*c;
        jacob[j][j] += 1.0;
    }
    dec(jacob,n,aux,p);
    if (!(*jacdfdp)(n,m,par,y,x,fp)) {
        save[-3]=5.0;
        goto Finish;
    }
    if (npar > m) inimat(1,n,m+1,npar,fp,0.0);
    /* prediction */
    for (l=0; l<=k-1; l++)
        for (j=k-1; j>=l; j--)
            elmvec(j*nnpar+n6+1,j*nnpar+n6+nnpar,nnpar,
                    y,y,1.0);
    /* correction */
    for (j=1; j<=npar; j++) {
        j5n=(j+5)*n;
        dupvec(1,n,j5n,y0,y);
        for (i=1; i<=n; i++)
            df[i]=h*(fp[i][j]+matvec(1,n,i,fy,y0))-
                    y[nnpar+j5n+i];
```

```
                    sol(jacob,n,p,df);
                    for (l=0; l<=k; l++) {
                        i=l*nnpar+j5n;
                        elmvec(i+1,i+n,-i,y,df,a[l]);
                    }
                }
            }
            while (x >= t) {
                /* calculate a row of the jacobian matrix and an
                    element of the residual vector */
                tobsdif=(tobs[ii]-x)/h;
                cobsii=cobs[ii];
                res[ii]=peideinterpol(cobsii,n,k,tobsdif,y)-obs[ii];
                if (!clean) {
                    for (i=1; i<=npar; i++)
                        yp[ii][i]=peideinterpol(cobsii+(i+5)*n,nnpar,k,
                                                        tobsdif,y);
                    /* introducing break-points */
                    if (bp[jj] != ii) {
                    } else if (first && fabs(res[ii]) < eps1) {
                        (*nbp)--;
                        for (i=jj; i<=(*nbp); i++) bp[i]=bp[i+1];
                        bp[*nbp+1]=0;
                    } else {
                        extra++;
                        if (first) par[m+jj]=obs[ii];
                        /* introducing a jacobian row and a residual
                            vector element for continuity requirements */
                        yp[nobs+jj][m+jj] = -weight;
                        mulrow(1,npar,nobs+jj,ii,yp,yp,weight);
                        res[nobs+jj]=weight*(res[ii]+obs[ii]-par[m+jj]);
                    }
                }
                if (ii == nobs)
                    goto Finish;
                else {
                    t=tobs[ii+1];
                    if (bp[jj] == ii && jj < *nbp) jj++;
                    hmax=t-tobs[ii];
                    hmin=hmax*in[1];
                    ii++;
                }
            }
            /* break-points introduce new initial values for y & yp */
            if (extra > 0) {
                for (i=1; i<=n; i++) {
                    y[i]=peideinterpol(i,n,k,tobsdif,y);
                    for (j=1; j<=npar; j++)
                        y[i+(j+5)*n]=peideinterpol(i+(j+5)*n,nnpar,
                                                        k,tobsdif,y);
                }
                for (l=1; l<=extra; l++) {
                    cobsii=cobs[bp[npar-m+l]];
                    y[cobsii]=par[npar+l];
                    for (i=1; i<=npar+extra; i++) y[cobsii+(5+i)*n]=0.0;
                    inivec(1+nnpar+(l+5)*n,nnpar+(l+6)*n,y,0.0);
                    y[cobsii+(5+npar+l)*n]=1.0;
                }
                npar += extra;
                extra=0;
                x=tobs[ii-1];
                /* evaluate jacobian */
                evaluate=0;
                decompose=evaluated=1;
                if (!(*jacdfdy)(n,m,par,y,x,fy)) {
                    save[-3]=4.0;
                    goto Finish;
                }
                nnpar=n*npar;
                goto Newstart;
            }
        }
    }
    Finish:
```

```
    if (save[-2] > *max) *max=save[-2];
    if (!first) (*monitor)(1,ncol,nrow,par,res,weight,*nis);
    free_integer_vector(p,1);
    free_real_vector(delta,1);
    free_real_vector(lastdelta,1);
    free_real_vector(df,1);
    free_real_vector(y0,1);
    free_real_matrix(jacob,1,n,1);
    return (save[-1] <= 40.0 && save[-3] == 0.0);
}

void peidereset(int n, int k, float hmin, float hmax, float hold,
        float xold, float y[], float save[], float *ch, float *x,
        float *h, int *decompose)
{
    /* this function is internally used by PEIDEFUNCT of PEIDE */

    int i,j;
    float c;

    if (*ch < hmin/hold)
        *ch = hmin/hold;
    else
        if (*ch > hmax/hold) *ch = hmax/hold;
    *x = xold;
    *h = hold*(*ch);
    c=1.0;
    for (j=0; j<=k*n; j += n) {
        for (i=1; i<=n; i++) y[j+i]=save[j+i]*c;
        c *= (*ch);
    }
    *decompose = 1;
}

void peideorder(int n, int k, float eps, float a[], float save[],
        float *tol, float *tolup, float *toldwn, float *tolconv,
        float *a0, int *decompose)
{
    /* this function is internally used by PEIDEFUNCT of PEIDE */

    int i,j;
    float c;

    c=eps*eps;
    j=((k-1)*(k+8))/2-38;
    for (i=0; i<=k; i++) a[i]=save[i+j];
    j += k+1;
    *tolup = c*save[j];
    *tol = c*save[j+1];
    *toldwn = c*save[j+2];
    *tolconv = eps/(2*n*(k+2));
    *a0 = a[0];
    *decompose = 1;
}

void peidestep(int n, int k, int fails, float tolup, float toldwn,
        float tol, float error, float delta[], float lastdelta[],
        float y[], float ymax[], int *knew, float *chnew)
{
    /* this function is internally used by PEIDEFUNCT of PEIDE */

    int i;
    float a1,a2,a3,aa,s;

    if (k <= 1)
        a1=0.0;
    else {
        s=FLT_MIN;
        for (i=1; i<=n; i++) {
            aa=y[k*n+i]/ymax[i];
            s += aa*aa;
        }
        a1=0.75*pow(toldwn/s,0.5/k);
    }
```

```
    a2=0.80*pow(tol/error,0.5/(k+1));
    if (k >= 5 || fails != 0)
        a3=0.0;
    else {
        s=FLT_MIN;
        for (i=1; i<=n; i++) {
            aa=(delta[i]-lastdelta[i])/ymax[i];
            s += aa*aa;
        }
        a3=0.70*pow(tolup/s,0.5/(k+2));
    }
    if (a1 > a2 && a1 > a3) {
        *knew = k-1;
        *chnew = a1;
    } else if (a2 > a3) {
        *knew = k;
        *chnew = a2;
    } else {
        *knew = k+1;
        *chnew = a3;
    }
}

float peideinterpol(int startindex, int jump, int k, float tobsdif,
                            float y[])
{
    /* this function is internally used by PEIDEFUNCT of PEIDE */

    int i;
    float s,r;

    s=y[startindex];
    r=tobsdif;
    for (i=1; i<=k; i++) {
        startindex += jump;
        s += y[startindex]*r;
        r *= tobsdif;
    }
    return s;
}
```

6. Special Functions

6.1 Elementary functions

6.1.1 Hyperbolic functions

A. arcsinh

Computes the inverse hyperbolic sine of the argument x.
If $|x| \leq 10^{10}$ then we use the procedure *logoneplusx* by writing

$$arcsinh(x) = ln(x + \sqrt{(x*x+1)}) = ln(1 + x + x^2/(1+\sqrt{(1+x^2)})).$$

If $|x| > 10^{10}$ we use the formula

$$arcsinh(x) = sign(x) * (ln(2) + ln(|x|)).$$

Function Parameters:

float arcsinh (x)

arcsinh: delivers the inverse hyperbolic sine of the argument x;

x: float;
entry: the real argument of *arcsinh(x)*.

Function used: logoneplusx.

```
#include <math.h>

float arcsinh(float x)
{
    float logoneplusx(float);
    float y;

    if (fabs(x) > 1.0e10)
        return ((x > 0.0) ? 0.69314718055995+log(fabs(x)) :
                            -0.69314718055995+log(fabs(x)));
    else {
        y=x*x;
        return ((x == 0.0) ? 0.0 :    ((x > 0.0) ?
                  logoneplusx(fabs(x)+y/(1.0+sqrt(1.0+y))) :
                  -logoneplusx(fabs(x)+y/(1.0+sqrt(1.0+y)))));
    }
}
```

B. arccosh

Computes the inverse hyperbolic cosine of the argument x.
If $x = 1$ then the value 0 is delivered.
If $1 < x \leq 10^{10}$ then we use the formula

$$arccosh(x) = ln(x + \sqrt{(x*x-1)}).$$

If $x > 10^{10}$ we use the formula

$$arccosh(x) = ln(2) + ln(x).$$

If x is close to 1, say $x = 1 + y$, then it is advised to use the procedure *logoneplusx* by writing

$$arccosh(x) = ln(1 + y + \sqrt{(y*(y+2))}\ ).$$

For example, if $x = exp(t)$, $t > 0$, t is small, then $y = exp(t) - 1$ is available in good relative accuracy, $y = 2 * exp(t/2) * sinh(t/2)$.

Function Parameters:

float arccosh (*x*)

arccosh: delivers the inverse hyperbolic cosine of the argument *x*;

x: float;
entry: the real argument of *arccosh(x)*, $x \geq 1$.

```
#include <math.h>

float arccosh(float x)
{
    return ((x <= 1.0) ? 0.0 : ((x > 1.0e10) ?
                         0.69314718055995+log(x) :
                         log(x+sqrt((x-1.0)*(x+1.0)))));
}
```

C. arctanh

Computes the inverse hyperbolic tangent of the argument *x*.
If $|x| < 1$ then we use the procedure *logoneplusx* by writing

$$arctanh(x) = 0.5 * ln((1+x)/(1-x)) = 0.5 * ln(1 + 2*x/(1-x)).$$

If $|x| = 1$ then the value is $sign(x) * FLT\_MAX$, where FLT_MAX is a large number.

Function Parameters:

float arctanh (*x*)

arctanh: delivers the inverse hyperbolic tangent of the argument *x*;

x: float;
entry: the real argument of *arctanh(x)*.

Function used: logoneplusx.

```
#include <float.h>
#include <math.h>

float arctanh(float x)
{
    float logoneplusx(float);
    float ax;

    if (fabs(x) >= 1.0)
        return ((x > 0.0) ? FLT_MAX : -FLT_MAX);
    else {
        ax=fabs(x);
        return ((x == 0.0) ? 0.0 :    ((x > 0.0) ?
                    0.5*logoneplusx(2.0*ax/(1.0-ax)) :
                    -0.5*logoneplusx(2.0*ax/(1.0-ax))));
    }
}
```

6.1.2 Logarithmic functions

logoneplusx

Computes the function $ln(1 + x)$ for $x > -1$. For values of x near zero, loss of relative accuracy in the computation of $ln(1+x)$ by use of the formulae $z=1+x$, $ln(1+x)=ln(z)$ occurs (since $1+x \approx 1$ for small x); the use of *logoneplusx* avoids this loss [HaCL68].

For $x < -0.2929$ or $x > 0.4142$, $ln(1+x)$ is evaluated by use of the standard function *ln* directly; otherwise, a polynomial expression of the form

$$\ln(1+x) = \sum_{i=0}^{6} c_i \left(\frac{x}{x+2}\right)^{2i+1}$$

is used, and for small x loss of relative accuracy does not take place.

Function Parameters:

float logoneplusx (x)

logoneplusx: delivers the value of $ln(1+x)$;

x: float;
entry: the real argument of $ln(1+x)$, $x > -1$.

```
#include <math.h>

float logoneplusx(float x)
{
    float y,z;

    if (x == 0.0)
        return 0.0;
    else if (x < -0.2928 || x > 0.4142)
        return log(1.0+x);
    else {
        z=x/(x+2.0);
        y=z*z;
        return z*(2.0+y*(0.666666666663366+y*(0.400000001206045+y*
                  (0.285714091590488+y*(0.22223823332791+y*
                  (0.1811136267967+y*0.16948212488))))));
    }
}
```

6.2 Exponential integral

6.2.1 Exponential integral

A. ei

Calculates the exponential integral

$$E_i(x) = \int_{-\infty}^{x} \frac{e^t}{t} dt$$

where the integral is to be interpreted as the Cauchy principal value. When $x > 0$, the related function

$$E_1(x) = \int_{x}^{\infty} \frac{e^{-t}}{t} dt$$

may be obtained from that of $E_i(x)$ by use of the relationship $E_1(x) = -E_i(-x)$. For $x=0$ the integral is undefined and the procedure will cause overflow.

The exponential integral is computed by means of the rational Chebyshev approximations given in [AbS65, CodT68, CodT69]. Only ratios of polynomials with equal degree l are considered.

Function Parameters:

float ei (*x*)

ei: delivers the value of the exponential integral;

x: float;
   entry: the argument of the integral.

Functions used: chepolsum, pol, jfrac.

```
#include <math.h>

float ei(float x)
{
    float chepolsum(int, float, float []);
    float pol(int, float, float []);
    float jfrac(int, float [], float []);
    float p[8],q[8];

    if (x > 24.0) {
        p[0]= 1.00000000000058;      q[1]  =  1.99999999924131;
        p[1]=x-3.00000016782085;     q[2]  = -2.99996432944446;
        p[2]=x-5.00140345515924;     q[3]  = -7.90404992298926;
        p[3]=x-7.49289167792884;     q[4]  = -4.31325836146628;
        p[4]=x-3.08336269051763e1;   q[5]  =  2.95999399486831e2;
        p[5]=x-1.39381360364405;     q[6]  = -6.74704580465832;
        p[6]=x+8.91263822573708;     q[7]  =  1.04745362652468e3;
        p[7]=x-5.31686623494482e1;
        return exp(x)*(1.0+jfrac(7,q,p)/x)/x;
    } else if (x > 12.0) {
        p[0]= 9.99994296074708e-1;  q[1]  =  1.00083867402639;
        p[1]=x-1.95022321289660;     q[2]  = -3.43942266899870;
        p[2]=x+1.75656315469614;     q[3]  =  2.89516727925135e1;
        p[3]=x+1.79601688769252e1;   q[4]  =  7.60761148007735e2;
        p[4]=x-3.23467330305403e1;   q[5]  =  2.57776384238440e1;
        p[5]=x-8.28561994140641;     q[6]  =  5.72837193837324e1;
        p[6]=x-1.86545454883399e1;   q[7]  =  6.95000655887434e1;
        p[7]=x-3.48334653602853;
        return exp(x)*jfrac(7,q,p)/x;
    } else if (x > 6.0) {
        p[0]= 1.00443109228078;      q[1]  =  5.27468851962908e-1;
        p[1]=x-4.32531132878135e1;   q[2]  =  2.73624119889328e3;
        p[2]=x+6.01217990830080e1;   q[3]  =  1.43256738121938e1;
        p[3]=x-3.31842531997221e1;   q[4]  =  1.00367439516726e3;
        p[4]=x+2.50762811293561e1;   q[5]  = -6.25041161671876;
        p[5]=x+9.30816385662165;     q[6]  =  3.00892648372915e2;
        p[6]=x-2.19010233854880e1;   q[7]  =  3.93707701852715;
        p[7]=x-2.18086381520724;
        return exp(x)*jfrac(7,q,p)/x;
    } else if (x > 0.0) {
        float t,r,x0,xmx0;
        p[0]= -1.95773036904548e8;   q[0]  = -8.26271498626055e7;
        p[1]=  3.89280421311201e6;   q[1]  =  8.91925767575612e7;
        p[2]= -2.21744627758845e7;   q[2]  = -2.49033375740540e7;
        p[3]= -1.19623669349247e5;   q[3]  =  4.28559624611749e6;
        p[4]= -2.49301393458648e5;   q[4]  = -4.83547436162164e5;
        p[5]= -4.21001615357070e3;   q[5]  =  3.57300298058508e4;
        p[6]= -5.49142265521085e2;   q[6]  = -1.60708926587221e3;
        p[7]= -8.66937339951070;     q[7]  =  3.41718750000000e1;
        x0=0.372507410781367;
        t=x/3.0-1.0;
        r=chepolsum(7,t,p)/chepolsum(7,t,q);
        xmx0=(x-409576229586.0/1099511627776.0)-
                0.767177250199394e-12;
        if (fabs(xmx0) > 0.037)
            t=log(x/x0);
```

```
        else {
            float z,z2;
            p[0] =  0.837207933976075e1;
            p[1] = -0.652268740837103e1;
            p[2] =  0.569955700306720;
            q[0] =  0.418603966988037e1;
            q[1] = -0.465669026080814e1;
            q[2] =  0.1e1;
            z=xmx0/(x+x0);
            z2=z*z;
            t=z*pol(2,z2,p)/pol(2,z2,q);
        }
        return t+xmx0*r;
    } else if (x > -1.0) {
        float y;
        p[0] = -4.41785471728217e4;  q[0]=7.65373323337614e4;
        p[1] =  5.77217247139444e4;  q[1]=3.25971881290275e4;
        p[2] =  9.93831388962037e3;  q[2]=6.10610794245759e3;
        p[3] =  1.84211088668000e3;  q[3]=6.35419418378382e2;
        p[4] =  1.01093806161906e2;  q[4]=3.72298352833327e1;
        p[5] =  5.03416184097568;    q[5]=1.0;
        y = -x;
        return log(y)-pol(5,y,p)/pol(5,y,q);
    } else if (x > -4.0) {
        float y;
        p[0]=8.67745954838444e-8;  q[0]=1.0;
        p[1]=9.99995519301390e-1;  q[1]=1.28481935379157e1;
        p[2]=1.18483105554946e1;   q[2]=5.64433569561803e1;
        p[3]=4.55930644253390e1;   q[3]=1.06645183769914e2;
        p[4]=6.99279451291003e1;   q[4]=8.97311097125290e1;
        p[5]=4.25202034768841e1;   q[5]=3.14971849170441e1;
        p[6]=8.83671808803844;     q[6]=3.79559003762122;
        p[7]=4.01377664940665e-1;  q[7]=9.08804569188869e-2;
        y = -1.0/x;
        return -exp(x)*pol(7,y,p)/pol(5,y,q);
    } else {
        float y;
        p[0] = -9.99999999998447e-1;  q[0]=1.0;
        p[1] = -2.66271060431811e1;   q[1]=2.86271060422192e1;
        p[2] = -2.41055827097015e2;   q[2]=2.92310039388533e2;
        p[3] = -8.95927957772937e2;   q[3]=1.33278537748257e3;
        p[4] = -1.29885688756484e3;   q[4]=2.77761949509163e3;
        p[5] = -5.45374158883133e2;   q[5]=2.40401713225909e3;
        p[6] = -5.66575206533869;     q[6]=6.31657483280800e2;
        y = -1.0/x;
        return -exp(x)*y*(1.0+y*pol(6,y,p)/pol(5,y,q));
    }
}
```

B. eialpha

Calculates a sequence of integrals [AbS65] of the form

$$\alpha_i(x) = \int_1^\infty e^{-xt} t^i \, dt \qquad (x > 0; \quad i = 0,...,n)$$

by use of the recursion

$$\alpha_0(x) = x^{-1}, \quad \alpha_i(x) = \alpha_0(x) + (i/x)\alpha_{i-1}(x) \qquad (i=1,...,n).$$

Function Parameters:

void eialpha (*x,n,alpha*)

x: float;
    entry: the real *x* in the integrand;

n: int;

entry: the integer n in the integrand;
alpha: float *alpha[0:n]*;
exit: the value of the integral is stored in *alpha[i]*, i=0,...,n.

```
#include <math.h>

void eialpha(float x, int n, float alpha[])
{
    int k;
    float a,b,c;

    c=1.0/x;
    a=exp(-x);
    b=alpha[0]=a*c;
    for (k=1; k<=n; k++) alpha[k]=b=(a+k*b)*c;
}
```

C. enx

Calculates a sequence of integrals of the form

$$\alpha_n(x) = \int_1^\infty \frac{e^{-xt}}{t^n}\,dt \qquad (x > 0;\quad n = n1,...,n2)\ . \tag{1}$$

The value of $\alpha_{n0}(x)$ where $n0 = \lfloor x \rfloor$ is first computed using (a) a call of *ei* if $n0 = 1$, (b) a Taylor series expansion in powers of $x - n0$ if $n0 \le 10$, and (c) (calling *nonexpenx* for this purpose) the relationship $\alpha_n(x) = e^{-x}\alpha'_n(x)$, where

$$\alpha'_n(x) = \frac{1}{x+}\,\frac{n}{1+}\,\frac{1}{x+}\,\frac{n+1}{1+}\cdots \tag{2}$$

with $n=n0$ if $n0 > 10$. Thereafter, the recursion

$$\alpha_{n+1}(x) = n^{-1}\{e^{-x} - x\alpha_n(x)\}$$

is used in a forward direction if $n0 < n2$, and in a backward direction if $n0 > n1$ to compute the required values of the functions (1). The successive convergence $C_r(x)$ of expansion (2) are computed until $|1 - C_r(x)/C_{r+1}(x)| \le \delta$, where δ is the value of the machine precision.

Function Parameters:

void enx (*x,n1,n2,a*)

x: float;
entry: the real positive x in the integrand;
n1,n2: int;
entry: lower and upper bound, respectively, of the integer n in the integrand;
a: float *a[n1:n2]*;
exit: the value of the integral is stored in *a[i]*, i=$n1$,...,$n2$.

Functions used: ei, nonexpenx.

```
#include <math.h>

void enx(float x, int n1, int n2, float a[])
{
    if (x <= 1.5) {
        float ei(float);
      int i;
```

```
            float w,e;
            w = -ei(-x);
            if (n1 == 1) a[1]=w;
            if (n2 > 1) e=exp(-x);
            for (i=2; i<=n2; i++) {
                w=(e-x*w)/(i-1);
                if (i >= n1) a[i]=w;
            }
        } else {
            int i,n;
            float w,e,an;
            n=ceil(x);
            if (n <= 10) {
                float f,w1,t,h,p[20];
                p[2] =0.37534261820491e-1;  p[11]=0.135335283236613;
                p[3] =0.89306465560228e-2;  p[12]=0.497870683678639e-1;
                p[4] =0.24233983686581e-2;  p[13]=0.183156388887342e-1;
                p[5] =0.70576069342458e-3;  p[14]=0.673794699908547e-2;
                p[6] =0.21480277819013e-3;  p[15]=0.247875217666636e-2;
                p[7] =0.67375807781018e-4;  p[16]=0.911881965554516e-3;
                p[8] =0.21600730159975e-4;  p[17]=0.335462627902512e-3;
                p[9] =0.70411579854292e-5;  p[18]=0.123409804086680e-3;
                p[10]=0.23253026570282e-5;  p[19]=0.453999297624848e-4;
                f=w=p[n];
                e=p[n+9];
                w1=t=1.0;
                h=x-n;
                i=n-1;
                do {
                    f=(e-i*f)/n;
                    t = -h*t/(n-i);
                    w1=t*f;
                    w += w1;
                    i--;
                } while (fabs(w1) > 1.0e-15*w);
            } else {
                float *allocate_real_vector(int, int);
                void free_real_vector(float *, int);
                void nonexpenx(float, int, int, float []);
                float *b;
                b=allocate_real_vector(n,n);
                nonexpenx(x,n,n,b);
                w=b[n]*exp(-x);
                free_real_vector(b,n);
            }
            if (n1 == n2 && n1 == n)
                a[n]=w;
            else {
                e=exp(-x);
                an=w;
                if (n <= n2 && n >= n1) a[n]=w;
                for (i=n-1; i>=n1; i--) {
                    w=(e-i*w)/x;
                    if (i <= n2) a[i]=w;
                }
                w=an;
                for (i=n+1; i<=n2; i++) {
                    w=(e-x*w)/(i-1);
                    if (i >= n1) a[i]=w;
                }
            }
        }
}
```

D. nonexpenx

Calculates a sequence of integrals of the form

$$\alpha_n(x) = e^x \int_1^\infty \frac{e^{-xt}}{t^n}\, dt \qquad (x > 0;\quad n = n1,...,n2)\ . \tag{1}$$

The value of $\alpha_{n0}(x)$ where $n0 = \lfloor x \rfloor$ is first computed using the methods (a) and (b) described in the documentation to *enx* if $n0 \le 10$ (calling *enx* for this purpose) and the continued fraction expansion (2) of that documentation if $n0 > 10$. Thereafter, the recursion

$$\alpha_{n+1}(x) = n^{-1}\{1 - x\alpha_n(x)\}$$

is used as described in the documentation to *enx* to compute the required values of the functions (1). See [AbS65, CodT68, CodT69, G61, G73].

Function Parameters:

void nonexpenx (*x,n1,n2,a*)

x: float;
  entry: the real positive *x* in the integrand;

n1,n2: int;
  entry: lower and upper bound, respectively, of the integer *n* in the integrand;

a: float *a[n1:n2]*;
  exit: the value of the integral is stored in *a[i]*, *i=n1,...,n2*.

Function used: enx.

```
#include <math.h>

void nonexpenx(float x, int n1, int n2, float a[])
{
	int i,n;
	float w,an;

	n = (x <= 1.5) ? 1 : ceil(x);
	if (n <= 10) {
		float *allocate_real_vector(int, int);
		void free_real_vector(float *, int);
		void enx(float, int, int, float []);
		float *b;
		b=allocate_real_vector(n,n);
		enx(x,n,n,b);
		w=b[n]*exp(x);
		free_real_vector(b,n);
	} else {
		int k,k1;
		float ue,ve,we,we1,uo,vo,wo,wo1,r,s;
		ue=1.0;
		ve=we=1.0/(x+n);
		we1=0.0;
		uo=1.0;
		vo = -n/(x*(x+n+1.0));
		wo1=1.0/x;
		wo=vo+wo1;
		w=(we+wo)/2.0;
		k1=1;
		k=k1;
		while (wo-we > 1.0e-15*w && we > we1 && wo < wo1) {
			we1=we;
			wo1=wo;
			r=n+k;
			s=r+x+k;
			ue=1.0/(1.0-k*(r-1.0)*ue/((s-2.0)*s));
			uo=1.0/(1.0-k*r*uo/(s*s-1.0));
			ve *= (ue-1.0);
			vo *= (uo-1.0);
			we += ve;
			wo += vo;
			w=(we+wo)/2.0;
			k1++;
```

```
                k=k1;
            }
        }
        an=w;
        if (n <= n2 && n >= n1) a[n]=w;
        for (i=n-1; i>=n1; i--) {
            w=(1.0-i*w)/x;
            if (i <= n2) a[i]=w;
        }
        w=an;
        for (i=n+1; i<=n2; i++) {
            w=(1.0-x*w)/(i-1);
            if (i >= n1) a[i]=w;
        }
}
```

6.2.2 Sine and cosine integral

A. sincosint

Calculates the sine integral *si(x)* and cosine integral *ci(x)* defined by

$$si(x) = \int_0^x \frac{\sin(t)}{t} dt , \qquad ci(x) = \gamma + \ln|x| + \int_0^x \frac{\cos(t)-1}{t} dt .$$

If $|x| \leq 4$ then *si(x)* is approximated by means of a Chebyshev series of the form

$$x \sum_{j=0}^{10} a_j \; T_{2j}(x)$$

and *ci(x)* by a similar series of the form

$$\gamma + \ln|x| - x^2 \sum_{j=0}^{10} a_j \; T_{2j}(x) .$$

For values of x outside this range, the functions *f(x)* and *g(x)* occurring in the formulae

$$si(x) = \tfrac{1}{2}\pi \text{sign}(x) - f(x)\cos(x) - g(x)\sin(x)$$
$$ci(x) = f(x)\sin(x) - g(x)\cos(x)$$

are evaluated by means of a call of *sincosfg*, and the values of *si(x)* and *ci(x)* are obtained by use of these relationships [AbS65, Bul67].

When using the procedure *sincosint* for large values of x, the relative accuracy mainly depends on the accuracy of the functions *sin(x)* and *cos(x)*.

Function Parameters:

void sincosint (*x,si,ci*)

x: float;
entry: the real argument of *si(x)* and *ci(x)*;
si: float *;
exit: the value of *si(x)*;
ci: float *;
exit: the value of *ci(x)*.

Functions used: sincosfg, chepolsum.

```
#include <math.h>

void sincosint(float x, float *si, float *ci)
{
    void sincosfg(float, float *, float *);
    float chepolsum(int, float, float []);
    float absx,z,f,g;

    absx=fabs(x);
    if (absx <= 4.0) {
        float a[11],z2;
        a[0] =2.7368706803630e0;    a[1] = -1.1106314107894e0;
        a[2] =1.4176562194666e-1;   a[3] = -1.0252652579174e-2;
        a[4] =4.6494615619880e-4;   a[5] = -1.4361730896642e-5;
        a[6] =3.2093684948229e-7;   a[7] = -5.4251990770162e-9;
        a[8] =7.1776288639895e-11;  a[9] = -7.6335493723482e-13;
        a[10]=6.6679958346983e-15;
        z=x/4.0;
        z2=z*z;
        g=z2+z2-1.0;
        *si = z*chepolsum(10,g,a);
        a[0] =2.9659610400727e0;    a[1] = -9.4297198341830e-1;
        a[2] =8.6110342738169e-2;   a[3] = -4.7776084547139e-3;
        a[4] =1.7529161205146e-4;   a[5] = -4.5448727803752e-6;
        a[6] =8.7515839180060e-8;   a[7] = -1.2998699938109e-9;
        a[8] =1.5338974898831e-11;  a[9] = -1.4724256070277e-13;
        a[10]=1.1721420798429e-15;
        *ci = 0.577215664901533+log(absx)-z2*chepolsum(10,g,a);
    } else {
        float cx,sx;
        sincosfg(x,&f,&g);
        cx=cos(x);
        sx=sin(x);
        *si = 1.570796326794897;
        if (x < 0.0) *si = -(*si);
        *si -= f*cx+g*sx;
        *ci = f*sx-g*cx;
    }
}
```

B. sincosfg

Evaluates the functions *f(x)*, *g(x)* related to the sine and cosine integrals [AbS65, Bul67] by means of the formulae

$$f(x) = ci(x)\sin(x) - \{si(x) - \pi/2\}\cos(x)$$
$$g(x) = -ci(x)\cos(x) - \{si(x) - \pi/2\}\sin(x).$$

When $|x| \le 4$ then the functions *si(x)* and *ci(x)* are evaluated by means of a call of *sincosint*, and the values of *f(x)* and *g(x)* are obtained by use of the above relationships. When $|x| > 4$ then *f(x)* is approximated by use of a Chebyshev series of the form

$$\frac{1}{x}\sum_{k=0}^{21} a_i\, T_i(\frac{8}{|x|} - 1)$$

and *g(x)* by a use of a similar series of the form

$$\frac{1}{x^2}\sum_{k=0}^{23} a_i\, T_i(\frac{8}{|x|} - 1)\ .$$

Function Parameters:

void sincosfg (x,f,g)

x: float;
entry: the real argument of $f(x)$ and $g(x)$;
f: float *;
exit: the value of $f(x)$;
g: float *;
exit: the value of $g(x)$.

Functions used: sincosint, chepolsum.

```
#include <math.h>

void sincosfg(float x, float *f, float *g)
{
    void sincosint(float, float *, float *);
    float chepolsum(int, float, float []);
    float absx,si,ci;

    absx=fabs(x);
    if (absx <= 4.0) {
        float cx,sx;
        sincosint(x,&si,&ci);
        cx=cos(x);
        sx=sin(x);
        si -= 1.570796326794897;
        *f = ci*sx-si*cx;
        *g = -ci*cx-si*sx;
    } else {
        float a[24];
        a[0] =  9.6578828035185e-1;  a[1]  = -4.3060837778597e-2;
        a[2] = -7.3143711748104e-3;  a[3]  =  1.4705235789868e-3;
        a[4] = -9.8657685732702e-5;  a[5]  = -2.2743202204655e-5;
        a[6] =  9.8240257322526e-6;  a[7]  = -1.8973430148713e-6;
        a[8] =  1.0063435941558e-7;  a[9]  =  8.0819364822241e-8;
        a[10]= -3.8976282875288e-8;  a[11] =  1.0335650325497e-8;
        a[12]= -1.4104344875897e-9;  a[13] = -2.5232078399683e-10;
        a[14]=  2.5699831325961e-10; a[15] = -1.0597889253948e-10;
        a[16]=  2.8970031570214e-11; a[17] = -4.1023142563083e-12;
        a[18]= -1.0437693730018e-12; a[19] =  1.0994184520547e-12;
        a[20]= -5.2214239401679e-13; a[21] =  1.7469920787829e-13;
        a[22]= -3.8470012979279e-14;
        *f = chepolsum(22,8.0/absx-1.0,a)/x;
        a[0] =  2.2801220638241e-1;  a[1]  = -2.6869727411097e-2;
        a[2] = -3.5107157280958e-3;  a[3]  =  1.2398008635186e-3;
        a[4] = -1.5672945116862e-4;  a[5]  = -1.0664141798094e-5;
        a[6] =  1.1170629343574e-5;  a[7]  = -3.1754011655614e-6;
        a[8] =  4.4317473520398e-7;  a[9]  =  5.5108696874463e-8;
        a[10]= -5.9243078711743e-8;  a[11] =  2.2102573381555e-8;
        a[12]= -5.0256827540623e-9;  a[13] =  3.1519168259424e-10;
        a[14]=  3.6306990848979e-10; a[15] = -2.2974764234591e-10;
        a[16]=  8.5530309424048e-11; a[17] = -2.1183067724443e-11;
        a[18]=  1.7133662645092e-12; a[19] =  1.7238877517248e-12;
        a[20]= -1.2930281366811e-12; a[21] =  5.7472339223731e-13;
        a[22]= -1.8415468268314e-13; a[23] =  3.5937256571434e-14;
        *g = 4.0*chepolsum(23,8.0/absx-1.0,a)/absx/absx;
    }
}
```

6.3 Gamma function

A. recipgamma

Calculates the reciprocal of the gamma function for arguments in the range [0.5,1.5]; moreover, odd and even parts are delivered.

recipgamma computes the values of the functions

$$\Gamma'(x) = \{\Gamma(1 - x)\}^{-1}$$

where

$$\Gamma(x) = \int_0^\infty t^{x-1} e^{-t}\, dt\ ,$$

and

$$\Gamma'_0(x) = (2+x)^{-1}\{\Gamma'(x) - \Gamma'(-x)\}$$
$$\Gamma'_e(x) = ½\{\Gamma'(x) + \Gamma'(-x)\}$$

for $-0.5 \le x \le 0.5$.

The functions $\Gamma'_0(x)$, $\Gamma'_e(x)$ are evaluated by use of truncated Chebyshev series, and the function $\Gamma'(x)$ recovered from them by use of the formula

$$\Gamma'(x) = \Gamma'_e(x) + (1 + x/2)\Gamma'_0(x).$$

Function Parameters:

float recipgamma (*x,odd,even*)

recipgamma: delivers $1/\Gamma(1-x)$;

x: float;
  entry: this argument should satisfy $-0.5 \le x \le 0.5$;
    (actually the gamma function Γ is calculated for $1-x$, i.e. if one wants to calculate $1/\Gamma(1)$, one has to set x to 0);

odd: float *;
  exit: the odd part of $1/\Gamma(1-x)$ divided by $(2x)$;
    i.e. $(1/\Gamma(1-x) - 1/\Gamma(1+x)) / (2x)$;

even: float *;
  exit: the even part of $1/\Gamma(1-x)$ divided by 2;
    i.e. $(1/\Gamma(1-x) + 1/\Gamma(1+x)) / 2$.

```
float recipgamma(float x, float *odd, float *even)
{
    int i;
    float alfa,beta,x2,b[13];

    b[1] = -0.283876542276024;   b[2]  = -0.076852840844786;
    b[3] =  0.001706305071096;   b[4]  =  0.001271927136655;
    b[5] =  0.000076309597586;   b[6]  = -0.000004971736704;
    b[7] = -0.000000865920800;   b[8]  = -0.000000033126120;
    b[9] =  0.000000001745136;   b[10] =  0.000000000242310;
    b[11]=  0.000000000009161;   b[12] = -0.000000000000170;
    x2=x*x*8.0;
    alfa = -0.000000000000001;
    beta=0.0;
    for (i=12; i>=2; i -= 2) {
        beta = -(alfa*2.0+beta);
        alfa = -beta*x2-alfa+b[i];
    }
    *even=(beta/2.0+alfa)*x2-alfa+0.921870293650453;
    alfa = -0.000000000000034;
    beta=0.0;
    for (i=11; i>=1; i -= 2) {
        beta = -(alfa*2.0+beta);
        alfa = -beta*x2-alfa+b[i];
    }
    *odd=(alfa+beta)*2.0;
    return (*odd)*x+(*even);
```

```
}
```

B. gamma

Computes the value of the gamma function at x

$$\Gamma(x) = \int_0^\infty t^{x-1} e^{-t}\, dt \,.$$

We distinguish between the following cases for the argument x:

$x < 0.5$:
In this case the formula $\Gamma(x) * \Gamma(1-x) = \pi/\sin(\pi * x)$ is used. However the sine function is not calculated directly on the argument $\pi * x$ but on the argument $\pi*(x \bmod 0.5)$, in this way a big decrease of precision is avoided. The precision here depends strongly on the precision of the sine function.

$0.5 \le x \le 1.5$:
Here the procedure *recipgamma* is called; moreover, $\Gamma(1) = 1$.

$1.5 < x \le 22$:
The recursion formula $\Gamma(1+x) = x * \Gamma(x)$ is used. The precision depends on the number of recursions needed. The upper bound of 22 has been chosen because now it is assured that for all integer arguments for which the value of the gamma function is representable (and this is the case for all integer arguments in the range [1,22]), this value is obtained, i.e. $\Gamma(i) = 1 * 2 * ... * (i\text{-}1)$.

$x > 22$:
Now the procedures *loggamma* and *exp* are used. The precision strongly depends on the precision of the exponential function, and no bound for the error can be given.

Function Parameters:

float gamma (x)

gamma: delivers the value of the gamma function at x;

x: float;
entry: the argument; if one of the following three conditions is fulfilled then overflow will occur:
1. the argument is too large;
2. the argument is a non-positive integer;
3. the argument is too close to a large (in absolute value) non-positive integer.

Functions used: recipgamma, loggamma.

```
#include <math.h>

float gamma(float x)
{
    float recipgamma(float, float *, float *);
    float loggamma(float);
    int inv;
    float y,s,f,g,odd,even;

    if (x < 0.5) {
        y=x-floor(x/2.0)*2;
        s = 3.14159265358979;
        if (y >= 1.0) {
            s = -s;
            y=2.0-y;
```

```
        }
        if (y >= 0.5) y=1.0-y;
        inv=1;
        x=1.0-x;
        f=s/sin(3.14159265358979*y);
    } else
        inv=0;
    if (x > 22.0)
        g=exp(loggamma(x));
    else {
        s=1.0;
        while (x > 1.5) {
            x=x-1.0;
            s *= x;
        }
        g=s/recipgamma(1.0-x,&odd,&even);
    }
    return (inv ? f/g : g);
}
```

C. loggamma

Computes the natural logarithm of the gamma function at x: $ln(\Gamma(x))$.
We distinguish between the following cases for the argument x (in most cases nothing is said about precision, as this highly depends on the precision of the natural logarithm):

$0 < x < 1$:
Here the recursion formula *loggamma(x) = loggamma(1+x) - ln(x)* is used.

$1 \leq x \leq 2$:
On the interval the truncated Chebyshev series for the function *loggamma(x) / ((x-1)*(x-2))* is used.

$2 < x \leq 13$:
The recursion formula *loggamma(x) = loggamma(1-x) + ln(x)* is used.

$13 < x \leq 22$:
As for $x < 1$ the formula *loggamma(x) = loggamma(1+x) - ln(x)* is used.

$x > 22$:
In this case *loggamma* is calculated by use of the asymptotic expansion for *loggamma(x) - (x-0.5) * ln(x)*.

Function Parameters:

float loggamma (x)

loggamma: delivers the value of the natural logarithm of the gamma function at x;
x: float;
entry: this argument must be positive.

```
#include <math.h>

float loggamma(float x)
{
    int i;
    float r,x2,y,f,u0,u1,u,z,b[19];

    if (x > 13.0) {
        r=1.0;
        while (x <= 22.0) {
            r /= x;
            x += 1.0;
        }
        x2 = -1.0/(x*x);
        r=log(r);
        return log(x)*(x-0.5)-x+r+0.918938533204672+
```

```
                (((0.595238095238095e-3*x2+0.793650793650794e-3)*x2+
                0.277777777777778e-2)*x2+0.833333333333333e-1)/x;
    } else {
        f=1.0;
        u0=u1=0.0;
        b[1]  = -0.0761141616704358;  b[2]  = 0.0084323249659328;
        b[3]  = -0.0010794937263286;  b[4]  = 0.0001490074800369;
        b[5]  = -0.0000215123998886;  b[6]  = 0.0000031979329861;
        b[7]  = -0.0000004851693012;  b[8]  = 0.0000000747148782;
        b[9]  = -0.0000000116382967;  b[10] = 0.0000000018294004;
        b[11] = -0.0000000002896918;  b[12] = 0.0000000000461570;
        b[13] = -0.0000000000073928;  b[14] = 0.0000000000011894;
        b[15] = -0.0000000000001921;  b[16] = 0.0000000000000311;
        b[17] = -0.0000000000000051;  b[18] = 0.0000000000000008;
        if (x < 1.0) {
            f=1.0/x;
            x += 1.0;
        } else
            while (x > 2.0) {
                x -= 1.0;
                f *= x;
            }
        f=log(f);
        y=x+x-3.0;
        z=y+y;
        for (i=18; i>=1; i--) {
            u=u0;
            u0=z*u0+b[i]-u1;
            u1=u;
        }
        return (u0*y+0.491415393029387-u1)*(x-1.0)*(x-2.0)+f;
    }
}
```

D. incomgam

Computes the incomplete gamma functions based on Padé approximations [AbS65, Lu70]. *incomgam* evaluates the functions

$$\gamma(a,x) = \int_0^x t^{a-1} e^{-t}\, dt\,, \qquad \Gamma(a,x) = \int_x^\infty t^{a-1} e^{-t}\, dt$$

to a prescribed relative accuracy ε.

If (a) $a,x < 3$ or (b) $x < a$ and $a \geq 3$, $\gamma(a,x)$ is computed by use of a continued fraction derived from a power series in ascending powers of x for this function, and $\Gamma(a,x)$ is determined from the relationship $\Gamma(a,x) = \Gamma(a) - \gamma(a,x)$. If neither of the above conditions holds, $\Gamma(a,x)$ is computed by use of a continued fraction derived from an asymptotic series in descending powers of x for this function, and the relationship $\gamma(a,x) = \Gamma(a) - \Gamma(a,x)$ is used.

The relative accuracy of the results depends not only on the quantity ε, but also on the accuracy of the functions *exp* and *gamma*. Especially for large values of x and a, the desired accuracy cannot be guaranteed.

Function Parameters:

void incomgam (*x,a,klgam,grgam,gam,eps*)

x: float;
entry: the independent argument x, $x \geq 0$;

a: float;
entry: the independent parameter a, $a > 0$;

klgam: float *;
    exit: the integral $\gamma(a,x)$ is delivered in *klgam*;
grgam: float *;
    exit: the integral $\Gamma(a,x)$ is delivered in *grgam*;
gam: float;
    entry: the value of $\Gamma(a)$;
        for this expression, the procedure *gamma* may be used;
eps: float;
    entry: the desired relative accuracy (value of ε above);
        the value of *eps* should not be smaller than the machine accuracy.

```
#include <float.h>
#include <math.h>

void incomgam(float x, float a, float *klgam, float *grgam,
                  float gam, float eps)
{
    int n;
    float c0,c1,c2,d0,d1,d2,x2,ax,p,q,r,s,r1,r2,scf;

    s=exp(-x+a*log(x));
    scf=FLT_MAX;
    if (x <= ((a < 3.0) ? 1.0 : a)) {
        x2=x*x;
        ax=a*x;
        d0=1.0;
        p=a;
        c0=s;
        d1=(a+1.0)*(a+2.0-x);
        c1=((a+1.0)*(a+2.0)+x)*s;
        r2=c1/d1;
        n=1;
        do {
            p += 2.0;
            q=(p+1.0)*(p*(p+2.0)-ax);
            r=n*(n+a)*(p+2.0)*x2;
            c2=(q*c1+r*c0)/p;
            d2=(q*d1+r*d0)/p;
            r1=r2;
            r2=c2/d2;
            c0=c1;
            c1=c2;
            d0=d1;
            d1=d2;
            if (fabs(c1) > scf || fabs(d1) > scf) {
                c0 /= scf;
                c1 /= scf;
                d0 /= scf;
                d1 /= scf;
            }
            n++;
        } while (fabs((r2-r1)/r2) > eps);
        *klgam = r2/a;
        *grgam = gam-(*klgam);
    } else {
        c0=a*s;
        c1=(1.0+x)*c0;
        q=x+2.0-a;
        d0=x;
        d1=x*q;
        r2=c1/d1;
        n=1;
        do {
            q += 2.0;
            r=n*(n+1-a);
            c2=q*c1-r*c0;
            d2=q*d1-r*d0;
            r1=r2;
```

```
                r2=c2/d2;
                c0=c1;
                c1=c2;
                d0=d1;
                d1=d2;
                if (fabs(c1) > scf || fabs(d1) > scf) {
                    c0 /= scf;
                    c1 /= scf;
                    d0 /= scf;
                    d1 /= scf;
                }
                n++;
            } while (fabs((r2-r1)/r2) > eps);
            *grgam = r2/a;
            *klgam = gam-(*grgam);
        }
}
```

E. incbeta

The incomplete beta function is defined as

$$B_x(p,q) = \int_0^x t^{p-1}(1-t)^{q-1}\,dt ,$$

$p > 0,\ q > 0,\ 0 \le x \le 1$, and the incomplete beta function ratio is $I_x(p,q) = B_x(p,q)\,/\,B_1(p,q)$. *incbeta* computes $I_x(p,q)$ for $0 \le x \le 1,\ p > 0,\ q > 0$ by use of the continued fraction corresponding to formula 26.5.8 in [AbS65], see also [G64, G67]. If $x > 0.5$ then the relation $I_x(p,q)=1-I_{1-x}(q,p)$ is used. The value of the continued fraction is approximated by the convergent $C_r(x)$, where r is the smallest integer for which $|1 - C_{r-1}(x)/C_r(x)| \le \varepsilon$, ε being a small positive real number supplied by the user.

It is advised to use in *incbeta* only small values of p and q, say $0<p\le 5,\ 0<q\le 5$. For other ranges of the parameters p and q, the procedures *ibpplusn* and *ibqplusn* can be used. *incbeta* satisfies *incbeta*=x if $x=0$ or $x=1$, whatever p and q. There is no control on the parameters x, p, q for their intended ranges.

Function Parameters:

float incbeta (*x,p,q,eps*)

incbeta: delivers the value of $I_x(p,q)$;

x: float;
entry: this argument should satisfy: $0 \le x \le 1$;

p: float;
entry: the parameter p, $p > 0$;

q: float;
entry: the parameter q, $q > 0$;

eps: float;
entry: the desired relative accuracy (value of ε above);
the value of *eps* should not be smaller than the machine accuracy.

Function used: gamma.

```
#include <math.h>

float incbeta(float x, float p, float q, float eps)
{
```

```
    float gamma(float);
    int m,n,neven,recur;
    float g,f,fn,fn1,fn2,gn,gn1,gn2,dn,pq;

    if (x == 0.0 || x == 1.0)
        return x;
    else {
        if (x > 0.5) {
            f=p;
            p=q;
            q=f;
            x=1.0-x;
            recur=1;
        } else
            recur=0;
        g=fn2=0.0;
        m=0;
        pq=p+q;
        f=fn1=gn1=gn2=1.0;
        neven=0;
        n=1;
        do {
            if (neven) {
                m++;
                dn=m*x*(q-m)/(p+n-1.0)/(p+n);
            } else
                dn = -x*(p+m)*(pq+m)/(p+n-1.0)/(p+n);
            g=f;
            fn=fn1+dn*fn2;
            gn=gn1+dn*gn2;
            neven=(!neven);
            f=fn/gn;
            fn2=fn1;
            fn1=fn;
            gn2=gn1;
            gn1=gn;
            n++;
        } while (fabs((f-g)/f) > eps);
        f=f*pow(x,p)*pow(1.0-x,q)*gamma(p+q)/gamma(p+1.0)/gamma(q);
        if (recur) f=1.0-f;
        return f;
    }
}
```

F. ibpplusn

The incomplete beta function is defined as

$$B_x(p,q) = \int_0^x t^{p-1}(1-t)^{q-1}\,dt ,$$

$p > 0$, $q > 0$, $0 \le x \le 1$, and the incomplete beta function ratio is $I_x(p,q) = B_x(p,q) / B_1(p,q)$. *ibpplusn* computes $I_x(p+n,q)$ for $n=0,1,...,nmax$, $0 \le x \le 1$, $p > 0$, $q > 0$ (see [G64, G67]). In [G64] the procedure *ibpplusn* is called "incomplete beta q fixed". There is no control on the parameters *x, p, q, nmax* for their intended ranges.

Function Parameters:

void ibpplusn (*x,p,q,nmax,eps,i*)

x: float;
entry: this argument should satisfy: $0 \le x \le 1$;

p: float;
entry: the parameter p, $p > 0$; it is advised to take $0 < p \le 1$;

q: float;

entry: the parameter q, $q > 0$;

nmax: int;

entry: *nmax* indicates the maximum number of function values $I_x(p+n,q)$ to be generated;

eps: float;

entry: the desired relative accuracy (value of ε above); the value of *eps* should not be smaller than the machine accuracy;

i: float *i[0:nmax]*;

exit: $i[n] = I_x(p+n,q)$ for n=0,1,...,*nmax*.

Functions used: ixqfix, ixpfix.

```
void ibpplusn(float x, float p, float q, int nmax, float eps,
                  float i[])
{
    void ixqfix(float, float, float, int, float, float []);
    void ixpfix(float, float, float, int, float, float []);
    int n;

    if (x == 0.0 || x == 1.0)
        for (n=0; n<=nmax; n++) i[n]=x;
    else {
        if (x <= 0.5)
            ixqfix(x,p,q,nmax,eps,i);
        else {
            ixpfix(1.0-x,q,p,nmax,eps,i);
            for (n=0; n<=nmax; n++) i[n]=1.0-i[n];
        }
    }
}
```

G. ibqplusn

The incomplete beta function is defined as

$$B_x(p,q) = \int_0^x t^{p-1}(1-t)^{q-1}\,dt ,$$

$p > 0,\ q > 0,\ 0 \le x \le 1$, and the incomplete beta function ratio is $I_x(p,q) = B_x(p,q) / B_1(p,q)$. *ibqplusn* computes $I_x(p,q+n)$ for $n=0,1,...,nmax,\ 0 \le x \le 1,\ p > 0,\ q > 0$ (see [G64, G67]). In [G64] the procedure *ibqplusn* is called "incomplete beta p fixed". There is no control on the parameters *x, p, q, nmax* for their intended ranges.

Function Parameters:

void ibqplusn (*x,p,q,nmax,eps,i*)

x: float;

entry: this argument should satisfy: $0 \le x \le 1$;

p: float;

entry: the parameter p, $p > 0$;

q: float;

entry: the parameter q, $q > 0$; it is advised to take $0 < q \le 1$;

nmax: int;

entry: *nmax* indicates the maximum number of function values $I_x(p,q+n)$ to be generated;

eps: float;
entry: the desired relative accuracy (value of ε above);
the value of *eps* should not be smaller than the machine accuracy;
i: float *i[0:nmax]*;
exit: $i[n] = I_x(p,q+n)$ for n=0,1,...,*nmax*.

Functions used: ixqfix, ixpfix.

```
void ibqplusn(float x, float p, float q, int nmax, float eps,
               float i[])
{
    void ixqfix(float, float, float, int, float, float []);
    void ixpfix(float, float, float, int, float, float []);
    int n;

    if (x == 0.0 || x == 1.0)
        for (n=0; n<=nmax; n++) i[n]=x;
    else {
        if (x <= 0.5)
            ixpfix(x,p,q,nmax,eps,i);
        else {
            ixqfix(1.0-x,q,p,nmax,eps,i);
            for (n=0; n<=nmax; n++) i[n]=1.0-i[n];
        }
    }
}
```

H. ixqfix

The four auxiliary procedures *ixqfix, ixpfix, forward* and *backward* are for procedures *incbeta, ibpplusn* and *ibqplusn.*

These auxiliary procedures are not described here. More information can be found in [G64] of *ibqplusn,* where the procedures *forward* and *backward* have the same name, while *ixqfix* and *ixpfix* are called "Isubx q fixed" and "Isubx p fixed", respectively. In the procedure *backward* we changed the starting value *nu* for the backward recurrence algorithm. The new value of *nu* is more realistic. Its computation is based on some asymptotic estimations. Also the initial value r=0 is changed into r=x.

Functions used: incbeta, forward, backward.

```
#include <math.h>

void ixqfix(float x, float p, float q, int nmax, float eps, float i[])
{
    float *allocate_real_vector(int, int);
    void free_real_vector(float *, int);
    float incbeta(float, float, float, float);
    void forward(float, float, float, float, float, int, float []);
    void backward(float, float, float, float, int, float, float []);
    int m,mmax;
    float s,iq0,iq1,q0,*iq;

    m=floor(q);
    s=q-m;
    q0 = (s > 0.0) ? s : s+1.0;
    mmax = (s > 0.0) ? m : m-1;
    iq0=incbeta(x,p,q0,eps);
    if (mmax > 0) iq1=incbeta(x,p,q0+1.0,eps);
    iq=allocate_real_vector(0,mmax);
    forward(x,p,q0,iq0,iq1,mmax,iq);
    backward(x,p,q,iq[mmax],nmax,eps,i);
    free_real_vector(iq,0);
```

```
}
```

I. ixpfix

See the documentation of *ixqfix*.

Functions used: incbeta, forward, backward.

```
#include <math.h>

void ixpfix(float x, float p, float q, int nmax, float eps, float i[])
{
    float *allocate_real_vector(int, int);
    void free_real_vector(float *, int);
    float incbeta(float, float, float, float);
    void forward(float, float, float, float, float, int, float []);
    void backward(float, float, float, float, int, float, float []);
    int m,mmax;
    float s,p0,i0,i1,iq0,iq1,*ip;

    m=floor(p);
    s=p-m;
    p0 = (s > 0.0) ? s : s+1.0;
    mmax = (s > 0.0) ? m : m-1;
    i0=incbeta(x,p0,q,eps);
    i1=incbeta(x,p0,q+1.0,eps);
    ip=allocate_real_vector(0,mmax);
    backward(x,p0,q,i0,mmax,eps,ip);
    iq0=ip[mmax];
    backward(x,p0,q+1.0,i1,mmax,eps,ip);
    iq1=ip[mmax];
    free_real_vector(ip,0);
    forward(x,p,q,iq0,iq1,nmax,i);
}
```

J. forward

See the documentation of *ixqfix*.

```
void forward(float x, float p, float q, float i0, float i1,
             int nmax, float i[])
{
    int m,n;
    float y,r,s;

    i[0]=i0;
    if (nmax > 0) i[1]=i1;
    m=nmax-1;
    r=p+q-1.0;
    y=1.0-x;
    for (n=1; n<=m; n++) {
        s=(n+r)*y;
        i[n+1]=((n+q+s)*i[n]-s*i[n-1])/(n+q);
    }
}
```

K. backward

See the documentation of *ixqfix*.

```
#include <math.h>

void backward(float x, float p, float q, float i0, int nmax,
```

```
                float eps, float i[])
{
    float *allocate_real_vector(int, int);
    void free_real_vector(float *, int);
    int m,n,nu,finish;
    float r,pq,y,logx,*iapprox;

    iapprox=allocate_real_vector(0,nmax);
    i[0]=i0;
    if (nmax > 0) {
        for (n=1; n<=nmax; n++) iapprox[n]=0.0;
        pq=p+q-1.0;
        logx=log(x);
        r=nmax+(log(eps)+q*log(nmax))/logx;
        nu=floor(r-q*log(r)/logx);
        while (1) {
            n=nu;
            r=x;
            while (1) {
                y=(n+pq)*x;
                r=y/(y+(n+p)*(1.0-r));
                if (n <= nmax) i[n]=r;
                n--;
                if (n < 1) break;
            }
            r=i0;
            for (n=1; n<=nmax; n++) r = i[n] *= r;
            finish=1;
            for (n=1; n<=nmax; n++)
                if (fabs((i[n]-iapprox[n])/i[n]) > eps) {
                    for (m=1; m<=nmax; m++) iapprox[m]=i[m];
                    nu += 5;
                    finish=0;
                    break;
                }
            if (finish) break;
        }
    }
    free_real_vector(iapprox,0);
}
```

6.4 Error function

A. errorfunction

Computes the error function *erf(x)* and complementary error function *erfc(x)* for a real argument, i.e.

$$erf(x) = \frac{2}{\sqrt{\pi}}\int_0^x e^{-t^2}dt \qquad \text{and} \qquad erfc(x) = \frac{2}{\sqrt{\pi}}\int_x^\infty e^{-t^2}dt .$$

When $x > 26$ then $erf(x) = 1$ and $erfc(x) = 0$; when $x < -5.5$ then $erf(x) = -1$ and $erfc(x) = 2$. Over the range $|x| \le 0.5$, $erf(x)$ is approximated by a Chebyshev rational function of the form

$$x \sum_{k=0}^{3} n_k x^{2k} \Big/ \sum_{k=0}^{3} d_k x^{2k} ,$$

and the relationship $erfc(x) = 1 - erf(x)$ is used. If $0.5 < x$ or $-5.5 \le x < 0.5$ then the function $nonexperfc(x) = exp(x^2) * erfc(|x|)$ is evaluated by means of a call of *nonexperfc*, the value of *erfc(|x|)* is recovered by multiplication by $exp(x^2)$, and the relationship *erf(|x|) = 1 - erfc(|x|)* is used.; if *x* lies in the second of the above ranges the further relationships

$$erf(x) = -erfc(|x|), \quad erfc(x) = 2 - erfc(|x|)$$

are used.

Function Parameters:

void errorfunction (*x,erf,erfc*)

x: float;
  entry: the real argument of *erf(x)* and *erfc(x)*;
erf: float *;
  exit: the value of *erf(x)*;
erfc: float *;
  exit: the value of *erfc(x)*.

Function used: nonexperfc.

```
#include <math.h>

void errorfunction(float x, float *erf, float *erfc)
{
    if (x > 26.0) {
        *erf = 1.0;
        *erfc = 0.0;
        return;
    } else if (x < -5.5) {
        *erf = -1.0;
        *erfc = 2.0;
        return;
    } else {
        float nonexperfc(float);
        float absx,c,p,q;
        absx=fabs(x);
        if (absx <= 0.5) {
            c=x*x;
            p=((-0.356098437018154e-1*c+0.699638348861914e1)*c+
                    0.219792616182942e2)*c+0.242667955230532e3;
            q=((c+0.150827976304078e2)*c+0.911649054045149e2)*c+
                    0.215058875869861e3;
            *erf = x*p/q;
            *erfc = 1.0-(*erf);
        } else {
            *erfc = exp(-x*x)*nonexperfc(absx);
            *erf = 1.0-(*erfc);
            if (x < 0.0) {
                *erf = -(*erf);
                *erfc = 2.0-(*erfc);
            }
        }
    }
}
```

B. nonexperfc

The error function *erf(x)* and complementary error function *erfc(x)* for a real argument are given by

$$erf(x) = \frac{2}{\sqrt{\pi}}\int_0^x e^{-t^2}\,dt \qquad \text{and} \qquad erfc(x) = \frac{2}{\sqrt{\pi}}\int_x^\infty e^{-t^2}\,dt\,.$$

nonexperfc computes $exp(x^2)*erfc(x)$, i.e.

$$nonexperfc(x) = \frac{2e^{x^2}}{\sqrt{\pi}} \int_x^\infty e^{-t^2} dt .$$

When $|x| \leq 0.5$ then the function *erfc(x)* is evaluated by means of a call of *errorfunction*, and the value of *nonexperfc(x)* is recovered by multiplication by $exp(x^2)$. If $0.5<|x|< 4$ then $nonexperfc(|x|)$ is approximated by a Chebyshev rational approximation (see [Cod69]) of the form

$$\sum_{k=0}^{4} n_k |x|^k \Big/ \sum_{k=0}^{4} d_k |x|^k ;$$

if, in addition, x is negative, the relationship

$$nonexperfc(x) = 2 * exp(x^2) - nonexperfc(|x|) \qquad (1)$$

is used. If $|x| \geq 4$ then *nonexperfc(x)* is approximated by a Chebyshev rational approximation (see [Cod69]) of the form

$$\frac{1}{|x|} \left\{ \frac{1}{\sqrt{\pi}} + \frac{1}{x^2} \sum_{k=0}^{3} n_k x^{-2k} \Big/ \sum_{k=0}^{3} d_k x^{-2k} \right\}$$

and again if x is negative then relationship (1) is used.

Function Parameters:

float nonexperfc (*x*)

nonexperfc: delivers the value of $exp(x^2)*erfc(x)$;

x: float;
entry: the real argument of *nonexperfc*.

Function used: errorfunction.

```
#include <math.h>

float nonexperfc(float x)
{
    void errorfunction(float, float *, float *);
    float absx,erf,erfc,c,p,q;

    absx=fabs(x);
    if (absx <= 0.5) {
        errorfunction(x,&erf,&erfc);
        return exp(x*x)*erfc;
    } else if (absx < 4.0) {
        c=absx;
        p=((((((-0.136864857382717e-6*c+0.564195517478974e0)*c+
                0.721175825088309e1)*c+0.431622272220567e2)*c+
                0.152989285046940e3)*c+0.339320816734344e3)*c+
                0.451918953711873e3)*c+0.300459261020162e3;
        q=((((((c+0.127827273196294e2)*c+0.770001529352295e2)*c+
                0.277585444743988e3)*c+0.638980264465631e3)*c+
                0.931354094850610e3)*c+0.790950925327898e3)*c+
                0.300459260956983e3;
        return ((x > 0.0) ? p/q : exp(x*x)*2.0-p/q);
    } else {
        c=1.0/x/x;
```

```
        p=(((0.223192459734185e-1*c+0.278661308609648e0)*c+
                0.226956593539687e0)*c+0.494730910623251e-1)*c+
                0.299610707703542e-2;
        q=(((c+0.198733201817135e1)*c+0.105167510706793e1)*c+
                0.191308926107830e0)*c+0.106209230528468e-1;
        c=(c*(-p)/q+0.564189583547756)/absx;
        return ((x > 0.0) ? c : exp(x*x)*2.0-c);
    }
}
```

C. inverseerrorfunction

Evaluates the inverse error function $y(x)$, where

$$x = \frac{2}{\sqrt{\pi}}\int_0^{y(x)} e^{-t^2}\,dt ,$$

the values of x and 1-x being supplied (for values of x near 1, loss of accuracy results from the formation of 1-x; this loss is avoided by defining y as a function of 1-x for such values of x, provided the supplied value of 1-x is sufficiently accurate).

When $|x| \le 0.8$, a telescoped power series is used to evaluate $y(x)$: y satisfies the equation

$$\left(\frac{dy}{dx}\right)^{-1} = 2\int_0^x y(t)\,dt - \frac{2}{\sqrt{\pi}}$$

so that if

$$y(x) = \sum_{n=0}^{\infty} C_n x^{2n-1} ,$$

then the $\{C_n\}$ may be extracted from the equation

$$1 + \left\{\sum_{m=1}^{\infty}(2m-1)C_m x^{2m-2}\right\}\left\{\sum_{n=1}^{\infty}\frac{C_n x^{2n}}{n} - \frac{2}{\sqrt{\pi}}\right\} = 0 ,$$

and from the $\{C_n\}$ a Chebyshev series (see [Str68]) of the form

$$y(x) \approx x\left\{\xi_0 + \sum_{n=1}^{22}\xi_n T_n\left(\frac{x^2}{0.32} - 1\right)\right\} ,$$

valid for $|x| \le 0.8$ may be obtained. Over the range $0.8 < |x| \le 0.9975$, an expansion of the form $y(x) = \beta(x)R(x)$, where

$$\beta(x) = \sqrt{-\ln(1-x^2)}, \qquad R(x) = \mathrm{sign}(x)\sum_{n=0}^{j}\lambda_n T_n(D_1\beta(x) + D_2)$$

with j=15 is used; over the range $2.5*10^{-3} > 1-x \geq 0.5*10^{-15}$ a similar expansion with j=23 is used, and over the range $0.5*10^{-15} > 1-x$, yet another similar expansion with j=12 is used.

Function Parameters:

void inverseerrorfunction (*x,oneminx,inverf*)

x: float;
entry: the argument of *inverseerrorfunction(x)*; it is necessary that $-1 < x < 1$; if $|x| > 0.8$ then the value of x is not used in the procedure;

oneminx: float;
entry: if $|x| \leq 0.8$ then the value of *oneminx* is not used in the procedure; if $|x| > 0.8$ then *oneminx* has to contain the value of $1-|x|$; in the case that $|x|$ is in the neighborhood of 1, cancellation of digits take place in the calculation of $1-|x|$; if the value $1-|x|$ is known exactly from another source then *oneminx* has to contain this value, which will give better results;

inverf: float *;
exit: the result of the procedure.

Function used: chepolsum.

```
#include <float.h>
#include <math.h>

void inverseerrorfunction(float x, float oneminx, float *inverf)
{
    float chepolsum(int, float, float []);
    float absx,p,betax,a[24];

    absx=fabs(x);
    if (absx > 0.8 && oneminx > 0.2) oneminx=0.0;
    if (absx <= 0.8) {
        a[0]  = 0.992885376618941;  a[1]  = 0.120467516143104;
        a[2]  = 0.016078199342100;  a[3]  = 0.002686704437162;
        a[4]  = 0.000499634730236;  a[5]  = 0.000098898218599;
        a[6]  = 0.000020391812764;  a[7]  = 0.000004327271618;
        a[8]  = 0.000000938081413;  a[9]  = 0.000000206734720;
        a[10] = 0.000000046159699;  a[11] = 0.000000010416680;
        a[12] = 0.000000002371501;  a[13] = 0.000000000543928;
        a[14] = 0.000000000125549;  a[15] = 0.000000000029138;
        a[16] = 0.000000000006795;  a[17] = 0.000000000001591;
        a[18] = 0.000000000000374;  a[19] = 0.000000000000088;
        a[20] = 0.000000000000021;  a[21] = 0.000000000000005;
        *inverf = chepolsum(21,x*x/0.32-1.0,a)*x;
    } else if (oneminx >= 25.0e-4) {
        a[0]  =  0.912158803417554;  a[1]  = -0.016266281867664;
        a[2]  =  0.000433556472949;  a[3]  =  0.000214438570074;
        a[4]  =  0.000002625751076;  a[5]  = -0.000003021091050;
        a[6]  = -0.000000012406062;  a[7]  =  0.000000062406609;
        a[8]  = -0.000000000540125;  a[9]  = -0.000000001423208;
        a[10] =  0.000000000034384;  a[11] =  0.000000000033584;
        a[12] = -0.000000000001458;  a[13] = -0.000000000000810;
        a[14] =  0.000000000000053;  a[15] =  0.000000000000020;
        betax=sqrt(-log((1.0+absx)*oneminx));
        p = -1.54881304237326*betax+2.56549012314782;
        p=chepolsum(15,p,a);
        *inverf = (x < 0.0) ? -betax*p : betax*p;
    } else if (oneminx >= 5.0e-16) {
        a[0]  =  0.956679709020493;  a[1]  = -0.023107004309065;
        a[2]  = -0.004374236097508;  a[3]  = -0.000576503422651;
        a[4]  = -0.000010961022307;  a[5]  =  0.000025108547025;
        a[6]  =  0.000010562336068;  a[7]  =  0.000002754412330;
        a[8]  =  0.000000432484498;  a[9]  = -0.000000020530337;
        a[10] = -0.000000043891537;  a[11] = -0.000000017684010;
        a[12] = -0.000000003991289;  a[13] = -0.000000000186932;
        a[14] =  0.000000000272923;  a[15] =  0.000000000132817;
```

```
        a[16] =  0.000000000031834;  a[17] =  0.000000000001670;
        a[18] = -0.000000000002036;  a[19] = -0.000000000000965;
        a[20] = -0.000000000000220;  a[21] = -0.000000000000010;
        a[22] =  0.000000000000014;  a[23] =  0.000000000000006;
        betax=sqrt(-log((1.0+absx)*oneminx));
        p = -0.559457631329832*betax+2.28791571626336;
        p=chepolsum(23,p,a);
        *inverf = (x < 0.0) ? -betax*p : betax*p;
    } else if (oneminx >= FLT_MIN) {
        a[0]  =  0.988575064066189;  a[1]  =  0.010857705184599;
        a[2]  = -0.001751165102763;  a[3]  =  0.000021196993207;
        a[4]  =  0.000015664871404;  a[5]  = -0.000000519041687;
        a[6]  = -0.000000037135790;  a[7]  =  0.000000001217431;
        a[8]  = -0.000000000176812;  a[9]  = -0.000000000011937;
        a[10] =  0.000000000000380;  a[11] = -0.000000000000066;
        a[12] = -0.000000000000009;
        betax=sqrt(-log((1.0+absx)*oneminx));
        p = -9.19999235883015/sqrt(betax)+2.79499082012460;
        p=chepolsum(12,p,a);
        *inverf = (x < 0.0) ? -betax*p : betax*p;
    } else
        *inverf = (x > 0.0) ? 26.0 : -26.0;
}
```

D. fresnel

Evaluates the Fresnel integrals

$$S(x) = \int_0^x \sin(\frac{\pi}{2}t^2)\,dt, \qquad C(x) = \int_0^x \cos(\frac{\pi}{2}t^2)\,dt\ .$$

If $|x| \leq 1.2$ then $S(x)$ is approximated by means of a Chebyshev rational function (see [Cod68]) of the form

$$x^3 \sum_{k=0}^{j} n_k x^{4k} \;/\; \sum_{k=0}^{j} d_k x^{4k}$$

and $C(x)$ by a similar function of the form

$$x \sum_{k=0}^{j} n_k x^{4k} \;/\; \sum_{k=0}^{j} d_k x^{4k}$$

where j=4. When $1.2 < |x| \leq 1.6$, similar approximations with j=5 are used. Over the range $1.6<|x|$, the functions $f(x)$ and $g(x)$ occurring in the formulae

$$S(x) = ½ - f(x)\cos(\pi x^2/2) - g(x)\sin(\pi x^2/2)$$
$$C(x) = ½ + f(x)\sin(\pi x^2/2) - g(x)\cos(\pi x^2/2)$$

are evaluated by means of a call of *fg*, and the values of $S(x)$ and $C(x)$ are obtained by use of these relationships.

Function Parameters:

void fresnel (x,c,s)

x: float;
  entry: the real argument of $C(x)$ and $S(x)$;
c: float *;

exit: the value of $C(x)$;

s: float *;

exit: the value of $S(x)$.

Function used: fg.

```
#include <math.h>

void fresnel(float x, float *c, float *s)
{
    void fg(float, float *, float *);
    float absx,x3,x4,a,p,q,f,g,c1,s1;

    absx=fabs(x);
    if (absx <= 1.2) {
        a=x*x;
        x3=a*x;
        x4=a*a;
        p=(((5.47711385682687e-6*x4-5.28079651372623e-4)*x4+
                1.76193952543491e-2)*x4-1.99460898826184e-1)*x4+1.0;
        q=(((1.18938901422876e-7*x4+1.55237885276994e-5)*x4+
                1.09957215025642e-3)*x4+4.72792112010453e-2)*x4+1.0;
        *c = x*p/q;
        p=(((6.71748466625141e-7*x4-8.45557284352777e-5)*x4+
                3.87782123463683e-3)*x4-7.07489915144523e-2)*x4+
                5.23598775598299e-1;
        q=(((5.95281227678410e-8*x4+9.62690875939034e-6)*x4+
                8.17091942152134e-4)*x4+4.11223151142384e-2)*x4+1.0;
        *s = x3*p/q;
    } else if (absx <= 1.6) {
        a=x*x;
        x3=a*x;
        x4=a*a;
        p=((((-5.68293310121871e-8*x4+1.02365435056106e-5)*x4-
                6.71376034694922e-4)*x4+1.91870279431747e-2)*x4-
                2.07073360335324e-1)*x4+1.00000000000111e0;
        q=((((4.41701374065010e-10*x4+8.77945377892369e-8)*x4+
                1.01344630866749e-5)*x4+7.88905245052360e-4)*x4+
                3.96667496952323e-2)*x4+1.0;
        *c = x*p/q;
        p=((((-5.76765815593089e-9*x4+1.28531043742725e-6)*x4-
                1.09540023911435e-4)*x4+4.30730526504367e-3)*x4-
                7.37766914010191e-2)*x4+5.23598775598344e-1;
        q=((((2.05539124458580e-10*x4+5.03090581246612e-8)*x4+
                6.87086265718620e-6)*x4+6.18224620195473e-4)*x4+
                3.53398342767472e-2)*x4+1.0;
        *s = x3*p/q;
    } else if (absx < 1.0e15) {
        fg(x,&f,&g);
        a=x*x;
        a=(a-floor(a/4.0)*4.0)*1.57079632679490;
        c1=cos(a);
        s1=sin(a);
        a = (x < 0.0) ? -0.5 : 0.5;
        *c = f*s1-g*c1+a;
        *s = -f*c1-g*s1+a;
    } else
        *c = *s = ((x > 0.0) ? 0.5 : -0.5);
}
```

E. fg

Evaluates the functions $f(x)$, $g(x)$ related to the Fresnel integrals by means of the formulae

$$f(x) = \{½ - S(x)\}\cos(\pi x^2/2) - \{½ - C(x)\}\sin(\pi x^2/2)$$
$$g(x) = \{½ - C(x)\}\cos(\pi x^2/2) + \{½ - S(x)\}\sin(\pi x^2/2).$$

When $|x| \le 1.6$ the functions $S(x)$ and $C(x)$ are evaluated by means of a call of *fresnel*, and the values of $f(x)$ and $g(x)$ are obtained by use of the above relationships. When $1.6<|x|\le 1.9$, $f(x)$ is approximated by use of a Chebyshev rational function (see [Cod68]) of the form

$$\frac{1}{x}\sum_{k=0}^{i} n_k x^{-4k} \Big/ \sum_{k=0}^{i} d_k x^{-4k}$$

and $g(x)$ by use of a similar function of the form

$$\frac{1}{x^3}\sum_{k=0}^{j} n_k x^{-4k} \Big/ \sum_{k=0}^{j} d_k x^{-4k}$$

with i=4 and j=5. When $|x| \ge 2.4$, similar expansions with i=j=5 are used. When $|x| > 2.4$, the approximating functions are

$$\frac{1}{x}\left\{\frac{1}{\pi} + \frac{1}{x^4}\sum_{k=0}^{5} n_k x^{-4k} \Big/ \sum_{k=0}^{5} d_k x^{-4k}\right\}$$

for $f(x)$ and

$$\frac{1}{x^3}\left\{\frac{1}{\pi^2} + \frac{1}{x^4}\sum_{k=0}^{6} n_k x^{-4k} \Big/ \sum_{k=0}^{6} d_k x^{-4k}\right\}$$

for $g(x)$.

Function Parameters:

void fg (*x*,*f*,*g*)

x: float;
 entry: the real argument of $f(x)$ and $g(x)$;
f: float *;
 exit: the value of $f(x)$;
g: float *;
 exit: the value of $g(x)$.

Function used: fresnel.

```
#include <math.h>

void fg(float x, float *f, float *g)
{
    void fresnel(float, float *, float *);
    float absx,c,s,c1,s1,a,xinv,x3inv,c4,p,q;

    absx=fabs(x);
    if (absx <= 1.6) {
        fresnel(x,&c,&s);
        a=x*x*1.57079632679490;
        c1=cos(a);
```

```
        s1=sin(a);
        a = (x < 0.0) ? -0.5 : 0.5;
        p=a-c;
        q=a-s;
        *f = q*c1-p*s1;
        *g = p*c1+q*s1;
    } else if (absx <= 1.9) {
        xinv=1.0/x;
        a=xinv*xinv;
        x3inv=a*xinv;
        c4=a*a;
        p=(((1.35304235540388e1*c4+6.98534261601021e1)*c4+
                4.80340655577925e1)*c4+8.03588122803942e0)*c4+
                3.18309268504906e-1;
        q=(((6.55630640083916e1*c4+2.49561993805172e2)*c4+
                1.57611005580123e2)*c4+2.55491618435795e1)*c4+1.0;
        *f = xinv*p/q;
        p=((((2.05421432498501e1*c4+1.96232037971663e2)*c4+
                1.99182818678903e2)*c4+5.31122813480989e1)*c4+
                4.44533827550512e0)*c4+1.01320618810275e-1;
        q=((((1.01379483396003e3*c4+3.48112147856545e3)*c4+
                2.54473133181822e3)*c4+5.83590575716429e2)*c4+
                4.53925019673689e1)*c4+1.0;
        *g = x3inv*p/q;
    } else if (absx <= 2.4) {
        xinv=1.0/x;
        a=xinv*xinv;
        x3inv=a*xinv;
        c4=a*a;
        p=((((7.17703249365140e2*c4+3.09145161574430e3)*c4+
                1.93007640786716e3)*c4+3.39837134926984e2)*c4+
                1.95883941021969e1)*c4+3.18309881822017e-1;
        q=((((3.36121699180551e3*c4+1.09334248988809e4)*c4+
                6.33747155851144e3)*c4+1.08535067500650e3)*c4+
                6.18427138172887e1)*c4+1.0;
        *f = xinv*p/q;
        p=((((3.13330163068756e2*c4+1.59268006085354e3)*c4+
                9.08311749529594e2)*c4+1.40959617911316e2)*c4+
                7.11205001789783e0)*c4+1.01321161761805e-1;
        q=((((1.15149832376261e4*c4+2.41315567213370e4)*c4+
                1.06729678030581e4)*c4+1.49051922797329e3)*c4+
                7.17128596939302e1)*c4+1.0;
        *g = x3inv*p/q;
    } else {
        xinv=1.0/x;
        a=xinv*xinv;
        x3inv=a*xinv;
        c4=a*a;
        p=((((2.61294753225142e4*c4+6.13547113614700e4)*c4+
                1.34922028171857e4)*c4+8.16343401784375e2)*c4+
                1.64797712841246e1)*c4+9.67546032967090e-2;
        q=((((1.37012364817226e6*c4+1.00105478900791e6)*c4+
                1.65946462621853e5)*c4+9.01827596231524e3)*c4+
                1.73871690673649e2)*c4+1.0;
        *f = (c4*(-p)/q+0.318309886183791)*xinv;
        p=(((((1.72590224654837e6*c4+6.66907061668636e6)*c4+
                1.77758950838030e6)*c4+1.35678867813756e5)*c4+
                3.87754141746378e3)*c4+4.31710157823358e1)*c4+
                1.53989733819769e-1;
        q=(((((1.40622441123580e8*c4+9.38695862531635e7)*c4+
                1.62095600500232e7)*c4+1.02878693056688e6)*c4+
                2.69183180396243e4)*c4+2.86733194975899e2)*c4+1.0;
        *g = (c4*(-p)/q+0.101321183642338)*x3inv;
    }
}
```

6.5 Bessel functions of integer order

6.5.1 Bessel functions J and Y

A. bessj0

Computes the value of the ordinary Bessel function of the first kind of order zero $J_0(x)$. When $|x| < 8$, $J_0(x)$ is evaluated by use of a Chebyshev polynomial approximation [Cle62] in the variable $2(x^2/16 - 1)$, and $|x| \geq 8$ by use of the formula

$$J_0(x) = \sqrt{\frac{2}{\pi |x|}} \left\{ P_0(x)\cos(x - \frac{9\pi}{4}) - Q_0(x)\sin(x - \frac{9\pi}{4}) \right\}$$

(*besspq0* being called to evaluate the functions P_0 and Q_0).

Function Parameters:

float bessj0 (x)

bessj0: delivers the ordinary Bessel function of the first kind of order zero with argument x;

x: float;
  entry: the argument of the Bessel function.

Function used: besspq0.

```
#include <math.h>

float bessj0(float x)
{
    if (x == 0.0) return 1.0;
    if (fabs(x) < 8.0) {
        int i;
        float z,z2,b0,b1,b2;
        static float ar[15]={-0.75885e-15, 0.4125321e-13,
            -0.194383469e-11, 0.7848696314e-10, -0.267925353056e-8,
            0.7608163592419e-7, -0.176194690776215e-5,
            0.324603288210051e-4, -0.46062616620628e-3,
            0.48191800694676e-2, -0.34893769411409e-1,
            0.158067102332097, -0.37009499387265, 0.265178613203337,
            -0.872344235285222e-2};
        x /= 8.0;
        z=2.0*x*x-1.0;
        z2=z+z;
        b1=b2=0.0;
        for (i=0; i<=14; i++) {
            b0=z2*b1-b2+ar[i];
            b2=b1;
            b1=b0;
        }
        return z*b1-b2+0.15772797147489;
    } else {
        void besspq0(float, float *, float *);
        float c,cosx,sinx,p0,q0;
        x=fabs(x);
        c=0.797884560802865/sqrt(x);
        cosx=cos(x-0.706858347057703e1);
        sinx=sin(x-0.706858347057703e1);
        besspq0(x,&p0,&q0);
        return c*(p0*cosx-q0*sinx);
    }
}
```

B. bessj1

Computes the value of the ordinary Bessel function of the first kind of order one $J_1(x)$. When $|x| < 8$, $J_1(x)$ is evaluated by use of a Chebyshev polynomial approximation [Cle62] in the variable $2(x^2/16 - 1)$, and $|x| \geq 8$ by use of the formula

$$J_1(x) = \text{sign}(x)\sqrt{\frac{2}{\pi|x|}}\left\{P_1(x)\sin(x-\frac{9\pi}{4}) + Q_1(x)\cos(x-\frac{9\pi}{4})\right\}$$

(*besspq1* being called to evaluate the functions P_1 and Q_1).

Function Parameters:

float bessj1 (x)

bessj1: delivers the ordinary Bessel function of the first kind of order one with argument x;

x: float;
entry: the argument of the Bessel function.

Function used: besspq1.

```
#include <math.h>

float bessj1(float x)
{
    if (x == 0.0) return 1.0;
    if (fabs(x) < 8.0) {
        int i;
        float z,z2,b0,b1,b2;
        static float ar[15]={-0.19554e-15, 0.1138572e-13,
            -0.57774042e-12, 0.2528123664e-10, -0.94242129816e-9,
            0.2949707007278e-7, -0.76175878054003e-6,
            0.158870192399321e-4, -0.260444389348581e-3,
            0.324027018268386e-2, -0.291755248061542e-1,
            0.177709117239728e0, -0.661443934134543e0,
            0.128799409885768e1, -0.119180116054122e1};
        x /= 8.0;
        z=2.0*x*x-1.0;
        z2=z+z;
        b1=b2=0.0;
        for (i=0; i<=14; i++) {
            b0=z2*b1-b2+ar[i];
            b2=b1;
            b1=b0;
        }
        return x*(z*b1-b2+0.648358770605265);
    } else {
        void besspq1(float, float *, float *);
        int sgnx;
        float c,cosx,sinx,p1,q1;
        sgnx = (x > 0.0) ? 1 : -1;
        x=fabs(x);
        c=0.797884560802865/sqrt(x);
        cosx=cos(x-0.706858347057703e1);
        sinx=sin(x-0.706858347057703e1);
        besspq1(x,&p1,&q1);
        return sgnx*c*(p1*sinx+q1*cosx);
    }
}
```

C. bessj

Generates an array of ordinary Bessel functions of the first kind of order k, $J_k(x)$,

k=0,...,n.

The method used [G67] is suitable for the determination of a sequence of numbers $f_0, f_1, \ldots$ which satisfy a recursion of the form

$$f_{k+1} + a_k f_k + b_k f_{k-1} = 0 \tag{1}$$

and decrease in magnitude with k more rapidly than any other solution of this difference equation, and in addition satisfy a relationship of the form

$$\sum_{k=0}^{\infty} \lambda_k f_k = s .$$

Setting $r_k = f_{k+1} / f_k$, formally

$$r_{k-1} = \frac{-b_k}{a_k -} \frac{b_{k+1}}{a_{k+1} -} \cdots$$

and with

$$s_{k-1} = \left\{ \sum_{m=k}^{\infty} \lambda_m f_m \right\} / f_{k-1} ,$$

$$s_{k-1} = r_{k-1}(\lambda_k + s_k).$$

By definition $f_0 = s/(\lambda_0+s_0)$. For suitably large ν, the numbers

$$r_{k-1}^{(\nu)} = \frac{-b_k}{a_k -} \frac{b_{k+1}}{a_{k+1} -} \cdots \frac{b_\nu}{a_\nu} ,$$

$$s_{k-1}^{(\nu)} = \sum_{m=k+1}^{\nu} \lambda_m r_k^{(\nu)} r_{k+1}^{(\nu)} r_{m-1}^{(\nu)}$$

are computed by use of the recursions

$$r_\nu^{(\nu)} = 0, \quad r_{k-1}^{(\nu)} = -b_k/(a_k + r_k^{(\nu)})$$

$$s_\nu^{(\nu)} = 0, \quad s_{k-1}^{(\nu)} = r_{k-1}^{(\nu)}(\lambda_k + s_k^{(\nu)})$$

for $k=\nu,\nu-1,\ldots,1$. Thereafter the required numbers $f_0, f_1, \ldots, f_n$ are recovered by use of the relationships

$$f_0^{(\nu)} = s / (\lambda_0 + s_0^{(\nu)}), \quad f_k^{(\nu)} = r_{k-1}^{(\nu)} f_{k-1}^{(\nu)}, \qquad (k=1,\ldots,n).$$

The functions $J_{a+k}(x)$ satisfy the recursion

$$J_{a+k+1}(x) - \{2(a+k)/x\} J_{a+k}(x) + J_{a+k-1}(x) = 0$$

and also the relationship

$$J_a(x) + \sum_{m=1}^{\infty} \frac{(a+2m)\Gamma(a+m)}{m!\,\Gamma(a+1)} J_{a+2m}(x) = \frac{(x/2)^2}{\Gamma(a+1)} .$$

Thus when $a = 0$, $a_k = 2k/x$, $b_k = -1$ and $\lambda_0 = 1$, $\lambda_{2m-1} = 0$, $\lambda_{2m} = 2$ (m=1,2,...), $s = 1$ may be taken in the above.

Function Parameters:

void bessj (x,n,j)

x: float;
entry: the argument of the Bessel functions;

n: int;
entry: the upper bound of the indices of array j; $n \geq 0$;

j: float $j[0:n]$;
exit: $j[k]$ is the ordinary Bessel function of the first kind of order k and argument x.

Function used: start.

```
#include <math.h>

void bessj(float x, int n, float j[])
{
    if (x == 0.0) {
        j[0]=1.0;
        for (; n>=1; n--) j[n]=0.0;
    } else {
        int start(float, int, int);
        int l,m,nu,signx;
        float x2,r,s;
        signx = (x > 0.0) ? 1 : -1;
        x=fabs(x);
        r=s=0.0;
        x2=2.0/x;
        l=0;
        nu=start(x,n,0);
        for (m=nu; m>=1; m--) {
            r=1.0/(x2*m-r);
            l=2-l;
            s=r*(l+s);
            if (m <= n) j[m]=r;
        }
        j[0]=r=1.0/(1.0+s);
        for (m=1; m<=n; m++) r = j[m] *= r;
        if (signx < 0.0)
            for (m=1; m<=n; m += 2) j[m] = -j[m];
    }
}
```

D. bessy01

Computes the ordinary Bessel functions of the second kind of orders zero and one: $Y_0(x)$ and $Y_1(x)$ for x>0. When $x < 8$, $Y_0(x)$ and $Y_1(x)$ are determined from relationships of the form

$$Y_0(x) = (2/\pi)ln(x)J_0(x) + y_0(x)$$
$$Y_1(x) = (2/\pi)ln(x)J_1(x) - 2/(\pi x) + xy_1(x)$$

where $y_0(x)$ and $y_1(x)$ are approximated by use of Chebyshev polynomials [Cle62]. When $x \geq 8$, the relationships

$$Y_0(x) = (2/(\pi x))^{1/2}\{P_0(x)\sin(x - 9\pi/4) + Q_0(x)\cos(x - 9\pi/4)\}$$
$$Y_1(x) = (2/(\pi x))^{1/2}\{Q_1(x)\sin(x - 9\pi/4) - P_1(x)\cos(x - 9\pi/4)\}$$

are used.

Function Parameters:

void bessy01 (*x,y0,y1*)

x: float;
  entry: the argument of the Bessel function; $x > 0$;
y0: float *;
  exit: *y0* has the value of the ordinary Bessel function of the second kind of order zero with argument *x*;
y1: float *;
  exit: *y1* has the value of the ordinary Bessel function of the second kind of order one with argument *x*.

Functions used: bessj0, bessj1, besspq0, besspq1.

```
#include <math.h>

void bessy01(float x, float *y0, float *y1)
{
    if (x < 8.0) {
        int i;
        float bessj0(float);
        float bessj1(float);
        float z,z2,c,lnx,b0,b1,b2;
        static float ar1[15]={0.164349e-14, -0.8747341e-13,
            0.402633082e-11, -0.15837552542e-9, 0.524879478733e-8,
            -0.14407233274019e-6, 0.32065325376548e-5,
            -0.563207914105699e-4, 0.753113593257774e-3,
            -0.72879624795521e-2, 0.471966895957634e-1,
            -0.177302012781143, 0.261567346255047,
            0.179034314077182, -0.274474305529745};
        static float ar2[15]={0.42773e-15, -0.2440949e-13,
            0.121143321e-11, -0.5172121473e-10, 0.187547032473e-8,
            -0.5688440039919e-7, 0.141662436449235e-5,
            -0.283046401495148e-4, 0.440478629867099e-3,
            -0.51316411610611e-2, 0.423191803533369e-1,
            -0.226624991556755, 0.675615780772188,
            -0.767296362886646, -0.128697384381350};
        c=0.636619772367581;
        lnx=c*log(x);
        c /= x;
        x /= 8.0;
        z=2.0*x*x-1.0;
        z2=z+z;
        b1=b2=0.0;
        for (i=0; i<=14; i++) {
            b0=z2*b1-b2+ar1[i];
            b2=b1;
            b1=b0;
        }
        *y0 = lnx*bessj0(8.0*x)+z*b1-b2-0.33146113203285e-1;
        b1=b2=0.0;
        for (i=0; i<=14; i++) {
            b0=z2*b1-b2+ar2[i];
            b2=b1;
            b1=b0;
        }
        *y1 = lnx*bessj1(8.0*x)-c+x*(z*b1-b2+0.2030410588593425e-1);
    } else {
        void besspq0(float, float *, float *);
        void besspq1(float, float *, float *);
        float c,cosx,sinx,p0,q0,p1,q1;
        c=0.797884560802865/sqrt(x);
```

```
        besspq0(x,&p0,&q0);
        besspq1(x,&p1,&q1);
        x -= 0.706858347057703e1;
        cosx=cos(x);
        sinx=sin(x);
        *y0 = c*(p0*sinx+q0*cosx);
        *y1 = c*(q1*sinx-p1*cosx);
    }
}
```

E. bessy

Generates an array of ordinary Bessel functions of the second kind of order k, $Y_k(x)$, $k=0,...,n$, for $x > 0$.

The functions $Y_0(x)$ and (if $n > 0$) $Y_1(x)$ are evaluated by means of a call of *bessy01*, and the recursion

$$Y_{k+1}(x) = (2k/x)Y_k(x) - Y_{k-1}(x)$$

is used.

Function Parameters:

void bessy (x,n,y)

x: float;
    entry: this argument should satisfy $x > 0$;
n: int;
    entry: the upper bound of the indices of array y; $n \geq 0$;
y: float $y[0:n]$;
    exit: $y[k]$ is the value of the ordinary Bessel function of the second kind of order k and argument x, $k=0,...,n$.

Function used: bessy01.

```
#include <math.h>

void bessy(float x, int n, float y[])
{
    void bessy01(float, float *, float *);
    int i;
    float y0,y1,y2;

    bessy01(x,&y0,&y1);
    y[0]=y0;
    if (n > 0) y[1]=y1;
    x=2.0/x;
    for (i=2; i<=n; i++) {
        y[i]=y2=(i-1)*x*y1-y0;
        y0=y1;
        y1=y2;
    }
}
```

F. besspq0

This procedure is an auxiliary procedure for the computation of the ordinary Bessel functions of order zero for large values of their argument.

besspq0 computes the values of the functions $P_0(x)$, $Q_0(x)$ occurring in the formulae

$$J_0(x) = (2/(\pi x))^{1/2}\{P_0(x)\cos(x - \pi/4) - Q_0(x)\sin(x - \pi/4)\}$$
$$Y_0(x) = (2/(\pi x))^{1/2}\{P_0(x)\sin(x - \pi/4) + Q_0(x)\cos(x - \pi/4)\}.$$

When $|x| < 8$, the above equations are used in the form

$$P_0(x) = (\pi x/2)^{1/2}\{Y_0(x)\sin(x - \pi/4) + J_0(x)\cos(x - \pi/4)\}$$

$$Q_0(x) = (\pi x/2)^{1/2}\{Y_0(x)\cos(x - \pi/4) - J_0(x)\sin(x - \pi/4)\}$$

(*bessj0* and *bessy0* being used to evaluate the functions J_0 and Y_0) and when $|x| \geq 8$, $P_0(x)$ and $Q_0(x)$ are evaluated by use of Chebyshev polynomial approximations [Cle62] in the variable $2(128/x^2 - 1)$.

Function Parameters:

void besspq0 (*x,p0,q0*)

x: float;
  entry: this argument should satisfy $x > 0$;
p0: float *;
  exit: the value of $P_0(x)$;
q0: float *;
  exit: the value of $Q_0(x)$.

Functions used: bessj0, bessy01.

```
#include <math.h>

void besspq0(float x, float *p0, float *q0)
{
    if (x < 8.0) {
        float bessj0(float);
        void bessy01(float, float *, float *);
        float b,cosx,sinx,j0x,y0;
        b=sqrt(x)*1.25331413731550;
        bessy01(x,&y0,&j0x);
        j0x=bessj0(x);
        x -= 0.785398163397448;
        cosx=cos(x);
        sinx=sin(x);
        *p0 = b*(y0*sinx+j0x*cosx);
        *q0 = b*(y0*cosx-j0x*sinx);
    } else {
        int i;
        float x2,b0,b1,b2,y;
        static float ar1[11]={-0.10012e-15, 0.67481e-15, -0.506903e-14,
            0.4326596e-13, -0.43045789e-12, 0.516826239e-11,
            -0.7864091377e-10, 0.163064646352e-8, -0.5170594537606e-7,
            0.30751847875195e-5, -0.536522046813212e-3};
        static float ar2[10]={-0.60999e-15, 0.425523e-14,
            -0.3336328e-13, 0.30061451e-12, -0.320674742e-11,
            0.4220121905e-10, -0.72719159369e-9, 0.1797245724797e-7,
            -0.74144984110606e-6, 0.683851994261165e-4};
        y=8.0/x;
        x=2.0*y*y-1.0;
        x2=x+x;
        b1=b2=0.0;
        for (i=0; i<=10; i++) {
            b0=x2*b1-b2+ar1[i];
            b2=b1;
            b1=b0;
        }
        *p0 = x*b1-b2+0.99946034934752;
        b1=b2=0.0;
        for (i=0; i<=9; i++) {
            b0=x2*b1-b2+ar2[i];
            b2=b1;
            b1=b0;
        }
        *q0 = (x*b1-b2-0.015555854605337)*y;
    }
}
```

G. besspq1

This procedure is an auxiliary procedure for the computation of the ordinary Bessel functions of order one for large values of their argument.

besspq1 computes the values of the functions $P_1(x)$, $Q_1(x)$ occurring in the formulae

$$J_1(x) = (2/(\pi x))^{1/2}\{P_1(x)\cos(x - 3\pi/4) - Q_1(x)\sin(x - 3\pi/4)\}$$
$$Y_1(x) = (2/(\pi x))^{1/2}\{P_1(x)\sin(x - 3\pi/4) + Q_1(x)\cos(x - 3\pi/4)\}.$$

When $|x| < 8$, the above equations are used in the form

$$P_1(x) = (\pi x/2)^{1/2}\{Y_1(x)\sin(x - 3\pi/4) + J_1(x)\cos(x - 3\pi/4)\}$$
$$Q_1(x) = (\pi x/2)^{1/2}\{Y_1(x)\cos(x - 3\pi/4) - J_1(x)\sin(x - 3\pi/4)\}$$

(*bessj1* and *bessy1* being used to evaluate the functions J_1 and Y_1) and when $|x| \geq 8$, $P_1(x)$ and $Q_1(x)$ are evaluated by use of Chebyshev polynomial approximations [Cle62] in the variable $2(128/x^2 - 1)$.

Function Parameters:

void besspq1 (*x,p1,q1*)

x: float;
entry: this argument should satisfy $x > 0$;
p1: float *;
exit: the value of $P_1(x)$;
q1: float *;
exit: the value of $Q_1(x)$.

Functions used: bessj1, bessy01.

```
#include <math.h>

void besspq1(float x, float *p1, float *q1)
{
    if (x < 8.0) {
        float bessj1(float);
        void bessy01(float, float *, float *);
        float b,cosx,sinx,j1x,y1;
        b=sqrt(x)*1.25331413731550;
        bessy01(x,&j1x,&y1);
        j1x=bessj1(x);
        x -= 0.785398163397448;
        cosx=cos(x);
        sinx=sin(x);
        *p1 = b*(j1x*sinx-y1*cosx);
        *q1 = b*(j1x*cosx+y1*sinx);
    } else {
        int i;
        float x2,b0,b1,b2,y;
        static float ar1[11]={0.10668e-15, -0.72212e-15, 0.545267e-14,
            -0.4684224e-13, 0.46991955e-12, -0.570486364e-11,
            0.881689866e-10, -0.187189074911e-8, 0.6177633960644e-7,
            -0.39872843004889e-5, 0.89898983308594e-3};
        static float ar2[11]={-0.10269e-15, 0.65083e-15, -0.456125e-14,
            0.3596777e-13, -0.32643157e-12, 0.351521879e-11,
            -0.4686363688e-10, 0.82291933277e-9, -0.2095978138408e-7,
            0.91386152579555e-6, -0.96277235491571e-4};
        y=8.0/x;
        x=2.0*y*y-1.0;
        x2=x+x;
        b1=b2=0.0;
        for (i=0; i<=10; i++) {
            b0=x2*b1-b2+ar1[i];
            b2=b1;
            b1=b0;
```

```
        }
        *p1 = x*b1-b2+1.0009030408600137;
        b1=b2=0.0;
        for (i=0; i<=10; i++) {
            b0=x2*b1-b2+ar2[i];
            b2=b1;
            b1=b0;
        }
        *q1 = (x*b1-b2+0.46777787069535e-1)*y;
    }
}
```

6.5.2 Bessel functions I and K

A. bessi0

Computes the value of the modified Bessel function of the first kind of order zero $I_0(x)$. For $|x| \leq 15$, a Chebyshev rational function approximation [Bla74] of the form

$$\sum_{k=0}^{14} n_k x^{2k} \;/\; \sum_{k=0}^{3} d_k x^{2k}$$

is used. When $|x| > 15$, the function $I_0'(x) = e^{-x}I_0(x)$ is evaluated by means of a call of *nonexpbessi0*, and the relationship $I_0(x) = e^{x}I_0'(x)$ is used.

Function Parameters:

float bessi0 (x)

bessi0: delivers the modified Bessel function of the first kind of order zero with argument x;

x: float;
entry: the argument of the Bessel function.

Function used: nonexpbessi0.

```
#include <math.h>

float bessi0(float x)
{
    if (x == 0.0) return 1.0;
    if (fabs(x) <= 15.0) {
        float z,denominator,numerator;
        z=x*x;
        numerator=(z*(z*(z*(z*(z*(z*(z*(z*(z*(z*(z*(z*(z*(z*
            0.210580722890567e-22+0.380715242345326e-19)+
            0.479440257548300e-16)+0.435125971262668e-13)+
            0.300931127112960e-10)+0.160224679395361e-7)+
            0.654858370096785e-5)+0.202591084143397e-2)+
            0.463076284721000e0)+0.754337328948189e2)+
            0.830792541809429e4)+0.571661130563785e6)+
            0.216415572361227e8)+0.356644482244025e9)+
            0.144048298227235e10);
        denominator=(z*(z*(z-0.307646912682801e4)+
            0.347626332405882e7)-0.144048298227235e10);
        return -numerator/denominator;
    } else {
        float nonexpbessi0(float);
        return exp(fabs(x))*nonexpbessi0(x);
    }
}
```

B. bessi1

Computes the value of the modified Bessel function of the first kind of order one $I_1(x)$. For $|x| \leq 15$, a Chebyshev rational function approximation [Bla74] of the form

$$\sum_{k=0}^{14} n_k x^{2k} \Big/ \sum_{k=0}^{3} d_k x^{2k}$$

for the function $x^{-1}I_1(x)$ is used, and the value of $I_1(x)$ is recovered. When $|x| > 15$, the function $I_1'(x) = e^{-x}I_1(x)$ is evaluated by means of a call of *nonexpbessi1*, and the relationship
$I_1(x) = e^x I_1'(x)$ is used.

Function Parameters:

float bessi1 (x)

bessi1: delivers the modified Bessel function of the first kind of order one with argument x;

x: float;
entry: the argument of the Bessel function.

Function used: nonexpbessi1.

```
#include <math.h>

float bessi1(float x)
{
    if (x == 0.0) return 0.0;
    if (fabs(x) <= 15.0) {
        float z,denominator,numerator;
        z=x*x;
        denominator=z*(z-0.222583674000860e4)+0.136293593052499e7;
        numerator=(z*(z*(z*(z*(z*(z*(z*(z*(z*(z*(z*(z*(z*
            0.207175767232792e-26+0.257091905584414e-23)+
            0.306279283656135e-20)+0.261372772158124e-17)+
            0.178469361410091e-14)+0.963628891518450e-12)+
            0.410068906847159e-9)+0.135455228841096e-6)+
            0.339472890308516e-4)+0.624726195127003e-2)+
            0.806144878821295e0)+0.682100567980207e2)+
            0.341069752284422e4)+0.840705772877836e5)+
            0.681467965262502e6);
        return x*(numerator/denominator);
    } else {
        float nonexpbessi1(float);
        return exp(fabs(x))*nonexpbessi1(x);
    }
}
```

C. bessi

Generates an array of modified Bessel functions of the first kind of order one $I_j(x)$, $j=0,...,n$.

The functions $I_j'(x) = e^{-|x|}I_j(x)$, $j=0,...,n$, are first evaluated by means of a call of *nonexpbessi*, and the required values of $I_j(x)$ are recovered by multiplication by e^x.

Function Parameters:

void bessi (x,n,i)

x: float;
 entry: the argument of the Bessel functions;
n: int;
 entry: the upper bound of the indices of the array i;
i: float *i[0:n]*;
 exit: *i[j]* contains the value of the modified Bessel function of the first kind of order j, j=0,...,n.

Function used: nonexpbessi.

```
#include <math.h>

void bessi(float x, int n, float i[])
{
    if (x == 0.0) {
        i[0]=1.0;
        for (; n>=1; n--) i[n]=0.0;
    } else {
        void nonexpbessi(float, int, float []);
        float expx;
        expx=exp(fabs(x));
        nonexpbessi(x,n,i);
        for (; n>=0; n--) i[n] *= expx;
    }
}
```

D. bessk01

Computes the modified Bessel functions of the third kind of orders zero and one: $K_0(x)$ and $K_1(x)$ for x>0.

For $0 < x < 1.5$, $K_0(x)$ and $K_1(x)$ are computed by use of truncated versions of the Taylor series expansions [AbS65]

$$I_0(x) = \sum_{\nu=0}^{\infty}\left(\frac{x^2}{4}\right)^{\nu} / (\nu!)^2$$

$$K_0(x) = -\{\ln(\frac{x}{2}) + \gamma\} I_0(x) + \sum_{\nu=1}^{\infty}\{\sum_{\tau=1}^{\nu}\frac{1}{\tau}\}(\frac{x^2}{4})^{\nu} / (\nu!)^2$$

$$I_1(x) = \frac{x}{2}\sum_{\nu=0}^{\infty}(\frac{x^2}{4})^{\nu} / \{\nu!(\nu+1)!\}$$

$$K_1(x) = \frac{1}{x} + \frac{\gamma x}{2} - \frac{x}{4} + \ln(\frac{x}{2})I_1(x) -$$

$$\frac{x}{4}\sum_{\nu=1}^{\infty}\left\{\frac{1}{\nu+1} - 2\gamma + 2\sum_{\tau=1}^{\nu}\frac{1}{\tau}\right\}(\frac{x^2}{4})^{\nu} \;/\; \{\nu!(\nu+1)!\} .$$

For $x \geq 1.5$, the functions $K_j'(x) = e^x K_j(x)$, j=0,1, are evaluated by a call of *nonexpbessk01*, and the relationship $K_j(x)=e^{-x}K_j'(x)$ is used.

Function Parameters:

void bessk01 (*x,k0,k1*)

x: float;
  entry: the argument of the Bessel functions; $x > 0$;
k0: float *;
  exit: *k0* has the value of the modified Bessel function of the third kind of order zero with argument *x*;
k1: float *;
  exit: *k1* has the value of the modified Bessel function of the third kind of order one with argument *x*.

Function used: nonexpbessk01.

```
#include <math.h>

void bessk01(float x, float *k0, float *k1)
{
    if (x <= 1.5) {
        int k;
        float c,d,r,sum0,sum1,t,term,t0,t1;
        sum0=d=log(2.0/x)-0.5772156649015328606;
        sum1 = c = -1.0-2.0*d;
        r=term=1.0;
        t=x*x/4.0;
        k=1;
        do {
            term *= t*r*r;
            d += r;
            c -= r;
            r=1.0/(k+1);
            c -= r;
            t0=term*d;
            t1=term*c*r;
            sum0 += t0;
            sum1 += t1;
            k++;
        } while (fabs(t0/sum0)+fabs(t1/sum1) > 1.0e-15);
        *k0 = sum0;
        *k1 = (1.0+t*sum1)/x;
    } else {
        void nonexpbessk01(float, float *, float *);
        float expx;
        expx=exp(-x);
        nonexpbessk01(x,k0,k1);
        *k1 *= expx;
        *k0 *= expx;
    }
}
```

E. bessk

Generates an array of modified Bessel functions of the third kind of order j, $K_j(x)$, j=0,...,n, for $x > 0$.

The functions $K_0(x)$ and $K_1(x)$ are first evaluated by means of a call of *bessk01*, and the recursion [AbS65]

$$K_{j+1}(x) = K_{j-1}(x) + (2j/x)K_j(x)$$

is then used.

Function Parameters:

void bessk (x,n,k)

x: float;
  entry: the argument of the Bessel functions; $x > 0$;
n: int;
  entry: the upper bound of the indices of array k; $n \geq 0$;
k: float $k[0:n]$;
  exit: $k[j]$ is the value of the modified Bessel function of the third kind of order j with argument x, j=0,...,n.

Function used: bessk01.

```
#include <math.h>

void bessk(float x, int n, float k[])
{
    void bessk01(float, float *, float *);
    int i;
    float k0,k1,k2;

    bessk01(x,&k0,&k1);
    k[0]=k0;
    if (n > 0) k[1]=k1;
    x=2.0/x;
    for (i=2; i<=n; i++) {
        k[i]=k2=k0+x*(i-1)*k1;
        k0=k1;
        k1=k2;
    }
}
```

F. nonexpbessi0

Computes the value of the modified Bessel function of the first kind of order zero multiplied by $e^{-|x|}$.

nonexpbessi0 evaluates the function $I_0'(x) = e^{-|x|}I_0(x)$. When $|x| \leq 15$, the function $I_0(x)$ is evaluated by means of a call of *bessi0*, and the function I_0' is computed by use of its defining relationship. When $|x| > 15$, a Chebyshev rational function approximation [Bla74] of the form

$$\sum_{k=0}^{4} n_k T_k(\frac{30}{x} - 1) \; / \; \sum_{k=0}^{3} d_k T_k(\frac{30}{x} - 1)$$

for the function $x^{1/2}I_0'(x)$ is used, and the value of $I_0'(x)$ is recovered.

Function Parameters:

float nonexpbessi0 (x)

nonexpbessi0: delivers the modified Bessel function of the first kind of order zero with argument x, multiplied by $e^{-|x|}$;

x: float;
entry: the argument of the Bessel function.

Function used: bessi0.

```
#include <math.h>

float nonexpbessi0(float x)
{
	if (x == 0.0) return 1.0;
	if (fabs(x) <= 15.0) {
		float bessi0(float);
		return exp(-fabs(x))*bessi0(x);
	} else {
		int i;
		float sqrtx,br,br1,br2,z,z2,numerator,denominator;
		static float ar1[4]={0.2439260769778, -0.115591978104435e3,
			0.784034249005088e4, -0.143464631313583e6};
		static float ar2[4]={1.0, -0.325197333369824e3,
			0.203128436100794e5, -0.361847779219653e6};
		x=fabs(x);
		sqrtx=sqrt(x);
		br1=br2=0.0;
		z=30.0/x-1.0;
		z2=z+z;
		for (i=0; i<=3; i++) {
			br=z2*br1-br2+ar1[i];
			br2=br1;
			br1=br;
		}
		numerator=z*br1-br2+0.346519833357379e6;
		br1=br2=0.0;
		for (i=0; i<=3; i++) {
			br=z2*br1-br2+ar2[i];
			br2=br1;
			br1=br;
		}
		denominator=z*br1-br2+0.865665274832055e6;
		return (numerator/denominator)/sqrtx;
	}
}
```

G. nonexpbessi1

Computes the value of the modified Bessel function of the first kind of order one multiplied by $e^{-|x|}$.

nonexpbessi1 evaluates the function $I_1'(x) = \text{sign}(x)e^{-|x|}I_1(x)$. When $|x| \leq 15$, the function $I_1(x)$ is evaluated by means of a call of *bessi1*, and the function I_1' is computed by use of its defining relationship. When $|x| > 15$, a Chebyshev rational function approximation [Bla74] of the form

$$\sum_{k=0}^{4} n_k T_k(\frac{30}{x} - 1) \;/\; \sum_{k=0}^{3} d_k T_k(\frac{30}{x} - 1)$$

for the function $x^{1/2}I_1'(x)$ is used, and the value of $I_1'(x)$ is recovered.

Function Parameters:

float nonexpbessi1 (x)

nonexpbessi1: delivers the modified Bessel function of the first kind of order one with argument x, multiplied by $e^{-|x|}$;

x: float;
entry: the argument of the Bessel function.

Function used: bessi1.

```
#include <math.h>

float nonexpbessi1(float x)
{
    if (x == 0.0) return 0.0;
    if (fabs(x) > 15.0) {
        int i,signx;
        float br,br1,br2,z,z2,sqrtx,numerator,denominator;
        static float ar1[4]={0.1494052814740e1, -0.362026420242263e3,
            0.220549722260336e5, -0.408928084944275e6};
        static float ar2[4]={1.0, -0.631003200551590e3,
            0.496811949533398e5, -0.100425428133695e7};
        signx = (x > 0.0) ? 1 : -1;
        x=fabs(x);
        sqrtx=sqrt(x);
        z=30.0/x-1.0;
        z2=z+z;
        br1=br2=0.0;
        for (i=0; i<=3; i++) {
            br=z2*br1-br2+ar1[i];
            br2=br1;
            br1=br;
        }
        numerator=z*br1-br2+0.102776692371524e7;
        br1=br2=0.0;
        for (i=0; i<=3; i++) {
            br=z2*br1-br2+ar2[i];
            br2=br1;
            br1=br;
        }
        denominator=z*br1-br2+0.26028876789105e7;
        return ((numerator/denominator)/sqrtx)*signx;
    } else {
        float bessi1(float);
        return exp(-fabs(x))*bessi1(x);
    }
}
```

H. nonexpbessi

Generates an array of modified Bessel functions of the first kind of order one $I_j(x)$, multiplied by $e^{-|x|}$, j=0,...,n.

nonexpbessi evaluates the functions $I_j'(x) = e^{-|x|}I_j(x)$, j=0,...,n. The method used [G67] is that described in the documentation to *bessj*. The functions $I_{a+k}'(x)$ satisfy the recursion

$$I_{a+k+1}'(x) + \{2(a+k)/x\}I_{a+k}'(x) - I_{a+k-1}(x) = 0$$

and the relationship

$$I_a(x) + 2\sum_{k=1}^{\infty} \frac{(a+k)\Gamma(2a+k)}{k!\,\Gamma(2a+1)} I_{a+k}'(x) = \frac{(x/2)^a}{\Gamma(a+1)}$$

when x=0, a_k and b_k in formula (1) of the documentation to *bessj* are $a_k = 2k/x$, $b_k = -1$ and $\lambda_k = 1$, s=1 may be taken in the further formulae of that documentation.

Function Parameters:

void nonexpbessi (x,n,i)

x: float;
entry: the argument of the Bessel functions;

n: int;
entry: the upper bound of the indices of the array *i*, $n \geq 0$;

i: float *i[0:n]*;
exit: *i[j]* contains the value of the modified Bessel function of the first kind of order *j*, multiplied by $e^{-|x|}$, $j=0,...,n$.

Function used: start.

```
#include <math.h>

void nonexpbessi(float x, int n, float i[])
{
    if (x == 0.0) {
        i[0]=1.0;
        for (; n>=1; n--) i[n]=0.0;
    } else {
        int start(float, int, int);
        int k,negative;
        float x2,r,s;
        negative = (x < 0.0);
        x=fabs(x);
        r=s=0.0;
        x2=2.0/x;
        k=start(x,n,1);
        for (; k>=1; k--) {
            r=1.0/(r+x2*k);
            s=r*(2.0+s);
            if (k <= n) i[k]=r;
        }
        i[0]=r=1.0/(1.0+s);
        if (negative)
            for (k=1; k<=n; k++) r = i[k] *= (-r);
        else
            for (k=1; k<=n; k++) r = i[k] *= r;
    }
}
```

I. nonexpbessk01

Computes the modified Bessel functions of the third kind of orders zero and one, $K_0(x)$ and $K_1(x)$ for $x>0$, multiplied by e^x.

nonexpbessk01 evaluates the functions $K_j'(x) = e^x K_j(x)$ for $x > 0$, $j=0,1$. For $0<x\leq 1.5$, the functions $K_0(x)$ and $K_1(x)$ are computed by a call of *bessk01* and the K_j' are evaluated by use of their defining relationship. For $1.5<x\leq 5$, the trapezoidal rule (see [Hu64] of *bessi0*) is used to evaluate the integrals (see [AbS65] of *bessi0*)

$$K_j'(x) = \sqrt{\frac{\pi}{2x}}\,\frac{\mu^{j+1/2}}{\Gamma(j+1/2)}\int_{-\infty}^{\infty} e^{-\mu t^2} t^{2j}\left(1+\frac{t^2\mu}{2x}\right)dt \qquad (j=0,1)$$

with $\mu = 2/5$. For $x > 5$, truncated Chebyshev expansions (see [Cle62, Lu69]) of the form

$$K_j'(x) = \sqrt{\frac{\pi}{2x}} \sum_{n=0} d_n T_n(\frac{10}{x} - 1)$$

are used.

Function Parameters:

void nonexpbessk01 $(x,k0,k1)$

x: float;
entry: the argument of the Bessel functions; $x > 0$;

$k0$: float *;
exit: $k0$ has the value of the modified Bessel function of the third kind of order zero with argument x, multiplied by e^x;

$k1$: float *;
exit: $k1$ has the value of the modified Bessel function of the third kind of order one with argument x, multiplied by e^x.

Function used: bessk01.

```
#include <math.h>

void nonexpbessk01(float x, float *k0, float *k1)
{
    if (x <= 1.5) {
        void bessk01(float, float *, float *);
        float expx;
        expx=exp(x);
        bessk01(x,k0,k1);
        *k0 *= expx;
        *k1 *= expx;
    } else if (x <= 5.0) {
        int i,r;
        float t2,s1,s2,term1,term2,sqrtexpr,exph2,x2;
        static float fac[20]={0.90483741803596, 0.67032004603564,
            0.40656965974060, 0.20189651799466, 0.82084998623899e-1,
            0.27323722447293e-1, 0.74465830709243e-2,
            0.16615572731739e-2, 0.30353913807887e-3,
            0.45399929762485e-4, 0.55595132416500e-5,
            0.55739036926944e-6, 0.45753387694459e-7,
            0.30748798795865e-8, 0.16918979226151e-9,
            0.76218651945127e-11, 0.28111852987891e-12,
            0.84890440338729e-14, 0.2098791048793e-15,
            0.42483542552916e-17};
        s1=0.5;
        s2=0.0;
        r=0.0;
        x2=x+x;
        exph2=1.0/sqrt(5.0*x);
        for (i=0; i<=19; i++) {
            r += 1.0;
            t2=r*r/10.0;
            sqrtexpr=sqrt(t2/x2+1.0);
            term1=fac[i]/sqrtexpr;
            term2=fac[i]*sqrtexpr*t2;
            s1 += term1;
            s2 += term2;
        }
        *k0 = exph2*s1;
        *k1 = exph2*s2*2.0;
    } else {
        int r,i;
        float br,br1,br2,cr,cr1,cr2,ermin1,erplus1,er,f0,f1,
                expx,y,y2;
        static float dr[14]={0.27545e-15, -0.172697e-14,
                0.1136042e-13, -0.7883236e-13, 0.58081063e-12,
                -0.457993633e-11, 0.3904375576e-10, -0.36454717921e-9,
                0.379299645568e-8, -0.450473376411e-7,
                0.63257510850049e-6, -0.11106685196665e-4,
```

```
                0.26953261276272e-3, -0.11310504646928e-1};
        y=10.0/x-1.0;
        y2=y+y;
        r=30;
        br1=br2=cr1=cr2=erplus1=er=0.0;
        for (i=0; i<=13; i++) {
            r -= 2;
            br=y2*br1-br2+dr[i];
            cr=cr1*y2-cr2+er;
            ermin1=r*dr[i]+erplus1;
            erplus1=er;
            er=ermin1;
            br2=br1;
            br1=br;
            cr2=cr1;
            cr1=cr;
        }
        f0=y*br1-br2+0.9884081742308258;
        f1=y*cr1-cr2+er/2.0;
        expx=sqrt(1.5707963267949/x);
        *k0 = f0 *=expx;
        *k1 = (1.0+0.5/x)*f0+(10.0/x/x)*expx*f1;
    }
}
```

J. nonexpbessk

Generates an array of modified Bessel functions of the third kind of order j, $K_j(x)$, multiplied by e^x, j=0,...,n, for $x > 0$.

nonexpbessk computes the values of the functions $K_j'(x) = e^x K_j(x)$, j=0,...,n, where $K_j(x)$ is a modified Bessel function of the third kind of order j, for $x > 0$.

The functions $K_0'(x)$ and $K_1'(x)$ are first evaluated by means of a call of *nonexpbessk01*, and the recursion [AbS65]

$$K_{j+1}'(x) = K_{j-1}'(x) + (2j/x)K_j'(x)$$

is then used.

Function Parameters:

void nonexpbessk (x,n,k)

x: float;
  entry: the argument of the Bessel functions; $x > 0$;

n: int;
  entry: the upper bound of the indices of array k; $n \geq 0$;

k: float $k[0:n]$;
  exit: $k[j]$ is value of the modified Bessel function of the third kind of order j multiplied by e^x, j=0,...,n.

Function used: nonexpbessk01.

```
#include <math.h>

void nonexpbessk(float x, int n, float k[])
{
    void nonexpbessk01(float, float *, float *);
    int i;
    float k0,k1,k2;

    nonexpbessk01(x,&k0,&k1);
    k[0]=k0;
    if (n > 0) k[1]=k1;
    x=2.0/x;
    for (i=2; i<=n; i++) {
```

```
            k[i]=k2=k0+x*(i-1)*k1;
            k0=k1;
            k1=k2;
        }
}
```

6.6 Bessel functions of real order

6.6.1 Bessel functions J and Y

A. bessjaplusn

Calculates the Bessel functions of the first kind of order $a+k$, $J_{a+k}(x)$, $(0 \leq k \leq n,\ 0 \leq a < 1)$. If $x=0$ then $J_a(x)$ equals 0 if $a>0$ and equals 1 if $a=0$, and $J_{a+k}(x)=0$ for $k=1,...,n$. If $a=½$, the functions $J_{a+k}(x)$ are evaluated by means of a call of *spherbessj*; otherwise the method used is that described in the documentation to *bessj*. Now a_k and b_k in formula (1) of that documentation are

$$a_k = 2(a+k)/x,\ \ b_k = -1,$$

and

$$\lambda_0 = 1,\ \lambda_{2m-1} = 0,\ \lambda_{2m} = (a+2m)\Gamma(a+m)/\{m!\Gamma(a+m)\}\ \ (m=1,2,...),$$

$s=(x/2)^a/\Gamma(a+1)$ may be taken in the further formulae of that documentation see [G67].

Function Parameters:

void bessjaplusn *(a,x,n,ja)*

a: float;
 entry: the noninteger part of the order; $0 \leq a < 1$;
x: float;
 entry: the argument value, $x \geq 0$;
n: int;
 entry: the upper bound of the indices of the array *ja*;
ja: float *ja[0:n]*;
 exit: *ja[k]* is assigned the value of the Bessel function of the first kind $J_{a+k}(x)$, $0 \leq k \leq n$.

Functions used: bessj, spherbessj, gamma, start.

```
#include <math.h>

void bessjaplusn(float a, float x, int n, float ja[])
{
    if (x == 0.0) {
        ja[0] = (a == 0.0) ? 1.0 : 0.0;
        for (; n>=1; n--) ja[n]=0.0;
    } else if (a == 0.0) {
        void bessj(float, int, float []);
        bessj(x,n,ja);
    } else if (a == 0.5) {
        void spherbessj(float, int, float []);
        float s;
        s=sqrt(x)*0.797884560802865;
        spherbessj(x,n,ja);
        for (; n>=0; n--) ja[n] *= s;
    } else {
        float gamma(float);
        int start(float, int, int);
        int k,m,nu;
```

```
        float a2,x2,r,s,l,labda;
        l=1.0;
        nu=start(x,n,0);
        for (m=1; m<=nu; m++) l=l*(m+a)/(m+1);
        r=s=0.0;
        x2=2.0/x;
        k = -1;
        a2=a+a;
        for (m=nu+nu; m>=1; m--) {
            r=1.0/(x2*(a+m)-r);
            if (k == 1)
                labda=0.0;
            else {
                l=l*(m+2)/(m+a2);
                labda=l*(m+a);
            }
            s=r*(labda+s);
            k = -k;
            if (m <= n) ja[m]=r;
        }
        ja[0]=r=1.0/gamma(1.0+a)/(1.0+s)/pow(x2,a);
        for (m=1; m<=n; m++) r = ja[m] *= r;
    }
}
```

B. bessya01

Computes the Bessel functions of the second kind (also called Neumann's functions) of order a and a+1: $Y_a(x)$ and $Y_{a+1}(x)$ for x>0, a≥0.

For $x < 3$, the above functions are evaluated by use of truncated Taylor series (see [T76a]). For x≥3, the functions $P_\alpha(x)$, $Q_\alpha(x)$ occurring in the formula

$$Y_\alpha(x) = (2/(\pi x))^{1/2}\{P_\alpha(x)\sin(x - (\alpha+½)\pi/2) + Q_\alpha(x)\cos(x - (\alpha+½)\pi/2)\}$$

are evaluated for $\alpha=a$, $a+1$ by means of a call of *besspqa01*; the values of $Y_a(x)$, $Y_{a+1}(x)$ are then recovered.

Function Parameters:

void bessya01 (*a,x,ya,ya1*)

a: float;
entry: the order;
x: float;
entry: this argument should satisfy $x > 0$;
ya: float *;
exit: the Neumann function of order a and argument x;
ya1: float *;
exit: the Neumann function of order a+1.

Functions used: bessy01, recipgamma, besspqa01.

```
#include <math.h>

void bessya01(float a, float x, float *ya, float *ya1)
{
    if (a == 0.0) {
        void bessy01(float, float *, float *);
        bessy01(x,ya,ya1);
    } else {
        int n,na,rec,rev;
        float b,c,d,e,f,g,h,p,pi,q,r,s;
        pi=4.0*atan(1.0);
        na=floor(a+0.5);
        rec = (a >= 0.5);
```

```
        rev = (a < -0.5);
        if (rev || rec) a -= na;
        if (a == -0.5) {
            p=sqrt(2.0/pi/x);
            f=p*sin(x);
            g = -p*cos(x);
        } else if (x < 3.0) {
            float recipgamma(float, float *, float *);
            b=x/2.0;
            d = -log(b);
            e=a*d;
            c = (fabs(a) < 1.0e-8) ? 1.0/pi : a/sin(a*pi);
            s = (fabs(e) < 1.0e-8) ? 1.0 : sinh(e)/e;
            e=exp(e);
            g=recipgamma(a,&p,&q)*e;
            e=(e+1.0/e)/2.0;
            f=2.0*c*(p*e+q*s*d);
            e=a*a;
            p=g*c;
            q=1.0/g/pi;
            c=a*pi/2.0;
            r = (fabs(c) < 1.0e-8) ? 1.0 : sin(c)/c;
            r *= pi*c*r;
            c=1.0;
            d = -b*b;
            *ya = f+r*q;
            *ya1 = p;
            n=1;
            do {
                f=(f*n+p+q)/(n*n-e);
                c=c*d/n;
                p /= (n-a);
                q /= (n+a);
                g=c*(f+r*q);
                h=c*p-n*g;
                *ya += g;
                *ya1 += h;
                n++;
            } while (fabs(g/(1.0+fabs(*ya)))+fabs(h/(1.0+fabs(*ya1))) >
                            1.0e-15);
            f = -(*ya);
            g = -(*ya1)/b;
        } else {
            void besspqa01(float,float,float *,float *,float *,float *);
            b=x-pi*(a+0.5)/2.0;
            c=cos(b);
            s=sin(b);
            d=sqrt(2.0/x/pi);
            besspqa01(a,x,&p,&q,&b,&h);
            f=d*(p*s+q*c);
            g=d*(h*s-b*c);
        }
        if (rev) {
            x=2.0/x;
            na = -na-1;
            for (n=0; n<=na; n++) {
                h=x*(a-n)*f-g;
                g=f;
                f=h;
            }
        } else if (rec) {
            x=2.0/x;
            for (n=1; n<=na; n++) {
                h=x*(a+n)*g-f;
                f=g;
                g=h;
            }
        }
        *ya = f;
        *ya1 = g;
    }
}
```

C. bessyaplusn

Generates an array of Bessel functions of the second kind of order $a+n$, $Y_{a+n}(x)$, $n=0,...,nmax$, for $x > 0$, $a \geq 0$.

The values of the functions $Y_a(x)$, $Y_{a+1}(x)$ are first obtained by means of a call of *bessya01*, and the recursion

$$Y_{a+n+1}(x) = -Y_{a+n-1}(x) + \{2(n+a)/x\}Y_{a+n}(x)$$

is then used.

Function Parameters:

void bessyaplusn *(a,x,nmax,jan)*

a: float;
 entry: the order;
x: float;
 entry: the argument value, $x > 0$;
nmax: int;
 entry: the upper bound of the indices of the array *yan*; $nmax \geq 0$;
yan: float *yan[0:nmax]*;
 exit: the values of the Bessel functions of the second kind of order $a+n$, for argument x are assigned to *yan[n]*, $0 \leq n \leq nmax$.

Function used: bessya01.

```
#include <math.h>

void bessyaplusn(float a, float x, int nmax, float yan[])
{
    void bessya01(float, float, float *, float *);
    int n;
    float y1;

    bessya01(a,x,&yan[0],&y1);
    a -= 1.0;
    x=2.0/x;
    if (nmax > 0) yan[1]=y1;
    for (n=2; n<=nmax; n++) yan[n] = -yan[n-2]+(a+n)*x*yan[n-1];
}
```

D. besspqa01

This procedure is an auxiliary procedure for the computation of the Bessel functions for large values of their argument.

besspqa01 evaluates the functions $P_\alpha(x)$, $Q_\alpha(x)$ occurring in the formulae

$$J_\alpha(x) = (2/(\pi x))^{1/2}\{P_\alpha(x)\cos(x - (\alpha+½)\pi/2) - Q_\alpha(x)\sin(x - (\alpha+½)\pi/2)\}$$
$$Y_\alpha(x) = (2/(\pi x))^{1/2}\{P_\alpha(x)\sin(x - (\alpha+½)\pi/2) + Q_\alpha(x)\cos(x - (\alpha+½)\pi/2)\}$$

for $\alpha = a, a+1$ ($a \geq 0$ and $x > 0$).

If $x < 3$ then the formulae

$$P_a(x) = (\pi x/2)^{1/2}\{Y_a(x)\sin(x - (a+½)\pi/2) + J_a(x)\cos(x - (a+½)\pi/2)\}$$
$$Q_a(x) = (\pi x/2)^{1/2}\{Y_a(x)\cos(x - (a+½)\pi/2) - J_a(x)\sin(x - (a+½)\pi/2)\}$$
$$P_{a+1}(x) = (\pi x/2)^{1/2}\{J_{a+1}(x)\sin(x - (a+½)\pi/2) - Y_{a+1}(x)\cos(x - (a+½)\pi/2)\}$$
$$Q_{a+1}(x) = (\pi x/2)^{1/2}\{J_{a+1}(x)\cos(x - (a+½)\pi/2) + Y_{a+1}(x)\sin(x - (a+½)\pi/2)\}$$

are used, *bessjaplusn* and *bessya01* being called to evaluate the functions $J_a,...,Y_{a+1}$. When $x \geq 3$, Chebyshev expansions are used (see [T76a, W45]).

Function Parameters:

void besspqa01 $(a,x,pa,qa,pa1,qa1)$

a: float;
entry: the order;
x: float;
entry: this argument should satisfy $x > 0$;
pa: float *;
exit: the value of $P_a(x)$;
qa: float *;
exit: the value of $Q_a(x)$;
$pa1$: float *;
exit: the value of $P_{a+1}(x)$;
$qa1$: float *;
exit: the value of $Q_{a+1}(x)$.

Functions used: besspq0, besspq1, bessjaplusn, bessya01.

```
#include <math.h>

void besspqa01(float a, float x, float *pa, float *qa, float *pa1,
               float *qa1)
{
    if (a == 0.0) {
        void besspq0(float, float *, float *);
        void besspq1(float, float *, float *);
        besspq0(x,pa,qa);
        besspq1(x,pa1,qa1);
    } else {
        int n,na,rec,rev;
        float b,pi,p0,q0;
        pi=4.0*atan(1.0);
        rev = (a < -0.5);
        if (rev) a = -a-1.0;
        rec = (a >= 0.5);
        if (rec) {
            na=floor(a+0.5);
            a -= na;
        }
        if (a == -0.5) {
            *pa = *pa1 = 1.0;
            *qa = *qa1 = 0.0;
        } else if (x >= 3.0) {
            float c,d,e,f,g,p,q,r,s,temp;
            c=0.25-a*a;
            b=x+x;
            f=r=1.0;
            g = -x;
            s=0.0;
            temp=x*cos(a*pi)/pi*1.0e15;
            e=temp*temp;
            n=2;
            do {
                d=(n-1+c/n);
                p=(2*n*f+b*g-d*r)/(n+1);
                q=(2*n*g-b*f-d*s)/(n+1);
                r=f;
                f=p;
                s=g;
                g=q;
                n++;
            } while ((p*p+q*q)*n*n < e);
            e=f*f+g*g;
            p=(r*f+s*g)/e;
            q=(s*f-r*g)/e;
            f=p;
```

```
            g=q;
            n--;
            while (n > 0) {
                r=(n+1)*(2.0-p)-2.0;
                s=b+(n+1)*q;
                d=(n-1+c/n)/(r*r+s*s);
                p=d*r;
                q=d*s;
                e=f;
                f=p*(e+1.0)-g*q;
                g=q*(e+1.0)+p*g;
                n--;
            }
            f += 1.0;
            d=f*f+g*g;
            *pa = f/d;
            *qa = -g/d;
            d=a+0.5-p;
            q += x;
            *pa1 = ((*pa)*q-(*qa)*d)/x;
            *qa1 = ((*qa)*q+(*pa)*d)/x;
        } else {
            void bessjaplusn(float, float, int, float []);
            void bessya01(float, float, float *, float *);
            float c,s,chi,ya,ya1,ja[2];
            b=sqrt(pi*x/2.0);
            chi=x-pi*(a/2.0+0.25);
            c=cos(chi);
            s=sin(chi);
            bessya01(a,x,&ya,&ya1);
            bessjaplusn(a,x,1,ja);
            *pa = b*(ya*s+c*ja[0]);
            *qa = b*(c*ya-s*ja[0]);
            *pa1 = b*(s*ja[1]-c*ya1);
            *qa1 = b*(c*ja[1]+s*ya1);
        }
        if (rec) {
            x=2.0/x;
            b=(a+1.0)*x;
            for (n=1; n<=na; n++) {
                p0=(*pa)-(*qa1)*b;
                q0=(*qa)+(*pa1)*b;
                *pa = *pa1;
                *pa1 = p0;
                *qa = *qa1;
                *qa1 = q0;
                b += x;
            }
        }
        if (rev) {
            p0 = *pa1;
            *pa1 = *pa;
            *pa = p0;
            q0 = *qa1;
            *qa1 = *qa;
            *qa = q0;
        }
    }
}
```

E. besszeros

Calculates the first *n* zeros of a Bessel function of the first or the second kind or its derivative.

besszero computes the first *n* zeros of either 1) $J_a(z)$ or 2) $Y_a(z)$ or 3) $dJ_a(z)/dz$ or 4) $dY_a(z)/dz$, ($a \geq 0$). Which of the above set of zeros is derived is determined by the value of the parameter *d* upon call: thus 1 for the zeros of $J_a(z)$, and so on.

Each zero is obtained by use of an initial approximation derived from an asymptotic expansion [AbS65], and subsequent improvement by use of a Newton-Raphson process [T76b, T78].

If $a < 3$, then with $\mu=4a^2$, the s-th zero of $J_a(z)$ or $Y_a(z)$ is given by

$$z_{a,s} \approx \beta - (\mu-1)/(8\beta) - 4(\mu-1)(7\mu-31)/\{3(8\beta)^2\} - 32(\mu-1)(83\mu^2-982\mu+3779)/\{15(8\beta)^5\} \ldots$$

where $\beta = (s + a/2 - 1/4)\pi$ for $J_a(z)$ and $\beta = (s + a/2 - 3/4)\pi$ for $Y_a(z)$.
Similarly the s-th zero of $dJ_a(z)/dz$ or $dY_a(z)/dz$ is given by

$$z'_{a,s} \approx \beta' - (\mu+3)/(8\beta') - 4(7\mu^2+82\mu-9)/\{3(8\beta')^3\} - 32(83\mu^3+2075\mu^2-3039\mu+3537)/\{15(8\beta')^5\} \ldots$$

where $\beta' = (s + a/2 - 3/4)\pi$ for $dJ_a(z)/dz$ and $\beta' = (s + a/2 - 1/4)\pi$ for $dY_a(z)/dz$.

If $a \geq 3$, then with $w(u)$ defined by

$$(2/3)(-u)^{3/2} = \{w(u)^2 - 1\}^{1/2} - \text{arc}\cos(w(u)^{-1})$$

i.e. $w(u) = 1/\cos\phi(u)$ where

$$\{\tan\phi(u)\} - \phi(u) = (2/3)(-u)^{3/2}$$

the s-th zero of $J_a(z)$ or $Y_a(z)$ is given by

$$z_{a,s} \approx aw[-a^{2/3}f\{3\pi(4s-2k-1)/8\}]$$

where

$$f\{x\} \approx x^{2/3}(1 + (5/48)x^{-2} - (5/36)x^{-4} + \ldots)$$

where k=0 for $J_a(z)$ and k=1 for $Y_a(z)$. Similarly, the s-th zero of $dJ_a(z)/dz$ or $dY_a(z)/dz$ is given by

$$z'_{a,s} \approx aw[-a^{2/3}g\{3\pi(4s-2k-1)/8\}]$$

where

$$g\{x\} \approx x^{2/3}(1 - (7/48)x^{-2} + (35/288)x^{-4} + \ldots)$$

where now k=1 for $dJ_a(z)/dz$ and k=0 for $dY_a(z)/dz$.

Function Parameters:

void besszeros (a,n,z,d)

a: float;
entry: the order of the Bessel function, $a \geq 0$;
n: int;
entry: the number of zeros to be evaluated, $n \geq 1$;
z: float $z[1:n]$;
exit: $z[j]$ is the j-th zero of the selected Bessel function;
d: int;
entry: the choice of d determines the type of the Bessel function of which the zeros are computed:
if d=1 then $J_a(z)$;
if d=2 then $Y_a(z)$;
if d=3 then $dJ_a(z)/dz$;
if d=4 then $dY_a(z)/dz$.

Function used: besspqa01.

```
#include <math.h>

void besszeros(float a, int n, float z[], int d)
{
    void besspqa01(float, float, float *, float *, float *, float *);
    int j,s;
    float aa,a2,b,bb,c,chi,co,mu,mu2,mu3,mu4,p,pi,pa,pa1,p0,p1,pp1,
          q,qa,qa1,q1,qq1,ro,si,t,tt,u,v,w,x,xx,x4,y,yy,fi;
```

```
pi=4.0*atan(1.0);
aa=a*a;
mu=4.0*aa;
mu2=mu*mu;
mu3=mu*mu2;
mu4=mu2*mu2;
if (d < 3) {
    p=7.0*mu-31.0;
    p0=mu-1.0;
    p1=4.0*(253.0*mu2-3722.0*mu+17869.0)/15.0/p*p0;
    q1=8.0*(83.0*mu2-982.0*mu+3779.0)/5.0/p;
} else {
    p=7.0*mu2+82.0*mu-9.0;
    p0=mu+3.0;
    p1=(4048.0*mu4+131264.0*mu3-221984.0*mu2-
        417600.0*mu+1012176.0)/60.0/p;
    q1=1.6*(83.0*mu3+2075.0*mu2-3039.0*mu+3537.0)/p;
}
t = (d == 1 || d == 4) ? 0.25 : 0.75;
tt=4.0*t;
if (d < 3) {
    pp1=5.0/48.0;
    qq1 = -5.0/36.0;
} else {
    pp1 = -7.0/48.0;
    qq1=35.0/288.0;
}
y=3.0*pi/8.0;
bb = (a >= 3.0) ? pow(a,-2.0/3.0) : 0.0;
for (s=1; s<=n; s++) {
    if (a == 0.0 && s == 1 && d == 3) {
        x=0.0;
        j=0;
    } else {
        if (s >= 3.0*a-8.0) {
            b=(s+a/2.0-t)*pi;
            c=1.0/b/b/64.0;
            x=b-1.0/b/8.0*(p0-p1*c)/(1.0-q1*c);
        } else {
            if (s == 1)
                x = ((d == 1) ? -2.33811 : ((d == 2) ? -1.17371 :
                        ((d == 3) ? -1.01879 : -2.29444)));
            else {
                x=y*(4.0*s-tt);
                v=1.0/x/x;
                x = -pow(x,2.0/3.0)*(1.0+v*(pp1+qq1*v));
            }
            u=x*bb;
            yy=2.0/3.0*pow(-u,1.5);
            if (yy == 0.0)
                fi=0.0;
            else if (yy > 1.0e5)
                fi=1.570796;
            else {
                float r,p,pp;
                if (yy <1.0) {
                    p=pow(3.0*yy,1.0/3.0);
                    pp=p*p;
                    p *= (1.0+pp*(-210.0+pp*(27.0-2.0*pp))/1575.0);
                } else {
                    p=1.0/(yy+1.570796);
                    pp=p*p;
                    p=1.570796-p*(1.0+pp*(2310.0+pp*(3003.0+pp*
                        (4818.0+pp*(8591.0+pp*16328.0))))/3465.0);
                }
                pp=(yy+p)*(yy+p);
                r=(p-atan(p+yy))/pp;
                fi=p-(1.0+pp)*r*(1.0+r/(p+yy));
            }
            v=fi;
            w=1.0/cos(v);
            xx=1.0-w*w;
            c=sqrt(u/xx);
            x=w*(a+c/a/u*((d < 3) ?
```

```
                -5.0/48.0/u-c*(-5.0/24.0/xx+1.0/8.0) :
                7.0/48.0/u+c*(-7.0/24.0/xx+3.0/8.0)));
        }
        j=0;
        do {
            xx=x*x;
            x4=xx*xx;
            a2=aa-xx;
            besspqa01(a,x,&pa,&qa,&pa1,&qa1);
            chi=x-pi*(a/2.0+0.25);
            si=sin(chi);
            co=cos(chi);
            ro = ((d == 1) ? (pa*co-qa*si)/(pa1*si+qa1*co) :
                    ((d == 2) ? (pa*si+qa*co)/(qa1*si-pa1*co) :
                    ((d == 3) ? a/x-(pa1*si+qa1*co)/(pa*co-qa*si) :
                                a/x-(qa1*si-pa1*co)/(pa*si+qa*co))));
            j++;
            if (d < 3) {
                u=ro;
                p=(1.0-4.0*a2)/6.0/x/(2.0*a+1.0);
                q=(2.0*(xx-mu)-1.0-6.0*a)/3.0/x/(2.0*a+1.0);
            } else {
                u = -xx*ro/a2;
                v=2.0*x*a2/(aa+xx)/3.0;
                w=a2*a2*a2;
                q=v*(1.0+(mu2+32.0*mu*xx+48.0*x4)/32.0/w);
                p=v*(1.0+(-mu2+40.0*mu*xx+48.0*x4)/64.0/w);
            }
            w=u*(1.0+p*ro)/(1.0+q*ro);
            x += w;
        } while (fabs(w/x) > 1.0e-13 && j < 5);
      }
      z[s]=x;
    }
}
```

F. start

This is an auxiliary procedure which computes a starting value of an algorithm used in several Bessel function procedures.

Certain stable methods for evaluating functions $f_0, f_1, ..., f_n$ which satisfy a three term recurrence relationship of the form $f_{k+1} + a_k f_k + b_k f_{k-1} = 0$ require a value of ν to be determined such that the value of the continued fraction

$$r_{k-1} = \frac{-b_k}{a_k -} \frac{b_{k+1}}{a_{k+1} -} \cdots \tag{1}$$

and that of its convergent

$$r_{k-1}^{(\nu)} = \frac{-b_k}{a_k -} \frac{b_{k+1}}{a_{k+1} -} \cdots \frac{b_\nu}{a_\nu}$$

should agree. (The theory of such methods [see T76b], which involve backward recursion, is described in outline in the documentation to *bessj* which concerns the case in which $f_k = J_k(x)$; the same methods are also implemented by *bessjaplusn* (for which $f_k = J_{a+k}(x)$), by *nonexpbessi* (for which $f_k = I_k(x)$), by *nonexpbessiaplusn* (for which $f_k = e^{-x} I_{a+k}(x)$), by *spherbessj* (for which $f_k=(\pi/(2x))^{1/2} J_{k+1/2}(x)$), and by *nonexpspherbessi* (for which $f_k=e^{-x}(\pi/(2x))^{1/2} I_{k+1/2}(x)$).)

The above requirement is equivalent to the condition that the tail

$$t(\nu) = \frac{-b_{\nu+1}}{a_{\nu+1}-} \frac{b_{\nu+2}}{a_{\nu+2}-} \cdots$$

of expansion (1) should be negligible. For the special cases considered, $t(\nu)$ represents the ratio of two Bessel functions of contiguous orders. Estimates of $t(\nu)$ may be obtained by the use of such formulae as

$$J_r(x) \approx \{2\pi r \tanh(\alpha)\}^{-1/2} \exp[r\{\tanh(\alpha)-\alpha\}]$$

where $r=x \cosh(\alpha)$, and

$$I_r(x) \approx (2\pi r)^{-1/2}(1+z^2)^{-1/4} e^{r\eta(z)}$$

where $z=x/r$ and

$$\eta(z) = (1+z^2)^{1/2} + \ln[z/\{1+(1+z^2)^{1/2}\}].$$

Function Parameters:

int start (x,n,t)

start: a starting value for the Miller algorithm for computing an array of Bessel functions;

x: float;

entry: the argument of the Bessel functions, $x > 0$;

n: int;

entry: the number of Bessel functions to be computed, $n \geq 0$;

t: int;

entry: the type of Bessel function in question,

t=0 corresponds to ordinary Bessel functions;

t=1 corresponds to modified Bessel functions.

```
#include <math.h>

int start(float x, int n, int t)
{
    int s;
    float p,q,r,y;

    s=2*t-1;
    p=36.0/x-t;
    r=n/x;
    if (r > 1.0 || t == 1) {
        q=sqrt(r*r+s);
        r=r*log(q+r)-q;
    } else
        r=0.0;
    q=18.0/x+r;
    r = (p > q) ? p : q;
    p=sqrt(2.0*(t+r));
    p=x*((1.0+r)+p)/(1.0+p);
    y=0.0;
    q=y;
    do {
        y=p;
        p /= x;
        q=sqrt(p*p+s);
        p=x*(r+q)/log(p+q);
        q=y;
    } while (p > q || p < q-1.0);
    return ((t == 1) ? floor(p+1.0) : -floor(-p/2.0)*2);
}
```

6.6.2 Bessel functions I and K

A. bessiaplusn

Generates an array of modified Bessel functions of the first kind of order $a+j$, $I_{a+j}(x)$, ($0 \leq j \leq n$, $0 \leq a < 1$).

When $x=0$ the above functions are evaluated directly; when $a=0$ or $a=0.5$ the procedures *bessi* or *nonexpspherbessi*, as is appropriate, is used. Otherwise the functions

$$e^{-|x|} I_{a+j}(x) \quad (j=0,...,n)$$

are evaluated by means of a call of *nonexpbessiaplusn*, and the values of the functions $I_{a+j}(x)$ $(j=0,...,n)$ are recovered by multiplication by $e^{|x|}$.

Function Parameters:

void bessiaplusn (*a,x,n,ia*)

a: float;
entry: the noninteger part of the order of the Bessel functions; $0 \leq a < 1$;
x: float;
entry: the argument value of the Bessel functions, $x \geq 0$;
n: int;
entry: the upper bound of the indices of the array *ia*; $n \geq 0$;
ia: float *ia[0:n]*;
exit: *ia[j]* is assigned the value of the modified Bessel function of the first kind of order $a+j$ and argument x, $I_{a+j}(x)$, $0 \leq j \leq n$.

Functions used: nonexpbessiaplusn, bessi, nonexpspherbessi.

```
#include <math.h>

void bessiaplusn(float a, float x, int n, float ia[])
{
    if (x == 0.0) {
        ia[0] = (a == 0.0) ? 1.0 : 0.0;
        for (; n>=1; n--) ia[n]=0.0;
    } else if (a == 0.0) {
        void bessi(float, int, float []);
        bessi(x,n,ia);
    } else if (a == 0.5) {
        void nonexpspherbessi(float, int, float []);
        float c;
        c=0.797884560802865*sqrt(fabs(x))*exp(fabs(x));
        nonexpspherbessi(x,n,ia);
        for (; n>=0; n--) ia[n] *= c;
    } else {
        void nonexpbessiaplusn(float, float, int, float[]);
        float expx;
        expx=exp(fabs(x));
        nonexpbessiaplusn(a,x,n,ia);
        for (; n>=0; n--) ia[n] *= expx;
    }
}
```

B. besska01

Computes the modified Bessel functions of the third kind of order a and $a+1$: $K_a(x)$ and $K_{a+1}(x)$ for $x>0$, $a \geq 0$.

For $0 < x < 1$, $K_a(x)$ and $K_{a+1}(x)$ are computed by using Taylor series (see [T75]). For

$x \geq 1$ the procedure calls for *nonexpbesska01*.

Function Parameters:

void besska01 (*a,x,ka,ka1*)

a: float;
  entry: the order;

x: float;
  entry: this argument should satisfy $x > 0$;

ka: float *;
  exit: the value of the modified Bessel function of the third kind of order a and argument x;

ka1: float *;
  exit: the value of the modified Bessel function of the third kind of order a+1 and argument x.

Functions used: bessk01, recipgamma, nonexpbesska01.

```
#include <math.h>

void besska01(float a, float x, float *ka, float *ka1)
{
    if (a == 0.0) {
        void bessk01(float, float *, float *);
        bessk01(x,ka,ka1);
    } else {
        int n,na,rec,rev;
        float f,g,h,pi;
        pi=4.0*atan(1.0);
        rev = (a < -0.5);
        if (rev) a = -a-1.0;
        rec = (a >= 0.5);
        if (rec) {
            na=floor(a+0.5);
            a -= na;
        }
        if (a == 0.5)
            f=g=sqrt(pi/x/2.0)*exp(-x);
        else if (x < 1.0) {
            float recipgamma(float, float *, float *);
            float a1,b,c,d,e,p,q,s;
            b=x/2.0;
            d = -log(b);
            e=a*d;
            c=a*pi;
            c = (fabs(c) < 1.0e-15) ? 1.0 : c/sin(c);
            s = (fabs(e) < 1.0e-15) ? 1.0 : sinh(e)/e;
            e=exp(e);
            a1=(e+1.0/e)/2.0;
            g=recipgamma(a,&p,&q)*e;
            *ka = f = c*(p*a1+q*s*d);
            e=a*a;
            p=0.5*g*c;
            q=0.5/g;
            c=1.0;
            d=b*b;
            *ka1 = p;
            n=1;
            do {
                f=(f*n+p+q)/(n*n-e);
                c=c*d/n;
                p /= (n-a);
                q /= (n+a);
                g=c*(p-n*f);
                h=c*f;
                *ka += h;
```

```
                *ka1 += g;
                n++;
            } while (h/(*ka)+fabs(g)/(*ka1) > 1.0e-15);
            f=(*ka);
            g=(*ka1)/b;
        } else {
            void nonexpbesska01(float, float, float *, float *);
            float expon;
            expon=exp(-x);
            nonexpbesska01(a,x,ka,ka1);
            f=expon*(*ka);
            g=expon*(*ka1);
        }
        if (rec) {
            x=2.0/x;
            for (n=1; n<=na; n++) {
                h=f+(a+n)*x*g;
                f=g;
                g=h;
            }
        }
        if (rev) {
            *ka1 = f;
            *ka = g;
        } else {
            *ka = f;
            *ka1 = g;
        }
    }
}
```

C. besskaplusn

Generates an array of modified Bessel functions of the third kind of order $a+n$, $K_{a+n}(x)$, $n=0,...,nmax$, for $x > 0$, $a \geq 0$.

The values of the functions $K_a(x)$, $K_{a+1}(x)$ are first obtained by means of a call of *besska01*, and the recursion

$$K_{a+n+1}(x) = K_{a+n-1}(x) + \{2(n+a)/x\}K_{a+n}(x)$$

is then used.

Function Parameters:

void besskaplusn (*a,x,nmax,kan*)

a: float;
  entry: the order; $a \geq 0$;
x: float;
  entry: the argument value, $x > 0$;
nmax: int;
  entry: the upper bound of the indices of the array *kan*; $nmax \geq 0$;
kan: float *kan[0:nmax]*;
  exit: the values of the modified Bessel functions of the third kind of order $a+n$, for argument x are assigned to *kan[n]*, $0 \leq n \leq nmax$.

Function used: besska01.

```
#include <math.h>

void besskaplusn(float a, float x, int nmax, float kan[])
{
    void besska01(float, float, float *, float *);
    int n;
    float k1;
```

```
    besska01(a,x,&kan[0],&k1);
    a -= 1.0;
    x=2.0/x;
    if (nmax > 0) kan[1]=k1;
    for (n=2; n<=nmax; n++) kan[n]=kan[n-2]+(a+n)*x*kan[n-1];
}
```

D. nonexpbessiaplusn

Generates an array of modified Bessel functions of the first kind of order $a+j$, $I_{a+j}(x)$, ($0 \le j \le n$, $0 \le a < 1$), multiplied by the factor e^{-x}. Thus, apart from the exponential factor the array entries are the same as those computed by *bessiaplusn*.

nonexpbessiaplusn computes the values of the functions $e^{-x}I_{a+j}(x)$ ($j=0,...,n$). When $x=0$, the above functions are evaluated directly; when $a=0$ or $a=0.5$ the procedure *nonexpbessi* or *nonexpspherbessi*, as is appropriate, is used. Otherwise the above functions are evaluated by the methods described in the documentation to *nonexpbessi*, where now $a_k=2(a+k)/x$, $b_k=-1$, $\lambda_0=1$, $\lambda_k=2(a+k)\Gamma(a+2k)/\{k!\Gamma(2a+1)\}$ and $s=(x/2)^a/\Gamma(a+1)$. See [G67].

Function Parameters:

void nonexpbessiaplusn *(a,x,n,ia)*

a: float;
entry: the noninteger part of the order $a+n$; $0 \le a < 1$;
x: float;
entry: the argument of the Bessel functions, $x \ge 0$;
n: int;
entry: the upper bound of the indices of the array *ia*; $n \ge 0$;
ia: float *ia[0:n]*;
exit: *ia[j]* is assigned the value of the modified Bessel function of the first kind of order $a+j$ and argument x, multiplied by e^{-x}, $0 \le j \le n$.

Functions used: nonexpbessi, nonexpspherbessi, gamma, start.

```
#include <math.h>

void nonexpbessiaplusn(float a, float x, int n, float ia[])
{
    if (x == 0.0) {
        ia[0] = (a == 0.0) ? 1.0 : 0.0;
        for (; n>=1; n--) ia[n]=0.0;
    } else if (a == 0.0) {
        void nonexpbessi(float, int, float []);
        nonexpbessi(x,n,ia);
    } else if (a == 0.5) {
        void nonexpspherbessi(float, int, float []);
        float c;
        c=0.797884560802865*sqrt(x);
        nonexpspherbessi(x,n,ia);
        for (; n>=0; n--) ia[n] *= c;
    } else {
        float gamma(float);
        int start(float, int, int);
        int m,nu;
        float r,s,labda,l,a2,x2;
        a2=a+a;
        x2=2.0/x;
        l=1.0;
        nu=start(x,n,1);
        r=s=0.0;
        for (m=1; m<=nu; m++) l=l*(m+a2)/(m+1);
        for (m=nu; m>=1; m--) {
```

```
            r=1.0/(x2*(a+m)+r);
            l=l*(m+1)/(m+a2);
            labda=l*(m+a)*2.0;
            s=r*(labda+s);
            if (m <= n) ia[m]=r;
        }
        ia[0]=r=1.0/(1.0+s)/gamma(1.0+a)/pow(x2,a);
        for (m=1; m<=n; m++) r = ia[m] *= r;
    }
}
```

E. nonexpbesska01

Computes the modified Bessel functions of the third kind of order a and a+1, $K_a(x)$ and $K_{a+1}(x)$, multiplied by the factor e^x, for x>0, a≥0. Thus, apart from the exponential factor, the functions are the same as those computed by *besska01*.

nonexpbesska01 evaluates the functions $K_\alpha'(x) = e^x K_\alpha(x)$ ($\alpha=a,a+1$). For $0<x<1$, the procedure *besska01* is called. For $x \geq 1$ the Bessel functions are computed by a Miller algorithm for confluent hypergeometric functions (see [T75]).

Function Parameters:

void nonexpbesska01 (*a,x,ka,ka1*)

a: float;
 entry: the order;
x: float;
 entry: this argument should satisfy $x > 0$;
ka: float *;
 exit: the value of the modified Bessel function of the third kind of order a and argument x;
ka1: float *;
 exit: the value of the modified Bessel function of the third kind of order a+1 and argument x.

Functions used: besska01, nonexpbessk01.

```
#include <math.h>

void nonexpbesska01(float a, float x, float *ka, float *ka1)
{
    if (a == 0.0) {
        void nonexpbessk01(float, float *, float *);
        nonexpbessk01(x,ka,ka1);
    } else {
        int n,na,rec,rev;
        float f,g,h,pi;
        pi=4.0*atan(1.0);
        rev = (a < -0.5);
        if (rev) a = -a-1.0;
        rec = (a >= 0.5);
        if (rec) {
            na=floor(a+0.5);
            a -= na;
        }
        if (a == -0.5)
            f=g=sqrt(pi/x/2.0);
        else if (x < 1.0) {
            void besska01(float, float, float *, float *);
            float expon;
            expon=exp(x);
            besska01(a,x,ka,ka1);
```

```
            f=expon*(*ka);
            g=expon*(*ka1);
        } else {
            float b,c,e,p,q;
            c=0.25-a*a;
            b=x+x;
            g=1.0;
            f=0.0;
            e=cos(a*pi)/pi*x*1.0e15;
            n=1;
            do {
                h=(2.0*(n+x)*g-(n-1+c/n)*f)/(n+1);
                f=g;
                g=h;
                n++;
            } while (h*n < e);
            p=q=f/g;
            e=b-2.0;
            do {
                p=(n-1+c/n)/(e+(n+1)*(2.0-p));
                q=p*(1.0+q);
                n--;
            } while (n > 0);
            f=sqrt(pi/b)/(1.0+q);
            g=f*(a+x+0.5-p)/x;
        }
        if (rec) {
            x=2.0/x;
            for (n=1; n<=na; n++) {
                h=f+(a+n)*x*g;
                f=g;
                g=h;
            }
        }
        if (rev) {
            *ka1 = f;
            *ka = g;
        } else {
            *ka = f;
            *ka1 = g;
        }
    }
}
```

F. nonexpbesskaplusn

Generates an array of modified Bessel functions of the third kind of order $a+n$, $K_{a+n}(x)$, n=0,...,$nmax$, multiplied by the factor e^x, for $x > 0$, $a \geq 0$. Thus, apart from the exponential factor, the functions are the same as those computed by *besskaplusn*

nonexpbesskaplusn computes the values of the functions $K_{a+n}'(x) = e^x K_{a+n}(x)$ (n=0,...,$nmax$). The values of the functions $K_a'(x)$, $K_{a+1}'(x)$ are first obtained by means of a call of *nonexpbesska01*, and the recursion

$$K_{a+n+1}'(x) = K_{a+n-1}'(x) + \{2(n+a)/x\}K_{a+n}'(x)$$

is then used.

Function Parameters:

void nonexpbesskaplusn (*a,x,nmax,kan*)

a: float;
  entry: the order; $a \geq 0$;

x: float;
  entry: the argument value, $x > 0$;

nmax: int;
  entry: the upper bound of the indices of the array *kan*; $nmax \geq 0$;

kan: float *kan[0:nmax]*;
exit: the values of the modified Bessel functions of the third kind of order $a+n$, for argument x multiplied by e^x are assigned to *kan[n]*, $0 \leq n \leq nmax$.

Function used: nonexpbesska01.

```
#include <math.h>

void nonexpbesskaplusn(float a, float x, int nmax, float kan[])
{
    void nonexpbesska01(float, float, float *, float *);
    int n;
    float k1;

    nonexpbesska01(a,x,&kan[0],&k1);
    a -= 1.0;
    x=2.0/x;
    if (nmax > 0) kan[1]=k1;
    for (n=2; n<=nmax; n++) kan[n]=kan[n-2]+(a+n)*x*kan[n-1];
}
```

6.6.3 Spherical Bessel functions

A. spherbessj

Calculates the spherical Bessel functions $J_{k+0.5}'(x) = (\pi/(2x))^{1/2} J_{k+0.5}(x)$, $k=0,...,n$, where $J_{k+0.5}(x)$ denotes the Bessel function of the first kind of order $k+0.5$, for $x \geq 0$.
The method used [G67] is that described in the documentation to *bessj*, with $a = 0.5$ in formula (1). Since

$$J_{0.5}'(x) = x^{-1}\sin(x),\ \lambda_0 = 1,\ \lambda_k = 0\ (k=1,2,...),$$

$s = x^{-1}\sin(x)$ may be taken in the further formulae of that documentation, which in consequence become a little simpler.

Function Parameters:

void spherbessj (*x,n,j*)

x: float;
entry: the value of the argument; $x \geq 0$;
n: int;
entry: the upper bound of the indices of array *j*; $n \geq 0$;
j: float *j[0:n]*;
exit: *j[k]* is the value of the spherical Bessel function $J_{k+0.5}'(x)$, $k=0,...,n$.

Function used: start.

```
#include <math.h>

void spherbessj(float x, int n, float j[])
{
    if (x == 0.0) {
        j[0]=1.0;
        for (; n>=1; n--) j[n]=0.0;
    } else if (n == 0) {
        float x2;
        if (fabs(x) < 0.015) {
            x2=x*x/6.0;
            j[0]=1.0+x2*(x2*0.3-1.0);
        } else
```

```
            j[0]=sin(x)/x;
    } else {
        int start(float, int, int);
        int m;
        float r,s;
        r=0.0;
        m=start(x,n,0);
        for (; m>=1; m--) {
            r=1.0/((m+m+1)/x-r);
            if (m <= n) j[m]=r;
        }
        if (x < 0.015) {
            s=x*x/6.0;
            j[0]=r=s*(s*0.3-1.0)+1.0;
        } else
            j[0]=r=sin(x)/x;
        for (m=1; m<=n; m++) r = j[m] *= r;
    }
}
```

B. spherbessy

Calculates the spherical Bessel functions $Y_{j+0.5}'(x) = (\pi/(2x))^{1/2}Y_{j+0.5}(x)$, $j=0,...,n$, where $Y_{j+0.5}(x)$ denotes the Bessel function of the third kind of order $j+0.5$, for $x > 0$.
The recursion [AbS65]

$$Y_{0.5}'(x) = -x^{-1}\cos(x),$$
$$Y_{1.5}'(x) = -x^{-2}\cos(x) - x^{-1}\sin(x)$$
$$Y_{j+0.5}'(x) = ((2j-1)/x)Y_{j-0.5}'(x) + Y_{j-1.5}'(x) \qquad (j=2,...,n)$$

is used.

Function Parameters:

void spherbessy (x,n,y)

x: float;
entry: the argument of the Bessel functions; $x > 0$;
n: int;
entry: the upper bound of the indices of array y; $n \geq 0$;
y: float $y[0:n]$;
exit: $y[j]$ is the value of the spherical Bessel function $Y_{j+0.5}'(x)$, $j=0,...,n$.

```
#include <math.h>

void spherbessy(float x, int n, float y[])
{
    if (n == 0)
        y[0] = -cos(x)/x;
    else {
        int i;
        float yi,yi1,yi2;
        yi2 = y[0] = -cos(x)/x;
        yi1=y[1]=(yi2-sin(x))/x;
        for (i=2; i<=n; i++) {
            y[i] = yi = -yi2+(i+i-1)*yi1/x;
            yi2=yi1;
            yi1=yi;
        }
    }
}
```

C. spherbessi

Calculates the modified spherical Bessel functions $I_{j+0.5}'(x) = (\pi/(2x))^{1/2} I_{j+0.5}(x)$, $j=0,...,n$, where $I_{j+0.5}(x)$ denotes the modified Bessel function of the first kind of order $j+0.5$, for $x \geq 0$.

The functions $I_{j+0.5}''(x) = e^{-x} I_{j+0.5}'(x)$, $j=0,...,n$, are first evaluated by means of a call of *nonexpspherbessi*; the values of the functions $I_{j+0.5}'(x)$ are then recovered by multiplication by e^x.

Function Parameters:

void spherbessi (x,n,i)

x: float;
  entry: the argument of the Bessel functions; $x \geq 0$;
n: int;
  entry: the upper bound of the indices of array i; $n \geq 0$;
i: float *i[0:n]*;
  exit: *i[j]* is the value of the modified spherical Bessel function $I_{j+0.5}'(x)$, $j=0,...,n$.

Function used: nonexpspherbessi.

```
#include <math.h>

void spherbessi(float x, int n, float i[])
{
    if (x == 0.0) {
        i[0]=1.0;
        for (; n>=1; n--) i[n]=0.0;
    } else {
        void nonexpspherbessi(float, int, float []);
        float expx;
        expx=exp(x);
        nonexpspherbessi(x,n,i);
        for (; n>=0; n--) i[n] *= expx;
    }
}
```

D. spherbessk

Calculates the modified spherical Bessel functions $K_{j+0.5}'(x) = (\pi/(2x))^{1/2} K_{j+0.5}(x)$, $j=0,...,n$, where $K_{j+0.5}(x)$ denotes the modified Bessel function of the third kind of order $j+0.5$, for $x > 0$.

The functions $K_{j+0.5}''(x) = e^x K_{j+0.5}'(x)$, $j=0,...,n$, are first evaluated by means of a call of *nonexpspherbessk*; the functions $K_{j+0.5}'(x)$ are then recovered by multiplication by e^{-x}.

Function Parameters:

void spherbessk (x,n,k)

x: float;
  entry: the argument value; $x > 0$;
n: int;
  entry: the upper bound of the indices of array k; $n \geq 0$;
k: float *k[0:n]*;
  exit: *k[j]* is the value of the modified spherical Bessel function $K_{j+0.5}'(x)$, $j=0,...,n$.

Function used: nonexpspherbessk.

```
#include <math.h>

void spherbessk(float x, int n, float k[])
{
      void nonexpspherbessk(float, int, float []);
      float expx;
      expx=exp(-x);
      nonexpspherbessk(x,n,k);
      for (; n>=0; n--) k[n] *= expx;
}
```

E. nonexpspherbessi

Calculates the modified spherical Bessel functions multiplied by e^{-x}. *nonexpspherbessi* evaluates $I_{j+0.5}''(x) = e^{-x}(\pi/(2x))^{1/2} I_{j+0.5}(x)$, $j=0,...,n$, where $I_{j+0.5}(x)$ denotes the modified Bessel function of the first kind of order $j+0.5$, for $x \geq 0$.

The ratio of two subsequent elements is computed using a backward recurrence formula according to Miller's method (see [G67]). Since the zeroth element is known to be $(1-e^{-2x})/(2x)$, the other elements follow immediately. The starting value is computed by *start*.

Function Parameters:

void nonexpspherbessi (x,n,i)

x: float;
entry: the argument of the Bessel functions; $x \geq 0$;

n: int;
entry: the upper bound of the indices of array i; $n \geq 0$;

i: float $i[0:n]$;
exit: $i[j]$ is the value of the modified spherical Bessel function $I_{j+0.5}''(x)$, $j=0,...,n$.

Function used: start.

```
#include <math.h>

void nonexpspherbessi(float x, int n, float i[])
{
      if (x == 0.0) {
          i[0]=1.0;
          for (; n>=1; n--) i[n]=0.0;
      } else {
          int start(float, int, int);
          int m;
          float x2,r;
          x2=x+x;
          i[0] = x2 = ((x == 0.0) ? 1.0 : ((x2 < 0.7) ?
                          sinh(x)/(x*exp(x)) : (1.0-exp(-x2))/x2));
          if (n != 0) {
              r=0.0;
              m=start(x,n,1);
              for (; m>=1; m--) {
                  r=1.0/((m+m+1)/x+r);
                  if (m <= n) i[m]=r;
              }
              for (m=1; m<=n; m++) x2 = i[m] *= x2;
          }
      }
}
```

F. nonexpspherbessk

Calculates the modified spherical Bessel functions multiplied by e^x. *nonexpspherbessk* evaluates the functions $K_{j+0.5}''(x) = e^x(\pi/(2x))^{1/2}K_{j+0.5}(x)$, $j=0,...,n$, where $K_{j+0.5}(x)$ denotes the modified Bessel function of the third kind of order $j+0.5$, for $x > 0$.

The recursion

$$K_{0.5}''(x) = \pi/(2x),$$
$$K_{1.5}''(x) = (1 + 1/x)K_{0.5}''(x)$$
$$K_{j+0.5}''(x) = ((2j-1)/x)K_{j-0.5}''(x) + K_{j-1.5}''(x) \qquad (j=2,...,n)$$

is used.

Function Parameters:

void nonexpspherbessk (x,n,k)

x: float;
entry: the argument of the Bessel functions; $x > 0$;
n: int;
entry: the upper bound of the indices of array k; $n \geq 0$;
k: float $k[0:n]$;
exit: $k[j]$ is the value of the modified spherical Bessel function $K_{j+0.5}''(x)$, $j=0,...,n$.

```
void nonexpspherbessk(float x, int n, float k[])
{
    int i;
    float ki,ki1,ki2;
    x=1.0/x;
    k[0]=ki2=x*1.5707963267949;
    if (n != 0) {
        k[1]=ki1=ki2*(1.0+x);
        for (i=2; i<=n; i++) {
            k[i]=ki=ki2+(i+i-1)*x*ki1;
            ki2=ki1;
            ki1=ki;
        }
    }
}
```

6.6.4 Airy functions

A. airy

Evaluates the functions $e^{-expon(z)}Ai(z)$ and $e^{expon(z)}Bi(z)$, where the Airy functions Ai and Bi (see [AbS65, Gor69]) may be defined by the formulae

$$Ai(z) = c_1f(z) - c_2g(z), \qquad Bi(z) = 3^{1/2}\{c_1f(z) + c_2g(z)\}$$

where

$$f(z) = \sum_{k=0}^{\infty} \frac{f_k z^{3k}}{(3k)!}, \qquad g(z) = \sum_{k=0}^{\infty} \frac{g_k z^{3k+1}}{(3k+1)!}$$

$$f_k = \prod_{\tau=0}^{k-1} (3\tau + 1), \qquad g_k = \prod_{\tau=0}^{k-1} (3\tau + 2)$$

$$c_1 = 3^{-2/3}\Gamma(\frac{2}{3}), \qquad c_2 = 3^{-1/3}\Gamma(\frac{1}{3})$$

and *expon(z)* = 0 for $z < 9$, *expon(z)* = $(2/3)z^{3/2}$ for $z \geq 9$. *airy* also evaluates the derivatives of these functions:

$$e^{-expon(z)}dAi(z)/dz \text{ and } e^{expon(z)}dBi(z)/dz.$$

When $-5 \leq z \leq 8$, the above defining formulae, and those obtained by differentiation are used to compute the Airy functions and their derivatives.

When $z \geq 8$, approximate versions of the formulae

$$Ai(z) = \frac{1}{2\sqrt{\pi}z^{1/4}} e^{-\frac{3}{2}z^{3/2}} \int_0^\infty \frac{\rho(x)}{1 + (3x/2z^{3/2})} dx$$

$$Bi(z) = \frac{1}{\sqrt{\pi}z^{1/4}} e^{\frac{3}{2}z^{3/2}} \int_0^\infty \frac{\rho(x)}{1 - (3x/2z^{3/2})} dx$$

where

$$\rho(x) = \frac{1}{\sqrt{\pi}2^{11/6}3^{2/3}} e^{-x} Ai[(3x/2)^{2/3}]$$

are used. These versions have the form

$$Ai(z) = \frac{1}{2\sqrt{\pi}z^{1/4}} e^{-\frac{3}{2}z^{3/2}} \sum_{i=1}^{n} \frac{w_i}{1 + (3x_i/2z^{3/2})}$$

$$Bi(z) = \frac{1}{\sqrt{\pi}z^{1/4}} e^{\frac{3}{2}z^{3/2}} \sum_{i=1}^{n} \frac{w_i}{1 - (3x_i/2z^{3/2})}$$

where

$$\int_0^\infty x^k \rho(x)\,dx = \frac{\Gamma(3k+½)}{54^k k!\,\Gamma(k+½)} = \sum_{i=1}^{n} x_i^k w_i \qquad (k=0,...,2n-1).$$

When $z < 0$, similar use is made of the formulae

$$Ai(-z) = \frac{1}{\sqrt{\pi}\,z^{1/4}} \int_0^\infty \frac{\cos(\beta-\frac{\pi}{4}) + (\frac{x}{\beta})\sin(\beta-\frac{\pi}{4})}{1+(x/\beta)^2}\,\rho(x)\,dx$$

$$Bi(-z) = \frac{1}{\sqrt{\pi}\,z^{1/4}} \int_0^\infty \frac{(\frac{x}{\beta})\cos(\beta-\frac{\pi}{4}) - \sin(\beta-\frac{\pi}{4})}{1+(x/\beta)^2}\,\rho(x)\,dx$$

where $\beta = 2z^{3/2}/3$. In both of the above cases the derivatives of Ai and Bi are obtained by differentiation of their approximation formulae.

Function Parameters:

void airy (*z,ai,aid,bi,bid,expon,first*)

z: float;
entry: the real argument of the Airy functions;
ai: float *;
exit: the value of the Airy function given by $e^{-expon(z)}Ai(z)$;
aid: float *;
exit: the value of the derivative of the Airy function given by $e^{-expon(z)}Ai(z)$;
bi: float *;
exit: the value of the Airy function given by $e^{expon(z)}Bi(z)$;
bid: float *;
exit: the value of the derivative of the Airy function given by $e^{expon(z)}Bi(z)$;
expon: float *;
exit: if $z < 9$ then 0 else $(2/3)z^{3/2}$;
first: int;
entry: the value of *first* should be zero unless the procedure is called for the first time; if *first* is nonzero then two internal static arrays are built up.

```
#include <math.h>

void airy(float z, float *ai, float *aid, float *bi, float *bid,
          float *expon, int first)
{
    int n,l;
    float s,t,u,v,sc,tc,uc,vc,x,k1,k2,k3,k4,c,zt,si,co,expzt,sqrtz,wwl,
          pl,pl1,pl2,pl3;
    static float c1,c2,sqrt3,sqrt1opi,pio4,xx[11],ww[11];

    if (first) {
        sqrt3=1.73205080756887729;
        sqrt1opi=0.56418958354775629;
```

```
        pio4=0.78539816339744831;
        c1=0.355028053887817;
        c2=0.258819403792807;
        xx[1] =1.4083081072180964e1;
        xx[2] =1.0214885479197331e1;
        xx[3] =7.4416018450450930;
        xx[4] =5.3070943061781927;
        xx[5] =3.6340135029132462;
        xx[6] =2.3310652303052450;
        xx[7] =1.3447970842609268;
        xx[8] =6.4188858369567296e-1;
        xx[9] =2.0100345998121046e-1;
        xx[10]=8.0594359172052833e-3;
        ww[1] =3.1542515762964787e-14;
        ww[2] =6.6394210819584921e-11;
        ww[3] =1.7583889061345669e-8;
        ww[4] =1.3712392370435815e-6;
        ww[5] =4.4350966639284350e-5;
        ww[6] =7.1555010917718255e-4;
        ww[7] =6.4889566103335381e-3;
        ww[8] =3.6440415875773282e-2;
        ww[9] =1.4399792418590999e-1;
        ww[10]=8.1231141336261486e-1;
    }
    *expon=0.0;
    if (z >= -5.0 && z <= 8.0) {
        u=v=t=uc=vc=tc=1.0;
        s=sc=0.5;
        n=3;
        x=z*z*z;
        while (fabs(u)+fabs(v)+fabs(s)+fabs(t) > 1.0e-18) {
            u=u*x/(n*(n-1));
            v=v*x/(n*(n+1));
            s=s*x/(n*(n+2));
            t=t*x/(n*(n-2));
            uc += u;
            vc += v;
            sc += s;
            tc += t;
            n += 3;
        }
        *bi=sqrt3*(c1*uc+c2*z*vc);
        *bid=sqrt3*(c1*z*z*sc+c2*tc);
        if (z < 2.5) {
            *ai=c1*uc-c2*z*vc;
            *aid=c1*sc*z*z-c2*tc;
            return;
        }
    }
    k1=k2=k3=k4=0.0;
    sqrtz=sqrt(fabs(z));
    zt=0.666666666666667*fabs(z)*sqrtz;
    c=sqrt1opi/sqrt(sqrtz);
    if (z < 0.0) {
        z = -z;
        co=cos(zt-pio4);
        si=sin(zt-pio4);
        for (l=1; l<=10; l++) {
            wwl=ww[l];
            pl=xx[l]/zt;
            pl2=pl*pl;
            pl1=1.0+pl2;
            pl3=pl1*pl1;
            k1 += wwl/pl1;
            k2 += wwl*pl/pl1;
            k3 += wwl*pl*(1.0+pl*(2.0/zt+pl))/pl3;
            k4 += wwl*(-1.0-pl*(1.0+pl*(zt-pl))/zt)/pl3;
        }
        *ai=c*(co*k1+si*k2);
        *aid=0.25*(*ai)/z-c*sqrtz*(co*k3+si*k4);
        *bi=c*(co*k2-si*k1);
        *bid=0.25*(*bi)/z-c*sqrtz*(co*k4-si*k3);
    } else {
        if (z < 9.0)
```

```
            expzt=exp(zt);
        else {
            expzt=1.0;
            *expon=zt;
        }
        for (l=1; l<=10; l++) {
            wwl=ww[l];
            pl=xx[l]/zt;
            pl1=1.0+pl;
            pl2=1.0-pl;
            k1 += wwl/pl1;
            k2 += wwl*pl/(zt*pl1*pl1);
            k3 += wwl/pl2;
            k4 += wwl*pl/(zt*pl2*pl2);
        }
        *ai=0.5*c*k1/expzt;
        *aid=(*ai)*(-0.25/z-sqrtz)+0.5*c*sqrtz*k2/expzt;
        if (z >= 8.0) {
            *bi=c*k3*expzt;
            *bid=(*bi)*(sqrtz-0.25/z)-c*k4*sqrtz*expzt;
        }
    }
}
```

B. airyzeros

Computes the zeros and associated values of the Airy functions *Ai(x)* and *Bi(x)*, and their derivatives.

Denoting by a_s, a_s', b_s, b_s' the *s*-th zero of *Ai(x), dAi(x)/dx, Bi(x), dBi(x)/dx* respectively (see [AbS65]):

$$a_s \approx -f\{3\pi(4s-1)/8\}, \quad a_s' \approx -g\{3\pi(4s-3)/8\}$$
$$b_s \approx -f\{3\pi(4s-3)/8\}, \quad b_s' \approx -g\{3\pi(4s-1)/8\}$$

where the functions *f* and *g* are defined in the documentation of *airy*. The appropriate member of the above set of approximations is used as the first iterate in a quadratic interpolation process for determining the zeros in question. The values of the Airy functions (and the associated values delivered in the array *vai*) are calculated by means of the procedure *airy*.

Function Parameters:

float airyzeros (*n,d,zai,vai*)

airyzeros: delivers the *n*-th zero of the selected Airy function;

d: int;
entry: an integer which selects the required Airy function, *d* = 0, 1, 2 or 3;

zai: float *zai[1:n]*;
exit: *zai[j]* contains the *j*-th zero of the selected Airy function:
if *d*=0 then *Ai(x)*,
if *d*=1 then *(d/dx)Ai(x)*,
if *d*=2 then *Bi(x)*,
if *d*=3 then *(d/dx)Bi(x)*;

vai: float *vai[1:n]*;
exit: *vai[j]* contains the value at *x* = *zai[j]* of the following function:
if *d*=0 then *(d/dx)Ai(x)*,
if *d*=1 then *Ai(x)*,
if *d*=2 then *(d/dx)Bi(x)*,
if *d*=3 then *Bi(x)*.

Function used: airy.

```
#include <math.h>

float airyzeros(int n, int d, float zai[], float vai[])
{
    void airy(float,float *,float *,float *,float *,float *,int);
    int a,found,i;
    float c,e,r,zaj,zak,vaj,daj,kaj,zz;

    a=((d == 0) || (d == 2));
    r = (d == 0 || d == 3) ? -1.17809724509617 : -3.53429173528852;
    airy(0.0,&zaj,&vaj,&daj,&kaj,&zz,1);
    for (i=1; i<=n; i++) {
        r += 4.71238898038469;
        zz=r*r;
        zaj = (i == 1 && d == 1) ? -1.01879297 :
               ((i == 1 && d == 2) ? -1.17371322 :
               pow(r,0.666666666666667)*
               (a ? -(1.0+(5.0/48.0-(5.0/36.0-(77125.0/82944.0-
               (108056875.0/6967296.0-(162375596875.0/334430208.0)/
               zz)/zz)/zz)/zz)/zz) : -(1.0-(7.0/48.0-(35.0/288.0-
               (181223.0/207360.0-(18683371.0/1244160.0-
               (91145884361.0/191102976.0)/zz)/zz)/zz)/zz)/zz)));
        if (d <= 1.0)
            airy(zaj,&vaj,&daj,&c,&e,&zz,0);
        else
            airy(zaj,&c,&e,&vaj,&daj,&zz,0);
        found=(fabs(a ? vaj : daj) < 1.0e-12);
        while (!found) {
            if (a) {
               kaj=vaj/daj;
               zak=zaj-kaj*(1.0+zaj*kaj*kaj);
            } else {
               kaj=daj/(zaj*vaj);
               zak=zaj-kaj*(1.0+kaj*(kaj*zaj+1.0/zaj));
            }
            if (d <= 1)
               airy(zak,&vaj,&daj,&c,&e,&zz,0);
            else
               airy(zak,&c,&e,&vaj,&daj,&zz,0);
            found=(fabs(zak-zaj) < 1.0e-14*fabs(zak) ||
                   fabs(a ? vaj : daj) < 1.0e-12);
            zaj=zak;
        }
        vai[i]=(a ? daj : vaj);
        zai[i]=zaj;
    }
    return zai[n];
}
```

7. Interpolation and Approximation

7.1 Real data in one dimension

7.1.1 Interpolation with general polynomials

newton

Calculates the coefficients of the Newton polynomial through given interpolation points and corresponding function values.

newton computes the divided differences $\delta_x^j f(x_0)$, j=0,...,n, derived from the function values f_i and their associated arguments x_i, i=0,...,n, (the x_i being assumed distinct). The $\{\delta_x^j f(x_0)\}$ are computed by use of the defining recursion

$$\delta_i^0 = f_i \qquad (i=0,...,n)$$

$$\delta_i^{j+1} = (\delta_{i+1}^j - \delta_i^j) / (x_{i+j+1} - x_i) \qquad (j=0,...,n-1;\ i=0,...,n-j-1)$$

when $\delta_x^j f(x_0) = \delta_0^j$.

Function Parameters:

void newton (n,x,f)

n: int;
entry: the degree of the polynomial;
x: float $x[0:n]$;
entry: the interpolation points;
f: float $f[0:n]$;
entry: the function values at the interpolation points;
exit: the coefficients of the Newton polynomial.

```
void newton(int n, float x[], float f[])
{
    int k,i,im1;
    float xim1,fim1;

    im1=0;
    for (i=1; i<=n; i++) {
        fim1=f[im1];
        xim1=x[im1];
        for (k=i; k<=n; k++) f[k]=(f[k]-fim1)/(x[k]-xim1);
        im1=i;
    }
}
```

7.1.2 Approximation in infinity norm

A. ini

Selects a set of integers out of a given set of integers. With integers m,n ($m > n$), *ini* determines a sequence of integer approximations $s(j)$, j=0,...,n; $s(j) > s(j-1)$, to the points in the range [0,m] at which $T_n((2x-m)/m)$, where $T_n(y) = \cos\{n \arccos(y)\}$ being a Chebyshev polynomial and $(2x-m)/m$ maps $x\in[-1,1]$ onto [0,m], assumes its maximal absolute values.

ini is an auxiliary procedure used in *minmaxpol*.

Function Parameters:

void ini (n,m,s)

n,m: int;
  entry: the number of points to be selected equals n+1; the reference set contains the numbers 0,1,...,m, ($m \geq n$);
s: int $s[0:n]$;
  exit: the selected integers are delivered in s.

```
#include <math.h>

void ini(int n, int m, int s[])
{
    int i,j,k,l;
    float pin2,temp;

    pin2=atan(1.0)*2.0/n;
    k=1;
    l=n-1;
    j=s[0]=0;
    s[n]=m;
    while (k < l) {
        temp=sin(k*pin2);
        i=temp*temp*m;
        j = s[k] = ((i <= j) ? j+1 : i);
        s[l]=m-j;
        l--;
        k++;
    }
    if (l*2 == n) s[l]=m/2;
}
```

B. sndremez

Exchanges at most n+1 numbers with numbers out of a reference set (see [Me64]). Given a sequence of real numbers g_i, i=0,...,m, which contains a subsequence $g_{s(j)}$, j=0,1,...,n<m; $s(j) > s(j-1)$, for which a)

$$|g_{s(j)}| = G, j=0,...,n,$$

and b)

$$\text{sign } g_{s(j)} = -\text{sign } g_{s(j-1)}, j=1,...,n,$$

a sequence of integers $S(j)$, j=0,1,...,n;

$$S(j) > S(j-1);\ S(0) \geq 0,\ S(n) \leq m,$$

is determined such that A)

$$|g_{S(j)}| \geq G, j=0,...,n,$$

and B)

$$\text{sign } g_{S(j)} = -\text{sign } g_{S(j-1)}, j=1,...,n,$$

and C) $|g_{S(j)}| > G$ for at least one j in the range $0 \leq j \leq n$ if $|g_i| > G$ for at least one i in the range $0 \leq i \leq n$.

The process used is as follows: let

$$H(1) = \max|g_i| \text{ for } s(0) < i < s(1)$$

and $|g_{I(1)}| = H(1)$ ($H(1)$ is zero if $s(0) = s(1)$ and $I(1)$ is then undefined; a similar convention is adopted below). If $H(1) \leq G$ then

$$S(0) = s(0),\ S(1) = s(1);$$

otherwise if

$$\text{sign } g_{I(1)} = \text{sign } g_{s(0)}$$

then

$$S(0) = I(1),\ S(1)=s(1),$$

whereas if

$$\text{sign } g_{I(1)} \neq \text{sign } g_{s(0)},$$

then

$$S(0) = s(0),\ S(1) = I(1).$$

Let

$$H(2) = \max|g_i| \text{ for } s(1) < i < s(2) \text{ and } |g_{I(2)}| = H(2).$$

If $H(2) \leq G$ then $S(2) = s(2)$; otherwise if sign $g_{I(2)}$ = sign $g_{S(1)}$ then $S(1)$ is replaced by $I(2)$ and $S(2) = s(2)$, whereas if sign $g_{I(2)} \neq$ sign $g_{S(1)}$ then $S(2) = I(2)$. The numbers g_i, $s(2)<i<s(3)$, are inspected, $S(2)$ is possibly replaced, and $S(3)$ is determined as before. In this way $S(0),...,S(n)$ are determined. Let

$$H' = \max|g_i| \text{ for } 0 \leq i < s(0) \text{ with } |g_{I'}|=H',$$

and

$$H'' = \max|g_i| \text{ for } s(n) < i \leq m, \text{ with } |g_{I''}| = H''.$$

If H', $H'' < G$, the numbers $S(j)$ are left as they stand.
If not, then if $H' > H''$ then a) if sign $g_{I'}$ = sign $g_{S(0)}$ then a') if $H'>|g_{s(0)}|$, $S(0)$ is replaced by I' and a") if sign $g_{I''} = g_{s(n)}$ and $H'' > |g_{s(n)}|$, $S(n)$ is replaced by I'';
b) if sign $g_{I'} \neq$ sign $g_{S(0)}$ then b') if $H' > |g_{s(n)}|$, $S(k)$ is replaced by $S(k-1)$, k=n,...,1, and $S(0)$ is replaced by I' and b") (with the new value of $S(n)$) if sign $g_{I''}$ = sign $g_{S(n)}$ and $H'' > |g_{s(n)}|$, $S(n)$ is replaced by I''. If $H'' \geq H'$ then similar modifications take place.

sndremez is an auxiliary procedure used in *minmaxpol*.

Function Parameters:

void sndremez (*n,m,s,g,em*)

n,m: int;
entry: the number of points to be exchanged is smaller than or equal to n+1;
the reference set contains the numbers 0,1,...,m, ($m \geq n$);
s: int *s[0:n]*;
entry: in *s* one must give n+1 (strictly) monotone increasing numbers out of 0,1,...,m;
exit: n+1 (strictly) monotone increasing numbers out of the numbers 0,1,...,m;
g: float *g[0:m]*;
entry: in array *g[0:m]* one must give the function values;
em: float *em[0:1]*;
entry: $0 < em[0] \leq g[i]$, i=0,...,m;
exit: $em[1]$ = infinity norm of array *g[0:m]*.

Function used: infnrmvec.

```
#include <math.h>

void sndremez(int n, int m, int s[], float g[], float em[])
{
    float infnrmvec(int, int, int *, float []);
    int s0,sn,sjp1,i,j,k,up,low,nm1;
    float max,msjp1,hi,hj,he,abse,h,temp1,temp2;

    s0=sjp1=s[0];
    he=em[0];
    low=s0+1;
    max=msjp1=abse=fabs(he);
    nm1=n-1;
```

```
    for (j=0; j<=nm1; j++) {
        up=s[j+1]-1;
        h=infnrmvec(low,up,&i,g);
        if (h > max) max=h;
        if (h > abse)
            if (he*g[i] > 0.0) {
                s[j] = (msjp1 < h) ? i : sjp1;
                sjp1=s[j+1];
                msjp1=abse;
            } else {
                s[j]=sjp1;
                sjp1=i;
                msjp1=h;
            }
        else {
            s[j]=sjp1;
            sjp1=s[j+1];
            msjp1=abse;
        }
        he = -he;
        low=up+2;
    }
    sn=s[n];
    s[n]=sjp1;
    hi=infnrmvec(0,s0-1,&i,g);
    hj=infnrmvec(sn+1,m,&j,g);
    if (j > m) j=m;
    if (hi > hj) {
        if (hi > max) max=hi;
        temp1 = (g[i] == 0.0) ? 0.0 : ((g[i] > 0.0) ? 1.0 : -1.0);
        temp2 = (g[s[0]]==0.0) ? 0.0 : ((g[s[0]]>0.0) ? 1.0 : -1.0);
        if (temp1 == temp2) {
            if (hi > fabs(g[s[0]])) {
                s[0]=i;
                if (g[j]/g[s[n]] > 1.0) s[n]=j;
            }
        }
        else {
            if (hi > fabs(g[s[n]])) {
                s[n] = (g[j]/g[s[nm1]] > 1.0) ? j : s[nm1];
                for (k=nm1; k>=1; k--) s[k]=s[k-1];
                s[0]=i;
            }
        }
    } else {
        if (hj > max) max=hj;
        temp1 = (g[j] == 0.0) ? 0.0 : ((g[j] > 0.0) ? 1.0 : -1.0);
        temp2 = (g[s[n]]==0.0) ? 0.0 : ((g[s[n]]>0.0) ? 1.0 : -1.0);
        if (temp1 == temp2) {
            if (hj > fabs(g[s[n]])) {
                s[n]=j;
                if (g[i]/g[s[0]] > 1.0) s[0]=i;
            }
        } else
            if (hj > fabs(g[s[0]])) {
                s[0] = (g[i]/g[s[1]] > 1.0) ? i : s[1];
                for (k=1; k<=nm1; k++) s[k]=s[k+1];
                s[n]=j;
            }
    }
    em[1]=max;
}
```

C. minmaxpol

Calculates the coefficients of the polynomial (as a sum of powers) which approximates a function, given for discrete arguments, in such a way that the infinity norm of the error vector is minimized.

With y_j (j=0,...,m) being a sequence of real arguments (y_{j-1} < y_j; j=1,...,m) and $f(y_j)$

corresponding function values, *minmaxpol* determines the coefficients $\{c_i\}$ of that polynomial

$$\sum_{i=0}^{n-1} c_i y^i$$

for which

$$\max \left| f(y_j) - \sum_{i=0}^{n-1} c_i y_j^i \right| \qquad (1)$$

is a minimum as j ranges over the values $0,...,m$ $(m > n)$.

The method used [Me64] involves the iterative construction of polynomials

$$\sum_{i=0}^{n-1} c_{k,i} y^i$$

for which

$$f(y_{s(k,j)}) - \sum_{i=0}^{n-1} c_{k,i} y_{s(k,j)}^i = (-1)^j \lambda_k \qquad (j=0,...,n) \qquad (2)$$

where $s(k,j)$, $j=0,...,n$, is a strictly increasing sequence selected from $0,1,...,m$. The discrepancies

$$g_{k,j}' = f(y_j) - \sum_{i=0}^{n-1} c_{k,i} y_j^i$$

at all points $j=0,...,m$ are then inspected, and from these a sequence $g_{k+1,j} = g'_{k,s(k+1,j)}$, $j=0,...,n$, is constructed which a) possesses the alternating property sign $g_{k+1,j} = -\text{sign } g_{k+1,j-1}$, $j=1,...,n$, and b) for which $|g_{k+1,j}| \geq |g_{k,j}|$, $j=0,...,n$, with (unless the process is to be terminated) $|g_{k+1,j}| > |g_{k,j}|$ for at least one j (for the details of this construction, see the documentation to *sndremez*). The coefficients $\{c_{k+1,i}\}$ in the succeeding polynomial are determined (by use of Newton's interpolation formula, and subsequent reduction to the given polynomial form) from condition (2) with k replaced by $k+1$.

Initially $s(0,j)$, $j=0,...,n$, is a sequence of integer approximations to the points at which $T_n((2x-m)/m)$, where $T_n(y) = \cos\{n \arccos(y)\}$ being a Chebyshev polynomial (see the documentation of *ini*) assumes its maximal absolute values in the range $0 \leq x \leq m$. The procedure *ini* is used to set the $s(0,j)$.

Function Parameters:

void minmaxpol (*n,m,y,fy,co,em*)

n: int;
entry: the degree of the approximating polynomial; $n \geq 0$;

m: int;
entry: the number of reference function values is $m+1$;
y,fy: float *y[0:m], fy[0:m]*;
entry: *fy[i]* is the function values at *y[i]*, i=0,...,m;
co: float *co[0:n]*;
exit: the coefficients of the approximating polynomial (*co[i]* is the coefficient of y^i);
em: float *em[0:3]*;
entry:
em[2]: maximum allowed number of iterations, say 10*n+5;
exit:
em[0]: the difference of the given function and the polynomial in the first approximation point;
em[1]: the infinity norm of the error of approximation over the discrete interval;
em[3]: the number of iterations performed.

Functions used: elmvec, dupvec, newton, pol, newgrn, ini, sndremez.

```
#include <math.h>

void minmaxpol(int n, int m, float y[], float fy[], float co[],
               float em[])
{
    int *allocate_integer_vector(int, int);
    float *allocate_real_vector(int, int);
    void free_integer_vector(int *, int);
    void free_real_vector(float *, int);
    void elmvec(int, int, int, float [], float [], float);
    void dupvec(int, int, int, float [], float []);
    float pol(int, float, float []);
    void newton(int, float [], float []);
    void newgrn(int, float [], float []);
    void ini(int, int, int []);
    void sndremez(int, int, int [], float [], float []);
    int np1,k,pomk,count,cnt,j,mi,*s,sjm1,sj,s0,up;
    float e,abse,abseh,*x,*b,*coef,*g;

    s=allocate_integer_vector(0,n+1);
    x=allocate_real_vector(0,n+1);
    b=allocate_real_vector(0,n+1);
    coef=allocate_real_vector(0,n+1);
    g=allocate_real_vector(0,m);

    np1=n+1;
    ini(np1,m,s);
    mi=em[2];
    abse=0.0;
    count=1;
    do {
        pomk=1;
        for (k=0; k<=np1; k++) {
            x[k]=y[s[k]];
            coef[k]=fy[s[k]];
            b[k]=pomk;
            pomk = -pomk;
        }
        newton(np1,x,coef);
        newton(np1,x,b);
        em[0]=e=coef[np1]/b[np1];
        elmvec(0,n,0,coef,b,-e);
        newgrn(n,x,coef);
        s0=sjm1=s[0];
        g[s0]=e;
        for (j=1; j<=np1; j++) {
            sj=s[j];
            up=sj-1;
```

```
            for (k=sjm1+1; k<=up; k++) g[k]=fy[k]-pol(n,y[k],coef);
            g[sj] = e = -e;
            sjm1=sj;
        }
        for (k=s0-1; k>=0; k--) g[k]=fy[k]-pol(n,y[k],coef);
        for (k=sj+1; k<=m; k++) g[k]=fy[k]-pol(n,y[k],coef);
        sndremez(np1,m,s,g,em);
        abseh=abse;
        abse=fabs(e);
        cnt=count;
        count++;
    } while (count <= mi && abse > abseh);
    em[2]=mi;
    em[3]=cnt;
    dupvec(0,n,0,co,coef);
    free_integer_vector(s,0);
    free_real_vector(x,0);
    free_real_vector(b,0);
    free_real_vector(coef,0);
    free_real_vector(g,0);
}
```

Worked Examples

Examples for chapter 1 procedures

Functions tested: **hshcomcol, hshcomprd**

Example:

The matrix

$$\begin{pmatrix} 3 & 4i \\ 4i & 5 \end{pmatrix}$$

is reduced to upper triangular form by successive calls of *hshcomcol* and *hshcomprd.*

```
#include <stdio.h>

void main ()
{
    float **allocate_real_matrix(int, int, int, int);
    void free_real_matrix(float **, int, int, int);
    int hshcomcol(int, int, int, float **, float **, float,
                  float *, float *, float *, float *);
    void hshcomprd(int, int, int, int, int, float **,
                   float **, float **, float **, float);
    float k,c,s,t;
    float **ar,**ai;

    ar=allocate_real_matrix(1,2,1,2);
    ai=allocate_real_matrix(1,2,1,2);
    ar[1][1]=3.0;
    ar[1][2]=ar[2][1]=0.0;
    ar[2][2]=5.0;
    ai[1][1]=0.0;
    ai[1][2]=ai[2][1]=4.0;
    ai[2][2]=0.0;
    if (hshcomcol(1,2,1,ar,ai,25.0e-28,&k,&c,&s,&t))
        hshcomprd(1,2,2,2,1,ar,ai,ar,ai,t);
    printf("After using hshcomcol and hshcomprd:\n"
        " %-3.1f+%-3.1f*I  %-3.1f+%-3.1f*I\n"
        " %-3.1f+%-3.1f*I  %-3.1f+%-3.1f*I\n"
        "k, c, s, t\n %6.1f  %6.1f  %6.1f  %6.1f",
        ar[1][1],ai[1][1],ar[1][2],ai[1][2],
        ar[2][1],ai[2][1],ar[2][2],ai[2][2],k,c,s,t);
    free_real_matrix(ar,1,2,1);
    free_real_matrix(ai,1,2,1);
}
```

Output:

```
After using hshcomcol and hshcomprd:
 8.0+0.0*I  0.0+1.6*I
 0.0+4.0*I  6.2+0.0*I
k, c, s, t
    5.0    -1.0    -0.0    40.0
```

Function tested: **elmcomcol**

Example:

With the matrix

$$A = \begin{pmatrix} 1+2i & -9+2i \\ -1+2i & -1-2i \end{pmatrix}$$

this example sets $A_{k,2} = A_{k,2} + (1\text{-}4i)A_{k,1}$, for k=1,2.

```
#include <stdio.h>

void main ()
{
    float **allocate_real_matrix(int, int, int, int);
    void free_real_matrix(float **, int, int, int);
    void elmcomcol(int, int, int, int, float **, float **,
                        float **, float **, float, float);
    float **ar,**ai;

    ar=allocate_real_matrix(1,2,1,2);
    ai=allocate_real_matrix(1,2,1,2);
    ar[1][1]=1.0;  ar[1][2] = -9.0;   ar[2][1]=ar[2][2] = -1.0;
    ai[1][1]=ai[1][2]=ai[2][1]=2.0;   ai[2][2] = -2.0;
    elmcomcol(1,2,2,1,ar,ai,ar,ai,1,-4);
    printf("Matrix after elimination:\n"
        "  %-3.1f+%-3.1f*I  %-3.1f+%-3.1f*I\n"
        " %-3.1f+%-3.1f*I  %-3.1f+%-3.1f*I\n",
        ar[1][1],ai[1][1],ar[1][2],ai[1][2],
        ar[2][1],ai[2][1],ar[2][2],ai[2][2]);
    free_real_matrix(ar,1,2,1);
    free_real_matrix(ai,1,2,1);
}
```

Output:

```
Matrix after elimination:
  1.0+2.0*I  0.0+0.0*I
 -1.0+2.0*I  6.0+4.0*I
```

Function tested: **rotcomcol**

Example:

The matrix

$$\begin{pmatrix} 4+3i & 5 \\ -5 & 4-3i \end{pmatrix}$$

is post multiplied by the rotation matrix

$$\begin{pmatrix} 0.08-0.06i & -0.1 \\ 0.1 & 0.08+0.06i \end{pmatrix}.$$

```
#include <stdio.h>

void main ()
{
    float **allocate_real_matrix(int, int, int, int);
```

```
    void free_real_matrix(float **, int, int, int);
    void rotcomcol(int, int, int, int, float **, float **,
                   float, float, float);
    int i,j;
    float **ar,**ai;

    ar=allocate_real_matrix(1,2,1,2);
    ai=allocate_real_matrix(1,2,1,2);
    ar[1][1]=4.0;  ar[1][2]=5.0;  ar[2][1] = -5.0;   ar[2][2]=4.0;
    ai[1][1]=3.0;  ai[1][2]=ai[2][1]=0.0;     ai[2][2] = -3.0;
    rotcomcol(1,2,1,2,ar,ai,0.08,0.06,-0.1);
    printf("After postmultiplication:\n"
        " %+3.1f%+3.1f*I  %+3.1f%+3.1f*I\n"
        " %+3.1f%+3.1f*I  %+3.1f%+3.1f*I\n",
        ar[1][1],ai[1][1],ar[1][2],ai[1][2],
        ar[2][1],ai[2][1],ar[2][2],ai[2][2]);
    free_real_matrix(ar,1,2,1);
    free_real_matrix(ai,1,2,1);
}
```

Output:

```
After postmultiplication:
 +1.0+0.0*I  -0.0-0.0*I
 +0.0-0.0*I  +1.0+0.0*I
```

Function tested: **comabs**

Example:

Compute $|0.3+0.4i|$.

```
#include <stdio.h>

void main ()
{
    float comabs(float, float);

    printf("The modulus of .3+.4*i equals %-4.2f",comabs(0.3,0.4));
}
```

Output:

```
The modulus of .3+.4*i equals 0.50
```

Function tested: **comsqrt**

Example:

Compute the value of $(-3+4i)^{1/2}$ with positive real part.

```
#include <stdio.h>

void main ()
{
    void comsqrt(float, float, float *, float *);
    float r,i;

    comsqrt(-3.0,4.0,&r,&i);
    printf("The square root of -3+4*i is %-4.2f+%-4.2f*i",r,i);
}
```

Output:

```
The square root of -3+4*i is 1.00+2.00*i
```

Function tested: **carpol**

Example:

Compute the value of r, $\cos\phi$ and $\sin\phi$, where $re^{i\phi} = 0.3+0.4i$.

```
#include <stdio.h>

void main ()
{
    void carpol(float, float, float *, float *, float *);
    float r,c,s;

    carpol(0.3,0.4,&r,&c,&s);
    printf("The polar coordinates of 0.3+0.4*i are \n"
           " modulus:  %-4.2f\n cosine of argument:  %-4.2f\n"
           " sine of argument:  %-4.2f",r,c,s);
}
```

Output:

```
The polar coordinates of 0.3+0.4*i are
 modulus:  0.50
 cosine of argument:  0.60
 sine of argument:  0.80
```

Function tested: **commul**

Example:

Compute the value of (0.1+0.2i)(0.3+0.4i).

```
#include <stdio.h>

void main ()
{
    void commul(float, float, float, float, float *, float *);
    float r,i;

    commul(0.1,0.2,0.3,0.4,&r,&i);
    printf("(.1+.2i)*(.3+.4*i) = %-5.2f+%-4.2f*i",r,i);
}
```

Output:

```
(.1+.2i)*(.3+.4*i) = -0.05+0.10*i
```

Function tested: **comdiv**

Example:

Compute the value of (-0.05+0.1i)/(0.1+0.2i).

```
#include <stdio.h>

void main ()
{
    void comdiv(float, float, float, float, float *, float *);
    float r,i;

    comdiv(-0.05,0.1,0.1,0.2,&r,&i);
    printf("(-.05+.1*i)/(.1+.2*i) = %-4.2f+%-4.2f*i",r,i);
}
```

Output:

```
(-.05+.1*i)/(.1+.2*i) = 0.30+0.40*i
```

Functions tested: **lngintadd, lngintsubtract, lngintmult, lngintdivide, lngintpower**

Example:

Compute the sum, difference, product, and quotient and remainder of $u = 3370707070$ and $v = 444$, and also compute v^4.

```
#include <stdio.h>

void main ()
{
    void lngintadd(int [], int [], int []);
    void lngintsubtract(int [], int [], int []);
    void lngintmult(int [], int [], int []);
    void lngintdivide(int [], int [], int [], int []);
    void lngintpower(int [], int, int []);
    int i,u[100],v[100],r1[100],r2[100];

    u[0]=5;
    u[1]=33;
    u[2]=u[3]=u[4]=u[5]=70;
    v[0]=2;
    v[1]=4;
    v[2]=44;
    printf("\nInput numbers:\n");
    for (i=1; i<=u[0]; i++) printf("%4d",u[i]);
    printf("\n");
    for (i=1; i<=v[0]; i++) printf("%4d",v[i]);
    printf("\n\nadd: ");

    lngintadd(u,v,r1);
    for (i=1; i<=r1[0]; i++) printf("%4d",r1[i]);
    printf("\nsubtract: ");
   lngintsubtract(u,v,r1);
    for (i=1; i<=r1[0]; i++) printf("%4d",r1[i]);
    printf("\nmultiple: ");
    lngintmult(u,v,r1);
    for (i=1; i<=r1[0]; i++) printf("%4d",r1[i]);
    printf("\ndivide:  Quotient: ");
    lngintdivide(u,v,r1,r2);
    for (i=1; i<=r1[0]; i++) printf("%4d",r1[i]);
    printf("  Remainder: ");
    for (i=1; i<=r2[0]; i++) printf("%4d",r2[i]);
    printf("\npower: ");
    lngintpower(v,4,r1);
    for (i=1; i<=r1[0]; i++) printf("%4d",r1[i]);
}
```

Output:

```
Input numbers:
  33  70  70  70  70
   4  44

add:   33  70  70  75  14
subtract:   33  70  70  66  26
multiple:    1  49  65  93  93  90  80
divide:  Quotient:     7  59  16  82  Remainder:     2  62
power:    3  88  62  60  24  96
```

Examples for chapter 2 procedures

Function tested: **derpol**

Example:

Compute the derivatives $d^jp(x)/dx^j$, for j=0,...,3, of

$$p(x) = -1 + x - 2x^2 + 3x^3$$

at the point x = 1.

```
#include <stdio.h>

void main ()
{
    void derpol(int, int, float, float []);
    float a[4] = {-1.0, 1.0, -2.0, 3.0};

    derpol(3,3,1.0,a);
    printf("The 0-th until and including the 3rd derivatives:"
           "  %-4.0f%-4.0f%-4.0f%-4.0f",a[0],a[1],a[2],a[3]);
}
```

Output:

```
The 0-th until and including the 3rd derivatives:  1   6   14  18
```

Function tested: **allortpol**

Example:

Compute the Laguerre polynomials $L_i(x)$, for i=0,...,5, which satisfy the recursion

$$L_0(x) = 1, \qquad L_1(x) = x - 1$$

$$L_{k+1}(x) = (x-2k-1)L_k(x) - k^2L_{k-1}(x) \qquad (k=1,2,...)$$

when x = 0.

```
#include <stdio.h>

void main ()
{
    void allortpol(int, float, float [], float [], float []);
    int i;
    float b[5],c[5],p[6];

    b[0]=1.0;
    for (i=1; i<=4; i++) {
        b[i]=2*i+1;
        c[i]=i*i;
    }
    allortpol(5,0.0,b,c,p);
    printf("ALLORTPOL delivers:  %-6.1f%-6.1f%-6.1f%-6.1f%-6.1f%-6.1f",
           p[0],p[1],p[2],p[3],p[4],p[5]);
}
```

Output:

```
ALLORTPOL delivers:  1.0   -1.0  2.0   -6.0  24.0  -120.0
```

Function tested: **chepolsum**

Example:

Compute the polynomial

$$1 + (1/2)T_1(x) + (1/4)T_2(x)$$

for i=-1,0,1, where $T_1(x)$ and $T_2(x)$ are the Chebyshev polynomials of first and second degree, respectively.

```
#include <stdio.h>

void main ()
{
    float chepolsum(int, float, float []);
    float a[3] = {1.0, 0.5, 0.25};

    printf("CHEPOLSUM delivers:  %-6.2f%-6.2f%-6.2f",
            chepolsum(2,-1.0,a),chepolsum(2,0.0,a),chepolsum(2,1.0,a));
}
```

Output:

```
CHEPOLSUM delivers:  0.75  0.75  1.75
```

Function tested: **oddchepolsum**

Example:

Compute the polynomial

$$1 + (1/2)T_1(x) + (1/5)T_3(x)$$

for i=-1,0,1, where $T_1(x)$ and $T_3(x)$ are the Chebyshev polynomials of first and third degree, respectively.

```
#include <stdio.h>

void main ()
{
    float oddchepolsum(int, float, float []);
    float a[2] = {0.5, 0.2};

    printf("ODDCHEPOLSUM delivers:  %-7.2f%-7.2f%-7.2f",
            oddchepolsum(1,-1.0,a),oddchepolsum(1,0.0,a),
            oddchepolsum(1,1.0,a));
}
```

Output:

```
ODDCHEPOLSUM delivers:  -0.70  -0.00  0.70
```

Functions tested: **chepol, allchepol**

Example:

The Chebyshev polynomials of the first kind of degrees 0,1,2 are evaluated at -1,0,1.

```
#include <stdio.h>

void main ()
{
    void allchepol(int, float, float []);
    float chepol(int, float);
    float t[3];

    allchepol(2,-1.0,t);
    printf("Delivers:\n %4.0f%4.0f%4.0f",t[0],t[1],t[2]);
    allchepol(2,0.0,t);
    printf("\n %4.0f%4.0f%4.0f",t[0],t[1],t[2]);
    allchepol(2,1.0,t);
    printf("\n %4.0f%4.0f%4.0f",t[0],t[1],t[2]);
    printf("\n\n %-6.0f%-6.0f%-6.0f",
```

```
            chepol(2,-1.0),chepol(2,0.0),chepol(2,1.0));
}
```

Output:

```
Delivers:
    1  -1   1
    1   0  -1
    1   1   1

 1      -1     1
```

Function tested: **fouser**

Example:

Evaluate $0.5 + \cos\theta + \sin\theta$ for $\theta = 0, \pi/2$ and π.

```
#include <math.h>
#include <stdio.h>

void main ()
{
    float fouser(int, float, float []);
    float pi, a[2] = {0.5, 1.0};

    pi=atan(1.0)*4.0;
    printf("FOUSER delivers:  %-7.2f%-7.2f%-7.2f",
            fouser(1,0.0,a),fouser(1,pi/2.0,a),fouser(1,pi,a));
}
```

Output:

```
FOUSER delivers:  1.50   1.50   -0.50
```

Function tested: **jfrac**

Example:

Compute the value of the *n*-th convergent of the continued fraction

$$1 + \frac{1}{2+}\,\frac{1}{2+}\,\ldots$$

for $n = 7,...,10$.

```
#include <stdio.h>

void main ()
{
    float fouser(int, float, float []);
    int i;
    float p[11],q[11];

    for (i=1; i<=10; i++) {
        p[i]=1.0;
        q[i]=2.0;
    }
    q[0]=1.0;
    printf("JFRAC delivers:  %11.7f%11.7f%11.7f%11.7f",
            jfrac(7,p,q),jfrac(8,p,q),jfrac(9,p,q),jfrac(10,p,q));
}
```

Output:

```
JFRAC delivers:     1.4142157  1.4142132  1.4142136  1.4142135
```

Functions tested: **chspol, polchs**

Example:

Convert the power sum $1+2x+3x^2$ into a sum of Chebyshev polynomials $b_0+b_1T_1(x)+b_2T_2(x)$ (by means of a call of *polchs*) and, as a check, transform the latter representation back into the original power sum (by means of a call of *chspol*).

```
#include <stdio.h>

void main ()
{
    void polchs(int, float []);
    void chspol(int, float []);
    float a[3] = {1.0, 2.0, 3.0};

    printf("          a[0]  a[1]  a[2]\ninput:"
           "    %-6.2f%-6.2f%-6.2f",a[0],a[1],a[2]);
    polchs(2,a);
    printf("\npolchs:  %-6.2f%-6.2f%-6.2f",a[0],a[1],a[2]);
    chspol(2,a);
    printf("\nchspol:  %-6.2f%-6.2f%-6.2f",a[0],a[1],a[2]);
}
```

Output:

```
          a[0]  a[1]  a[2]
input:    1.00  2.00  3.00
polchs:   2.50  2.00  1.50
chspol:   1.00  2.00  3.00
```

Functions tested: **polshtchs, shtchspol**

Example:

Convert the power sum $1+2x+3x^2$ into a sum of shifted Chebyshev polynomials $b_0+b_1T_1'(x)+b_2T_2'(x)$ (by means of a call of *polshtchs*) and, as a check, transform the latter representation back into the original power sum (by means of a call of *shtchspol*).

```
#include <stdio.h>

void main ()
{
    void polshtchs(int, float []);
    void shtchspol(int, float []);
    float a[3];

    a[0]=1.0;  a[1]=2.0;  a[2]=3.0;
    printf("             a[0]  a[1]  a[2]\ninput:"
           "       %-6.2f%-6.2f%-6.2f",a[0],a[1],a[2]);
    polshtchs(2,a);
    printf("\npolshtchs:  %-6.2f%-6.2f%-6.2f",a[0],a[1],a[2]);
    shtchspol(2,a);
    printf("\nshtchspol:  %-6.2f%-6.2f%-6.2f",a[0],a[1],a[2]);
}
```

Output:

```
             a[0]  a[1]  a[2]
input:       1.00  2.00  3.00
polshtchs:   3.12  2.50  0.38
shtchspol:   1.00  2.00  3.00
```

Functions tested: **newgrn, grnnew**

Example:

Convert the power sum $1+2x+3x^2$ into its Newton sum of the form

$$\delta_0^{\,0}+(x-x_0)\delta_0^{\,1}+(x-x_0)(x-x_1)\delta_0^{\,2}$$

(by means of a call of *newgrn*) and, as a check, transform the latter representation back into the original power sum (by means of a call of *grnnew*).

```
#include <stdio.h>

void main ()
{
    void grnnew(int, float [], float []);
    void newgrn(int, float [], float []);
    float x[2] = {1.0, 2.0},  a[3] = {1.0, 2.0, 3.0};

    printf("            a[0]    a[1]    a[2]\ninput:"
           "  %7.2f%7.2f%7.2f",a[0],a[1],a[2]);
    grnnew(2,x,a);
    printf("\ngrnnew: %7.2f%7.2f%7.2f",a[0],a[1],a[2]);
    newgrn(2,x,a);
    printf("\nnewgrn: %7.2f%7.2f%7.2f",a[0],a[1],a[2]);
}
```

Output:

```
            a[0]    a[1]    a[2]
input:      1.00    2.00    3.00
grnnew:     6.00   11.00    3.00
newgrn:     1.00    2.00    3.00
```

Function tested: **lintfmpol**

Example:

Convert the power sum $1+2x+3x^2$ into its power sum in y with $x = 2y + 3$, and, as a check, transform the latter representation back into the original power sum (both transformations are carried out by means of a call of *lintfmpol*).

```
#include <stdio.h>

void main ()
{
    void lintfmpol(float, float, int, float []);
    float a[3] = {1.0, 2.0, 3.0};

    printf("               a[0]    a[1]    a[2]\ninput:"
           "     %7.2f%7.2f%7.2f",a[0],a[1],a[2]);
    lintfmpol(2.0,3.0,2,a);
    printf("\nlintfmpol: %7.2f%7.2f%7.2f   (power sum in y)",
           a[0],a[1],a[2]);
    lintfmpol(0.5,-1.5,2,a);
    printf("\nlintfmpol: %7.2f%7.2f%7.2f   (power sum in x)",
           a[0],a[1],a[2]);
}
```

Output:

```
               a[0]    a[1]    a[2]
input:         1.00    2.00    3.00
lintfmpol:    34.00   40.00   12.00   (power sum in y)
lintfmpol:     1.00    2.00    3.00   (power sum in x)
```

Function tested: **intchs**

Example:

Determine the coefficients $\{b_j\}$ in the expansion

$$\int_0^x \{1 + \frac{1}{2}T_1(x) + \frac{1}{5}T_2(x) + \frac{1}{10}T_3(x)\}\,dx = \sum_{j=1}^{4} b_j T_j(x)$$

where $\{T_j(x)\}$ are Chebyshev polynomials.

```
#include <math.h>
#include <stdio.h>

void main ()
{
    void intchs(int, float [], float []);
    float b[5], a[4] = {1.0, 0.5, 0.2, 0.1};

    intchs(3,a,b);
    printf("INTCHS delivers:%8.4f%8.4f%8.4f%8.4f",b[1],b[2],b[3],b[4]);
}
```

Output:

```
INTCHS delivers:  0.9000  0.1000  0.0333  0.0125
```

Examples for chapter 3 procedures

Functions tested: **determ, gsselm**

Example:

Compute the determinant of the 4×4 Hilbert matrix A for which $A_{i,j} = 1/(i+j-1)$, $i,j=1,...,4$.

```
#include <stdio.h>

void main ()
{
    float **allocate_real_matrix(int, int, int, int);
    void free_real_matrix(float **, int, int, int);
    void gsselm(float **, int, float [], int [], int []);
    float determ(float **, int, int);
    int i,j,ri[5],ci[5];
    float d,**a,aux[8];

    a=allocate_real_matrix(1,4,1,4);
    for (i=1; i<=4; i++)
        for (j=1; j<=4; j++) a[i][j]=1.0/(i+j-1);
    aux[2]=1.0e-5;
    aux[4]=8;
    gsselm(a,4,aux,ri,ci);
    d = (aux[3] == 4) ? determ(a,4,aux[1]) : 0.0;
    printf("Determinant = %e",d);
    free_real_matrix(a,1,4,1);
}
```

Output:

```
Determinant = 1.65344e-07
```

Function tested: **decsol**

Example:

Solve the 4×4 linear system of equations $Ax=b$, where A is the Hilbert matrix, for which $A_{i,j}=1/(i+j-1)$, $i,j=1,...,4$, and $b_i = 1/(2+i)$, $i=1,...,4$, (so that the solution is the third unit vector: $(0,0,1,0)^T$).

```
#include <stdio.h>

void main ()
{
    float **allocate_real_matrix(int, int, int, int);
    void free_real_matrix(float **, int, int, int);
    void decsol(float **, int, float [], float []);
    int i,j;
    float **a,b[5],aux[4];

    a=allocate_real_matrix(1,4,1,4);
    for (i=1; i<=4; i++) {
        for (j=1; j<=4; j++) a[i][j]=1.0/(i+j-1);
        b[i]=a[i][3];
    }
    aux[2]=1.0e-5;
    decsol(a,4,aux,b);
    printf("Solution: %e  %e  %e  %e\n",b[1],b[2],b[3],b[4]);
    printf("Sign(Det) =%3.0f\nNumber of eliminations =%3.0f\n",
           aux[1],aux[3]);
    free_real_matrix(a,1,4,1);
}
```

Output:

```
Solution: 0.00000e+00  0.00000e+00  1.00000e+00  -0.00000e+00
Sign(Det) =  1
Number of eliminations =  4
```

Function tested: **gsssol**

Example:

Solve the 4×4 linear system of equations $Ax=b$, where A is the Hilbert matrix, for which $A_{i,j}=1/(i+j-1)$, $i,j=1,...,4$, and $b_i = 1/(2+i)$, $i=1,...,4$, (so that the solution is the third unit vector: $(0,0,1,0)^T$).

```
#include <stdio.h>

void main ()
{
    float **allocate_real_matrix(int, int, int, int);
    void free_real_matrix(float **, int, int, int);
    void gsssol(float **, int, float [], float []);
    int i,j;
    float **a,b[5],aux[8];

    a=allocate_real_matrix(1,4,1,4);
    for (i=1; i<=4; i++) {
        for (j=1; j<=4; j++) a[i][j]=1.0/(i+j-1);
        b[i]=a[i][3];
    }
    aux[2]=1.0e-5;
    aux[4]=8;
    gsssol(a,4,aux,b);
    printf("Solution: %e  %e  %e  %e\n",b[1],b[2],b[3],b[4]);
    printf("Sign(Det) =%3.0f\nNumber of eliminations =%3.0f\n"
           "Max(abs(a[i,j])) = %e\nUpper bound growth = %e\n",
```

```
            aux[1],aux[3],aux[5],aux[7]);
    free_real_matrix(a,1,4,1);
}
```

Output:

```
Solution: 0.00000e+00  1.78814e-07  1.000000e+00  -0.00000e+00
Sign(Det) =  1
Number of eliminations =  4
Max(abs(a[i,j])) = 1.00000e+00
Upper bound growth = 1.59619e+00
```

Function tested: **gsssolerb**

Example:

Solve the 4×4 linear system of equations $Ax=b$, where A is the Hilbert matrix, for which $A_{i,j}=1/(i+j-1)$, $i,j=1,...,4$, and $b_i = 1/(2+i)$, $i=1,...,4$, (so that the solution is the third unit vector: $(0,0,1,0)^T$), and provide an upper bound for the relative error in x.

```
#include <stdio.h>

void main ()
{
    float **allocate_real_matrix(int, int, int, int);
    void free_real_matrix(float **, int, int, int);
    void gsssolerb(float **, int, float [], float []);
    int i,j;
    float **a,b[5],aux[12];

    a=allocate_real_matrix(1,4,1,4);
    for (i=1; i<=4; i++) {
        for (j=1; j<=4; j++) a[i][j]=1.0/(i+j-1);
        b[i]=a[i][3];
    }
    aux[0]=aux[2]=aux[6]=1.0e-14;
    aux[4]=8;
    gsssolerb(a,4,aux,b);
    printf("Solution: %e  %e  %e  %e\n",b[1],b[2],b[3],b[4]);
    printf("Sign(Det) =%3.0f\nNumber of eliminations =%3.0f\n"
            "Max(abs(a[i,j])) = %e\nUpper bound growth = %e\n"
            "1-norm of the inverse matrix = %e\n"
            "Upper bound rel. error in the calc. solution = %e",
            aux[1],aux[3],aux[5],aux[7],aux[9],aux[11]);
    free_real_matrix(a,1,4,1);
}
```

Output:

```
Solution: 0.00000e+00  1.78814e-07  1.000000e+00  -0.00000e+00
Sign(Det) =  1
Number of eliminations =  4
Max(abs(a[i,j])) = 1.00000e+00
Upper bound growth = 1.59619e+00
1-norm of the inverse matrix = 1.36200e+04
Upper bound rel. error in the calc. solution = 2.77896e-08
```

Function tested: **decinv**

Example:

Compute the inverse of the 4×4 matrix

$$\begin{pmatrix} 4 & 2 & 4 & 1 \\ 30 & 20 & 45 & 12 \\ 20 & 15 & 36 & 10 \\ 35 & 28 & 70 & 20 \end{pmatrix}.$$

```
#include <stdio.h>

void main ()
{
    float **allocate_real_matrix(int, int, int, int);
    void free_real_matrix(float **, int, int, int);
    void decinv(float **, int, float []);
    int i;
    float **a,aux[4];

    a=allocate_real_matrix(1,4,1,4);
    a[1][1]=4.0;   a[1][2]=2.0;   a[1][3]=4.0;   a[1][4]=1.0;
    a[2][1]=30.0;  a[2][2]=20.0;  a[2][3]=45.0;  a[2][4]=12.0;
    a[3][1]=20.0;  a[3][2]=15.0;  a[3][3]=36.0;  a[3][4]=10.0;
    a[4][1]=35.0;  a[4][2]=28.0;  a[4][3]=70.0;  a[4][4]=20.0;
    aux[2]=1.0e-5;
    decinv(a,4,aux);
    printf("Calculated inverse:\n");
    for (i=1; i<=4; i++)
       printf(" %4.0f%4.0f%4.0f%4.0f\n",a[i][1],a[i][2],a[i][3],a[i][4]);
    printf("\naux[1] =%3.0f\naux[3] =%3.0f\n",aux[1],aux[3]);
    free_real_matrix(a,1,4,1);
}
```

Output:

```
Calculated inverse:
    4  -2   4  -1
  -30  20 -45  12
   20 -15  36 -10
  -35  28 -70  20

aux[1] =  1
aux[3] =  4
```

Function tested: **gssinv**

Example:

Compute the inverse of the 4×4 matrix

$$\begin{pmatrix} 4 & 2 & 4 & 1 \\ 30 & 20 & 45 & 12 \\ 20 & 15 & 36 & 10 \\ 35 & 28 & 70 & 20 \end{pmatrix}.$$

```
#include <stdio.h>

void main ()
{
    float **allocate_real_matrix(int, int, int, int);
    void free_real_matrix(float **, int, int, int);
    void gssinv(float **, int, float []);
    int i;
    float **a,aux[10];

    a=allocate_real_matrix(1,4,1,4);
```

```
    a[1][1]=4.0;  a[1][2]=2.0;  a[1][3]=4.0;  a[1][4]=1.0;
    a[2][1]=30.0; a[2][2]=20.0; a[2][3]=45.0; a[2][4]=12.0;
    a[3][1]=20.0; a[3][2]=15.0; a[3][3]=36.0; a[3][4]=10.0;
    a[4][1]=35.0; a[4][2]=28.0; a[4][3]=70.0; a[4][4]=20.0;
    aux[2]=1.0e-5;
    aux[4]=8;
    gssinv(a,4,aux);
    printf("Calculated inverse:\n");
    for (i=1; i<=4; i++)
        printf(" %4.0f%4.0f%4.0f%4.0f\n",a[i][1],a[i][2],a[i][3],a[i][4]);
    printf("\nAUX elements:\n%e  %e  %e  %e  %e\n",
            aux[1],aux[3],aux[5],aux[7],aux[9]);
    free_real_matrix(a,1,4,1);
}
```

Output:

```
Calculated inverse:
    4  -2   4  -1
  -30  20 -45  12
   20 -15  36 -10
  -35  28 -70  20

AUX elements:
1.00000e+00  4.00000e+00  7.00000e+01  1.12529e+02  1.54996e+02
```

Function tested: **gssinverb**

Example:

Compute the inverse of the 4×4 matrix

$$\begin{pmatrix} 4 & 2 & 4 & 1 \\ 30 & 20 & 45 & 12 \\ 20 & 15 & 36 & 10 \\ 35 & 28 & 70 & 20 \end{pmatrix}$$

and obtain an error bound for the computed inverse.

```
#include <stdio.h>

void main ()
{
    float **allocate_real_matrix(int, int, int, int);
    void free_real_matrix(float **, int, int, int);
    void gssinverb(float **, int, float []);
    int i;
    float **a,aux[12];

    a=allocate_real_matrix(1,4,1,4);
    a[1][1]=4.0;  a[1][2]=2.0;  a[1][3]=4.0;  a[1][4]=1.0;
    a[2][1]=30.0; a[2][2]=20.0; a[2][3]=45.0; a[2][4]=12.0;
    a[3][1]=20.0; a[3][2]=15.0; a[3][3]=36.0; a[3][4]=10.0;
    a[4][1]=35.0; a[4][2]=28.0; a[4][3]=70.0;    a[4][4]=20.0;
    aux[0]=aux[2]=aux[6]=1.0e-14;
    aux[4]=8;
    gssinverb(a,4,aux);
    printf("Calculated inverse:\n");
    for (i=1; i<=4; i++)
        printf(" %4.0f%4.0f%4.0f%4.0f\n",a[i][1],a[i][2],a[i][3],a[i][4]);
    printf("\nAUX elements:\n%e\n%e\n%e\n%e\n%e\n%e\n",
            aux[1],aux[3],aux[5],aux[7],aux[9],aux[11]);
    free_real_matrix(a,1,4,1);
}
```

Output:

```
Calculated inverse:
    4  -2   4  -1
  -30  20 -45  12
   20 -15  36 -10
  -35  28 -70  20

AUX elements:
1.00000e+00
4.00000e+00
7.00000e+01
1.12529e+02
1.54996e+02
2.22941e-08
```

Function tested: **gssitisol**

Example:

Solve the 4×4 linear system of equations $Ax=b$, where $A_{i,j}=840/(i+j-1)$, $i,j=1,...,4$, and b is the third column of A (so that the solution is the third unit vector: $(0,0,1,0)^T$).

```
#include <stdio.h>

void main ()
{
    float **allocate_real_matrix(int, int, int, int);
    void free_real_matrix(float **, int, int, int);
    void gssitisol(float **, int, float [], float []);
    int i,j;
    float **a,b[5],aux[14];

    a=allocate_real_matrix(1,4,1,4);
    for (i=1; i<=4; i++) {
        for (j=1; j<=4; j++) a[i][j]=840/(i+j-1);
        b[i]=a[i][3];
    }
    aux[2]=aux[10]=1.0e-5;
    aux[4]=8;
    aux[12]=5.0;
    gssitisol(a,4,aux,b);
    printf("Solution: %e  %e  %e  %e\n",b[1],b[2],b[3],b[4]);
    printf("Sign(Det) =%3.0f\nNumber of eliminations =%3.0f\n"
           "Max(abs(a[i,j])) = %e\nUpper bound growth = %e\n"
           "Norm last correction vector = %e\n"
           "Norm residual vector = %e\n",
           aux[1],aux[3],aux[5],aux[7],aux[11],aux[13]);
    free_real_matrix(a,1,4,1);
}
```

Output:

```
Solution: -5.68989e-16  6.71685e-15  1.00000e+00  1.08316e-14
Sign(Det) =  1
Number of eliminations =  4
Max(abs(a[i,j])) = 8.40000e+02
Upper bound growth = 1.34080e+03
Norm last correction vector = 3.57885e-10
Norm residual vector = 1.32202e-11
```

Function tested: **gssitisolerb**

Example:

Solve the 4×4 linear system of equations $Ax=b$, where $A_{i,j}=840/(i+j-1)$, $i,j=1,...,4$, and

b is the third column of A (so that the solution is the third unit vector: $(0,0,1,0)^T$), and provide a realistic upper bound for the relative error in it.

```
#include <stdio.h>

void main ()
{
    float **allocate_real_matrix(int, int, int, int);
    void free_real_matrix(float **, int, int, int);
    void gssitisolerb(float **, int, float [], float []);
    int i,j;
    float **a,b[5],aux[14];

    a=allocate_real_matrix(1,4,1,4);
    for (i=1; i<=4; i++) {
        for (j=1; j<=4; j++) a[i][j]=840/(i+j-1);
        b[i]=a[i][3];
    }
    aux[0]=aux[2]=aux[10]=1.0e-14;
    aux[4]=8;
    aux[6]=aux[8]=0.0;
    aux[12]=5.0;
    gssitisolerb(a,4,aux,b);
    printf("Solution: %e  %e  %e  %e\n",b[1],b[2],b[3],b[4]);
    printf("Sign(Det) =%3.0f\nNumber of eliminations =%3.0f\n"
           "Max(abs(a[i,j])) = %e\nUpper bound growth = %e\n"
           "Norm calculated inverse matrix : = %e\n"
           "Upper bound for the relative error : = %e\n"
           "Norm residual vector = %e\n",
           aux[1],aux[3],aux[5],aux[7],aux[9],aux[11],aux[13]);
    free_real_matrix(a,1,4,1);
}
```

Output:

```
Solution: 2.67937e-15  -3.38878e-14  1.00000e+00  -5.68474e-14
Sign(Det) =  1
Number of eliminations =  4
Max(abs(a[i,j])) = 8.40000e+02
Upper bound growth = 1.34080e+03
Norm calculated inverse matrix : = 1.62142e+01
Upper bound for the relative error : = 1.10328e-09
Norm residual vector = 6.80441e-11
```

Functions tested: **chldec2, chlsol2, chlinv2**

Example:

Solve the 4×4 linear system of equations $Ax=b$, where A is the symmetric positive definite coefficient matrix (the Pascal matrix of order 4) for which

$$A_{i,j} = \binom{i+j-2}{j-1}, \qquad i,j=1,...,4$$

and $b_i = 2^i$, i=1,...,4, and obtain the inverse of A.

```
#include <math.h>
#include <stdio.h>

void main ()
{
    float *allocate_real_vector(int, int);
    float **allocate_real_matrix(int, int, int, int);
    void free_real_vector(float *, int);
    void free_real_matrix(float **, int, int, int);
    void chldec2(float **, int, float []);
```

```
    void chlsol2(float **, int, float []);
    void chlinv2(float **, int);
    int i,j;
    float **pascal2,*b,*aux;

    pascal2=allocate_real_matrix(1,4,1,4);
    b=allocate_real_vector(1,4);
    aux=allocate_real_vector(2,3);
    for (j=1; j<=4; j++) {
        pascal2[1][j]=1.0;
        for (i=2; i<=j; i++)
            pascal2[i][j] = (i == j) ?
                    pascal2[i-1][j]*2.0 : pascal2[i][j-1]+pascal2[i-1][j];
        b[j]=pow(2.0,j);
    }
    aux[2]=1.0e-11;
    chldec2(pascal2,4,aux);
    if (aux[3] == 4) {
        chlsol2(pascal2,4,b);
        chlinv2(pascal2,4);
    } else
        printf("Matrix not positive definite");
    printf("Solution with CHLDEC2 and CHLSOL2:\n %e  %e  %e  %e\n",
            b[1],b[2],b[3],b[4]);
    printf("\nInverse matrix with CHLINV2:\n");
    for (i=1; i<=4; i++) {
        for (j=1; j<=4; j++)
            if (j < i)
                printf("           ");
            else
                printf("%11.5f",pascal2[i][j]);
        printf("\n");
    }
    free_real_matrix(pascal2,1,4,1);
    free_real_vector(b,1);
    free_real_vector(aux,2);
}
```

Output:

```
Solution with CHLDEC2 and CHLSOL2:
 0.00000e+00  4.00000e+00  -4.00000e+00  2.00000e+00

Inverse matrix with CHLINV2:
    4.00000   -6.00000    4.00000   -1.00000
              14.00000  -11.00000    3.00000
                         10.00000   -3.00000
                                     1.00000
```

Functions tested: **chldec1, chlsol1, chlinv1**

Example:

Solve the 4×4 linear system of equations $Ax=b$, where A is the symmetric positive definite coefficient matrix (the Pascal matrix of order 4) for which

$$A_{i,j} = \binom{i+j-2}{j-1}, \qquad i,j=1,...,4$$

and $b_i = 2^i$, $i=1,...,4$, and obtain the inverse of A.

```
#include <math.h>
#include <stdio.h>

void main ()
{
    float *allocate_real_vector(int, int);
```

```
    void free_real_vector(float *, int);
    void chlsol1(float [], int, float []);
    void chlinv1(float [], int);
    int i,j,jj;
    float *pascal1,*b,*aux;

    pascal1=allocate_real_vector(1,((4+1)*4)/2);
    b=allocate_real_vector(1,4);
    aux=allocate_real_vector(2,3);

    jj=1;
    for (j=1; j<=4; j++) {
        pascal1[jj]=1.0;
        for (i=2; i<=j; i++)
            pascal1[jj+i-1] = (i == j) ?
                    pascal1[jj+i-2]*2.0 : pascal1[jj+i-2]+pascal1[jj+i-j];
        b[j]=pow(2.0,j);
        jj += j;
    }
    aux[2]=1.0e-11;
    chldec1(pascal1,4,aux);
    if (aux[3] == 4) {
        chlsol1(pascal1,4,b);
        chlinv1(pascal1,4);
    } else
        printf("Matrix not positive definite");
    printf("Solution with CHLDEC1 and CHLSOL1:\n %e  %e  %e  %e\n",
            b[1],b[2],b[3],b[4]);
    printf("\nInverse matrix with CHLINV1:\n");
    for (i=1; i<=4; i++) {
        for (j=1; j<=4; j++)
            if (j < i)
                printf("           ");
            else
                printf("%11.5f",pascal1[((j-1)*j)/2+i]);
        printf("\n");
    }
    free_real_vector(pascal1,1);
    free_real_vector(b,1);
    free_real_vector(aux,2);
}
```

Output:

```
Solution with CHLDEC1 and CHLSOL1:
 0.00000e+00  4.00000e+00  -4.00000e+00  2.00000e+00

Inverse matrix with CHLINV1:
    4.00000   -6.00000    4.00000   -1.00000
              14.00000  -11.00000    3.00000
                         10.00000   -3.00000
                                     1.00000
```

Functions tested: **chldecsol2, chldeterm2, chldecinv2**

Example:

Solve the 4×4 linear system of equations $Ax=b$, where A is the symmetric positive definite coefficient matrix (the Pascal matrix of order 4) for which

$$A_{i,j} = \binom{i+j-2}{j-1}, \qquad i,j=1,...,4$$

and $b_i = 2^i$, i=1,...,4, and obtain the determinant and the inverse of A.

```
#include <math.h>
#include <stdio.h>
```

```
void main ()
{
    float *allocate_real_vector(int, int);
    float **allocate_real_matrix(int, int, int, int);
    void free_real_vector(float *, int);
    void free_real_matrix(float **, int, int, int);
    void chldecsol2(float **, int, float [], float []);
    float chldeterm2(float **, int);
    void chldecinv2(float **, int, float []);
    int i,j;
    float determinant,**pascal2,*b,*aux;

    pascal2=allocate_real_matrix(1,4,1,4);
    b=allocate_real_vector(1,4);
    aux=allocate_real_vector(2,3);

    for (j=1; j<=4; j++) {
        pascal2[1][j]=1.0;
        for (i=2; i<=j; i++)
            pascal2[i][j] = (i == j) ?
                    pascal2[i-1][j]*2.0 : pascal2[i][j-1]+pascal2[i-1][j];
        b[j]=pow(2.0,j);
    }
    aux[2]=1.0e-11;
    chldecsol2(pascal2,4,aux,b);
    if (aux[3] == 4)
        determinant=chldeterm2(pascal2,4);
    else
        printf("Matrix not positive definite");
    printf("Solution with CHLDECSOL2:\n %e  %e  %e  %e\n",
            b[1],b[2],b[3],b[4]);
    printf("\nDeterminant with CHLDETERM2: %e\n",determinant);
    for (j=1; j<=4; j++) {
        pascal2[1][j]=1.0;
        for (i=2; i<=j; i++)
            pascal2[i][j] = (i == j) ?
                    pascal2[i-1][j]*2.0 : pascal2[i][j-1]+pascal2[i-1][j];
    }
    chldecinv2(pascal2,4,aux);
    printf("\nInverse matrix with CHLDECINV2:\n");
    for (i=1; i<=4; i++) {
        for (j=1; j<=4; j++)
            if (j < i)
                printf("           ");
            else
                printf("%11.5f",pascal2[i][j]);
        printf("\n");
    }

    free_real_matrix(pascal2,1,4,1);
    free_real_vector(b,1);
    free_real_vector(aux,2);
}
```

Output:

```
Solution with CHLDECSOL2:
 0.00000e+00  4.00000e+00  -4.00000e+00  2.00000e+00

Determinant with CHLDETERM2: 1.00000e+00

Inverse matrix with CHLDECINV2:
    4.00000   -6.00000    4.00000   -1.00000
              14.00000  -11.00000    3.00000
                         10.00000   -3.00000
                                     1.00000
```

Functions tested: **chldecsol1, chldeterm1, chldecinv1**

Example:

Solve the 4×4 linear system of equations $Ax=b$, where A is the symmetric positive definite coefficient matrix (the Pascal matrix of order 4) for which

$$A_{i,j} = \binom{i+j-2}{j-1}, \qquad i,j=1,...,4$$

and $b_i = 2^i$, i=1,...,4, and obtain the determinant and the inverse of A.

```
#include <math.h>
#include <stdio.h>

void main ()
{
    float *allocate_real_vector(int, int);
    void free_real_vector(float *, int);
    void chldecsol1(float [], int, float [], float []);
    float chldeterm1(float [], int);
    void chldecinv1(float [], int, float []);
    int i,j,jj;
    float determinant,*pascal1,*b,*aux;

    pascal1=allocate_real_vector(1,((4+1)*4)/2);
    b=allocate_real_vector(1,4);
    aux=allocate_real_vector(2,3);

    jj=1;
    for (j=1; j<=4; j++) {
        pascal1[jj]=1.0;
        for (i=2; i<=j; i++)
            pascal1[jj+i-1] = (i == j) ?
                    pascal1[jj+i-2]*2.0 : pascal1[jj+i-2]+pascal1[jj+i-j];
        b[j]=pow(2.0,j);
        jj += j;
    }
    aux[2]=1.0e-11;
    chldecsol1(pascal1,4,aux,b);
    if (aux[3] == 4)
        determinant=chldeterm1(pascal1,4);
    else
        printf("Matrix not positive definite");
    printf("Solution with CHLDECSOL1:\n %e  %e  %e  %e\n",
          b[1],b[2],b[3],b[4]);
    printf("\nDeterminant with CHLDETERM1: %e\n",determinant);
    jj=1;
    for (j=1; j<=4; j++) {
        pascal1[jj]=1.0;
        for (i=2; i<=j; i++)
            pascal1[jj+i-1] = (i == j) ?
                    pascal1[jj+i-2]*2.0 : pascal1[jj+i-2]+pascal1[jj+i-j];
        jj += j;
    }
    chldecinv1(pascal1,4,aux);
    printf("\nInverse matrix with CHLDECINV1:\n");
    for (i=1; i<=4; i++) {
        for (j=1; j<=4; j++)
            if (j < i)
                printf("           ");
            else
                printf("%11.5f",pascal1[((j-1)*j)/2+i]);
        printf("\n");
    }

    free_real_vector(pascal1,1);
    free_real_vector(b,1);
    free_real_vector(aux,2);
}
```

Output:

```
Solution with CHLDECSOL1:
 0.00000e+00  4.00000e+00  -4.00000e+00  2.00000e+00

Determinant with CHLDETERM1: 1.00000e+00

Inverse matrix with CHLDECINV1:
    4.00000    -6.00000     4.00000    -1.00000
              14.00000   -11.00000     3.00000
                          10.00000    -3.00000
                                       1.00000
```

Function tested: **determsym2**

Example:

Evaluate the determinant of the matrix

$$\begin{pmatrix} -3 & -3 & -18 & -30 & 18 \\ -3 & -1 & -4 & -48 & 8 \\ -18 & -4 & -6 & -274 & 6 \\ -30 & -48 & -274 & 119 & 19 \\ 18 & 8 & 6 & 19 & 216 \end{pmatrix}.$$

```
#include <stdio.h>
void main ()
{
    float **allocate_real_matrix(int, int, int, int);
    void free_real_matrix(float **, int, int, int);
    void decsym2(float **, int, float, int [], int [], float []);
    float detmsym2(float [], int, int []);
    int i,j,aux[6],p[6];
    float tol,determinant,**a,detaux[6];

    a=allocate_real_matrix(1,5,1,5);
    a[1][1]=a[1][2] = -3.0; a[1][3] = -18.0;
    a[1][4] = -30.0;  a[1][5]=18.0;
    a[2][2] = -1.0;    a[2][3] = -4.0;     a[2][4] = -48.0;  a[2][5]=8.0;
    a[3][3] = -6.0;    a[3][4] = -274.0;  a[3][5]=6.0;
    a[4][4]=119.0; a[4][5]=19.0;    a[5][5]=216.0;
    for (i=1; i<=5; i++)
       for (j=i+1; j<=5; j++) a[j][i]=a[i][j];
    tol=1.0e-6;
    decsym2(a,5,tol,aux,p,detaux);
    if (aux[2] == 1)
       printf("\nThe matrix is symmetric.");
    else
       printf("The matrix is asymmetric, results are meaningless.");
    determinant=determsym2(detaux,5,aux);
    printf("\nThe determinant of the matrix :%8.2f\n",determinant);
    free_real_matrix(a,1,5,1);
}
```

Output:

```
The matrix is symmetric.
The determinant of the matrix :  168.00
```

Function tested: **decsolsym2**

Example:

Solve the system of equations

$$\begin{pmatrix} -3 & -3 & -18 & -30 & 18 \\ -3 & -1 & -4 & -48 & 8 \\ -18 & -4 & -6 & -274 & 6 \\ -30 & -48 & -274 & 119 & 19 \\ 18 & 8 & 6 & 19 & 216 \end{pmatrix} \begin{pmatrix} x_1 \\ x_2 \\ x_3 \\ x_4 \\ x_5 \end{pmatrix} = \begin{pmatrix} 327 \\ 291 \\ 1290 \\ 275 \\ 1720 \end{pmatrix}.$$

```
#include <stdio.h>

void main ()
{
    float **allocate_real_matrix(int, int, int, int);
    void free_real_matrix(float **, int, int, int);
    void decsolsym2(float **, int, float [], float, int []);
    int i,j,aux[6];
    float tol,**a,b[6];

    a=allocate_real_matrix(1,5,1,5);
    a[1][1]=a[1][2] = -3.0;    a[1][3] = -18.0;
   a[1][4] = -30.0;     a[1][5]=18.0;
    a[2][2] = -1.0;     a[2][3] = -4.0;     a[2][4] = -48.0;     a[2][5]=8.0;
    a[3][3] = -6.0;     a[3][4] = -274.0;  a[3][5]=6.0;
    a[4][4]=119.0;      a[4][5]=19.0;   a[5][5]=216.0;
    b[1]=327.0;     b[2]=291.0;     b[3]=1290.0; b[4]=275.0;    b[5]=1720.0;
    for (i=1; i<=5; i++)
        for (j=i+1; j<=5; j++) a[j][i]=a[i][j];
    tol=1.0e-6;
    decsolsym2(a,5,b,tol,aux);
    if (aux[2] == 1)
        printf("\nThe matrix is symmetric.");
    else
        printf("The matrix is asymmetric, results are meaningless.");
    printf("\nInertia : %2d,%2d,%2d\n",aux[3],aux[4],aux[5]);
    printf("\nThe computed solution:\n%10.5f%10.5f%10.5f%10.5f%10.5f\n",
            b[1],b[2],b[3],b[4],b[5]);
    free_real_matrix(a,1,5,1);
}
```

Output:

```
The matrix is symmetric.
Inertia :  3, 2, 0

The computed solution:
  -7.00000  -2.00000  -1.00000  -4.00000   9.00000
```

Functions tested: **lsqortdec, lsqsol, lsqdglinv**

Example:

Derive the least squares solution of the system of equations

$$\begin{aligned} -2x_1 + x_2 &= 0 \\ -x_1 + x_2 &= 1 \\ x_1 + x_2 &= 2 \\ 2x_1 + x_2 &= 2 \\ x_1 + 2x_2 &= 3 \end{aligned}$$

```
#include <stdio.h>

void main ()
```

```
{
    float **allocate_real_matrix(int, int, int, int);
    void free_real_matrix(float **, int, int, int);
    void lsqortdec(float **, int, int, float [], float [], int []);
    void lsqsol(float **, int, int, float [], int [], float []);
    void lsqdglinv(float **, int, float [], int [], float []);
    float vecvec(int, int, int, float [], float []);
    int i,piv[3];
    float sum,temp,**a,**c,b[6],x[6],diag[3],aid[3],aux[6];

    a=allocate_real_matrix(1,5,1,2);
    c=allocate_real_matrix(1,5,1,2);
    aux[2]=1.0e-6;
    a[1][1]=c[1][1] = -2.0;
    a[2][1]=c[2][1] = -1.0;
    a[3][1]=c[3][1]=1.0;
    a[4][1]=c[4][1]=2.0;
    a[5][1]=c[5][1]=1.0;
    a[1][2]=a[2][2]=a[3][2]=a[4][2]=c[1][2]=c[2][2]=c[3][2]=c[4][2]=1.0;
    a[5][2]=c[5][2]=2.0;
    b[1]=x[1]=0.0;
    b[2]=x[2]=1.0;
    b[3]=x[3]=b[4]=x[4]=2.0;
    b[5]=x[5]=3.0;
    lsqortdec(a,5,2,aux,aid,piv);
    if (aux[3] == 2) {
        lsqsol(a,5,2,aid,piv,x);
        lsqdglinv(a,2,aid,piv,diag);
        sum=0.0;
        for (i=1; i<=5; i++) {
            temp=b[i]-c[i][1]*x[1]-c[i][2]*x[2];
            sum += temp*temp;
        }
        printf("Aux[2, 3, 5] =  %e   %e   %e\n"
                "LSQ solution :  %e   %e\n"
                "Residue (delivered) :  %e\n"
                "Residue (checked)   :  %e\n"
                "Diagonal of inverse M'M :  %e   %e\n",
                aux[2],aux[3],aux[5],x[1],x[2],sqrt(vecvec(3,5,0,x,x)),
                sqrt(sum),diag[1],diag[2]);
    }
    free_real_matrix(a,1,5,1);
    free_real_matrix(c,1,5,1);
}
```

Output:

```
Aux[2, 3, 5] =  1.000000e-06   2.00000e+00   3.31662e+00
LSQ solution :  5.00000e-01   1.25000e+00
Residue (delivered) :  5.00000e-01
Residue (checked)   :  5.00000e-01
Diagonal of inverse M'M :  9.52381e-02   1.30952e-01
```

Function tested: **lsqortdecsol**

Example:

Derive the least squares solution of the system of equations

$$-2x_1 + x_2 = 0$$
$$-x_1 + x_2 = 1$$
$$x_1 + x_2 = 2$$
$$2x_1 + x_2 = 2$$
$$x_1 + 2x_2 = 3$$

```
#include <stdio.h>

void main ()
{
```

```
    float **allocate_real_matrix(int, int, int, int);
    void free_real_matrix(float **, int, int, int);
    void lsqortdecsol(float **, int, int, float [], float [], float []);
    float vecvec(int, int, int, float [], float []);
    int i;
    float sum,temp,**a,**c,b[6],x[6],diag[3],aux[6];

    a=allocate_real_matrix(1,5,1,2);
    c=allocate_real_matrix(1,5,1,2);
    aux[2]=1.0e-6;
    a[1][1]=c[1][1] = -2.0;
    a[2][1]=c[2][1] = -1.0;
    a[3][1]=c[3][1]=1.0;
    a[4][1]=c[4][1]=2.0;
    a[5][1]=c[5][1]=1.0;
    a[1][2]=a[2][2]=a[3][2]=a[4][2]=c[1][2]=c[2][2]=c[3][2]=c[4][2]=1.0;
    a[5][2]=c[5][2]=2.0;
    b[1]=x[1]=0.0;
    b[2]=x[2]=1.0;
    b[3]=x[3]=b[4]=x[4]=2.0;
    b[5]=x[5]=3.0;
    lsqortdecsol(a,5,2,aux,diag,x);
    if (aux[3] == 2) {
        sum=0.0;
        for (i=1; i<=5; i++) {
            temp=b[i]-c[i][1]*x[1]-c[i][2]*x[2];
            sum += temp*temp;
        }
        printf("Aux[2, 3, 5] =  %e   %e   %e\n"
               "LSQ solution :  %e   %e\n"
               "Residue (delivered) :  %e\n"
               "Residue (checked)   :  %e\n"
               "Diagonal of inverse M'M :  %e   %e\n",
               aux[2],aux[3],aux[5],x[1],x[2],sqrt(vecvec(3,5,0,x,x)),
               sqrt(sum),diag[1],diag[2]);
    }
    free_real_matrix(a,1,5,1);
    free_real_matrix(c,1,5,1);
}
```

Output:

```
Aux[2, 3, 5] =  1.000000e-06   2.00000e+00   3.31662e+00
LSQ solution :  5.00000e-01   1.25000e+00
Residue (delivered) :  5.00000e-01
Residue (checked)   :  5.00000e-01
Diagonal of inverse M'M :  9.52381e-02   1.30952e-01
```

Function tested: **lsqinv**

Example:

Computes the inverse T of $S'S$, where S' is the transpose of S given by

$$S' = \begin{pmatrix} -2 & -1 & 1 & 2 & 1 \\ 1 & 1 & 1 & 1 & 2 \end{pmatrix}.$$

```
#include <stdio.h>

void main ()
{
    float **allocate_real_matrix(int, int, int, int);
    void free_real_matrix(float **, int, int, int);
    void lsqortdec(float **, int, int, float [], float [], int []);
    void lsqinv(float **, int, float [], int []);
    float tammat(int, int, int, int, float **, float **);
```

```
    float matmat(int, int, int, int, float **, float **);
    int i,j,piv[3];
    float **a,**c,**t,aid[3],aux[6];

    a=allocate_real_matrix(1,5,1,2);
    c=allocate_real_matrix(1,5,1,2);
    t=allocate_real_matrix(1,2,1,2);
    aux[2]=1.0e-6;
    a[1][1]=c[1][1] = -2.0;
    a[2][1]=c[2][1] = -1.0;
    a[3][1]=c[3][1]=1.0;
    a[4][1]=c[4][1]=2.0;
    a[5][1]=c[5][1]=1.0;
    a[1][2]=a[2][2]=a[3][2]=a[4][2]=c[1][2]=c[2][2]=c[3][2]=c[4][2]=1.0;
    a[5][2]=c[5][2]=2.0;
    lsqortdec(a,5,2,aux,aid,piv);
    if (aux[3] == 2) {
        lsqinv(a,2,aid,piv);
        t[1][1]=a[1][1];
        t[2][2]=a[2][2];
        t[2][1]=t[1][2]=a[1][2];
        for (j=1; j<=2; j++)
            for (i=1; i<=5; i++) a[i][j]=matmat(1,2,i,j,c,t);
        printf("Aux[2, 3, 5] =  %e   %e   %e\n"
               "\nInverse:\n   %e   %e\n  %e    %e\n"
               "\nCheck:  S' * (S * T) :\n   %e   %e\n  %e   %e\n",
               aux[2],aux[3],aux[5],t[1][1],t[1][2],t[2][1],t[2][2],
               tammat(1,5,1,1,c,a),tammat(1,5,1,2,c,a),
               tammat(1,5,2,1,c,a),tammat(1,5,2,2,c,a));
    }
    free_real_matrix(a,1,5,1);
    free_real_matrix(c,1,5,1);
    free_real_matrix(t,1,2,1);
}
```

Output:

```
Aux[2, 3, 5] =  1.000000e-06   2.00000e+00   3.31662e+00

Inverse:
   9.52381e-02   -2.38095e-02
  -2.38095e-02    1.30952e-01

Check:  S' * (S * T) :
   1.00000e+00   2.98023e-08
  -7.45058e-09   1.000000e+00
```

Functions tested: **lsqdecomp, lsqrefsol**

Example:

Minimize $\|b-Ax\|_E$, where $x \epsilon R^3$ and

$$A = \begin{pmatrix} 1 & 0 & 8 \\ 0 & 3 & 2 \\ 1 & 2 & 10^{-5} \\ 0 & 0 & 0 \end{pmatrix}, \qquad b = \begin{pmatrix} 25 \\ 12 \\ 5.00003 \\ 1 \end{pmatrix}$$

subject to the constraint

$$x_1 + 1000x_2 + 5x_3 = 2016 .$$

```
#include <stdio.h>

void main ()
{
```

```
    float **allocate_real_matrix(int, int, int, int);
    void free_real_matrix(float **, int, int, int);
    void lsqdecomp(float **, int, int, int, float [], float [], int []);
    void lsqrefsol(float **, float **, int, int, int, float [], float [],
                   int [], float [], float *, float [], float []);
    int n,m,n1,i,j,ci[4];
    float ldx,**a,**qr,aux[8],b[6],res[6],aid[4],x[4];

    a=allocate_real_matrix(1,5,1,3);
    qr=allocate_real_matrix(1,5,1,3);
    n=5;  m=3;  n1=1;
    a[1][1]=1.0;  a[1][2]=1000.0;  a[1][3]=5.0;
    a[2][1]=1.0;  a[2][2]=0.0;     a[2][3]=8.0;
    a[3][1]=0.0;  a[3][2]=3.0;     a[3][3]=2.0;
    a[4][1]=1.0;  a[4][2]=2.0;     a[4][3]=1.0e-5;
    a[5][1]=a[5][2]=a[5][3]=0.0;
    b[1]=2016.0;  b[2]=25.0;  b[3]=12.0;  b[4]=5.00003;  b[5]=1.0;
    aux[2]=1.0e-6;
    aux[6]=5.0;
    for (i=1; i<=5; i++)
       for (j=1; j<=3; j++) qr[i][j]=a[i][j];
    lsqdecomp(qr,n,m,n1,aux,aid,ci);
    lsqrefsol(a,qr,n,m,n1,aux,aid,ci,b,&ldx,x,res);
    printf("The solution vector:\n   %e   %e   %e\n"
          "\nThe residual vector:\n %e\n  %e\n  %e\n  %e\n"
          "Number of iterations: %3.0f\n"
          "Norm last correction of x:  %e\n",
          x[1],x[2],x[3],res[2],res[3],res[4],res[5],aux[7],ldx);
    free_real_matrix(a,1,5,1);
    free_real_matrix(qr,1,5,1);
}
```

Output:

```
The solution vector:
   1.00000e+00   2.00000e+00   3.00000e+00

The residual vector:
 -2.25629e-09
  9.10484e-09
  2.28816e-09
  1.00000e+00
Number of iterations:   2
Norm last correction of x:  3.87508e-08
```

Function tested: **solovr**

Example:

Determine that $x \epsilon R^5$ with minimum $\|x\|_2$ which minimizes $\|Ax-b\|_2$, where

$$A = \begin{pmatrix} 22 & 10 & 2 & 3 & 7 \\ 14 & 7 & 10 & 0 & 8 \\ -1 & 13 & -1 & -11 & 3 \\ -3 & -2 & 13 & -2 & 4 \\ 9 & 8 & 1 & -2 & 4 \\ 9 & 1 & -7 & 5 & -1 \\ 2 & -6 & 6 & 5 & 1 \\ 4 & 5 & 0 & -2 & 2 \end{pmatrix}, \qquad b = \begin{pmatrix} -1 \\ 2 \\ 1 \\ 4 \\ 0 \\ -3 \\ 1 \\ 0 \end{pmatrix}.$$

```
#include <stdio.h>

void main ()
{
    float **allocate_real_matrix(int, int, int, int);
    void free_real_matrix(float **, int, int, int);
    int solovr(float **, int, int, float [], float []);
    int i;
    float **a,b[9],em[8];

    a=allocate_real_matrix(1,8,1,5);
    a[1][1]=22;    a[1][2]=a[2][3]=10.0; a[1][3]=a[7][1]=a[8][5]=2.0;
    a[1][4]=a[3][5]=3.0; a[1][5]=a[2][2]=7.0; a[2][1]=14.0; a[2][5]=8.0;
    a[2][4]=a[8][3]=0.0; a[3][1]=a[3][3]=a[6][5] = -1.0; a[3][2]=13.0;
    a[3][4] = -11.0; a[4][1] = -3.0;
    a[4][2]=a[4][4]=a[5][4]=a[8][4] = -2.0;
    a[4][3]=13.0; a[4][5]=a[5][5]=a[8][1]=4.0; a[5][1]=a[6][1]=9.0;
    a[5][2]=8.0; a[5][3]=a[6][2]=a[7][5]=1.0; a[6][3] = -7.0;
    a[6][4]=a[7][4]=a[8][2]=5.0; a[7][2] = -6.0; a[7][3]=6.0;
    b[1] = -1.0; b[2]=2.0; b[3]=b[7]=1.0; b[4]=4.0; b[5]=b[8]=0.0;
    b[6] = -3.0;
    em[0]=1.0e-14; em[2]=1.0e-12; em[4]=80.0; em[6]=1.0e-10;
    i=solovr(a,8,5,b,em);
    printf("Number of singular values not found : %2d\n"
        "Norm : %e\nMaximal neglected subdiagonal element : %e\n"
        "Number of iterations : %3.0f\nRank : %3.0f\n\nSolution vector"
        "\n %13.6e\n %13.6e\n %13.6e\n %13.6e\n %13.6e\n",
        i,em[1],em[3],em[5],em[7],b[1],b[2],b[3],b[4],b[5]);
    free_real_matrix(a,1,8,1);
}
```

Output:

```
Number of singular values not found :  0
Norm : 4.40000e+01
Maximal neglected subdiagonal element : 2.75552e-16
Number of iterations :   5
Rank :   3

Solution vector
  -8.33333e-02
  -1.25858e-17
   2.50000e-01
  -8.33333e-02
   8.33333e-02
```

Function tested: **solund**

Example:

Determine that $x \epsilon \mathrm{R}^8$ with minimum $\|x\|_2$ which minimizes $\|A^T x-b\|_2$, where

$$A = \begin{pmatrix} 22 & 10 & 2 & 3 & 7 \\ 14 & 7 & 10 & 0 & 8 \\ -1 & 13 & -1 & -11 & 3 \\ -3 & -2 & 13 & -2 & 4 \\ 9 & 8 & 1 & -2 & 4 \\ 9 & 1 & -7 & 5 & -1 \\ 2 & -6 & 6 & 5 & 1 \\ 4 & 5 & 0 & -2 & 2 \end{pmatrix}, \qquad b = \begin{pmatrix} -1 \\ 2 \\ 1 \\ 4 \\ 0 \end{pmatrix}.$$

```
#include <stdio.h>

void main ()
{
    float **allocate_real_matrix(int, int, int, int);
    void free_real_matrix(float **, int, int, int);
    int solund(float **, int, int, float [], float []);
    int i;
    float **a,b[9],em[8];

    a=allocate_real_matrix(1,8,1,5);
    a[1][1]=22;     a[1][2]=a[2][3]=10.0; a[1][3]=a[7][1]=a[8][5]=2.0;
    a[1][4]=a[3][5]=3.0; a[1][5]=a[2][2]=7.0; a[2][1]=14.0; a[2][5]=8.0;
    a[2][4]=a[8][3]=0.0; a[3][1]=a[3][3]=a[6][5] = -1.0; a[3][2]=13.0;
    a[3][4] = -11.0; a[4][1] = -3.0;
    a[4][2]=a[4][4]=a[5][4]=a[8][4] = -2.0;
    a[4][3]=13.0; a[4][5]=a[5][5]=a[8][1]=4.0; a[5][1]=a[6][1]=9.0;
    a[5][2]=8.0; a[5][3]=a[6][2]=a[7][5]=1.0; a[6][3] = -7.0;
    a[6][4]=a[7][4]=a[8][2]=5.0; a[7][2] = -6.0; a[7][3]=6.0;
    b[1] = -1.0; b[2]=2.0; b[3]=1.0; b[4]=4.0; b[5]=0.0;
    em[0]=1.0e-14; em[2]=1.0e-12; em[4]=80.0; em[6]=1.0e-10;
    i=solund(a,8,5,b,em);
    printf("Number of singular values not found : %2d\n"
        "Norm : %e\nMaximal neglected subdiagonal element : %e\n"
        "Number of iterations : %3.0f\nRank : %3.0f\n\nSolution vector"
        "\n %13.6e\n %13.6e\n %13.6e\n %13.6e\n %13.6e\n %13.6e"
        "\n %13.6e\n %13.6e\n",i,em[1],em[3],em[5],em[7],
        b[1],b[2],b[3],b[4],b[5],b[6],b[7],b[8]);
    free_real_matrix(a,1,8,1);
}
```

Output:

```
Number of singular values not found :  0
Norm : 4.40000e+01
Maximal neglected subdiagonal element : 2.75552e-16
Number of iterations :    5
Rank :   3

Solution vector
   1.64103e-02
   1.48077e-02
  -4.83974e-02
   1.00000e-02
  -6.79487e-03
   1.16026e-02
   3.00000e-02
  -8.39744e-03
```

Function tested: **homsol**

Example:

With the 8×5 matrix A of rank 3 given by

$$A = \begin{pmatrix} 22 & 10 & 2 & 3 & 7 \\ 14 & 7 & 10 & 0 & 8 \\ -1 & 13 & -1 & -11 & 3 \\ -3 & -2 & 13 & -2 & 4 \\ 9 & 8 & 1 & -2 & 4 \\ 9 & 1 & -7 & 5 & -1 \\ 2 & -6 & 6 & 5 & 1 \\ 4 & 5 & 0 & -2 & 2 \end{pmatrix}$$

the vectors $v_j \epsilon R^5$ $(j=4,5)$ defining the complementary column subspace in which $x \epsilon R^5$ must lie for the condition $Ax=0$ to hold, and the vectors u_j^T $(j=4,5)$ defining the complementary row subspace in which $u^T \epsilon R^8$ must lie for the condition $x^T A=0$ to hold are determined.

```
#include <stdio.h>

void main ()
{
    float **allocate_real_matrix(int, int, int, int);
    void free_real_matrix(float **, int, int, int);
    int homsol(float **, int, int, float **, float []);
    int i,j;
    float **a,**v,em[8];

    a=allocate_real_matrix(1,8,1,5);
    v=allocate_real_matrix(1,5,1,5);
    a[1][1]=22;    a[1][2]=a[2][3]=10.0; a[1][3]=a[7][1]=a[8][5]=2.0;
    a[1][4]=a[3][5]=3.0; a[1][5]=a[2][2]=7.0; a[2][1]=14.0; a[2][5]=8.0;
    a[2][4]=a[8][3]=0.0; a[3][1]=a[3][3]=a[6][5] = -1.0; a[3][2]=13.0;
    a[3][4] = -11.0; a[4][1] = -3.0;
    a[4][2]=a[4][4]=a[5][4]=a[8][4] = -2.0;
    a[4][3]=13.0; a[4][5]=a[5][5]=a[8][1]=4.0; a[5][1]=a[6][1]=9.0;
    a[5][2]=8.0; a[5][3]=a[6][2]=a[7][5]=1.0; a[6][3] = -7.0;
    a[6][4]=a[7][4]=a[8][2]=5.0; a[7][2] = -6.0; a[7][3]=6.0;
    em[0]=1.0e-14; em[2]=1.0e-12; em[4]=80.0; em[6]=1.0e-10;
    i=homsol(a,8,5,v,em);
    printf("Number of singular values not found : %2d\n"
        "Norm : %e\nMaximal neglected subdiagonal element : %e\n"
        "Number of iterations : %3.0f\nRank : %3.0f\n",
        i,em[1],em[3],em[5],em[7]);
    for (j=em[7]+1; j<=5; j++) {
        printf("\nColumn number : %2d\n",j);
        for (i=1; i<=5; i++)
            printf(" %13.6e %13.6e\n",a[i][j],v[i][j]);
            printf(" %13.6e\n %13.6e\n %13.6e\n",
                    a[6][j],a[7][j],a[8][j]);
    }
    free_real_matrix(a,1,8,1);
    free_real_matrix(v,1,5,1);
}
```

Output:

```
Number of singular values not found :  0
Norm : 4.40000e+01
Maximal neglected subdiagonal element : 2.75552e-16
Number of iterations :    5
Rank :   3
```

```
Column number :  4
   3.47086e-01   4.19095e-01
  -6.07234e-01  -4.40509e-01
   1.22075e-01   5.20045e-02
   6.18826e-01  -6.76059e-01
  -4.63444e-03  -4.12977e-01
   3.34099e-01
  -3.35284e-02
  -1.35472e-02

Column number :  5
  -2.55331e-01  -0.00000e+00
  -1.73598e-01   4.18548e-01
  -2.20812e-01   3.48790e-01
   4.11655e-02   2.44153e-01
   9.20442e-01  -8.02217e-01
  -2.88960e-02
   6.13276e-02
  -4.90581e-02
```

Function tested: **psdinv**

Example:

Compute the generalized inverse of the 8×5 matrix given by

$$\begin{pmatrix} 22 & 10 & 2 & 3 & 7 \\ 14 & 7 & 10 & 0 & 8 \\ -1 & 13 & -1 & -11 & 3 \\ -3 & -2 & 13 & -2 & 4 \\ 9 & 8 & 1 & -2 & 4 \\ 9 & 1 & -7 & 5 & -1 \\ 2 & -6 & 6 & 5 & 1 \\ 4 & 5 & 0 & -2 & 2 \end{pmatrix}.$$

```
#include <stdio.h>

void main ()
{
    float **allocate_real_matrix(int, int, int, int);
    void free_real_matrix(float **, int, int, int);
    int psdinv(float **, int, int, float []);
    int i,j;
    float **a,em[8];

    a=allocate_real_matrix(1,8,1,5);
    a[1][1]=22;    a[1][2]=a[2][3]=10.0; a[1][3]=a[7][1]=a[8][5]=2.0;
    a[1][4]=a[3][5]=3.0; a[1][5]=a[2][2]=7.0; a[2][1]=14.0; a[2][5]=8.0;
    a[2][4]=a[8][3]=0.0; a[3][1]=a[3][3]=a[6][5] = -1.0; a[3][2]=13.0;
    a[3][4] = -11.0; a[4][1] = -3.0;
    a[4][2]=a[4][4]=a[5][4]=a[8][4] = -2.0;
    a[4][3]=13.0; a[4][5]=a[5][5]=a[8][1]=4.0; a[5][1]=a[6][1]=9.0;
    a[5][2]=8.0; a[5][3]=a[6][2]=a[7][5]=1.0; a[6][3] = -7.0;
    a[6][4]=a[7][4]=a[8][2]=5.0; a[7][2] = -6.0; a[7][3]=6.0;
    em[0]=1.0e-14; em[2]=1.0e-12; em[4]=80.0; em[6]=1.0e-10;
    i=psdinv(a,8,5,em);
    printf("Number of singular values not found : %2d\n"
        "Norm : %e\nMaximal neglected subdiagonal element : %e\n"
        "Number of iterations : %3.0f\nRank : %3.0f\n",
        i,em[1],em[3],em[5],em[7]);
```

```
    printf("\nTranspose of pseudo-inverse, first three columns:\n");
    for (i=1; i<=8; i++)
        printf(" %13.6e %13.6e %13.6e\n",a[i][1],a[i][2],a[i][3]);
    printf("\nLast two columns:\n");
    for (i=1; i<=8; i++)
        printf("              %13.6e %13.6e\n",a[i][4],a[i][5]);
    free_real_matrix(a,1,8,1);
}
```

Output:

```
Number of singular values not found :  0
Norm : 4.40000e+01
Maximal neglected subdiagonal element : 2.75552e-16
Number of iterations :   5
Rank :   3

Transpose of pseudo-inverse, first three columns:
   2.11298e-02   4.61538e-03  -2.10737e-03
   9.31090e-03   2.21154e-03   2.05288e-02
  -1.10978e-02   2.74038e-02  -3.88622e-03
  -7.91667e-03  -5.00000e-03   3.37500e-02
   5.51282e-03   9.80769e-03  -8.97436e-04
   1.43189e-02  -2.59615e-03  -2.01362e-02
   4.89583e-03  -1.50000e-02   1.53125e-02
   1.50641e-03   7.40385e-03  -1.69872e-03

Last two columns:
               7.60417e-03   3.80609e-03
              -2.08333e-04   1.00160e-02
              -2.76042e-02   4.20673e-03
              -5.41667e-03   1.04167e-02
              -5.00000e-03   3.20513e-03
               1.28125e-02  -6.20994e-03
               1.23958e-02   2.60417e-03
              -5.00000e-03   1.60256e-03
```

Functions tested: **solbnd, decbnd, determbnd**

Example:

Solve the system of equations

$$\begin{aligned} 2x_1 - x_2 \qquad\qquad &= 1 \\ -x_1 + 2x_2 - x_3 \qquad &= 0 \\ - x_2 + 2x_3 - x_4 \quad &= 0 \\ - x_3 + 2x_4 - x_5 &= 0 \\ - x_4 + 2x_5 &= 1 \end{aligned}$$

and evaluate the determinant of the coefficient matrix.

```
#include <stdio.h>

void main ()
{
    void decbnd(float [], int, int, int, float [], float [], int []);
    void solbnd(float [], int, int, int, float [], int [], float []);
    float determbnd(float [], int, int, int, int);
    int i,rowind[6];
    float band[14],mult[5],right[6],aux[6];

    for (i=1; i<=13; i++)
        band[i] = (((i+1)/3)*3 < i) ? 2.0 : -1.0;
    right[1]=right[5]=1.0;
    right[2]=right[3]=right[4]=0.0;
    aux[2]=1.0e-12;
    decbnd(band,5,1,1,aux,mult,rowind);
    if (aux[3] == 5) {
        solbnd(band,5,1,1,mult,rowind,right);
```

```
        printf("Delivers: %8.4f %8.4f %8.4f %8.4f %8.4f\n"
               "Determinant is  %e\n",right[1],right[2],right[3],
               right[4],right[5],determbnd(band,5,1,1,aux[1]));
    }
}
```

Output:

```
Delivers:   1.0000   1.0000   1.0000   1.0000   1.0000
Determinant is  6.00000e+00
```

Function tested: **decsolbnd**

Example:

Solve the system of equations

$$\begin{aligned} 2x_1 - x_2 &= 1 \\ -x_1 + 2x_2 - x_3 &= 0 \\ - x_2 + 2x_3 - x_4 &= 0 \\ - x_3 + 2x_4 - x_5 &= 0 \\ - x_4 + 2x_5 &= 1 \end{aligned}$$

and evaluate the determinant of the coefficient matrix.

```
#include <stdio.h>

void main ()
{
    void decsolbnd(float [], int, int, int, float [], float []);
    float determbnd(float [], int, int, int, int);
    int i;
    float band[14],right[6],aux[6];

    for (i=1; i<=13; i++)
        band[i] = (((i+1)/3)*3 < i) ? 2.0 : -1.0;
    right[1]=right[5]=1.0;
    right[2]=right[3]=right[4]=0.0;
    aux[2]=1.0e-12;
    decsolbnd(band,5,1,1,aux,right);
    if (aux[3] == 5)
        printf("Delivers: %8.4f %8.4f %8.4f %8.4f %8.4f\n"
               "Determinant is  %e\n",right[1],right[2],right[3],
               right[4],right[5],determbnd(band,5,1,1,aux[1]));
}
```

Output:

```
Delivers:   1.0000   1.0000   1.0000   1.0000   1.0000
Determinant is  6.00000e+00
```

Function tested: **decsoltri**

Example:

Solve the system of equations $Ax=b$, where A is a 30×30 tridiagonal matrix for which $A_{i+1,i} = 2*i$, $A_{i,i} = i+10$, $A_{i,i+1} = i$, and b is the second column of A (the solution is given by the second unit vector).

```
#include <math.h>
#include <stdio.h>

void main ()
{
    void decsoltri(float [], float [], float [], int,
```

```
                    float [], float []);
    float vecvec(int, int, int, float [], float []);
    int i;
    float d[31],sub[31],super[31],b[31],aux[6];

    for (i=1; i<=30; i++) {
        sub[i]=i*2;
        super[i]=i;
        d[i]=i+10;
        b[i]=0.0;
    }
    b[1]=1.0;  b[2]=12.0;  b[3]=4.0;
    aux[2]=1.0e-6;
    decsoltri(sub,d,super,30,aux,b);
    b[2]--;
    printf("AUX[3] and AUX[5]:  %e   %e\n"
            "Error in the solution:  %e\n",
            aux[3],aux[5],sqrt(vecvec(1,30,0,b,b)));
}
```

Output:

```
AUX[3] and AUX[5]:  3.00000e+01   1.24000e+02
Error in the solution:  0.00000e+00
```

Function tested: **soltripiv**

Example:

Solve the two systems of equations $Ax=b1$, $Ax=b2$, where A is a 30×30 tridiagonal matrix for which $A_{i+1,i} = 2*i$, $A_{i,i} = i+10$, $A_{i,i+1} = i$, and $b1$, $b2$ are the second and third column of A, respectively, (the solutions of the systems are given by the second and third unit vector, respectively).

```
#include <stdio.h>

void main ()
{
    void dectripiv(float [], float [], float [], int,
                    float [], float [], int []);
    void soltripiv(float [], float [], float [], int,
                    float [], int [], float []);
    float vecvec(int, int, int, float [], float []);
    int i,piv[30];
    float d[31],sub[31],super[31],aid[31],b1[31],b2[31],aux[6];

    for (i=1; i<=30; i++) {
        sub[i]=i*2;
        super[i]=i;
        d[i]=i+10;
        b1[i]=b2[i]=0.0;
    }
    b1[1]=1.0;  b1[2]=12.0;  b1[3]=4.0;
    b2[2]=2.0;  b2[3]=13.0;  b2[4]=6.0;
    aux[2]=1.0e-6;
    dectripiv(sub,d,super,30,aid,aux,piv);
    soltripiv(sub,d,super,30,aid,piv,b1);
    soltripiv(sub,d,super,30,aid,piv,b2);
    b1[2]--;  b2[3]--;
    printf("AUX[3] and AUX[5]:  %e   %e\n"
            "Error in b1:  %e\nError in b2:  %e\n",aux[3],aux[5],
            sqrt(vecvec(1,30,0,b1,b1)),sqrt(vecvec(1,30,0,b2,b2)));
}
```

Output:

```
AUX[3] and AUX[5]:  3.00000e+01   1.24000e+02
Error in b1:  0.00000e+00
```

```
Error in b2:  0.00000e+00
```

Function tested: **decsoltripiv**

Example:

Solve the system of equations $Ax=b$, where A is a 30×30 tridiagonal matrix for which $A_{i+1,i} = 2*i$, $A_{i,i} = i+10$, $A_{i,i+1} = i$, and b is the second column of A (the solution is given by the second unit vector).

```
#include <math.h>
#include <stdio.h>

void main ()
{
    void decsoltripiv(float [], float [], float [], int,
                      float [], float []);
    float vecvec(int, int, int, float [], float []);
    int i;
    float d[31],sub[31],super[31],b[31],aux[6];

    for (i=1; i<=30; i++) {
        sub[i]=i*2;
        super[i]=i;
        d[i]=i+10;
        b[i]=0.0;
    }
    b[1]=1.0;  b[2]=12.0;  b[3]=4.0;
    aux[2]=1.0e-6;
    decsoltripiv(sub,d,super,30,aux,b);
    b[2]--;
    printf("AUX[3] and AUX[5]:  %e   %e\n"
           "Error in the solution:  %e\n",
           aux[3],aux[5],sqrt(vecvec(1,30,0,b,b)));
}
```

Output:

```
AUX[3] and AUX[5]:  3.00000e+01   1.24000e+02
Error in the solution:  0.00000e+00
```

Functions tested: **chlsolbnd, chldecbnd, chldetermbnd**

Example:

Solve the system of equations

$$\begin{aligned} 2x_1 - x_2 &= 1 \\ -x_1 + 2x_2 - x_3 &= 0 \\ - x_2 + 2x_3 - x_4 &= 0 \\ - x_3 + 2x_4 - x_5 &= 0 \\ - x_4 + 2x_5 &= 1 \end{aligned}$$

and evaluate the determinant of the coefficient matrix.

```
#include <stdio.h>

void main ()
{
    float *allocate_real_vector(int, int);
    void free_real_vector(float *, int);
    void chldecbnd(float [], int, int, float []);
    void chlsolbnd(float [], int, int, float []);
    float chldetermbnd(float [], int, int);
    int i;
```

```
    float *symband,*right,*aux;

    symband=allocate_real_vector(1,9);
    right=allocate_real_vector(1,5);
    aux=allocate_real_vector(2,3);

    for (i=1; i<=9; i++)
        symband[i] = ((i/2)*2 < i) ? 2.0 : -1.0;
    right[1]=right[5]=1.0;
    right[2]=right[3]=right[4]=0.0;
    aux[2]=1.0e-12;
    chldecbnd(symband,5,1,aux);
    if (aux[3] == 5) {
        chlsolbnd(symband,5,1,right);
        printf("Delivers: %8.4f %8.4f %8.4f %8.4f %8.4f\n"
               "Determinant is  %e\n",right[1],right[2],right[3],
               right[4],right[5],chldetermbnd(symband,5,1));
    }
    free_real_vector(symband,1);
    free_real_vector(right,1);
    free_real_vector(aux,2);
}
```

Output:

```
Delivers:   1.0000   1.0000   1.0000   1.0000   1.0000
Determinant is  6.00000e+00
```

Functions tested: **chldecsolbnd, chldetermbnd**

Example:

Solve the system of equations

$$\begin{aligned} 2x_1 - x_2 \quad\quad\quad\quad &= 1 \\ -x_1 + 2x_2 - x_3 \quad\quad &= 0 \\ - x_2 + 2x_3 - x_4 \quad &= 0 \\ - x_3 + 2x_4 - x_5 &= 0 \\ - x_4 + 2x_5 &= 1 \end{aligned}$$

and evaluate the determinant of the coefficient matrix.

```
#include <stdio.h>

void main ()
{
    float *allocate_real_vector(int, int);
    void free_real_vector(float *, int);
    void chldecsolbnd(float [], int, int, float [], float []);
    float chldetermbnd(float [], int, int);
    int i;
    float *symband,*right,*aux;

    symband=allocate_real_vector(1,9);
    right=allocate_real_vector(1,5);
    aux=allocate_real_vector(2,3);

    for (i=1; i<=9; i++)
        symband[i] = ((i/2)*2 < i) ? 2.0 : -1.0;
    right[1]=right[5]=1.0;
    right[2]=right[3]=right[4]=0.0;
    aux[2]=1.0e-12;
    chldecsolbnd(symband,5,1,aux,right);
    if (aux[3] == 5) {
        printf("Delivers: %8.4f %8.4f %8.4f %8.4f %8.4f\n"
               "Determinant is  %e\n",right[1],right[2],right[3],
               right[4],right[5],chldetermbnd(symband,5,1));
    }
    free_real_vector(symband,1);
```

```
    free_real_vector(right,1);
    free_real_vector(aux,2);
}
```

Output:

```
Delivers:   1.0000   1.0000   1.0000   1.0000   1.0000
Determinant is  6.00000e+00
```

Function tested: **decsolsymtri**

Example:

Solve the system of equations $Ax=b$, where A is a 100×100 symmetric tridiagonal matrix for which $A_{i+1,i} = A_{i,i+1} = 2*i$, $A_{i,i} = i$, and b is the second column of A (the solution is given by the second unit vector).

```
#include <math.h>
#include <stdio.h>

void main ()
{
    void decsolsymtri(float [], float [], int, float [], float []);
    float vecvec(int, int, int, float [], float []);
    int i;
    float d[101],co[101],b[101],aux[6];

    for (i=1; i<=100; i++) {
        d[i]=i;
        co[i]=i*2;
        b[i]=0.0;
    }
    b[1]=b[2]=2.0;  b[3]=4.0;
    aux[2]=1.0e-6;
    decsolsymtri(d,co,100,aux,b);
    b[2]--;
    printf("AUX[3] and AUX[5]:  %e   %e\n"
           "Error in the solution:  %e\n",
           aux[3],aux[5],sqrt(vecvec(1,100,0,b,b)));
}
```

Output:

```
AUX[3] and AUX[5]:  1.00000e+02   4.93000e+02
Error in the solution:  0.00000e+00
```

Function tested: **conjgrad**

Example:

Solve the system of equations $Ax=b$, $b \in \mathrm{R}^{13}$, where

$$A_{i+1,i} = A_{i,i+1} = -1,\ A_{i,i} = 2,\ A_{i,j} = 0 \text{ for } |i-j|>1,$$

and $b_1 = 1$, $b_i = 0$ (i=2,...,12), $b_{13} = 4$, using x=(0,0,...,0) as an initial approximation to the solution.

```
#include <stdio.h>

void a(float x[], float b[])
{
    int i;
    b[0]=2.0*x[0]-x[1];
    for (i=1; i<=11; i++) b[i] = -x[i-1]+2.0*x[i]-x[i+1];
    b[12]=2.0*x[12]-x[11];
}
```

```
int moveon(int iter, float norm)
{
   return (iter<20 && norm>1.0e-10);
}

void main ()
{
   void conjgrad(void (*)(float [], float []), float [], float [],
                 int, int, int (*)(int, float), int *, float *);
   int it,i;
   float no,x[13],b[13];

   for (i=0; i<=12; i++) x[i]=b[i]=0.0;
   b[0]=1.0;  b[12]=4.0;
   conjgrad(a,x,b,0,12,moveon,&it,&no);
   printf("Delivers:\nIT = %3d,   NO = %e\n"
          "\n     X          R\n\n",it,no);
   for (i=0; i<=12; i++) printf(" %+e  %+e\n",x[i],b[i]);
}
```

Output:

```
Delivers:
IT =  13,    NO = 1.83603e-14

     X             R

 +1.21429e+00  +3.17607e-08
 +1.42857e+00  +6.44881e-08
 +1.64286e+00  -4.73635e-09
 +1.85714e+00  +4.73824e-08
 +2.07143e+00  +2.98148e-09
 +2.28571e+00  +9.84739e-09
 +2.50000e+00  +6.16825e-08
 +2.71429e+00  -2.00927e-08
 +2.92857e+00  +1.94551e-08
 +3.14286e+00  -8.59740e-10
 +3.35714e+00  +5.69948e-08
 +3.57143e+00  -6.92489e-10
 +3.78571e+00  -5.46159e-08
```

Function tested: **eqilbrcom**

Example:

Equilibrate the 3×3 matrix

$$\begin{pmatrix} 1 & 0 & 1024i \\ 0 & 1 & 0 \\ i/1024 & 0 & 2 \end{pmatrix}.$$

```
#include <math.h>
#include <stdio.h>

void main ()
{
   float **allocate_real_matrix(int, int, int, int);
   void free_real_matrix(float **, int, int, int);
   void inimat(int, int, int, int, float **, float);
   void eqilbrcom(float **, float **, int, float [], float [], int []);
   int inter[4];
   float **a1,**a2,em[8],d[4];

   a1=allocate_real_matrix(1,3,1,3);
   a2=allocate_real_matrix(1,3,1,3);
```

```
    em[0]=5.0e-7;
    em[6]=10.0;
    inimat(1,3,1,3,a1,0.0);
    inimat(1,3,1,3,a2,0.0);
    a1[1][1]=a1[2][2]=1.0;
    a1[3][3]=2.0;
    a2[1][3]=pow(2.0,10.0);
    a2[3][1]=pow(2.0,-10.0);
    eqilbrcom(a1,a2,3,em,d,inter);
    printf("Equilibrated matrix:\n %2.0f %2.0f %2.0f\n"
           " %2.0f %2.0f  I\n %2.0f  I %2.0f\n"
           "\nEM[7]:       %2.0f\nD[1:3]:       %2.0f %5.0f %5.0f\n"
           "INTER[1:3]: %2d %5d %5d\n",a1[1][1],a1[1][2],
           a1[1][3],a1[2][1],a1[2][2],a1[3][1],a1[3][3],em[7],
           d[1],d[2],d[3],inter[1],inter[2],inter[3]);
    free_real_matrix(a1,1,3,1);
    free_real_matrix(a2,1,3,1);
}
```

Output:

```
Equilibrated matrix:
  1  0  0
  0  1  I
  0  I  2

EM[7]:        4
D[1:3]:       1  1024     1
INTER[1:3]:   2     0     0
```

Function tested: **hshhrmtri**

Example:

Reduce the 4×4 complex matrix

$$A = \begin{pmatrix} 3 & 1 & 0 & 2i \\ 1 & 3 & -2i & 0 \\ 0 & 2i & 1 & 1 \\ -2i & 0 & 1 & 1 \end{pmatrix}$$

to real tridiagonal form, and estimate the value of $\|A\|$.

```
#include <stdio.h>

void main ()
{
    float **allocate_real_matrix(int, int, int, int);
    void free_real_matrix(float **, int, int, int);
    void hshhrmtri(float **, int, float [], float [], float [],
                        float [], float [], float []);
    void inimat(int, int, int, int, float **, float);
    float **a,d[5],b[5],bb[5],tr[4],ti[4],em[2];

    a=allocate_real_matrix(1,4,1,4);
    inimat(1,4,1,4,a,0.0);
    a[1][1]=a[2][2]=3.0;
    a[1][2]=a[3][3]=a[3][4]=a[4][4]=1.0;
    a[3][2]=2.0;
    a[4][1] = -2.0;
    em[0]=1.0e-6;
    hshhrmtri(a,4,d,b,bb,em,tr,ti);
    printf("HSHHRMTRI delivers\n\nD[1:4]:  %7.3f %7.3f %7.3f %7.3f\n"
           "B[1:3]:  %7.3f %7.3f %7.3f\n"
           "BB[1:3]: %7.3f %7.3f %7.3f\n"
```

```
            "EM[1]:   %7.3f\n",
            d[1],d[2],d[3],d[4],b[1],b[2],b[3],bb[1],bb[2],bb[3],em[1]);
    free_real_matrix(a,1,4,1);
}
```

Output:

```
HSHHRMTRI delivers

D[1:4]:     3.000   1.400   2.600   1.000
B[1:3]:     2.236   0.800   2.236
BB[1:3]:    5.000   0.640   5.000
EM[1]:      6.000
```

Functions tested: **valsymtri, vecsymtri**

Example:

Compute the largest and second largest eigenvalues, and their corresponding eigenvectors, of the 4×4 matrix A for which

$$A_{i,i}=2 \ (i=1,2,3,4),\ A_{i,i-1}=A_{i-1,i}=-1 \ (i=2,3,4),\ A_{i,j}=0 \ (|i-j|\geq 2).$$

```
#include <stdio.h>

void main ()
{
    float **allocate_real_matrix(int, int, int, int);
    void free_real_matrix(float **, int, int, int);
    void valsymtri(float [], float [], int, int, int,
                        float [], float []);
    void vecsymtri(float [], float [], int, int, int,
                        float [], float **, float []);
    float b[5],d[5],bb[4],val[3],em[10],**vec;

    vec=allocate_real_matrix(1,4,1,2);
    em[0]=1.0e-6;  em[1]=4.0;  em[2]=1.0e-5;
   em[4]=1.0e-3;   em[6]=1.0e-6;   em[8]=5.0;
    d[1]=d[2]=d[3]=d[4]=2.0;
    b[4]=0.0;  b[1]=b[2]=b[3] = -1.0;
    bb[1]=bb[2]=bb[3]=1.0;
    valsymtri(d,bb,4,1,2,val,em);
    vecsymtri(d,b,4,1,2,val,vec,em);
    printf("The eigenvalues:\n  %12.5e   %12.5e\n\nThe eigenvectors:\n"
            "  %12.5e   %12.5e\n  %12.5e   %12.5e\n  %12.5e   %12.5e\n"
            "  %12.5e   %12.5e\n"
            "\nEM[7] = %e\nEM[3] =%3.0f\nEM[5] =%3.0f\nEM[9] =%3.0f\n",
            val[1],val[2],vec[1][1],vec[1][2],vec[2][1],vec[2][2],
            vec[3][1],vec[3][2],vec[4][1],vec[4][2],
            em[7],em[3],em[5],em[9]);
    free_real_matrix(vec,1,4,1);
}
```

Output:

```
The eigenvalues:
    3.6180e+00      2.6180e+00

The eigenvectors:
   -3.7175e-01      6.0150e-01
    6.0150e-01     -3.7175e-01
   -6.0150e-01     -3.7175e-01
    3.7175e-01      6.0150e-01

EM[7]  = 1.27244e-05
EM[3]  = 21
EM[5]  =  1
EM[9]  =  6
```

Function tested: **eigsym1**

Example:

Compute the two largest eigenvalues and corresponding eigenvectors of the 4×4 matrix A for which $A_{i,j} = 1/(i+j-1)$, $i,j=1,2,3,4$.

```
#include <stdio.h>

void main ()
{
    float *allocate_real_vector(int, int);
    float **allocate_real_matrix(int, int, int, int);
    void free_real_vector(float *, int);
    void free_real_matrix(float **, int, int, int);
    void eigsym1(float [], int, int, float [], float **, float []);
    int i,j;
    float a[11],val[3],em[10],**vec;

    vec=allocate_real_matrix(1,4,1,2);
    em[0]=1.0e-6;  em[2]=1.0e-5;  em[4]=1.0e-3;
    em[6]=1.0e-5;  em[8]=5.0;
    for (i=1; i<=4; i++)
        for (j=i; j<=4; j++) a[(j*j-j)/2+i]=1.0/(i+j-1);
    eigsym1(a,4,2,val,vec,em);
    printf("The eigenvalues:\n  %12.5e   %12.5e\n\nThe eigenvectors:\n"
           "  %12.5e   %12.5e\n  %12.5e   %12.5e\n  %12.5e   %12.5e\n"
           "  %12.5e   %12.5e\n\nEM[1] = %e\n"
           "EM[7] = %e\nEM[3] =%3.0f\nEM[5] =%3.0f\nEM[9] =%3.0f\n",
           val[1],val[2],vec[1][1],vec[1][2],vec[2][1],vec[2][2],
           vec[3][1],vec[3][2],vec[4][1],vec[4][2],
           em[1],em[7],em[3],em[5],em[9]);
    free_real_matrix(vec,1,4,1);
}
```

Output:

```
The eigenvalues:
    1.5002e+00     1.6914e-01

The eigenvectors:
   -7.9261e-01     5.8208e-01
   -4.5192e-01    -3.7050e-01
   -3.2242e-01    -5.0958e-01
   -2.5216e-01    -5.1405e-01

EM[1] = 2.08333e+00
EM[7] = 6.25799e-06
EM[3] = 27
EM[5] =  1
EM[9] =  1
```

Function tested: **symeigimp**

Example:

Determine approximations to the eigenvalues and eigenvectors of the matrix

$$\begin{pmatrix} 6 & 4 & 4 & 1 \\ 4 & 6 & 1 & 4 \\ 4 & 1 & 6 & 4 \\ 1 & 4 & 4 & 6 \end{pmatrix}.$$

```
#include <stdio.h>

void main ()
{
    float **allocate_real_matrix(int, int, int, int);
    void free_real_matrix(float **, int, int, int);
    void symeigimp(int, float **, float **, float [],
                        float [], float [], float []);
    int qrisym(float **, int, float [], float []);
    int i,j;
    float **a,**x,val[5],lbound[5],ubound[5],em[6],aux[6];

    a=allocate_real_matrix(1,4,1,4);
    x=allocate_real_matrix(1,4,1,4);
    a[1][1]=a[2][2]=a[3][3]=a[4][4]=6.0;
    a[1][2]=a[2][1]=a[3][1]=a[1][3]=4.0;
    a[4][2]=a[2][4]=a[3][4]=a[4][3]=4.0;
    a[1][4]=a[4][1]=a[3][2]=a[2][3]=1.0;
    for (i=1; i<=4; i++)
        for (j=i; j<=4; j++) x[i][j]=x[j][i]=a[i][j];
    em[0]=1.0e-6;  em[4]=100.0;   em[2]=1.0e-5;
    qrisym(x,4,val,em);
    aux[0]=0.0;    aux[4]=10.0;   aux[2]=1.0e-6;
    symeigimp(4,a,x,val,lbound,ubound,aux);
    printf("\nThe exact eigenvalues are:  -1,  5,  5,  15\n\n"
        "The computed eigenvalues:\n %12.6e\n %12.6e\n %12.6e\n"
        " %12.6e\n\n Lowerbounds   Upperbounds\n %e   %e\n"
        " %e   %e\n %e   %e\n %e   %e\n"
        "\nNumber of iterations =%3.0f\n"
        "Infinity norm of A   =%3.0f\n"
        "Maximum absolute element of residu = %e",
        val[1],val[2],val[3],val[4],
        lbound[1],ubound[1],lbound[2],ubound[2],
        lbound[3],ubound[3],lbound[4],ubound[4],
        aux[5],aux[1],aux[3]);
    free_real_matrix(a,1,4,1);
    free_real_matrix(x,1,4,1);
}
```

Output:

```
The exact eigenvalues are:  -1,  5,  5,  15

The computed eigenvalues:
 -1.00000e+00
  5.00000e+00
  5.00000e+00
  1.50000e+01

 Lowerbounds   Upperbounds
 1.54543e-13   1.54543e-13
 1.13840e-06   1.13840e-06
 6.67992e-07   6.67992e-07
 4.05009e-14   4.05009e-14

Number of iterations =  1
Infinity norm of A   = 15
Maximum absolute element of residu = 3.57628e-07
```

Functions tested: **comvalqri, comveches**

Example:

Determine all eigenvalues and corresponding eigenvectors of the 4×4 matrix

$$\begin{pmatrix} -1 & -1 & -1 & -1 \\ 1 & 0 & 0 & 0 \\ 0 & 1 & 0 & 0 \\ 0 & 0 & 1 & 0 \end{pmatrix}.$$

```
#include <stdio.h>

void main ()
{
    float **allocate_real_matrix(int, int, int, int);
    void free_real_matrix(float **, int, int, int);
    int comvalqri(float **, int, float [], float [], float []);
    void comveches(float **, int, float, float,
                        float [], float [], float []);
    int i,j,m,k;
    float **a,re[5],im[5],u[5],v[5],em[10];

    a=allocate_real_matrix(1,4,1,4);
    em[0]=1.0e-6;  em[2]=1.0e-6;  em[1]=4.0;  em[4]=40.0;
    em[6]=1.0e-5;  em[8]=5.0;
    for (i=1; i<=4; i++)
        for (j=1; j<=4; j++)
            a[i][j] = (i == 1) ? -1.0 : ((i-j == 1) ? 1.0 : 0.0);
    m=comvalqri(a,4,em,re,im);
    printf("The number of not calculated eigenvalues: %2d\n"
            "\nThe eigenvalues and eigenvectors:\n",m);
    for (j=m+1; j<=4; j++) {
        for (i=1; i<=4; i++)
            for (k=1; k<=4; k++)
                a[i][k] = (i == 1) ? -1.0 : ((i-k == 1) ? 1.0 : 0.0);
        comveches(a,4,re[j],im[j],em,u,v);
        printf("\n %e    %e\n\n"
            "          %12.6e   %12.6e\n          %12.6e   %12.6e\n"
            "          %12.6e   %12.6e\n          %12.6e   %12.6e\n",
            re[j],im[j],u[1],v[1],u[2],v[2],u[3],v[3],u[4],v[4]);
    }
    printf("\nEM[3] = %e\nEM[7] = %e\nEM[5] = %3.0f\nEM[9] = %3.0f\n",
            em[3],em[7],em[5],em[9]);
    free_real_matrix(a,1,4,1);
}
```

Output:

```
The number of not calculated eigenvalues:  0

The eigenvalues and eigenvectors:

 3.09017e-01   9.51057e-01

          -2.93894e-01   -4.04508e-01
          -4.75527e-01    1.54509e-01
           7.03717e-07    5.00000e-01
           4.75528e-01    1.54509e-01

 3.09017e-01   -9.51057e-01

          -2.93894e-01    4.04508e-01
          -4.75527e-01   -1.54509e-01
           7.03717e-07   -5.00000e-01
           4.75528e-01   -1.54509e-01
```

```
 -8.09017e-01   5.87785e-01

           2.93892e-01    4.04509e-01
           1.28674e-06   -5.00000e-01
          -2.93893e-01    4.04508e-01
           4.75527e-01   -1.54508e-01

 -8.09017e-01  -5.87785e-01

           2.93892e-01   -4.04509e-01
           1.28674e-06    5.00000e-01
          -2.93893e-01   -4.04508e-01
           4.75527e-01    1.54508e-01

EM[3] = 5.08525e-11
EM[7] = 7.60844e-07
EM[5] =   8
EM[9] =   1
```

Function tested: **reaeig3**

Example:

Compute all eigenvalues and corresponding eigenvectors of the 4×4 matrix

$$\begin{pmatrix} 1 & 1 & 1 & 1 \\ 1/2 & 1/3 & 1/4 & 1/5 \\ 1/3 & 1/4 & 1/5 & 1/6 \\ 1/4 & 1/5 & 1/6 & 1/7 \end{pmatrix}.$$

```
#include <stdio.h>

void main ()
{
    float **allocate_real_matrix(int, int, int, int);
    void free_real_matrix(float **, int, int, int);
    int reaeig3(float **, int, float [], float [], float **);
    int i,j,m;
    float **a,**vec,val[5],em[6];

    a=allocate_real_matrix(1,4,1,4);
    vec=allocate_real_matrix(1,4,1,4);
    for (i=1; i<=4; i++)
        for (j=1; j<=4; j++)
            a[i][j] = (i == 1) ? 1.0 : 1.0/(i+j-1);
    em[0]=1.0e-6;
    em[2]=1.0e-5;
    em[4]=40.0;
    m=reaeig3(a,4,em,val,vec);
    printf("The number of not calculated eigenvalues: %3.0f\n\n"
        "The eigenvalues and corresponding eigenvectors:\n",m);
    for (i=m+1; i<=4; i++)
        printf("\n %12.6e   %12.6e\n                %12.6e"
               "\n               %12.6e\n                %12.6e\n",
               val[i],vec[1][i],vec[2][i],vec[3][i],vec[4][i]);
    printf("\nEM[1] = %e\nEM[3] = %e\nEM[5] = %3.0f\n",
            em[1],em[3],em[5]);
    free_real_matrix(a,1,4,1);
    free_real_matrix(vec,1,4,1);
}
```

Output:

```
The number of not calculated eigenvalues:   0
```

```
The eigenvalues and corresponding eigenvectors:

  1.88663e+00     1.00000e+00
                  3.94224e-01
                  2.77320e-01
                  2.15088e-01

 -1.98015e-01     1.00000e+00
                 -7.38848e-01
                 -3.11624e-01
                 -1.47542e-01

 -1.22829e-02    -4.63474e-01
                  1.00000e+00
                 -1.54255e-01
                 -3.76579e-01

 -1.44118e-04     1.09571e-01
                 -6.20841e-01
                  1.00000e+00
                 -4.88746e-01

EM[1] = 4.00000e+00
EM[3] = 7.25585e-08
EM[5] =    3
```

Function tested: **eighrm**

Example:

Compute the largest eigenvalue and corresponding eigenvector of the 4×4 matrix

$$\begin{pmatrix} 3 & 1 & 0 & 2i \\ 1 & 3 & -2i & 0 \\ 0 & 2i & 1 & 1 \\ -2i & 0 & 1 & 1 \end{pmatrix}.$$

```
#include <stdio.h>

void main ()
{
    float **allocate_real_matrix(int, int, int, int);
    void free_real_matrix(float **, int, int, int);
    void sclcom(float **, float **, int, int, int);
    void inimat(int, int, int, int, float **, float);
    void eighrm(float **, int, int, float [], float **, float **,
                float []);
    int i;
    float **a,**vecr,**veci,val[2],em[10];

    a=allocate_real_matrix(1,4,1,4);
    vecr=allocate_real_matrix(1,4,1,1);
    veci=allocate_real_matrix(1,4,1,1);

    inimat(1,4,1,4,a,0.0);
    a[1][1]=a[2][2]=3.0;
    a[1][2]=a[3][3]=a[3][4]=a[4][4]=1.0;
    a[3][2]=2.0;  a[4][1] = -2.0;
    em[0]=5.0e-6;
    em[2]=1.0e-5;
    em[4]=0.01;
    em[6]=1.0e-5;
    em[8]=5.0;
    eighrm(a,4,1,val,vecr,veci,em);
    sclcom(vecr,veci,4,1,1);
```

```
    printf("Largest eigenvalue: %e\n"
           "\nCorresponding eigenvector:\n",val[1]);
    for (i=1; i<=4; i++)
       printf(" %5.2f %6.3f*I\n",vecr[i][1],veci[i][1]);
    printf("\nEM[1]:  %e\nEM[3]:  %e\nEM[5]:  %e\nEM[7]:  %e"
           "\nEM[9]:  %e\n",em[1],em[3],em[5],em[7],em[9]);
    free_real_matrix(a,1,4,1);
    free_real_matrix(vecr,1,4,1);
    free_real_matrix(veci,1,4,1);
}
```

Output:

```
Largest eigenvalue: 4.82843e+00

Corresponding eigenvector:
   1.00  0.000*I
   1.00  0.000*I
   0.00  0.414*I
   0.00 -0.414*I

EM[1]:  6.00000e+00
EM[3]:  1.60000e+01
EM[5]:  1.00000e+00
EM[7]:  5.30322e-06
EM[9]:  1.00000e+00
```

Function tested: **qrihrm**

Example:

Compute all eigenvalues and corresponding eigenvectors of the 4×4 matrix

$$\begin{pmatrix} 3 & 1 & 0 & 2i \\ 1 & 3 & -2i & 0 \\ 0 & 2i & 1 & 1 \\ -2i & 0 & 1 & 1 \end{pmatrix}.$$

```
#include <stdio.h>

void main ()
{
    float **allocate_real_matrix(int, int, int, int);
    void free_real_matrix(float **, int, int, int);
    void sclcom(float **, float **, int, int, int);
    void inimat(int, int, int, int, float **, float);
    int qrihrm(float **, int, float [], float **, float **, float []);
    int i;
    float **a,**vr,**vi,val[5],em[6];

    a=allocate_real_matrix(1,4,1,4);
    vr=allocate_real_matrix(1,4,1,4);
    vi=allocate_real_matrix(1,4,1,4);

    inimat(1,4,1,4,a,0.0);
    a[1][1]=a[2][2]=3.0;
    a[3][2]=2.0;  a[4][1] = -2.0;
    a[1][2]=a[3][3]=a[3][4]=a[4][4]=1.0;
    em[0]=em[2]=5.0e-5;
    em[4]=20.0;
    printf("QRIHRM: %2d\n",qrihrm(a,4,val,vr,vi,em));
    sclcom(vr,vi,4,2,3);
    printf("\nEigenvalues:\n VAL[1]: %7.3f\n VAL[2]: %7.3f\n"
        " VAL[3]: %7.3f\n VAL[4]: %7.3f\n"
        "\nEigenvectors corresponding to\n   VAL[2]   ,    VAL[3]\n",
```

```
        val[1],val[2],val[3],val[4]);
    for (i=1; i<=4; i++)
        printf(" %3.0f%3.0f*I  ,  %3.0f%3.0f*I\n",
        vr[i][2],vi[i][2],vr[i][3],vi[i][3]);
    printf("\nEM[1]:  %e\nEM[3]:  %e\nEM[5]:  %e\n",
            em[1],em[3],em[5]);
    free_real_matrix(a,1,4,1);
    free_real_matrix(vr,1,4,1);
    free_real_matrix(vi,1,4,1);
}
```

Output:

```
QRIHRM:  0

Eigenvalues:
 VAL[1]:   4.828
 VAL[2]:   4.000
 VAL[3]:   0.000
 VAL[4]:  -0.828

Eigenvectors corresponding to
   VAL[2]   ,    VAL[3]
   1  0*I   ,    0 -1*I
  -1 -0*I   ,    0  1*I
   0 -1*I   ,    1 -0*I
   0 -1*I   ,    1  0*I

EM[1]:  6.00000e+00
EM[3]:  5.63109e-07
EM[5]:  4.00000e+00
```

Function tested: **valqricom**

Example:

Determine the roots of the polynomial

$$x^4 + (4+2i)x^3 + (5+6i)x^2 + (2+6i)x + 2i$$

by computing the eigenvalues of the companion matrix

$$\begin{pmatrix} -4-2i & -5-6i & -2-6i & -2i \\ 1 & 0 & 0 & 0 \\ 0 & 1 & 0 & 0 \\ 0 & 0 & 1 & 0 \end{pmatrix}.$$

```
#include <stdio.h>
void main ()
{
    float **allocate_real_matrix(int, int, int, int);
    void free_real_matrix(float **, int, int, int);
    void inimat(int, int, int, int, float **, float);
    int valqricom(float **, float **, float [], int, float [],
                  float [], float []);
    int i;
    float **a1,**a2,b[4],val1[5],val2[5],em[6];

    a1=allocate_real_matrix(1,4,1,4);
    a2=allocate_real_matrix(1,4,1,4);
    inimat(1,4,1,4,a1,0.0);
    inimat(1,4,1,4,a2,0.0);
    a1[1][1] = -4.0;  a1[1][2] = -5.0;
    a1[1][3] = a2[1][1] = a2[1][4] = -2.0;
    a2[1][2] = a2[1][3] = -6.0;
    b[1]=b[2]=b[3]=1.0;
```

```
    em[0]=5.0e-6;  em[1]=27.0;  em[2]=1.0e-6;  em[4]=15.0;
    printf("VALQRICOM: %2d\n",valqricom(a1,a2,b,4,em,val1,val2));
    printf("\n Eigenvalues:\n    Real part     Imaginary part\n");
    for (i=1; i<=4; i++)
       printf(" %12.4e    %12.4e\n",val1[i],val2[i]);
    printf("\nEM[3]: %e\nEM[5]: %3.0f\n",em[3],em[5]);
    free_real_matrix(a1,1,4,1);
    free_real_matrix(a2,1,4,1);
}
```

Output:

```
VALQRICOM:  0

 Eigenvalues:
    Real part      Imaginary part
   -9.989e-01       -1.001e+00
   -1.001e+00       -9.994e-01
   -1.005e+00       -3.410e-03
   -9.953e-01        3.406e-03

EM[3]: 8.90246e-06
EM[5]:   8
```

Function tested: **qricom**

Example:

Compute the eigenvalues and eigenvectors of the matrix

$$\begin{pmatrix} -4-2i & -5-6i & -2-6i & -2i \\ 1 & 0 & 0 & 0 \\ 0 & 1 & 0 & 0 \\ 0 & 0 & 1 & 0 \end{pmatrix}.$$

```
#include <stdio.h>

void main ()
{
    float **allocate_real_matrix(int, int, int, int);
    void free_real_matrix(float **, int, int, int);
    void sclcom(float **, float **, int, int, int);
    int qricom(float **, float **, float [], int, float [],
                  float [], float [], float **, float **);
    void inimat(int, int, int, int, float **, float);
    int i;
    float **a1,**a2,**vec1,**vec2,b[4],val1[5],val2[5],em[6];

    a1=allocate_real_matrix(1,4,1,4);
    a2=allocate_real_matrix(1,4,1,4);
    vec1=allocate_real_matrix(1,4,1,4);
    vec2=allocate_real_matrix(1,4,1,4);
    inimat(1,4,1,4,a1,0.0);
    inimat(1,4,1,4,a2,0.0);
    a1[1][1] = -4.0;  a1[1][2] = -5.0;
    a1[1][3] = a2[1][1] = a2[1][4] = -2.0;
    a2[1][2] = a2[1][3] = -6.0;
    b[1]=b[2]=b[3]=1.0;
    em[0]=5.0e-6;  em[1]=27.0;  em[2]=1.0e-6;  em[4]=15.0;
    printf("QRICOM: %2d\n",qricom(a1,a2,b,4,em,val1,val2,vec1,vec2));
    printf("\n Eigenvalues:\n    Real part     Imaginary part\n");
    for (i=1; i<=4; i++)
       printf(" %12.4e    %12.4e\n",val1[i],val2[i]);
    sclcom(vec1,vec2,4,1,4);
    printf("\nFirst eigenvector:\n    Real part     Imaginary part\n");
```

```
    for (i=1; i<=4; i++)
        printf(" %12.4e    %12.4e\n",vec1[i][1],vec2[i][1]);
    printf("\nEM[3]: %e\nEM[5]: %3.0f\n",em[3],em[5]);
    free_real_matrix(a1,1,4,1);
    free_real_matrix(a2,1,4,1);
    free_real_matrix(vec1,1,4,1);
    free_real_matrix(vec2,1,4,1);
}
```

Output:

```
QRICOM:  0

 Eigenvalues:
    Real part     Imaginary part
   -1.001e+00      -9.994e-01
   -9.989e-01      -1.001e+00
   -1.005e+00      -3.410e-03
   -9.953e-01       3.406e-03

First eigenvector:
    Real part     Imaginary part
    1.000e+00      -3.585e-08
   -5.003e-01       4.994e-01
    8.288e-04      -4.997e-01
    2.492e-01       2.504e-01

EM[3]: 8.90247e-06
EM[5]:   8
```

Function tested: **eigcom**

Example:

Compute the eigenvalues and eigenvectors of the matrix

$$\begin{pmatrix} 1+3i & 2+i & 3+2i & 1+i \\ 3+4i & 1+2i & 2+i & 4+3i \\ 2+3i & 1+5i & 3+i & 5+2i \\ 1+2i & 3+i & 1+4i & 5+3i \end{pmatrix}.$$

```
#include <stdio.h>

void main ()
{
    float *allocate_real_vector(int, int);
    float **allocate_real_matrix(int, int, int, int);
    void free_real_vector(float *, int);
    void free_real_matrix(float **, int, int, int);
    int eigcom(float **, float **, int, float [], float [],
                  float [], float **, float **);
    int i;
    float **ar,**ai,**vr,**vi,*valr,*vali,em[8];

    valr=allocate_real_vector(1,4);
    vali=allocate_real_vector(1,4);
    ar=allocate_real_matrix(1,4,1,4);
    ai=allocate_real_matrix(1,4,1,4);
    vr=allocate_real_matrix(1,4,1,4);
    vi=allocate_real_matrix(1,4,1,4);

    ar[1][1]=ar[1][4]=ar[2][2]=ar[3][2]=ar[4][1]=ar[4][3]=
    ai[1][2]=ai[1][4]=ai[2][3]=ai[3][3]=ai[4][2]=1.0;
    ar[1][2]=ar[2][3]=ar[3][1]=ai[1][3]=ai[2][2]=ai[3][4]=ai[4][1]=2.0;
    ar[1][3]=ar[2][1]=ar[3][3]=ar[4][2]=
```

```
    ai[1][1]=ai[2][4]=ai[3][1]=ai[4][4]=3.0;
    ar[2][4]=ai[2][1]=ai[4][3]=4.0;
    ar[3][4]=ar[4][4]=ai[3][2]=5.0;
    em[0]=5.0e-6;  em[2]=1.0e-5;  em[4]=10.0;  em[6]=10.0;
    printf("EIGCOM: %2d\n",eigcom(ar,ai,4,em,valr,vali,vr,vi));
    printf("\n Eigenvalues:\n");
    for (i=1; i<=4; i++)
        printf(" %12.5e%12.5e * I\n",valr[i],vali[i]);
    printf("\nFirst eigenvector:\n");
    for (i=1; i<=4; i++)
        printf(" %12.5e%12.5e\n",vr[i][1],vi[i][1]);
    printf("\nEM[1]: %6.2f\nEM[3]:   %e\nEM[5]: %3.0f\nEM[7]: %3.0f\n",
            em[1],em[3],em[5],em[7]);
    free_real_vector(valr,1);
    free_real_vector(vali,1);
    free_real_matrix(ar,1,4,1);
    free_real_matrix(ai,1,4,1);
    free_real_matrix(vr,1,4,1);
    free_real_matrix(vi,1,4,1);
}
```

Output:

```
EIGCOM:  0

 Eigenvalues:
  -3.3710e+00 -7.7045e-01 * I
   9.7837e+00  9.3225e+00 * I
   1.3657e+00 -1.4011e+00 * I
   2.2217e+00  1.8490e+00 * I

First eigenvector:
  -5.0610e-01  5.8345e-01
   1.0000e+00 -3.5252e-11
   5.1832e-01 -7.1466e-01
  -5.5348e-01  1.8756e-02

EM[1]:  15.30
EM[3]:   2.12406e-05
EM[5]:   5
EM[7]:   4
```

Function tested: **qzival**

Example:

With

$$A = \begin{pmatrix} 2 & 3 & -3 & 4 \\ 1 & -1 & 5 & 1 \\ 0 & 2 & 6 & 8 \\ 1 & 1 & 0 & 4 \end{pmatrix}, \qquad B = \begin{pmatrix} 1 & 5 & 9 & 0 \\ 2 & 6 & 10 & 2 \\ 3 & 7 & 11 & -1 \\ 4 & 8 & 12 & 3 \end{pmatrix}$$

determine complex numbers $\alpha^{(j)}$ and real numbers $\beta^{(j)}$ such that $\beta^{(j)}A-\alpha^{(j)}B$ is singular (j=1,...,4) and evaluate the quotients $\lambda^{(j)}=\alpha^{(j)}/\beta^{(j)}$.

```
#include <stdio.h>

void main ()
{
    float **allocate_real_matrix(int, int, int, int);
    void free_real_matrix(float **, int, int, int);
    void qzival(int, float **, float **, float [], float [],
                float [], int [], float []);
    int k,iter[5];
```

```
    float **a,**b,alfr[5],alfi[5],beta[5],em[2];

    a=allocate_real_matrix(1,4,1,4);
    b=allocate_real_matrix(1,4,1,4);
    a[1][1]=2.0;  a[1][2]=3.0;        a[1][3] = -3.0;   a[1][4]=4.0;
    a[2][1]=1.0;  a[2][2] = -1.0;     a[2][3]=5.0;      a[2][4]=1.0;
    a[3][1]=0.0;  a[3][2]=2.0;        a[3][3]=6.0;      a[3][4]=8.0;
    a[4][1]=1.0;  a[4][2]=1.0;        a[4][3]=0.0;      a[4][4]=4.0;
    b[1][1]=1.0;  b[1][2]=5.0;        b[1][3]=9.0;      b[1][4]=0.0;
    b[2][1]=2.0;  b[2][2]=6.0;        b[2][3]=10.0;     b[2][4]=2.0;
    b[3][1]=3.0;  b[3][2]=7.0;        b[3][3]=11.0;     b[3][4] = -1.0;
    b[4][1]=4.0;  b[4][2]=8.0;        b[4][3]=12.0;     b[4][4]=3.0;
    em[0]=1.0e-35;
    em[1]=1.0e-6;
    qzival(4,a,b,alfr,alfi,beta,iter,em);
    for (k=1; k<=4; k++)
        printf("ITER[%1d]=%3d\n",k,iter[k]);
    printf("\nALFA(real part)    ALFA(imaginary part)        BETA\n");
    for (k=1; k<=4; k++)
        printf(" %12.6e  %16.6e  %21.6e\n",alfr[k],alfi[k],beta[k]);
    printf("\nLAMBDA(real part)  LAMBDA(imaginary part)\n");
    for (k=1; k<=4; k++)
        if (beta[k] == 0.0)
            printf("  INFINITE             INDEFINITE\n");
        else
            printf(" %12.6e  %16.6e\n",alfr[k]/beta[k],alfi[k]/beta[k]);
    free_real_matrix(a,1,4,1);
    free_real_matrix(b,1,4,1);
}
```

Output:

```
ITER[1]=  0
ITER[2]=  0
ITER[3]=  0
ITER[4]=  9

ALFA(real part)    ALFA(imaginary part)        BETA
 -4.43471e+00       0.00000e+00            0.00000e+00
  2.96138e-01       0.00000e+00           -1.47027e-01
 -1.67632e-01       5.34577e-01            1.69637e+00
 -1.33137e-04      -4.24572e-04            1.34729e-03

LAMBDA(real part)  LAMBDA(imaginary part)
  INFINITE          INDEFINITE
 -2.01418e+00      -0.00000e+00
 -9.88182e-02       3.15130e-01
 -9.88182e-02      -3.15130e-01
```

Function tested: **qzi**

Example:

With

$$A = \begin{pmatrix} 2 & 3 & -3 & 4 \\ 1 & -1 & 5 & 1 \\ 0 & 2 & 6 & 8 \\ 1 & 1 & 0 & 4 \end{pmatrix}, \qquad B = \begin{pmatrix} 1 & 5 & 9 & 0 \\ 2 & 6 & 10 & 2 \\ 3 & 7 & 11 & -1 \\ 4 & 8 & 12 & 3 \end{pmatrix}$$

determine complex numbers $\alpha^{(j)}$ and real numbers $\beta^{(j)}$ such that $\beta^{(j)}A-\alpha^{(j)}B$ is singular (j=1,...,4), and evaluate the quotients $\lambda^{(j)}=\alpha^{(j)}/\beta^{(j)}$, and also determine the components of normalized eigenvectors $x^{(j)}$ such that $\beta^{(j)}Ax^{(j)}=\alpha^{(j)}Bx^{(j)}$.

```
#include <stdio.h>

void main ()
{
    float **allocate_real_matrix(int, int, int, int);
    void free_real_matrix(float **, int, int, int);
    void qzi(int, float **, float **, float **, float [],
             float [], float [], int [], float []);
    int k,l,iter[5];
    float **a,**b,**x,alfr[5],alfi[5],beta[5],em[2];

    a=allocate_real_matrix(1,4,1,4);
    b=allocate_real_matrix(1,4,1,4);
    x=allocate_real_matrix(1,4,1,4);
    a[1][1]=2.0;  a[1][2]=3.0;     a[1][3] = -3.0;   a[1][4]=4.0;
    a[2][1]=1.0;  a[2][2] = -1.0;  a[2][3]=5.0;      a[2][4]=1.0;
    a[3][1]=0.0;  a[3][2]=2.0;     a[3][3]=6.0;      a[3][4]=8.0;
    a[4][1]=1.0;  a[4][2]=1.0;     a[4][3]=0.0;      a[4][4]=4.0;
    b[1][1]=1.0;  b[1][2]=5.0;     b[1][3]=9.0;      b[1][4]=0.0;
    b[2][1]=2.0;  b[2][2]=6.0;     b[2][3]=10.0;     b[2][4]=2.0;
    b[3][1]=3.0;  b[3][2]=7.0;     b[3][3]=11.0;     b[3][4] = -1.0;
    b[4][1]=4.0;  b[4][2]=8.0;     b[4][3]=12.0;     b[4][4]=3.0;
    for (k=1; k<=4; k++)
        for (l=1; l<=4; l++)  x[k][l] = (k == l) ? 1.0 : 0.0;
    em[0]=1.0e-35;
    em[1]=1.0e-6;
    qzi(4,a,b,x,alfr,alfi,beta,iter,em);
    for (k=1; k<=4; k++)
        printf("ITER[%1d]=%3d\n",k,iter[k]);
    printf("\nEigenvectors:\n");
    for (k=1; k<=4; k++)
        printf(" %12.6e  %12.6e  %12.6e  %12.6e\n",
               x[k][1],x[k][2],x[k][3],x[k][4]);
    printf("\nALFA(real part)    ALFA(imaginary part)      BETA\n");
    for (k=1; k<=4; k++)
        printf(" %12.6e  %16.6e  %21.6e\n",alfr[k],alfi[k],beta[k]);
    printf("\nLAMBDA(real part)  LAMBDA(imaginary part)\n");
    for (k=1; k<=4; k++)
        if (beta[k] == 0.0)
            printf("  INFINITE          INDEFINITE\n");
        else
            printf(" %12.6e  %16.6e\n",
                   alfr[k]/beta[k],alfi[k]/beta[k]);
    free_real_matrix(a,1,4,1);
    free_real_matrix(b,1,4,1);
    free_real_matrix(x,1,4,1);
}
```

Output:

```
ITER[1]=  0
ITER[2]=  0
ITER[3]=  0
ITER[4]=  9

Eigenvectors:
 -5.00000e-01   1.00000e+00  -6.29479e-01   6.51869e-01
  1.00000e+00  -3.82291e-02   1.00000e+00   6.19239e-09
 -5.00000e-01  -3.04666e-02   1.65899e-01   1.09309e-01
  5.78136e-08  -7.63352e-01  -5.84739e-01   1.77444e-01

ALFA(real part)    ALFA(imaginary part)      BETA
 -4.43471e+00       0.00000e+00            0.00000e+00
  2.96138e-01       0.00000e+00           -1.47027e-01
 -1.67632e-01       5.34577e-01            1.69637e+00
 -1.33137e-04      -4.24572e-04            1.34729e-03

LAMBDA(real part)  LAMBDA(imaginary part)
  INFINITE          INDEFINITE
 -2.01418e+00      -0.00000e+00
 -9.88182e-02       3.15130e-01
 -9.88182e-02      -3.15130e-01
```

Function tested: **qrisngvaldec**

Example:

Compute the singular value decomposition of the 6×5 matrix A for which

$$A_{i,j} = 1/(i+j-1).$$

```
#include <stdio.h>

void main ()
{
    float **allocate_real_matrix(int, int, int, int);
    void free_real_matrix(float **, int, int, int);
    int qrisngvaldec(float **, int, int, float [], float **, float []);
    int i,j;
    float **a,**v,val[5],em[8];

    a=allocate_real_matrix(1,6,1,5);
    v=allocate_real_matrix(1,5,1,5);
    for (i=1; i<=6; i++)
        for (j=1; j<=5; j++) a[i][j]=1.0/(i+j-1);
    em[0]=1.0e-6;  em[2]=1.0e-5;  em[4]=25.0;  em[6]=1.0e-5;
    i=qrisngvaldec(a,6,5,val,v,em);
    printf("Number of singular values not found : %2d\n"
        "Infinity norm : %e\nMax neglected subdiagonal element : %e\n"
        "Number of iterations : %3.0f\nNumerical rank : %3.0f\n"
        "\nSingular values :\n",
        i,em[1],em[3],em[5],em[7]);
    for (i=1; i<=5; i++)
        printf("  %12.6e\n",val[i]);
    printf("\nMatrix U, first 3 columns :\n");
    for (i=1; i<=6; i++)
        printf("  %12.6e  %12.6e  %12.6e\n",a[i][1],a[i][2],a[i][3]);
    printf("\n           Last 2 columns :\n");
    for (i=1; i<=6; i++)
        printf("           %12.6e  %12.6e\n",a[i][4],a[i][5]);
    free_real_matrix(a,1,6,1);
    free_real_matrix(v,1,5,1);
}
```

Output:

```
Number of singular values not found :  0
Infinity norm : 2.28333e+00
Max neglected subdiagonal element : 2.11744e-05
Number of iterations :   3
Numerical rank :   5

Singular values :
   1.59212e+00
   2.24496e-01
   1.36102e-02
   1.25732e-04
   1.41708e-05

Matrix U, first 3 columns :
  -7.54979e-01   6.10111e-01  -2.33268e-01
  -4.39093e-01  -2.26021e-01   7.05900e-01
  -3.17031e-01  -3.73070e-01   2.11268e-01
  -2.49995e-01  -3.95578e-01  -1.47776e-01
  -2.07050e-01  -3.84833e-01  -3.68053e-01
  -1.76997e-01  -3.64582e-01  -4.95334e-01

           Last 2 columns :
            9.87459e-03  -2.57538e-02
           -6.88724e-03   2.67619e-01
           -3.89435e-01  -5.02005e-01
            8.52031e-01  -1.34877e-01
           -2.25330e-01   7.40444e-01
           -2.67327e-01  -3.30543e-01
```

Functions tested: **zerpol, bounds**

Example:

Determine the zeros of the polynomial

$$z^7 - 3z^6 - 3z^5 + 25z^4 - 46z^3 + 38z^2 - 12z$$

by means of a call of *zerpol*, and derive the real and imaginary parts of the centers and corresponding radii of discs containing the true zeros of the above polynomial.

```
#include <stdio.h>

void main ()
{
    int zerpol(int, float [], float [], float [], float [], float []);
    void bounds(int, float [], float [], float [], float,
                float, float [], float [], float []);
    int i,j;
    float a[8],d[8],re[8],im[8],em[5],recentre[8],imcentre[8],bound[8];

    a[7]=1.0;   a[6]=a[5] = -3.0;  a[4]=25.0;  a[3] = -46.0;
    a[2]=38.0;  a[1] = -12.0;      a[0]=0.0;
    em[0]=1.0e-6;  em[1]=40.0;
    i=zerpol(7,a,em,re,im,d);
    printf("Coefficients of polynomial:\n  ");
    for (j=7; j>=0; j--) printf("%5.0f",a[j]);
    printf("\n\nNumber of not found zeros: %2d\n"
           "Fail indication:           %3.0f\n"
           "Number of new starts:      %3.0f\n"
           "Number of iterations:      %3.0f\n\nZeros:\n",
           i,em[2],em[3],em[4]);
    for (j=i+1; j<=7; j++)
        if (im[j] == 0.0)
           printf(" %12.6e\n",re[j]);
        else
           printf(" %12.6e  %12.6e\n",re[j],im[j]);
    if (i == 0) {
        bounds(7,a,re,im,0.0,0.0,recentre,imcentre,bound);
        printf("\nReal and imaginary part of centre + radius\n");
        for (j=1; j<=7; j++)
           printf("  %12.6e  %12.6e  %12.6e\n",
                  recentre[j],imcentre[j],bound[j]);
    }
}
```

Output:

```
Coefficients of polynomial:
     1   -3   -3   25  -46   38  -12    0

Number of not found zeros:  0
Fail indication:            0
Number of new starts:       0
Number of iterations:      11

Zeros:
  1.99994e+00
 -3.00000e+00
  9.99974e-01  -9.99995e-01
  9.99974e-01   9.99995e-01
  9.93030e-01
  1.00708e+00
  0.00000e+00

Real and imaginary part of centre + radius
   1.99994e+00   0.00000e+00   7.00949e-05
  -3.00000e+00   0.00000e+00   1.07083e-06
   9.99974e-01  -9.99995e-01   3.35562e-05
   9.99974e-01   9.99995e-01   3.35562e-05
   1.00005e+00   0.00000e+00   1.49372e-02
   1.00005e+00   0.00000e+00   1.49372e-02
```

```
   0.00000e+00   0.00000e+00   0.00000e+00
```

Function tested: **allzerortpol**

Example:

Determine the roots of the Chebyshev polynomial $T_3(x)$ for which

$$T_0(x)=1, \quad T_1(x)=x, \quad T_{k+1}(x)=xT_k(x)-2^{-k}T_{k-1}(x), \; k=1,2.$$

```
#include <stdio.h>

void main ()
{
    void allzerortpol(int, float [], float [], float [], float []);
    float b[4],c[4],zer[4],em[6];

    em[0]=em[2]=1.0e-6;  em[4]=15.0;
    b[2]=b[1]=b[0]=0.0;
    c[0]=0.0;  c[1]=0.5;  c[2]=0.25;
    allzerortpol(3,b,c,zer,em);
    printf("The three zeros:\n %13.6e\n %13.6e\n %13.6e\n\n"
            "EM[1]: %5.2f\nEM[3]: %9.3e\nEM[5]: %2.0f\n",
            zer[1],zer[2],zer[3],em[1],em[3],em[5]);
}
```

Output:

```
The three zeros:
  -8.66026e-01
   8.66025e-01
 -1.000000e-06

EM[1]:  1.50
EM[3]:  7.07e-07
EM[5]:  1
```

Function tested: **lupzerortpol**

Example:

Determine the two smaller and two larger roots of the Laguerre polynomial $L_3(x)$ for which

$$L_0(x)=1, \quad L_1(x)=x-1, \quad L_{k+1}(x)=(x-2k-1)L_k(x)-k^2L_{k-1}(x), \; k=1,2.$$

```
#include <stdio.h>

void main ()
{
    void lupzerortpol(int, int, float [], float [], float [], float []);
    int i;
    float b[4],c[4],zer[3],em[7];

    em[0]=em[2]=1.0e-6;  em[4]=45.0;  em[6]=1.0;
    for (i=0; i<=2; i++) {
        b[i]=2*i+1;
        c[i]=i*i;
    }
    lupzerortpol(3,2,b,c,zer,em);
    printf("The two lower zeros:\n %13.6e\n %13.6e\n\n"
            "EM[1]: %5.2f\nEM[3]: %9.3e\nEM[5]: %3.0f\n",
            zer[1],zer[2],em[1],em[3],em[5]);
    em[6]=0.0;
    for (i=0; i<=2; i++) {
        b[i] = -2*i-1;
        c[i]=i*i;
    }
```

```
    lupzerortpol(3,2,b,c,zer,em);
    printf("\nThe two upper zeros:\n %13.6e\n %13.6e\n\n"
           "EM[1]: %5.2f\nEM[3]: %9.3e\nEM[5]: %3.0f\n",
           -zer[1],-zer[2],em[1],em[3],em[5]);
}
```

Output:

```
The two lower zeros:
   4.15775e-01
   2.29428e+00

EM[1]:  9.00
EM[3]:  4.79e-08
EM[5]:  10

The two upper zeros:
   6.28995e+00
   2.29428e+00

EM[1]:  9.00
EM[3]:  3.67e-10
EM[5]:  12
```

Function tested: **selzerortpol**

Example:

Determine the third root of the Legendre polynomial $P_4(x)$ for which

$$P_0(x)=1, \quad P_1(x)=x, \quad P_{k+1}(x)=xP_k(x)-\{k^2/(4k^2-1)\}P_{k-1}(x),\ k=1,2,3.$$

```
#include <stdio.h>

void main ()
{
    void selzerortpol(int, int, int, float [], float [],
                           float [], float []);
    int i;
    float b[5],c[5],zer[4],em[6];

    em[0]=em[2]=1.0e-6;
    for (i=0; i<=3; i++) {
       b[i]=0.0;
       c[i]=i*i/(4.0*i*i-1.0);
    }
    selzerortpol(4,3,3,b,c,zer,em);
    printf("The third zero:\n %13.6e\n\nEM[1]: %5.2f\nEM[5]: %3.0f\n",
           zer[3],em[1],em[5]);
}
```

Output:

```
The third zero:
  -3.39981e-01

EM[1]:  1.33
EM[5]:  10
```

Function tested: **alljaczer**

Example:

Compute the roots of the Jacobi polynomial $P_3^{(-1/2,-1/2)}(x)$.

```
#include <stdio.h>
```

```
void main ()
{
    void alljaczer(int, float, float, float []);
    float x[4];

    alljaczer(3,-0.5,-0.5,x);
    printf("Delivers:\n  %13.6e %13.6e %13.6e\n",x[1],x[2],x[3]);
}
```

Output:

```
Delivers:
   -8.66025e-01   0.00000e+00   8.66025e-01
```

Function tested: **alllagzer**

Example:

Compute the roots of the Laguerre polynomial $L_3^{(-1/2)}(x)$.

```
#include <stdio.h>

void main ()
{
    void alllagzer(int, float, float []);
    float x[4];

    alllagzer(3,-0.5,x);
    printf("Delivers:\n  %13.6e %13.6e %13.6e\n",x[1],x[2],x[3]);
}
```

Output:

```
Delivers:
    5.52534e+00   1.78449e+00   1.90164e-01
```

Function tested: **comkwd**

Example:

Compute the roots of x^2 - $2(-0.1+0.3i)x$ - $(0.11+0.02i)$.

```
#include <stdio.h>

void main ()
{
    void comkwd(float, float, float, float,
                float *, float *, float *, float *);
    float gr,gi,kr,ki;

    comkwd(-0.1,0.3,0.11,0.02,&gr,&gi,&kr,&ki);
    printf("x**2-2(-0.1+0.3*i)*x-(0.11+0.02*i) has roots\n"
           " %6.2f+%4.2f*i\n %6.3f+%4.2f*i\n",gr,gi,kr,ki);
}
```

Output:

```
x**2-2(-0.1+0.3*i)*x-(0.11+0.02*i) has roots
  -0.30+0.40*i
  0.100+0.20*i
```

Examples for chapter 4 procedures

Function tested: **euler**

Example:

Apply the Euler transformation to the series

$$\sum_{i=1}^{\infty} \frac{(-1)^{i-1}}{i^2} .$$

```
#include <stdio.h>

float a(int i)
{
    return (pow(-1,i)/((i+1)*(i+1)));
}

void main ()
{
    float euler(float (*)(int), float, int);

    printf("Delivers:  %13.6e\n",euler(a,1.0e-6,100));
}
```

Output:

```
Delivers:    8.22467e-01
```

Function tested: **sumposseries**

Example:

Evaluate the sum

$$\sum_{i=1}^{\infty} \frac{1}{i^2} .$$

```
#include <stdio.h>

float ai(float i)
{
    return 1.0/(i*i);
}

void main ()
{
    float sumposseries(float (*)(float), int, float, int, int, int);

    printf("SUMPOSSERIES delivers:  %e\n",
            sumposseries(ai,100,1.0e-6,8,100,10));
}
```

Output:

```
SUMPOSSERIES delivers:  1.64493e+00
```

Function tested: **qadrat**

Example:

Evaluate the integral

$$\int_0^{\pi} \sin(x)\,dx \;.$$

```
#include <math.h>
#include <stdio.h>

float a(float x)
{
    return (sin(x));
}

void main ()
{
    float qadrat(float *, float, float, float (*)(float), float []);
    float t,q,e[4];

    e[1]=e[2]=1.0e-6;
    q=qadrat(&t,0.0,3.141592653589,a,e);
    printf("Delivers:  %13.6e  %3.0f\n",q,e[3]);
}
```

Output:

```
Delivers:    2.00000e+00    0
```

Function tested: **integral**

Example:

Evaluate the integral

$$\int_{\alpha}^{\beta} \frac{10}{x^2}\,dx$$

with, in succession, α = -1 and ß = -2, -4, -20, -100.

```
#include <stdio.h>

float fx(float x)
{
    return 10.0/(x*x);
}

void main ()
{
    float integral(float, float, float (*)(float), float [],
                   int, int);
    int ua,ub,i;
    float e[7],a;
    static float b[4]={2.0, 4.0, 20.0, 100.0};

    printf("INTEGRAL delivers:\n");
    ua=1;
    e[1]=e[2]=1.0e-6;
    for (i=0; i<=3; i++) {
        ub=(b[i] < 50.0);
        a=integral(-1.0,-b[i],fx,e,ua,ub);
        printf(" %e  %3.0f   %e   %3.0f  %3.0f\n",
               a,e[3],e[4],e[5],e[6]);
        ua=0;
```

```
    }
}
```

Output:

```
INTEGRAL delivers:
 -5.00000e+00    0   -5.00000e+00    -2    2
 -7.50000e+00    0   -7.50000e+00    -4    1
 -9.50000e+00    0   -9.50000e+00   -20    0
 -9.99999e+00    1   -9.99999e+00     0    0
```

Function tested: **tricub**

Example:

Evaluate the double integral

$$\iint_{\Omega} \cos(x)\cos(y)\,dxdy$$

over the triangle Ω in the x-y plane with vertices $(0,0)$, $(0,\pi/2)$, $(\pi/2,\pi/2)$.

```
#include <math.h>
#include <stdio.h>

float e(float x, float y)
{
    return cos(x)*cos(y);
}

void main ()
{
    float tricub(float, float, float, float, float, float,
                    float (*)(float,float), float, float);
    int i;
    float pi,acc;

    printf("TRICUB delivers:\n");
    pi=3.14159265359;
    acc=1.0;
    for (i=0; i<=5; i++) {
        acc *= 1.0e-1;
        printf(" %9.1e   %e\n",acc,
                tricub(0.0,0.0,0.0,pi/2.0,pi/2.0,pi/2.0,e,acc,acc));
    }
}
```

Output:

```
TRICUB delivers:
     1e-01   5.00640e-01
   1.0e-02   5.00640e-01
   1.0e-03   5.00640e-01
   1.0e-04   4.99991e-01
   1.0e-05   4.99999e-01
   1.0e-06   4.99999e-01
```

Function tested: **reccof**

Example:

Evaluate the coefficients c_1, c_2 in the recursion

$$C_{k+1}(x) = xC_k(x) - c_kC_{k-1}(x) \qquad (k=1,2)$$

for the Chebyshev polynomials of the second kind which are orthogonal over [-1,1] with respect to the weight function $w(x)=(1-x^2)^{1/2}$.

```
#include <stdio.h>

float a(float x)
{
   return sqrt(1.0-x*x);
}

void main ()
{
   void reccof(int, int, float *, float (*)(float), float [],
                  float [], float [], int);
   float x,b[3],c[3],l[3];

   reccof(2,200,&x,a,b,c,l,1);
   printf("Delivers: %7.3f %7.3f\n",c[1],c[2]);
}
```

Output:

```
Delivers:   0.250   0.250
```

Function tested: **gsswtssym**

Example:

Compute the weights w_i which render the formula

$$\pi \sum_{i=1}^{5} w_i f(z_i) = \int_{-1}^{1} \frac{1}{\sqrt{(1-x^2)}} f(x)\,dx$$

where

$$z_i = \cos\{\frac{2(n-i)-1}{2n}\pi\} \qquad i=1,...,n/2, \qquad n=5$$

exact for all polynomials of degree < 10 by means of a call of *gsswtssym* with c_1=1/2, c_k=1/4 (k=2,3,...).

```
#include <math.h>
#include <stdio.h>

void main ()
{
   void gsswtssym(int, float [], float [], float []);
   float pi,zer[3],w[4],c[5];

   pi=4.0*atan(1.0);
   c[1]=0.5;  c[2]=0.25; c[3]=0.25; c[4]=0.25;
   zer[1]=cos(0.9*pi);
   zer[2]=cos(0.7*pi);
   gsswtssym(5,zer,c,w);
   printf("Results:\n %7.3f %7.3f %7.3f %7.3f %7.3f",
             w[1]*pi,w[2]*pi,w[3]*pi,w[2]*pi,w[1]*pi);
}
```

Output:

```
Results:
   0.628   0.628   0.628   0.628   0.628
```

Function tested: **gssjacwghts**

Example:

Evaluate the integral

$$\int_{-1}^{1}(1-x)(1+x)^2 e^x dx$$

by use of a fifth order Gauss-Jacobi quadrature formula. The exact value is *2e - 10/e*. (The error is printed out.)

```
#include <math.h>
#include <stdio.h>

void main ()
{
    void gssjacwghts(int, float, float, float [], float []);
    int n;
    float alfa,beta,ind,x[6],w[6];

    alfa=1.0;
    beta=2.0;
    n=5;
    ind=0.0;
    gssjacwghts(n,alfa,beta,x,w);
    for (n=1; n<=5; n++) ind += w[n]*exp(x[n]);
    printf("Delivers:  %13.6e\n",ind-2.0*exp(1.0)+10.0/exp(1.0));
}
```

Output:

```
Delivers:    1.01034e-07
```

Function tested: **gsslagwghts**

Example:

Evaluate the integral

$$\int_{0}^{\infty}\sin(x)e^{-x} dx$$

by use of a ten point Gauss-Laguerre quadrature formula. The exact value is *0.5*. (The error is printed out.)

```
#include <math.h>
#include <stdio.h>

void main ()
{
    void gsslagwghts(int, float, float [], float []);
    int n;
    float ind,x[11],w[11];

    gsslagwghts(10,0.0,x,w);
    ind=0.0;
    for (n=10; n>=1; n--) ind += w[n]*sin(x[n]);
    printf("Delivers:  %13.6e\n",ind-0.5);
}
```

Output:

```
Delivers:    1.78814e-07
```

Function tested: **jacobnnf**

Example:

Compute approximations to the partial derivatives $\partial f_i/\partial x_j$ $(i,j=1,2)$ of the components of the function $f(x)$ given by

$$f_1 = x_1^3 + x_2 \qquad\qquad f_2 = 10x_2$$

at the point $x = (2,1)$.

```
#include <stdio.h>

void f1(int n, float x[], float f[])
{
    f[1]=x[1]*x[1]*x[1]+x[2];
    f[2]=x[2]*10.0;
}

float di(int i)
{
    return ((i == 1) ? 1.0e-5 : 1.0);
}

void main ()
{
    float **allocate_real_matrix(int, int, int, int);
    void free_real_matrix(float **, int, int, int);
    void jacobnnf(int, float [], float [], float **,
                  float (*)(int), void (*)(int, float[], float[]));
    int i;
    float **jac,x[3],f[3];

    jac=allocate_real_matrix(1,2,1,2);
    x[1]=2.0;  x[2]=1.0;
    f1(2,x,f);
    jacobnnf(2,x,f,jac,di,f1);
    printf("The calculated jacobian is:\n %6.1f %6.1f\n %6.1f %6.1f\n",
           jac[1][1],jac[1][2],jac[2][1],jac[2][2]);
    free_real_matrix(jac,1,2,1);
}
```

Output:

```
The calculated jacobian is:
  12.0     1.0
   0.0    10.0
```

Function tested: **jacobnmf**

Example:

Compute approximations to the partial derivatives $\partial f_i/\partial x_j$ $(i=1,2,3;\ j=1,2)$ of the components of the function $f(x)$ given by

$$f_1 = x_1^3 + x_2, \qquad f_2 = 10x_2 + x_1x_2, \qquad f_3 = x_1x_2$$

at the point $x = (2,1)$.

```
#include <stdio.h>

void f1(int n, int m, float x[], float f[])
{
    f[1]=x[1]*x[1]*x[1]+x[2];
    f[2]=x[2]*10.0+x[2]*x[1]*x[1];
    f[3]=x[1]*x[2];
```

```
}

float di(int i)
{
    return ((i == 2) ? 1.0 : 1.0e-5);
}

void main ()
{
    float **allocate_real_matrix(int, int, int, int);
    void free_real_matrix(float **, int, int, int);
    void jacobnmf(int, int, float [], float [], float **,
                  float (*)(int), void (*)(int, int, float[], float[]));
    int i;
    float **jac,x[3],f[4];

    jac=allocate_real_matrix(1,3,1,2);
    x[1]=2.0;  x[2]=1.0;
    f1(3,2,x,f);
    jacobnmf(3,2,x,f,jac,di,f1);
    printf("The calculated jacobian is:\n"
        "  %7.1f  %7.1f\n  %7.1f  %7.1f\n  %7.1f  %7.1f\n",
        jac[1][1],jac[1][2],jac[2][1],jac[2][2],jac[3][1],jac[3][2]);
    free_real_matrix(jac,1,3,1);
}
```

Output:

```
The calculated jacobian is:
     12.0      1.0
      4.0     14.0
      1.0      2.0
```

Function tested: **jacobnbndf**

Example:

Compute approximations to the partial derivatives

$$\partial f_i/\partial x_j, \qquad i=1,...,5;\ j=\max(1,i-1),...,\min(5,i+1)$$

of the components of the function $f(x)$ given by

$$f_1 = (3-2x_1)x_1 - 2x_2 + 1$$
$$f_i = (3-2x_i)x_i - 2x_{i+1} - x_{i-1} + 1 \qquad (i=2,3,4)$$
$$f_5 = 4 - x_4 - 2x_5$$

at the point $x = (-1,-1,-1,-1,-1)$.

```
#include <stdio.h>

int func(int n, int l, int u, float x[], float f[])
{
    int i;

    for (i=l; i<=((u == 5) ? 4 : u); i++) {
        f[i]=(3.0-2.0*x[i])*x[i]+1.0-2.0*x[i+1];
        if (i != 1) f[i] -= x[i-1];
    }
    if (u == 5) f[5]=4.0-x[4]-x[5]*2.0;
}

float di(int i)
{
    return ((i == 5) ? 1.0 : 1.0e-6);
}

void main ()
{
    void jacobnbndf(int, int, int, float [], float [],
                    float [], float (*)(int),
```

```
                     int (*)(int, int, int, float[], float[]));
    int i;
    float jac[14],x[6],f[6];

    for (i=1; i<=5; i++) x[i] = -1.0;
    func(5,1,5,x,f);
    jacobnbndf(5,1,1,x,f,jac,di,func);
    printf("The calculated tridiagonal jacobian is:\n"
        "   %4.0f  %4.0f\n   %4.0f  %4.0f  %4.0f\n"
        "         %4.0f  %4.0f  %4.0f\n"
        "               %4.0f  %4.0f  %4.0f\n"
        "                     %4.0f  %4.0f\n",
        jac[1],jac[2],jac[3],jac[4],jac[5],jac[6],jac[7],jac[8],
        jac[9],jac[10],jac[11],jac[12],jac[13]);
}
```

Output:

```
The calculated tridiagonal jacobian is:
     7     -2
    -1      7     -2
           -1      7     -2
                  -1      7     -2
                         -1     -2
```

Examples for chapter 5 procedures

Function tested: **zeroin**

Example:

Determine a zero of $e^{-3x}(x-1)+x^3$ in the interval [0,1].

```
#include <math.h>
#include <stdio.h>

float fx(float  x)
{
    return exp(-x*3.0)*(x-1.0)+x*x*x;
}

float tolx(float  x)
{
    return fabs(x)*1.0e-6+1.0e-6;
}

void main ()
{
    int zeroin(float *, float *, float (*)(float), float (*)(float));
    float x,y;

    x=0.0;
    y=1.0;
    if (zeroin(&x,&y,fx,tolx))
        printf("Calculated zero: %e\n",x);
    else
        printf("No zero found.");
}
```

Output:

```
Calculated zero: 4.89703e-01
```

Function tested: **zeroinrat**

Example:

Determine a zero of $e^{-3x}(x-1)+x^3$ in the interval [0,1].

```
#include <math.h>
#include <stdio.h>

float fx(float  x)
{
    return exp(-x*3.0)*(x-1.0)+x*x*x;
}

float tolx(float  x)
{
    return fabs(x)*1.0e-6+1.0e-6;
}

void main ()
{
    int zeroinrat(float *, float *, float (*)(float), float (*)(float));
    float x,y;

    x=0.0;
    y=1.0;
    if (zeroinrat(&x,&y,fx,tolx))
        printf("Calculated zero: %e\n",x);
    else
        printf("No zero found.");
}
```

Output:

```
Calculated zero: 4.89703e-01
```

Function tested: **zeroinder**

Example:

Determine a zero of $e^{-3x}(x-1)+x^3$ in the interval [0,1].

```
#include <math.h>
#include <stdio.h>

float f(float  x)
{
    return exp(-x*3.0)*(x-1.0)+x*x*x;
}

float df(float  x)
{
    return exp(-x*3.0)*(-3.0*x+4.0)+3.0*x*x;
}

float tolx(float  x)
{
    return fabs(x)*1.0e-6+1.0e-6;
}

void main ()
{
    int zeroinder(float *, float *, float (*)(float),
                      float (*)(float), float (*)(float));
    float x,y;

    x=0.0;
    y=1.0;
    if (zeroinder(&x,&y,f,df,tolx))
        printf("Calculated zero and function value:\n  %e     %e"
               "\nOther straddling approximation and function value:"
               "\n  %e    %e\n",x,f(x),y,f(y));
```

```
    else
        printf("No zero found.");
}
```

Output:

```
Calculated zero and function value:
  4.89703e-01     1.64360e-08
Other straddling approximation and function value:
  4.89701e-01    -1.92349e-06
```

Function tested: **quanewbnd1**

Example:

Solve the system of equations $f(x) = 0$ $(f,x\epsilon R^n)$ where

$$f^{(1)} = (3 - 2x^{(1)})x^{(1)} - 2x^{(2)} + 1$$
$$f^{(k)} = (3 - 2x^{(k)})x^{(k)} - 2x^{(k+1)} - x^{(k-1)} + 1 \qquad (k=2,...,n-1)$$
$$f^{(n)} = (3 - 2x^{(n)})x^{(n)} - x^{(n-1)} + 1$$

with initial approximation $x_0^{(k)} = -1$ $(k=1,...,n)$, for $n = 600$.

```
#include <stdio.h>

int fun(int n, int l, int u, float x[], float f[])
{
    int i;
    float x1,x2,x3;

    x1 = (l == 1) ? 0.0 : x[l-1];
    x2=x[l];
    x3 = (l == n) ? 0.0 : x[l+1];
    for (i=l; i<=u; i++) {
        f[i]=(3.0-2.0*x2)*x2+1.0-x1-x3*2.0;
        x1=x2;
        x2=x3;
        x3 = (i <= n-2) ? x[i+2] : 0.0;
    }
    return (1);
}

void main ()
{
    float *allocate_real_vector(int, int);
    void free_real_vector(float *, int);
    void quanewbnd1(int, int, int, float [], float [],
                    int (*)(int, int, int, float[], float[]),
                    float [], float []);
    int i;
    float *x,*f,in[6],out[6];

    x=allocate_real_vector(1,600);
    f=allocate_real_vector(1,600);
    for (i=1; i<=600; i++) x[i] = -1.0;
    in[0]=1.0e-6;  in[1]=in[2]=in[3]=1.0e-5;  in[4]=20000.0;
    in[5]=0.001;
    quanewbnd1(600,1,1,x,f,fun,in,out);
    printf("Norm Residual vector: %e\n"
           "Length of last step:  %e\n"
           "Number of function component evaluations: %6.0f\n"
           "Number of iterations: %3.0f\nReport: %3.0f\n",
           out[2],out[1],out[3],out[4],out[5]);
    free_real_vector(x,1);
    free_real_vector(f,1);
}
```

Output:

```
Norm Residual vector: 2.39722e-06
```

```
Length of last step:  2.70116e-05
Number of function component evaluations:   5998
Number of iterations:   6
Report:   0
```

Function tested: **minin**

Example:

Determine an approximation to the point in the interval [1,4] at which the function

$$f(x) = \sum_{i=1}^{20} \left(\frac{2i-5}{x-i^2} \right)^2$$

assumes a minimum value.

```
#include <stdio.h>
#include <math.h>

float f(float x)
{
    int i;
    float s,temp;

    s=0.0;
    for (i=1; i<=20; i++) {
        temp=(i*2-5)/(x-i*i);
        s += temp*temp;
    }
    return s;
}

float tol(float x)
{
    return (fabs(x)*1.0e-6+1.0e-6);
}

void main ()
{
    float minin(float *, float *, float *, float (*)(float),
                float (*)(float));
    float m,x,a,b;

    a=1.0+tol(1.0);
    b=4.0-tol(4.0);
    m=minin(&x,&a,&b,f,tol);
    printf("Minimum is  %e\nFor x is  %e\nin the interval with "
           "endpoints  %e   %e",m,x,a,b);
}
```

Output:

```
Minimum is  3.67670e+00
For x is  3.02290e+00
in the interval with endpoints  3.02290e+00   3.02291e+00
```

Function tested: **mininder**

Example:

Determine an approximation to the point in the interval [1.01,3.99] at which the function

$$f(x) = \sum_{i=1}^{20} \left(\frac{2i-5}{x-i^2} \right)^2$$

assumes a minimum value. The derivative of this function is

$$\frac{df(x)}{dx} = -2 \sum_{i=1}^{20} \frac{(2i-5)^2}{(x-i^2)^3} .$$

```
#include <stdio.h>
#include <math.h>

float f(float x)
{
    int i;
    float s,temp;

    s=0.0;
    for (i=1; i<=20; i++) {
        temp=(i*2-5)/(x-i*i);
        s += temp*temp;
    }
    return s;
}

float df(float x)
{
    int i;
    float s,temp1,temp2;

    s=0.0;
    for (i=1; i<=20; i++) {
        temp1=i*2-5;
        temp2=x-i*i;
        s += (temp1*temp1)/(temp2*temp2*temp2);
    }
    return -s*2.0;
}

float tol(float x)
{
    return (fabs(x)*1.0e-6+1.0e-6);
}

void main ()
{
    float mininder(float *, float *, float (*)(float),
                        float (*)(float), float (*)(float));
    float m,x,y;

    x=1.01;
    y=3.99;
    m=mininder(&x,&y,f,df,tol);
    printf("Minimum is  %e\nFor x is  %e  and y is  %e\n",m,x,y);
}
```

Output:

```
Minimum is  3.67670e+00
For x is  3.02291e+00  and y is  3.02292e+00
```

Function tested: **praxis**

Example:

Calculate the minimum of the function

$$f(x) = 100(x_2-x_1^2)^2 + (1-x_1)^2$$

using (-1.2,1) as an initial estimate.

```
#include <stdio.h>

float f(int n, float x[])
{
    float temp;

    temp=x[2]-x[1]*x[1];
    return temp*temp*100.0+(1.0-x[1])*(1.0-x[1]);
}

void main ()
{
    void praxis(int, float [], float (*)(int, float[]),
                float [], float []);
    float x[3],in[10],out[7];

    in[0]=1.0e-6;  in[1]=in[2]=1.0e-6;  in[5]=250.0;
    in[6]=in[7]=in[8]=in[9]=1.0;
    x[1] = -1.2;  x[2]=1.0;
    praxis(2,x,f,in,out);
    if (out[1] == 0.0) printf("Normal Termination\n\n");
    printf("Minimum is  %e\nFor x is  %e  %e\n"
           "The initial function value was  %e\n"
           "The number of function evaluations needed was %4.0f\n"
           "The number of line searches was %4.0f\n"
           "The step size in the last iteration step was  %e\n",
           out[2],x[1],x[2],out[3],out[4],out[5],out[6]);
}
```

Output:

```
Normal Termination

Minimum is  0.00000e+00
For x is  1.00000e+00  1.00000e+00
The initial function value was  2.42000e+01
The number of function evaluations needed was  180
The number of line searches was   68
The step size in the last iteration step was  4.28428e-08
```

Functions tested: **rnk1min, flemin**

Example:

Determine the value of $x = (x_1,x_2)$ yielding a minimum of the function

$$f(x) = 100(x_2-x_1^2)^2 + (1-x_1)^2$$

twice: firstly by use of *rnk1min*, and secondly by *flemin*. In both cases the initial approximation is taken to be (-1.2,1).

```
#include <stdio.h>

float rosenbrock(int n, float x[], float g[])
{
    float temp;

    temp=x[2]-x[1]*x[1];
    g[1]=(-temp*400.0+2.0)*x[1]-2.0;
    g[2]=temp*200.0;
    return temp*temp*100.0+(1.0-x[1])*(1.0-x[1]);
}
```

```
void main ()
{
    float rnk1min(int, float [], float [], float [],
                   float (*)(int, float[], float[]),
                   float [], float []);
    float flemin(int, float [], float [], float [],
                  float (*)(int, float[], float[]), float [], float []);
    int again;
    float f,x[3],g[3],h[4],in[9],out[5];

    in[0]=1.0e-6;  in[1]=in[2]=1.0e-5;  in[3]=1.0e-4;  in[4]=1.0e-5;
    in[5] = -10.0;  in[6]=1.0;  in[7]=100.0;  in[8]=0.01;
    x[1] = -1.2;  x[2]=1.0;
    again=1;
    f=rnk1min(2,x,g,h,rosenbrock,in,out);
    while (1) {
        printf("\nLeast value:  %e\nx:  %e  %e\n"
            "Gradient:  %e  %e\n"
            "Metric:      %e  %e\n                          %e\n"
            "OUT: %e\n     %e\n     %e\n     %e\n     %e\n",
            f,x[1],x[2],g[1],g[2],h[1],h[2],h[3],out[0],out[1],out[2],
            out[3],out[4]);
        if (again) {
            x[1] = -1.2;  x[2]=1.0;
            again=0;
            f=flemin(2,x,g,h,rosenbrock,in,out);
        } else
            break;
    }
}
```

Output:

```
Least value:  5.68434e-14
x:  1.00000e+00  1.00000e+00
Gradient:  4.76860e-07  -1.13687e-11
Metric:     4.89787e-01  9.77621e-01
                         1.95649e+00
OUT: 5.21398e-07
     4.76860e-07
     5.60000e+01
     1.10000e+01
     6.00000e+00

Least value:  1.42109e-14
x:  1.000000e+00  1.000000e+00
Gradient:  -2.38413e-07  -2.84217e-12
Metric:     4.90739e-01  9.80828e-01
                         1.96534e+00
OUT: 2.61484e-07
     2.38413e-07
     4.30000e+01
     6.00000e+00
     0.00000e+00
```

Function tested: **marquardt**

Example:

Determine the parameters p_1, p_2 and p_3 of best fit of the function

$$g(x,p) = p_1 + p_2*exp(p_3x)$$

when $x=x_i$, to data readings y_i (i=1,...,b), where $x_1,...,x_6$ are -5, -3, -1, 1, 3, 5 and $y_1,...,y_6$ are 127.0, 151.0, 379.0, 421.0, 460.0, 426.0.

The components of the residual vector are

$$p_1 + p_2*exp(p_3x_i) - y_i \qquad (i=1,...,6).$$

The elements of the Jacobian matrix are

$$\partial g/\partial p_1 = 1, \quad \partial g/\partial p_2 = exp(p_3x_i), \quad \partial g/\partial p_3 = p_2*x_i*exp(p_3x_i), \qquad (i=1,...,6).$$

```
#include <math.h>
#include <stdio.h>

float x[7],y[7];

int expfunct(int m, int n,float par[], float rv[])
{
    int i;

    for (i=1; i<=m; i++) {
        if (par[3]*x[i] > 680.0) return 0;
        rv[i]=par[1]+par[2]*exp(par[3]*x[i])-y[i];
    }
    return 1;
}

void jacobian(int m, int n, float par[], float rv[], float **jac)
{
    int i;
    float ex;

    for (i=1; i<=m; i++) {
        jac[i][1]=1.0;
        jac[i][2]=ex=exp(par[3]*x[i]);
        jac[i][3]=x[i]*par[2]*ex;
    }
}

void main ()
{
    float **allocate_real_matrix(int, int, int, int);
    void free_real_matrix(float **, int, int, int);
    void marquardt(int, int, float [], float [], float **,
                        int (*)(int, int, float[], float[]),
                        void (*)(int, int, float[], float[], float **),
                        float [], float []);
    float in[7],out[8],rv[7],par[4],**jjinv;

    jjinv=allocate_real_matrix(1,3,1,3);
    in[0]=1.0e-6;  in[3]=1.0e-4;  in[4]=1.0e-1;  in[5]=75.0;
    in[6]=1.0e-2;
    x[1] = -5.0;  x[2] = -3.0;  x[3] = -1.0;  x[4]=1.0;
    x[5]=3.0;  x[6]=5.0;
    y[1]=127.0;  y[2]=151.0;  y[3]=379.0;  y[4]=421.0;
    y[5]=460.0;  y[6]=426.0;
    par[1]=580.0;  par[2] = -180.0;  par[3] = -0.160;
    marquardt(6,3,par,rv,jjinv,expfunct,jacobian,in,out);
    printf("Parameters:\n  %9.4e  %9.4e  %9.4e\n\nOUT:\n"
        " %14.6e\n %14.6e\n %14.6e\n %14.6e\n %14.6e\n %14.6e\n"
        " %14.6e\n\nLast residual vector:\n"
        " %6.1f  %6.1f  %6.1f  %6.1f  %6.1f  %6.1f\n",
        par[1],par[2],par[3],out[7],out[2],out[6],out[3],out[4],
        out[5],out[1],rv[1],rv[2],rv[3],rv[4],rv[5],rv[6]);
    free_real_matrix(jjinv,1,3,1);
}
```

Output:

```
Parameters:
   5.232e+02   -1.568e+02   -1.998e-01

OUT:
    7.22150e+07
    1.15716e+02
    1.72813e-03
    1.65459e+02
    2.30000e+01
    2.20000e+01
    0.00000e+00

Last residual vector:
  -29.6    86.6   -47.3   -26.2   -22.9    39.5
```

Function tested: **gssnewton**

Example:

Determine the parameters p_1, p_2 and p_3 of best fit of the function

$$g(x,p) = p_1 + p_2*exp(p_3x)$$

when $x=x_i$, to data readings y_i (i=1,...,b), where $x_1,...,x_6$ are -5, -3, -1, 1, 3, 5 and $y_1,...,y_6$ are 127.0, 151.0, 379.0, 421.0, 460.0, 426.0.

The components of the residual vector are

$$p_1 + p_2*exp(p_3x_i) - y_i \qquad (i=1,...,6).$$

The elements of the Jacobian matrix are

$$\partial g_i/\partial p_1 = 1, \quad \partial g_i/\partial p_2 = exp(p_3x_i), \quad \partial g_i/\partial p_3 = p_2*x_i*exp(p_3x_i), \qquad (i=1,...,6).$$

```
#include <math.h>
#include <stdio.h>

float x[7],y[7];

int expfunct(int m, int n,float par[], float g[])
{
    int i;

    for (i=1; i<=m; i++) {
        if (par[3]*x[i] > 680.0) return 0;
        g[i]=par[1]+par[2]*exp(par[3]*x[i])-y[i];
    }
    return 1;
}

void jacobian(int m, int n, float par[], float g[], float **jac)
{
    int i;
    float ex;

    for (i=1; i<=m; i++) {
        jac[i][1]=1.0;
        jac[i][2]=ex=exp(par[3]*x[i]);
        jac[i][3]=x[i]*par[2]*ex;
    }
}

void main ()
{
    float **allocate_real_matrix(int, int, int, int);
    void free_real_matrix(float **, int, int, int);
    void gssnewton(int, int, float [], float [], float **,
                   int (*)(int, int, float[], float[]),
                   void (*)(int, int, float[], float[], float **),
                   float [], float []);
    float in[8],out[10],g[7],par[4],**v;

    v=allocate_real_matrix(1,3,1,3);
    in[0]=1.0e-6;  in[1]=in[2]=1.0e-6;  in[5]=75.0;  in[4]=1.0e-6;
    in[6]=14.0;  in[7]=1.0;
    x[1] = -5.0;  x[2] = -3.0;  x[3] = -1.0;  x[4]=1.0;
    x[5]=3.0;  x[6]=5.0;
    y[1]=127.0;  y[2]=151.0;  y[3]=379.0;  y[4]=421.0;
    y[5]=460.0;  y[6]=426.0;
    par[1]=580.0;  par[2] = -180.0;  par[3] = -0.160;
    gssnewton(6,3,par,g,v,expfunct,jacobian,in,out);
    printf("Parameters:\n  %9.4e   %9.4e   %9.4e\n\nOUT:\n"
        " %14.6e\n %14.6e\n %14.6e\n %14.6e\n %14.6e\n %14.6e\n"
        " %14.6e\n %14.6e\n %14.6e\n\nLast residual vector:\n"
        " %6.1f  %6.1f  %6.1f  %6.1f  %6.1f  %6.1f\n",
        par[1],par[2],par[3],out[6],out[2],out[3],out[4],out[5],
        out[1],out[7],out[8],out[9],g[1],g[2],g[3],g[4],g[5],g[6]);
    free_real_matrix(v,1,3,1);
}
```

Output:

```
Parameters:
   5.233e+02   -1.570e+02   -1.996e-01

OUT:
    6.88262e-06
    1.15716e+02
    1.65459e+02
    3.40000e+01
    1.10000e+01
    3.00000e+00
    1.10000e+01
    3.00000e+00
    2.33968e+03

Last residual vector:
  -29.6    86.6   -47.3   -26.2   -22.9    39.5
```

Function tested: **rk1**

Example:

Compute the solution at $x = 1$ of the differential equation

$$dy/dx = -y$$

with initial condition $y(0) = 1$.

```
#include <math.h>
#include <stdio.h>

float fxy(float x, float y)
{
    return -y;
}

void main ()
{
    void rk1(float *, float, float, float *, float,
                float (*)(float, float), float [], float [], int);
    int first;
    float x,y,d[5],e[3];

    e[1]=e[2]=1.0e-4;
    first=1;
    rk1(&x,0.0,1.0,&y,1.0,fxy,e,d,first);
    printf("RK1 delivers:\n  x = %e\n  y = %e      yexact =  %e",
                x,y,exp(-x));
}
```

Output:

```
RK1 delivers:
  x = 1.00000e+00
  y = 3.67877e-01      yexact =  3.67879e-01
```

Function tested: **rke**

Example:

Compute the solution at $t = 1$ and $t = -1$ of the system

$$dx/dt = y - z$$
$$dy/dt = x^2 + 2y + 4t$$
$$dz/dt = x^2 + 5x + z^2 + 4t,$$

with $x = y = 0$ and $z = 2$ at $t = 0$.

```
#include <math.h>
#include <stdio.h>

void rhs(int n, float t, float y[])
{
    float xx,yy,zz;

    xx=y[1];
    yy=y[2];
    zz=y[3];
    y[1]=yy-zz;
    y[2]=xx*xx+2.0*yy+4.0*t;
    y[3]=xx*(xx+5.0)+2.0*zz+4.0*t;
}

void info(int n, float t, float te, float y[], float data[])
{
    float et,t2,aex,aey,aez,rex,rey,rez;

    if (t == te) {
        et=exp(t);
        t2=2.0*t;
        rex = -et*sin(t2);
        aex=rex-y[1];
        rex=fabs(aex/rex);
        rey=et*et*(8.0+2.0*t2-sin(2.0*t2))/8.0-t2-1.0;
        rez=et*(sin(t2)+2.0*cos(t2))+rey;
        aey=rey-y[2];
        rey=fabs(aey/rey);
        aez=rez-y[3];
        rez=fabs(aez/rez);
        printf("\nT = %2.0f\n"
            "Relative and absolute errors in x, y and z:\n"
            "  RE(X)     RE(Y)     RE(Z)     AE(X)     AE(Y)     AE(Z)\n"
            " %7.2e  %7.2e  %7.2e  %7.2e  %7.2e  %7.2e\n"
            "Number of integration steps performed : %3.0f\n"
            "Number of integration steps skipped   : %3.0f\n"
            "Number of integration steps rejected  : %3.0f\n",
            t,rex,rey,rez,fabs(aex),fabs(aey),fabs(aez),
            data[4],data[6],data[5]);
    }
}

void main ()
{
    void rke(float *, float *, int, float [],
             void (*)(int, float, float[]), float [], int,
             void (*)(int, float, float, float [], float []));
    float t,te,y[4],data[7];

    te=1.0;
    while (1) {
        y[1]=y[2]=0.0;
        y[3]=2.0;
        t=0.0;
        data[1]=data[2]=1.0e-5;
        rke(&t,&te,3,y,rhs,data,1,info);
        if (te != 1.0) break;
        te = -1.0;
    }
}
```

Output:

```
T =  1
Relative and absolute errors in x, y and z:
  RE(X)     RE(Y)     RE(Z)     AE(X)     AE(Y)     AE(Z)
 4.8e-07  1.5e-06  1.4e-06  1.2e-06  1.3e-05  1.2e-05
Number of integration steps performed :   9
Number of integration steps skipped   :   0
Number of integration steps rejected  :   5

T = -1
```

```
Relative and absolute errors in x, y and z:
  RE(X)     RE(Y)     RE(Z)     AE(X)     AE(Y)     AE(Z)
 4.5e-07  0.0e+00  2.2e-07  1.5e-07  0.0e+00  8.9e-08
Number of integration steps performed :  10
Number of integration steps skipped   :   0
Number of integration steps rejected  :   7
```

Function tested: **rk4a**

Example:

The solution of the differential equation

$$dy/dx = 1 - 2(x^2+y), \quad x \geq 0,$$
$$y = 0 \text{ at } x = 0,$$

is represented by the parabola $y = x(1-x)$. Find the value of x for which the curve of the solution intersects the line $y+x=0$.

```
#include <math.h>
#include <stdio.h>

float b(float x, float y)
{
   return x+y;
}

float fxy(float x, float y)
{
   return 1.0-2.0*(x*x+y);
}

void main ()
{
   void rk4a(float *, float, float (*)(float, float),
             float *, float, float (*)(float, float),
             float [], float [], int, int, int);
   float x,y,d[5],e[6];

   e[0]=e[1]=e[2]=e[3]=e[4]=e[5]=1.0e-4;
   rk4a(&x,0.0,b,&y,0.0,fxy,e,d,1,1,1);
   printf("x =  %e         Exactly : 2.00000\ny = %e\n"
             "y-x*(1-x)  = %e\n",x,y,y-x*(1-x));
}
```

Output:

```
x =  1.99995e+00        Exactly : 2.00000
y = -2.00006e+00
y-x*(1-x) = -2.10879e-04
```

Function tested: **rk4na**

Example:

Obtain the period of the solution of the van der Pol equation

$$dx_1/dt = x_2$$
$$dx_2/dt = 10(1-x_1^2)x_2-x_1, \quad t \geq 0.$$

```
#include <stdio.h>

float b(int n, float x[])
{
   return x[2];
}
```

```
float fxj(int n, int k, float x[])
{
    return ((k == 1) ? x[2] : 10.0*(1.0-x[1]*x[1])*x[2]-x[1]);
}

void main ()
{
    void rk4na(float [], float [], float (*)(int, float[]),
                  float (*)(int, int, float[]), float [], float [],
                  int, int, int, int);
    int j,first;
    float x0,e[8],xa[3],x[3],d[6];

    for (j=0; j<=5; j++) e[j]=0.1e-4;
    e[6]=e[7]=1.0e-4;
    printf("VAN DER POL\n\nEPS = %e\n\n"
           "The values of x[0],x[1],x[2],p:\n",e[0]);
    x0=xa[0]=xa[2]=0.0;
    xa[1]=2.0;
    printf(" %8.5f %8.5f %8.5f %8.5f\n",xa[0],xa[1],xa[2],x0);
    first=1;
    for (j=1; j<=4; j++) {
        rk4na(x,xa,b,fxj,e,d,first,2,0,1);
        x0=x[0]-x0;
        printf(" %8.5f %8.5f %8.5f %8.5f\n",x[0],x[1],x[2],x0);
        first=0;
        x0=x[0];
    }
}
```

Output:

```
VAN DER POL

EPS = 1.000000e-05

The values of x[0],x[1],x[2],p:
   0.00000   2.00000   0.00000   0.00000
   9.32386  -2.01429   0.00000   9.32386
  18.86305   2.01429   0.00000   9.53919
  28.40223  -2.01429  -0.00000   9.53918
  37.94142   2.01429   0.00000   9.53919
```

Function tested: **rk5na**

Example:

The van der Pol equation in the phase plane

$$dx_1/dx_0 = (10(1-x_0^2)x_1-x_0)/x_1$$

can be integrated by *rk5na*. The starting values are $x_0 = 2$, $x_1 = 0$. The integration proceeds until the next zero of x_1, then it continues until the next zero and so on until the fourth zero is encountered.

```
#include <math.h>
#include <stdio.h>

float b(int n, float x[])
{
    return x[1];
}

float fxj(int n, int k, float x[])
{
    return ((k == 0) ? x[1] : 10.0*(1.0-x[0]*x[0])*x[1]-x[0]);
}

void main ()
{
```

```
    void rk5na(float [], float [], float (*)(int, float[]),
                float (*)(int, int, float[]), float [], float [],
                int, int, int, int);
    int j,first;
    float e[6],xa[2],x[2],d[5];

    for (j=0; j<=3; j++) e[j]=1.0e-4;
    e[4]=e[5]=1.0e-5;
    d[1]=0.0;
    xa[0]=2.0;  xa[1]=0.0;
    j=0;
    printf("Results:\n     x[0]         x[1]          s\n"
           " %9.5f   %9.5f   %9.5f\n",xa[0],xa[1],fabs(d[1]));
    do {
        first=(j == 0);
        rk5na(x,xa,b,fxj,e,d,first,1,1,0);
        j++;
        printf(" %9.5f   %9.5f   %9.5f\n",x[0],x[1],fabs(d[1]));
    } while (j < 4);
}
```

Output:

```
Results:
     x[0]         x[1]          s
   2.00000      0.00000      0.00000
  -2.01429      0.00000     29.38738
   2.01429      0.00000     58.78843
  -2.01429     -0.00000     88.18951
   2.01429     -0.00000    117.59058
```

Function tested: **multistep**

Example:

Compute the solution at $x = 1$ and at $x = 10$ of the differential equations

$$dy_1/dx = 0.04(1 - y_1 - y_2) - y_1(10^4 y_2 + 3*10^7 y_1)$$
$$dy_2/dx = 3*10^7 y_1^2$$

with the initial conditions $y_1 = 0$ and $y_2 = 0$ at $x = 0$.

```
#include <stdio.h>

void der(float f[], int n, float x, float y[])
{
    float r;

    f[2]=r=3.0e7*y[1]*y[1];
    f[1]=0.04*(1-y[1]-y[2])-1.0e4*y[1]*y[2]-r;
}

int avail(int n, float x, float y[], float **jac)
{
    float r;

    jac[2][1]=r=6.0e7*y[1];
    jac[1][1] = -0.04-1.0e4*y[2]-r;
    jac[1][2] = -0.04-1.0e4*y[1];
    jac[2][2]=0.0;
    return 1;
}

void out(float h, int k, int n, float x, float y[])
{
    return;
}

void main ()
{
    float *allocate_real_vector(int, int);
```

```
    float **allocate_real_matrix(int, int, int, int);
    void free_real_vector(float *, int);
    void free_real_matrix(float **, int, int, int);
    int multistep(float *, float, float [], float, float,
                float [], float, int *, float [],
                void (*)(float [], int, float, float[]),
                int (*)(int, float, float [], float **), float **, int,
                int, void (*)(float, int, int, float, float []));
    int i,first;
    float x,xend,hmin,eps,y[13],ymax[3],*d,**jac;

    d=allocate_real_vector(-40,12);
    jac=allocate_real_matrix(1,2,1,2);
    hmin=1.0e-6;
    eps=1.0e-6;
    first=1;
    x=0.0;
    y[1]=y[2]=0.0;
    ymax[1]=0.0001;
    ymax[2]=1.0;
    printf("Delivers with hmin = %5.1e  and eps = %5.1e\n",hmin,eps);
    for (i=1; i<=10; i+=9) {
        xend=i;
        multistep(&x,xend,y,hmin,5,ymax,eps,&first,d,
                  der,avail,jac,1,2,out);
        printf("    %e    %e\n",y[1],y[2]);
    }
    free_real_vector(d,-40);
    free_real_matrix(jac,1,2,1);
}
```

Output:

```
Delivers with hmin = 1.0e-06  and eps = 1.0e-06
    3.07463e-05    3.35095e-02
    1.62340e-05    1.58613e-01
```

Function tested: **diffsys**

Example:

Solve the system of differential equations

$$y_1' = y_2$$
$$y_2' = y_1 + 2y_4 - \mu_1(y_1+\mu)/s_1 - \mu(y_1-\mu_1)/s_2$$
$$y_3' = y_4$$
$$y_4' = y_3 + 2y_2 - \mu_1 y_3/s_1 - \mu y_3/s_2$$

where

$$s_1 = \{(y_1+\mu)^2 + y_3^2\},\quad s_2 = \{(y_1-\mu_1)^2 + y_3^2\},\quad \mu_1 = 1-\mu.$$

This problem arises from the restricted problem of three bodies. When $\mu = 1/82.45$, the solution is a closed orbit with period $t = 6.192169331396$.

```
#include <math.h>
#include <stdio.h>

int passes,k;

void der(int n, float x, float y[], float dy[])
{
    float mu,mu1,y1,y2,y3,y4,s1,s2;

    mu=1.0/82.45;
    mu1=1.0-mu;
    passes++;
    y1=y[1];
    y2=dy[1]=y[2];
    y3=y[3];
```

```
    y4=dy[3]=y[4];
    s1=(y1+mu)*(y1+mu)+y3*y3;
    s2=(y1-mu1)*(y1-mu1)+y3*y3;
    s1 *= sqrt(s1);
    s2 *= sqrt(s2);
    dy[2]=y1+2.0*y4-mu1*(y1+mu)/s1-mu*(y1-mu1)/s2;
    dy[4]=y3-2.0*y2-mu1*y3/s1-mu*y3/s2;
}

void out(int n, float x, float xe, float y[], float s[])
{
    k++;
    if (x >= xe)
        printf(" %3d   %4d    %e   %e\n",k,passes,y[1],y[3]);
}

void main ()
{
    int i;
    float x,xe,tol,h0,y[5],s[5];

    printf("Results with DIFFSYS are :\n"
           "  K   DER.EV.      Y[1]           Y[3]\n");
    tol=1.0e-2;
    for (i=1; i<=2; i++) {
        tol *= 1.0e-2;
        passes=k=0;
        x=0.0;
        xe=6.192169331396;
        y[1]=1.2;
        y[2]=y[3]=0.0;
        y[4] = -1.04935750983;
        s[1]=s[2]=s[3]=s[4]=0.0;
        h0=0.2;
        diffsys(&x,xe,4,y,der,tol,tol,s,h0,out);
    }
}
```

Output:

```
Results with DIFFSYS are :
  K   DER.EV.      Y[1]           Y[3]
  30   2585     1.32403e+00   -3.19608e-02
  33   3414     1.20036e+00   -1.22485e-04
```

Function tested: **ark**

Example:

Compute the values of:

(1) $y(1)$ and $y(2)$ of the initial value problem

$$dy/dx = y - 2x/y, \quad y(0) = 1$$

and

(2) $u(0.6,0)$ of the Cauchy problem

$$du/dt = 0.5*du/dx, \quad u(0,x) = exp(-x^2).$$

```
#include <math.h>
#include <stdio.h>

void der1(int *m0, int *m, float *t, float v[])
{
    v[1] -= 2.0*(*t)/v[1];
}

void der2(int *m0, int *m, float *t, float v[])
{
    int j;
    float v1,v2,v3;
```

```
    v2=v[*m0];
    (*m0)++;
    (*m)--;
    v3=v[*m0];
    for (j=(*m0); j<=(*m); j++) {
        v1=v2;
        v2=v3;
        v3=v[j+1];
        v[j]=250.0*(v3-v1)/3.0;
    }
}

void out1(int *m0, int *m, float *t, float *te, float y[],
            float data[])
{
    if (*t == *te) {
        if (*t == 1.0)
            printf("\nProblem 1\n\n"
                "  x   integration steps  y(computed)     y(exact)\n");
        printf("%2.0f        %3.0f            %e   %e\n",
                *t,data[8],y[1],sqrt(2.0*(*t)+1));
        *te = 2.0;
    }
}

void out2(int *m0, int *m, float *t, float *te, float u[],
            float data[])
{
    if (fabs((*t)-0.6) < 1.0e-5)
        printf("\n\nProblem 2\n\n"
            " derivative calls   u(.6,0)(computed)    u(.6,0)exact\n"
            "     %4.0f              %e        %e\n",
            data[1]*data[8],u[0],exp(-0.09));
}

void main ()
{
    float *allocate_real_vector(int, int);
    void free_real_vector(float *, int);
    void ark(float *, float *, int *, int *, float [],
            void (*)(int *, int *, float *, float[]), float [],
            void (*)(int *, int *, float *, float *,
                        float [], float []));
    int m0,m,i;
    static float dat1[13]={3.0, 3.0, 1.0, 1.0, 1.0e-3, 1.0e-6,
                1.0e-6, 0.0, 0.0, 0.0, 1.0, 1.0, 0.5, 1.0/6.0};
    static float dat2[14]={4.0, 3.0, 0.0, 500.0/3.0, 0.0,
            -1.0, -1.0, 0.0, 0.0, 0.0, 1.0, 0.5, 1.0/6.0, 1.0/24.0};
    float t,te,y[2],*u,data[15];

    u=allocate_real_vector(-150,150);
    for (i=1; i<=13; i++) data[i]=dat1[i-1];
    t=0.0;
    y[1]=1.0;
    te=1.0;
    m0=m=1;
    ark(&t,&te,&m0,&m,y,der1,data,out1);
    for (i=1; i<=14; i++) data[i]=dat2[i-1];
    data[3]=sqrt(8.0);
    data[5]=data[3]/data[4];
    m0 = -150;
    m=150;
    t=0.0;
    u[0]=1.0;
    for (i=1; i<=m; i++) u[i]=u[-i]=exp(-(0.003*i)*(0.003*i));
    te=0.6;
    ark(&t,&te,&m0,&m,u,der2,data,out2);
    free_real_vector(u,-150);
}
```

Output:

```
Problem 1
```

```
 x   integration steps  y(computed)     y(exact)
 1          38          1.73205e+00   1.73205e+00
 2          55          2.23610e+00   2.23607e+00


Problem 2

 derivative calls    u(.6,0)(computed)    u(.6,0)exact
     144                9.13933e-01        9.13931e-01
```

Function tested: **efrk**

Example:

Consider the system of differential equations:

$$dy_1/dx = -y_1 + y_1y_2 + 0.99y_2$$
$$dy_2/dx = -1000*(-y_1 + y_1y_2 + y_2)$$

with the initial conditions $y_1 = 1$ and $y_2 = 0$ at $x = 0$. The solution at $x = 50$ is approximately $y_1 = 0.7658783202487$ and $y_2 = 0.4337103535768$. The following program shows some different calls of the procedure *efrk*.

Note that the computations are carried out in double precision by using the macro statement: `#define float double.`
From these results it appears that calls with *thirdorder* equals nonzero in *efrk* are less advisable.

```
#define float double

#include <math.h>
#include <stdio.h>

unsigned int passes;

void der(int m0, int m, float x, float y[])
{
    float y1,y2;

    y1=y[1];
    y2=y[2];
    y[1]=(y1+0.99)*(y2-1.0)+0.99;
    y[2]=1000.0*((1.0+y1)*(1.0-y2)-1.0);
    passes++;
}

void out(int m0, int m, float x, float xe, float y[],
            float *sigma, float *phi, float *diameter, int k,
            float *step, int r, int l)
{
    float s;

    s=(-1000.0*y[1]-1001.0+y[2])/2.0;
    *sigma=fabs(s-sqrt(s*s+10.0*(y[2]-1.0)));
    *diameter=2.0*(*step)*fabs(1000.0*
                    (1.99*y[2]-2.0*y[1]*(1.0-y[2])));
    if (x == 50.0)
        printf(" %2d  %2d  %4d  %5u   %e   %e\n",
                r,l,k,passes,y[1],y[2]);
}

void main ()
{
    void efrk(float *, float, int, int, float [], float *,
            float *, float *, void (*)(int, int, float, float[]),
            int *, float *, float, float, float [], int, float,
            void (*)(int, int, float, float, float [], float *,
                        float *, float *, int, float *, int, int));
    int k,r,l,i;
```

```
    float x,xe,sigma,phi,step,diameter,y[3],beta[7];

    printf("The results with EFRK are:\n\n"
        "  R   L    K   DER.EV.     Y[1]            Y[2]\n");
    phi=4.0*atan(1.0);
    beta[0]=beta[1]=1.0;
    for (r=1; r<=3; r++)
        for (l=1; l<=3; l++) {
            for (k=2; k<=r; k++) beta[k]=beta[k-1]/k;
            for (i=1; i<=2; i++) {
                step = ((i == 1) ? 1.0 : 0.1);
                passes=k=0;
                x=y[2]=0.0;
                y[1]=1.0;
                out(1,2,x,xe,y,&sigma,&phi,&diameter,k,&step,r,l);
                efrk(&x,50.0,1,2,y,&sigma,&phi,&diameter,der,&k,
                        &step,r,l,beta,r>=3,1.0e-3,out);
            }
            printf("\n");
        }
}
```

Output:

```
The results with EFRK are:

  R   L    K   DER.EV.     Y[1]            Y[2]
  1   1   237    474    7.65813e-01    4.33689e-01
  1   1   501   1002    7.65848e-01    4.33701e-01

  1   2    52    156    7.65571e-01    4.33615e-01
  1   2   501   1503    7.65848e-01    4.33701e-01

  1   3    52    208    7.65571e-01    4.33615e-01
  1   3   500   2000    7.65849e-01    4.33701e-01

  2   1  3315   9945    7.65878e-01    4.33710e-01
  2   1  1050   3150    7.65878e-01    4.33710e-01

  2   2   174    696    7.65878e-01    4.33710e-01
  2   2   501   2004    7.65878e-01    4.33709e-01

  2   3   161    805    7.65878e-01    4.33710e-01
  2   3   501   2505    7.65878e-01    4.33710e-01

  3   1  7005  28020    7.65878e-01    4.33713e-01
  3   1  3253  13012    7.65878e-01    4.33711e-01

  3   2  3733  18665    7.65878e-01    4.33711e-01
  3   2  3732  18660    7.65878e-01    4.33711e-01

  3   3  5392  32352    7.65878e-01    4.33711e-01
  3   3  5391  32346    7.65878e-01    4.33711e-01
```

Function tested: **efsirk**

Example:

Solve the differential equation:

$$dy/dx = -e^x\{y - ln(x)\} + 1/x$$

with the initial value $y(0.01) = ln(0.01)$ over the two ranges [0.01,0.4] and [0.4,8]. The analytic solution is $y(x) = ln(x)$.

The above equation is written in autonomous form (taking $y^{(2)}(x) = x$) as

$$dy^{(1)}/dx = -\ exp(y^{(2)})\{y^{(1)} - ln(y^{(2)})\} + 1/y^{(2)}$$
$$dy^{(2)}/dx = 1.$$

The Jacobian matrix J_n for this set of equations has elements

$$J_n^{(1,1)} = -exp(y^{(2)}),\quad J_n^{(1,2)} = -\{y^{(1)} - ln(y^{(2)}) - 1/y^{(2)}\}exp(y^{(2)}) - 1/(y^{(2)})^2$$

$J_n^{(2,1)} = J_n^{(2,2)} = 0$, and its eigenvalue with largest absolute value is $\delta_n = -exp(y^{(2)})$.

```
#include <math.h>
#include <stdio.h>

float lnx;

void der(int m, float y[], float *delta)
{
    float y2;

    y2=y[2];
    *delta = -exp(y2);
    lnx=log(y2);
    y[1]=(y[1]-lnx)*(*delta)+1.0/y2;
    y[2]=1.0;
}

void jac(int m, float **j, float y[], float *delta)
{
    float y2;

    y2=y[2];
    j[1][1]=(*delta);
    j[1][2]=(y[1]-lnx-1.0/y2)*(*delta)-1.0/(y2*y2);
    j[2][1]=j[2][2]=0.0;
}

void outp(float x, float xe, int m, float y[],
            float delta, float **j, int n)
{
    float y1;

    if (x == xe) {
        y1=y[1];
        lnx=log(x);
        printf("\n N =%3d    X =%4.1f    Y(X) = %7.5f"
            "     DELTA = %4.2f\n ABS.ERR. = %7.2e"
            "   REL.ERR. = %7.2e\n",
            n,x,y1,delta,fabs(y1-lnx),fabs((y1-lnx)/lnx));
    }
}

void main ()
{
    void efsirk(float *, float, int, float [], float *,
            void (*)(int, float[], float *),
            void (*)(int, float **, float [], float *),
            float **, int *, float, float, float, float, int,
            void (*)(float, float, int, float [],
                                float, float **, int));
    int n;
    float x,xe,delta,y[3],**j;

    j=allocate_real_matrix(1,2,1,2);
    printf("EFSIRK delivers:\n");
    xe=0.4;
    x=0.01;
    y[1]=log(0.01);
    y[2]=x;
    efsirk(&x,xe,2,y,&delta,der,jac,j,&n,1.0e-2,1.0e-2,0.005,1.5,
            0,outp);
    xe=8.0;
    efsirk(&x,xe,2,y,&delta,der,jac,j,&n,1.0e-2,1.0e-2,0.005,1.5,
            0,outp);
    free_real_matrix(j,1,2,1);
}
```

Output:

```
EFSIRK delivers:
```

```
N = 10     X = 0.4     Y(X) = -0.91093          DELTA = -1.44
ABS.ERR. = 5.4e-03     REL.ERR. = 5.8e-03

N = 92     X = 8.0     Y(X) = 2.07639     DELTA = -2962.96
ABS.ERR. = 3.1e-03     REL.ERR. = 1.5e-03
```

Function tested: **eferk**

Example:

Consider the system of differential equations:

$$dy_1/dx = -y_1 + y_1y_2 + 0.99y_2$$
$$dy_2/dx = -1000*(-y_1 + y_1y_2 + y_2)$$

with the initial conditions $y_1 = 1$ and $y_2 = 0$ at $x = 0$. The solution at $x = 50$ is approximately $y_1 = 0.7658783202487$ and $y_2 = 0.4337103535768$. The following program shows some different calls of the procedure *eferk*.

```
#include <math.h>
#include <stdio.h>

int passes,pasjac;

void der(int m, float y[])
{
    float y1,y2;

    y1=y[1];
    y2=y[2];
    y[1]=(y1+0.99)*(y2-1.0)+0.99;
    y[2]=1000.0*((1.0+y1)*(1.0-y2)-1.0);
    passes++;
}

void jacobian(int m, float **j, float y[], float *sigma)
{
    j[1][1]=y[2]-1.0;
    j[1][2]=0.99+y[1];
    j[2][1]=1000.0*(1.0-y[2]);
    j[2][2] = -1000.0*(1.0+y[1]);
    *sigma=fabs(j[2][2]+j[1][1]-sqrt((j[2][2]-j[1][1])*
                (j[2][2]-j[1][1])+4.0*j[2][1]*j[1][2]))/2.0;
    pasjac++;
}

void out(float x, float xe, int m, float y[], float **j, int k)
{
    if (x == 50.0)
        printf("%3d  %4d    %4d     %e   %e\n",
               k,passes,pasjac,y[1],y[2]);
}

void main ()
{
    float **allocate_real_matrix(int, int, int, int);
    void free_real_matrix(float **, int, int, int);
    void eferk(float *, float, int, float [], float *, float,
            void (*)(int, float[]), float **,
            void (*)(int, float **, float [], float *),
            int *, int, int, float, float, float, float, int,
            void (*)(float, float, int, float [], float **, int));
    int i,k;
    float x,xe,sigma,phi,tol,y[3],**j;

    j=allocate_real_matrix(1,2,1,2);
    printf("The results with EFERK are:\n\n"
        " K   DER.EV. JAC.EV.      Y[1]           Y[2]\n");
    phi=4.0*atan(1.0);
    tol=1.0;
```

```
    for (i=1; i<=4; i++) {
        passes=pasjac=0;
        x=y[2]=0.0;
        y[1]=1.0;
        xe=50.0;
        eferk(&x,xe,2,y,&sigma,phi,der,j,jacobian,&k,1,1,tol,tol,
                1.0e-6,50.0,0,out);
        tol *= 1.0e-1;
    }
    free_real_matrix(j,1,2,1);
}
```

Output:

```
The results with EFERK are:

 K   DER.EV. JAC.EV.     Y[1]          Y[2]
164   328     164     7.65878e-01   4.33834e-01
171   342     171     7.65879e-01   4.33487e-01
195   390     195     7.65878e-01   4.33596e-01
265   530     265     7.65878e-01   4.33713e-01
```

Function tested: **liniger1vs**

Example:

Consider the system of differential equations:

$$dy_1/dx = -y_1 + y_1y_2 + 0.99y_2$$

$$dy_2/dx = -1000*(-y_1 + y_1y_2 + y_2)$$

with the initial conditions $y_1 = 1$ and $y_2 = 0$ at $x = 0$. The solution at $x = 50$ is approximately $y_1 = 0.7658783202487$ and $y_2 = 0.4337103535768$. The following program shows some integration of this problem with variable and constant stepsizes.

```
#include <math.h>
#include <stdio.h>

void f(int m, float a[], float *sigma)
{
    float a1,a2;

    a1=a[1];
    a2=a[2];
    a[1]=(a1+0.99)*(a2-1.0)+0.99;
    a[2]=1000.0*((1.0+a1)*(1.0-a2)-1.0);
}

void jacobian(int m, float **j, float y[], float *sigma)
{
    j[1][1]=y[2]-1.0;
    j[1][2]=0.99+y[1];
    j[2][1]=1000.0*(1.0-y[2]);
    j[2][2] = -1000.0*(1.0+y[1]);
    *sigma=fabs(j[2][2]+j[1][1]-sqrt((j[2][2]-j[1][1])*
                (j[2][2]-j[1][1])+4.0*j[2][1]*j[1][2]))/2.0;
}

void out(float x, float xe, int m, float y[],
            float sigma, float **j, float info[])
{
    if (x == 50.0)
        printf("%3.0f %3.0f %3.0f %3.0f %3.0f %3.0f  %5.1e  %e   %e\n",
                info[1],info[2],info[3],info[4],info[5],info[6],info[9],
                y[1],y[2]);
}

void main ()
{
    float **allocate_real_matrix(int, int, int, int);
```

```
    void free_real_matrix(float **, int, int, int);
    void liniger1vs(float *, float, int, float [], float *,
                void (*)(int, float[], float *), float **,
                void (*)(int, float **, float [], float *),
                int, float, float, float, float,  float [],
                void (*)(float, float, int, float [], float,
                         float **, float []));
    int i,itmax;
    float x,sigma,reta,y[3],**j,info[10];

    j=allocate_real_matrix(1,2,1,2);
    printf("The results with LINIGER1VS are:\n\n");
    reta=1.0;
    for (i=1; i<=3; i++) {
        reta *= 1.0e-2;
        x=y[2]=0.0;
        y[1]=1.0;
        liniger1vs(&x,50.0,2,y,&sigma,f,j,jacobian,10,0.1,50.0,
                   reta,reta,info,out);
    }
    printf("\n");
    reta = -1.0;
    for (i=1; i<=3; i++) {
        reta *= 1.0e-2;
        x=y[2]=0.0;
        y[1]=1.0;
        liniger1vs(&x,50.0,2,y,&sigma,f,j,jacobian,10,0.1,1.0,
                   reta,reta,info,out);
    }
    free_real_matrix(j,1,2,1);
}
```

Output:

```
The results with LINIGER1VS are:

 17  21   8   2   1   2  2e-02  7.72046e-01   4.35681e-01
 13  25  23   2   0   2  2e-02  7.67399e-01   4.34198e-01
163 326 163   2   0   2  2e-02  7.65986e-01   4.33745e-01

 50  52   1   0  50   2  0e+00  7.66670e-01   4.33081e-01
 50 104   3   0  50   3  0e+00  7.66183e-01   4.33811e-01
 50 152  12   0  50   4  0e+00  7.66185e-01   4.33809e-01
```

Function tested: **liniger2**

Example:

Consider the system of differential equations:

$$dy_1/dx = -y_1 + y_1y_2 + 0.99y_2$$

$$dy_2/dx = -1000*(-y_1 + y_1y_2 + y_2)$$

with the initial conditions $y_1 = 1$ and $y_2 = 0$ at $x = 0$. The solution at $x = 50$ is approximately $y_1 = 0.7658783202487$ and $y_2 = 0.4337103535768$. The following program shows some different calls of *liniger2*.

```
#include <math.h>
#include <stdio.h>

int passes,pasjac;

float f(int m, float y[], int i, float *sigma1, float *sigma2)
{
    if (i == 1)
        return (y[1]+0.99)*(y[2]-1.0)+0.99;
    else {
        passes++;
        return 1000.0*((1.0+y[1])*(1.0-y[2])-1.0);
    }
```

```
}

void jacobian(int m, float **j, float y[],
                    float *sigma1, float *sigma2)
{
    j[1][1]=y[2]-1.0;
    j[1][2]=0.99+y[1];
    j[2][1]=1000.0*(1.0-y[2]);
    j[2][2] = -1000.0*(1.0+y[1]);
    *sigma1=fabs(j[2][2]+j[1][1]-sqrt((j[2][2]-j[1][1])*
                    (j[2][2]-j[1][1])+4.0*j[2][1]*j[1][2]))/2.0;
    pasjac++;
}

int evaluate1(int i)
{
    return (i == 1);
}

int evaluate2(int i)
{
    return 1;
}

void out(float x, float xe, int m, float y[],
            float sigma1, float sigma2, float **j, int k)
{
    if (x == 50.0)
        printf("%3d  %4d    %4d     %e   %e\n",
                k,passes,pasjac,y[1],y[2]);
}

void main ()
{
    float **allocate_real_matrix(int, int, int, int);
    void free_real_matrix(float **, int, int, int);
    void liniger2(float *, float, int, float [], float *, float *,
            float (*)(int, float[], int, float *, float *),
            int (*)(int), float **,
            void (*)(int, float **, float [], float *, float *),
            int *, int, float, float, float,
            void (*)(float, float, int, float [], float, float,
                        float **, int));
    int i,k,itmax;
    float x,sigma1,sigma2,step,y[3],**j;

    j=allocate_real_matrix(1,2,1,2);
    printf("The results with LINIGER2 (second order) are:\n"
        " K   DER.EV. JAC.EV.     Y[1]          Y[2]\n");
    for (i=1; i<=2; i++) {
        step = (i == 1) ? 10.0 : 1.0;
        for (itmax=1; itmax<=3; itmax += 2) {
            passes=pasjac=0;
            x=y[2]=0.0;
            y[1]=1.0;
            sigma2=0.0;
            liniger2(&x,50.0,2,y,&sigma1,&sigma2,f,evaluate1,j,
                        jacobian,&k,itmax,step,1.0e-4,1.0e-4,out);
        }
    }
    printf("\nThe results with LINIGER2 (third order) are:\n"
        " K   DER.EV. JAC.EV.     Y[1]          Y[2]\n");
    for (i=1; i<=2; i++) {
        step = (i == 1) ? 10.0 : 1.0;
        for (itmax=1; itmax<=3; itmax += 2) {
            passes=pasjac=0;
            x=y[2]=0.0;
            y[1]=1.0;
            sigma2=0.0;
            liniger2(&x,50.0,2,y,&sigma1,&sigma2,f,evaluate2,j,
                        jacobian,&k,itmax,step,1.0e-4,1.0e-4,out);
        }
    }
    free_real_matrix(j,1,2,1);
```

```
}
```

Output:

```
The results with LINIGER2 (second order) are:
 K   DER.EV. JAC.EV.      Y[1]          Y[2]
  5     5       5     7.65612e-01   4.33970e-01
  5    15       5     7.65755e-01   4.33671e-01
 50    50      50     7.65865e-01   4.33710e-01
 50   101      50     7.65877e-01   4.33710e-01

The results with LINIGER2 (third order) are:
 K   DER.EV. JAC.EV.      Y[1]          Y[2]
  5     5       5     7.65612e-01   4.33970e-01
  5    15      15     7.65882e-01   4.33712e-01
 50    50      50     7.65865e-01   4.33710e-01
 50   101     101     7.65879e-01   4.33711e-01
```

Function tested: **gms**

Example:

Solve the system of differential equations:

$$dy^{(1)}/dx = -1000y^{(1)}(y^{(1)}+y^{(2)}-1.999987)$$
$$dy^{(2)}/dx = -2500y^{(2)}(y^{(1)}+y^{(2)}-2)$$

with the initial value $y^{(1)}(0) = y^{(2)}(0) = 1$ over the range [0,50]. The The Jacobian matrix J_n for this set of equations has elements

$$J_n^{(1,1)} = 1999.987 - 1000(2y^{(1)}+y^{(2)}), \quad J_n^{(1,2)} = -1000y^{(2)}$$
$$J_n^{(2,1)} = -2500y^{(2)}, \quad J_n^{(2,2)} = 2500(2-y^{(1)}-y^{(2)}).$$

```
#include <math.h>
#include <stdio.h>

void der(int r, float y[], float *delta)
{
    float y1,y2;

    y1=y[1];
    y2=y[2];
    y[1] = -1000.0*y1*(y1+y2-1.999987);
    y[2] = -2500.0*y2*(y1+y2-2.0);
}

void jac(int r, float **j, float y[], float *delta)
{
    float y1,y2;

    y1=y[1];
    y2=y[2];
    j[1][1] = 1999.987-1000.0*(2.0*y1+y2);
    j[1][2] = -1000.0*y1;
    j[2][1] = -2500.0*y2;
    j[2][2] = 2500.0*(2.0-y1-2.0*y2);
}

void outp(float x, float xe, int r, float y[], float delta,
            int n, int jev, int lu)
{
    float ye1,ye2;

    if (x == 50.0) {
        ye1=0.5976546988;
        ye2=1.4023434075;
        printf(" X = %2.0f     N = %4d     JEV = %3d     LU = %4d\n"
                " Y1 = %e     REL.ERR. = %5.2e\n"
                " Y2 = %e     REL.ERR. = %5.2e\n",
                x,n,jev,lu,y[1],fabs((y[1]-ye1)/ye1),
```

```
                y[2],fabs((y[2]-ye2)/ye2));
    }
}

void main ()
{
    void gms(float *, float, int, float [], float, float,
            float, float *, void (*)(int, float [], float *),
            void (*)(int, float **, float [], float *),
            float, float, int *, int *, int *, int, int,
            void (*)(float, float, int, float [], float,
                        int, int, int));
    int n,jev,lu;
    float x,y[3],delta;

    printf("The results with GMS are:\n");
    y[1]=y[2]=1.0;
    x=0.0;
    delta=0.0;
    gms(&x,50.0,2,y,0.01,0.001,0.5,&delta,der,jac,1.0e-5,1.0e-5,
            &n,&jev,&lu,0,0,outp);
}
```

Output:

```
The results with GMS are:
 X = 50     N = 1851     JEV = 608     LU = 1742
 Y1 = 5.97714e-01     REL.ERR. = 1.00e-04
 Y2 = 1.40228e+00     REL.ERR. = 4.2e-05
```

Function tested: **impex**

Example:

Consider the autonomous system of differential equations:

$$dy_1/dx = 0.2(y_2 - y_1)$$
$$dy_2/dx = 10y_1 - (60 - y_3/8)y_2 + y_3/8$$
$$dy_3/dx = 1$$

with initial conditions $y_1=y_2=y_3=0$ at $x=0$. The solution at several points in the interval [0,400] may be obtained by the following program. (The solution at $x=400$ is $y_1=22.24222011$, $y_2=27.11071335$)

```
#include <math.h>
#include <stdio.h>

int nfe,nje,point;
float print[6];

void f(float t, float y[], float f1[], int n)
{
    nfe++;
    f1[1]=0.2*(y[2]-y[1]);
    f1[2]=10.0*y[1]-(60.0-0.125*y[3])*y[2]+0.125*y[3];
    f1[3]=1.0;
}

int available(float t, float y[], float **a, int n)
{
    nje++;
    a[1][1] = -0.2;
    a[1][2] = 0.2;
    a[1][3]=a[3][1]=a[3][2]=a[3][3]=0.0;
    a[2][1]=10.0;
    a[2][2]=0.125*y[3]-60.0;
    a[2][3]=0.125*(1.0+y[2]);
    return 1;
}
```

```
void update(float sw[], float r1[], int n)
{
    int i;
    float s1,s2;

    for (i=1; i<=n; i++) {
        s1=1.0/sw[i];
        s2=fabs(r1[i]);
        if (s1 < s2) sw[i]=1.0/s2;
    }
}

void control(float *tp, float t, float h, float hnew, float **y,
             float err[], int n, float tend)
{
    int i;
    float c[6],*x,s,s2,s3,s4;

    x=allocate_real_vector(1,n);
    while (1) {
        s=(t-(*tp))/h;
        s2=s*s;
        s3=s2*s;
        s4=s3*s;
        c[3]=(s2-s)/2.0;
        c[4] = -s3/6.0+s2/2.0-s/3.0;
        c[5]=s4/24.0-s3/4.0+11.0*s2/24.0-s/4.0;
        for (i=1; i<=n; i++)
            x[i]=y[1][i]-s*y[2][i]+c[3]*y[3][i]+
                    c[4]*y[4][i]+c[5]*y[5][i];
        printf(" %6.2f  %7.2e  %e   %e   %4d  %3d\n",
            *tp,err[3],x[1],x[2],nfe,nje);
        if (*tp >= tend) break;
        point++;
        *tp = print[point];
        if (*tp > t) break;
    }
    free_real_vector(x,1);
}

void main ()
{
    void impex(int, float, float, float [],
               void (*)(float, float [], float [], int),
               int (*)(float, float [], float **, int),
               float, float, int, float, float [],
               void (*)(float [], float [], int), int *fail,
               void (*)(float *, float, float, float, float **,
                        float [], int, float));
    void dupvec(int, int, int, float [], float []);
    float vecvec(int, int, int, float [], float []);
    void elmvec(int, int, int, float [], float [], float);
    int n,fail,i,it;
    float t,tend,eps,hmax,l,h2,y[4],sw[4],f1[4],f2[4],z[4],x[4],
          n1,n2;

    printf("The results with IMPEX are:\n\n");
    n=3;
    nje=nfe=0;
    t=0.0;
    tend=400.0;
    eps=1.0e-5;
    hmax=400.0;
    y[1]=y[2]=y[3]=0.0;
    sw[1]=sw[2]=sw[3]=1.0;
    print[1]=0.1;  print[2]=1.0;  print[3]=10.0;  print[4]=100.0;
    print[5]=400.0;
    dupvec(1,n,0,z,y);
    for (i=1; i<=n; i++)
        x[i] = (y[i] == 0.0) ? eps : (1.0+eps)*y[i];
    n1=sqrt(vecvec(1,n,0,x,x))*eps;
    f(t,x,f1,n);
    for (it=1; it<=5; it++) {
        f(t,z,f2,n);
```

```
        elmvec(1,n,0,f2,f1,-1.0);
        n2=n1/sqrt(vecvec(1,n,0,f2,f2));
        dupvec(1,n,0,z,x);
        elmvec(1,n,0,z,f2,n2);
    }
    f(t,z,f2,n);
    elmvec(1,n,0,f2,f1,-1.0);
    l=sqrt(vecvec(1,n,0,f2,f2))/n1;
    h2=pow(eps*320.0,1.0/5.0)/(4.0*l);
    printf("EPS = %e\nInterval of integration = (%1.0f,%3.0f)\n"
        "Maximally allowed stepsize = %e\n\nLipschconst = %e\n"
        "Starting stepsize = %e\nFunctional eval = %2d\n\n"
        "     X     ERROR         Y[1]            Y[2]         NFE  NJE\n",
        eps,t,tend,hmax,l,h2,nfe);
    impex(n,t,tend,y,f,available,h2,hmax,0,eps,sw,update,&fail,
          control);
    printf("\nNumber of functional evaluations =%4d\n"
        "Number of Jacobian evaluations   = %3d\n",nfe,nje);
}
```

Output:

```
The results with IMPEX are:

EPS = 1.000000e-05
Interval of integration = (0,400)
Maximally allowed stepsize = 4.00000e+02

Lipschconst = 6.01554e+01
Starting stepsize = 1.31733e-03
Functional eval =  7

     X     ERROR        Y[1]           Y[2]          NFE  NJE
   0.00   0.0e+00  0.00000e+00   0.00000e+00     7    0
   0.10   1.4e-06  1.49652e-06   1.73902e-04    57    4
   1.00   1.8e-06  1.91084e-04   2.08362e-03    98    9
  10.00   3.5e-06  1.30151e-02   2.34488e-02   142   11
 100.00   3.8e-06  3.06299e-01   3.27548e-01   253   14
 400.00   1.7e-05  2.22389e+01   2.71076e+01   581   34

Number of functional evaluations = 581
Number of Jacobian evaluations   =  34
```

Function tested: **modifiedtaylor**

Example:

Solve the differential equation:

$$Du(t) = -e^t\{u(t) - ln(t)\} + 1/t \qquad (1)$$

where $D=d/dt$, with initial condition $u(0.01) = ln(0.01)$ (the analytic solution is $u(t) = ln(t)$) over the ranges $[0.01,e]$ and $[e,e^2]$, using a fourth order modified Taylor series method with third order exact stability polynomial with coefficients $\beta_j = 1/j!$ (j=0,...,3) $\beta_4 = 0.018455702$ (whose range constant is $\beta(4) = 6.025$). The required derivatives of $u(t)$ are given by

$$D^2u(t) = e^t\{ln(t) + 1/t - u(t) - Du(t)\} - 1/t^2$$
$$D^3u(t) = e^t\{ln(t) + 2/t - u(t) - 2Du(t) - D^2u(t) - 1/t^2\} + 2/t^3$$
$$D^4u(t) = D^3u(t) - 2(1 + 3/t)/t^3 + e^t[\{1 - (2 - 2/t)/t\}/t - Du(t) - 2D^2u(t) - D^3u(t)].$$

The spectral radius with respect to the eigenvalues lying in the left half-plane of the Jacobian matrix derived from equation (1) is simply e^t.

```
#include <math.h>
#include <stdio.h>

float expt,lnt,c0,c1,c2,c3;

void der(float t, int m0, int m, int i, float a[])
```

```
{
    if (i == 1) {
        expt=exp(t);
        lnt=log(t);
        c0=a[0];
        c1 = a[0] = -expt*c0+1.0/t+expt*lnt;
    }
    if (i == 2) c2=a[0]=expt*(lnt+1.0/t-c0-c1)-1.0/t/t;
    if (i == 3)
        c3=a[0]=expt*(lnt+2.0/t-c0-2.0*c1-c2-1.0/t/t)+2.0/t/t/t;
    if (i == 4)
        a[0]=c3-2.0*(1.0+3.0/t)/t/t/t+expt*((1.0-(2.0-2.0/t)/t)/t-
                c1-c2*2.0-c3);
}

float sigma(float t, int m0, int m)
{
    return exp(t);
}

float aeta(float t, int m0, int m)
{
    return 1.0e-5;
}

float reta(float t, int m0, int m)
{
    return 1.0e-4;
}

void op(float t, float te, int m0, int m, float u[],
            int k, float eta, float rho)
{
    if (t == te)
        printf("\nNumber of steps: %3d\n"
            "Solution:  T = %8.6f    U(T) = %8.6f\n",k,t,u[0]);
}

void main ()
{
    float *allocate_real_vector(int, int);
    void free_real_vector(float *, int);
    void modifiedtaylor(float *, float, int, int, float [],
            float (*)(float, int, int), float,
            void (*)(float, int, int, int, float []),
            int *, float [], float, int, float (*)(float, int, int),
            float (*)(float, int, int), float *, float *,
            void (*)(float, float, int, int, float [],
                    int, float, float));
    int j,k;
    float t,te,eta,rho,u[1],*data;

    data=allocate_real_vector(-2,4);
    printf("The results with MODIFIEDTAYLOR are:\n");
    data[-2]=4.0;  data[-1]=3.0;  data[0]=6.025;  data[1]=1.0;
    data[2]=0.5;  data[3]=1.0/6.0;  data[4]=0.018455702;
    t=u[0]=1.0e-2;
    k=0;
    for (j=1; j<=2; j++) {
        te = (j == 1) ? exp(1.0) : te*te;
        modifiedtaylor(&t,te,0,0,u,sigma,1.0e-4,der,&k,data,1.5,
                1,aeta,reta,&eta,&rho,op);
    }
    free_real_vector(data,-2);
}
```

Output:

```
The results with MODIFIEDTAYLOR are:

Number of steps:  46
Solution:  T = 2.718282    U(T) = 1.000028
```

```
Number of steps: 424
Solution:  T = 7.389056    U(T) = 1.999989
```

Function tested: **eft**

Example:

Solve the differential equation:

$$Du(t) = -e^t\{u(t) - ln(t)\} + 1/t \qquad (1)$$

where $D=d/dt$, with initial condition $u(0.01) = ln(0.01)$ (the analytic solution is $u(t) = ln(t)$) over the ranges $[0.01,e]$ and $[e,e^2]$, using a third order, first order consistent, exponentially fitted Taylor series method. The equation is solved eight times with different absolute and relative tolerances. The derivatives of (1) are as described in the test program of *modifiedtaylor*.

```
#include <math.h>
#include <stdio.h>

float expt,lnt,u0,u1,u2,accuracy;

void derivative(float t, int m0, int m, int i, float u[])
{
    if (i == 1) {
        expt=exp(t);
        lnt=log(t);
        u0=u[0];
        u1=u[0]=expt*(lnt-u0)+1.0/t;
    } else if (i == 2)
        u2=u[0]=expt*(lnt-u0-u1+1.0/t)-1.0/t/t;
    else
        u[0]=expt*(lnt-u0-2.0*u1-u2+2.0/t-1.0/t/t)+2.0/t/t/t;
}

float sigma(float t, int m0, int m)
{
    return exp(t);
}

float diameter(float t, int m0, int m)
{
    return 2.0*exp(2.0*t/3.0);
}

float aeta1(float t, int m0, int m)
{
    return accuracy/10.0;
}

float aeta2(float t, int m0, int m)
{
    return accuracy/10.0*((t < 3.0) ? 1.0 : exp(2.0*(t-3.0)));
}

float reta1(float t, int m0, int m)
{
    return accuracy;
}

float reta2(float t, int m0, int m)
{
    return accuracy*((t < 3.0) ? 1.0 : exp(2.0*(t-3.0)));
}

void out(float t, float te, int m0, int m, float u[],
           int k, float eta, float rho)
{
    if (t == te) printf(" %3d  %e  ",k,u[0]);
}
```

```
void main ()
{
    void eft(float *, float, int, int, float [],
            float (*)(float, int, int), float,
            float (*)(float, int, int),
            void (*)(float, int, int, int, float []),
            int *, float, int,
            float (*)(float, int, int), float (*)(float, int, int),
            float *, float *, float, float *,
            void (*)(float, float, int, int, float [],
                        int, float, float));
    int j,k,l;
    float t,te,te1,te2,eta,rho,pi,hs,u[1];

    printf("The results with EFT are:\n"
        "  K     U(TE1)         K     U(TE2)       RETA\n");
    pi=4.0*atan(1.0);
    te1=exp(1.0);
    te2=exp(2.0);
    accuracy=1.0;
    for (j=1; j<=4; j++) {
        accuracy *= 1.0e-1;
        t=0.01;
        u[0]=log(t);
        k=0;
        hs=0.0;
        for (l=1; l<=2; l++) {
            te = (l == 1) ? te1 : te2;
            eft(&t,te,0,0,u,sigma,pi,diameter,derivative,&k,
                1.5,2,aeta1,reta1,&eta,&rho,1.0e-4,&hs,out);
        }
        printf(" %6.1e\n",accuracy);
    }
    printf("\nWith relaxed accuracy conditions for t > 3 :\n"
        "  K     U(TE1)         K     U(TE2)       RETA\n");
    accuracy=1.0;
    for (j=1; j<=4; j++) {
        accuracy *= 1.0e-1;
        t=0.01;
        u[0]=log(t);
        k=0;
        hs=0.0;
        for (l=1; l<=2; l++) {
            te = (l == 1) ? te1 : te2;
            eft(&t,te,0,0,u,sigma,pi,diameter,derivative,&k,
                1.5,2,aeta2,reta2,&eta,&rho,1.0e-4,&hs,out);
        }
        printf(" %6.1e\n",accuracy);
    }
}
```

Output:

```
The results with EFT are:
  K     U(TE1)        K     U(TE2)        RETA
 15  1.00156e+00     42  2.00008e+00     1e-01
 22  1.00141e+00     52  2.00007e+00   1.0e-02
 37  1.00024e+00     94  2.00001e+00   1.0e-03
 60  1.00003e+00    174  2.00000e+00   1.0e-04

With relaxed accuracy conditions for t > 3 :
  K     U(TE1)        K     U(TE2)        RETA
 15  1.00156e+00     42  2.00008e+00     1e-01
 22  1.00141e+00     50  2.00005e+00   1.0e-02
 37  1.00024e+00     69  2.00002e+00   1.0e-03
 60  1.00003e+00    103  2.00000e+00   1.0e-04
```

Function tested: **rk2**

Example:

The van del Pol equation

$$d^2y/dx^2 = 10(1-y^2)(dy/dx) - y, \quad x \geq 0,$$
$$y = 2,\ dy/dx = 0, \quad x = 0,$$

can be integrated by *rk2*. At the points x=*9.32386578, 18.86305405, 28.40224162, 37.94142918* the derivative dy/dx vanishes.

Note that the computations are carried out in double precision by using the macro statement: `#define float double.`

```
#define float double

#include <math.h>
#include <stdio.h>

float fxyz(float x, float y, float z)
{
    return 10.0*(1.0-y*y)*z-y;
}

void main ()
{
    void rk2(float *, float, float, float *, float, float *, float,
            float (*)(float, float, float), float [], float [], int);
    int i,fi;
    float x,y,z,e[5],d[6],
            b[4]={9.32386578, 18.86305405, 28.40224162, 37.94142918};

    e[1]=e[2]=e[3]=e[4]=e[5]=1.0e-8;
    printf("RK2 delivers :\n");
    for (i=0; i<=3; i++) {
        fi=(b[i] < 10.0);
        rk2(&x,0.0,b[i],&y,2.0,&z,0.0,fxyz,e,d,fi);
        printf("  X = %+e  Y = %+e    DY/DX = %+e\n",x,y,z);
    }
}
```

Output:

```
RK2 delivers :
  X = +9.32387e+00  Y = -2.01429e+00    DY/DX = +7.55013e-08
  X = +1.88631e+01  Y = +2.01429e+00    DY/DX = -7.09799e-06
  X = +2.84022e+01  Y = -2.01429e+00    DY/DX = +1.27100e-05
  X = +3.79414e+01  Y = +2.01429e+00    DY/DX = -1.83014e-05
```

Function tested: **rk2n**

Example:

The second order differential equation

$$d^2y_1/dx^2 = -5(y_1 + dy_2/dx) + y_2,$$
$$d^2y_2/dx^2 = -5(y_2 + dy_1/dx) + y_1, \quad x \geq 0,$$
$$y_1 = dy_2/dx = 1, \quad y_2 = dy_1/dx = 0, \quad x = 0$$

with analytic solution

$$y_1 = (5/6)e^{-x} + e^{-2x} - (1/2)e^{-3x} - (1/3)e^{-4x},$$
$$y_2 = (5/6)e^{-x} - e^{-2x} + (1/2)e^{-3x} - (1/3)e^{-4x}$$

can be integrated by *rk2n* from 0 to 5 with 1,2,3,4 as reference points.

```
#include <stdio.h>

float fxyzj(int n, int j, float x, float y[], float z[])
{
    return -5.0*(y[j]+z[j])+((j == 1) ? y[2] : y[1]);
}
```

```
void main ()
{
    void rk2n(float *, float, float, float [], float [], float [],
            float [], float (*)(int, int, float, float [], float []),
            float [], float [], int, int);
    int k,fi;
    float b,x,expx,y[3],ya[3],z[3],za[3],e[9],d[8];

    printf("Results from RK2N :\n");
    for (k=1; k<=8; k++) e[k]=1.0e-5;
    ya[1]=za[2]=1.0;
    ya[2]=za[1]=0.0;
    b=1.0;
    do {
        fi=(b == 1.0);
        rk2n(&x,0.0,b,y,ya,z,za,fxyzj,e,d,fi,2);
        expx=exp(-x);
        ya[1] = -expx*(expx*(expx*(expx/3.0+0.5)-1.0)-5.0/6.0);
        ya[2] = -expx*(expx*(expx*(expx/3.0-0.5)+1.0)-5.0/6.0);
        za[1]=expx*(expx*(expx*(expx/0.75+1.5)-2.0)-5.0/6.0);
        za[2]=expx*(expx*(expx*(expx/0.75-1.5)+2.0)-5.0/6.0);
        printf("\n  X = %6.4f\nY[1]-YEXACT[1] = %+6.2e    "
            "Y[2]-YEXACT[2] = %+6.2e\nZ[1]-ZEXACT[1] = %+6.2e    "
            "Z[2]-ZEXACT[2] = %+6.2e\n",x,y[1]-ya[1],y[2]-ya[2],
            z[1]-za[1],z[2]-za[2]);
        b += 1.0;
    } while (b < 5.0);
}
```

Output:

```
Results from RK2N :

  X = 1.0000
Y[1]-YEXACT[1] = +6.0e-08    Y[2]-YEXACT[2] = +1.5e-08
Z[1]-ZEXACT[1] = -8.9e-08    Z[2]-ZEXACT[2] = -6.0e-08

  X = 2.0000
Y[1]-YEXACT[1] = -7.5e-08    Y[2]-YEXACT[2] = +7.5e-09
Z[1]-ZEXACT[1] = +3.4e-07    Z[2]-ZEXACT[2] = -6.7e-08

  X = 3.0000
Y[1]-YEXACT[1] = -8.6e-08    Y[2]-YEXACT[2] = -2.2e-08
Z[1]-ZEXACT[1] = +4.5e-07    Z[2]-ZEXACT[2] = +8.9e-08

  X = 4.0000
Y[1]-YEXACT[1] = -6.2e-08    Y[2]-YEXACT[2] = -4.7e-08
Z[1]-ZEXACT[1] = +3.4e-07    Z[2]-ZEXACT[2] = +2.0e-07
```

Function tested: **rk3**

Example:

Compute the solution of

$$y'' = xy, \quad y(0) = 0, \quad y'(0) = 1$$

at interval [0,1].

```
#include <math.h>
#include <stdio.h>

float fxy(float x, float y)
{
    return x*y;
}

void main ()
{
    void rk3(float *, float, float, float *, float, float *, float,
            float (*)(float, float), float [], float [], int);
```

```
    int n,i,fi;
    float x,y,z,e[6],d[6],x3,s,term,
            b[4]={0.25, 0.50, 0.75, 1.00};

    e[1]=e[3]=1.0e-6;
    e[2]=e[4]=1.0e-6;
    printf("RK3 delivers :\n");
    for (i=0; i<=3; i++) {
        fi=(b[i] < 0.3);
        rk3(&x,0.0,b[i],&y,0.0,&z,1.0,fxy,e,d,fi);
        x3=x*x*x;
        term=x;
        s=0.0;
        n=3;
        do {
            s += term;
            term=term*x3/n/(n+1);
            n += 3;
        } while (fabs(term) > 1.0e-14);
        printf("Y-YEXACT= %+e    X= %4.2f    Y= %8.6f\n",y-s,x,y);
    }
}
```

Output:

```
RK3 delivers :
Y-YEXACT= +0.00000e+00     X= 0.25     Y= 0.250326
Y-YEXACT= +5.96046e-08     X= 0.50     Y= 0.505224
Y-YEXACT= +0.00000e+00     X= 0.75     Y= 0.776633
Y-YEXACT= -1.19209e-07     X= 1.00     Y= 1.085340
```

Function tested: **rk3n**

Example:

Solve the second order differential equation

$$y_1'' = y_2, \quad y_2'' = -y_1$$

with $y_1(0) = y_2(0) = 1$, $y_1'(0) = y_2'(0) = 0$, the exact solutions of which are

$$y_1(x) = cosh(x/\sqrt{2})*cos(x/\sqrt{2}) + sinh(x/\sqrt{2})*sin(x/\sqrt{2})$$
$$y_2(x) = cosh(x/\sqrt{2})*cos(x/\sqrt{2}) - sinh(x/\sqrt{2})*sin(x/\sqrt{2}).$$

These equations can be integrated by *rk3n* because first derivatives are absent from them. Integration is carried out over the intervals [ν,ν+1], ν=0,1,...,4.

```
#include <math.h>
#include <stdio.h>

float fxyj(int n, int j, float x, float y[])
{
    return ((j == 1) ? y[2] : -y[1]);
}

void main ()
{
    void rk3n(float *, float, float, float [], float [], float [],
                float [], float (*)(int, int, float, float[]),
                float [], float [], int, int);
    int k,fi,i,n,j;
    float b,x,y[3],ya[3],z[3],e[9],d[8],x2,term;

    printf("Results from RK3N :\n");
    for (k=1; k<=8; k++) e[k]=1.0e-5;
    fi=1;
    y[1]=y[2]=1.0;
    z[1]=z[2]=0.0;
    b=0.0;
    do {
        b += 1.0;
```

```
        rk3n(&x,0.0,b,y,y,z,z,fxyj,e,d,fi,2);
        ya[1]=ya[2]=0.0;
        term=1.0;
        x2=x*x*0.5;
        n=1;
        do {
            for (i=1; i<=2; i++) {
                j=(i+n-2)/2;
                ya[i] += term*((j%2 == 0) ? 1 : -1);
            }
            term=term*x2/n/(n*2-1);
            n++;
        } while (fabs(term) > 1.0e-14);
        printf(" ABS(YEXACT[1]-Y[1])+ABS(YEXACT[2]-Y[2]) = %e\n",
                fabs(y[1]-ya[1])+fabs(ya[2]-y[2]));
        fi=0;
    } while (b < 5.0);
}
```

Output:

```
Results from RK3N :
 ABS(YEXACT[1]-Y[1])+ABS(YEXACT[2]-Y[2]) = 2.98023e-07
 ABS(YEXACT[1]-Y[1])+ABS(YEXACT[2]-Y[2]) = 3.57628e-07
 ABS(YEXACT[1]-Y[1])+ABS(YEXACT[2]-Y[2]) = 1.43051e-06
 ABS(YEXACT[1]-Y[1])+ABS(YEXACT[2]-Y[2]) = 4.29153e-06
 ABS(YEXACT[1]-Y[1])+ABS(YEXACT[2]-Y[2]) = 1.33514e-05
```

Function tested: **arkmat**

Example:

Solve the initial boundary value problem

$$\frac{\partial^2 u(t,x,y)}{\partial t^2} = \frac{\partial^2 u(t,x,y)}{\partial x^2} + \frac{\partial^2 u(t,x,y)}{\partial y^2} \tag{1}$$

on the domain $0 \le t \le 1$, $0 \le x \le \pi$, $0 \le y \le 1$, with initial conditions

$$u(0,x,y) = \sin(x)*\cos(\pi y/2), \quad (\partial/\partial t)u(t,x,y)_{t=0}=0 \quad (0 \le x \le \pi,\ 0 \le y \le 1)$$

and boundary values

$$u(t,x,0) = \sin(x)*\cos\{(1+\pi^2/4)t\}, \quad u(t,x,1) = 0 \quad (0 \le x \le \pi)$$

$$u(t,0,y) = u(t,\pi,y) = 0 \quad (0 \le y \le 1)$$

over the range $0 \le t \le 1$. The analytic solution of the above initial boundary value problem is

$$u(t,x,y) = \cos\{(1+\pi^2/4)t\}*\sin(x)*\cos(\pi y/2).$$

Equation (1) may be reduced to a form of the first order in t by setting

$$v(t,x,y) = Du(t,x,y)$$

where $D=\partial/\partial t$, when

$$Dv(t,x,y) = \frac{\partial^2 u(t,x,y)}{\partial x^2} + \frac{\partial^2 u(t,x,y)}{\partial y^2}.$$

The above first order form is solved by discretizing the variables x,y, and including the construction of the boundary values in the integration of the resulting set of first order differential equations in t, making use of the conditions

$$D^2u(t,x,1) = 0 \quad (0 \le x \le \pi)$$

$$D^2u(t,0,y) = D^2u(t,\pi,y) = 0 \quad (0 \le y \le 1)$$

and further use of the symmetry property (easily deduced from the form of the original equation and the initial and boundary conditions) that $u(t,x,-y) = u(t,x,y)$, to assist in the determination of $D^2u(t,x,0)$ $(0 \le x \le \pi)$.

Taking $\Delta x = \pi/(m-1)$, $\Delta y = 1/(n-1)$ and for $i=1,...,n$; $j=1,...,m$

$$U^{(i,j)}(t) = u(t,(j-1)\Delta x,(i-1)\Delta y)$$
$$U^{(n+i,j)}(t) = v(t,(j-1)\Delta x,(i-1)\Delta y)$$

the equations

$$DU^{(i,j)} = U^{(n+i,j)} \quad (i=1,...,n;\ j=1,...,m)$$
$$DU^{(n+i,j)} = \{U^{(i,j+1)} - 2U^{(i,j)} + U^{(i,j-1)}\}/(\Delta x)^2 + \{U^{(i+1,j)} - 2U^{(i,j)} + U^{(i-1,j)}\}/(\Delta y)^2 \quad (i=2,...,n-1;\ j=2,...,m-1)$$
$$DU^{(n+i,1)} = DU^{(n+i,m)} = 0 \quad (i=1,...,n)$$
$$DU^{(2n,j)} = 0 \quad (j=2,...,m-1)$$
$$DU^{(n+1,j)} = \{U^{(1,j+1)} - 2U^{(1,j)} + U^{(1,j-1)}\}/(\Delta x)^2 + 2\{U^{(2,j)} - U^{(1,j)}\}/(\Delta y)^2 \quad (j=2,...,m-1)$$

for the components of the $2n \times m$ matrix U are derived. The initial values for this set of equations are

$$U^{(i,j)} = \sin\{(j-1)\Delta x\}*\cos\{\pi(i-1)\Delta y/2\}, \quad U^{(n+1,j)} = 0 \quad (i=1,...,n;\ j=1,...,m).$$

The above set of equations with $m=n=10$ are solved by use of *arkmat* with parameters *type* = 3, *order* = 2 and constant step length in t equal to 0.1.

```
int tel;
float h1k,h2k,hpi,h1,h2;

#include <stdio.h>
#include <math.h>

void der(int m, int n, float t, float **u, float **du)
{
    int i,j;

    n=n/2;
    for (i=2; i<=n-1; i++)
        for (j=2; j<=m-1; j++) {
            du[i][j]=u[i+n][j];
            du[i+n][j]=(u[i][j+1]-2.0*u[i][j]+u[i][j-1])/h1k+
                        (u[i+1][j]-2.0*u[i][j]+u[i-1][j])/h2k;
        }
    for (j=1; j<=m; j += m-1) {
        inimat(n+1,n+n,j,j,du,0.0);
        for (i=1; i<=n; i++) du[i][j]=u[n+1][j];
    }
    for (i=1; i<=n; i += n-1)
        for (j=2; j<=m-1; j++) {
            du[i][j]=u[i+n][j];
            if (i == 1)
                du[n+1][j]=(u[1][j+1]-2.0*u[1][j]+u[1][j-1])/h1k+
                            (2.0*u[2][j]-2.0*u[1][j])/h2k;
            else
                du[2*n][j]=0.0;
        }
}

void out(float t, float te, int m, int n, float **u, int type,
            int order, float *spr)
{
    int i;

    tel++;
    if (t == te) {
        for (i=1; i<=10; i++)
            printf("%6.3f  %6.3f  %9.6f   %9.6f\n",
                (i-1)*h1,(i-1)*h2,u[i][i],
                sin(h1*(i-1))*cos(hpi*h2*(i-1))*cos(t*sqrt(1.0+hpi*hpi)));
        printf("\nThe number of integration steps: %2d\n"
            " Type is %1d   Order is %1d\n",tel,type,order);
    }
```

```
}

void main ()
{
    float **allocate_real_matrix(int, int, int, int);
    void free_real_matrix(float **, int, int, int);
    void arkmat(float *, float, int, int, float **,
        void (*)(int, int, float, float **, float **),
        int, int *, float *,
        void (*)(float, float, int, int, float **, int, int, float *));
    int i,j,n,m,typ,orde;
    float **u,t,te,cos1,spr;

    u=allocate_real_matrix(1,20,1,10);
    hpi=2.0*atan(1.0);
    h2=1.0/9.0;
    h1=(2.0*hpi)/9.0;
    n=m=10;
    h1k=h1*h1;
    h2k=h2*h2;
    tel=0;
    t=0.0;
    te=1.0;
    for (j=1; j<=m; j++) u[n][j]=sin(h1*(j-1));
    for (i=1; i<=n; i++) {
        cos1=cos(h2*hpi*(i-1));
        for (j=1; j<=m; j++) u[i][j]=u[n][j]*cos1;
    }
    inimat(n+1,n+n,1,m,u,0.0);
    typ=3;
    orde=2;
    spr=80.0;
    arkmat(&t,te,m,n+n,u,der,typ,&orde,&spr,out);
    free_real_matrix(u,1,20,1);
}
```

Output:

```
 0.000   0.000   0.000000   -0.000000
 0.349   0.111  -0.095201   -0.096735
 0.698   0.222  -0.170723   -0.173474
 1.047   0.333  -0.211983   -0.215398
 1.396   0.444  -0.213228   -0.216664
 1.745   0.556  -0.178920   -0.181802
 2.094   0.667  -0.122388   -0.124360
 2.443   0.778  -0.062138   -0.063139
 2.793   0.889  -0.016787   -0.017057
 3.142   1.000   0.000000   -0.000000

The number of integration steps: 10
 Type is 3   Order is 2
```

Function tested: **femlagsym**

Example:

Compute approximations to the solution of the boundary value problem

$$-D\{e^x Dy(x)\} + \cos(x)y(x) = e^x\{\sin(x) - \cos\{x]\} + \sin(2x)/2$$
$$y(0) = y(\pi) = 0$$

where $D=d/dx$, at the points $x_i = i\pi/n$, $i=0,...,n$, (approximating the solution on a uniform grid) for the six cases in which $n = 10, 20$ and the order of the method used is 2, 4, 6. The analytical solution to the above problem is $y(x) = \sin(x)$.

```
#include <math.h>
#include <stdio.h>

float r(float x)
```

```
{
    return cos(x);
}

float p(float x)
{
    return exp(x);
}

float f(float x)
{
    return exp(x)*(sin(x)-cos(x))+sin(2.0*x)/2.0;
}

void main ()
{
    void femlagsym(float [], float [], int, float (*)(float),
                   float (*)(float), float (*)(float), int, float []);
    int n,i,order;
    float pi,x[21],y[21],e[7],rho,d;

    printf("FEMLAGSYM delivers:\n");
    for (n=10; n<=20; n += 10) {
        e[1]=e[4]=1.0;
        e[2]=e[3]=e[5]=e[6]=0.0;
        pi=3.14159265358979;
        for (i=0; i<=n; i++) x[i]=pi*i/n;
        printf("N=%2d\n",n);
        for (order=2; order<=6; order += 2) {
            femlagsym(x,y,n,p,r,f,order,e);
            rho=0.0;
            for (i=0; i<=n; i++) {
                d=fabs(y[i]-sin(x[i]));
                if (rho < d) rho=d;
            }
            printf("     ORDER=%1d     MAX.ERROR= %7.3e\n",order,rho);
        }
    }
}
```

Output:

```
FEMLAGSYM delivers:
N=10
     ORDER=2     MAX.ERROR= 1.36e-02
     ORDER=4     MAX.ERROR= 7.56e-05
     ORDER=6     MAX.ERROR= 8.74e-08
N=20
     ORDER=2     MAX.ERROR= 3.41e-03
     ORDER=4     MAX.ERROR= 4.91e-06
     ORDER=6     MAX.ERROR= 1.77e-07
```

Function tested: **femlag**

Example:

Compute approximations to the solution of the boundary value problem

$$-D^2y(x) + e^xy(x) = sin(x)\{1+e^x\}$$

$$y(0) = y(\pi) = 0$$

where $D=d/dx$, at the points $x_i = i\pi/n$, $i=0,...,n$, (approximating the solution on a uniform grid) for the six cases in which $n = 10$, 20 and the order of the method used is 2, 4, 6. The analytical solution to the above problem is $y(x) = sin(x)$.

```
#include <math.h>
#include <stdio.h>

float r(float x)
{
```

```
    return exp(x);
}

float f(float x)
{
    return sin(x)*(1.0+exp(x));
}

void main ()
{
    void femlag(float [], float [], int, float (*)(float),
                        float (*)(float), int, float []);
    int n,i,order;
    float pi,x[21],y[21],e[7],rho,d;

    printf("FEMLAG delivers:\n");
    for (n=10; n<=20; n += 10) {
        e[1]=e[4]=1.0;
        e[2]=e[3]=e[5]=e[6]=0.0;
        pi=3.14159265358979;
        for (i=0; i<=n; i++) x[i]=pi*i/n;
        printf("N=%2d\n",n);
        for (order=2; order<=6; order += 2) {
            femlag(x,y,n,r,f,order,e);
            rho=0.0;
            for (i=0; i<=n; i++) {
                d=fabs(y[i]-sin(x[i]));
                if (rho < d) rho=d;
            }
            printf("     ORDER=%1d     MAX.ERROR= %7.3e\n",order,rho);
        }
    }
}
```

Output:

```
FEMLAG delivers:
N=10
     ORDER=2     MAX.ERROR= 1.60e-03
     ORDER=4     MAX.ERROR= 1.54e-05
     ORDER=6     MAX.ERROR= 1.19e-07
N=20
     ORDER=2     MAX.ERROR= 4.01e-04
     ORDER=4     MAX.ERROR= 9.93e-07
     ORDER=6     MAX.ERROR= 8.74e-08
```

Function tested: **femlagspher**

Example:

Solve the boundary value problem

$$-(y'x^{nc})'/x^{nc} + y = 1 - x^4 + (12 + 4*nc)x^2$$

$$0 < x < 1; \quad y'(0) = y(1) = 0.$$

The analytical solution is $y(x) = 1 - x^4$. We approximate the solution on a uniform grid, i.e. $x_i=i/n$, i=0,...,n, with n = 10, 20 and the order of the method used is 2 and 4.

```
#include <math.h>
#include <stdio.h>

int nc;

float r(float x)
{
    return 1.0;
}

float f(float x)
{
```

```
    return (12+4*nc)*x*x+1.0-x*x*x*x;
}

void main ()
{
    void femlagspher(float [], float [], int, int, float (*)(float),
                     float (*)(float), int, float []);
    int n,i,order;
    float x[21],y[21],e[7],rho,d;

    printf("FEMLAGSPHER delivers:\n");
    for (n=10; n<=20; n += 10)
        for (nc=0; nc<=2; nc++) {
            e[2]=e[4]=1.0;
            e[1]=e[3]=e[5]=e[6]=0.0;
            for (i=0; i<=n; i++) x[i]=(float)(i)/(float)(n);
            printf("N= %2d    NC=%2d\n",n,nc);
            for (order=2; order<=4; order += 2) {
                femlagspher(x,y,n,nc,r,f,order,e);
                rho=0.0;
                for (i=0; i<=n; i++) {
                    d=fabs(y[i]-1.0+x[i]*x[i]*x[i]*x[i]);
                    if (rho < d) rho=d;
                }
                printf("     ORDER=%1d    MAX.ERROR= %7.3e\n",order,rho);
            }
        }
}
```

Output:

```
FEMLAGSPHER delivers:
N= 10    NC= 0
     ORDER=2    MAX.ERROR= 4.37e-03
     ORDER=4    MAX.ERROR= 3.10e-06
N= 10    NC= 1
     ORDER=2    MAX.ERROR= 1.42e-02
     ORDER=4    MAX.ERROR= 5.51e-05
N= 10    NC= 2
     ORDER=2    MAX.ERROR= 2.46e-02
     ORDER=4    MAX.ERROR= 1.26e-04
N= 20    NC= 0
     ORDER=2    MAX.ERROR= 1.09e-03
     ORDER=4    MAX.ERROR= 2.98e-07
N= 20    NC= 1
     ORDER=2    MAX.ERROR= 3.53e-03
     ORDER=4    MAX.ERROR= 3.87e-06
N= 20    NC= 2
     ORDER=2    MAX.ERROR= 6.10e-03
     ORDER=4    MAX.ERROR= 9.48e-06
```

Function tested: **femlagskew**

Example:

Compute approximations to the solution of the boundary value problem

$$-y'' + \cos(x)y' + e^x y = \sin(x)\{1 + e^x\} + \{\cos(x)\}^2$$

$$0 < x < \pi; \quad y(0) = y(\pi) = 0.$$

The analytical solution is $y(x) = \sin(x)$. We approximate the solution on a uniform grid, i.e. $x_i = i\pi/n$, i=0,...,n, with n = 10, 20 and the order of the method used is 2, 4 and 6.

```
#include <math.h>
#include <stdio.h>

float q(float x)
{
    return cos(x);
}
```

```
float r(float x)
{
    return exp(x);
}

float f(float x)
{
    return sin(x)*(1.0+exp(x))+cos(x)*cos(x);
}

void main ()
{
    void femlagskew(float [], float [], int, float (*)(float),
                    float (*)(float), float (*)(float), int, float []);
    int n,i,order;
    float pi,x[21],y[21],e[7],rho,d;

    printf("FEMLAGSKEW delivers:\n");
    for (n=10; n<=20; n += 10) {
        e[1]=e[4]=1.0;
        e[2]=e[3]=e[5]=e[6]=0.0;
        pi=3.14159265358979;
        for (i=0; i<=n; i++) x[i]=pi*i/n;
        printf("N=%2d\n",n);
        for (order=2; order<=6; order += 2) {
            femlagskew(x,y,n,q,r,f,order,e);
            rho=0.0;
            for (i=0; i<=n; i++) {
                d=fabs(y[i]-sin(x[i]));
                if (rho < d) rho=d;
            }
            printf("     ORDER=%1d     MAX.ERROR= %7.3e\n",order,rho);
        }
    }
}
```

Output:

```
FEMLAGSKEW delivers:
N=10
     ORDER=2     MAX.ERROR= 2.95e-03
     ORDER=4     MAX.ERROR= 2.56e-05
     ORDER=6     MAX.ERROR= 1.42e-07
N=20
     ORDER=2     MAX.ERROR= 7.55e-04
     ORDER=4     MAX.ERROR= 1.65e-06
     ORDER=6     MAX.ERROR= 9.41e-08
```

Function tested: **femhermsym**

Example:

Compute approximations to the solution of the boundary value problem

$$y'''' - (\cos(x)y')' + e^x y = \sin(x)\{1 + e^x + 2\cos(x)\}$$

$$0 < x < \pi; \quad y(0) = y(\pi) = 0; \quad y'(0) = 1; \quad y'(\pi) = -1.$$

The analytical solution is $y(x) = \sin(x)$. We approximate the solution on a uniform grid, i.e. $x_i=i\pi/n$, i=0,...,n, with n = 5, 10 and the order of the method used is 4, 6 and 8.

```
#include <math.h>
#include <stdio.h>

float p(float x)
{
    return 1.0;
}

float q(float x)
{
```

```
    return cos(x);
}

float r(float x)
{
    return exp(x);
}

float f(float x)
{
    return sin(x)*(1.0+exp(x)+2.0*cos(x));
}

void main ()
{
    void femhermsym(float [], float [], int, float (*)(float),
            float (*)(float), float (*)(float), float (*)(float),
            int, float []);
    int n,i,order;
    float pi,x[11],y[19],e[5],rho1,rho2,d1,d2;

    printf("FEMHERMSYM delivers:\n");
    for (n=5; n<=10; n += 5) {
        e[1]=e[3]=0.0;  e[2]=1.0;  e[4] = -1.0;
        pi=3.14159265358979;
        for (i=0; i<=n; i++) x[i]=pi*i/n;
        printf("N=%2d\n",n);
        for (order=4; order<=8; order += 2) {
            femhermsym(x,y,n,p,q,r,f,order,e);
            rho1=rho2=0.0;
            for (i=1; i<=n-1; i++) {
                d1=fabs(y[2*i-1]-sin(x[i]));
                if (rho1 < d1) rho1=d1;
                d2=fabs(y[2*i]-cos(x[i]));
                if (rho2 < d2) rho2=d2;
            }
            printf("     ORDER=%1d\n"
                "            MAX. ABS(Y[2*I-1]-Y(X[I])) = %7.3e\n"
                "            MAX. ABS(Y[2*I]-Y'(X[I]))  = %7.3e\n",
                order,rho1,rho2);
        }
    }
}
```

Output:

```
FEMHERMSYM delivers:
N= 5
     ORDER=4
            MAX. ABS(Y[2*I-1]-Y(X[I])) = 4.82e-04
            MAX. ABS(Y[2*I]-Y'(X[I]))  = 4.54e-04
     ORDER=6
            MAX. ABS(Y[2*I-1]-Y(X[I])) = 7.55e-06
            MAX. ABS(Y[2*I]-Y'(X[I]))  = 2.82e-06
     ORDER=8
            MAX. ABS(Y[2*I-1]-Y(X[I])) = 4.81e-06
            MAX. ABS(Y[2*I]-Y'(X[I]))  = 4.74e-06
N=10
     ORDER=4
            MAX. ABS(Y[2*I-1]-Y(X[I])) = 3.00e-05
            MAX. ABS(Y[2*I]-Y'(X[I]))  = 3.67e-05
     ORDER=6
            MAX. ABS(Y[2*I-1]-Y(X[I])) = 2.89e-05
            MAX. ABS(Y[2*I]-Y'(X[I]))  = 3.25e-05
     ORDER=8
            MAX. ABS(Y[2*I-1]-Y(X[I])) = 2.21e-05
            MAX. ABS(Y[2*I]-Y'(X[I]))  = 2.46e-05
```

Function tested: **nonlinfemlagskew**

Example:

Compute approximations to the solution of the boundary value problem

$$(y'x^2)'/x^2 = exp(y) + exp(y') - exp(1-x^2) - exp(2x) - 6$$

$$0 < x < 1; \quad y'(0) = y(1) = 0;$$

The analytical solution is $y(x) = 1 - x^2$. We approximate the solution on a uniform grid, i.e. $x_i = i/n$, $i=0,...,n$.

```
#include <math.h>
#include <stdio.h>

int nc;

float f(float x, float y, float z)
{
    return exp(y)+exp(z)-exp(1.0-x*x)-exp(-2.0*x)-2.0-2*nc;
}

float fy(float x, float y, float z)
{
    return exp(y);
}

float fz(float x, float y, float z)
{
    return exp(z);
}

void main ()
{
    void nonlinfemlagskew(float [], float [], int,
                float (*)(float, float, float),
                float (*)(float, float, float),
                float (*)(float, float, float), int, float []);
    int n,i;
    float x[51],y[51],e[7],rho,d;

    printf("NONLINFEMLAGSKEW delivers:\n");
    for (nc=0; nc<=2; nc++)
        for (n=25; n<=50; n += 25) {
            e[2]=e[4]=1.0;
            e[1]=e[3]=e[5]=e[6]=0.0;
            for (i=0; i<=n; i++) {
                x[i]=(float)(i)/(float)(n);
                y[i]=0.0;
            }
            printf(" N=%2d     NC=%1d",n,nc);
            nonlinfemlagskew(x,y,n,f,fy,fz,nc,e);
            rho=0.0;
            for (i=0; i<=n; i++) {
                d=fabs(y[i]-1.0+x[i]*x[i]);
                if (rho < d) rho=d;
                }
            printf("     MAX.ERROR= %7.3e\n",rho);
        }
}
```

Output:

```
NONLINFEMLAGSKEW delivers:
 N=25     NC=0     MAX.ERROR= 2.47e-04
 N=50     NC=0     MAX.ERROR= 6.19e-05
 N=25     NC=1     MAX.ERROR= 1.41e-03
 N=50     NC=1     MAX.ERROR= 3.99e-04
 N=25     NC=2     MAX.ERROR= 2.44e-03
 N=50     NC=2     MAX.ERROR= 7.02e-04
```

Function tested: **richardson**

Example:

Solve the system of equations in the 144 unknowns $U_{j,l}$ (j,l=1,...,12)

$$U_{1,j} = 0, \quad U_{12,j} = \pi^2(j-1)^2h^2 \qquad (j=1,...,12)$$
$$U_{l,1} = 0, \quad U_{l,12} = \pi^2(l-1)^2h^2 \qquad (l=1,...,12)$$
$$U_{j,l-1} + U_{j-1,l} + U_{j+1,l} + U_{j,l+1} - 4U_{j,l} = 2h^2(x_l^2+y_j^2) \qquad (j,l=2,...,11)$$

where $x_l = (l-1)h$, $y_j = (j-1)h$ and $h = \pi/11$, by use of *richardson* with parameters a=0.163, b=7.83 and n=50. The $U_{j,l}$ are clearly approximations to the solution of the partial differential equation

$$\{(\partial/\partial x)^2 + (\partial/\partial y)^2\}\hat{u}(x,y) = 2(x^2+y^2)$$

with boundary values $\hat{u}(x,0) = 0$, $\hat{u}(x,\pi) = \pi^2x^2$ $(0 \le x \le \pi)$, $\hat{u}(0,y) = 0$, $\hat{u}(\pi,y) = \pi^2y^2$ $(0 \le y \le \pi)$ which has the analytical solution $\hat{u}(x,y) = x^2y^2$ (approximately $U_{j,l} = \hat{u}\{(l-1)h,(j-1)h\}$).

```
#include <stdio.h>

float h,h2;

void residual(int lj, int uj, int ll, int ul, float **u)
{
    int ujmin1,ulmin1,ljplus1,j,l;
    float u2,u1[12];

    ujmin1=uj-1;
    ulmin1=ul-1;
    ljplus1=lj+1;
    for (j=lj; j<=uj; j++) {
        u1[j]=u[j][ll];
        u[j][ll]=0.0;
    }
    for (l=ll+1; l<=ulmin1; l++) {
        u1[lj]=u[lj][l];
        u[lj][l]=0.0;
        for (j=ljplus1; j<=ujmin1; j++) {
            u2=u[j][l];
            u[j][l]=(4.0*u2-u1[j-1]-u1[j]-u[j+1][l]-u[j][l+1])+
                        2.0*((j*h)*(j*h)+(l*h)*(l*h))*h2;
            u1[j]=u2;
        }
        u[uj][l]=0.0;
    }
    for (j=lj; j<=uj; j++) u[j][ul]=0.0;
}

void out1(float **u, int lj, int uj, int ll, int ul, int *n,
            float discr[], int k, float rateconv, float domeigval)
{
    if (k == *n)
        printf("  K     DISCR[1]      DISCR[2]      RATECONV\n"
            " %2d  %e  %e  %e\n",k,discr[1],discr[2],rateconv);
}

void main ()
{
    float **allocate_real_matrix(int, int, int, int);
    void free_real_matrix(float **, int, int, int);
    void richardson(float **, int, int, int, int,
            int, void (*)(int, int, int, int, float **),
            float, float, int *, float [], int *, float *, float *,
            void (*)(float **, int, int, int, int, int *, float [],
                        int, float, float));
    int j,l,lj,uj,ll,ul,n,k;
    float pi,domeigval,rateconv,a,b,discr[3],**u;

    u=allocate_real_matrix(0,11,0,11);
    printf("RICHARDSON delivers:\n\n");
    pi=3.14159265358979;
    lj=0;  uj=11;  ll=0;  ul=11;  n=50;
```

```
    a=0.163;  b=7.83;
    h=pi/(uj-lj);
    h2=h*h;
    for (j=lj; j<=uj; j++)
        for (l=ll; l<=ul; l++)
            u[j][l] = (j==lj || j==uj || l==ll || l==ul) ?
                            (j*h)*(j*h)*(l*h)*(l*h) : 1.0;
    richardson(u,lj,uj,ll,ul,1,residual,a,b,&n,discr,&k,&rateconv,
                &domeigval,out1);
    free_real_matrix(u,0,11,0);
}
```

Output:

```
RICHARDSON delivers:

  K     DISCR[1]      DISCR[2]      RATECONV
 50  1.40216e-04  4.62895e-05  2.92251e-01
```

Function tested: **elimination**

Example:

Solve the system of equations described in the documentation of the test program for *richardson*. *richardson* is first called precisely as in that documentation but with the value allocated to *a* being 0.326. *elimination* is then called. Note that the computations are carried out in double precision by using the macro statement: #define float double.

```
#define float double

#include <stdio.h>

int nn;
float h,h2,d1,d2;

void residual(int lj, int uj, int ll, int ul, float **u)
{
    int ujmin1,ulmin1,ljplus1,j,l;
    float u2,u1[12];

    ujmin1=uj-1;
    ulmin1=ul-1;
    ljplus1=lj+1;
    for (j=lj; j<=uj; j++) {
        u1[j]=u[j][ll];
        u[j][ll]=0.0;
    }
    for (l=ll+1; l<=ulmin1; l++) {
        u1[lj]=u[lj][l];
        u[lj][l]=0.0;
        for (j=ljplus1; j<=ujmin1; j++) {
            u2=u[j][l];
            u[j][l]=(4.0*u2-u1[j-1]-u1[j]-u[j+1][l]-u[j][l+1])+
                            2.0*((j*h)*(j*h)+(l*h)*(l*h))*h2;
            u1[j]=u2;
        }
        u[uj][l]=0.0;
    }
    for (j=lj; j<=uj; j++) u[j][ul]=0.0;
}

void out2(float **u, int lj, int uj, int ll, int ul, int *p,
            float discr[], int k, float rateconv, float domeigval)
{
    if (k == *p)
        printf(" %2d  %e   %e   %e\n",k,discr[1],discr[2],rateconv);
}

void out1(float **u, int lj, int uj, int ll, int ul, int *n,
```

```
          float discr[], int k, float rateconv, float domeigval)
{
   if (k == 0)
      d1=d2=1.0;
   else {
      d2=d1;
      d1=domeigval;
      *n = (fabs((d1-d2)/d2) < 1.0e-4) ? k : nn;
      if (k == *n)
         printf("  K    DISCR[1]       DISCR[2]       RATECONV\n"
            " %2d  %e   %e   %e\n",k,discr[1],discr[2],rateconv);
   }
}

void main ()
{
   float **allocate_real_matrix(int, int, int, int);
   void free_real_matrix(float **, int, int, int);
   void richardson(float **, int, int, int, int,
         int, void (*)(int, int, int, int, float **),
         float, float, int *, float [], int *, float *, float *,
         void (*)(float **, int, int, int, int, int *, float [],
                  int, float, float));
   void elimination(float **, int, int, int, int,
         void (*)(int, int, int, int, float **),
         float, float, int *, float [], int *, float *, float *,
         void (*)(float **, int, int, int, int, int *, float [],
                  int, float, float));
   int j,l,lj,uj,ll,ul,n,p,k;
   float pi,domeigval,rateconvr,rateconve,rateconv,a,b,discr[3],**u;

   u=allocate_real_matrix(0,11,0,11);
   printf("RICHARDSON and ELIMINATION deliver:\n\n");
   pi=3.14159265358979;
   lj=0;  uj=11;  ll=0;  ul=11;  n=50;
   a=0.326;  b=7.83;
   h=pi/(uj-lj);
   h2=h*h;
   for (j=lj; j<=uj; j++)
      for (l=ll; l<=ul; l++)
         u[j][l] = (j==lj || j==uj || l==ll || l==ul) ?
                     (j*h)*(j*h)*(l*h)*(l*h) : 1.0;
   nn=n;
   richardson(u,lj,uj,ll,ul,1,residual,a,b,&n,discr,&k,&rateconv,
               &domeigval,out1);
   rateconvr=rateconv;
   printf("\n dominant eigenvalue:  %e\n\n",domeigval);
   elimination(u,lj,uj,ll,ul,residual,a,b,&p,discr,&k,&rateconv,
               &domeigval,out2);
   rateconve=rateconv;
   nn=n+p;
   printf("\nTotal number of iterations: %2d\n"
      "Rate of convergence with respect to\n"
      "   the zeroth iterand of RICHARDSON:   %e\n",
      nn,(n*rateconvr+p*rateconve)/nn);
   free_real_matrix(u,0,11,0);
}
```

Output:

```
RICHARDSON and ELIMINATION deliver:

  K    DISCR[1]       DISCR[2]       RATECONV
 38  1.15236e-01   2.05237e-02   2.16041e-01

 dominant eigenvalue:  1.62087e-01

  7  5.17283e-05   1.79648e-05   1.05355e+00

Total number of iterations: 45
Rate of convergence with respect to
   the zeroth iterand of RICHARDSON:   3.46320e-01
```

Function tested: **peide**

Example:

The following initial value problem arising from biochemistry

$$y_1(0) = 1.0, \quad y_2(0) = 0.0 \qquad (1)$$

$$dy_1/dt = -(1-y_2)y_1 + q_2 y_2 \qquad (2)$$

$$dy_2/dt = q_1\{(1-y_2)y_1 - (q_2+q_3)y_2\}$$

is obtained. Setting $q_i = exp(p_i)$, i=1,2,3, equations (2) become

$$dy_1/dt = -(1-y_2)y_1 + exp(p_2)y_2 = f_1(t,y,p) \qquad (3)$$

$$dy_2/dt = exp(p_1)\{(1-y_2)y_1 - (exp(p_2) + exp(p_3))y_2\} = f_2(t,y,p).$$

Data have been obtained from formulae (1,2) with $q_1=1000$, $q_2=0.99$, $q_3=0.01$ (i.e. $p_1=6.907755$, $p_2=-0.01005034$, $p_3=-4.6051702$) by tabulating the values of the function $y_2(t)$ obtained for $t^{(i)}=0.0002i$ (i=1,...,10), $t^{(i)}=0.02(i-10)$ (i=11,...,15), $t^{(16)}=1.0$, $t^{(17)}=2.0$, $t^{(i)}=5.0(i-17)$ (i=18,...,23). The initial guess for the determination of the vector q with the above numerical components $\{q_i\}$ is taken to be $q^{(0)}$, where $q_1^{(0)}=1600$, $q_2^{(0)}=0.8$, $q_3^{(0)}=1.2$ (so that $p^{(0)}$ has the components $p_1^{(0)}=ln(1600)\approx7.377759$, $p_2^{(0)}=ln(0.8)\approx-0.2231436$, $p_3^{(0)}=0.1823216$). The superscripts i=17, 19 and 21 specify the selected break-points.

From equation (3), we have

$$\partial f_1/\partial y_1 = y_2-1, \quad \partial f_1/\partial y_2 = y_1+exp(p_2) \qquad (4)$$

$$\partial f_2/\partial y_1 = exp(p_1)(1-y_2), \quad \partial f_2/\partial y_2 = -exp(p_1)\{y_1+exp(p_2)+exp(p_3)\}$$

$$\partial f_1/\partial p_1 = \partial f_1/\partial p_3 = 0, \quad \partial f_1/\partial p_2 = exp(p_2)y_2 \qquad (5)$$

$$\partial f_2/\partial p_1 = exp(p_1)\{(1-y_2)y_1 - (exp(p_2)+exp(p_3))y_2\}$$

$$\partial f_2/\partial p_2 = -exp(p_1+p_2)y_2, \quad \partial f_2/\partial p_3 = -exp(p_1+p_3)y_2.$$

The initial values (1) are not functions of p.

$$ymax = (1,1) \qquad (6)$$

is taken to be an estimate of $(\max|y_1(t,p)|, \max|y_2(t,p)|)$ over the range $0 \le t \le 30$.

The central terms in formulae (3) are evaluated by use of the procedure *deriv*, the elements of the Jacobian matrix given by formulae (4) are evaluated by use of the procedure *jacdfdy*, those of the Jacobian matrix given by formulae (5) by use of the procedure *jacdfdp*, the initial values (1) and estimates (6) are inserted by use of the procedure *callystart*, the required data for the computations are inserted by use of the procedure *data*, and the procedure *monitor* has been used to monitor the computations.

```
#include <math.h>
#include <stdio.h>

void communication(int post, float fa, int n, int m, int nobs,
        int nbp, float par[], float res[], int bp[], float **jtjinv,
        float in[], float out[], int weight, int nis)
{
    float vecvec(int, int, int, float [], float []);
    int i,j;
    float c,*conf;

    conf=allocate_real_vector(1,m);
    if (post == 5) {
        printf("\nThe first residual vector\n  I       RES[I]\n");
        for (i=1; i<=nobs; i++) printf(" %2d   %+e\n",i,res[i]);
    } else if (post == 3) {
        printf("\nThe Euclidean norm of the residual vector : %e\n"
               "Calculated parameters:\n",
               sqrt(vecvec(1,nobs,0,res,res)));
        for (i=1; i<=m; i++) printf("   %+e\n",par[i]);
        printf("Number of integration steps performed: %d\n",nis);
    } else if (post == 4) {
        if (nbp == 0)
```

```
            printf("Minimization is started without break-points\n");
         else {
            printf("Minimization is started with weight =%3d\n"
               "The extra parameters are the observations:\n",
                  weight);
            for (i=1; i<=nbp; i++) printf("  %5d\n",bp[i]);
         }
         printf("Starting values of the parameters:\n");
         for (i=1; i<=m; i++) printf("  %+e\n",par[i]);
         printf("Rel. tol. for the Eucl. norm of the res. vector:"
            " %e\nAbs. tol. for the Eucl. norm of the res. vector:"
            " %e\nRelative starting value of lambda: %e\n",
            in[3],in[4],in[6]);
      } else if (post == 1) {
         printf("\nStarting values of the parameters:\n");
         for (i=1; i<=m; i++) printf("  %+e\n",par[i]);
         printf("Numer of equations    : %2d\n"
            "Number of observations: %2d\n\n"
            "Machine precision                          : %6.2e\n"
            "Relative local error bound for integration: %6.2e\n"
            "Relative tolerance for residual            : %6.2e\n"
            "Absolute tolerance for residual            : %6.2e\n"
            "Maximum number of integrations to perform : %3.0f\n"
            "Relative starting value of lambda          : %6.2e\n"
            "Relative minimal steplength                : %6.2e\n",
            n,nobs,in[0],in[2],in[3],in[4],in[5],in[6],in[1]);
         if (nbp == 0)
            printf("There are no break-points\n");
         else {
            printf("Break-points are the observations:");
            for (i=1; i<=nbp; i++) printf("  %3d",bp[i]);
         }
         printf("\nThe alpha-point of the f-distribution: %5.2f\n",
               fa);
      } else if (post == 2) {
         if (out[1] == 0.0)
            printf("\nNormal termination of the process\n");
         else if (out[1] == 1.0)
            printf("Number of integrations allowed was exceeded\n");
         else if (out[1] == 2.0)
            printf("Minimal steplength was decreased four times\n");
         else if (out[1] == 3.0)
            printf("A call of deriv delivered false\n");
         else if (out[1] == 4.0)
            printf("A call of jacdfdy delivered false\n");
         else if (out[1] == 5.0)
            printf("A call of jacdfdp delivered false\n");
         else if (out[1] == 6.0)
            printf("Precision asked for may not be attained\n");
         if (nbp == 0)
            printf("Last integration was performed "
                  "without break-points\n");
         else {
            printf("The process stopped with Break-points:");
            for (i=1; i<=nbp; i++) printf("  %3d",bp[i]);
         }
         printf("\nEucl. norm of the last residual vector : %e\n"
               "Eucl. norm of the first residual vector: %e\n"
               "Number of integrations performed        : %2.0f\n"
               "Last improvement of the euclidean norm : %e\n"
               "Condition number of J'*J                : %e\n"
               "Local error bound was exceeded (maxim.): %2.0f\n",
               out[2],out[3],out[4],out[6],out[7],out[5]);
         printf("\n   Parameters    Confidence Interval\n");
         for (i=1; i<=m; i++) {
            conf[i]=sqrt(m*fa*jtjinv[i][i]/(nobs-m))*out[2];
            printf("  %+e       %+e\n",par[i],conf[i]);
         }
         c = (nobs == m) ? 0.0 : out[2]*out[2]/(nobs-m);
         printf("\nCorrelation matrix          Covariance matrix\n");
         for (i=1; i<=m; i++) {
            for (j=1; j<=m; j++) {
               if (i == j)
                  printf("                         ");
```

```
                if (i > j)
                    printf("%9.3e  ",
                        jtjinv[i][j]/sqrt(jtjinv[i][i]*jtjinv[j][j]));
                else
                    printf("%9.3e  ",jtjinv[i][j]*c);
            }
            printf("\n");
        }
        printf("\nThe last residual vector\n  I       RES[I]\n");
        for (i=1; i<=nobs; i++) printf(" %2d   %+e\n",i,res[i]);
    }
    free_real_vector(conf,1);
}

int jacdfdp(int n, int m, float par[], float y[], float x,
                float **fp)
{
    float y2;

    y2=y[2];
    fp[1][1]=fp[1][3]=0.0;
    fp[1][2]=y2*exp(par[2]);
    fp[2][1]=exp(par[1])*(y[1]*(1.0-y2)-
                (exp(par[2])+exp(par[3]))*y2);
    fp[2][2] = -exp(par[1]+par[2])*y2;
    fp[2][3] = -exp(par[1]+par[3])*y2;
    return 1;
}

void data(int nobs, float tobs[], float obs[], int cobs[])
{
    int i;
    static float a[23]={0.0002, 0.0004, 0.0006, 0.0008, 0.001,
            0.0012, 0.0014, 0.0016, 0.0018, 0.002, 0.02, 0.04, 0.06,
            0.08, 0.1, 1.0, 2.0, 5.0, 10.0, 15.0, 20.0, 25.0, 30.0};
    static float b[23]={0.1648, 0.2753, 0.3493, 0.3990, 0.4322,
            0.4545, 0.4695, 0.4795, 0.4862, 0.4907, 0.4999, 0.4998,
            0.4998, 0.4998, 0.4998, 0.4986, 0.4973, 0.4936, 0.4872,
            0.4808, 0.4743, 0.4677, 0.4610};

    tobs[0]=0.0;
    for (i=1; i<=nobs; i++) {
        tobs[i]=a[i-1];
        obs[i]=b[i-1];
        cobs[i]=2;
    }
    printf("\nThe observations were:\n"
        "  I    TOBS[I]    COBS[I]    OBS[I]\n");
    for (i=1; i<=nobs; i++)
        printf(" %2d   %7.4f      %1d        %6.4f\n",
                i,tobs[i],cobs[i],obs[i]);
}

void callystart(int n, int m, float par[], float y[], float ymax[])
{
    y[1]=ymax[1]=ymax[2]=1.0;
    y[2]=0.0;
}

int deriv(int n, int m, float par[], float y[], float x,
            float df[])
{
    float y2;

    y2=y[2];
    df[1] = -(1.0-y2)*y[1]+exp(par[2])*y2;
    df[2]=exp(par[1])*((1.0-y2)*y[1]-(exp(par[2])+exp(par[3]))*y2);
    return 1;
}

int jacdfdy(int n, int m, float par[], float y[], float x,
                float **fy)
{
    fy[1][1] = -1.0+y[2];
```

```
    fy[1][2]=exp(par[2])+y[1];
    fy[2][1]=exp(par[1])*(1.0-y[2]);
    fy[2][2] = -exp(par[1])*(exp(par[2])+exp(par[3])+y[1]);
    return 1;
}

void monitor(int post, int ncol, int nrow, float par[],
        float res[], int weight, int nis)
{
}

void main ()
{
    float **allocate_real_matrix(int, int, int, int);
    void free_real_matrix(float **, int, int, int);
    void peide(int, int, int, int *, float [], float [],
            int [], float **, float [], float [],
            int (*)(int,int,float [],float [],float,float []),
            int (*)(int,int,float [],float [],float,float **),
            int (*)(int,int,float [],float [],float,float **),
            void (*)(int,int,float [],float [],float[]),
            void (*)(int,float [],float [],int[]),
            void (*)(int,int,int,float [],float [],int,int));
    int m,n,nobs,nbp,bp[4];
    float fa,par[7],res[27],**jtjinv,in[7],out[8];

    jtjinv=allocate_real_matrix(1,3,1,3);
    printf("          E S C E P - problem\n");
    m=3;  n=2;  nobs=23;  nbp=3;
    par[1]=log(1600.0);  par[2]=log(0.8);  par[3]=log(1.2);
    in[0]=1.0e-14;  in[3]=in[4]=1.0e-4;  in[5]=50.0;  in[6]=1.0e-2;
    in[1]=1.0e-4;  in[2]=1.0e-5;
    bp[1]=17;  bp[2]=19;  bp[3]=21;
    fa=4.94;
    communication(1,fa,n,m,nobs,nbp,par,res,bp,jtjinv,in,out,0,0);
    peide(n,m,nobs,&nbp,par,res,bp,jtjinv,in,out,deriv,jacdfdy,
            jacdfdp,callystart,data,monitor);
    communication(2,fa,n,m,nobs,nbp,par,res,bp,jtjinv,in,out,0,0);
    free_real_matrix(jtjinv,1,3,1);
}
```

Output:

```
        E S C E P - problem

Starting values of the parameters:
  +7.37776e+00
  -2.23144e-01
  +1.82322e-01
Numer of equations     :  2
Number of observations: 23

Machine precision                              : 1.00e-14
Relative local error bound for integration: 1.00e-05
Relative tolerance for residual               : 1.00e-04
Absolute tolerance for residual               : 1.00e-04
Maximum number of integrations to perform :  50
Relative starting value of lambda            : 1.00e-02
Relative minimal steplength                  : 1.00e-04
Break-points are the observations:   17   19   21
The alpha-point of the f-distribution:  4.94

The observations were:
  I   TOBS[I]   COBS[I]   OBS[I]
  1   0.0002     2        0.1648
  2   0.0004     2        0.2753
  3   0.0006     2        0.3493
  4   0.0008     2        0.3990
  5   0.0010     2        0.4322
  6   0.0012     2        0.4545
  7   0.0014     2        0.4695
  8   0.0016     2        0.4795
  9   0.0018     2        0.4862
```

```
 10    0.0020    2     0.4907
 11    0.0200    2     0.4999
 12    0.0400    2     0.4998
 13    0.0600    2     0.4998
 14    0.0800    2     0.4998
 15    0.1000    2     0.4998
 16    1.0000    2     0.4986
 17    2.0000    2     0.4973
 18    5.0000    2     0.4936
 19   10.0000    2     0.4872
 20   15.0000    2     0.4808
 21   20.0000    2     0.4743
 22   25.0000    2     0.4677
 23   30.0000    2     0.4610

Normal termination of the process
Last integration was performed without break-points

Eucl. norm of the last residual vector : 1.43818e-04
Eucl. norm of the first residual vector: 1.33107e+00
Number of integrations performed       : 12
Last improvement of the euclidean norm : 2.65941e-05
Condition number of J'*J               : 2.57723e+02
Local error bound was exceeded (maxim.): 37

   Parameters     Confidence Interval
  +6.90767e+00       +3.21391e-04
  -1.00394e-02       +1.69753e-04
  -4.60523e+00       +1.95162e-03

Correlation matrix             Covariance matrix
                         6.97e-09   1.42e-09  -9.19e-09
 3.85e-01                           1.94e-09  -1.43e-08
-2.17e-01  -6.39e-01                           2.57e-07

The last residual vector
  I       RES[I]
  1   +1.66893e-06
  2   -2.91467e-05
  3   +2.79844e-05
  4   -3.89516e-05
  5   +3.05176e-05
  6   +3.08454e-05
  7   -2.03252e-05
  8   -3.99351e-06
  9   +1.04308e-05
 10   +1.37687e-05
 11   -5.12600e-05
 12   +2.36630e-05
 13   -1.37091e-06
 14   -2.63751e-05
 15   -5.13792e-05
 16   +2.25008e-05
 17   +6.94096e-05
 18   -1.31726e-05
 19   +2.15769e-05
 20   -1.93715e-05
 21   -3.47793e-05
 22   -2.28882e-05
 23   +1.80006e-05
```

Examples for chapter 6 and 7 procedures

Function tested: **ei**

Example:

Evaluate

$$\int_x^\infty \frac{1}{te^t}dt$$

for $x = 0.5$.

```
#include <stdio.h>

void main ()
{
    float ei(float);

    printf("EI delivers:  %e\n",-ei(-0.5));
}
```

Output:

```
EI delivers:  5.59774e-01
```

Function tested: **eialpha**

Example:
Evaluate

$$\int_1^\infty \frac{t^i}{e^{xt}}dt$$

for $x = 0.25$ and i=0,1,...,5.

```
#include <stdio.h>

void main ()
{
    void eialpha(float, int, float []);
    int k;
    float a[6];

    printf("EIALPHA delivers:\n");
    eialpha(0.25,5,a);
    for (k=0; k<=5; k++) {
        printf(" %2d   %e\n",k,a[k]);
    }
}
```

Output:

```
EIALPHA delivers:
  0   3.11520e+00
  1   1.55760e+01
  2   1.27723e+02
  3   1.53580e+03
  4   2.45758e+04
  5   4.91520e+05
```

Functions tested: **enx, nonexpenx**

Example:
Evaluate in succession: a)

$$\int_1^\infty \frac{1}{e^{xt}t^n}\,dt$$

for $x = 1.1$ and $n = 40, 41, 42$, by means of a call of *enx*; b)

$$e^x \int_1^\infty \frac{t^n}{e^{xt}}\,dt$$

for $x = 50.1$ and $n = 1$, by means of a call of *nonexpenx*.

```
#include <stdio.h>

void main ()
{
    float *allocate_real_vector(int, int);
    void free_real_vector(float *, int);
    void enx(float, int, int, float []);
    void nonexpenx(float, int, int, float []);
    int i;
    float b[2],*a;

    printf("ENX and NONEXPENX deliver:\n");
    a=allocate_real_vector(40,42);
    enx(1.1,40,42,a);
    for (i=40; i<=42; i++) printf(" E(%2d,1.1) = %e\n",i,a[i]);
    nonexpenx(50.1,1,1,b);
    printf("\nEXP(50.1)*E(1,50.1) = %e\n",b[1]);
    free_real_vector(a,40);
}
```

Output:

```
ENX and NONEXPENX deliver:
 E(40,1.1) = 8.29521e-03
 E(41,1.1) = 8.09366e-03
 E(42,1.1) = 7.90166e-03

EXP(50.1)*E(1,50.1) = 1.95767e-02
```

Functions tested: **sincosint, sincosfg**

Example:

Evaluate

$$SI(x) = \int_0^x \frac{\sin(t)}{t}\,dt\,, \qquad CI(x) = \gamma + \ln|x| + \int_0^x \frac{\cos(t) - 1}{t}\,dt\,,$$

$$F(x) = CI(x)\sin(x) - \left(SI(x) - \frac{\pi}{2}\right)\cos(x)\,,$$

$$G(x) = -CI(x)\cos(x) - \left(SI(x) - \frac{\pi}{2}\right)\sin(x)$$

for $x = 1$.

```
#include <stdio.h>

void main ()
{
```

```
    void sincosint(float, float *, float *);
    void sincosfg(float, float *, float *);
    float si,ci,f,g;
    sincosint(1.0,&si,&ci);
    sincosfg(1.0,&f,&g);
    printf("SINCOSINT and SINCOSFG deliver:\n");
    printf(" SI(1) = %+e   CI(1) = %+e\n"
           "  F(1) = %+e    G(1) = %+e\n",si,ci,f,g);
}
```

Output:

```
SINCOSINT and SINCOSFG deliver:
 SI(1) = +9.46083e-01   CI(1) = +3.37404e-01
  F(1) = +6.21450e-01    G(1) = +3.43378e-01
```

Function tested: **recipgamma**

Example:

Evaluate

$\Gamma'(x) = \{\Gamma(1-x)\}^{-1}$, $\Gamma_0'(x) = (2+x)^{-1}\{\Gamma'(x) - \Gamma'(-x)\}$, $\Gamma_e'(x) = \{\Gamma'(x) + \Gamma'(-x)\}/2$ for $x = 0.4$ and $x = 0$.

```
#include <stdio.h>

void main ()
{
    float recipgamma(float, float *, float *);
    float x,odd,even;
    printf("RECIPGAMMA delivers:\n");
    x=recipgamma(0.4,&odd,&even);
    printf(" 0.4   %e   %e   %e\n",x,odd,even);
    x=recipgamma(0.0,&odd,&even);
    printf(" 0.0   %e   %e   %e\n",x,odd,even);
}
```

Output:

```
RECIPGAMMA delivers:
 0.4   6.71505e-01   -5.69444e-01   8.99283e-01
 0.0   1.00000e+00   -5.77216e-01   1.00000e+00
```

Function tested: **gamma**

Example:

Evaluate $\Gamma(x)$ for x = -8.5, 0.25, 1.5 and 22.

```
#include <stdio.h>

void main ()
{
    float gamma(float);
    int i;
    static float x[4]={-8.5, 0.25, 1.5, 22.0};
    printf("GAMMA delivers:\n");
    for (i=0; i<=3; i++)
        printf(" %6.2f   %+e\n",x[i],gamma(x[i]));
}
```

Output:

```
GAMMA delivers:
  -8.50   -2.63352e-05
```

```
   0.25   +3.62561e+00
   1.50   +8.86227e-01
  22.00   +5.10909e+19
```

Function tested: **loggamma**

Example:

Evaluate $ln\{\Gamma(x)\}$ for x = 0.25, 1.5, 12, 15 and 80.

```
#include <stdio.h>

void main ()
{
    float loggamma(float);
    int i;
    static float x[5]={0.25, 1.5, 12.0, 15.0, 80.0};
    printf("LOGGAMMA delivers:\n");
    for (i=0; i<=4; i++)
        printf(" %6.2f   %+e\n",x[i],loggamma(x[i]));
}
```

Output:

```
LOGGAMMA delivers:
   0.25   +1.28802e+00
   1.50   -1.20782e-01
  12.00   +1.75023e+01
  15.00   +2.51912e+01
  80.00   +2.69291e+02
```

Function tested: **incomgam**

Example:

Evaluate $\gamma(4,3)$ and $\Gamma(4,3)$.

```
#include <stdio.h>

void main ()
{
    float p,q;
    void incomgam(float, float, float *, float *, float, float);
    printf("INCOMGAM delivers:\n");
    incomgam(3.0,4.0,&p,&q,1.0*2.0*3.0,1.0e-6);
    printf("  KLGAM and GRGAM are : %e  %e\n",p,q);
}
```

Output:

```
INCOMGAM delivers:
  KLGAM and GRGAM are : 2.11661e+00  3.88339e+00
```

Function tested: **incbeta**

Example:

Evaluate $I_x(p,q)$ for x = 0.3, p = 1.4, q = 1.5.

```
#include <stdio.h>
void main ()
{
    float incbeta(float, float, float, float);
```

```
    printf("INCBETA delivers:  %e\n",incbeta(0.3,1.4,1.5,1.0e-6));
}
```

Output:

```
INCBETA delivers:  2.79116e-01
```

Function tested: **ibpplusn**

Example:
Evaluate $I_x(p+n,q)$ for $x = 0.3$, $p = 0.4$, $q = 1.5$, and $n = 0, 1, 2$.

```
#include <stdio.h>

void main ()
{
    void ibpplusn(float, float, float, int, float, float []);
    float isubx[3];

    ibpplusn(0.3,0.4,1.5,2,1.0e-6,isubx);
    printf("IBPPLUSN delivers:\n %e   %e   %e\n",
           isubx[0],isubx[1],isubx[2]);
}
```

Output:

```
IBPPLUSN delivers:
 7.21671e-01   2.79116e-01   9.89328e-02
```

Function tested: **ibqplusn**

Example:
Evaluate $I_x(p,q+n)$ for $x = 0.3$, $p = 1.4$, $q = 0.5$, and $n = 0, 1, 2$.

```
#include <stdio.h>

void main ()
{
    void ibqplusn(float, float, float, int, float, float []);
    float isubx[3];

    ibqplusn(0.3,1.4,0.5,2,1.0e-6,isubx);
    printf("IBQPLUSN delivers:\n %e   %e   %e\n",
           isubx[0],isubx[1],isubx[2]);
}
```

Output:

```
IBQPLUSN delivers:
 8.94496e-02   2.79116e-01   4.47287e-01
```

Functions tested: **errorfunction, nonexperfc**

Example:
Evaluate

$$erf(1) = \frac{2}{\sqrt{\pi}}\int_0^1 e^{-t^2}\,dt$$

$$erfc(1) = \frac{2}{\sqrt{\pi}}\int_1^\infty e^{-t^2}\,dt$$

$$nonexperfc(x) = \frac{2e^{x^2}}{\sqrt{\pi}}\int_x^\infty e^{-t^2}\,dt$$

for $x = 100$.

```
#include <stdio.h>

void main ()
{
    void errorfunction(float, float *, float *);
    float nonexperfc(float);
    float erf,erfc,p;

    errorfunction(1.0,&erf,&erfc);
    p=nonexperfc(100.0);
    printf("ERRORFUNCTION and NONEXPERFC deliver:\n\n"
            "ERF(1)   = %e\nERFC(1) = %e\nNONEXPERFC(100) = %e\n",
            erf,erfc,p);
}
```

Output:

```
ERRORFUNCTION and NONEXPERFC deliver:

ERF(1)   = 8.42701e-01
ERFC(1) = 1.57299e-01
NONEXPERFC(100) = 5.64161e-03
```

Function tested: **inverseerrorfunction**

Example:

Evaluate

$$x = \frac{2}{\sqrt{\pi}}\int_0^{y(x)} e^{-t^2}\,dt$$

for $x = 0.6$ and $1\text{-}10^{-30}$.

```
#include <stdio.h>

void main ()
{
    void inverseerrorfunction(float, float, float *);
    float inverf1,inverf2;

    inverseerrorfunction(0.6,0.0,&inverf1);
    inverseerrorfunction(1.0,1.0e-30,&inverf2);
    printf("INVERSEERRORFUNCTION delivers:\n\n"
            " X = 0.6,  INVERF = %e\n X = 1.0,  INVERF = %e\n",
```

```
                inverf1,inverf2);
}
```

Output:

```
INVERSEERRORFUNCTION delivers:

 X = 0.6,   INVERF = 5.95116e-01
 X = 1.0,   INVERF = 8.14862e+00
```

Functions tested: **fresnel, fg**

Example:

Evaluate

$$C(x) = \int_0^x \cos(\frac{\pi t^2}{2}) dt \ , \qquad S(x) = \int_0^x \sin(\frac{\pi t^2}{t}) \ dt \ ,$$

$$F(x) = \left(\frac{1}{2} - S(x)\right)\cos(\frac{\pi x^2}{2}) - \left(\frac{1}{2} - C(x)\right)\sin(\frac{\pi x^2}{2}) \ ,$$

$$G(x) = \left(\frac{1}{2} - C(x)\right)\cos(\frac{\pi x^2}{2}) + \left(\frac{1}{2} - S(x)\right)\sin(\frac{\pi x^2}{2})$$

for $x = 1$.

```
#include <stdio.h>

void main ()
{
    void fresnel(float, float *, float *);
    void fg(float, float *, float *);
    float c,s,f,g;

    fresnel(1.0,&c,&s);
    fg(1.0,&f,&g);
    printf("FRESNEL and FG deliver:\n\n C(1) = %e   S(1) = %e\n"
           " F(1) = %e   G(1) = %e\n",c,s,f,g);
}
```

Output:

```
FRESNEL and FG deliver:

 C(1) = 7.79893e-01   S(1) = 4.38259e-01
 F(1) = 2.79893e-01   G(1) = 6.17409e-02
```

Functions tested: **bessj0, bessj1, bessj**

Example:

Evaluate $J_k''(x)$ - $J_k'(x)$, k=0,1, where $J_k''(x)$ is $J_k(x)$ as evaluated by *bessj* (k=0,1) and $J_0'(x)$, $J_1'(x)$ are $J_0(x)$, $J_1(x)$ as evaluated by *bessj0* and *bessj1*, respectively, for x = 1, 5, 10, 25.

```
#include <stdio.h>
```

```
void main ()
{
    void bessj(float, int, float []);
    float bessj0(float);
    float bessj1(float);
    int i;
    float j[2];
    float x[4]={1.0, 5.0, 10.0, 25.0};

    for (i=0; i<=3; i++) {
        bessj(x[i],1,j);
        printf(" %6.1f   %+7.2e   %+7.2e\n",
               x[i],j[0]-bessj0(x[i]),j[1]-bessj1(x[i]));
    }
}
```

Output:

```
   1.0   -2.2e-08   +2.5e-08
   5.0   +3.0e-09   +3.0e-08
  10.0   +5.1e-08   +2.8e-08
  25.0   +9.8e-08   +4.3e-08
```

Function tested: **bessy01**

Example:

Evaluate $Y_0(1)$ and $Y_1(1)$.

```
#include <stdio.h>

void main ()
{
    void bessy01(float, float *, float *);
    float x,y0,y1;

    x=1.0;
    bessy01(x,&y0,&y1);
    printf("BESSY01 delivers:\n %3.1f   %e   %e\n",x,y0,y1);
}
```

Output:

```
BESSY01 delivers:
 1.0   8.82570e-02   -7.81213e-01
```

Function tested: **bessy**

Example:

Evaluate $Y_i(1)$ for $i = 0, 1, 2$.

```
#include <stdio.h>

void main ()
{
    void bessy(float, int, float []);
    float y[3];

    bessy(1.0,2,y);
    printf("BESSY delivers:\n  %e   %e   %e\n",y[0],y[1],y[2]);
}
```

Output:

```
BESSY delivers:
```

```
  8.82570e-02    -7.81213e-01    -1.65068e+00
```

Functions tested: **besspq0, besspq1**

Example:

Verify the relation

$$P_0(x)*P_1(x) + Q_0(x)*Q_1(x) = 1$$

for $x = 1, 3, 5, 10$.

```
#include <stdio.h>

void main ()
{
    void besspq0(float, float *, float *);
    void besspq1(float, float *, float *);
    int i;
    float p,q,r,s;
    float x[4]={1.0, 3.0, 5.0, 10.0};

    printf("BESSPQ0 and BESSPQ1 deliver:\n");
    for (i=0; i<=3; i++) {
        besspq0(x[i],&p,&q);
        besspq1(x[i],&r,&s);
        printf("    %7.2e",fabs(p*r+q*s-1.0));
    }
}
```

Output:

```
BESSPQ0 and BESSPQ1 deliver:
   2.2e-08    1.2e-07    2.8e-08    4.8e-08
```

Functions tested: **bessi, bessk**

Example:

Verify the relation

$$x*\{I_{n-1}(x)*K_{n-1}(x) + I_n(x)*K_{n-1}(x)\} = 1$$

for $x = 1,...,20$ and $n = 1,...,5$.

```
#include <stdio.h>

void main ()
{
    void bessi(float, int, float []);
    void bessk(float, int, float []);
    int j,n;
    float x,i[6],k[6];

    printf("BESSI and BESSK deliver:");
    for (j=1; j<=20; j++) {
        x=j;
        printf("\n %3.0f",x);
        bessi(x,5,i);
        bessk(x,5,k);
        for (n=1; n<=5; n++)
            printf(" %+9.2e",x*(i[n]*k[n-1]+i[n-1]*k[n])-1.0);
    }
}
```

Output:

```
BESSI and BESSK deliver:
   1  -2.0e-08  -3.1e-09  +3.9e-08  +6.5e-09  -3.6e-08
```

```
 2  +1.6e-08  +2.6e-08  +4.1e-08  -2.9e-08  +1.2e-08
 3  -3.5e-08  -5.1e-09  +8.0e-08  +3.6e-08  +1.1e-08
 4  +7.0e-08  +5.2e-08  -2.2e-08  -8.9e-08  -8.1e-08
 5  +6.4e-08  +8.1e-08  +1.3e-07 +1.00e-07  +8.0e-08
 6  +1.4e-07  +1.8e-07  +1.8e-07  +2.0e-07  +1.7e-07
 7  +1.1e-07  +1.7e-07  +2.1e-07  +2.3e-07  +1.8e-07
 8  +4.4e-08  +4.6e-08  +2.3e-08  +1.8e-09  +2.4e-08
 9  -9.2e-08  -4.3e-08  -3.4e-08  -3.5e-08  -5.1e-08
10  +1.9e-08  -1.9e-08  -1.6e-08  +6.4e-08  +8.2e-08
11  +9.3e-08  +2.3e-08  -3.2e-08  -1.3e-08  -9.4e-09
12  +6.7e-08  +5.8e-08  +3.2e-08  -2.2e-08  -3.7e-08
13  +5.7e-08  +5.4e-08  +8.8e-08  +7.1e-08  +1.1e-07
14  -8.3e-08  -8.8e-08  -3.4e-08  +9.2e-09  +2.3e-08
15  -7.5e-08  -5.9e-08  -8.2e-08  -8.5e-08  -7.0e-08
16  +1.1e-07  +1.3e-07  +4.5e-08  +3.8e-08  +1.2e-07
17  -3.2e-08  -6.5e-08  -1.2e-07  -1.5e-07  -1.8e-07
18  -2.1e-08  +2.4e-08  -5.8e-09  -3.6e-08  -4.9e-08
19  -8.1e-08  -9.6e-08  -1.0e-07  -1.2e-07  -1.1e-07
20  +1.9e-08  +3.9e-10  +9.1e-09  +5.1e-08  +7.0e-08
```

Function tested: **bessk01**

Example:

Evaluate $K_0(x)$ and $K_1(x)$ for $x = 0.5$, 1.5 and 2.5.

```
#include <stdio.h>

void main ()
{
    void bessk01(float, float *, float *);
    int j;
    float x,k0,k1;

    printf("BESSK01 delivers:\n");
    x=0.5;
    for (j=1; j<=3; j++) {
        bessk01(x,&k0,&k1);
        printf(" %4.1f   %e   %e\n",x,k0,k1);
        x += 1.0;
    }
}
```

Output:

```
BESSK01 delivers:
  0.5   9.24419e-01   1.65644e+00
  1.5   2.13806e-01   2.77388e-01
  2.5   6.23475e-02   7.38908e-02
```

Function tested: **nonexpbessk01**

Example:

Evaluate $e^xK_0(x)$ and $e^xK_1(x)$ for $x = 0.5$, 1.0, 1.5, 2.0 and 2.5.

```
#include <stdio.h>

void main ()
{
    void nonexpbessk01(float, float *, float *);
    int j;
    float x,k0,k1;

    printf("NONEXPBESSK01 delivers:\n");
    x=0.5;
    for (j=1; j<=5; j++) {
```

```
        nonexpbessk01(x,&k0,&k1);
        printf(" %4.1f   %e   %e\n",x,k0,k1);
        x += 0.5;
    }
}
```

Output:

```
NONEXPBESSK01 delivers:
  0.5   1.52411e+00   2.73101e+00
  1.0   1.14446e+00   1.63615e+00
  1.5   9.58210e-01   1.24317e+00
  2.0   8.41568e-01   1.03348e+00
  2.5   7.59549e-01   9.00175e-01
```

Function tested: **nonexpbessk**

Example:

Evaluate $e^x K_i(x)$ for x = 0.5, 1.0, 1.5, 2.0 and i = 0, 1, 2.

```
#include <stdio.h>

void main ()
{
    void nonexpbessk(float, int, float []);
    int j;
    float x,k[3];

    printf("NONEXPBESSK delivers:\n");
    x=0.5;
    for (j=1; j<=4; j++) {
        nonexpbessk(x,2,k);
        printf(" %4.1f   %e   %e   %e\n",x,k[0],k[1],k[2]);
        x += 0.5;
    }
}
```

Output:

```
NONEXPBESSK delivers:
  0.5   1.52411e+00   2.73101e+00   1.24481e+01
  1.0   1.14446e+00   1.63615e+00   4.41677e+00
  1.5   9.58210e-01   1.24317e+00   2.61576e+00
  2.0   8.41568e-01   1.03348e+00   1.87505e+00
```

Function tested: **bessjaplusn**

Example:

Evaluate $J_{a+i}(x)$ for x = 2, a = 0.78 and i = 0, 1, 2.

```
#include <stdio.h>

void main ()
{
    void bessjaplusn(float, float, int, float []);
    int n;
    float a,x,ja[3];

    x=2.0;
    a=0.78;
    n=2;
    bessjaplusn(a,x,n,ja);
    printf("BESSJAPLUSN delivers:\n"
           " X = %2.0f    A = %4.2f    N = %d\n"
           " %e   %e   %e\n",x,a,n,ja[0],ja[1],ja[2]);
```

```
}
```

Output:

```
BESSJAPLUSN delivers:
 X =  2    A = 0.78    N = 2
 5.73061e-01   4.15295e-01   1.66163e-01
```

Function tested: **besspqa01**

Example:

Verify the relation

$$P_a(x)*P_{a+1}(x) + Q_a(x)*Q_{a+1}(x) = 1$$

for $a = 0$ and $x = 1, 3, 5, 10, 15, 20, 50$.

```
#include <math.h>
#include <stdio.h>

void main ()
{
    void besspqa01(float, float, float *, float *, float *, float *);
    int i;
    float p,q,r,s;
    static float x[7]={1.0, 3.0, 5.0, 10.0, 15.0, 20.0, 50.0};

    printf("BESSPQA01 delivers:\n");
    for (i=0; i<=6; i++) {
        besspqa01(0.0,x[i],&p,&q,&r,&s);
        printf(" %5.0f    %e\n",x[i],fabs(p*r+q*s-1.0));
    }
}
```

Output:

```
BESSPQA01 delivers:
     1   2.18651e-08
     3   1.17937e-07
     5   2.78083e-08
    10   4.84759e-08
    15   1.29200e-08
    20   4.45196e-08
    50   2.90220e-08
```

Function tested: **besszeros**

Example:

Compute the first two zeros of $Y_{3.14}(z)$.

```
#include <stdio.h>

void main ()
{
    void besszeros(float, int, float [], int);
    int n,d;
    float a,z[3];

    a=3.14;
    n=d=2;
    besszeros(a,n,z,d);
    printf("The first two zeros of the Bessel function Y of "
           "the order 3.14:\n  %e   %e\n",z[1],z[2]);
}
```

Output:

```
The first two zeros of the Bessel function Y of the order 3.14:
  4.68478e+00   8.27659e+00
```

Functions tested: **spherbessi, nonexpspherbessi**

Example:

Compute

$$\sqrt{\frac{\pi}{2x}}I_{j+½}(x) \qquad \text{and} \qquad e\left(\frac{1}{e^x}\sqrt{\frac{\pi}{2x}}I_{j+½}(x)\right)$$

for $x = 1$ and $j = 0,...,3$.

```
#include <math.h>
#include <stdio.h>

void main ()
{
    void spherbessi(float, int, float []);
    void nonexpspherbessi(float, int, float []);
    int n;
    float x,expx,i1[4],i2[4];

    printf("SPHERBESSI and NONEXPSPHERBESSI deliver:\n");
    x=1.0;
    expx=exp(x);
    n=3;
    spherbessi(x,n,i1);
    nonexpspherbessi(x,n,i2);
    for (n=0; n<=3; n++)
        printf(" %d   %e  %e\n",n,i1[n],i2[n]*expx);
}
```

Output:

```
SPHERBESSI and NONEXPSPHERBESSI deliver:
  0   1.17520e+00  1.17520e+00
  1   3.67879e-01  3.67879e-01
  2   7.15629e-02  7.15629e-02
  3   1.00651e-02  1.00651e-02
```

Functions tested: **spherbessk, nonexpspherbessk**

Example:

Compute

$$\sqrt{\frac{\pi}{2x}}K_{j+½}(x) \qquad \text{and} \qquad \frac{1}{e^2}\left(e^x\sqrt{\frac{\pi}{2x}}K_{j+½}(x)\right)$$

for $x = 2$ and $j = 0,...,3$.

```
#include <math.h>
#include <stdio.h>

void main ()
```

```
{
    void spherbessk(float, int, float []);
    void nonexpspherbessk(float, int, float []);
    int n;
    float x,expx,k1[4],k2[4];

    printf("SPHERBESSK and NONEXPSPHERBESSK deliver:\n");
    x=2.0;
    expx=exp(-x);
    n=3;
    spherbessk(x,n,k1);
    nonexpspherbessk(x,n,k2);
    for (n=0; n<=3; n++)
        printf("  %d   %e  %e\n",n,k1[n],k2[n]*expx);
}
```

Output:

```
SPHERBESSK and NONEXPSPHERBESSK deliver:
  0   1.06292e-01  1.06292e-01
  1   1.59438e-01  1.59438e-01
  2   3.45449e-01  3.45449e-01
  3   1.02306e+00  1.02306e+00
```

Function tested: **airy**

Example:

Compute *Ai(x), dAi(x)/dx, Bi(x)* and *dBi(x)/dx* for $x = 9.654894$.

```
#include <math.h>
#include <stdio.h>

void main ()
{
    void airy(float, float *, float *, float *, float *,
              float *, int);
    float a,b,c,d,e;

    airy(9.654894,&a,&b,&c,&d,&e,1);
    printf("AIRY delivers:\n"
           " AI  (9.654894) = %+e\n AID(9.654894) = %+e\n"
           " BI  (9.654894) = %+e\n BID(9.654894) = %+e\n",
           a*exp(-e),b*exp(-e),c*exp(e),d*exp(e));
}
```

Output:

```
AIRY delivers:
 AI  (9.654894) = +3.28735e-10
 AID(9.654894) = -1.02980e-09
 BI  (9.654894) = +1.55839e+08
 BID(9.654894) = +4.80104e+08
```

Function tested: **airyzeros**

Example:

Compute the *i*-th zero of *Bi(x)* and the value of *dBi(x)/dx* at this point for $i = 3$.

```
#include <stdio.h>

void main ()
{
    float airyzeros(int, int, float [], float []);
    float b,zbi[4],vbid[4];
```

```
    b=airyzeros(3,2,zbi,vbid);
    printf("AIRYZEROS delivers:\n"
           " The third zero of BI(X) is  %e\n"
           " The value of (D/DX)BI(X) in this point is  %e\n",
           b,vbid[3]);
}
```

Output:

```
AIRYZEROS delivers:
 The third zero of BI(X) is  -4.83074e+00
 The value of (D/DX)BI(X) in this point is  8.36986e-01
```

Function tested: **newton**

Example:

Compute the divided differences $\delta_x^j f(x_0)$, j=0,1,2, from the function values $f_0=1$, $f_1=f_2=0$ and associated arguments $x_0=0$, $x_1=1/2$. $x_2=1$.

```
#include <stdio.h>

void main ()
{
    void newton(int, float [], float []);
    float x[3],f[3];

    x[0]=0.0;  x[1]=0.5;  x[2]=1.0;
    f[0]=1.0;  f[1]=f[2]=0.0;
    newton(2,x,f);
    printf("The Newton coefficients are:\n %e   %e   %e",
           f[0],f[1],f[2]);
}
```

Output:

```
The Newton coefficients are:
 1.00000e+00   -2.00000e+00   2.00000e+00
```

Function tested: **ini**

Example:

Determine integer approximations $s(j)$, j=0,1,2, to the points in the range [0,20] at which $T_2((x-10)/10)$ assumes its maximal absolute values.

```
#include <stdio.h>

void main ()
{
    void ini(int, int, int []);
    int s[3];

    ini(2,20,s);
    printf("INI selects out of 0,1,...,20 the numbers:\n"
           "  %4d %4d %4d\n",s[0],s[1],s[2]);
}
```

Output:

```
INI selects out of 0,1,...,20 the numbers:
     0   10   20
```

Function tested: **sndremez**

Example:

Given the numbers g_i, i=0,...,7, in order: 10, 12, -15, -10, -14, 15, 10, 11 and $s(j)$, j=0,1,2, in order: 0, 3, 6 (so that $g_{s(0)} = 10$, $g_{s(1)} = -10$, $g_{s(2)} = 10$) selects an integer sequence $S(j)$, j=0,1,2, such that the numbers $g_{S(j)}$ have the properties described in the documentation to *sndremez*.

```
#include <stdio.h>

void main ()
{
    int s[3];
    float em[2],g[8];

    printf("SNDREMEZ delivers:\n\n");
    g[0]=10.0;  g[1]=12.0;  g[2] = -15.0;  g[3] = -10.0;
    g[4] = -14.0;  g[5]=15.0;  g[6]=10.0;  g[7]=11.0;
    em[0]=10.0;  s[0]=0.0;  s[1]=3.0;  s[2]=6.0;
    printf("The numbers:\n S[J]:  %d  %d  %d\n G[S[j]]:"
           "  %4.0f %4.0f %4.0f\n\n",
           s[0],s[1],s[2],g[s[0]],g[s[1]],g[s[2]]);
    sndremez(2,7,s,g,em);
    printf("are exchanged with:\n S[J]:"
           "  %d  %d  %d\n G[S[j]]:  %4.0f %4.0f %4.0f\n\n"
           "The reference set of function values is:\n"
           " %4.0f %4.0f %4.0f %4.0f %4.0f %4.0f %4.0f %4.0f\n",
           s[0],s[1],s[2],g[s[0]],g[s[1]],g[s[2]],
           g[0],g[1],g[2],g[3],g[4],g[5],g[6],g[7]);
}
```

Output:

```
SNDREMEZ delivers:

The numbers:
 S[J]:  0  3  6
 G[S[j]]:    10  -10   10

are exchanged with:
 S[J]:  0  2  5
 G[S[j]]:    10  -15   15

The reference set of function values is:
   10   12  -15  -10  -14   15   10   11
```

Function tested: **minmaxpol**

Example:

Determine c_0 and c_1 such that

$$\max |f(y_j) - c_0 - c_1 y_j|$$

where $y_j = -1+0.01j$ and $f(y) = 1/(10-y)$, is a minimum as j ranges over the values 0,...,200.

```
#include <stdio.h>

float f(float x)
{
    return 1.0/(x-10.0);
}

void compute(int n, float a, float b, float (*f)(float))
{
    int k,l,m;
    float r,idm,*coef,em[4],*y,*fy;
```

```
    float *allocate_real_vector(int, int);
    void free_real_vector(float *, int);
    void minmaxpol(int, int, float [], float [], float [], float []);

    coef=allocate_real_vector(0,n);
    em[2]=10*n+5;
    m=100*n+10;
    y=allocate_real_vector(0,m);
    fy=allocate_real_vector(0,m);
    idm=(b-a)/m;
    r=y[0]=a;
    fy[0]=f(r);
    r=y[m]=b;
    fy[m]=f(r);
    l=m-1;
    for (k=1; k<=l; k++) {
        r=y[k]=a+k*idm;
        fy[k]=f(r);
    }
    minmaxpol(n,m,y,fy,coef,em);
    printf(" COEF:    ");
    for (k=0; k<=n; k++) printf("  %e",coef[k]);
    printf("\n EM[0:3]:  %e   %e   %4.0f   %3.0f\n",
            em[0],em[1],em[2],em[3]);
    free_real_vector(coef,0);
    free_real_vector(y,0);
    free_real_vector(fy,0);
}

void main ()
{
    int n;

    printf("MINMAXPOL delivers:\n");
    n=1;
    printf(" Degree = %d\n",n);
    compute(n,-1.0,1.0,f);
}
```

Output:

```
MINMAXPOL delivers:
 Degree = 1
 COEF:      -1.00504e-01  -1.01010e-02
 EM[0:3]:  -5.06310e-04   5.06310e-04     15     3
```

Appendix A: References

[AbS65] M. Abramowitz, I.A. Stegun, *Handbook of Mathematical Functions*, Dover Publications, Inc., 1965.

[Ad67] D.A. Adams, "A Stopping Criterion for Polynomial Root Finding", *CACM* 10, No. 10, October 1967, 655-658.

[AHU74] A.V. Aho, J.E. Hopcroft, J.D. Ullman, *The Design and Analysis of Computer Algorithms*, Addison-Wesley, 1974.

[Bab72] I. Babuska, "Numerical Stability in Problems of Linear Algebra", *SIAM J. Numer. Anal.* 9, 1972, 53-77.

[Bak77] M. Bakker, "Galerkin Methods in Spherical Regions", Report NW 50, Mathematical Centre, Amsterdam, 1977.

[BakH76] M. Bakker, P.W. Hemker, P.J. van der Houwen, S.J. Polak, M. van Veldhuizen, "Colloquium on Discretization Methods", MC Syllabus NR 27, Mathematisch Centrum, Amsterdam, 1976.

[Be72] P.A. Beentjes, "An Algol 60 Version of Stabilized Runge Kutta Methods", Report NR 23, Mathematical Centre, Amsterdam, 1972.

[Be74] P.A. Beentjes, "Some Special Formulas of the England Class of Fifth Order Runge-Kutta Schemes", Report NW 24, Mathematical Centre, Amsterdam, 1974.

[BeDHKW73] P.A. Beentjes, K. Dekker, H.C. Hemker, S.P.N. van Kampen, G.M. Willems, "Colloquium Stiff Differential Equations 2", MC Syllabus NR 15.2, Mathematical Centre, Amsterdam, 1973.

[BeDHV74] P.A. Beentjes, K. Dekker, H.C. Hemker, M.V. Veldhuizen, "Colloquium Stiff Differential Equations 3", MC Syllabus NR 15.3, Mathematical Centre, Amsterdam, 1974.

[BjG67] A. Bjorck, G.H. Golub, "Iterative Refinement of Least Squares Solutions by Householder Transformation", *BIT* 7, 1967, 322-337.

[Bla74] J.M. Blair, "Rational Chebyshev Approximations for the Modified Bessel Functions $I_0(x)$ and $I_1(x)$", *Math. Comp.* 28, 1974, 581-583.

[Bre73] R.P. Brent, *Algorithms for Minimization without Derivatives*, Prentice-Hall, 1973.

[Bro71] C.G. Broyden, "The Convergence of an Algorithm for Solving Sparse Nonlinear Systems", *Math. Comp.* 25, No.114, 1971, 285-294.

[Bul67] R. Bulirsch, "Numerical Calculation of the Sine, Cosine and Fresnel Integrals", *Numer. Math.* 9, 1967, 380-385.

[BulS65] R. Bulirsch, J. Stoer, "Numerical Treatment of Ordinary Differential Equations by Extrapolation Methods", *Numer. Math.* 8, 1965, 1-13.

[BunK77] J.R. Bunch, L. Kaufman, "Some Stable Methods for Calculating Inertia and Solving Symmetric Linear Systems", *Math. Comp.* 31, 1977, 163-180.

[BunKP76] J.R. Bunch, L. Kaufman, B.N. Parlett, "Decomposition of a Symmetric Matrix", *Numerische Mathematik* 27, 1976, 95-109.

[Bus72a] J.C.P. Bus, "Linear Systems with Calculation of Error Bounds and Iterative Refinement", LR 3.4.19, Mathematical Centre, Amsterdam, 1972.

[Bus72b] J.C.P. Bus, "Minimization of Functions of Several Variables", NR 29, Mathematical Centre, Amsterdam, 1972.

[Bus76] J.C.P. Bus (ed.), "Colloquium Numerieke Programmatuur", MC Syllabus NR 29.1a, 29.1b, Mathematical Centre, Amsterdam, 1976.

[BusD75] J.C.P. Bus, T.J. Dekker, "Two Efficient Algorithms with Guaranteed Convergence for Finding a Zero of a Function", *ACM Transactions on Mathematical Software*, 1975, 330-345.

[BusDK75] J.C.P. Bus, B. van Domselaar, J. Kok, "Nonlinear Least Squares Estimation", Report NW 17, Mathematical Centre, Amsterdam, 1975.

[Busi71] P.A. Businger, "Monitoring the Numerical Stability of Gaussian Elimination", *Numer. Math.* 16, 1971, 360-361.

[BusiG65] P. Businger, G.H. Golub, "Linear Least Squares Solution by Householder Transformations", *Numerische Mathematik* 7, 1965, 269-276.

[Cle62] C.W. Clenshaw, "Chebyshev Series for Mathematical Functions", *Mathematical Tables* Vol. 5, Nat. Physical Lab., Her Majesty's Stationery Office, London, 1962.

[Cod68] W.J. Cody, "Chebyshev Approximations for the Fresnel Integrals", *Math. Comp.* 22, 1968, 450-453.

[Cod69] W.J. Cody, "Rational Chebyshev Approximations for the Error

Function", *Math. Comp.* 23, 1969, 631-637.

[CodT68] W.J. Cody, H.C. Thacher, Jr., "Rational Chebyshev Approximations for the Exponential Integral ei(x)", *Math. Comp.* 22, 1968, 641-649.

[CodT69] W.J. Cody, H.C. Thacher, Jr., "Chebyshev Approximations for the Exponential Integral ei(x)", *Math. Comp.* 23, 1969, 289-303.

[CooHHS] T.M.T. Coolen, P.W. Hemker, P.J. van der Houwen, E. Slagt, "Algol 60 Procedures for Initial and Boundary Value Problems", MC Syllabus NR 20, Mathematical Centre, Amsterdam, 1973.

[Dan69] J.W. Daniel, "Summation of a Series of Positive Terms by Condensation Transformations", *Math. Comp.* 23, 1969, 91-96.

[Dav59] W.C. Davidon, "Variable Metric Methods for Minimization", Argonne National Lab., Report 5990, 1959.

[De72] K. Dekker, "An Algol 60 Version of Exponentially Fitted Runge-Kutta Methods", Report NR 25, Mathematical Centre, Amsterdam, 1972.

[Dek66] T.J. Dekker, "Newton-Laguerre Iteration", Mathematisch Centrum MR 82, Amsterdam, 1966.

[Dek68] T.J. Dekker, "ALGOL 60 Procedures in Numerical Algebra, Part 1", Mathematical Centre Tract 22, Mathematisch Centrum, Amsterdam, 1968.

[Dek71] T.J. Dekker, "Numerical Algebra", MC Syllabus NR 12, Mathematical Centre, Amsterdam, 1971.

[DekHH72] T.J. Dekker, P.W. Hemker, P.J. van der Houwen, "Colloquium Stiff Differential Equations 1", MC Syllabus NR 15.1, Mathematical Centre, Amsterdam, 1972.

[DekHo68] T.J. Dekker, W. Hoffmann, "ALGOL 60 Procedures in Numerical Algebra, Part 2", Mathematical Centre Tract 23, Mathematisch Centrum, Amsterdam, 1968.

[DekR71] T.J. Dekker, C.J. Roothart, "Introduction to Numerical Analysis", Report CR 24, Mathematical Centre, Amsterdam, 1971.

[Eng12] R. England, "Error Estimates for Runge-Kutta Type Solutions to Systems of Ordinary Differential Equations", *Computer Journal* 12, 1969, 166-169.

[Fle63] R. Fletcher, "A New Approach to Variable Metric Algorithms", *Computer Journal* 6, 1963, 163-168.

[FoP68] L. Fox, I.B. Parker, *Chebyshev Polynomials in Numerical Analysis*, Oxford University Press, 1968.

[Fox72] P. Fox, "A Comparative Study of Computer Programs for Integrating Differential Equations", *CACM* 15, 1972, 941-948.

[Fr61] J.G. Francis, "The QR Transformation, Parts 1 and 2", *Computer Journal* 4, 1961, 265-271, 332-345.

[G61] W. Gautschi, "Recursive Computation of Certain Integrals", *JACM* 8, 1961, 21-40.

[G64] W. Gautschi, "Incomplete Beta Function Ratios", *CACM* 7, 1964, 143-144.

[G67] W. Gautschi, "Computational Aspects of Three-term Recurrence Relations", *SIAM Review* 9, 1967, 24-82.

[G68a] W. Gautschi, "Construction of Gauss-Christoffel Formulas", *Math. Comp.* 22, 1968, 251-270.

[G68b] W. Gautschi, "Gaussian Quadrature Formulas", *CACM* 11, No. 6, 1968, 432-436.

[G70] W. Gautschi, "Generation of Gaussian Quadrature Rules and Orthogonal Polynomials", in "Colloquium Approximatietheorie", MC Syllabus NR 14, Mathematisch Centrum, Amsterdam, 1970.

[G73] W. Gautschi, "Exponential Integrals", *CACM* 16, 1973, 761-763.

[Ge69] W.M. Gentleman, "An Error Analysis of Goertzel's (Watt's) Method for Computing Fourier Coefficients", *Comput. J.*, Vol. 12, 1969, 160-165.

[GolP67] A.A. Goldstein, J.F. Price, "An Effective Algorithm for Minimization", *Numer. Math.* 10, 1967, 184-189.

[GolW69] G.H. Golub, J.H. Welsch, "Calculation of Gauss Quadrature Rules", *Math. Comp.* 23, 1969, 221-230.

[Gor69] R.G. Gordon, "Evaluation of Airy Functions", *The Journal of Chemical Physics* 51, 1969, 23-24.

[Gr67] J.L. Greenstadt, "On the Relative Efficiencies of Gradient Methods", *Math. Comp.* 21, 1967, 360-367.

[GrK69] R.T. Gregory, D.L. Karney, *A Collection of Matrices for Testing Computational Algorithms*, Wiley-Interscience, 1969.

[H73] R.W. Hamming, *Numerical Methods for Scientists and Engineers*, McGraw-Hill, 1973.

[HaCL68] J.F. Hart, E.W. Cheney, C.L. Lawson, H.J. Maehly, C.K. Mesztenyi, J.R. Rice, H.C. Thacher, Jr., C. Witzgall, *Computer Approximations*, Wiley, New York, 1968.

[Har61] H.O. Hartley, "The Modified Gauss-Newton Method", *Technometrics* 3, 1961, 269-280.

[He71] P.W. Hemker, "An Algol 60 Procedure for the Solution of Stiff Differential Equations", Report MR 128, Mathematical Centre, Amsterdam, 1971.

[He73] P.W. Hemker, "A Sequence of Nested Cubature Rules", Report NW 3, Mathematical Centre, Amsterdam, 1973.

[He75] P.W. Hemker, "Galerkin's Method and Lobatto Points", Report 24, Mathematisch Centrum, Amsterdam, 1975.

[He80] P.W. Hemker (ed.), "NUMAL, Numerical Procedures in ALGOL 60", MC Syllabus, 47.1-47.7, Mathematisch Centrum, Amsterdam, 1980.

[HEFS72] T.E. Hull, W.H. Enright, B.M. Fellen, A.E. Sedgwick, "Comparing Numerical Methods for Ordinary Differential Equations", *SIAM Journal on Numerical Analysis* 9, 1972, 603-637.

[Hu64] D.B. Hunter, "The Calculation of Certain Bessel Functions", *Math. Comp.* 18, 1964, 123-128.

[K69] D.E. Knuth, *The Art of Computer Programming, Seminumerical Algorithms*, Vol. 2, Addison-Wesley Pub. Co., 1969.

[K73] D.E. Knuth, *The Art of Computer Programming, Vol. 3, Sorting and Searching*, Addison-Wesley, 1973.

[L57] C. Lanczos, *Applied Analysis*, Prentice Hall, 1957.

[Li73] B. Lindberg, "IMPEX 2, A Procedure for the Solution of Systems of Stiff Differential Equations", Report TRITA-NA-7303, Royal Institute of Technology, Stockholm, 1973.

[LiW70] W. Liniger, R.A. Willoughby, "Efficient Integration Methods for Stiff Systems of Ordinary Differential Equations", *SIAM J. Numer. Anal.* 7, 1970, 47-66.

[Lu69] Y.L. Luke, *The Special Functions and their Approximations*, 2 Vols., Academic Press, 1969.

[Lu70] Y.L. Luke, "Evaluation of the Gamma Function by means of Padé Approximations", *SIAM J. Math. Anal.* 1, 1970, 266-281.

[M63] D.W. Marquardt, "An Algorithm for Least-Squares Estimation of Nonlinear Parameters", *J. SIAM* 11, 1963, 431-441.

[Me64] G. Meinardus, *Approximation of Function and their Numerical Treatment*, Springer Tracts in Natural Philosophy, Vol. 4, 1964.

[MS71] C.B. Moler, G.W. Stewart, "An algorithm for the Generalized Matrix Eigenvalue Problem Ax=λBx", report STAN-CS-232-71, Stanford University.

[Mu66] D.J. Mueller, "Householder's Method for Complex Matrices and Eigensystems of Hermitian Matrices", *Numerische Mathematik* 8, 1966, 72-92.

[N73] A.C.R. Newbery, "Error Analysis for Fourier Series Evaluation", *Math. Comp.*, Vol. 26, 1973, 923-924.

[Os60] E.E. Osborne, "On Preconditioning of Matrices", *J. Assoc. Comput. Mach.* 7, 1960, 338-354.

[PaR69] B.N. Parlett, C. Reinsch, "Balancing a Matrix for Calculation of Eigenvalues and Eigenvectors", *Numerische Mathematik* 13, 1969, 293-304.

[PeW71] G. Peters, J.H. Wilkinson, "Practical Problems Arising in the Solution of Polynomial Equations", *Journal of the Institute of Mathematics and Its Applications.* 8, No. 1, 1971, 16-35.

[Po70] M.J.D. Powell, "Rank One Methods for Unconstrained Optimization", in *Integer and Nonlinear Programming*, edited by J. Abadie, North-Holland, 1970.

[Pr92] W.H. Press, S.A. Teukolsky, W.T. Vetterling, B.P. Flannery, *Numerical Recipes in C*, second edition, Cambridge University Press, 1992.

[R71] J.K. Reid, "On the Method of Conjugate Gradients for the Solution of Large Sparse Systems of Linear Equations", in *Large Sparse Sets of Linear Equations*, edited by J.K. Reid, Academic Press, 1971.

[Re67] C. Reinsch, "A note on Trigonometric Interpolation", Report No. 6709, Mathematics Department, Technische Universität München, 1967.

[Re71] C.H. Reinsch, "A Stable, Rational QR Algorithm for the Computation of the Eigenvalues of an Hermitian, Tridiagonal

Matrix", *Math. Comp.* 25, 1971, 591-597.

[Rey77] Th.H.P. Reymer, "Berekening Van Nulpunten Van Reele Polynomen En Foutgrenzen Voor Deze Nulpunten", Doctoraal Scriptie UVA, April 1977.

[Rie76] H.J.J. te Riele (ed.), "Colloquium Numerieke Programmatuur", MC Syllabus NR 29.2, Mathematical Centre, Amsterdam, 1977.

[Riv74] T.J. Rivlin, *The Chebyshev Polynomials*, Wiley, 1974.

[RoF72] C.J. Roothart, H. Fiolet, "Quadrature Procedures", Report MR 137, Mathematical Centre, Amsterdam, 1972.

[ShT74] M. Shaw and J. Traub, "On the Number of Multiplications for the Evaluation of a Polynomial and some of its Derivatives", *J. Assoc. Comput. Mach.*, 1974, Vol. 21, No. 1, 161-167.

[Sp67] H. Spaeth, "The Damped Taylor's Series Method for Minimizing a Sum of Squares and for Solving Systems of Nonlinear Equations", *CACM* 10, 1967, 726-728.

[St72] J. Stoer, *Einfuehrung in die Numerische Mathematik I*, Heidelberger Taschenbuecher 105, Springer-Verlag, 1972.

[StF73] G. Strang, G.J. Fix, *An Analysis of the Finite Element Method*, Prentice-Hall, 1973.

[Str68] A.J. Strecok, "On the Calculation of the Inverse of the Error Function", *Math. Comp.* 22, 1968, 144-158.

[T75] N.M. Temme, "On the Numerical Evaluation of the Ordinary Bessel Function of the Third Kind", *J. Comput. Phys.* 19, 1975, 324-337.

[T76a] N.M. Temme, "On the Numerical Evaluation of the Ordinary Bessel Function of the Second Kind", *J. Comput. Phys.* 21, 1976, 343-350.

[T76b] N.M. Temme, "Speciale Functies", in "Colloquium Numerieke Programmatuur", J.C.P. Bus (ed.), MC Syllabus NR 29.1b, Mathematical Centre, Amsterdam, 1976.

[T78] N.M. Temme, "An Algorithm with Algol 60 Implementation for the Calculation of the Zeros of a Bessel Function", Report TW 179, Mathematical Centre, Amsterdam, 1978.

[Vh68] P.J. van der Houwen, "Finite Difference Methods for Solving Partial Differential Equations", Mathematical Centre Tract 20, Mathematical Centre, Amsterdam, 1968.

[Vh70a] P.J. van der Houwen, "One-Step Methods for Linear Initial Value

Problems I, Polynomial Methods", Report TW 119, Mathematical Centre, Amsterdam, 1970.

[Vh70b] P.J. van der Houwen, "One-Step Methods for Linear Initial Value Problems II, Polynomial Methods", Report TW 122, Mathematical Centre, Amsterdam, 1970.

[Vh71] P.J. van der Houwen, "Stabilized Runge Kutta Method with Limited Storage Requirements", Report TW 124, Mathematical Centre, Amsterdam, 1971.

[Vh72] P.J. van der Houwen, "One-Step Methods with Adaptive Stability Functions for the Integration of Differential Equations", Lecture Notes of the Conference on Numerische Insbesondere Approximationstheoretische Behandlung Von Funktional-Gleichungen, Oberwolfach, December 3-12, 1972.

[VhBDS71] P.J. van der Houwen, P. Beentjes, K. Dekker, E. Slagt, "One Step Methods for Linear Initial Value Problems III, Numerical Examples", Report TW 130, Mathematical Centre, Amsterdam, 1971.

[VhK71] P.J. van der Houwen, J. Kok, "Numerical Solution of a Minimax Problem", Report TW 123, Mathematical Centre, Amsterdam, 1971.

[VhV74] P.J. van der Houwen, J.G. Verwer, "Generalized Linear Multistep Methods 1, Development of Algorithms with Zero-Parasitic Roots", Report NW 10, Mathematisch Centrum, Amsterdam, 1974.

[Vn75] B. van Domselaar, "Nonlinear Parameter Estimation in Initial Value Problems", Report NW 18, Mathematical Centre, Amsterdam, 1975.

[Vnw65] A. van Wijngaarden, "Course Scientific Computing B, Process Analysis", Mathematisch Centrum CR 18, Amsterdam, 1965.

[Vr66] J.M. Varah, "Eigenvectors of a Real Matrix by Inverse Iteration", Technical Report No. CS 34, Stanford University, 1966.

[Vw74] J.G. Verwer, "Generalized Linear Multistep Methods 2, Numerical Applications", Report NW 12, Mathematisch Centrum, Amsterdam, 1974.

[W45] G.N. Watson, *A Treatise on the Theory of Bessel Functions*, Cambridge University Press, 1945.

[Wi63] J.H. Wilkinson, *Rounding Errors in Algebraic Processes*, London,

1963.

[Wi65] J.H. Wilkinson, *The Eigenvalue Problem*, Oxford, 1965.

[WiR71] J.H. Wilkinson, C. Reinsch, *Handbook of Automatic Computation, Vol. 2, Linear Algebra*, Springer-Verlag, 1971.

[Wo74] H. Wozniakowski, "Rounding Error Analysis for the Evaluation of a Polynomial and some of its Derivatives", *SIAM J. Numer. Anal.*, Vol. 11, No. 4, 1974, 780-787.

[Wy81] P. Wynn, "NUMAL in FORTRAN", IIMAS, Universidad Nacional Autónoma de México, Comunicaciones técnicas Nos. 48.0-48.11, 1981.

[Z64] J.A. Zonneveld, "Automatic Numerical Integration", Mathematical Centre Tract 8, Mathematisch Centrum, Amsterdam, 1964.

Appendix B: Prototype Declarations

```
float absmaxmat(int, int, int, int, int *, int *, float **);
void airy(float, float *, float *, float *, float *,
       float *, int);
float airyzeros(int, int, float [], float []);
void allchepol(int, float, float []);
void alljaczer(int, float, float, float []);
void alllagzer(int, float, float []);
int **allocate_integer_matrix(int, int, int, int);
int *allocate_integer_vector(int, int);
float **allocate_real_matrix(int, int, int, int);
float *allocate_real_vector(int, int);
void allortpol(int, float, float [], float [], float []);
void allortpolsym(int, float, float [], float []);
void allzerortpol(int, float [], float [], float [], float []);
float arccosh(float);
float arcsinh(float);
float arctanh(float);
void ark(float *, float *, int *, int *, float [],
    void (*)(int *, int *, float *, float[]), float [],
    void (*)(int *, int *, float *, float *,
       float [], float []));
void arkmat(float *, float, int, int, float **,
       void (*)(int, int, float, float **, float **),
       int, int *, float *,
       void (*)(float, float, int, int, float **, int, int, float *));
void backward(float, float, float, float, int, float, float []);
void bakcomhes(float **, float **, float [], float [], float [],
       float **, float **, int, int, int);
void bakhrmtri(float **, int, int, int, float **,
       float **, float [], float []);
void baklbr(int, int, int, float [], int [], float **);
void baklbrcom(int, int, int, float [], int [], float **, float **);
void bakreahes1(float **, int, int [], float []);
void bakreahes2(float **, int, int, int, int [], float **);
void baksymtri1(float [], int, int, int, float **);
void baksymtri2(float **, int, int, int, float **);
void besspqa01(float, float, float *, float *, float *, float *);
void bessi(float, int, float []);
float bessi0(float);
float bessi1(float);
void bessiaplusn(float, float, int, float []);
void bessj(float, int, float []);
float bessj0(float);
float bessj1(float);
void bessjaplusn(float, float, int, float []);
void bessk(float, int, float []);
void bessk01(float, float *, float *);
void besska01(float, float, float *, float *);
void besskaplusn(float, float, int, float []);
void besspq0(float, float *, float *);
void besspq1(float, float *, float *);
void bessy(float, int, float []);
void bessy01(float, float *, float *);
void bessya01(float, float, float *, float *);
void bessyaplusn(float, float, int, float []);
void besszeros(float, int, float [], int);
void bounds(int, float [], float [], float [], float,
       float, float [], float [], float []);
void carpol(float, float, float *, float *, float *);
float chepol(int, float);
float chepolsum(int, float, float []);
void chldec1(float [], int, float []);
void chldec2(float **, int, float []);
void chldecbnd(float [], int, int, float []);
void chldecinv1(float [], int, float []);
void chldecinv2(float **, int, float []);
void chldecsol1(float [], int, float [], float []);
void chldecsol2(float **, int, float [], float []);
void chldecsolbnd(float [], int, int, float [], float []);
```

```
float chldeterm1(float [], int);
float chldeterm2(float **, int);
float chldetermbnd(float [], int, int);
void chlinv1(float [], int);
void chlinv2(float **, int);
void chlsol1(float [], int, float []);
void chlsol2(float **, int, float []);
void chlsolbnd(float [], int, int, float []);
void chsh2(float, float, float, float, float *, float *, float *);
void chspol(int, float []);
void colcst(int, int, int, float **, float);
float comabs(float, float);
void comcolcst(int, int, int, float **, float **, float, float);
void comdiv(float, float, float, float, float *, float *);
int comeig1(float **, int, float [], float [], float [], float **);
int comeigval(float **, int, float [], float [], float []);
float comeucnrm(float **, float **, int, int);
void comfouser1(int, float, float [], float [], float *, float *);
void comfouser2(int, float, float [], float [], float *, float *);
void comfouser(int, float, float [], float *, float *);
void comkwd(float, float, float, float,
       float *, float *, float *, float *);
void commatvec(int, int, int, float **, float **,
       float [], float [], float *, float *);
void commul(float, float, float, float, float *, float *);
void comrowcst(int, int, int, float **, float **, float, float);
void comscl(float **, int, int, int, float []);
void comsqrt(float, float, float *, float *);
int comvalqri(float **, int, float [], float [], float []);
void comveches(float **, int, float, float,
       float [], float [], float []);
void conjgrad(void (*)(float [], float []), float [], float [],
       int, int, int (*)(int, float), int *, float *);
float cosser(int, float, float []);
void davupd(float [], int, float [], float [], float, float);
void dec(float **, int, float [], int []);
void decbnd(float [], int, int, int, float [], float [], int []);
void decinv(float **, int, float []);
void decsol(float **, int, float [], float []);
void decsolbnd(float [], int, int, int, float [], float []);
void decsolsym2(float **, int, float [], float, int []);
void decsolsymtri(float [], float [], int, float [], float []);
void decsoltri(float [], float [], float [], int,
       float [], float []);
void decsoltripiv(float [], float [], float [], int,
       float [], float []);
void decsym2(float **, int, float, int [], int [], float []);
void decsymtri(float [], float [], int, float []);
void dectri(float [], float [], float [],    int, float []);
void dectripiv(float [], float [], float [], int,
       float [], float [], int []);
void derpol(int, int, float, float []);
float determ(float **, int, int);
float determbnd(float [], int, int, int, int);
float determsym2(float [], int, int []);
void diffsys(float *, float, int, float [],
       void (*)(int, float, float [], float []),
       float, float, float [], float,
       void (*)(int, float, float, float [], float []));
void dupcolvec(int, int, int, float **, float []);
void dupmat(int, int, int, int, float **, float **);
void duprowvec(int, int, int, float **, float []);
void dupvec(int, int, int, float [], float []);
void dupveccol(int, int, int, float [], float **);
void dupvecrow(int, int, int, float [], float **);
void eferk(float *, float, int, float [], float *, float,
       void (*)(int, float[]), float **,
       void (*)(int, float **, float [], float *),
       int *, int, int, float, float, float, float, int,
       void (*)(float, float, int, float [], float **, int));
void efrk(float *, float, int, int, float [], float *,
       float *, float *, void (*)(int, int, float, float[]),
       int *, float *, float, float, float [], int, float,
       void (*)(int, int, float, float, float [], float *,
```

```
            float *, float *, int, float *, int, int));
void efsirk(float *, float, int, float [], float *,
        void (*)(int, float[], float *),
        void (*)(int, float **, float [], float *),
        float **, int *, float, float, float, float, int,
        void (*)(float, float, int, float [],
            float, float **, int));
void eft(float *, float, int, int, float [],
    float (*)(float, int, int), float,
    float (*)(float, int, int),
    void (*)(float, int, int, int, float []),
    int *, float, int,
    float (*)(float, int, int), float (*)(float, int, int),
    float *, float *, float, float *,
    void (*)(float, float, int, int, float [],
        int, float, float));
float ei(float);
void eialpha(float, int, float []);
int eigcom(float **, float **, int, float [], float [],
        float [], float **, float **);
void eighrm(float **, int, int, float [], float **, float **,
        float []);
void eigsym1(float [], int, int, float [], float **, float []);
void eigsym2(float **, int, int, float [], float **, float []);
int eigvalcom(float **, float **, int, float [],
        float [], float []);
void eigvalhrm(float **, int, int, float [], float []);
void eigvalsym1(float [], int, int, float [], float []);
void eigvalsym2(float **, int, int, float [], float []);
void elimination(float **, int, int, int, int,
        void (*)(int, int, int, int, float **),
        float, float, int *, float [], int *, float *, float *,
        void (*)(float **, int, int, int, int, int *, float [],
            int, float, float));
void elmcol(int, int, int, int, float **, float **, float);
void elmcolrow(int, int, int, int, float **, float **, float);
void elmcolvec(int, int, int, float **, float [], float);
void elmcomcol(int, int, int, int, float **, float **,
        float **, float **, float, float);
void elmcomrowvec(int, int, int, float **, float **,
        float [], float [], float, float);
void elmcomveccol(int, int, int, float [], float [],
        float **, float **, float, float);
void elmrow(int, int, int, int, float **, float **, float);
void elmrowcol(int, int, int, int, float **, float **, float);
void elmrowvec(int, int, int, float **, float [], float);
void elmvec(int, int, int, float [], float [], float);
void elmveccol(int, int, int, float [], float **, float);
void elmvecrow(int, int, int, float [], float **, float);
void enx(float, int, int, float []);
void eqilbr(float **, int, float [], float [], int []);
void eqilbrcom(float **, float **, int, float [], float [], int []);
void erbelm(int, float [], float);
void errorfunction(float, float *, float *);
float euler(float (*)(int), int *, float, int);
void femhermsym(float [], float [], int, float (*)(float),
        float (*)(float), float (*)(float), float (*)(float),
        int, float []);
void femlag(float [], float [], int, float (*)(float),
        float (*)(float), int, float []);
void femlagskew(float [], float [], int, float (*)(float),
        float (*)(float), float (*)(float), int, float []);
void femlagspher(float [], float [], int, int, float (*)(float),
        float (*)(float), int, float []);
void femlagsym(float [], float [], int, float (*)(float),
        float (*)(float), float (*)(float), int, float []);
void fg(float, float *, float *);
float flemin(int, float [], float [], float [],
        float (*)(int, float[], float[]), float [], float []);
void fleupd(float [], int, float [], float [], float, float);
void forward(float x, float p, float q, float i0, float i1,
        int nmax, float i[]);
float fouser(int, float, float []);
float fouser1(int, float, float [], float []);
```

```
float fouser2(int, float, float [], float []);
void free_integer_matrix(int **, int, int, int);
void free_integer_vector(int *, int);
void free_real_matrix(float **, int, int, int);
void free_real_vector(float *, int);
void fresnel(float, float *, float *);
void fulmatvec(int, int, int, int, float **, float [], float []);
void fulsymmatvec(int, int, int, int, float [], float [], float []);
void fultamvec(int, int, int, int, float **, float [], float []);
float gamma(float);
void gms(float *, float, int, float [], float, float,
    float, float *, void (*)(int, float [], float *),
    void (*)(int, float **, float [], float *),
    float, float, int *, int *, int *, int, int,
    void (*)(float, float, int, float [], float,
        int, int, int));
void grnnew(int, float [], float []);
void gssitisolerb(float **, int, float [], float []);
void gsselm(float **, int, float [], int [], int []);
void gsserb(float **, int, float [], int [], int []);
void gssinv(float **, int, float []);
void gssinverb(float **, int, float []);
void gssitisol(float **, int, float [], float []);
void gssjacwghts(int, float, float, float [], float []);
void gsslagwghts(int, float, float [], float []);
void gssnewton(int, int, float [], float [], float **,
        int (*)(int, int, float[], float[]),
        void (*)(int, int, float[], float[], float **),
            float [], float []);
void gssnri(float **, int, float [], int [], int []);
void gsssol(float **, int, float [], float []);
void gsssolerb(float **, int, float [], float []);
void gsswts(int, float [], float [], float [], float []);
void gsswtssym(int, float [], float [], float []);
void hestgl2(int, float **, float **);
void hestgl3(int, float **, float **, float **);
int homsol(float **, int, int, float **, float []);
void homsolsvd(float **, float [], float **, int, int);
void hsh2col(int, int, int, int, float, float, float **, float **);
void hsh2row2(int, int, int, int, float, float,
        float **, float **);
void hsh2row3(int, int, int, int, int, float, float,
        float **, float **, float **);
void hsh3col(int, int, int, int, float, float, float,
        float **, float **);
void hsh3row2(int, int, int, float, float, float,
        float **, float **);
void hsh3row3(int, int, int, int, float, float, float,
        float **, float **, float **);
void hshcolmat(int, int, int, int, int, float, float **, float **);
void hshcoltam(int, int, int, int, int, float, float **, float **);
int hshcomcol(int, int, int, float **, float **, float,
        float *, float *, float *, float *);
void hshcomhes(float **, float **, int, float [], float [],
        float [], float [], float []);
void hshcomprd(int, int, int, int, int, float **,
        float **, float **, float **, float);
void hshdecmul(int, float **, float **, float);
void hshhrmtri(float **, int, float [], float [], float [],
        float [], float [], float []);
void hshhrmtrival(float **, int, float [], float [], float []);
void hshreabid(float **, int, int, float [], float [], float []);
void hshrowmat(int, int, int, int, int, float, float **, float **);
void hshrowtam(int, int, int, int, int, float, float **, float **);
void hshvecmat(int, int, int, int, float, float [], float **);
void hshvectam(int, int, int, int, float, float [], float **);
void ibpplusn(float, float, float, int, float, float []);
void ibqplusn(float, float, float, int, float, float []);
void ichcol(int, int, int, int, float **);
void ichrow(int, int, int, int, float **);
void ichrowcol(int, int, int, int, float **);
void ichseq(int, int, int, int, float []);
void ichseqvec(int, int, int, int, float []);
void ichvec(int, int, int, float []);
```

```
void impex(int, float, float, float [],
       void (*)(float, float [], float [], int),
       int (*)(float, float [], float **, int),
       float, float, int, float, float [],
       void (*)(float [], float [], int), int *,
       void (*)(float *, float, float, float, float **,
           float [], int, float));
float incbeta(float, float, float, float);
void incomgam(float, float, float *, float *, float, float);
float infnrmcol(int, int, int, int *, float **);
float infnrmmat(int, int, int, int, int *, float **);
float infnrmrow(int, int, int, int *, float **);
float infnrmvec(int, int, int *, float []);
void ini(int, int, int []);
void inimat(int, int, int, int, float **, float);
void inimatd(int, int, int, float **, float);
void inisymd(int, int, int, float [], float);
void inisymrow(int, int, int, float [], float);
void inivec(int, int, float [], float);
void intchs(int, float [], float []);
float integral(float, float, float (*)(float), float [],
       int, int);
void inv(float **, int, int []);
float inv1(float **, int, int [], int [], int);
void inverseerrorfunction(float, float, float *);
void itisol(float **, float **, int, float [],
       int [], int [], float []);
void itisolerb(float **, float **, int, float [],
       int [], int [], float []);
void ixpfix(float, float, float, int, float, float []);
void ixqfix(float, float, float, int, float, float []);
void jacobnbndf(int, int, int, float [], float [],
       float [], float (*)(int),
       int (*)(int, int, int, float[], float[]));
void jacobnmf(int, int, float [], float [], float **,
       float (*)(int), void (*)(int, int, float[], float[]));
void jacobnnf(int, float [], float [], float **,
       float (*)(int), void (*)(int, float[], float[]));
float jfrac(int, float [], float []);
void linemin(int, float [], float [], float, float *, float [],
       float (*)(int, float[], float[]), float, float *,
       float, float *, int *, int, float []);
void liniger1vs(float *, float, int, float [], float *,
       void (*)(int, float[], float *), float **,
       void (*)(int, float **, float [], float *),
       int, float, float, float, float,  float [],
       void (*)(float, float, int, float [], float,
           float **, float []));
void liniger2(float *, float, int, float [], float *, float *,
       float (*)(int, float[], int, float *, float *),
       int (*)(int), float **,
       void (*)(int, float **, float [], float *, float *),
       int *, int, float, float, float,
       void (*)(float, float, int, float [], float, float,
           float **, int));
void lintfmpol(float, float, int, float []);
void lngintadd(int [], int [], int []);
void lngintdivide(int [], int [], int [], int []);
void lngintmult(int [], int [], int []);
void lngintpower(int [], int, int []);
void lngintsubtract(int [], int [], int []);
float loggamma(float);
float logoneplusx(float);
void lsqdecomp(float **, int, int, int, float [], float [], int []);
void lsqdglinv(float **, int, float [], int [], float []);
void lsqinv(float **, int, float [], int []);
void lsqortdec(float **, int, int, float [], float [], int []);
void lsqortdecsol(float **, int, int, float [], float [], float []);
void lsqrefsol(float **, float **, int, int, int, float [], float [],
       int [], float [], float *, float [], float []);
void lsqsol(float **, int, int, float [], int [], float []);
void lupzerortpol(int, int, float [], float [], float [], float []);
void marquardt(int, int, float [], float [], float **,
       int (*)(int, int, float[], float[]),
```

```
        void (*)(int, int, float[], float[], float **),
            float [], float []);
float matmat(int, int, int, int, float **, float **);
float mattam(int, int, int, int, float **, float **);
float matvec(int, int, int, float **, float []);
int maxelmrow(int, int, int, int, float **, float **, float);
void mergesort(float [], int [], int, int);
float minin(float *, float *, float *, float (*)(float),
        float (*)(float));
float mininder(float *, float *, float (*)(float),
        float (*)(float), float (*)(float));
void minmaxpol(int, int, float [], float [], float [], float []);
void modifiedtaylor(float *, float, int, int, float [],
        float (*)(float, int, int), float,
        void (*)(float, int, int, int, float []),
        int *, float [], float, int, float (*)(float, int, int),
        float (*)(float, int, int), float *, float *,
        void (*)(float, float, int, int, float [],
            int, float, float));
void mulcol(int, int, int, int, float **, float **, float);
void mulrow(int, int, int, int, float **, float **, float);
int multistep(float *, float, float [], float, float,
        float [], float, int *, float [],
        void (*)(float [], int, float, float[]),
        int (*)(int, float, float [], float **), float **, int,
        int, void (*)(float, int, int, float, float []));
void mulvec(int, int, int, float [], float [], float);
void nonexpbessi(float, int, float []);
float nonexpbessi0(float);
float nonexpbessi1(float);
void nonexpbessiaplusn(float, float, int, float []);
void nonexpbessk(float, int, float []);
void nonexpbessk01(float, float *, float *);
void nonexpbesska01(float, float, float *, float *);
void nonexpbesskaplusn(float, float, int, float []);
void nonexpspherbessi(float, int, float []);
void nonexpspherbessk(float, int, float []);
void newgrn(int, float [], float []);
void newton(int, float [], float []);
void nonexpenx(float, int, int, float []);
float nonexperfc(float);
void nonlinfemlagskew(float [], float [], int,
            float (*)(float, float, float),
            float (*)(float, float, float),
            float (*)(float, float, float), int, float []);
void norderpol(int, int, float, float []);
float oddchepolsum(int, float, float []);
float onenrmcol(int, int, int, float **);
float onenrminv(float **, int);
float onenrmmat(int, int, int, int, int *, float **);
float onenrmrow(int, int, int, float **);
float onenrmvec(int, int, float []);
void orthog(int, int, int, float **);
float ortpol(int, float, float [], float []);
float ortpolsym(int, float, float []);
void peide(int, int, int, int *, float [], float [],
        int [], float **, float [], float [],
        int (*)(int,int,float [],float [],float,float []),
        int (*)(int,int,float [],float [],float,float **),
        int (*)(int,int,float [],float [],float,float **),
        void (*)(int,int,float [],float [],float[]),
        void (*)(int,float [],float [],int[]),
        void (*)(int,int,int,float [],float [],int,int));
float pol(int, float, float []);
void polchs(int, float []);
void polshtchs(int, float []);
void praxis(int, float [], float (*)(int, float[]),
        float [], float []);
void pretfmmat(float **, int, int, float []);
int psdinv(float **, int, int, float []);
void psdinvsvd(float **, float [], float **, int, int, float []);
void psttfmmat(float **, int, float **, float []);
float qadrat(float *, float, float, float (*)(float), float []);
int qricom(float **, float **, float [], int, float [],
```

```
        float [], float [], float **, float **);
int qrihrm(float **, int, float [], float **, float **, float []);
int qrisngval(float **, int, int, float [], float []);
int qrisngvalbid(float [], float [], int, float []);
int qrisngvaldec(float **, int, int, float [], float **, float []);
int qrisngvaldecbid(float [], float [], int, int, float **,
            float **, float []);
int qrisym(float **, int, float [], float []);
int qrisymtri(float **, int, float [], float [], float [], float []);
int qrivalhrm(float **, int, float [], float []);
int qrivalsym1(float [], int, float [], float []);
int qrivalsym2(float **, int, float [], float []);
int qrivalsymtri(float [], float [], int, float []);
void quanewbnd(int, int, int, float [], float [], float [],
        int (*)(int, int, int, float[], float[]),
        float [], float []);
void quanewbnd1(int, int, int, float [], float [],
        int (*)(int, int, int, float[], float[]),
        float [], float []);
void qzi(int, float **, float **, float **, float [],
        float [], float [], int [], float []);
void qzival(int, float **, float **, float [], float [],
        float [], int [], float []);
int reaeig1(float **, int, float [], float [], float **);
int reaeig3(float **, int, float [], float [], float **);
int reaeigval(float **, int, float [], float []);
int reaqri(float **, int, float [], float [], float **);
void reascl(float **, int, int, int);
int reavalqri(float **, int, float [], float []);
void reaveches(float **, int, float, float [], float []);
void reccof(int, int, float *, float (*)(float), float [],
        float [], float [], int);
float recipgamma(float, float *, float *);
void resvec(int, int, int, int, float **, float [], float [], float);
void richardson(float **, int, int, int, int,
        int, void (*)(int, int, int, int, float **),
        float, float, int, float [], int *, float *, float *,
        void (*)(float **, int, int, int, int, int, float [],
            int, float, float));
void rk1(float *, float, float, float *, float,
        float (*)(float, float), float [], float [], int);
void rk2(float *, float, float, float *, float, float *, float,
        float (*)(float, float, float), float [], float [], int);
void rk2n(float *, float, float, float [], float [], float [],
        float [], float (*)(int, int, float, float [], float []),
        float [], float [], int, int);
void rk3(float *, float, float, float *, float, float *, float,
        float (*)(float, float), float [], float [], int);
void rk3n(float *, float, float, float [], float [], float [],
        float [], float (*)(int, int, float, float[]),
        float [], float [], int, int);
void rk4a(float *, float, float (*)(float, float),
        float *, float, float (*)(float, float),
        float [], float [], int, int, int);
void rk4na(float [], float [], float (*)(int, float[]),
        float (*)(int, int, float[]), float [], float [],
        int, int, int, int);
void rk5na(float [], float [], float (*)(int, float[]),
        float (*)(int, int, float[]), float [], float [],
        int, int, int, int);
void rke(float *, float *, int, float [],
        void (*)(int, float, float[]), float [], int,
        void (*)(int, float, float, float [], float []));
float rnk1min(int, float [], float [], float [],
        float (*)(int, float[], float[]),
        float [], float []);
void rnk1upd(float [], int, float [], float);
void rotcol(int, int, int, int, float **, float, float);
void rotcomcol(int, int, int, int, float **, float **,
        float, float, float);
void rotcomrow(int, int, int, float **, float **,
        float, float, float);
void rotrow(int, int, int, int, float **, float, float);
void rowcst(int, int, int, float **, float);
```

```
void rowperm(int [], int, int, int, float **);
float scaprd1(int, int, int, int, int, float [], float []);
void sclcom(float **, float **, int, int, int);
void selzerortpol(int, int, int, float [], float [],
          float [], float []);
float seqvec(int, int, int, int, float [], float []);
void shtchspol(int, float []);
void sincosfg(float, float *, float *);
void sincosint(float, float *, float *);
float sinser(int, float, float []);
void sndremez(int, int, int [], float [], float []);
void sol(float **, int, int [], float []);
void solbnd(float [], int, int, int, float [], int [], float []);
void solelm(float **, int, int [], int [], float []);
int solovr(float **, int, int, float [], float []);
void solsvdovr(float **, float [], float **, int, int,
      float [], float []);
void solsvdund(float **, float [], float **, int, int,
      float [], float []);
void solsym2(float **, int, float [], int [], float []);
void solsymtri(float [], float [], int, float []);
void soltri(float [], float [], float [], int, float []);
void soltripiv(float [], float [], float [], int,
      float [], int [], float []);
int solund(float **, int, int, float [], float []);
void spherbessi(float, int, float []);
void spherbessj(float, int, float []);
void spherbessk(float, int, float []);
void spherbessy(float, int, float []);
int start(float, int, int);
float sumortpol(int, float, float [], float [], float []);
float sumortpolsym(int, float, float [], float []);
float sumposseries(float (*)(float), int, float, int, int, int);
void symeigimp(int, float **, float **, float [], float [],
      float [], float []);
float symmatvec(int, int, int, float [], float []);
void symresvec(int, int, int, int, float [], float [], float [], float);
void system_error(char []);
float tammat(int, int, int, int, float **, float **);
float tamvec(int, int, int, float **, float []);
void taypol(int, int, float, float []);
void tfmprevec(float **, int);
void tfmreahes(float **, int, float [], int []);
void tfmsymtri1(float [], int, float [], float [], float [],
      float []);
void tfmsymtri2(float **, int, float [], float [], float [],
      float []);
float tricub(float, float, float, float, float, float,
      float (*)(float,float), float, float);
int valqricom(float **, float **, float [], int, float [],
      float [], float []);
void valsymtri(float [], float [], int, int, int,
      float [], float []);
void vecperm(int [], int, int, float []);
void vecsymtri(float [], float [], int, int, int,
      float [], float **, float []);
float vecvec(int, int, int, float [], float []);
int zeroin(float *, float *, float (*)(float), float (*)(float));
int zeroinder(float *, float *, float (*)(float),
      float (*)(float), float (*)(float));
int zeroinrat(float *, float *, float (*)(float), float (*)(float));
int zerpol(int, float [], float [], float [], float [], float []);
```

Appendix C: Procedure Descriptions

INIVEC initializes a vector with a constant.
INIMAT initializes a matrix with a constant.
INIMATD initializes a (co)diagonal of a matrix.
INISYMD initializes a (co)diagonal of a symmetric matrix, whose upper triangle is stored columnwise in a one-dimensional array.
INISYMROW initializes a row of a symmetric matrix, whose upper triangle is stored columnwise in a one-dimensional array.
DUPVEC copies a vector into another vector.
DUPVECROW copies a row vector into a vector.
DUPROWVEC copies a vector into a row vector.
DUPVECCOL copies a column vector into a vector.
DUPCOLVEC copies a vector into a column vector.
DUPMAT copies a matrix into another matrix.
MULVEC stores a constant multiplied by a vector into a vector.
MULROW stores a constant multiplied by a row vector into a row vector.
MULCOL stores a constant multiplied by a solumn vector into a column vector.
COLCST multiplies a column vector by a constant.
ROWCST multiplies a row vector by a constant.
VECVEC forms the scalar product of a vector and a vector.
MATVEC forms the scalar product of a row vector and a vector.
TAMVEC forms the scalar product of a column vector and a vector.
MATMAT forms the scalar product of a row vector and a column vector.
TAMMAT forms the scalar product of a column vector and a column vector.
MATTAM forms the scalar product of a row vector and a row vector.
SEQVEC forms the scalar product of two vectors given in one-dimensional arrays, where the mutual spacings between the indices of the first vector change linearly.
SCAPRD1 forms the scalar product of two vectors given in one-dimensional arrays, where the spacings of both vectors are constant.
SYMMATVEC forms the scalar product of a vector and a row of a symmetric matrix, whose upper triangle is given columnwise in a one-dimensional array.
FULMATVEC forms the product A*B of a given matrix A and a vector

B.

FULTAMVEC forms the product A^T*B, where A^T is the transpose of a given matrix A and B is a vector.

FULSYMMATVEC forms the product A*B, where A is a symmetric matrix whose upper triangle is stored columnwise in a one-dimensional array and B is a vector.

RESVEC calculates the residual vector A*B+X*C, where A is a given matrix, B and C are vectors and X is a scalar.

SYMRESVEC calculates the residual vector A*B+X*C, where A is a symmetric matrix whose upper triangle is stored columnwise in a one-dimensional array, B and C are vectors and X is a scalar.

HSHVECMAT premultiplies a matrix by a Householder matrix, the vector defining this Householder matrix being given in a one-dimensional array.

HSHCOLMAT premultiplies a matrix by a Householder matrix, the vector defining this Householder matrix being given as a column in a two-dimensional array.

HSHROWMAT premultiplies a matrix by a Householder matrix, the vector defining this Householder matrix being given as a row in a two-dimensional array.

HSHVECTAM postmultiplies a matrix by a Householder matrix, the vector defining this Householder matrix being given in a one-dimensional array.

HSHCOLTAM postmultiplies a matrix by a Householder matrix, the vector defining this Householder matrix being given as a column in a two-dimensional array.

HSHROWTAM postmultiplies a matrix by a Householder matrix, the vector defining this Householder matrix being given as a row in a two-dimensional array.

ELMVEC adds a constant times a vector to a vector.

ELMCOL adds a constant times a column vector to a column vector.

ELMROW adds a constant times a row vector to a row vector.

ELMVECCOL adds a constant times a column vector to a vector.

ELMCOLVEC adds a constant times a vector to a column vector.

ELMVECROW adds a constant times a row vector to a vector.

ELMROWVEC adds a constant times a vector to a row vector.

ELMCOLROW adds a constant times a row vector to a column vector.

ELMROWCOL adds a constant times a column vector to a row vector.

MAXELMROW adds a constant times a row vector to a row vector, it also delivers the subscript of an element of the new row vector which is of maximum absolute value.

ICHVEC interchanges two vectors given in a one-dimensional array.
ICHCOL interchanges two columns of a matrix.
ICHROW interchanges two rows of a matrix.
ICHROWCOL interchanges a row and a column of a matrix.
ICHSEQVEC interchanges a row and a column of an upper triangular matrix, which is stored columnwise in a one-dimensional array.
ICHSEQ interchanges two columns of an upper triangular matrix, which is stored columnwise in a one-dimensional array.
ROTCOL replaces two column vectors X and Y by two vectors CX+SY and CY-SX.
ROTROW replaces two row vectors X and Y by two vectors CX+SY and CY-SX.
INFNRMVEC calculates the infinity norm of a vector.
INFNRMROW calculates the infinity norm of a row vector.
INFNRMCOL calculates the infinity norm of a column vector.
INFNRMMAT calculates the infinity norm of a matrix.
ONENRMVEC calculates the 1-norm of a vector.
ONENRMROW calculates the 1-norm of a row vector.
ONENRMCOL calculates the 1-norm of a column vector.
ONENRMMAT calculates the 1-norm of a matrix.
ABSMAXMAT calculates the modulus of the largest element of a matrix and delivers the indices of the maximal element.
REASCL normalizes the columns of a two-dimensional array.
COMCOLCST multiplies a complex column vector by a complex number.
COMROWCST multiplies a complex row vector by a complex number.
COMMATVEC forms the scalar product of a complex row vector and a complex vector.
HSHCOMCOL transforms a complex vector into a vector proportional to a unit vector.
HSHCOMPRD premultiplies a complex matrix with a complex Householder matrix.
ELMCOMVECCOL adds a complex number times a complex column vector to a complex vector.
ELMCOMCOL adds a complex number times a complex column vector to a complex column vector.
ELMCOMROWVEC adds a complex number times a complex vector to a complex row vector.
ROTCOMCOL replaces two complex column vectors X and Y by two complex vectors CX+SY and CY-SX.
ROTCOMROW replaces two complex row vectors X and Y by two complex vectors CX+SY and CY-SX.

CHSH2 finds a complex rotation matrix.
COMEUCNRM calculates the Euclidean norm of a complex matrix with some lower diagonals.
COMSCL normalizes real and complex eigenvectors.
SCLCOM normalizes the columns of a complex matrix.
COMABS calculates the modulus of a complex number.
COMSQRT calculates the square root of a complex number.
CARPOL transforms the Cartesian coordinates of a complex number into polar coordinates.
COMMUL calculates the product of two complex numbers.
COMDIV calculates the quotient of two complex numbers.
LNGINTADD computes the sum of long nonnegative integers.
LNGINTSUBTRACT computes the difference of long nonnegative integers.
LNGINTMULT computes the product of long nonnegative integers.
LNGINTDIVIDE computes the quotient with remainder of long nonnegative integers.
LNGINTPOWER computes U^P, where U is a long nonnegative integer and P is the positive (single-length) exponent.
POL evaluates a polynomial.
TAYPOL evaluates the first k terms of a Taylor series.
NORDERPOL evaluates the first k normalized derivatives of a polynomial.
DERPOL evaluates the first k derivatives of a polynomial.
ORTPOL evaluates the value of an n-degree orthogonal polynomial, given by a set of recurrence coefficients.
ORTPOLSYM evaluates the values of an n-degree orthogonal polynomial, given by a set of recurrence coefficients.
ALLORTPOL evaluates the value of all orthogonal polynomials up to a given degree, given a set of recurrence coefficients.
ALLORTPOLSYM evaluates the values of all orthogonal polynomials up to a given degree, given a set of recurrence coefficients.
SUMORTPOL evaluates a finite series expressed in orthogonal polynomials, given by a set of recurrence coefficients.
SUMORTPOLSYM evaluates a finite series expressed in orthogonal polynomials, given by a set of recurrence coefficients.
CHEPOLSUM evaluates a finite sum of Chebyshev polynomials.
ODDCHEPOLSUM evaluates a finite sum of Chebyshev polynomials of odd degree.
CHEPOL evaluates a Chebyshev polynomial.
ALLCHEPOL evaluates all Chebyshev polynomials up to a certain

degree.

SINSER evaluates a sine series.

COSSER evaluates a cosine series.

FOUSER evaluates a fourier series with equal sine and cosine coefficients.

FOUSER1 evaluates a fourier series.

FOUSER2 evaluates a fourier series.

COMFOUSER evaluates a complex fourier series with real coefficients.

COMFOUSER1 evaluates a complex fourier series.

COMFOUSER2 evaluates a complex fourier series.

JFRAC calculates a terminating continued fraction.

POLCHS transforms a polynomial from power sum into Chebyshev sum form.

CHSPOL transforms a polynomial from Chebyshev sum into power sum form.

POLSHTCHS transforms a polynomial from power sum into shifted Chebyshev sum form.

SHTCHSPOL transforms a polynomial from shifted Chebyshev sum form into power sum form.

GRNNEW transforms a polynomial from power sum into Newton sum form.

NEWGRN transforms a polynomial from Newton sum into power sum form.

LINTFMPOL transforms a polynomial in X into a polynomial in Y (Y=A*X+B).

INTCHS computes the indefinite integral of a given Chebyshev series.

DEC performs a triangular decomposition with partial pivoting.

GSSELM performs a triangular decomposition with a combination of partial and complete pivoting.

ONENRMINV calculates the 1-norm of the inverse of a matrix whose triangularly decomposed form is delivered by GSSELM.

ERBELM calculates a rough upper bound for the error in the solution of a system of linear equations whose matrix is triangularly decomposed by GSSELM.

GSSERB performs a triangular decomposition of the matrix of a system of linear equations and calculates an upper bound for the relative error in the solution of that system.

GSSNRI performs a triangular decomposition and calculates the 1-norm of the inverse matrix.

DETERM calculates the determinant of a triangularly decomposed matrix.

SOL solves a system of linear equations whose matrix has been triangularly decomposed by DEC.

DECSOL solves a system of linear equations whose order is small relative to the number of binary digits in the number representation.

SOLELM solves a system of linear equations whose matrix has been triangularly decomposed by GSSELM or GSSERB.

GSSSOL solves a system of linear equations.

GSSSOLERB calculates the inverse of a matrix and 1-norm, an upper bound for the error in the inverse matrix is also given.

INV calculates the inverse of a matrix that has been triangularly decomposed by DEC.

DECINV calculates the inverse of a matrix whose order is small relative to the number of binary digits in the number representation.

INV1 calculates the inverse of a matrix that has been triangularly decomposed by GSSELM or GSSERB, the 1-norm of the inverse matrix might also be calculated.

GSSINV calculates the inverse of a matrix.

GSSINVERB calculates the inverse of a matrix and 1-norm, an upper bound for the error in the inverse matrix is also given.

ITISOL solves a system of linear equations whose matrix has been triangularly decomposed by GSSELM or GSSERB, this solution is improved iteratively.

GSSITISOL solves a system of linear equations and the solution is improved iteratively.

ITISOLERB solves a system of linear equations whose matrix has been triangularly decomposed by GSSNRI, this solution is improved iteratively and an upper bound for the error in the solution is calculated.

GSSITISOLERB solves a system of linear equations, this solution is improved iteratively and an upper bound for the error in the solution is calculated.

CHLDEC2 calculates the Cholesky decomposition of a positive definite symmetric matrix whose upper triangle is given in a two-dimensional array.

CHLDEC1 calculates the Cholesky decomposition of a positive definite symmetric matrix whose upper triangle is given columnwise in a one-dimensional array.

CHLDETERM2 calculates the determinant of a positive definite symmetric matrix, the Cholesky decomposition being given in a two-dimensional array.

CHLDETERM1 calculates the determinant of a positive definite symmetric matrix, the Cholesky decomposition being given

	columnwise in a one-dimensional array.
CHLSOL2	solves a system of linear equations if the coefficient matrix has been decomposed by CHLDEC2 or CHLDECSOL2.
CHLSOL1	solves a system of linear equations if the coefficient matrix has been decomposed by CHLDEC1 or CHLDECSOL1.
CHLDECSOL2	solves a positive definite symmetric system of linear equations by Cholesky's square root method, the coefficient matrix should be given in the upper triangle of a two-dimensional array.
CHLDECSOL1	solves a positive definite symmetric system of linear equations by Cholesky's square root method, the coefficient matrix should be given columnwise in a one-dimensional array.
CHLINV2	calculates the inverse of a positive definite symmetric matrix, if the matrix has been decomposed by CHLDEC2 or CHLDECSOL2.
CHLINV1	calculates the inverse of a positive definite symmetric matrix, if the matrix has been decomposed by CHLDEC1 or CHLDECSOL1.
CHLDECINV2	calculates the inverse of a positive definite symmetric matrix by Cholesky's square root method, the coefficient matrix given in the upper triangle of a two-dimensional array.
CHLDECINV1	calculates the inverse of a positive definite symmetric matrix by Cholesky's square root method, the coefficient matrix given columnwise in a one-dimensional array.
DECSYM2	calculates the symmetric decomposition of a symmetric matrix.
DETERMSYM2	calculates the determinant of a symmetric matrix, the symmetric decomposition being given.
SOLSYM2	solves a symmetric system of linear equations if the coefficient matrix has been decomposed by DECSYM2 or DECSOLSYM2.
DECSOLSYM2	solves a symmetric system of linear equations by symmetric decomposition.
LSQORTDEC	delivers the Householder triangularization with column interchanges of a matrix of a linear least squares problem.
LSQDGLINV	calculates the diagonal elements of the inverse of M^TM, where M is the coefficient matrix of a linear least squares problem.

LSQSOL solves a linear least squares problem if the coefficient matrix has been decomposed by LSQORTDEC.

LSQORTDECSOL solves a linear least squares problem by Householder triangularization with column interchanges and calculates the diagonal of the inverse of M^TM, where M is the coefficient matrix.

LSQINV calculates the inverse of the matrix S^TS, where S is the coefficient matrix of a linear least squares problem.

LSQDECOMP computes the QR decomposition of a linear least squares problem with linear constraints.

LSQREFSOL solves a linear least squares problem with linear constraints if the coefficient matrix has been decomposed by LSQDECOMP.

SOLSVDOVR solves an overdetermined system of linear equations, multiplying the right hand side by the pseudo-inverse of the given matrix.

SOLOVR calculates the singular values decomposition and solves an overdetermined system of linear equations.

SOLSVDUND solves an underdetermined system of linear equations, multiplying the right hand side by the pseudo-inverse of the given matrix.

SOLUND calculates the singular values decomposition and solves an underdetermined system of linear equations.

HOMSOLSVD solves the homogeneous system of linear equations AX=0 and X^TA=0, where A denotes a matrix and X a vector (the singular value decomposition being given).

HOMSOL solves the homogeneous system of linear equations AX=0 and X^TA=0, where A denotes a matrix and X a vector.

PSDINVSVD calculates the pseudo-inverse of a matrix (the singular value decomposition being given).

PSDINV calculates the pseudo-inverse of a matrix.

DECBND performs a triangular decomposition of a band matrix, using partial pivoting.

DETERMBND calculates the determinant of a band matrix.

SOLBND solves a system of linear equations, the matrix being decomposed by DECBND.

DECSOLBND solves a system of linear equations by Gaussian elimination with partial pivoting if the coefficient matrix is in band form and is stored rowwise in a one-dimensional array.

DECTRI performs a triangular decomposition of a tridiagonal matrix.

DECTRIPIV performs a triangular decomposition of a tridiagonal matrix, using partial pivoting.

SOLTRI solves a tridiagonal system of linear equations, the triangular decomposition being given.

DECSOLTRI solves a tridiagonal system of linear equations and performs the triangular decomposition without pivoting.

SOLTRIPIV solves a tridiagonal system of linear equations, the triangular decomposition being given.

DECSOLTRIPIV solves a tridiagonal system of linear equations and performs the triangular decomposition with partial pivoting.

CHLDECBND performs the Cholesky decomposition of a positive definite symmetric band matrix.

CHLDETERMBND calculates the determinant of a positive definite symmetric band matrix.

CHLSOLBND solves a positive definite symmetric linear system, the triangular decomposition being given.

CHLDECSOLBND solves a positive definite symmetric linear system and performs the triangular decomposition by Cholesky's method.

DECSYMTRI performs the triangular decomposition of a symmetric tridiagonal matrix.

SOLSYMTRI solves a symmetric tridiagonal system of linear equations, the triangular decomposition being given.

DECSOLSYMTRI solves a symmetric tridiagonal system of linear equations and performs the tridiagonal decomposition.

CONJGRAD solves a positive definite symmetric system of linear equations by the method of conjugate gradients.

EQILBR equilibrates a matrix by means of a diagonal similarity transformation.

BAKLBR performs the back transformation corresponding to EQILBR.

EQILBRCOM equilibrates a complex matrix.

BAKLBRCOM transforms the eigenvectors of a complex equilibrated (by EQILBRCOM) matrix into the eigenvectors of the original matrix.

TFMSYMTRI2 transforms a real symmetric matrix into a similar tridiagonal one by means of Householder's transformation.

BAKSYMTRI2 performs the back transformation corresponding to TFMSYMTRI2.

TFMPREVEC in combination with TFMSYMTRI2 calculates the transforming matrix.

TFMSYMTRI1 transforms a real symmetric matrix into a similar tridiagonal one by means of Householder's transformation.

BAKSYMTRI1 performs the back transformation corresponding to TFMSYMTRI1.

TFMREAHES transforms a matrix into a similar upper Hessenberg matrix by means of Wilkinson's transformation.

BAKREAHES1 performs the back transformation (on a vector) corresponding to TFMREAHES.

BAKREAHES2 performs the back transformation (on columns) corresponding to TFMREAHES.

HSHHRMTRI transforms a Hermitian matrix into a similar real symmetric tridiagonal matrix.

HSHHRMTRIVAL delivers the main diagonal elements and the squares of the codiagonal elements of a Hermitian tridiagonal matrix which is unitary similar with a given Hermitian matrix.

BAKHRMTRI performs the back transformation corresponding to HSHHRMTRI.

HSHCOMHES transforms a complex matrix by means of Householder's transformation followed by a complex diagonal transformation into a similar unitary upper Hessenberg matrix with a real nonnegative subdiagonal.

BAKCOMHES performs the back transformation corresponding to HSHCOMHES.

HSHREABID transforms a matrix to bidiagonal form, by premultiplying and postmultiplying with orthogonal matrices.

PSTTFMMAT calculates the postmultiplying matrix from the data generated by HSHREABID.

PRETFMMAT calculates the premultiplying matrix from the data generated by HSHREABID.

VALSYMTRI calculates all, or some consecutive, eigenvalues of a symmetric tridiagonal matrix by means of linear interpolation using a Sturm sequence.

VECSYMTRI calculates eigenvectors of a symmetric tridiagonal matrix by means of inverse iteration.

QRIVALSYMTRI calculates the eigenvalues of a symmetric tridiagonal matrix by means of QR iteration.

QRISYMTRI calculates the eigenvalues and eigenvectors of a symmetric tridiagonal matrix by means of QR iteration.

EIGVALSYM2 calculates all (or some) eigenvalues of a symmetric matrix using linear interpolation of a function derived from a Sturm sequence.

EIGSYM2 calculates eigenvalues and eigenvectors by means of

inverse iteration.

EIGVALSYM1 calculates all (or some) eigenvalues of a symmetric matrix using linear interpolation of a function derived from a Sturm sequence.

EIGSYM1 calculates eigenvalues and eigenvectors by means of inverse iteration.

QRIVALSYM2 calculates the eigenvalues of a symmetric matrix by means of QR iteration.

QRISYM calculates all eigenvalues and eigenvectors of a symmetric matrix by means of QR iteration.

QRIVALSYM1 calculates the eigenvalues of a symmetric matrix by means of QR iteration.

MERGESORT delivers a permutation of indices corresponding to sorting the elements of a given vector into non-decreasing order.

VECPERM permutes the elements of a given vector according to a given permutation of indices.

ROWPERM permutes the elements of a given row according to a given permutation of indices.

ORTHOG orthogonalizes some adjacent matrix columns according to the modified Gram-Schmidt method.

SYMEIGIMP improves an approximation of a real symmetric eigensystem and calculates error bounds for the eigenvalues.

REAVALQRI calculates the eigenvalues of a real upper Hessenberg matrix, provided that all eigenvalues are real, by means of single QR iteration.

REAVECHES calculates an eigenvector corresponding to a given real eigenvalue of a real upper Hessenberg matrix by means of inverse iteration.

REAQRI calculates all eigenvalues and eigenvectors of a real upper Hessenberg matrix, provided that all eigenvalues are real, by means of single QR iteration.

COMVALQRI calculates the real and complex eigenvalues of a real upper Hessenberg matrix by means of double QR iteration.

COMVECHES calculates the eigenvector corresponding to a given complex eigenvalue of a real upper Hessenberg matrix by means of inverse iteration.

REAEIGVAL calculates the eigenvalues of a matrix, provided that all eigenvalues are real.

REAEIG1 calculates the eigenvalues and eigenvectors of a matrix, provided that they are all real.

REAEIG3 calculates the eigenvalues and eigenvectors of a matrix,

provided that they are all real.
COMEIGVAL calculates the eigenvalues of a matrix.
COMEIG1 calculates the eigenvalues and eigenvectors of a matrix.
EIGVALHRM calculates the eigenvalues of a complex Hermitian matrix.
EIGHRM calculates the eigenvalues and eigenvectors of a complex Hermitian matrix.
QRIVALHRM calculates the eigenvalues of a complex Hermitian matrix.
QRIHRM calculates the eigenvalues and eigenvectors of a complex Hermitian matrix.
VALQRICOM calculates the eigenvalues of a complex upper Hessenberg matrix with a real subdiagonal.
QRICOM calculates the eigenvalues and eigenvectors of a complex upper Hessenberg matrix.
EIGVALCOM calculates the eigenvalues of a complex matrix.
EIGCOM calculates the eigenvalues and eigenvectors of a complex matrix.
QZIVAL computes generalized eigenvalues by means of QZ iteration.
QZI computes generalized eigenvalues and eigenvectors by means of QZ iteration.
HSHDECMUL is an auxiliary procedure for the computation of generalized eigenvalues.
HESTGL3 is an auxiliary procedure for the computation of generalized eigenvalues.
HESTGL2 is an auxiliary procedure for the computation of generalized eigenvalues.
HSH2COL is an auxiliary procedure for the computation of generalized eigenvalues.
HSH3COL is an auxiliary procedure for the computation of generalized eigenvalues.
HSH2ROW3 is an auxiliary procedure for the computation of generalized eigenvalues.
HSH2ROW2 is an auxiliary procedure for the computation of generalized eigenvalues.
HSH3ROW3 is an auxiliary procedure for the computation of generalized eigenvalues.
HSH3ROW2 is an auxiliary procedure for the computation of generalized eigenvalues.
QRISNGVALBID calculates the singular values of a bidiagonal matrix.
QRISNGVALDECBID calculates the singular values decomposition of a matrix of which the bidiagonal and the pre- and postmultiplying matrices are given.
QRISNGVAL calculates the singular values of a given matrix.

QRISNGVALDEC calculates the singular values decomposition UDV^T, with U and V orthogonal and D positive diagonal.

ZERPOL calculates all roots of a polynomial with real coefficients by Laguerre's method.

BOUNDS calculates the error in approximated zeros of a polynomial with real coefficients.

ALLZERORTPOL calculates all zeros of an orthogonal polynomial.

LUPZERORTPOL calculates a number of adjacent upper or lower zeros of an orthogonal polynomial.

SELZERORTPOL calculates a number of adjacent zeros of an orthogonal polynomial.

ALLJACZER calculates the zeros of a Jacobian polynomial.

ALLLAGZER calculates the zeros of a Laguerre polynomial.

COMKWD calculates the roots of a quadratic equation with complex coefficients.

EULER performs the summation of an alternating infinite series.

SUMPOSSERIES performs the summation of an infinite series with positive monotonically decreasing terms using the van Wijngaarden transformation.

QADRAT computes the definite integral of a function of one variable over a finite interval.

INTEGRAL calculates the definite integral of a function of one variable over a finite or infinite interval or over a number of consecutive intervals.

TRICUB computes the definite integral of a function of two variables over a triangular domain.

RECCOF calculates recurrence coefficients of an orthogonal polynomial, a weight function being given.

GSSWTS calculates the Gaussian weights of a weight function, the recurrence coefficients being given.

GSSWTSSYM calculates the Gaussian weights of a weight function, the recurrence coefficients being given.

GSSJACWGHTS computes the abscissae and weights for Gauss-Jacobi quadrature.

GSSLAGWGHTS computes the abscissae and weights for Gauss-Lagrange quadrature.

JACOBNNF calculates the Jacobian matrix of an n-dimensional function of n variables using forward differences.

JACOBNMF calculates the Jacobian matrix of an n-dimensional function of m variables using forward differences.

JACOBNBNDF calculates the Jacobian matrix of an n-dimensional function of n variables, if the Jacobian is known to be a band matrix.

ZEROIN finds (in a given interval) a zero of a function of one variable.

ZEROINRAT finds (in a given interval) a zero of a function of one variable.

ZEROINDER finds (in a given interval) a zero of a function of one variable using values of the function and of its derivative.

QUANEWBND solves a system of non-linear equations of which the Jacobian (being a band matrix) is given.

QUANEWBND1 solves a system of non-linear equations of which the Jacobian is a band matrix.

MININ minimizes a function of one variable in a given interval.

MININDER minimizes a function of one variable in a given interval, using values of the function and of its derivative.

LINEMIN minimizes a function of several variables in a given direction.

RNK1UPD adds a rank-1 matrix to a symmetric matrix.

DAVUPD adds a rank-2 matrix to a symmetric matrix.

FLEUPD adds a rank-2 matrix to a symmetric matrix.

PRAXIS minimizes a function of several variables.

RNK1MIN minimizes a function of several variables.

FLEMIN minimizes a function of several variables.

MARQUARDT calculates the least squares solution of an overdetermined system of non-linear equations with Marquardt's method.

GSSNEWTON calculates the least squares solution of an overdetermined system of non-linear equations with the Gauss-Newton method.

RK1 solves a single first order differential equation by means of a fifth order Runge-Kutta method.

RKE solves a system of first order differential equations (initial value problem) by means of a fifth order Runge-Kutta method.

RK4A solves a single first order differential equation by means of a fifth order Runge-Kutta method; the integration is terminated as soon as a condition on X and Y, which is supplied by the user, is satisfied.

RK4NA solves a system of first order differential equations (initial value problem) by means of a fifth order Runge-Kutta method; the integration is terminated as soon as a condition on X[0], X[1],..., X[n], supplied by the user, is satisfied.

RK5NA solves a system of first order differential equations (initial value problem) by means of a fifth order Runge-Kutta method; the arc length is introduced as an integration variable; the integration is terminated as soon as a condition on X[0], X[1],..., X[n], supplied by the user, is satisfied.

MULTISTEP solves a system of first order differential equations (initial value problem) by means of a variable order multistep method Adams-Moulton, Adams-Bashforth or Gear's method; the order of accuracy is automatic, up to fifth order; this method is suitable for stiff systems.

DIFFSYS solves a system of first order differential equations (initial value problem) by extrapolation, applied to low order results, a high order of accuracy is obtained; this method is suitable for smooth problems when high accuracy is required.

ARK solves a system of first order differential equations (initial value problem) by means of a stabilized Runge-Kutta method with limited storage requirements.

EFRK solves a system of first order differential equations (initial value problem) by means of a first, second or third order, exponentially fitted Runge-Kutta method; automatic step size control is not provided; this method can be used to solve stiff systems with known eigenvalue spectrum.

EFSIRK solves an autonomous system of first order differential equations (initial value problem) by means of a third order, exponentially fitted, semi-implicit Runge-Kutta method; this method can be used to solve stiff systems.

EFERK solves an autonomous system of first order differential equations (initial value problem) by means of an exponentially fitted, third order Runge-Kutta method; this method can be used to solve stiff systems with known eigenvalue spectrum.

LINIGER1VS solves an autonomous system of first order differential equations (initial value problem) by means of an implicit, exponentially fitted first order one-step method; this method can be used to solve stiff systems.

LINIGER2 solves an autonomous system of first order differential equations (initial value problem) by means of an implicit, exponentially fitted first order one-step method; automatic step size control is not provided; this method can be used to solve stiff systems.

GMS solves an autonomous system of first order differential equations (initial value problem) by means of a third order multistep method; this method can be used to solve stiff systems.

IMPEX solves an autonomous system of first order differential equations (initial value problem) by means of the implicit midpoint rule with smoothing and extrapolation; this method is suitable for the integration of stiff differential equations.

MODIFIEDTAYLOR solves a system of first order differential equations (initial value problem) by means of a first, second or third order one-step Taylor method; this method can be used to solve large and sparse systems, provided higher order derivatives can easily be obtained.

EFT solves a system of first order differential equations (initial value problem) by means of a variable order Taylor method; this method can be used to solve stiff systems, with known eigenvalue spectrum, provided higher order derivatives can easily be obtained.

RK2 solves a single second order differential equation (initial value problem) by means of a fifth order Runge-Kutta method.

RK2N solves a system of second order differential equations (initial value problem) by means of a fifth order Runge-Kutta method.

RK3 solves a single second order differential equation (initial value problem) by means of a fifth order Runge-Kutta method; this method can only be used if the right hand side of the differential equation does not depend on y'.

RK3N solves a system of second order differential equations (initial value problem) by means of a fifth order Runge-Kutta method; this method can only be used if the right hand side of the differential equation does not depend on y'.

ARKMAT solves a system of first order differential equations (initial boundary-value problem) by means of a stabilized Runge-Kutta method, in particular suitable for systems arising from two-dimensional time-dependent partial differential equations.

FEMLAGSYM solves a linear two-point boundary-value problem for a second order self-adjoint differential equation by a Ritz-Galerkin method.

FEMLAG solves a linear two-point boundary-value problem for a second order self-adjoint differential equation by a Ritz-Galerkin method; the coefficient of y' is supposed to be unity.

FEMLAGSPHER solves a linear two-point boundary-value problem for a second order self-adjoint differential equation with spherical coordinates by a Ritz-Galerkin method.

FEMLAGSKEW solves a linear two-point boundary-value problem for a second order differential equation by a Ritz-Galerkin method.

FEMHERMSYM solves a linear two-point boundary-value problem for a fourth order self-adjoint differential equation with Dirichlet boundary conditions by a Ritz-Galerkin method.

NONLINFEMLAGSKEW solves a nonlinear two-point boundary-value problem for a second order differential equation with spherical coordinates by a Ritz-Galerkin method and Newton iteration.

RICHARDSON solves a system of linear equations with positive real eigenvalues (elliptic boundary value problem) by means of a non-stationary second order iterative method.

ELIMINATION solves a system of linear equations with positive real eigenvalues (elliptic boundary value problem) by means of a non-stationary second order iterative method, which is an acceleration of Richardson's method.

PEIDE estimates unknown parameters in a system of first order differential equations; the unknown variables may appear non-linearly both in the differential equations and its initial values; a set of observed values of some components of the solution of the differential equations must be given.

ARCSINH computes the inverse hyperbolic sine for a real argument.

ARCCOSH computes the inverse hyperbolic cosine for a real argument.

ARCTANH computes the inverse hyperbolic tangent for a real argument.

LOGONEPLUSX evaluates the logarithmic function *ln*(1+x).

EI calculates the exponential integral.

EIALPHA calculates a sequence of integrals of the form ($e^{-xt}t^n$ dt), from t=1 to t=infinity.

ENX calculates a sequence of exponential integrals E(n,x) = the integral from 1 to infinity of e^{-xt}/t^n dt.

NONEXPENX calculates a sequence of integrals e^x * E(n,x), see ENX.

SINCOSINT calculates the sine integral SI(x) and the cosine integral CI(x).

SINCOSFG is an auxiliary procedure for the sine and cosine integrals.

RECIPGAMMA calculates the reciprocal of the gamma function for arguments in the range [0.5,1.5]; moreover odd and even parts are delivered.

GAMMA calculates the gamma function.
LOGGAMMA calculates the natural logarithm of the gamma function for positive arguments.
INCOMGAM calculates the incomplete gamma functions.
INCBETA calculates the incomplete beta function I(x,p,q) for $0 \leq x \leq 1$, $p>0$, $q>0$.
IBPPLUSN calculates the incomplete beta function ratios I(x,p+n,q) for n=0,1,...,nmax, $0 \leq x \leq 1$, $p>0$, $q>0$.
IBQPLUSN calculates the incomplete beta function ratios I(x,p,q+n) for n=0,1,...,nmax, $0 \leq x \leq 1$, $p>0$, $q>0$.
IXQFIX is an auxiliary procedure for the computation of incomplete bessel functions.
IXPFIX is an auxiliary procedure for the computation of incomplete bessel functions.
FORWARD is an auxiliary procedure for the computation of incomplete bessel functions.
BACKWARD is an auxiliary procedure for the computation of incomplete bessel functions.
ERRORFUNNCTION computes the error function (erf) and complementary error function (erfc) for a real argument.
NONEXPERFC computes erfc(x) * exp(x^2); see ERRORFUNCTION.
INVERSEERRORFUNCTION calculates the inverse error function y = inverf(x).
FRESNEL calculates the fresnel integrals C(x) and S(x).
FG is an auxiliary procedure for the computation of fresnel integrals.
BESSJ0 calculates the ordinary bessel function of the first kind of order zero.
BESSJ1 calculates the ordinary bessel function of the first kind of order one.
BESSJ calculates the ordinary bessel functions of the first kind of order l, l=0,...,n.
BESSY01 calculates the ordinary bessel functions of the second kind of orders zero and one with argument x, $x>0$.
BESSY calculates the ordinary bessel functions of the second kind of order l, l=0,...,n, with argument x, $x>0$.
BESSPQ0 is an auxiliary procedure for the computation of the ordinary bessel functions of order zero for large values of their argument.
BESSPQ1 is an auxiliary procedure for the computation of the ordinary bessel functions of order one for large values of

their argument.

BESSI0 calculates the modified bessel function of the first kind of order zero.

BESSI1 calculates the modified bessel function of the first kind of order one.

BESSI calculates the modified bessel functions of the first kind of order l, l=0,...,n.

BESSK01 calculates the modified bessel functions of the third kind of orders zero and one with argument x, x>0.

BESSK calculates the modified bessel functions of the third kind of order l, l=0,...,n, with argument x, x>0.

NONEXPBESSI0 calculates the modified bessel function of the first kind of order zero; the result is multiplied by $e^{-|x|}$.

NONEXPBESSI1 calculates the modified bessel function of the first kind of order one; the result is multiplied by $e^{-|x|}$.

NONEXPBESSI calculates the modified bessel functions of the first kind of order l, l=0,...,n; the result is multiplied by $e^{-|x|}$.

NONEXPBESSK01 calculates the modified bessel functions of the third kind of orders zero and one with argument x, x>0; the result is multiplied by e^x.

NONEXPBESSK calculates the modified bessel functions of the third kind of order l, l=0,...,n, with argument x, x>0; the result is multiplied by e^x.

BESSJAPLUSN calculates the bessel functions of the first kind of order a+k, 0≤k≤n, 0≤a<1.

BESSYA01 calculates the bessel functions of the second kind (also called Neumann's functions) of order a and a+1, a≥0, and argument x>0.

BESSYAPLUSN calculates the bessel functions of the second kind of order a+n, n=0,...,nmax, a≥0, and argument x>0.

BESSPQA01 is an auxiliary procedure for the computation of the bessel functions for large values of their argument.

BESSZEROS calculates zeros of a bessel function (of first or second kind) and of its derivative.

START is an auxiliary procedure in bessel function procedures.

BESSIAPLUSN calculates the modified bessel functions of the first kind of order a+n, n=0,...,nmax, a≥0, and argument x≥0.

BESSKA01 calculates the modified bessel functions of the third kind of orders a and a+1, a≥0, and argument x>0.

BESSKAPLUSN calculates the modified bessel functions of the third kind of order a+n, n=0,...,nmax, a≥0, and argument x>0.

NONEXPBESSIAPLUSN calculates the modified bessel functions of the first kind of order a+n, n=0,...,nmax, a≥0 and argument x≥0, multiplied by e^{-x}.

NONEXPBESSKA01 calculates the modified bessel functions of the third kind of order a and a+1, a≥0 and argument x, x>0, multiplied by the factor e^{x}.

NONEXPBESSKAPLUSN calculates the modified bessel functions of the third kind of order a+n, n=0,...,nmax, a≥0 and argument x>0, multiplied by e^{x}.

SPHERBESSJ calculates the spherical bessel functions of the first kind.

SPHERBESSY calculates the spherical bessel functions of the third kind.

SPHERBESSI calculates the modified spherical bessel functions of the first kind.

SPHERBESSK calculates the modified spherical bessel functions of the third kind.

NONEXPSPHERBESSI calculates the modified spherical bessel functions of the first kind multiplied by e^{-x}.

NONEXPSPHERBESSK calculates the modified spherical bessel functions of the third kind multiplied by e^{x}.

AIRY evaluates the Airy functions AI(z) and BI(z) and their derivatives.

AIRYZEROS computes the zeros and associated values of the Airy functions AI(z) and BI(z) and their derivatives.

NEWTON calculates the coefficients of the Newton polynomial through given interpolation points and corresponding function values.

INI selects a subset of integers out of a given set of integers; it is an auxiliary procedure for MINMAXPOL.

SNDREMEZ exchanges at most n+1 numbers with numbers out of a reference set; it is an auxiliary procedure for MINMAXPOL.

MINMAXPOL calculates the coefficients of the polynomial that approximates a function, given for discrete arguments, such that the infinity norm of the error vector is minimized.

Appendix D: Memory Management Utilities

Dynamic storage allocation is effected in C by the use of manipulation procedures. For example, placed at the beginning of a procedure, the code

```
integer *p;
p=allocate_integer_vector(l,u);
```

reserves storage for an integer vector p of range [*l,u*]; at the end of the procedure the code

```
free_integer_vector(p,l);
```

frees the storage for alternative use. Similarly, the code

```
float **q;
q=allocate_real_matrix(lr,ur,lc,uc);
```

reserves storage for a real matrix q of range [*lr:ur,lc:uc*] (this is done by allocating pointers to an array of pointers in C). The code

```
free_real_matrix(q,lr,ur,lc);
```

frees storage. (In this usage, array elements are referred to in the form q[i][j].)

The following utilities are used in the library.

```
#include <stdio.h>
#include <alloc.h>

void system_error(char error_message[])
{
    void exit(int);

    printf("%s",error_message);
    exit(1);
}


int *allocate_integer_vector(int l, int u)
{
    /*  Allocates an integer vector of range [l..u].  */

    void system_error(char *);
    int *p;

    p=(int *)malloc((unsigned) (u-l+1)*sizeof(int));
    if (!p) system_error("Failure in allocate_integer_vector().");
    return p-l;
}


float *allocate_real_vector(int l, int u)
{
    /*  Allocates a real vector of range [l..u].  */

    void system_error(char *);
    float *p;

    p=(float *)malloc((unsigned) (u-l+1)*sizeof(float));
    if (!p) system_error("Failure in allocate_real_vector().");
```

```
    return p-1;
}


int **allocate_integer_matrix(int lr, int ur, int lc, int uc)
{
    /*  Allocates an integer matrix of range [lr..ur][lc..uc].  */

    void system_error(char *);
    int i, **p;

    p=(int **)malloc((unsigned) (ur-lr+1)*sizeof(int*));
    if (!p) system_error("Failure in allocate_integer_matrix().");
    p -= lr;

    for (i=lr; i<=ur; i++){
        p[i]=(int *)malloc((unsigned) (uc-lc+1)*sizeof(int));
        if (!p[i]) system_error("Failure in allocate_integer_matrix().");
        p[i] -= lc;
    }
    return p;
}


float **allocate_real_matrix(int lr, int ur, int lc, int uc)
{
    /*  Allocates a real matrix of range [lr..ur][lc..uc].  */

    void system_error(char *);
    int i;
    float **p;

    p=(float **)malloc((unsigned) (ur-lr+1)*sizeof(float*));
    if (!p) system_error("Failure in allocate_real_matrix().");
    p -= lr;

    for (i=lr; i<=ur; i++){
        p[i]=(float *)malloc((unsigned) (uc-lc+1)*sizeof(float));
        if (!p[i]) system_error("Failure in allocate_real_matrix().");
        p[i] -= lc;
    }
    return p;
}


void free_integer_vector(int *v, int l)
{
    /*  Frees an integer vector of range [l..u].  */

    free((char*) (v+l));
}


void free_real_vector(float *v, int l)
{
    /*  Frees a real vector of range [l..u].  */

    free((char*) (v+l));
}


void free_integer_matrix(int **m, int lr, int ur, int lc)
{
    /*  Frees an integer matrix of range [lr..ur][lc..uc].  */

    int i;

    for (i=ur; i>=lr; i--) free((char*) (m[i]+lc));
    free((char*) (m+lr));
```

```
}


void free_real_matrix(float **m, int lr, int ur, int lc)
{
    /*  Frees a real matrix of range [lr..ur][lc..uc].  */

    int i;

    for (i=ur; i>=lr; i--) free((char*) (m[i]+lc));
    free((char*) (m+lr));
}
```

Index of Procedures

Index of Worked Examples

T - #0295 - 101024 - C0 - 254/178/44 [46] - CB - 9780849373763 - Gloss Lamination